BASIC MACHINE TECHNOLOGY

BASIC MACHINE TECHNOLOGY

C. Thomas Olivo, Ed.D.

Bobbs-Merrill Educational Publishing, Indianapolis

The Bobbs-Merrill Company, Inc.
4300 W. 62nd Street
Indianapolis, Indiana 46268

First Edition
First Printing 1980

Cover and interior design by Alpha Technical Services, Inc.

Library of Congress Cataloging in Publication Data
Olivo, C. Thomas.
  Basic machine technology.
  Includes index.
  1. Machine-shop practice.   I. Title.
TJ1160.044   621.9   79-11958
ISBN 0-672-97171-2

No comprehensive writing is ever done without sacrifice and devotion. Thus, Basic Machine Technology is affectionately dedicated to my wife, Hilda G. Olivo. Over these many years of enquiry, study, research, and analysis, she has been an outstanding teacher, constant companion, and reviewer. These provided the continuous thread of understanding and support.

Contents

Part Four   DRILLING MACHINES: TECHNOLOGY AND PROCESSES

Part Five   LATHE TECHNOLOGY AND PROCESSES

Part Six   MILLING MACHINES: TECHNOLOGY AND PROCESSES

List Of Appendix Tables

ABOUT THE AUTHOR:

C. Thomas Olivo is an internationally-recognized industrial-vocational educator with extensive experience as a journeyman tool and die maker, machine trades teacher, and technical institute director. Dr. Olivo also served as Chief of Vocational Curriculum Development, New York State Education Department.

PREFACE

Machine technology is essential to each individual's existence and to the Nation's well-being. Transforming raw materials into goods, products, and services that are necessary to maintain high standards of healthful living, economic stability, and national security depends on *machine technology*.

Machine technology is used as a generic term. Broadly interpreted, it relates to the study of skills and technical knowledge. These are applied to design, measurement, forming, fabricating, assembling, inspecting, and other processes in machine and metal products manufacturing, machine and tool design, and other mechanical and related industrial occupations.

The clusters of job titles at many employment levels in the occupational constellation comprise a significantly high percent of the Nation's work force. Job levels within each cluster range the full spectrum of workers from beginning machine operators to highly skilled machine, tool, and industrial engineers and researchers. Predominantly, the greatest numbers are employed as production workers, semiskilled operatives, skilled craftspersons and technicians, and supervisors, middle management, and semiprofessional workers.

The worker at each level and in each occupation must master a varying combination of skills and accompanying technical knowledge or theory. In each instance, the skills and technical knowledge are interdependent. They combine the "why-to-do" technology with the "how-to-do" manipulation of tools, machines, instruments, materials, and the like. Skills must also be developed in dimensional measurement. Industrial drawings must be interpreted. Shop and laboratory sketches are to be made. Dimensions are to be calculated. Materials must be selected, layed out, machined, and fitted.

Equally important is the dependency of these skills on the worker's ability to relate them to mathematics, physical science, blueprint reading, drafting and design principles.

Basic Machine Technology is complemented with a *Student's Manual* (Workbook) and a *Teacher's Guide* and *Answer Book*. The textbook, student's manual, and teacher's guide are based on extensive labor market studies of job titles, specifications, and levels. Occupational and task analyses were made to inventory common skill and technology elements at each successive stage of development. Curriculum studies were conducted to establish student/learner needs. Functional, effective teaching methods and performance objectives were researched. Curriculum and instructional material resources were studied in actual teaching and learning situations over many years.

Institutional and on-the-job organizational patterns were diagnosed for effective preemployment, retraining, and upgrading programs. Comprehensive long-term analyses were made of scope, content, organization, difficulty index, and quality of resource materials used by individuals for improvement through self-study. Formalized programs and instruction provided by school systems, industry, the military, labor organizations, and other manpower development agencies were reviewed.

Another important dimension of *Basic Machine Technology* relates to the application of *SI Metrics* in American and Canadian industry, labor, military, vocational-technical schools, and other postsecondary institutional training programs. The amount and depth to which metric concepts, measurements, and procedures are interwoven into the units is based on the author's first-hand experiences over a number of years in assessing training programs in the United States, Canada, and abroad. Judgments on metrics in machine technology are made from these experiences and the vantage point of a journeyman tool and die maker, former machine trades teacher, technical institute director, and state director of industrial-technical education.

It is obvious that *Basic Machine Technology* is the end product of extensive study, analysis, and experience. The text covers basic hand tools, cutting tools, and machine tool setups and processes; measurement and layout instruments and tools, and common bench and assembly practices. Industrial safety is threaded in detail throughout the units. Essential blueprint reading and mathematics principles are applied directly to typical shop jobs. Drilling machines, floor grinders, power hacksaws and band cutoff machines, engine lathes, and horizontal milling machines are the fundamental machine tools that are treated.

The *Student's Manual* (Workbook) provides a tremendous range of test items. These simulate actual industrial conditions. Each unit provides for individualized directed learning experiences. The *Student's Manual* interweaves the hands-on development of skills and technology in layout, machining, fabricating, and assembly of a single workpiece or the production of multiple parts.

The *Teacher's Guide and Answer Book* provides suggestions for effective teaching, learner-directed activities and assignments, and the solutions to all shop problems in the *Student's Manual*. Detailed descriptions of the scope, organization, and applications of the textbook, *Student's Manuals,* and the *Teacher's Guide* follow in the *Introduction to Basic Machine Technology*.

C. Thomas Olivo

ACKNOWLEDGMENTS

A statement of grateful appreciation is made to the many teachers and directors in post secondary institutions offering machine and manufacturing technology curriculums; other occupationally competent teachers in vocational/industrial/technical schools providing instruction in the machine trades, and to the instructors in in-plant training programs who assisted. Recognition is made to teachers and instructors in similar institutional and industrial positions in Europe, the Middle East, Southeast Asia, and South America, who made on-site assessments, and curriculum and facilities planning possible.

Also recognized are those shop foremen and industrial management persons who permitted in-plant analyses of modern design, process, and control techniques.

The work and contribution of the leaders in the curriculum and instructional materials centers of the Vocational-Technical Education Consortium of States and other regional vocational education research centers is appreciated.

A particular word of commendation is made to the Canadian, British, German, Austrian and Iranian machine tool and instrument companies and institutions. They provided technical assistance, advice, relevant data and in-plant experience, particularly in relation to dimensional measurements in SI Metrics.

Through an acknowledgement of the participating companies that follows, comes a personal "thank you" to each individual who provided photos and other supporting technical documents for that company. The late Manley Hanson who served as managing director, Do All Company's Middle East Area, is singled out for the unusual technical resources provided while in overseas service, with Lincoln Peters, Public Relations Manager, who continued to provide outstanding assistance. A very special thanks goes to Fred Clarkson of the L. S. Starrett Company for supplying many additional illustrations and instruments to the publisher's staff during the final preparation of this text.

The author also expresses his sincere appreciation to the following companies, which generously supplied illustrations, technical data, or other resources and assistance during the preparation of *Basic Machine Technology*:

Acme-Cleveland Corporation
American Tool, Incorporated
B. C. Ames Co.
Armstrong Bros. Tool Co.
Bridgeport Machine Division of Textron Inc.
Brown & Sharpe Manufacturing Company
Buck Supreme Inc.
Burgmaster Division, Houdaille Industries, Inc.
Cincinnati Milacron, Inc.
Clausing Corporation
The Cooper Group
Cushman Industries, Inc.
The Desmond-Stephan Manufacturing Company
Do All Company
Enco Manufacturing Company
Federal Products Corporation
Hardinge Brothers, Inc.
Illinois/Eclipse: A Division of Illinois Tool Works Inc.
The Jacobs Manufacturing Company
Kasto-Racine, Inc.
Kearney & Trecker Corporation
Kennametal Inc.
LeBlond Machine Tool Company
The Lodge & Shipley Company
Morse Cutting Tools Division, Gulf & Western Manufacturing Company
W. H. Nichols Company
Norton Company, Abrasives Marketing Group
South Bend Lathe Company
Standard Tool Division/Lear Siegler, Inc.
The L. S. Starrett Company
TRW Geometric Tool Division
TRW Greenfield Tap & Die Division
Universal Vise & Tool Company
VR/Wesson Division of Fansteel
The Walton Company
Warner & Swasey Company, Turning Machine Division
J. H. Williams Division of TRW Inc.

Thanks is also expressed to Peter J. Olivo, District Manager, Chevron Hoffman Fuel Company, and former master toolmaker, Danbury, Connecticut, and Pat L. Musto, Machine Trades and Technology instructor and industrial consultant, Brooklyn, New York, for technical editing of the complete manuscript; Thomas P. Olivo, Vocational Curriculum Specialist, Vocational Education Media Center, Clemson University, South Carolina, for occupational analysis research and curriculum planning services; Dr. Leon Tunkel, Executive Director, National Occupational Competency Testing Institute, Albany, New York, for logistics support; and to Susan M. Ketcham of NOCTI for secretarial assistance.

C. Thomas Olivo

INTRODUCTION TO BASIC MACHINE TECHNOLOGY

SCOPE AND CONTENT

Basic Machine Technology has a companion *Student's Manual* and a *Teacher's Guide and Answer Book*. The textbook incorporates manipulative skills with the complementary technology. These must be mastered to become a productive worker with promotional capability for a cluster of job opportunities within the work force.

Basic Machine Technology is organized for persons who are to enter the field and others seeking to develop the skills and technology essential to advance in the occupational cluster. The organization and contents at this level deal with theory (why-to-do) and practice (how-to-do). These are related to fundamental measurements (using line-graduated and micrometer measuring tools) and to layout and other bench cutting, forming, and assembly processes.

The machining processes and technology are related to such basic machine tools as power hacksaws and horizontal band machines, drilling machines and accessories, engine lathes, and horizontal milling machines (including the dividing head). Both SI Metric and customary inch standard systems and units of measure are incorporated in all bench and machine tool setups, machining processes, and measurements.

RELATED MATHEMATICS, PHYSICAL SCIENCE, AND BLUEPRINT READING

Basic Machine Technology also incorporates fundamental principles of mathematics, physical science, and blueprint reading and sketching, and design. These are applied in functional ways in shop, laboratory, and design and development activities. The productive worker must be able at each level of job responsibility to use appropriate formulas and make computations. Design, production, and measurement information must be communicated and interpreted accurately from drawings, sketches, design data, and other technical directions.

Physical and chemical phenomena and principles (identified through occupational analyses) are related to layout, tool, cutter, and machine setups, and machining processes. Dimensional accuracy, machinability, surface finish, tool life, safety (to mention a few factors) all depend on a knowledge of basic science and practical applications. Each principle is interwoven into the theory and processes to which it applies.

INDUSTRIAL TERMS AND SAFETY

Shop and laboratory terms represent the technical language and method of communicating that is required and used in industry. Thus, each unit contains technical (occupational) terms. These are expressed and interpreted in the special way in which they are used on the job. The terms are also grouped and restated at the end of the unit. The purpose is to provide a review and for emphasis.

Worker safety and the protection and safe operation of a machine tool and accessory are covered within a unit in relation to a setup or process. CAUTIONS appear throughout the text. The CAUTIONS highlight the safe use of hand tools and orderly procedures for machine tool operations. These are necessary for personal and group safety and for the protection of sensitive and expensive instruments and equipment. Safety factors are again grouped at the end of the unit for further careful analysis and easy reference.

ILLUSTRATIONS AND SUMMARY OF EACH UNIT

Illustrations are included to reinforce the descriptive material. Other art work has been especially prepared as line drawings. These, too, strengthen and extend the written content. The illustrations show a condition, setup, mechanism, tool, or operation in a simplified graphic form.

The technical content of each unit is further crystallized. A series of condensed statements is brought together in the *Summary*.

REVIEW AND SELF-TEST

A series of self-test items are included at the end of each unit. These are designed to serve a two-fold purpose. The answers provide (1) a review of the technology and processes and (2) a measure of how well the content is mastered.

SIGNIFICANT TECHNICAL TABLES

Workers in machine technology and related occupations use industrial reference tables. The tables provide formulas, design, engineering and other important data. These are used for computing dimensions and values that are applied to hand and machine processes. The complete set of tables that are used within the text is contained in the appropriate *Appendix*. Parts of tables are given in some of the units. These provide examples of how the technical information is applied in actual shop practice.

BASIC MACHINE TECHNOLOGY STUDENT'S MANUAL

The *Student's Manual* is designed as a study, assignment, and test item workbook. The units in the *Student's Manual* parallel the content and experiences in the textbook. The *Student's Manual* contains test items that are designed to measure performance objectives. The test items complement the end-of-unit review and self-test items in the text, and may be used for both pretesting and post-testing.

Directed student learning activities are detailed in each unit. These are followed by *Student Test Items*. Both the units and the test items proceed in a logical learning progression.

BASIC MACHINE TECHNOLOGY TEACHER'S GUIDE AND ANSWER BOOK

The units in the *Teacher's Guide and Answer Book* parallel the instructional units in the textbook and *Student's Manual*. The organization of the teach-ing/learning content into units provides flexibility in subdividing the material into a series of lessons and in outlining courses and programs for full-time and part-time instruction.

Solutions to test items, step-by-step work processes, and other guidelines are provided. Answers to safe personal, machine, and tool safety practices are analyzed and stressed.

LEARNING PATTERN AND ORGANIZATION

This textbook and the *Student's Manual* and *Teacher's Guide and Answer Book* consist of a number of *Parts*. These represent major categories within the occupation. The *Parts* are subdivided into *Sections*. Each *Section* contains a grouping of significant related activities that bond together. The skills and technology that relate to each section are described in detail in what are called *Units*. The relationship to be found among the *Units, Sections,* and *Parts* is illustrated by reviewing the contents page of *Basic Machine Technology*.

Restated, *Basic Machine Technology* evolved from occupational, job, and task analyses. Labor market studies of skilled manpower projections and job and career ladders were made. Importantly, these were assessed against skill and technology needs of individuals; patterns of learning and teaching under different organizational systems and structures; the nature, scope and utilization of student and teacher resource materials; and performance measurement criteria, and evaluation and testing techniques.

BASIC MACHINE TECHNOLOGY IN SKILLED MANPOWER/WOMANPOWER DEVELOPMENT

The organization, content and format of *Basic Machine Technology* provides flexibility for use at many levels and under differing learning conditions. Effectiveness and efficiency in human resource development are incorporated in the materials. The series is planned for the following, as well as other, training applications:

- *Basic Machine Technology* may be used for full-time or part-time instruction in an institutional setting, cooperative work-experience program, as related instruction, or any combination of in-plant or institutional training program.

- The *Basic Machine Technology textbook* and *Student's Manual* are geared for use within technical institutes and junior colleges, and vocational industrial/technical schools that have machine technology, metal products manufacturing, mechanical technology, machine trades, and other related curriculums and programs.

- The *Basic Machine Technology textbook* and *Student's Manual* are adapted to inplant industrial training programs, cooperative training programs within union training centers, military occupational specialty programs, and other manpower/womanpower development agency training programs.

- The *Basic Machine Technology textbook* and *Student's Manual* provide resource materials for self-study by craftspersons, technicians, supporting personnel in related fields, training directors, management persons, and others who must have an understanding of the theory, processes, practices, and products associated with machine technology.

PART ONE

MEASUREMENT, LAYOUT TOOLS, AND MEASURING INSTRUMENTS

SECTION ONE

MEASUREMENT AND MEASUREMENT SYSTEMS

Accurate interpretation and measurement of dimensions are skills that every individual involved in machine technology must have. A thorough understanding of dimensions, units of measure, measurement systems, and measurement terms is necessary for proficiency in machine technology. This section presents the basic principles and practices that are foundational to accurate dimensional measurement.

Unit 1
THE UNIVERSAL LANGUAGE OF MEASUREMENT

SI METRIC UNITS OF MEASURE

There are seven basic units of measure. These are incorporated in a system of measurement known as *SI* metrics. *SI* is the abbreviation of the worldwide-accepted International System of Units (Système International d' Unités). Metric units are used in *SI*.

The seven basic units are supplemented by two *supplementary units* and other *derived units*. The basic, supplementary and derived units can be used to describe and measure every physical object and known scientific phenomenon.

The *basic units* deal with (1) *length*, (2) *mass*, (3) *time*, (4) *electric current*, (5) *amount of (molecular) substance*, (6) *temperature*, and (7) *light*. The accepted term for each quantity and the symbol of each basic unit are given in Table 1-1.

DIMENSIONAL MEASUREMENTS

Each measurement is written or expressed as a *dimension*. Whoever designs an object uses a dimension to describe or record a distance or a particular feature. *Dimensional measurements* are the foundation of machine technology and of all industry, business and commerce.

Dimensional measurement relates to straight and curved lines, angles, areas, and volumes. Plane surfaces, as well as regular and irregular-shaped objects, are included.

Dimensional measurements are necessary in designing, building, assembling, operating and maintaining all parts and mechanisms. Some measurements are computed; others are taken directly. Actual measurements may be made with direct-reading measuring rules and other measuring instruments. The degree of accuracy establishes the precision requirements of the measuring tool or instrument.

Measurement also requires the ability to compare values. Two examples are illustrated in Fig. 1-1. A measurement of one millionth part of an inch (0.000001″) is compared with one inch. This relationship is more clearly visualized in inch and mile units of measure. The 0.000001″ to 1″ measurement ratio is approximately the same as 1″ to 16 miles.

Similarly, one one-millionth part of one millimeter is shown in comparison to 25.4 millimeters (one inch). This value has the same relationship as 25.4 mm has to 25.4 kilometers.

There are three prime reasons for dimensional measurement:

● To describe a physical object
● To construct a physical object. (The object may be a single part, a component in a mechanism, or a complete unit of movable and stationary parts.)
● To control the way in which an object is produced by many different people. Controlled measurements are necessary in the manufacture of interchangeable parts.

Dimensional Measurements To Construct An Object

Regardless of how crudely or precisely an object is to be made, there must be some idea of size and form. In creative work and homecrafts these are translated directly by the artist or craftsperson from a concept. The object is shaped until it conforms to a desired size.

Measurements must be taken when a *single* part or

Table 1-1. BASIC UNITS OF MEASURE IN SI METRICS

Basic Units of Measure	Length	Mass	Time	Electric Current	Amount of Substance	Thermodynamic Temperature	Luminous Intensity
Quantity	meter	kilogram	second	ampere	mole	kelvin	candela
SI Symbol	m	kg	s	A	mol	K	cd

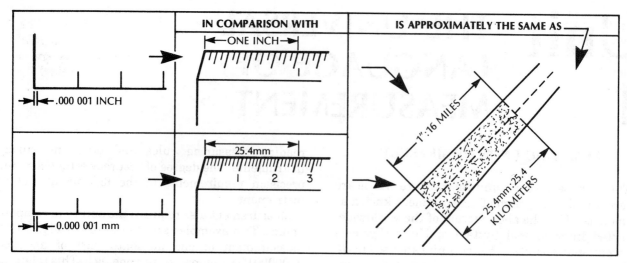

Fig. 1-1. Approximate comparisons of linear measurements.

unit must fit with or move in relation to another part. The *fitting* may be done on a trial-and-error basis for single rough parts. Simple, crude measurements may be taken. However, the dimensional measurements must be controlled when *a number* of the same part are to be produced. The reproduced parts must be of uniform size and shape.

The *function* of the object usually determines the precision of measurement and the surface finish. For instance, direct "rough" measurements may be taken to make one ornamental iron grille for a building. The precision of these measurements differs from those required to mass-produce fine parts for a microminiature device or instrument. The skill of the craftsperson is limited by the ability to measure accurately.

Dimensional Measurements To Control Manufacture

Outside of home arts and crafts, objects are usually mass-produced by others. Many of the parts are *interchangeable*. This requires controlling and measuring dimensions accurately. *Interchangeability* means that parts produced by one or more persons or organizations must interchange. Equally important is the fact that a standard of measurement is used. The standard must be accepted by all parties involved.

In the early years of such inventions as the printing press with movable type, steam engines, and textile machines, the measurements taken were crude by modern standards. Parts were hand-fitted, and surfaces were roughly machined. The efficiency of the machine was low.

Today's automotive, diesel and jet engines; high-speed motors and generators; nuclear power plants;

scientific instruments—in fact, all mass-produced units, machines, and other rotating and mating parts—could not function under the tolerances used years ago. Standards and measurements have become increasingly more precise.

Dimensional measurements must be communicated among a team. Each individual must speak the same language of measurement. Each must be able to translate numerical values. This must take place regardless of geographic location and country. The engineer and designer, the semiskilled worker, skilled craftsperson and technician, and the assembler all use dimensional measurement. Thus, measurement is the universal language for controlling manufacture.

Dimensional Measurements And Technological Progress

The transfer of knowledge in all branches of pure and applied science and among the various levels of craftspersons may be repeated by another only when specific measurements are given.

Technological progress depends on the capability to transfer new knowledge and expertise to the design and construction of new goods and products. This advancement requires group involvement. Each individual must be able to measure within prescribed limits of accuracy. Accuracy relates to a specific system of measurement. As stated before, measurements may be taken in one plant. They also may be spread out among many industries located in different parts of the world.

CHARACTERISTICS OF MEASUREMENT

The most widely used dimensional measurement is

length or linear distance. The term used is *linear measurement*. It means that a distance separating two points may be reproduced. Each linear measurement has two parts: a *multiplier* (Fig. 1-2A) and the *unit of length* (Fig. 1-2B). The multiplier may be a whole number, a fractional or decimal value, or a combination of the two.

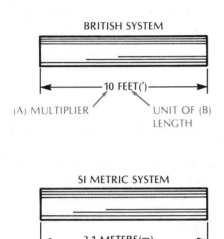

Fig. 1-2. Parts of a linear measurement.

Dimensions, Features And Measurements

Every line of measurement must have *direction*. The bases of linear measurements are pictured in Fig. 1-3. The measurement of an actual dimension begins at a *reference point*. The measurement terminates (ends) at the *measured point*. These two

points lie along the *line of measurement*. The two points are also called *references*. Two such points on one part of an object or between parts are called *features*. The features represent the physical characteristics of the part.

The notes on drawings, and lines that are used for laying out or describing features, are bounded by other lines, edges, ends, or plane surfaces. Three examples of the features of different parts are given in Fig. 1-4. *Features* are also bounded by areas. Measurements are taken from an end or an inside or an outside edge. Most measurement lines are edges that are formed by the intersection of two planes. Measurements are indicated on drawings as dimensions. A *dimension* really states the designer's concept of the exact size of the intended feature. By contrast, a *measurement* relates to the actual size. Two shop terms that are used to identify common features are *male* and *female*. These features appear in Fig. 1-5.

Measurements are also affected by temperature. This fact is particularly important with precise work. In such cases, a feature may measure larger or smaller than a required dimension. Measurement accuracy depends on the temperature of both the measuring instrument and the workpiece.

Precision, Tolerances And Accuracy

Three other dimension-related terms are discussed at this point. *Precision* refers to the degree of exactness. For instance, a part may be dimensioned to indicate a precision of $\pm 0.001''$ or ± 0.02mm. The actual measured dimension, if it *falls within these*

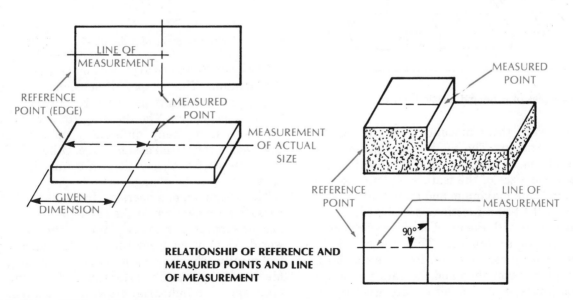

Fig. 1-3. Bases of linear measurements.

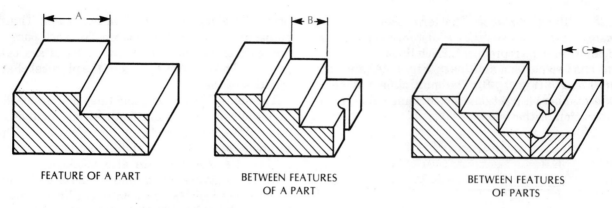

Fig. 1-4. Dimensions and workpiece features.

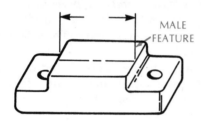

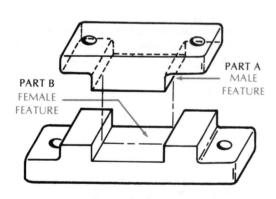

Fig. 1-5. Examples of male and female features.

MEASUREMENT AND INDUSTRIAL DRAWINGS

An object may be described by spoken words or with a photograph or some other form of a picture. The sketch of the die block in Fig. 1-6 is such a picture. These methods of communication do not, however, reveal:

- Exact dimensions
- Cutaway sections which show internal details
- Required machining
- Additional work processes.

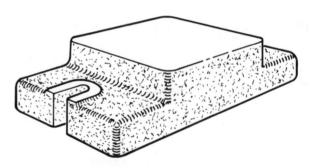

Fig. 1-6. Picture (sketch) of a die block.

The craftsperson needs to rely on precise technical information. The shape, dimensions, and other important specifications must be furnished on a drawing or sketch. The drawing conveys identical information to many different people. Some are involved in the initial design stage. Others deal with the same part or mechanism in a final inspection, assembly or testing stage.

The designers, engineers, technicians, and craftspersons all may work in the same plant. They may also be separated in many plants or throughout the world. All may not speak the same language, except technically. The drawing is the *universal language.* Measurement is an important part of the universal language of an industrial drawing. Measurement is required in the production of every part and device.

limits, may vary these amounts from the size as stated on the drawing. The ± 0.001″ or ± 0.02mm are known as *tolerances.* The tolerance gives the fineness to the range of sizes of the feature. Tolerances are needed to indicate maximum and minumum sizes of parts that will fit and function as designed.

Accuracy deals with the number of measurements that conform to a standard. This number is in comparison with those that vary from the standard. However, in common shop and laboratory practice the terms *precision* and *accuracy* are used interchangeably.

INDUSTRIAL DRAWINGS AND BLUEPRINT READING

An *original* drawing (referred to as a *master drawing*) is usually produced with the aid of drafting instruments. Simple parts sometimes are sketched freehand. Still other drawings are made in automated drafting machines. Statistical and design data provide input for a computer. The computer programs and controls automated drafting equipment. A master drawing may be produced.

Complicated shapes and circuits may also be drawn directly on a workpiece. These layouts guide the worker in performing hand or machine processes and assembly work.

Original drawings are usually filed as a record. Subsequent design changes are made on these drawings. Reproductions are generally made of original drawings for use in the shop and laboratory. The duplicate is referred to as a *print of the original*. While few of these prints are *blueprints*, the name still is applied generally among craftspersons.

By developing an ability to *read a blueprint* or sketch, the craftsperson is able to:
- Form a mental picture of the part
- Visualize and perform different production and assembly processes
- Make additional drawings that may be needed to machine a part
- Take necessary measurements.

Reading a print requires an understanding of basic principles of representation and dimensioning. Combined into what is called *drafting*, the principles include:
- The alphabet of lines
- Views that provide exact information about shape
- Dimensions and notes. These relate to measurements and accuracy
- Cutting planes and sections. These show cross-sectional details of special areas of a workpiece
- Shop sketching techniques.

Drawings and prints are required in the shop and laboratory. Some drawings conform to American standards. Others follow SI metrics. Two typical drawings are reproduced in Fig. 1-7. The placement of views (projection) follows American standards. However, two different dimensioning practices are used: inch standard and inch/millimeter. The drawings are reduced from the *scale* (proportional size) originally used. Fig. 1-7A has *dual dimensioning*. This is where inch standard and metric standard dimensions both are used. Fig. 1-7B provides complete information to produce the support block. American drafting standards and the inch standard unit of measurement are used in Fig. 1-7B.

The craftsperson must be able to work equally well in both systems. Subsequent hand and machine work units include new principles of blueprint reading with applications. These are intended to develop an ability to interpret how each operation is represented on a print. This technology is foundational to the actual taking of measurements and performing each hand and machine process.

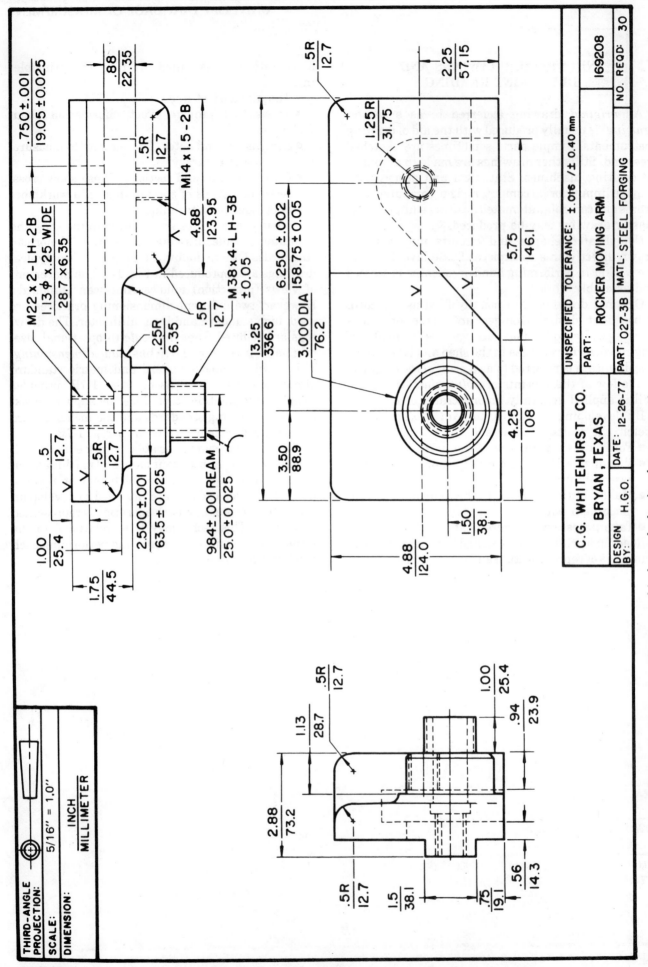

Fig. 1-7(A). A drawing dimensioned in both metric (millimeter) and inch standard units of measurement.

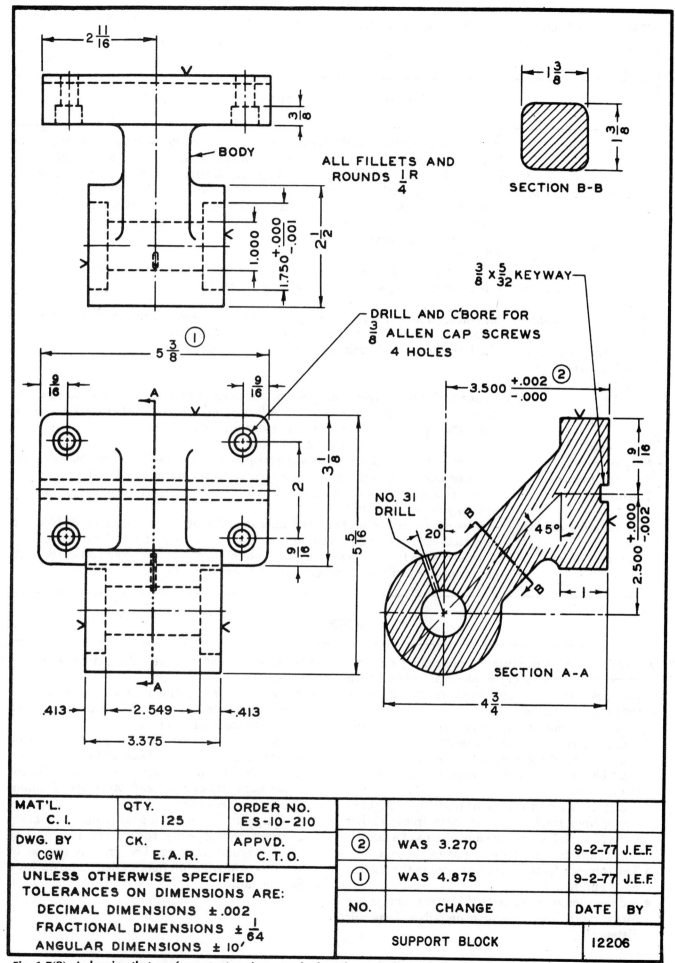

ALL FILLETS AND ROUNDS $\frac{1}{4}$ R

BODY

SECTION B-B

$\frac{3}{8} \times \frac{5}{32}$ KEYWAY

DRILL AND C'BORE FOR $\frac{3}{8}$ ALLEN CAP SCREWS 4 HOLES

NO. 31 DRILL

SECTION A-A

MAT'L. C. I.	QTY. 125	ORDER NO. ES-10-210				
DWG. BY CGW	CK. E. A. R.	APPVD. C. T. O.	②	WAS 3.270	9-2-77	J.E.F.
UNLESS OTHERWISE SPECIFIED TOLERANCES ON DIMENSIONS ARE: DECIMAL DIMENSIONS ±.002 FRACTIONAL DIMENSIONS ± $\frac{1}{64}$ ANGULAR DIMENSIONS ± 10'			①	WAS 4.875	9-2-77	J.E.F.
			NO.	CHANGE	DATE	BY
			SUPPORT BLOCK		12206	

Fig. 1-7(B). A drawing that conforms to American standards and uses inch standard units of measure.

MEASUREMENT TERMS

Interchangeability	The control of surface finish and all features of a part. Regardless of the geographic location of manufacture, each part meets dimensional specifications exactly. Each part may be interchanged with every other similar part to serve an intended function.
SI Metrics, Basic Units	Seven internationally accepted units of measure. All other measurements are derived from the basic units. The basic, supplementary and derived units of measure are adequate to measure all objects and physical phenomena.
Functions of dimensional measurements	● To describe a physical object ● To construct an object ● To control the manner in which others produce a physical object.
Dimensional measurement	The universal language of measuring a dimension.
Linear measurement	Measuring a length between two points along a line of measurement. Expressing a distance in terms of a unit of length and a multiplier.
Reference and measured points (references)	The beginning and termination points of a measurement.
Features	Two definable points from which measurements may be taken. These points establish the physical characteristics of a part. A feature may be bounded by lines, edges and ends. The features may be in the same or different planes.
Dimension	A statement of the exact size of a feature.
Measurement	The actual size of a feature. (A measurement may be affected by temperature change.)
Precision (accuracy)	A stated degree of conformance of a measurement to the standard.
Tolerance	A dimensional range within which a feature or part will perform a specific function and may be interchanged.
Graphic representation	A drawing of a part or mechanism. Dimensions and other specifications are included. Information in visual and written form from which a part may be produced.
Blueprint reading	A general shop term. The accurate interpretation of size, shape and other technical data contained on a reproduced copy of a drawing.

SUMMARY

● Measurement is essential in controlling size and shape when producing a physical object. Each measurement is a statement of the distance between two points.

● Each measurement begins at a reference point and ends at a measured point. Measurements are taken along a line of measurement. The measurement must be related to a measurable characteristic of the feature.

● A feature of a part may be bounded by straight or curved lines, inside or outside edges, areas, or other parts of a solid object. The dimensions of a feature may relate to one part or a number of parts.

● Two common features are described in the shop as male and female.

● A dimension represents the designer's concept of a *perfect part*. The dimension describes a specific distance.

● A measurement provides the actual size of a feature.

● Precision refers to the degree a measurement conforms to a stated standard. Tolerances specify a range of acceptable sizes.

● An *industrial drawing* graphically provides precise information about size, shape and other requirements. Measurements are taken based on the dimensions, tolerances, and other specifications found on a drawing.

● A *blueprint*, in shop language, refers to many different kinds of prints. The print is a reproduction of an original drawing. The print is needed by workers to provide full information for producing a part or unit.

UNIT 1 REVIEW AND SELF-TEST

1. a. Cite two examples of direct measurements.
   b. Give two other examples of computed measurements.
2. State three major functions of dimensional measurements.
3. Describe briefly the relationship between producing interchangeable parts and dimensional measurements.
4. Tell how each of the following measurement characteristics relates to dimensional measurement:
   a. features
   b. reference point
   c. measured point
5. Differentiate between the use of the following measurement terms: (a) *precision,* (b) *tolerance,* (c) *limits,* and (d) *accuracy.*
6. Give three reasons why a craftsperson must be able to read and interpret industrial drawings.

Unit 2 BASIC SYSTEMS OF MEASUREMENT

The circle is recorded in history as the first standard unit of measure. This important concept was noted in Chaldea before the year 4,000 B.C. As civilization advanced, the standard units evolved into systems of measurement. Each new standard has more adequately defined physical objects to ever higher levels of precision.

HISTORICAL CONTRIBUTIONS TO MEASUREMENT

Early Egyptian Measurements

The earliest recorded standard of linear measure was the *cubit*. The Royal Egyptian Cubit was defined as the length of the pharaoh's forearm. This extended from his elbow to the tip of his middle finger. A metal bar was marked to indicate this linear measurement. This bar set the *Royal Cubit* as a standard. The year was about 4,000 B.C. Multiples and subdivisions of the cubit were developed in Egypt between 4,000 and 2,000 B.C. The Egyptians added units such as the *span* of the outstretched hand (nearly nine inches), the *palm (almost three inches),* the *mile* (equal to 1,000 cubits), and others.

Southern European And British Measurements

Around 500 B.C., the Greeks introduced the *stadia*. The stadia was equal to 1/10th of the Egyptian *meridian mile*. A mile covered 1,000 *paces*. Later, in Rome, the *foot* was divided into twelve parts. Thus, the ancient Romans contributed the *inch*.

King Henry I of England established the *yard* in the twelfth century. An iron standard, known as the *iron ulna*, provided a physical unit of length. The *yard* represented the length from the point of King Henry's nose to the end of his thumb. In this same century, the English decreed the standard unit of length to be the inch. The inch equalled the length of *three barley corns laid end to end*. By 1558, Queen Elizabeth proclaimed the length of a special bronze bar to be the standard *yard* of England. Interestingly, this measure was continued for 300 years.

In the eighteenth century French scientists saw the need to develop a *constant standard*. The standard had to be universal and unchanging. By the closing years of the eighteenth century, the *meter* was defined. One meter was equal to one ten-millionth of the distance from the north pole to the equator along a line near Paris. In 1798 a physical standard was produced of the meter length. This became the master reference. It was used to produce duplicate *meter standards*. The meter was divided into units in multiples of ten for smaller measurements.

The International Prototype Meter

Almost 100 years later, in 1889, the meter was redefined. A platinumiridium alloy bar was made one meter in length. The bar was accepted by many nations who participated in the International Conference of Weights and Measures. Thirty similar standard meters were constructed and calibrated with each other. The one most nearly like the standard of 1798 (known as the *Meter of the Archives*) was selected as the *International Prototype Meter*. This particular standard meter bar is located in the Bureau of Weights and Measures, near Paris.

Legalized Use Of The Metric System In The United States

The United States Congress in 1866 legalized the use of the metric system. The yard was defined as being equal to 3600/3937 of a meter. In 1893 the meter was adopted as the reference standard for all length measurements in all measurement systems. As a participant to and as one of the signing nations at the International Conference on Weights and Measures, the United States in 1889 received one of the thirty prototype meter standards (#27). The next year prototype #21 was obtained.

The International Prototype Meter was established by the Mendenhall Act of 1893 as the legal standard for linear units of measure in the United States. The Act also defined the United States inch as being equal to 25.4000508 millimeters. This conversion factor was changed in 1959. It was then agreed that the *international inch* would be equal to 25.4

millimeters.

While the metric unit has been legal in the United States for more than 100 years, the meter has never been mandated as a unit of measure. Successive International Conferences on Weights and Measures have refined the meter. It now is defined in terms of a more precise, fixed value: the wavelength of a particular light. This new constant value makes it possible to accurately reproduce the meter in laboratories anywhere in the world.

THREE BASIC SYSTEMS OF MEASUREMENT

The British (United States) System Of Linear Measurement

The most important measurement at this point is that of *length*, Table 2-1. Units of measure for length are first considered in the British Imperial System. Units in the metric system follow.

Table 2-1. INCH STANDARD LINEAR UNITS OF MEASURE

International Inch(") = Smallest Unit of Measure	
12"	= 1 foot(')
3'	= 1 yard (yd)
5½ yds	= 1 rod
1760 yds	= 1 mile (mi)
5280 ft	= 1 mile (mi)

SI Metrics
1" = 25.4 millimeters (mm)

To repeat, the inch is the most widely used unit of linear measure in the British Imperial System. The inch is subdivided into a number of equal parts. Each part represents a common fraction or a decimal fraction equivalent. These are shown in Fig. 2-1.

Inch Common Fractions—The fractions most commonly used in steel rules represent halves, quarters, eighths, sixteenths, thirty-seconds, and sixty-fourths of an inch. The enlarged illustration in Fig. 2-1A shows how these fractions are related.

Inch Decimal Fractions—The inch is also divided into tenths, fiftieths and hundredths of an inch. In the shop and laboratory these parts are called *decimal fractions*, or *decimal divisions* (Fig. 2-1B.) Decimals are applied to both the British and the metric system basic units of measurement. Fraction-

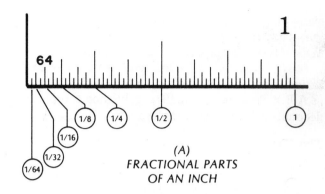

(A)
FRACTIONAL PARTS
OF AN INCH

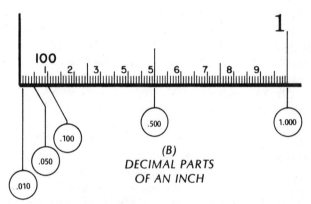

(B)
DECIMAL PARTS
OF AN INCH

Fig. 2-1. Fractional parts of an inch on steel rules.

al divisions that are smaller than 1/64" or 1/100" are not practical to measure with a rule.

A skilled craftsperson often must make a measurement to a closer degree of accuracy than is possible with a steel rule. When a linear measurement requires an accuracy to one one-thousandth part of an inch (.001") or finer (.0001"), measuring tools such as *micrometers* or *vernier calipers* may be used. Metric measurements of 0.02 mm and 0.002 mm are taken with similar metric graduated measuring tools.

The Decimal-Inch System

Back in the 1850s decimals were used in surveying work in the United States. The foot was divided into tenths and hundredths to simplify linear measurements. Up to the 1860s, measurements of ±1/64" were considered in industry to be practical and accurate. However, as new inventions required a number of movable parts, it was necessary to maintain closer tolerances. Accuracies of 1/1,000" and, later, 1/10,000" became accepted precision standards.

In the late 1920s the Ford Motor Company promoted a *decimal-inch system* (Fig. 2-1B). This system was based on the decimal division of the inch. Dimensions were given on drawings with subdivisions of the inch in multiples of ten. Parts were measured according to the decimal-inch system. The term *mil* was coined to express *thousandths*.

Ford's action was accepted by professional organizations such as the Society of Automotive Engineers (S.A.E.). This society in 1946 published a complete decimal-inch dimensioning system.

The Metric System

The units of linear measure in the metric system are similar to those in the decimal-inch system. The difference is in the use of the *meter* instead of the inch. The meter (m) is the standard unit of measure in the metric system. For all practical shop purposes, the *meter is equal to 39.37."* All other units in the metric system are multiples of the meter. The centimeter equals 0.01 meter or 0.3937". The millimeter equals 0.001 m or 0.03937."

Centi and *milli* are prefixes. A series of prefixes is used to denote a fixed relationship to the meter. Some prefixes, as shown in Table 2-2, indicate fractional values. Other prefixes show multiple values greater than one (1). The prefix *milli* (thousandth) is the standard unit of measure in *SI Metrics* (International System of Measurement). The prefix *kilo* means one thousand times; *hecto* means one hundred times; and *deka* means ten times. These prefixes and a number of others are applied to the seven basic units of measure in SI Metrics.

LIMITATIONS OF THE INCH AND METRIC MEASUREMENT SYSTEMS

The metric system has the advantage over the British inch system for ease of computation and communication. However, the inch system provides fractional parts of the basic units that are more functional. For example, there is a significant range of finely spaced divisions that can be read on steel rules graduated in inches and fractional parts.

Readings of 1/64" and 1/100" are common. Micrometer measurements in the British system may be taken with a wide range of discrimination: one one-thousandth and one ten-thousandth part of the inch.

By contrast, the millimeter (milli" = 1/1000; meter = 39.37") is 1/1000 of 39.37", or almost 0.040". The next smaller unit in the metric system, 0.1 mm, is one-tenth this size, or about 0.004". It is almost impossible to graduate or read such fractional measurements on a steel rule. Similarly, for micrometer measurements this same 0.1 mm (0.004") is too inaccurate for most shop measurements.

The next submultiple in SI Metrics is 0.01 mm (the equivalent of almost 0.0004"). A 0.01 mm measurement may require a closer tolerance and more expensive machining than is necessary. In practice, metric measurements are rounded off to the nearest whole or half millimeter value. This produces larger measurement errors than the inch system.

COMMON LINEAR MEASURING TOOLS

The degrees of precision to which parts may be measured are illustrated in Table 2-3. Note that linear measurements may be taken with accuracies of 1/64", 1/100" and 1/2 mm using steel rules. Micrometers, vernier tools and dial indicators are used on precision measurements. Repeating, these are held to one one-thousandth (0.001) and one ten-thousandth (0.0001) part of the inch, or 0.02 mm or 0.002 mm (almost 0.001" and 0.0001"), respectively.

Gage blocks provide accuracies in measuring to within 1/100,000th and 1/1,000,000th part of an inch. Similar accuracies in metric measurements may be made with metric gage blocks. Comparator instruments of high amplification and other measuring devices are used to measure to still finer degrees of accuracy.

Table 2-2. COMMON PREFIXES AND VALUES

Prefix	Multiples (Greater Than 1)				Fractional (Less Than 1)					
	tera	giga	mega	kilo	deci*	centi*	milli	micro	nano	pico
Quantity	1 trillion (10^{12})	1 billion (10^9)	1 million (10^6)	1 thousand (10^3)	$\frac{1}{10}$ (10^{-1})	$\frac{1}{100}$ (10^{-2})	$\frac{1}{1000}$ (10^{-3})	1 millionth (10^{-6})	1 billionth (10^{-9})	1 trillionth (10^{-12})
SI Symbol	T	G	M	k	d	c	m	u	n	p

*While not recommended in SI Metrics, these two prefixes are widely used in the machine, metal, and other industries.

Table 2-3. MEASURING TO VARIOUS DEGREES OF ACCURACY

Common Measuring Tools and Gage Blocks	Degree of Accuracy	
	(inch standard)	(metric standard)
 Steel Rule	$\dfrac{1}{64}$ ", $\dfrac{1}{100}$ "	$\dfrac{1}{2}$ mm
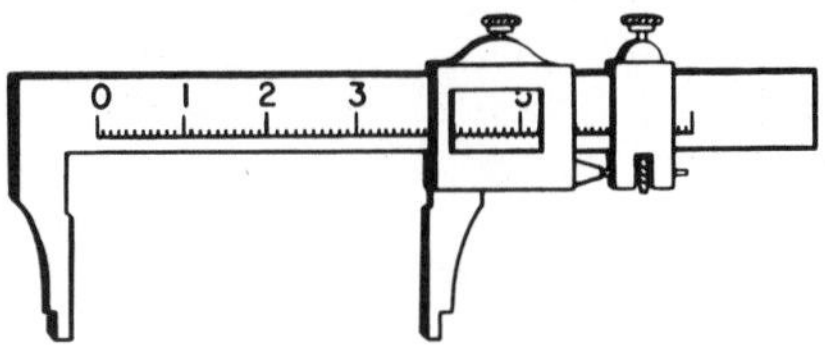 Vernier Caliper	$\dfrac{1}{100}$ ", (0.010") $\dfrac{1}{1,000}$ " (0.001")	$\dfrac{1}{2}$ mm, 0.02 mm
Micrometer · Dial Indicator	$\dfrac{1}{1,000}$ ", (0.001") $\dfrac{1}{10,000}$ ", (0.0001")	0.02 mm 0.002 mm
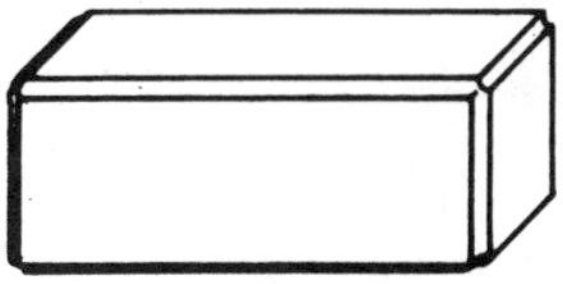 Gage Blocks	$\dfrac{1}{100,000}$ ", (0.00001") $\dfrac{1}{1,000,000}$ " (0.000001")	0.0002 mm 0.00002 mm

COMPARISON OF STEEL RULE SCALES

Three units of measure have been discussed in relation to the steel rule as a line-graduated measuring tool. These units are enlarged and are shown graphically in Fig. 2-2. Graduations that are commonly used in each system are identified. The readings on the decimal-inch scales combine the advantages of the precision of the fractional-inch scale and the readability of the metric scale.

CONVERSION TABLES

Most trade handbooks contain conversion tables for ready use by skilled craftspersons and technicians. Practical daily applications require the changing of fractional and metric values to decimal equivalents, and vise versa. Table 2-4 shows selected examples of the kind of information that is provided by conversion tables.

Conversion tables with accuracies of measurements rounded to five or six decimal places are widely used in shops and laboratories. Expanded tables now include fractional, decimal and metric unit equivalents.

REPRESENTATION OF AN OBJECT

As stated earlier, a working drawing contains full information about the shape, size and other specifications of a part. Examination of a drawing or sketch shows that it consists of a series of lines. These are of different shapes, types, thicknesses and lengths. The lines are used in combination with units of measure.

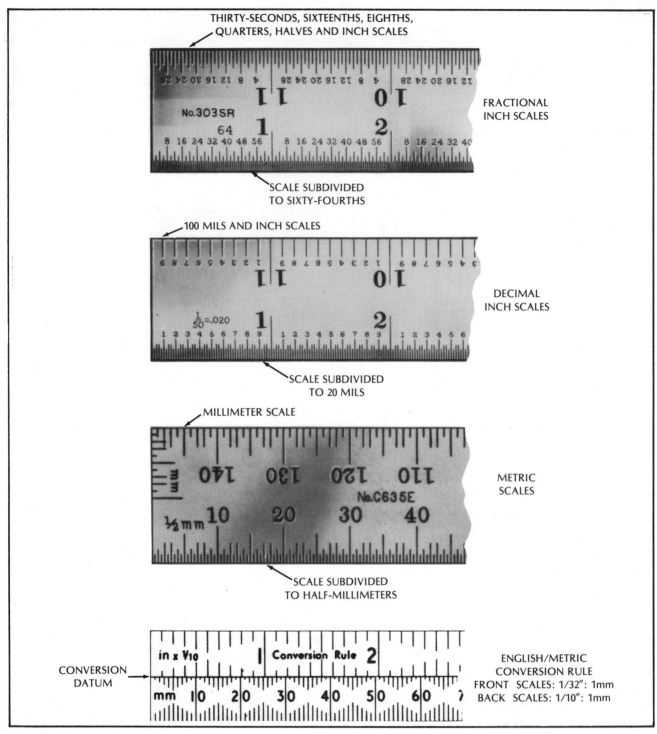

Fig. 2-2. Comparison of graduations on different steel rule scales. (Courtesy of The L.S. Starrett Company)

Table 2-4. PARTIAL CONVERSION TABLE OF STANDARD MEASUREMENT UNITS

Measurement Units			
Millimeter	Decimal Inch	Fractional Inch	Equivalent Decimal Inch
0.1			0.00394
0.2			0.00787
	0.01		0.01000
0.3			0.01181
0.397		1/64	0.01563
0.4			0.01575
0.5			0.01968
	0.02		0.02000
0.6			0.02362
0.7			0.02756
	0.03		0.03000
0.794		1/32	0.03125
0.8			0.03149
0.9			0.03543
1.0			0.03937
	0.04		0.04000
	0.09		0.09000
2.3			0.09055
2.381		3/32	0.09375
2.4			0.09448
2.5			0.09842
	0.1		0.10000
2.6			0.10236
2.7			0.10629
2.778		7/64	0.10938
2.8			0.11023
2.9			0.11417
25.3			0.99606
25.4	1.00		1.00000
25.5			1.00394

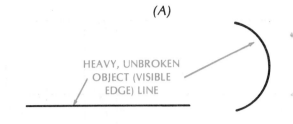

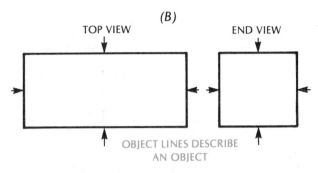

Fig. 2-3. Applications of object lines.

These units are associated with dimensions and the measurement of a feature.

The lines and dimensions make it possible for a worker to visualize and produce a part. An *alphabet of lines* is used as standard in drafting practice. This alphabet has been adopted by the American National Standards Institute (ANSI). The six basic types of lines in the alphabet are:

- Object lines
- Extension lines
- Dimension lines
- Hidden lines
- Center lines
- Projection lines

Object, extension and dimension lines are covered at this point. The other lines are treated in subsequent units.

Object Lines To Describe Features

An *object line* is used to represent the *shape* of a part. The object line (also known as a *visible edge line*) is a heavy, solid line (Fig. 2-3). This line clearly defines the outline of the part.

The thickness of an object line may vary. Comparative weight (thickness) lines are used. The

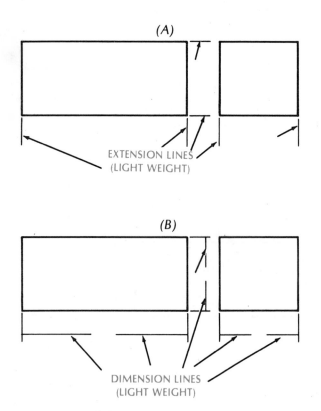

Fig. 2-4. Applications of extension and dimension lines.

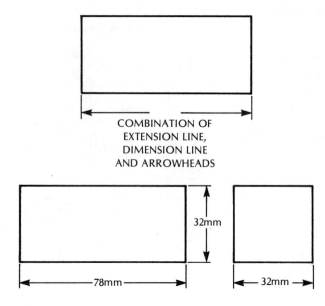

Fig. 2-5. Dimensioned (working) drawing.

weight depends on the size, shape and details of the part to be produced. A solid, heavy object line is shown in Fig. 2-3A. A number of object lines are combined in Fig. 2-3B. These show clearly a rectangular part that is square in cross section (a rectangular solid).

Extension And Dimension Lines

Once the physical shape is *read*, the worker still needs information about size. Thus, the drawing includes dimensions. Dimensioning requires two other types of lines: *extension* and *dimension* lines.

Extension lines are fine, solid lines like those in Fig. 2-4A. Extension lines start close to an object line and extend out from it a short distance.

A dimension line (Fig. 2-4B) connects the extension lines. The end points of a dimension are indicated by arrowheads. These are drawn on the dimension line to indicate the beginning and ending points of a dimension.

Fig. 2-5 shows a combination of object, extension and dimension lines and arrowheads. The drawing tells the worker that (1) the part is 32 mm square by 78 mm long, and (2) the ends are to be machined at right angles to the axis.

MEASUREMENT TERMS

Early Linear Units of Measure (4,000 B.C. to 500 B.C.)	Egyptian: Royal cubit, meridian mile, span, palm
	Greek: stadia, mile
	Roman: inch
Linear Units of Measure (1,200 A.D. to date)	British: yard, foot, inch
	French: meter, millimeter
	International: inch = 25.4 mm
International prototype meter	A standard linear unit of measure representing a constant value. The meter equals the wavelength of a specific light radiated under controlled conditions. A constant length designated as the meter by the International Conference of Weights and Measures.
Basic unit of measure (British Imperial System)	The international inch, which equals 3600/3937 of one meter
Decimal-inch system	Equal divisions of the standard inch in multiples of ten.
Metric system of linear measure	Equal divisions of the meter as the standard unit of linear measure in multiples of 10, 100, 1,000, etc.
Line-graduated measuring tools	Linear measurement tools on which equal divisions of the inch or meter are indicated by equally spaced lines. Measurements are read directly from such measuring tools.
Decimal Prefixes	Part of a system of values that is a multiple of ten and represents a quantity. Each prefix is attached to a unit of measure.
SI Metric Limitations	Measurements with a limited range of discrimination. Maximum degree of accuracy with direct scale and micrometer readings.
Object line	A solid, heavy line that represents the outline, or shape, of an object.
Extension line	A fine, unbroken line used in combination with a dimension line. Extension lines mark the beginning and ending points of a dimension.
Dimension line	A fine, unbroken line. The ends terminate in arrowheads. The arrowheads indicate the limits of a dimension.

SUMMARY

- Civilization's progress has been accompanied by continued refinements of measurement units. These are accepted universally as standards.
- Recorded units of measure dating from 4,000 B.C. to the last few years of the eighteenth century were based primarily on changing quantities.
- Constant quantities and conditions are the foundation of the seven basic units of measure in SI Metrics.
- The meter as the metric system unit was legalized in the United States as far back as 1866. It was never mandated.
- The International Inch was legally established in 1959 to equal 25.4 millimeters.
- The inch is subdivided into fractional parts: 1/64″, 1/32″, 1/16″, 1/8″, 1/4″ and 1/2″.
- The decimal inch incorporates desirable features of both metric and fractional inch measurements. The inch is divided into multiples of 10, 50 and 100.

- The SI Metric basic unit of linear measure is the meter. Subdivisions of the meter are designated by prefixes.
- The common and decimal subdivisions of the inch are more functional to read than are metric subdivisions. The inch subdivisions provide a better range of discrimination than do those of the metric system.
- The steel rule, vernier caliper and micrometer are practical basic linear measuring tools. The range of precision with these tools is from 1/64″ to 0.0001″ and from 1/2 mm to 0.02 mm.
- The shape of an object is shown on a drawing by solid, heavy object lines.
- A dimension is represented on a drawing by extension and dimension lines. Arrowheads locate the end points, or distance, referred to by a dimension.

---

1. Describe briefly how the *constant meter standard* of 1798 revolutionized linear measurements.
2. Identify the smallest practical unit of length that it is possible to measure with a steel rule in (a) the British/United States, (b) decimal-inch, and (c) metric systems.
3. List three instruments or other measuring tools that are used for layout and measuring to accuracies of 0.000 001″ (0.000 025 mm).

4. Name five basic American National Standards Insitute (ANSI) lines that are used by designers to represent the shape, features and dimensions of a part on an industrial blueprint.
5. a. Make a simple front and right-side view sketch of a 76 mm × 125 mm rectangular block (1½″ thick) with a drilled hole in the center. Dimension the part.
   b. Identify five basic ANSI lines on the sketch.

SECTION TWO

MEASUREMENT WITH LINE-GRADUATED INSTRUMENTS

The accuracy with which metal parts are shaped and dimensioned depends, in part, on the craftperson's ability to correctly manipulate and read a variety of semi-precision and precision measuring instruments. This section presents the principals of and correct practices for using semi-precision line-graduated measuring instruments and related transfer measuring tools.

Unit 3

DIMENSIONAL MEASUREMENT WITH LINE-GRADUATED RULES

The terms *scale* and *rule* are commonly used in industry. The difference may be explained by comparing an architectural drafting scale with a machinist's rule. Different divisions (graduations) on the drafting scale may represent different linear measurements. As examples, a drawing may show a scale of 1″ = 1′ or 1 cm = 1 m. By contrast, the graduations on the rule represent actual measurements. These may be in inches and common or decimal fractions. The graduations may also be in millimeters and fractional parts of millimeters in the metric system.

DISCRIMINATION AND MEASUREMENT SCALES

The terms *scale* and *graduation* are used interchangeably in the shop. There are usually four sets of graduations (scales) on a steel rule, one on each edge front and back. These scales are 1/64″, 1/32″, 1/16″ and 1/8″. The scale to use depends on the *discrimination* that is required. *Discrimination* refers to the degree to which a unit of length is divided and the accuracy

of the tool for taking measurements. Fig. 3-1 illustrates the discrimination among international inch, decimal inch, and metric type rules.

For instance, when an inch is divided into 64 equal parts, each division represents 1/64 of an inch. On a metric rule, the centimeter (10 mm) is divided into millimeters and half millimeters. The finest reading that can be taken on this rule has a discrimination of one-half millimeter. Thus, common metric rules have a coarser (less accurate) reading than the 1/64″ scale in the inch system.

ALIGNMENT OF GRADUATIONS

The type of rule and the scale to use depends on the nature of the measurement and the degree of precision required. Five different types of steel rules are illustrated in Fig. 3-2.

The reference points that were described earlier are important in measurement. Measurements with the steel rule must fall along the line of measurement. Either the end of the steel rule or a particular graduation must be *aligned* with the reference point

DISCRIMINATION

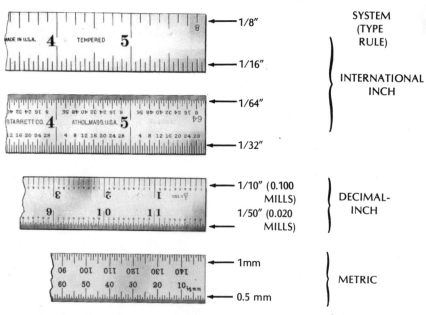

	SYSTEM (TYPE RULE)
1/8″	INTERNATIONAL INCH
1/16″	
1/64″	
1/32″	
1/10″ (0.100 MILLS)	DECIMAL-INCH
1/50″ (0.020 MILLS)	
1mm	METRIC
0.5 mm	

Fig. 3-1. Discriminations of three common types of steel rules. (Courtesy of The L.S. Starrett Company)

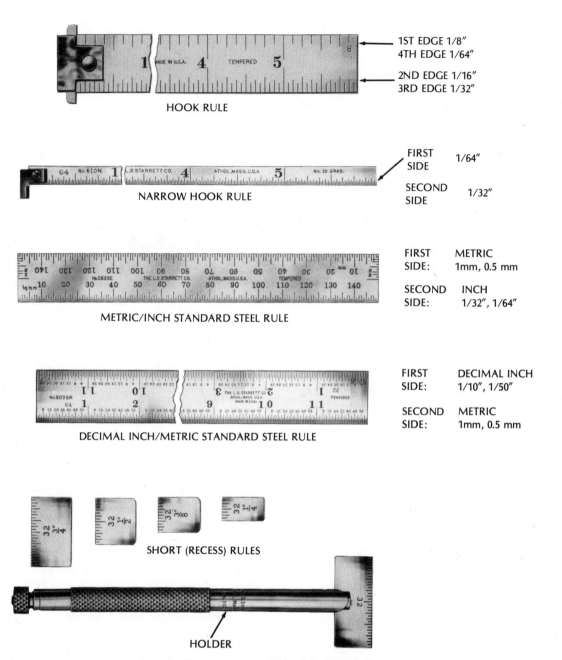

HOOK RULE

1ST EDGE 1/8″
4TH EDGE 1/64″

2ND EDGE 1/16″
3RD EDGE 1/32″

NARROW HOOK RULE

FIRST SIDE   1/64″

SECOND SIDE   1/32″

METRIC/INCH STANDARD STEEL RULE

FIRST SIDE: METRIC   1mm, 0.5 mm

SECOND SIDE: INCH   1/32″, 1/64″

DECIMAL INCH/METRIC STANDARD STEEL RULE

FIRST SIDE: DECIMAL INCH   1/10″, 1/50″

SECOND SIDE: METRIC   1mm, 0.5 mm

SHORT (RECESS) RULES

HOLDER

Fig. 3-2. Common types of steel rules. (Courtesy of The L.S. Starrett Company)

or edge. The measurement is then read at the *measured point.*

Aligned means that the edge or a graduation on a rule falls in the same plane as the measurement reference point or edge.

FLEXIBLE STEEL TAPES

Steel tapes (Fig. 3-3) are another form of line-graduated measuring rule. Steel tapes are more practical for making long linear measurements.

When using steel tapes for precision measure-

ments, be sure to keep the flexible tape on and flush against the line of measurement (Fig. 3-4). Compensation must also be made for expansion and contraction caused by differences in temperature.

Most *pocket size tapes* are designed with a square head of a specified size. This design permits the tape to be used to measure to a measuring point in a confined area.

ERRORS IN READING MEASUREMENTS

Measurements should be taken as accurately as

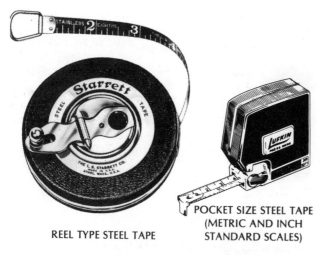

REEL TYPE STEEL TAPE

POCKET SIZE STEEL TAPE (METRIC AND INCH STANDARD SCALES)

Fig. 3-3. Flexible tapes are used for long measurements. (Courtesy of The L.S. Starrett Company, and The Cooper Group)

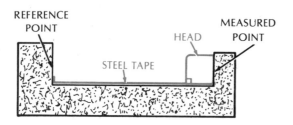

Fig. 3-4. Correct placement of a steel tape along the line of measurement.

possible using the *best scale. Best scale* means that the scale selected gives the most accurate reading depending on the requirements of the job. For example, a rule graduated in 20 mils, or 1/64″, or 1/2 mm is intended for measurements to that degree of accuracy.

Errors in reading measurements are caused by inaccurate *observation,* improper *manipulation,* and worker *bias.*

Observation Errors

Whenever possible, the steel rule should be held at right angles to the workpiece and flush along the line of measurement (Fig. 3-5A.) Under these conditions, the position of the observer does not influence the reading.

Parallax Errors—Parallax is one form of observation error. Parallax relates to the misreading of a measurement. The misreading is caused by the position or angle from which the graduated division is read in relation to the measuring point.

The preferred accurate method of measuring a feature is shown in Fig. 3-5A. The parallax condition

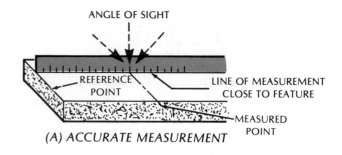

(A) ACCURATE MEASUREMENT

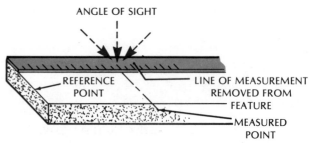

(B) MEASUREMENT ERROR DUE TO PARALLAX

Fig. 3-5. Accurate and inaccurate measurements.

as shown in Fig. 3-5B affects the reading, causing an *observational error.*

Errors In Reading Graduations—Fine graduations, particularly 1/64″ and 1/100″, are another possible source of observational error. In matching such fine graduations with a reference point on a workpiece, the exact dimension may be read incorrectly.

Dimensional Errors Caused By Manipulation

Errors in measurement also may be caused by improperly handling *(manipulating)* the steel rule, the workpiece, or both. Sometimes a rule is *cramped.* Cramping is the result of using excessive force. The rule is squeezed, causing it to bend. This condition is exaggerated in Fig. 3-6A.

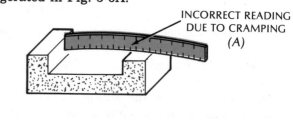

INCORRECT READING DUE TO CRAMPING (A)

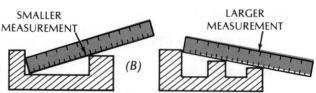

SMALLER MEASUREMENT   LARGER MEASUREMENT

(B)

Fig. 3-6. Common measurement errors caused by incorrect manipulation.

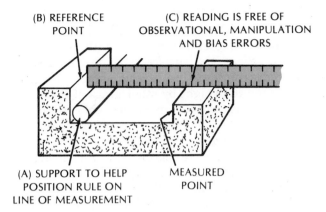

Fig. 3-7. Correct measurement technique.

Another common error is to take a measurement when the rule does not lie on the line of measurement. The two line drawings in Fig. 3-6B show the kind of errors that are caused by incorrect positioning of the rule in relation to the feature to be measured.

There is also the possibility of *slippage*. This is especially true if the part is heavy or if it takes a comparatively long time to make a measurement. A dimensional error is caused by such movement between the measuring tool and the workpiece.

Good measurement practice requires that the rule be positioned on the line of measurement and held against the reference point using a *light touch*. This light touch is developed with repeated practice.

Measurements Errors Caused by Worker Bias

Finally, measurements are affected by the worker's *bias*. *Bias* means that an individual unconsciously reads an error into a measurement. Sometimes this is caused by a person's natural tendency to move either the rule or the line of sight to avoid parallax. It is difficult to manipulate a workpiece by hand, keep the rule along the line of measurement and aligned at the reference point, and, at the same time, see the graduation at the measured point.

HOW TO ACCURATELY MEASURE WITH THE STEEL RULE

Step 1   Select a steel rule with graduations (scale) within the required limits of accuracy.

Step 2   Remove burrs from the workpiece. Clean the rule and work.
Determine whether the measurement is free of observational, manipulative and observer bias.

Step 3   Align the steel rule so the scale edge is along the line of measurement. See Fig. 3-7A. If necessary, use a support piece to ensure that the rule edge remains along the line of measurement.

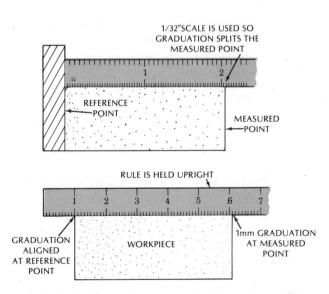

Fig. 3-8. Standard practices for taking accurate linear measurements.

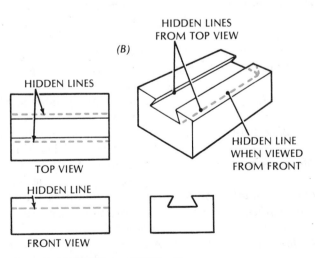

Fig. 3-9. Applications of hidden lines.

Step 4   Position the steel rule with the end held lightly against the reference point (Fig. 3-7B).

Step 5   Read the measurement at the measured point (Fig. 3-7C). For greater precision in measurement, or if a great number of measurements are required, use a magnifier.

Note: If the measurement point does not align with a graduation, turn the rule to another edge having a finer graduation.

Step 6   Recheck the measurement using standard practices as illustrated in Fig. 3-8.

REPRESENTING INVISIBLE EDGES ON A SHOP DRAWING

Many workpieces are designed with one or more "hidden" edges and surfaces. When drawings are prepared according to drafting standards, such edges are covered from sight by other areas of the object. However, the worker must be able to obtain information about hidden edges and surfaces from the drawing alone.

The hidden edges are also known as *invisible edges*. Invisible edges are represented by a series of medium-weight, small-dash lines. These are called *hidden lines*. An example of a hidden line is given in Fig. 3-9A. Fig. 3-9B shows an application of the hidden line.

SAFE PRACTICES IN DIMENSIONAL MEASUREMENT

- Remove burrs from a workpiece. Burrs cause inaccuracies in measurement and unsafe handling conditions. Burrs also affect the accurate fitting and movement of mating parts.
- Recheck all precision measurements.

- Consider the possibility of inaccurate measurement readings. These may result from observational, manipulative, and worker bias errors.
- Make it a practice to check and clean measuring tools and the workpiece surface before taking a measurement.

MEASUREMENT AND MEASURING TOOL TERMS

Scale	The graduations on a rule representing a specific linear dimension.
Steel rule	A common line-graduated measuring tool. A tool used in the shop and laboratory for taking and reading measurements directly from a workpiece.
Discrimination	The degree to which a unit of length is divided and a measurement accurately measures a dimension.
Aligned	In relation to a direct measurement, the edge of the steel rule falls on the line of measurement. Also, a graduation aligns with or *splits* the measured point.
Steel tape	A flexible steel rule of great length. A flexible rule that may be wound into a container or head.
Best scale	The most appropriate scale to use to obtain a measurement within a desired degree of accuracy.
Parallax error	A condition resulting in a dimensional reading error. An error caused by changing the viewing position of a measurement.
Measurement bias	Conditions affecting measurement accuracy. Measurement errors caused by slippage, reading into or assuming a measurement, or other personal judgment.
Observational error	An error caused by faulty reading (viewing) of a measurement.
Manipulative error	A measurement error caused by improper positioning or handling of a measuring tool or workpiece.
Hidden line	A medium-weight line of short dashes. A dash line used in drafting to represent an invisible edge or hidden surface.

SUMMARY

- A steel rule should be selected for linear measurements. The scales should be graduated to the degree of accuracy specified on a drawing. A second consideration is the measurement system. Parts with dimensions given in the metric system should be measured with metric scales.
- Decimal-inch system measurements should be taken with a rule having inch graduations subdivided into decimal values.
- Observational errors in measurement commonly result from parallax errors.
- Manipulative errors in measurement are largely caused by cramping a steel rule, holding the rule at an angle to the reference point, and not aligning the reading with the measuring point.
- Bias errors are personal errors introduced into a measurement by slippage or the tendency to "read into" a measurement.
- Workpieces must be clean and free of burrs, which may cause inaccurate measurements and the improper fitting of mating parts.
- All direct-reading measurements should be rechecked for accuracy.
- An invisible edge or hidden surface is represented on a drawing by a hidden line.

UNIT 3 REVIEW AND SELF-TEST

1. State what graduated scale on a steel rule should be used to measure each of the following dimensions:
   a. $3\frac{1}{64}''$; $2\frac{19}{32}''$
   b. 5.01"; 14.08"
   c. 75 mm; 10.7 mm
2. Indicate three conditions that produce errors in reading measurements on a steel rule.
3. Define the term *aligned* in relation to a direct measurement.
4. Draw a side view of a machined part that has a dovetail cut into the top face and a tapped hole at a right-angle to the top face.
5. State one general practice followed by the craftsperson to check a measurement that has been taken.
6. Give two reasons for removing burrs before measuring and/or fitting two mating parts.

Unit 4
LAYOUT AND TRANSFER MEASURING TOOLS

Layout is a common shop term. Layout refers to preparing a part so that work processes can be performed according to specific requirements. A layout identifies the actual features. Lines are scribed on a workpiece:

- To show boundaries for machining or performing hand operations,
- To indicate the dimensional limits to which material is to be removed or a part is to be formed, and
- To position a part for subsequent operations.

Layout requires the use of three groups of layout, measurement and inspection instruments:

- Group 1) The steel rule with adaptions. These make the steel rule more functional and precise for layout as well as measurement operations
- Group 2) Related tools for the transfer of measurements
- Group 3) Precision tools, gages and instruments.

The basic measuring and transfer tools in the first two groups are covered in this unit. Principles and applications of micrometers, verniers, dial indicators and gage blocks in group 3 are covered in Unit 5.

THE DEPTH GAGE

Many machined surfaces, like those shown in Fig. 4-1, require a depth measurement for a groove, hole, recess or step. Sometimes, it is not practical to use a steel rule. The rule may be too wide, or it may be easy to hold the rule on an angle and read it incorrectly.

One of the common adaptations of the steel rule is the addition of a *head*. The head spans a groove, slot or other indentation and rests on the reference surface. The steel rule slides through the head and may be made to *bottom* at the measured point. This combination of head and rule is called a *depth gage*.

The rules for this type of depth gage are narrow and flat. In other cases, rods are used for small holes where a measurement with a narrow rule is not possible. The rods may or may not be graduated. Examples of the narrow rule and rod types are illustrated in Fig. 4-2.

Some depth gages are marked for angles of 30°, 45° and 60° as shown in Fig. 4-3. The blade may be moved to the right or left. A depth gage with an angle adjustment is useful for making a linear measurement to a surface that is at 30°, 45°, or 60° to the reference surface. Depth gages, however, are *not* intended to be used to make angular measurements.

HOW TO MEASURE WITH A DEPTH GAGE

Step 1   Remove burrs and chips from the workpiece and wipe it clean.

Step 2   Select a depth gage with a blade size that will fit into the groove, hole or indentation. Then, loosen the clamping screw so the blade (rule) may be moved by a slight pressure.

Step 3   Hold the head firmly on the reference surface (Fig. 4-4).

Step 4   Slide the rule gently into the opening until it just *bottoms* (Fig. 4-5). Lock the head and blade.

Step 5   Release the pressure on the head. Then, try to lightly slide the head on the reference surface. *Feel* whether the end of the rule is bottomed correctly.

Step 6   Remove the depth gage and read the depth. If a nongraduated rod is used, place a steel rule next to the rod and read the measurement.

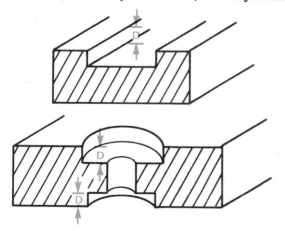

Fig. 4-1. Depth measurements.

28

CAUTION: In good shop practice, the measurement steps are repeated. The rule or rod is checked to see that it is properly bottomed and the measurement is read correctly.

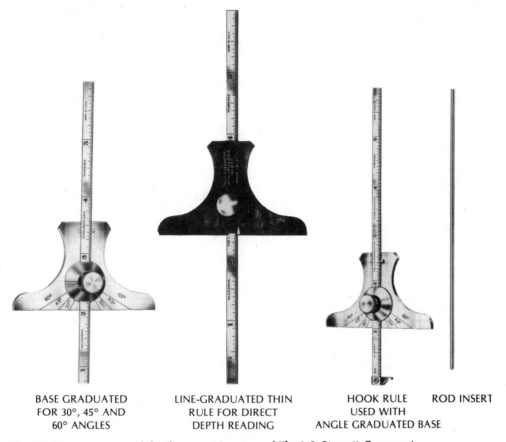

BASE GRADUATED
FOR 30°, 45° AND
60° ANGLES

LINE-GRADUATED THIN
RULE FOR DIRECT
DEPTH READING

HOOK RULE
USED WITH
ANGLE GRADUATED BASE

ROD INSERT

Fig. 4-2. Common types of depth gages. (Courtesy of The L.S. Starrett Company)

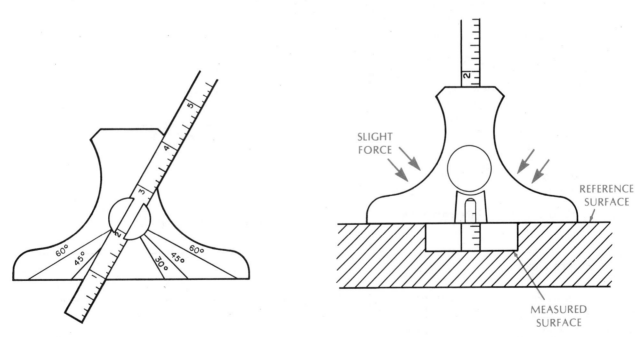

Fig. 4-3. Standard angular adjustments of a rule.

SLIGHT
FORCE

REFERENCE
SURFACE

MEASURED
SURFACE

Fig. 4-4. Head held against the reference surface.

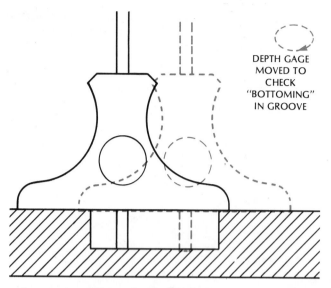

Fig. 4-5. Bottoming a depth gage blade.

THE COMBINATION SET

The *combination set* is an adaptation of the steel rule. The *combination set* includes a *square head*, *center head* and a *protractor head* (Fig. 4-6). The grooved rule is used with each head. The groove permits the rule to be moved into position and clamped.

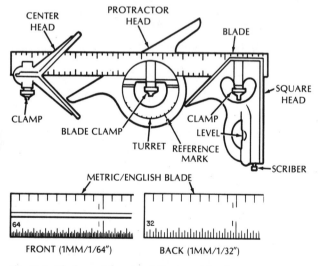

Fig. 4-6. Blade and heads of the combination set.

The Square Head

The *square head* provides a right angle reference. This addition extends the use of the steel rule to laying out right angle or parallel lines. Similarly, the square head and rule may be used to measure the

depth of a hole, indentation, or other linear distance.

In reading a measurement, the head covers a portion of one scale beyond the measured point. Although there are four scales on the steel rule, the locking groove on one side permits direct use of only two scales. The scale that provides the required discrimination should be selected for the head side (the *top* scale in Fig. 4-7). The dimension should be measured at the graduation on the head side (labeled "direct reading" in Fig. 4-7). Sighting to the scale on the opposite side (*bottom* scale in Fig. 4-7) often produces an observational error.

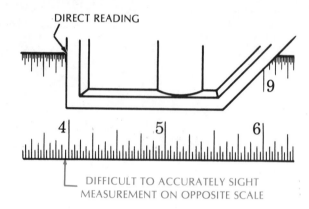

Fig. 4-7. Error in "sighting."

Another feature of the square head is that one of its surfaces is machined at a 45° angle. This angle permits the layout and measurement of many parts that require a 45° angle. This application is illustrated in Fig. 4-8.

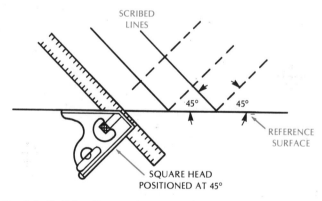

Fig. 4-8. Scribing lines at 45°.

HOW TO MEASURE AND LAY OUT WITH A SQUARE HEAD

Preparation

Step 1   Remove burrs and chips. Wipe the combination square, other tools, and the workpiece clean.

Step 2 Select the scale on the graduated blade that will provide the required accuracy of measurement.

Step 3 Turn the clamping nut so the blade may be moved with a slight force.

Measuring A Length

Step 1 Hold the head firmly against the reference surface.

Step 2 Move the blade until it exactly splits the measured point (Fig. 4-9). Read the dimension on the blade of the square head.

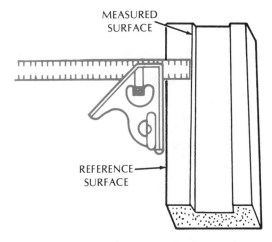

Fig. 4-9. Measuring with a square head.

Laying Out A Line

Step 1 Move the blade until it extends the required distance from the head. This measurement is read on the scale of the blade on the head side. Lock the blade and the square head.

Step 2 Position the head firmly against the reference surface. Scribe the required line.

Step 3 Recheck the dimension for accuracy.

THE CENTER HEAD

The *center head* is designed so that the measuring or layout surface of the rule falls along the centerline of a diameter. The rule may be moved in the center head and locked in any desired position.

The center head is used primarily with round stock and cylindrical surfaces. It provides a quick and satisfactory method of measuring a diameter. An application of the center head for laying out centerlines on a round object is shown in Fig. 4-10.

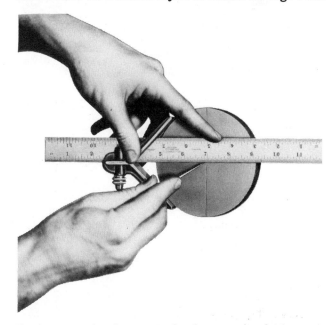

Fig. 4-10. Locating the center using the center head. (Courtesy of The L.S. Starrett Company)

HOW TO MEASURE AND LAY OUT WITH A CENTER HEAD

Laying Out A Center

Step 1 Remove burrs and chips, and wipe the center head and workpiece clean. Check the sharpness of the scriber point.

Step 2 Place the "V" legs of the center head against the outside diameter (surface). Hold it in this position.

Step 3 Scribe the centerline with a sharp scriber. The scriber should be held so the point is as close to the blade as possible (Fig. 4-11A).

Step 4 Turn either the round bar 90° or the center head over (Fig. 4-11B). Hold the center head firmly against the bar.

Step 5 Scribe the second centerline.

NOTE: Sometimes this process is repeated until four centerlines are scribed (Fig. 4-11C). The center is located at the center of the small area. The center of a *boss* (round raised feature) or other round section that is slightly irregular may be found in this same manner.

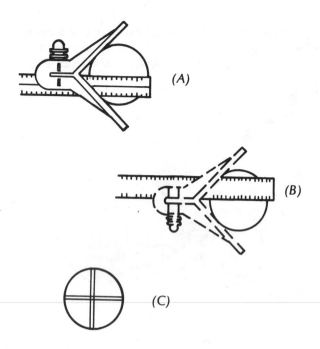

Fig. 4-11. Laying out a center.

Measuring A Diameter

Step 1   Slide the blade in the center head. Bring the end of the blade (rule) flush with the outside diameter of the workpiece.

Step 2   Read the graduation at the measured point. The measurement on the scale of the blade represents the diameter of the workpiece.

THE PROTRACTOR HEAD

The *protractor head* is used to measure angles to an accuracy of 1°. This head contains a turret that is graduated in degrees from 90° to 0° to 90°, a range of 180°.

The base of the protractor head is held against the reference surface (Fig. 4-12). The blade (steel rule) is

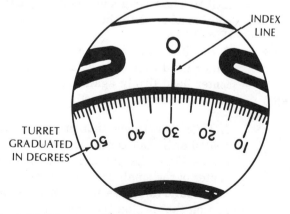

Fig. 4-12. Direct reading of an angle on a protractor head.

secured to the turret. The turret is turned until the included angle of the blade and the protractor head coincides with the angle to be measured. The number of degrees is read at the reference mark (index line) above the turret (Fig. 4-13).

The protractor head is also useful for measurement and layout work when dimensions are to be indicated on angular surfaces.

HOW TO MEASURE AND LAY OUT ANGLES WITH A PROTRACTOR HEAD

Measuring An Angle

Step 1   Remove burrs and chips. Wipe the protractor head, other layout tools, and the work.

Step 2   Insert the blade (rule) so it extends far enough to measure or lay out the required surface.

Step 3   Loosen the turret clamp to permit the turret to be rotated when a slight force is applied.

Step 4   Hold the protractor base firmly against the reference surface of the work (Fig. 4-13). Position the protractor head so that the blade and turret may be moved to the desired angle.

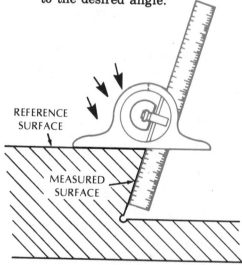

Fig. 4-13. Positioning the protractor base and blade.

Step 5   Bring the blade down gently to the workpiece and adjust it. The desired angle is reached when there is no light showing between the measured surface of the work and the blade.

Step 6   Tighten the clamping screw. Read the graduation on the turret that coincides with the reference mark on the protractor head.

Measurements On An Angular Surface

Step 1 Hold the protractor head securely against the reference surface.

Step 2 Extend the rule until the end of it splits the measured point.

Step 3 Read the measurement directly on one of the scales on the rule.

NOTE: More precise angles, requiring an accuracy within 1/12 of one degree, or 5 minutes (5′), may be measured with a vernier bevel protractor.

A sine bar (described in a later unit) is the most accurate of all angle layout tools. The accuracy of the sine bar accessory is limited by the precision of the measuring tools that are used.

A work surface is usually coated with a color-producing, fast-drying fluid. Layout lines may be scribed easily and are clearly visible on such a surface. The work must be free of oil, grease and burrs. All tools should be examined in advance for any imperfections that can impair their accuracy. This checking is standard practice for all bench work, machine processes, assembly operations and inspection.

TRANSFER INSTRUMENTS

The *caliper* is a tool that is applied to mechanically transfer a measurement. There are four basic forms of calipers. These are shown in Fig. 4-14. One of these calipers is called a *divider*. The other three calipers are the (a) *inside caliper,* (b) *outside caliper,* and (c) the *hermaphrodite caliper.* This last named caliper combines some of the features of the divider and the other calipers.

Technically speaking, calipers are used to *transfer* measurements. These originate at a reference point or surface and terminate at the measured point or surface. Still other calipers, like the slide caliper rule, are made with scales. Measurements are read directly. The slide caliper is treated later in this unit. The more precise vernier calipers are covered in another unit. The simplest caliper types are explained first.

MEASURING AND LAYING OUT WITH THE DIVIDER

The divider is a simple, accurate tool for transfering or laying out a linear dimension. *Parallax* in

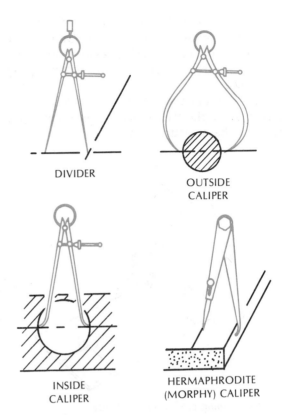

Fig. 4-14. Four basic caliper forms. (Courtesy of The L.S. Starrett Company)

measurement is at a minimum. The sharp points of the divider legs are accurately aligned with measurement lines. The divider legs may be set to a desired dimension with a rule or they may be adjusted to conform to a measurement. The distance between the divider points is then measured on a rule.

It is common practice when setting the divider legs for a measurement to place the point end of one divider leg on a graduation on a rule. Usually, the 1″ or 1 cm graduation on the metric scale is used. The distance to the point end of the other leg is read on the rule (Fig. 4-15). The difference between the scale readings represents the measurement.

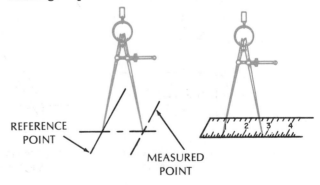

Fig. 4-15. Setting the divider legs.

Circles may be scribed by setting the divider legs at the distance representing the radius. One leg is set in the centerpoint. The other leg is swung around (Fig. 4-16). A full circle or an arc is scribed by applying a slight force.

When a number of parts have the same linear measurement, the divider needs to be set only once to duplicate the measurement.

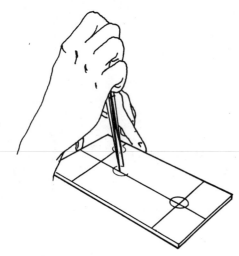

Fig. 4-16. Scribing circles with a divider.

INSIDE AND OUTSIDE CALIPERS

An important point is reemphasized for measurements that are made with inside and outside calipers. The actual line of measurement is established by *feel* in using the tool. For example, when the legs of an outside caliper are adjusted so they just slide over a piece of stock, the measurement taken is on the line of measurement (Fig. 4-17). If the outside caliper cannot slide over the workpiece with this *feel*, or if it is set too large, an inaccurate measurement results. This is to say that accuracy is developed by correct practice. Force applied to a caliper causes the legs to

spring, producing an inaccurate measurement.

Calipers are also used to transfer a measurement to or from a particular feature of the workpiece. This means that either the inside or outside caliper is *set* to a particular size. The setting may be done by measuring the leg distance with a steel rule, micrometer, or gage. A caliper leg may also be set against another surface feature. The degree of accuracy required determines the measuring tool to use. Three common caliper applications are illustrated in Fig. 4-18.

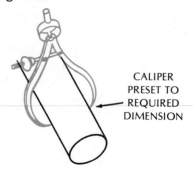

CALIPER PRESET TO REQUIRED DIMENSION

Fig. 4-17. Adjusting the outside caliper.

In turn, the caliper is then used to compare a feature against a specific dimension. An example is given in Fig. 4-19. The size is established by the feel of the caliper legs at the line of measurement. To

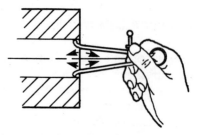

Fig. 4-19. Comparing the dimension of a feature to a preset caliper.

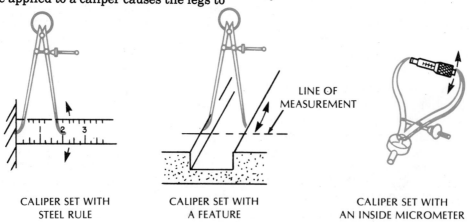

CALIPER SET WITH STEEL RULE

CALIPER SET WITH A FEATURE

LINE OF MEASUREMENT

CALIPER SET WITH AN INSIDE MICROMETER

Fig. 4-18. Common methods of setting calipers

repeat, if any force is applied, the caliper legs will *spring*. The lighter the worker feels the caliper measurement, the more accurate the measurement.

Calipers are usually *rocked* (Fig. 4-20). One leg is held lightly or rests on a surface. The other leg is moved in a series of arcs. This *rocking* movement helps to establish when the caliper is located at the line of measurement. It is important to be able to manipulate the caliper from a comfortable position.

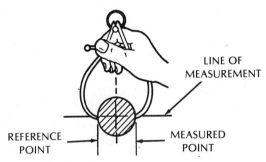

Fig. 4-20. The caliper is "rocked" to establish the line of measurement.

As a transfer tool, a caliper may be set according to the feature of the workpiece that is to be measured. The caliper setting then is measured with a measuring tool. The setting also may be compared with a similar feature or part having the desired measurement.

```
HOW TO SET AND MEASURE
WITH INSIDE AND OUTSIDE CALIPERS
```

Setting An Outside Caliper

Step 1   Hold the steel rule so that it rests on the fingertips of one hand.

Step 2   Hold the outside caliper in the other hand. It should be possible to comfortably turn the adjusting screw with the thumb and the first and middle fingers.

Step 3   Set one caliper leg against the end of the steel rule. Move the second leg so it is as near as possible to the line of measurement.

Step 4   Adjust the second leg until it splits the division of the required measurement on the steel rule. The setting is illustrated in Fig. 4-21.

Step 5   Recheck the measurement by holding the one leg against the end of the rule. The end point of the second leg should be at the measured point.

Fig. 4-21. Setting an outside caliper.

Setting An Inside Caliper

Step 1   Hold the steel rule vertically so the end rests on a plane surface (Fig. 4-22).

Step 2   Place the measuring point of one inside caliper leg on the flat plane surface and against the slide rule.

Step 3   Hold the caliper and turn the adjusting screw until the end point of the second leg splits the measured point. The positions of the hands, rule and caliper are shown in Fig. 4-22.

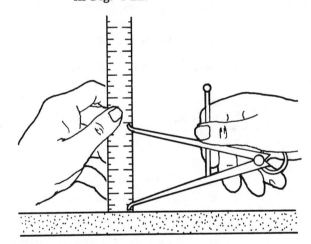

Fig. 4-22. Setting an inside caliper.

Step 4   Move the second leg of the caliper in a slight arc. This is to ensure that the legs are set along the line of measurement at the measured point.

Measuring A Feature With An Outside Caliper

CAUTION: Stop the machine before any steps are taken to caliper a surface. There must be no movement of a cutting tool or of the work.

Step 1   Hold the caliper near the head end (Fig. 4-23). This permits free movement and a sensitive feel as the caliper is moved on the workpiece.

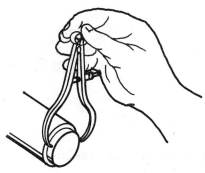

Fig. 4-23. Measuring an outside diameter with a caliper.

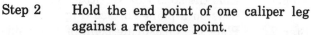

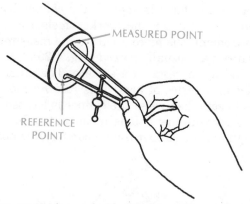

Fig. 4-24. Measuring an inside diameter with a caliper.

Step 2   Hold the end point of one caliper leg against a reference point.

Step 3   Adjust the second end point until the caliper slides easily on the work. Fig. 4-23 shows the correct position of the fingers and the caliper legs.

Step 4   Measure the distance between the two legs.

Step 5   Compare the measurement against the required dimension. Determine the difference in measurement.

Step 6   Move the cutting tool into or away from the workpiece.
NOTE: In turning operations, the adjustment is one half the difference of the measurement.

Step 7   Take the adjusted cut for a short distance (just enough for the caliper legs to measure the part). Stop the machine or hand operation.
NOTE: If a considerable amount of material is to be removed, it may require more than one cut. Also, for precision machining a series of *roughing* cuts is taken. A minimum of *stock* (material) is left for a final, or *finish*, cut.

Step 8   Recheck the depth of cut by resetting the caliper at the line of measurement.

Step 9   Repeat steps 1-8 until the required measurement is reached.

Measuring A Feature With An Inside Caliper

Step 1   Rest one leg of the inside caliper against the reference point.

Step 2   Move the end of the second leg in an arc so it swings across the line of measurement. (Fig. 4-24.)

Step 3   Keep adjusting the caliper until it just touches the measured point of the workpiece.

Step 4   Measure the distance across the caliper legs. Set the cutting tool according to the amount of material still to be removed. If the cut is too deep, move the cutting tool away from the workpiece.

THE HERMAPHRODITE CALIPER

The hermaphrodite caliper (Fig. 4-25) is a layout and measurement tool. It has one leg like that of a divider and one like that of a caliper. The caliper leg may be turned 180° for either inside or outside work. The caliper leg usually rests or rides on the reference surface. The divider leg is used to scribe lines parallel to a surface, find the center of a part, or take measurements.

HOW TO LAY OUT AND MEASURE WITH HERMAPHRODITE CALIPERS

Setting The Caliper

Step 1   Place the caliper leg against a reference point or surface.

Step 2   Adjust the divider point until it splits the measured point.

Scribing Parallel Lines

Step 1   Hold the caliper leg against the reference surface (Fig. 4-25A). Bring the divider leg perpendicular to the reference surface and to the line of measurement.

Step 2   Move the caliper and divider leg at the same time (Fig. 4-25B). Apply a slight force to the divider leg to scribe the parallel line.

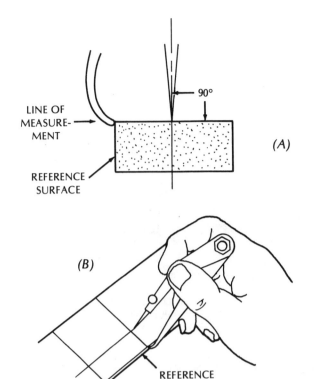

LINE OF MEASUREMENT

REFERENCE SURFACE

(A)

(B)

REFERENCE SURFACE

Fig. 4-25. Scribing a parallel line.

Scribing Arcs

Step 1   Set the divider leg at the required radius.

Step 2   Hold the caliper leg against the outside surface of the workpiece.

Step 3   Swing the divider (scriber) leg in an arc. To find the center of a round surface, turn the workpiece 90° and swing successive arcs. The center of the arcs is the center of the circular area.

Taking A Measurement

Step 1   Repeat the steps for adjusting the caliper legs between the reference surface and the point to be measured.

Step 2   Read the distance across the legs with a steel rule. The dimension may also be checked against a measurement that is to be duplicated.

SLIDE CALIPER RULE

The *slide caliper rule* combines the features of an inside and outside caliper and a rule in one instrument (Fig. 4-26). The slide caliper rule has the advantage of providing a direct reading. The measurement can be *fixed* (locked) at any position.

Locking prevents measurement changes caused by handling. The direct-reading feature eliminates the need to remember a measurement. This is an important feature when a number of cuts are to be taken or a series of operations performed.

However, the slide caliper rule has disadvantages that must be considered at all times:

• The range of use is less than with regular caliper and rule combinations

• The instrument cannot be adjusted to compensate for wear or the springing of the legs

• The range of discrimination is limited by the 1/64″ or 1/2 mm scale.

**HOW TO MEASURE WITH
THE SLIDE CALIPER RULE**

Step 1   Check the accuracy of the jaws by bringing them together. Note whether they are parallel and are not *sprung*. Check the measurement across the jaws with a micrometer or other gage.

Step 2   Release the lock so the jaw may be moved easily. Place one leg at the reference point.

Step 3   Adjust the second leg to the measured point on the line of measurement. Lock the movable jaw and frame. The correct positioning for an outside measurement is shown in Fig. 4-27A. The inside measurement position is shown in Fig. 4-27B.

Step 4   Read the measurement at the graduation on the rule. The measurement appears at either the "IN" (inside) or "OUT" (outside) index mark, depending on how the slide caliper is used.

Using The Slide Caliper Rule As A Gage

Step 1   Set the slide caliper rule to the required dimension. Lock it in this position.

Step 2   *Gage* the internal or external feature as shown in Fig. 4-27A and B. The dimension of the workpiece is correct when the feature *just fits* the caliper setting (the feel between the work and the instrument).

THE SURFACE GAGE

The *surface gage* is a layout and transfer measuring tool. The base is machined with a "V" groove to

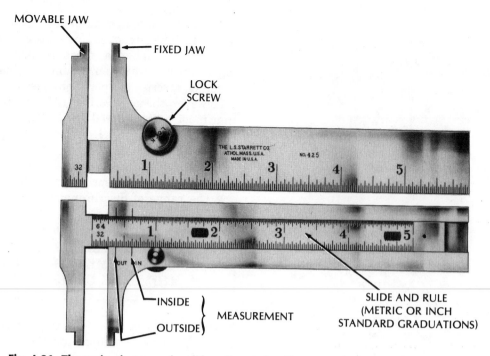

Fig. 4-26. The major features of a slide caliper rule. (Courtesy of The L.S. Starrett Company)

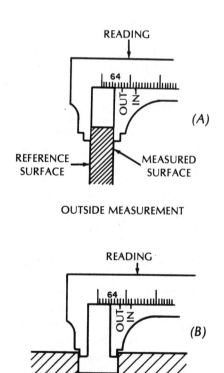

Fig. 4-27. Measuring with a slide caliper rule.

rest on a flat or round surface. An adjustable spindle is provided with a clamp to hold and position a scriber. These main parts are illustrated in Fig. 4-28.

The scriber is set at the approximate required linear measurement. It is then accurately raised and lowered to the required dimension by the adjusting screw. The base also has two *guide pins*. These are

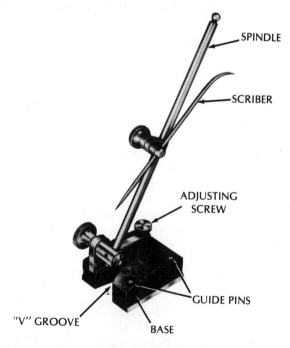

Fig. 4-28. The main parts of a surface gage. (Courtesy of The L.S. Starrett Company)

pushed down when the surface gage is to be used to scribe lines parallel to a reference surface.

When scribing a line, it is easy to spring the scriber if it extends too far from the spindle. For this reason, both the scriber and the spindle should be positioned close to the work.

HOW TO LAY OUT AND TAKE MEASUREMENTS WITH A SURFACE GAGE

Step 1   Check the base to be sure it is free of burrs and clean. The scriber point must be sharp.

Step 2   Position the spindle and scriber. They should be as close to the work as possible for the layout operation.

Step 3   Place the reference face of the combination square on a surface plate (Fig. 4-29). Firmly hold the square head. Slide the blade down until the end touches the surface plate.

Step 4   Hold the surface gage lightly on the reference surface (surface plate). Bring the surface gage close to the square and roughly set the scriber. Turn the adjusting screw to move the scriber point to the required measurement. These adjustments are illustrated in Fig. 4-29.

Step 5   Hold the workpiece securely. Position the surface gage near the work. Scribe the required line on the workpiece.

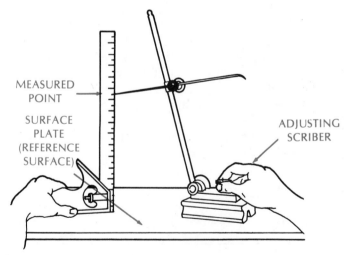

Fig. 4-29. Setting a surface gage to a specified height.

Scribing Lines Parallel To An Edge

Step 1   Lower the guide pins. Hold the surface

gage base against the reference surface (Fig. 4-30).

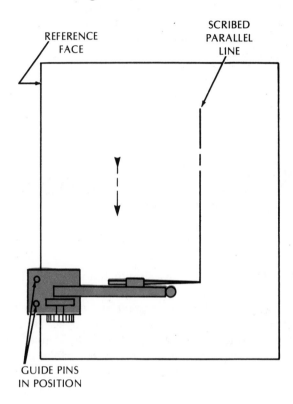

Fig. 4-30. The position of the surface gage for scribing parallel lines.

Step 2   Adjust the scriber to the required dimension.

Step 3   Position and lightly hold the surface gage. Guide it with the pins and move it along the reference surface. Apply a little force to scribe the required parallel line.

Taking a Measurement

Step 1   Position the surface gage and scriber with a minimum of overhang. Adjust the scriber to the approximate measured point.

Step 2   Turn the adjusting screw to position the scriber point exactly on the measured point.

Step 3   Determine the measurement. Read the surface gage measurement with a line-graduated rule. The measurement also may be compared against a gage or other similar part.

THE SURFACE PLATE

The *surface plate* is an auxiliary accessory. As an

accessory it is combined with other tools, measuring instruments and test equipment. It is used universally to provide a reference surface. The surface plate is essential to layout and measurement processes. Fig. 4-31 provides examples of popular manufactured surface plates.

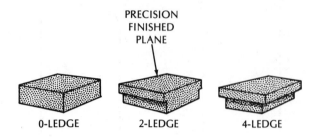

PRECISION
FINISHED
PLANE

0-LEDGE   2-LEDGE   4-LEDGE

Fig. 4-31. General styles of precision granite surface plates.

A surface plate may be a simple, flat plate that has been accurately machined. The term usually refers to a plate that has been machined and scraped to an extreme accuracy or ground to a fine surface finish. The underside is honeycombed with a number of webbed sections. These prevent the surface plate from *warping,* thus providing a permanent flat surface.

The *granite* surface plate is used extensively today. Such a surface plate has a number of advantages over the cast-iron type:
- High degree of surface flatness
- Highly stable. Maintains surface flatness because there is no warpage
- Capability to remain flat, even when *chipped.* Cast iron surface plates, by contrast, tend to *crater* and require rescraping.
- Simplified maintenance. The surfaces are rust-proof. They may be cleaned easily with soap and water, naphtha, and other readily available cleaners.

The size, construction and degree of accuracy of surface plates varies. The size naturally is governed by the nature of the layout, measurement, or assembly operations that are to be performed. The degree of precision required determines: (1) the accuracy to which the flat (plane) surface must be machined and scraped, and (2) the finish of the surface. The flat surface is used as a reference point or surface.

CAUTION: Care and judgment must be exercised in placing measuring tools, instruments and workpieces on the surface plate. All should be clean and burr-free. The surface plate also must be checked continuously for burrs and carefully wiped clean.

SAFE PRACTICES IN USING LAYOUT AND TRANSFER MEASURING TOOLS

- Examine the workpiece, layout tools and measuring instruments. Each item should be free of burrs.
- Use a wiping cloth to clean the workpiece and layout and measuring tools. All chips and foreign matter should be removed.
- Handle all tools and parts carefully. Avoid hitting tools or parts together or scraping them against each other.

- Recheck each layout and measurement before cutting or forming. Each dimension should meet the drawing or other technical specification.
- Use a light touch to feel the accuracy of a measurement at the line of measurement. Force applied to a layout or measuring tool may cause the legs to spread. This produces an inaccurate measurement.
- Replace each layout and measurement tool in a protective container or storage area.

TERMS APPLIED TO TRANSFER AND LAYOUT TOOLS

Layout — Scribing and identifying the location, shape and size of one or more features of a workpiece.

Transfer tools — A group of hand tools and instruments used to indirectly take, transfer, or assist in establishing a measurement. The measurement is usually read on another measuring tool.

Depth gage	A tool for measuring the depth of holes, grooves and slots. A plain or line-graduated depth measurement tool.
Combination set	Additions of square, center and protractor heads to a grooved, adjustable steel rule. These adaptations extend the use and accuracy of the steel rule for layout work and measurements.
Divider	An adjustable layout and measuring tool having two sharply pointed legs.
Caliper	A transfer measurement tool. The two bent legs are adjustable to the size of a feature. Inside calipers are used for internal measurements; outside calipers are used for external features.
Slide caliper rule	A line-graduated instrument which combines the features of an inside and outside caliper and a rule.
Feel (as applied to measurement and layout)	The handling of a measuring or layout tool at the line of measurement. The craftsperson *senses* the physical contact of the tool or instrument at the reference and measured points of the workpiece.
Surface gage	A layout tool having a base with locating pins, an adjustable spindle and a scriber. The surface gage may be set to a particular size for layout work or for the transfer of a measurement.
Surface plate	A true, flat metal or granite plate. A ribbed cast iron plate with a perfect plane as a reference surface. A reference plane used for layout, measurement and assembly operations which require a flat working surface.

SUMMARY

- The steel rule is one of the basic layout and measuring tools. Adaptations of the rule extend its use. The depth gage; combination square, center head and protractor; and slide caliper rule are commonly used in shops and laboratories.
- Layout refers to tools, techniques and processes of marking a work surface. Layout lines show the boundaries of a feature and serve as a guide for hand or machine operations.
- Layout work requires the use of layout and measuring tools, positioning and clamping devices, and other bench-work accessories.
- The accurate machining or fitting of parts or a feature depends on the ability of the craftsperson to *feel*. This is the point where the measuring tool accurately touches the workpiece. Contact is at the reference and measured points on the line of measurement.
- The three general types of calipers include the inside caliper, outside caliper, and the hermaphrodite *(morphy)* caliper. Of the three calipers, the *morphy* caliper may be used for *both* layout work *and* transferring measurements. Inside and outside calipers are used principally to measure internal and external features.
- The surface plate is an accessory used with measurement, layout and other hand tools. It provides a true, finely machined or scraped-plane layout surface.
- A slide caliper rule is used for inside or outside dimensions. The dimension may be preset or the jaws may be adjusted to a dimensional size. The jaws also may be locked at this position.
- Accuracy in layout work and measurement depends on feel and judgment. Tools and workpieces must be clean and free of burrs. They must be protected from damage or inaccuracies caused by forcing, hitting or scraping.

UNIT 4 REVIEW AND SELF-TEST

1. List three functions that are served by layout lines.
2. Identify the tool to use to take the following measurements:
   a. The diameter of a hole
   b. The distance from the face to the bottom of a hole
   c. The vertical height to the bottom of a dovetail slot
   d. A part that is machined with an angle to within 1°
   e. Across the diameter of a round bar of stock.
3. Tell what function transfer instruments serve.
4. Describe how a measurement on an inside caliper may be accurately read on a steel rule.
5. Explain the meaning of using a *scale that provides the necessary discrimination.*
6. Cite three examples of layout processes using an hermaphrodite caliper.
7. Describe briefly how to take an inside measurement with a slide caliper rule.
8. Identify three layout work processes requiring the use of a surface gage.
9. Cite three advantages of granite surface plates over cast iron surface plates.
10. State three safety precautions to take to ensure accuracy in laying out processes.

SECTION THREE

MEASUREMENTS WITH MICROMETERS, VERNIERS, AND INDICATORS

Dimensional measurements that must be more accurate than 1/100″ or 1/2 mm require the use of precision measuring instruments. Such instruments include standard and vernier micrometers; vernier calipers, depth gages, height gages, and bevel protractors; test and dial indicators; and gage blocks. The principles of and correct practices for manipulating and reading these instruments are presented in this section.

Unit 5

MICROMETER and VERNIER PRINCIPLES AND MEASUREMENTS

The line-graduated rule is one of the most widely used and practical measuring tools in industry. Measurements may be taken quickly and read directly. However, the accuracy is limited to 1/64″, 1/100″, or 1/2 mm, depending on whether the inch or metric scale is used.

Many parts or features of a part require a degree of accuracy that is ± 0.001″ (± 0.02 mm) or ± 0.0001″ (± 0.002 mm). The standard micrometer is widely used for precision measurements to within 0.001″ (or 0.02 mm for metric dimensioning). More precise measurements to within 0.0001″ or 0.002 mm are taken with a *vernier micrometer*.

Micrometers are made for direct hand use. They may also be designed as part of another measuring instrument. Of the many different types of micrometers, the three that are most commonly used are covered in this unit. Their functions parallel those of calipers.

The *outside micrometer* is treated first as the basic precision measuring instrument. This is followed by adding the vernier scale to extend the range of precision of the regular micrometer. The shape of the micrometer is then changed to permit taking inside measurements with the *inside micrometer*. The third group of micrometers relates to accurate depth measurements and the *depth micrometer*. The term used for *micrometer* and for micrometer measuring practices in the shop and laboratory is *mike*.

Among the advantages of the micrometer are the following:

• Easier and clearer readability than rules or verniers
• Consistently accurate measurements with close tolerances
• There is no observational error due to parallax
• Portable, comparatively easy to handle, fast operation, and relatively inexpensive
• Design features that include a built-in adjustment to correct for wear.

While today's micrometer includes many refinements, the basic shape and the principle of operation are similar to the early micrometer caliper. In 1877 J. R. Brown and Lucian Sharpe started producing the micrometer in the United States.

MAJOR DESIGN FEATURES

The major parts of the micrometer are identified in Fig. 5-1. The micrometer *anvil* and *spindle* are of a hardened alloy steel. The ends, or measuring surfaces, are finely lapped so they are perfectly parallel to each other. As illustrated, the thread is cut directly on the spindle.

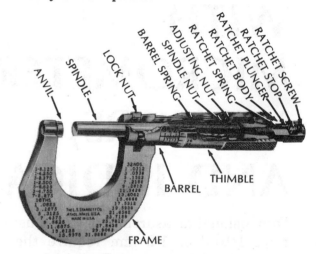

Fig. 5-1. Cutaway view of a 1-inch standard micrometer. (Courtesy of The L.S. Starrett Company)

The *ratchet stop* is attached to the end of the spindle. The spring action in the ratchet provides a constant amount of pressure (force). This action makes it possible to obtain the correct feel for all measurements.

Many micrometers are provided with a *spindle lock*. Rotating the lock nut in one direction *locks* the spindle from turning. Slight rotation in the opposite direction releases the pressure. The spindle is then free to turn. Outside micrometers are available in a wide range of sizes from 1/2″ to 60″ (12 mm to 1500 mm) capacity. The maximum range of measurements that may be taken with a micrometer is usually one inch or 25 millimeters.

The micrometer design provides for wear on the anvil and spindle faces and adjustments due to frame conditions. The thimble and spindle are made in two parts. These may be loosened and the thimble scale

---

CONTROL OF MEASUREMENT CONDITIONS

In every measurement there are four conditions that must be controlled:

- **First condition.** *Control of the part.* This requires manufacturing controls to produce precisely machined surfaces. Accurate measurements may be taken from such surfaces.
- **Second condition.** *Built-in design features.* Such features permit accurate readings to be taken consistently with the measuring instrument.
- **Third condition.** *The craftsperson* and expertise in taking and reading a measurement.
- **Fourth condition.** *Environment.* Precision measurements require adequate temperature controls of the workpiece and measuring instruments. Vibration, dust and other environmental conditions that impair the accuracy of a measurement must be eliminated.

The higher the degree of accuracy required, the closer the four conditions must be controlled.

---

turned and reset to zero at the index line. The two parts are then secured as a single part. There is also an adjusting nut. This may be turned to take up the clearance caused by the wearing of the threads.

MICROMETER PRINCIPLE

The *inch* and *metric* outside micrometers are used for similar measurements. The instruments are handled in the same manner. Direct readings are made on the *barrel* and *thimble scales*. All micrometers operate on the principle that a *circular movement* of a threaded spindle produces an *axial movement*. Thus, as the spindle assembly is turned, the end (face) of the spindle moves toward or away from the fixed anvil. The distance of movement per revolution depends on the *pitch* of the threaded spindle. The

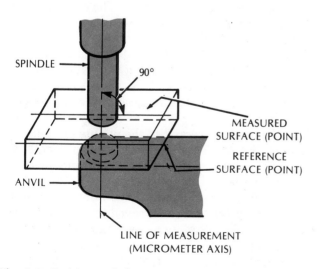

Fig. 5-2. Positions, of the micrometer and workpiece for correct measurement.

SPINDLE

90°

MEASURED SURFACE (POINT)

REFERENCE SURFACE (POINT)

ANVIL

LINE OF MEASUREMENT (MICROMETER AXIS)

pitch refers to the number of threads either per inch or in 25 millimeters.

The workpiece is calipered along its line of measurement. The actual measurement is the distance between the reference and measured points. Fig. 5-2 shows the position of the micrometer and workpiece in taking a correct measurement. The accuracy of the measurement depends on the feel. The feel relates to the contact between the instrument, work, and the operator in a controlled environment.

The Standard Inch Micrometer

The screw thread on the spindle of the standard inch micrometer has forty threads per inch. Its *pitch is* 1/40″, or 0.025″. As the spindle is turned one complete revolution, it moves 0.025″. These conditions are illustrated in Fig. 5-3A and B.

Attached to the spindle is the *thimble*. The *thimble* is also graduated. There are twenty-five graduations around the circumference (periphery) of the thimble. By dividing the movement of the spindle (0.025″) for each revolution by 25, the distance represented by each graduation on the thimble is 0.001″ (Fig. 5-4). For greater readability, the lengths of some of the lines that represent the graduations on the thimble and barrel are varied. For example, since every fourth graduation on the barrel represents 100 thousandths (0.025″ × 4), each hundred thousandth is represented by a longer graduation line. These lines are numbered 0, 1, 2, 3, etc., to 10. The graduations on the thimble are numbered 5, 10, 15, 20 and 0 (for 25). There usually is a smaller sized number for the intermediate graduations. Each graduation on the thimble represents one thousandth of an inch.

The micrometer measurement is represented by

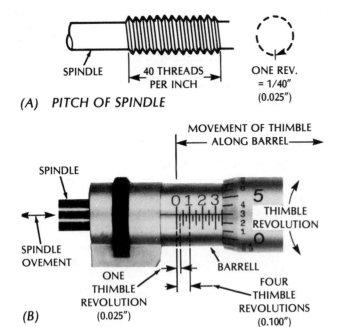

(A) PITCH OF SPINDLE

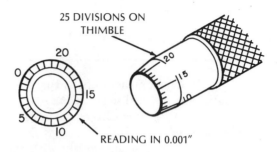

Fig. 5-4. The graduations on the thimble of a standard micrometer.

Fig. 5-3. The relationships between (A) the micrometer spindle pitch (40 threads per inch) and the spindle/thimble movement (1/40", or 0.025", per revolution), and (B) the spacing between graduations on the micrometer barrel (each represents 0.025").

Before proceeding with the steps required to take a micrometer measurement, a number of safe practices must be considered. These are taken to ensure that the micrometer is accurate, it conforms to the measurement standard, and the workpiece and all contact surfaces are properly prepared.

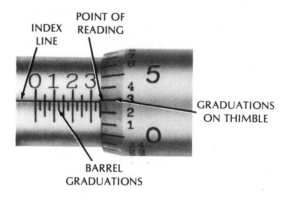

Fig. 5-5. Establishing a micrometer reading.

the reading on the barrel and the thimble. The reading is taken at the point where the index line on the barrel crosses a graduated (or fraction) line on the thimble. This is pointed out in Fig. 5-5. The barrel and thimble readings are added to obtain the micrometer measurement.

SAFE PRACTICES IN USING A MICROMETER OR VERNIER

- Stop the machine if a mounted piece of work is to be measured. Be sure there is no movement of the work.
- Check the workpiece. Remove burrs, dirt, abrasives, and other foreign particles with a clean cloth. Recheck the surface by slowly and lightly moving the hand across the faces to be measured.
- Clean the measuring surfaces of the micrometer. Bring the anvil and spindle faces lightly together on a piece of paper. Gently pull the paper through. Blow away any remaining paper fibres from the faces.
- Check the accuracy of the micrometer. The thimble reading should be *zero*. Any variation requires adjustment. The cap holding the thimble and spindle are unlocked with a micrometer spanner wrench. The thimble is moved gently

until a *zero reading* is obtained. The cap is then locked. The thimble is secured at *zero*.
- Turn the spindle carefully. If it must be moved a long distance, hold the frame in one hand. Gently roll the thimble along the palm of the other hand (Fig. 5-6). This action produces minimum wear on the threads and other bearing surfaces.
- Use the ratchet to equalize the minute pressure that is applied to obtain uniform measurements.
- Feel the measurement at the line of measurement. The feel is established as the anvil and spindle faces are brought to the reference and measured points and are moved on the work.
- Carefully place the micrometer, as well as all other measuring instruments, in a location where they will not be hit, dropped, or otherwise damaged.

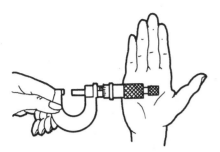

Fig. 5-6. Roll the thimble gently for long spindle movements.

HOW TO MEASURE WITH A MICROMETER

Taking a Measurement

Step 1   Adjust the micrometer opening so it is slightly larger than the feature to be measured.

Step 2   Hold the frame of the micrometer with the fingers of one hand (Fig. 5-7). Place the micrometer on the work so the anvil is held lightly against the reference point.

Step 3   Turn the ratchet (or the thimble directly if there is no ratchet) with the fingers of the other hand. Continue turning until the spindle face just touches the work (measured point).

Note: The axis of the micrometer spindle must be held perpendicular to the reference plane. The feel of the measurement then will be along the line of measurement at the reference points. The correct positions of the hands and micrometer are illustrated in Fig. 5-7.

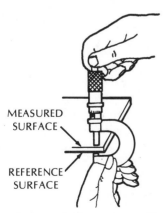

MEASURED
SURFACE

REFERENCE
SURFACE

Fig. 5-7. Correct finger and micrometer positions for taking an accurate measurement.

Reading a Measurement To Within 0.001″

Step 1   Lock the spindle with the lock nut. Note the last numeral that appears on the graduated barrel ("A" in Fig. 5-8).

Step 2   Add 0.025″ for each additional line which shows on the barrel ("B" in Fig. 5-8).

Step 3   Add the number of the graduation on the thimble. ("C" in Fig. 5-8).

Step 4   Read the micrometer measurement in thousandths. The measurement is the *sum* of the barrel and thimble readings ("A", "B", and "C" in Fig. 5-8).

Note: Where a micrometer larger than one inch is used, the total measurement equals the whole number and the decimal fraction read on the micrometer head. For example, with an 8″ micrometer, the reading would be 7″ plus whatever decimal is shown on the micrometer.

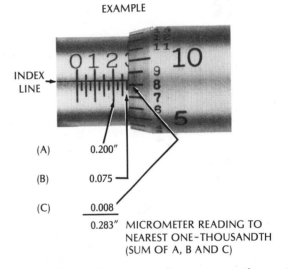

EXAMPLE

INDEX
LINE

(A)   0.200″

(B)   0.075″

(C)   0.008″

0.283″   MICROMETER READING TO
NEAREST ONE-THOUSANDTH
(SUM OF A, B AND C)

Fig. 5-8. Reading a micrometer to the nearest one-thousandth inch (0.001″).

Estimating to Finer Than 0.001″

Step 1   Estimate the distance between the last graduation on the thimble and the index line. For instance. in Fig. 5-9 the distance between "8" and "9" is about one third (0.0003″).

Step 2   Add this decimal fractional value ("D" in Fig. 5-9) to the *thousandths* reading (the sum of "A", "B", and "C" in Fig. 5-9, or 0.283″). The answer gives an approximate four-place decimal value.

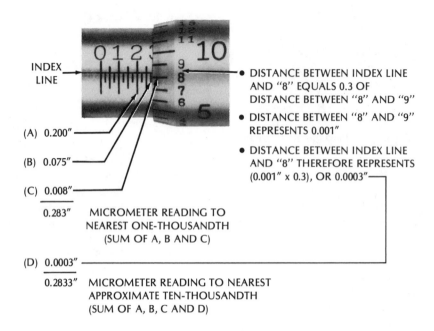

INDEX LINE

● DISTANCE BETWEEN INDEX LINE AND "8" EQUALS 0.3 OF DISTANCE BETWEEN "8" AND "9"

● DISTANCE BETWEEN "8" AND "9" REPRESENTS 0.001"

● DISTANCE BETWEEN INDEX LINE AND "8" THEREFORE REPRESENTS (0.001" x 0.3), OR 0.0003"

(A) 0.200"

(B) 0.075"

(C) 0.008"

0.283"   MICROMETER READING TO NEAREST ONE-THOUSANDTH (SUM OF A, B AND C)

(D) 0.0003"

0.2833"   MICROMETER READING TO NEAREST APPROXIMATE TEN-THOUSANDTH (SUM OF A, B, C AND D)

Fig. 5-9. Reading a nonvernier micrometer to the nearest approximate ten-thousandth inch (0.0001").

Through correct practice, micrometer measurements are read directly and almost automatically. Starting at the index line, the readings on the barrel and thimble are read as a continuous numerical value.

THE VERNIER MICROMETER AND 0.0001" (0.002 mm) MEASUREMENTS

The vernier micrometer (Fig. 5-10) is used to make measurements that are within the one *ten-thousandths of an inch* (0.0001") range or its metric system equivalent (0.002 mm). The vernier principle dates back to 1631 and is named for its inventor, Pierre Vernier. The addition of the vernier scale to the barrel of a micrometer extends its use in making more precise measurements.

The vernier principle is comparatively simple. There are ten graduations on the top of the micrometer barrel. These occupy the same space as nine graduations on the thimble. The difference between the width of one of the nine spaces on the thimble and one of the ten spaces on the barrel is one-tenth of one space.

Each graduation on the thimble represents one thousandth of an inch. Thus, the difference between a graduation on the barrel and the thimble is one-tenth of one thousandth, or one ten-thousandth part of an inch.

The vernier reading is established by sighting across the graduated lines on the barrel and the divisions on the thimble. The line that coincides represents the *ten-thousandths vernier reading*. This reading is added to the three-place (one-thousandth) reading. The result is a four-place (ten-thousandth) vernier micrometer reading.

HOW TO MEASURE WITH THE VERNIER MICROMETER (0.0001")

Standard Micrometer Measurement (0.001")

Step 1   Adjust the micrometer opening to accommodate the work size.

Step 2   Position the micrometer on the work. Bring the anvil and spindle faces into contact with the work. Lock the spindle.

Step 3   Read the one-thousandth measurement. (The sum of "A", "B," and "C" in Fig. 5-11)

Vernier Measurement (0.0001")

Step 1   Sight along the vernier lines on the barrel. Note the graduation that *aligns*

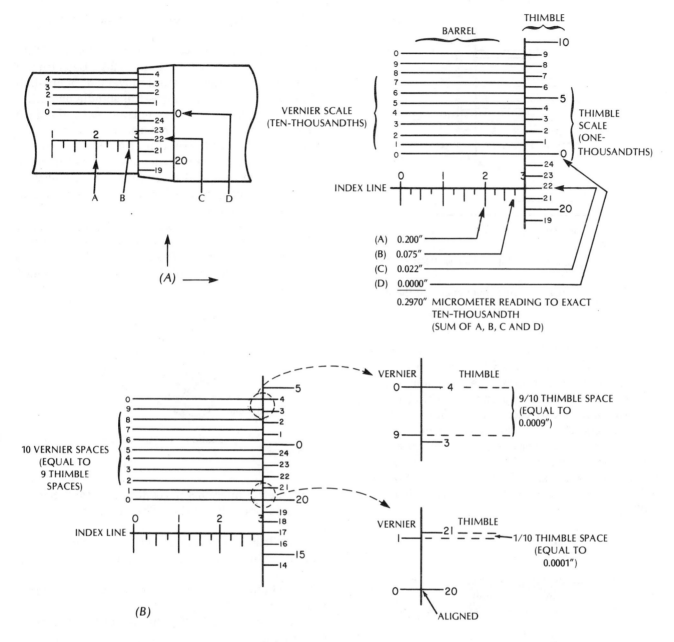

Fig. 5-10. Vernier micrometer scales and a sample measurement.

(coincides) with a graduation on the thimble. In Fig. 5-11 the vernier graduation cuts the 12 thimble graduation.

Step 2   Read the number of the vernier graduation.

Step 3   Combine the vernier reading ("D" in Fig. 5-11) with the thousandth reading. The answer is the four-place decimal reading called the *vernier, or ten-thousandths, reading*. This reading is 0.3813″ in Fig. 5-11.

METRIC MICROMETERS

The principles and the applications of metric micrometers are similar. The difference lies in the basic unit of measure and the discriminating ability of the instrument. For example, international-inch micrometers can be used to measure to one one-thousandth or one-ten-thousandth part of the inch.

By comparison, metric micrometers have a discrimination of one one-hundredth (0.01) and one one-thousandth (0.001) of a millimeter. The 0.01 mm

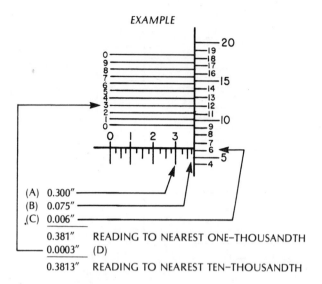

EXAMPLE

(A) 0.300″
(B) 0.075″
(C) 0.006″
0.381″ — READING TO NEAREST ONE-THOUSANDTH
0.0003″ (D)
0.3813″ — READING TO NEAREST TEN-THOUSANDTH

Fig. 5-11. How to read a vernier micrometer to the nearest ten-thousandth inch (0.0001″).

is equal to 0.0004″. This is normally a higher degree of accuracy than is required for some work. At the same time, it is a coarser measurement than is needed on parts where an accuracy within 0.0001″ is practical and most economical. The 0.001 mm measurement (equal to 0.00004″) is beyond the range of accuracy of most workpieces and the instrument itself. *For practical purposes,* the inch vernier micrometer has a discrimination value (0.0001″) that is *four times greater* than the metric micrometer (0.0004″).

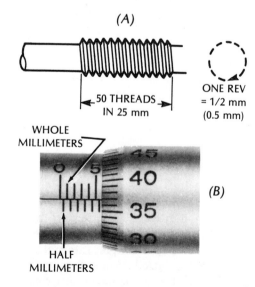

(A)

50 THREADS IN 25 mm

ONE REV = 1/2 mm (0.5 mm)

WHOLE MILLIMETERS

(B)

HALF MILLIMETERS

Fig. 5-12. Relationship between (A) the metric micrometer spindle pitch, the spindle/thimble movement, and (B) the spacing between graduations on the barrel.

Metric Micrometer Graduations and Readings

The screw of the metric micrometer has fifty threads in 25 millimeters (Fig. 5-12A). Each turn of the spindle produces an axial movement of one-half millimeter. Two turns moves the spindle one millimeter (almost 0.040″). In contrast with the inch micrometer, there are two rows of graduations on the metric micrometer barrel (Fig. 5-12B). The top row next to the index line represents whole millimeters. These start at zero, and every fifth one is numbered. The graduations on the lower row represent half millimeters (0.50 mm).

There are fifty graduations on the thimble (Fig. 5-13). One complete revolution of the thimble moves it 1/2 mm (or 0.50 mm). The difference between two graduations on the thimble is 1/50 of 0.50 mm, or 0.01 millimeter.

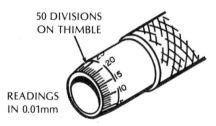

50 DIVISIONS ON THIMBLE

READINGS IN 0.01mm

Fig. 5-13. The graduations on the thimble of a metric micrometer.

The metric micrometer reading is established by three additions (Fig. 5-14): The values to be added are:

• The last whole number of millimeters that is visible in the top row near the thimble ("A" in Fig. 5-14).

• The half-millimeter graduation in the lower row that shows ("B" in Fig. 5-14).

• The graduation on the thimble at the index line ("C" in Fig. 5-14).

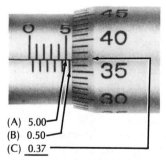

(A) 5.00
(B) 0.50
(C) 0.37
5.87 mm (READING TO NEAREST ONE ONE-HUNDREDTH)

Fig. 5-14. Taking a metric micrometer reading.

Like the inch micrometer, any fractional part that falls between two graduations may be estimated. This estimating adds to the accuracy of the measurement.

**HOW TO MEASURE WITH
THE METRIC MICROMETER**

Step 1   Adjust the micrometer opening to the size of the workpiece. Position the micrometer. Bring the faces into contact with the feature to be measured.

Step 2   Take the measurement. Follow the same precautions and steps as for the regular micrometer. Turn the lock nut when the measurement is reached.

Step 3   Note the number of graduations (whole millimeters) on the top row of the barrel ("A" in Fig. 5-14).

Step 4   Add 0.50 mm if one of the half-millimeter graduations is exposed on the lower row of the barrel ("B" in Fig. 5-14).

Step 5   Note the graduation on the thimble at the index line. Add this value ("C" in Fig. 5-14, or hundredths of a millimeter).

Step 6   Estimate (if needed) any fractional part between two divisions on the thimble at the index line. Add this as the estimated three-place millimeter value.

Step 7   Read the barrel and thimble readings directly. This represents the measurement in millimeters and fractional (decimal) parts of a millimeter.

Step 8   Repeat each step to recheck the feel of the measurement and the accuracy of the reading.

THE DEPTH MICROMETER

The depth micrometer combines the features of a micrometer head and a base. The same operations are performed as with a depth gage having a steel rule blade. The precision of the depth micrometer is within the 0.001″ range on the regular inch micrometer and 1/2 (0.50) mm on the metric micrometer head. While vernier graduations on the head would increase the possible range of accuracy to 0.0001″, such measurements would not be consistently reliable.

Two of the points that were emphasized with the depth gage must be observed with the depth micrometer:

● The base must be firmly held against the reference point. The position of the base should not be disturbed by the rod as it is bottomed at the measured

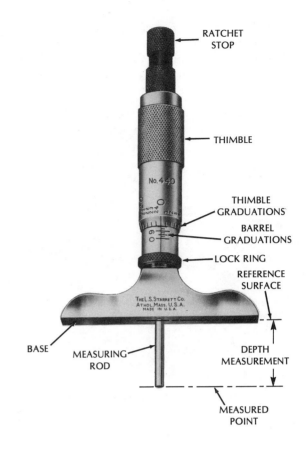

Fig. 5-15. Functional features of the depth micrometer. (Courtesy of The L.S. Starrett Company)

point (Fig. 5-15). The ratchet should be used to obtain an equal pressure for all measurements.

● The surface of the work on which the base rests must be square in relation to the line of measurement. The micrometer and work must be free from burrs, nicks and dirt.

The usual range of measurements for depth micrometers is one inch or 25 millimeters. This range is increased by the addition of interchangeable measuring rods. As with all measuring tools, care must be taken to properly *seat the measuring rods*. The accuracy of the zero setting of the depth micrometer may be easily checked by holding the base against a flat surface. The face (end) of the measuring rod is then brought into contact with the surface, using the ratchet (Fig. 5-16). Any variation from zero should be adjusted following the steps used with a regular micrometer. Gage blocks of specified sizes may be used to check the accuracy along the depth range of the instrument.

Another point to remember with the depth micrometer is that *it measures in reverse* to other micrometers. Fig. 5-17 illustrates the differences in reading a regular micrometer and a depth micrometer: (1) The *zero reading* of the depth micrometer appears when the thimble is at the *topmost position*. (2) The depth micrometer *graduations* are in *reverse order*. (3) The thimble reads *clockwise*.

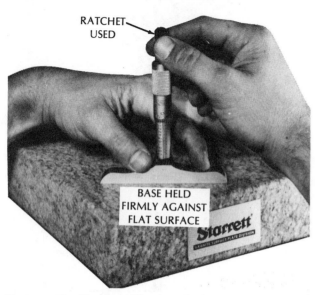

Fig. 5-16. Checking the calibration of a depth micrometer with a granite surface plate. (Courtesy of The L.S. Starrett Company)

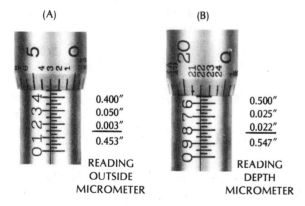

(A)

0.400″
0.050″
0.003″

0.453″

READING OUTSIDE MICROMETER

(B)

0.500″
0.025″
0.022″

0.547″

READING DEPTH MICROMETER

Fig. 5-17. Comparison of outside and depth micrometer graduations and reading techniques.

HOW TO MEASURE WITH THE DEPTH MICROMETER

Step 1   Prepare the workpiece. Select the appropriate length measuring rod. Prepare and test the depth micrometer for accuracy. Then, adjust the measuring rod so it is almost to the depth to be measured.

Step 2   Position and apply a slight pressure to hold the micrometer base against the reference surface.

Step 3   Lower the measuring rod. Use the ratchet to *bottom* the rod.

Step 4   Note the numbered graduation that appears in full at the index line on the barrel. The first number of the reading is one (1) less than the full numeral that is exposed. In the illustration (Fig. 5-18) this measurement is 0.700″.

Step 5   Note the last (0.025″) graduation that is covered (Fig. 5-18). This reading becomes the second and third decimal place reading (0.075″).

Step 6   Read the graduation on the thimble at the index line (Fig. 5-18). This represents thousandths of an inch (0.012″).

Step 7   Combine the readings at the index line of the graduations on the barrel and the spindle. The sum represents the depth in thousandths (Fig. 5-18).

Note: Where extension rods are used, the length of the rod is added to the measurement on the micrometer head.

DEPTH MEASUREMENTS IN METRIC

● Follow the same procedures with the metric depth micrometers as described for the inch-standard type. The measurements will be in millimeters and fractional parts of a millimeter.

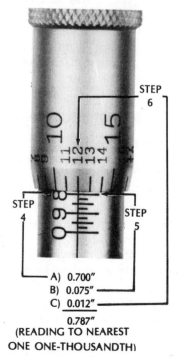

STEP 6

STEP 4

STEP 5

A) 0.700″
B) 0.075″
C) 0.012″

0.787″
(READING TO NEAREST ONE ONE-THOUSANDTH)

Fig. 5-18. How to read a depth micrometer to the nearest one-thousandth inch (0.001″).

INSIDE MICROMETERS

The *inside micrometer* (Fig. 5-19) consists of a micrometer head, a *chuck* and two contact points. Precision extension rods and a spacing collar extend the range of measurements possible with a single micrometer head. A handle is provided for use when small holes or distances are to be measured. A micrometer head with an extension rod set is illustrated in Fig. 5-20.

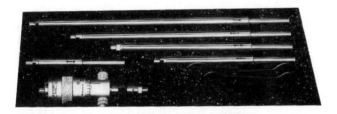

Fig. 5-19. An inside micrometer head with an extension rod inserted. (Courtesy of The L.S. Starrett Company)

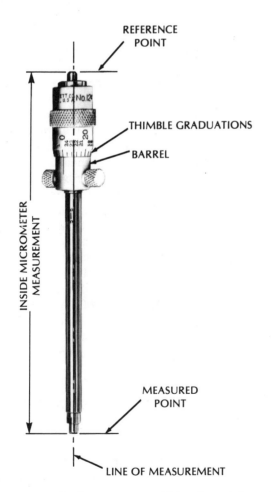

Fig. 5-20. An inside micrometer set, with head, extension rods, and wrenches.

The graduations on the micrometer head may be in thousandths of an inch or in hundredths of a millimeter, and finer. The instrument, like that of all other instruments, must be checked against an accurate standard and adjusted for accuracy.

Fig. 5-21. Taking an inside diameter measurement.

Application of the Inside Micrometer

A measurement is taken by closing the inside micrometer to almost the size of the feature to be measured. The micrometer thimble is turned until contact is established at the measured point along the line of measurement. The micrometer is removed from the work. The reading is made directly by combining the barrel and thimble readings. The features and correct alignment of an inside micrometer for taking an accurate inside measurement are shown in Fig. 5-21.

Because the inside micrometer is handled more, the thimble does not move freely on the barrel as do those of other micrometers. This added friction secures the measurement so that it will not change when the reading is being taken. Sometimes, the micrometer is set to the measurement of the workpiece and then compared with a required measurement. Such measurements guide the craftsperson on the amount of material to be removed and the depth to which a cut is to be set.

The same precautions for checking the work and instrument for accuracy and cleanliness must be taken with the inside micrometer. It is especially necessary to check the shoulders of the extension rods, to seat them accurately, and to lock the rods securely in place.

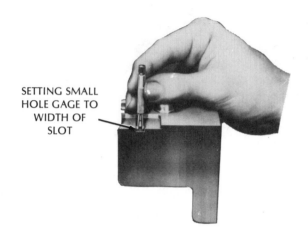

SETTING SMALL HOLE GAGE TO WIDTH OF SLOT

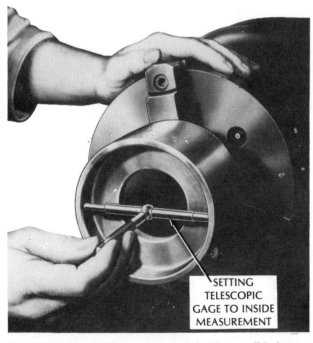

SETTING TELESCOPIC GAGE TO INSIDE MEASUREMENT

Fig. 5-22. Taking inside measurements with a small-hole gage (A) and a telescopic gage (B).

TELESCOPE AND SMALL-HOLE GAGES

The micrometer head size limits its use to minimum internal measurements of close to two

inches or 50 millimeters. For smaller measurements, it is necessary to use transfer measuring tools. Among these are the *telescope* and *small-hole gages*. Measurements with these gages extend down to 1/8" (3 mm). Applications of a small-hole and a telescope gage are given in Fig. 5-22.

The telescope gage (Fig 5-22B) consists of two tubes (legs) that are spring loaded. The legs are locked at the reference points of a measurement by a locking nut or screw in a handle. The measurement is then transferred to and read on a micrometer or other measuring instrument.

HOW TO MEASURE WITH THE INSIDE MICROMETER

Step 1   Prepare the workpiece. Select an inside micrometer of the correct size. Test it for accuracy. Adjust it to the approximate dimension to be measured.

Step 2   Place one leg of the micrometer head against the reference point (Fig. 5-21)

Step 3   Hold the rod end by the fingers of the other hand. Swing a series of arcs in the work to establish the line of measurement and the approximate measured point.

Step 4   Turn the thimble slowly, continuing to move the rod end until light contact is made. The measurement should then represent the correct measured distance.

Step 5   Remove the inside micrometer. Read the measurement at the index line on the barrel (0.125") and the thimble (0.020").

Step 6   Add the length of the extension rod. The total of barrel, thimble and extension rod represents the measurement. (In Fig. 5-21, the barrel reading is 0.125", the thimble reading is 0.020", and the extension rod length is 3.000." Added together, these produce a reading of 3.145".)

Step 7   Retrace the last four steps to recheck the accuracy of the measurement.

LINES USED ON PARTS DRAWINGS

The Centerline

The *centerline* is widely applied on drawings and in layout work. A *centerline* is used to locate the center of a circle, arc, angle, or other linear distance. The centerline also may show that certain features of a part are *symmetrical*. (Symmetrical means that every distance, line, and surface shape is the same on both sides of a centerline.)

A centerline is represented on a drawing by a series of lightweight, alternately long and short dashes (Fig. 5-23). Four common applications appear in Fig. 5-24. Note that the intersection of the centerlines marks the center. Circles and arcs are scribed around a center.

LINE COMBINATIONS

Thus far, five different types of lines have been presented. These are: object lines, hidden lines,

extension lines, dimension lines and centerlines. Most drawings consist of a combination of these lines. The combinations, with dimensions added, usually provide a full description of the shape and size of a part. The five types of lines are combined in Fig. 5-25 to represent an object.

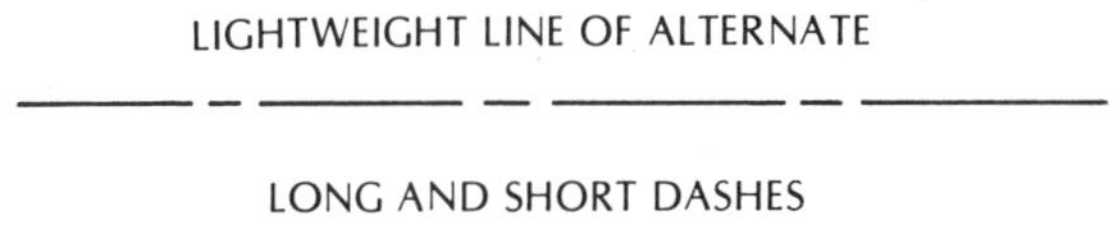

LIGHTWEIGHT LINE OF ALTERNATE

LONG AND SHORT DASHES

Fig. 5-23. Representing a centerline.

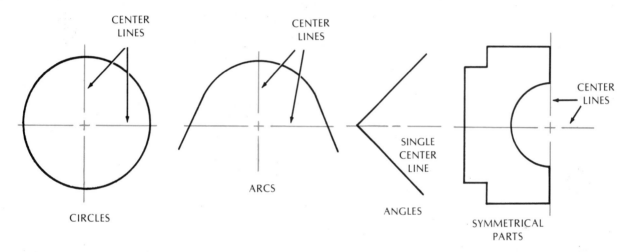

Fig. 5-24. Common applications of centerlines.

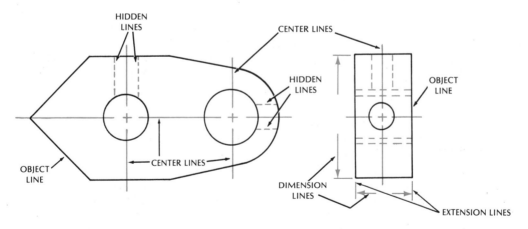

Fig. 5-25. Object, hidden, dimension, and extension lines and centerlines used in combination.

MEASUREMENT TERMS FOR MICROMETERS AND VERNIERS

Basic micrometers	Three basic types referred to as the *depth mike, inside mike* and *outside mike*. The name identifies the measurements that may be taken with each instrument.
Micrometer (Mike) reading	A precise measurement representing the distance between two reference points at the line of measurement. A direct reading which is the total of the barrel, thimble and vernier scale readings. Where applicable, the length of the extension rod is added.
Estimating a fractional part (0.0001″ or 0.002 mm)	Observing the fractional part of a 0.001″ or 0.02 mm division on the thimble at the index line. The addition of the fractional (distance) value to the whole number value on the thimble.
Index line	A horizontal line on the barrel. A fixed point from which micrometer measurements are read.
Telescoping gage	A nongraduated transfer measuring tool. A set of spring-loaded legs that are adjustable for internal measurements. The legs are locked at a correct measurement.
Scale (barrel and thimble)	The graduations on the barrel and thimble of the micrometer head. The barrel scale represents the pitch of the spindle thread. The thimble scale represents 1/25th or 1/50th of the spindle thread pitch (0.001″ or 0.02 mm, respectively).
Control conditions	Control of workpiece production, worker expertise, reliability of the instrument, and environment for precision measuring.
Micrometer head	A precisely machined and graduated adjustable measuring unit. A measuring tool that consistently produces precise measurements. The addition of a frame or base or extension rods extends the use. Inside, outside, depth, and other precise linear measuring instruments are examples.

SUMMARY

- Micrometers are basic measuring tools. Regular micrometers are graduated to measure within 0.001″ or 0.02 mm.
- A vernier scale may be added to the micrometer barrel. The vernier scale makes it possible to read measurements to within 0.0001″ or its metric equivalent.
- Provision is made in the design of the micrometer to compensate for wear of the contact measuring surfaces and the threaded spindle.
- Micrometer measurements are based on the principle that the rotary motion of a screw thread produces an axial linear movement.
- The pitch of the metric micrometer screw thread is 1/2 mm (approximately 0.020″) as contrasted with the 1/40″ pitch (0.025″) for the inch micrometer.
- The barrel of the metric micrometer has two scales: the upper scale (1 mm), and the lower scale (0.5 mm). The inch micrometer has a single row of graduations, each representing 0.025″.

- The accuracy of micrometer measurements (in fact, all measurements) depends on four controls:
  - (1) production of the workpiece
  - (2) accuracy built into the design features of the instrument
  - (3) the expertise of the craftsperson
  - (4) the environment.
- The thimble of the regular metric micrometer is graduated in two-hundredths (0.02) of a millimeter; the inch micrometer thimble, 0.001″.
- A regular micrometer measurement includes the reading on the barrel and thimble. Where extension rods are used, the additional length is added.
- Small inside measurements are usually transferred to a micrometer with small-hole gages or a telescoping gage.
- Layout and machine processes require the use of centerlines. Centerlines are applied to locate center points. Other part features are laid out from a center or centerline.

UNIT 5 REVIEW AND SELF-TEST

1. Cite three advantages of using micrometers over steel rules and verniers.
2. Tell why measurements taken with a micrometer with a ratchet stop are consistently more accurate than those made with a standard micrometer.
3. State three safe practices that must be observed when using a micrometer or vernier caliper.
4. Indicate how a 0.001″ micrometer may be used to estimate a measurement finer than 0.001″.
5. Describe briefly the *vernier principle* as applied to an outside micrometer.
6. Explain the statement: *In practical machin-*

ing, the inch standard vernier micrometer has a discrimination value that is four times greater than a 0.01mm metric micrometer.
7. State how the spindle graduations of a depth micrometer differ from those on an outside micrometer.
8. Explain how to take an accurate measurement with an inside micrometer.
9. Cite an example of where *telescope and small hole gages* must be used instead of inside micrometers or transfer measuring tools.
10. Describe a centerline as used on a drawing and state one of its functions.

Unit 6

VERNIER AND DIAL INDICATING INSTRUMENTS AND GAGE BLOCKS

Measurements in thousandths of an inch or 0.02 millimeters are still the most widely used in industry. Finer measurements are obtained by applying the vernier principle to metric and inch-standard micrometers and other instruments. The vernier graduations (scale) extend the range of fineness of measurement to 0.0001″ (one ten-thousandth of an inch) and 0.002 mm (the metric equivalent of 0.00008″).

Vernier measuring instruments incorporate the vernier principle with special features of measuring tools. The vernier measuring instruments described in this unit are:

- Vernier calipers (inside and outside)
- Vernier depth gage
- Vernier height gage
- Vernier bevel protractor.

The vernier principle is also applied to graduated handwheels and tables on machine tools. In each of these instances, the *vernier* increases the precision to which a machine or a workpiece may be set for machining operations.

Two additional groups of precision measuring and layout instruments also are covered in this unit:

- *Test indicators* and *dial indicators*—These pro-vide for accurate, direct measurements by *amplification*.
- *Gage blocks*—These are combined with vernier measuring and dial indicating instruments for both measurement and layout work. Gage blocks are made to extremely close tolerances. These blocks permit measurements to within one one-millionth part of an inch or equivalent metric values.

VERNIER INSTRUMENTS PARTS AND SCALES

In the mid 1800's, Joseph R. Brown (a cofounder with his father, of the Brown and Sharpe Manufacturing Company) applied the vernier principle to the slide caliper. The resultant series of measuring instruments are identified by names that incorporated the term *vernier* and the name of the type of instrument. The combination of the *vernier* and *height gage* is known as the *vernier height gage*. Similarly, the *vernier depth gage* and *vernier bevel protractor* incorporate the main features of these instruments.

A *vernier scale* generally is imprinted on the adjustable jaw of a measuring instrument. A *beam*

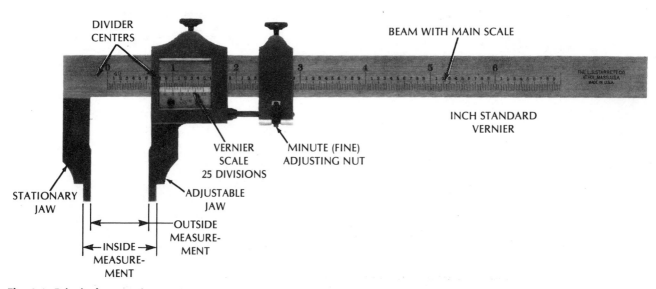

Fig. 6-1. **Principal parts of a vernier caliper. (Courtesy of The L.S. Starrett Company)**

with a stationary jaw contains the *main scale*. The vernier and main scales are designed in a straight line, as for the vernier caliper (Fig. 6-1), vernier height gage and vernier depth gage. These scales also may be included along a curved surface. The vernier bevel protractor is an example.

The readings on the main and vernier scales of angle or linear measuring instruments indicate a particular measurement.

Regardless of how a movable jaw, blade or rule is adjusted, the principle of measurement is the same. Also, the technique by which the instrument is brought into proper contact with surfaces that are to be measured or layed out is similar.

Many different types of vernier measuring instruments have been developed by changing the design of the frame (*beam*), the jaws, or other parts. For example, the bodies of some universal vernier bevel protractors are graduated in whole degrees (Fig. 6-2). The vernier scale on the turret, which holds and positions the measuring blade, is graduated in 5-minute (5') increments. The combination of the main and vernier scale readings (Fig. 6-1) provides direct angular measurements accurate to within 5'.

THE VERNIER PRINCIPLE AND VERNIER CALIPER MEASUREMENT

The vernier caliper differs from the vernier micrometer in construction and method of taking a measurement. The vernier caliper is used for internal (inside) and external (outside) linear measurements. An explanation of the main and vernier scales is made in relation to the inch standard. The same principle also applies to metric measurements in decimal parts of a millimeter. The common *25-division vernier* is used to describe the vernier principle.

Inch-Standard Vernier Caliper

The main scale, on the beam, is divided into inches (Fig. 6-3). Each inch is subdivided into 40 equal parts. The distance between graduations is 1/40", or 0.025". The second scale is the *vernier scale*. It is imprinted on the movable jaw and therefore moves along the main scale. The 25 divisions on the vernier scale correspond in length to 24 divisions on the main scale

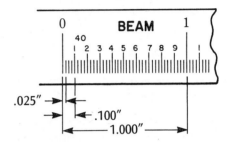

Fig. 6-3. Each inch on the main scale of an inch-standard vernier caliper is subdivided into 40 equal parts.

(Fig. 6-4). The difference between a main-scale graduation and a vernier-scale graduation is 1/25 of 0.025", or 0.001".

When the sliding leg of the vernier caliper is moved until the first vernier scale graduation coincides with the first graduation on the beam, the leg is opened 0.001". This reading is illustrated in Fig. 6-5. As this small movement is continued until the second

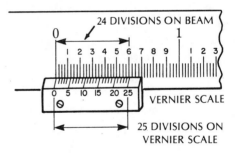

Fig. 6-4. The difference between main-scale graduations and vernier-scale graduations is 0.001".

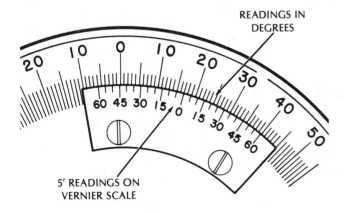

Fig. 6-2. Main and vernier scale graduations on a vernier bevel protractor.

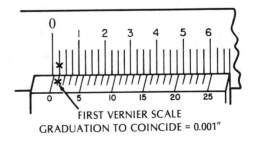

Fig. 6-5. Vernier scale reading of 0.001".

graduations coincide, the opening represents a distance of 0.002″. When the leg has moved 0.025″, the *zero index line* on the vernier scale and the first graduation (0.025″) on the beam coincide (Fig. 6-6). This represents a measurement of 0.025″.

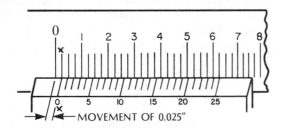

Fig. 6-6. Zero (0) on the vernier scale coincides with a 0.025″ graduation on the main scale.

A vernier caliper reading consists of (1) the number of inches (for dimensions larger than one inch) and (2) decimal parts of an inch. The measurement is read from the beam and vernier graduations. The decimal values on the beam (0.025″ graduations) and the 0.001″ graduations on the vernier scale are added to any whole inch values.

Two other features of vernier calipers are: (1) inside and outside measurements may be taken directly, and (2) an adjustment screw permits the movable jaw to be minutely adjusted to a measurement. The jaw is then locked in position by a clamping screw. Because a vernier instrument is often more difficult to use than a micrometer, the worker must establish which instrument is most practical for the job.

Metric Vernier Caliper Measurements

The applications of the metric and inch-standard vernier calipers are the same. The major difference is that measurements are taken in metric units. The beam of a metric vernier caliper is graduated in millimeters. A graduation on the beam, illustrated in Fig. 6-7, equals 1 mm. Every tenth graduation is marked with a numeral.

The illustration in Fig. 6-7 shows a beam with two scales. The top scale is offset horizontally from the bottom scale to compensate for the width of the vernier jaws. *Internal* (inside) measurements are read on the top scale. The bottom scale has the same graduations as the top. These are one millimeter apart. The bottom scale is used for direct *external* measurements. On the model in Fig. 6-7, inside and outside measurements are read on the same side. Vernier calipers are available with only inch-standard graduations. Others have one scale for metric and a second one for inch-standard measurements.

On the all-metric vernier caliper in Fig. 6-7, the top and bottom vernier scales are graduated in 0.02 mm. Every fifth graduation is numbered. The "10" graduation on the vernier scale represents 0.10 mm. "20" is 0.20 mm, etc.

The metric vernier caliper is read in the same manner as the inch-standard vernier. The decimal fraction of a millimeter on the vernier scale is added to the whole number of millimeters on the beam. The decimal value is established at the point where a vernier graduation coincides with a graduation on the beam.

The inside measurement indicated on the top and vernier scales in Fig. 6-7 is 78.08 mm. This measurement consists of a beam reading of 70 + 8, or 78.00 mm. The vernier reading is the "4" graduation × 0.02 mm or 0.08 mm. The correct measurement is 78.08 mm. The *external reading*, taken from the bottom scale, is 70.08 mm.

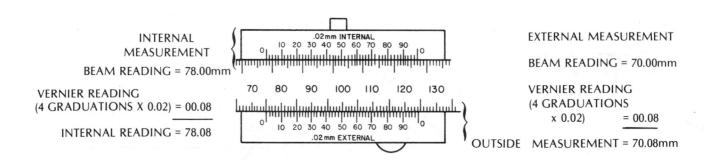

Fig. 6-7. Beam and vernier scales with metric graduations.

HOW TO READ A VERNIER CALIPER MEASUREMENT

Taking The Measurement

Step 1   Check the workpiece and the vernier caliper. They must be clean and free of burrs.

Step 2   Check the vernier caliper in the fully closed position. The reading should be zero (0).

Step 3   Loosen the clamping screws. Slide the movable jaw. The distance from the stationary jaw should be slightly smaller than the inside measurement to be taken. The jaw opening should be a little larger for an outside measurement.

Step 4   Bring the vernier caliper to the feature of the workpiece that is to be measured. Make "rough" adjustments, if needed, then, lock the clamping screw.

Step 5   Hold the solid leg to the reference point. The beam is brought to a position parallel to the line of measurement.

Step 6   Bring the movable jaw into position at the measured point. Turn the fine adjusting nut until the caliper jaw just slides over the measured point. Lock the movable jaw at this point. Recheck the feel of the measurement.
          Note: The correct position of a vernier caliper is shown in Fig. 6-8. The linear dimension is accurately measured along the line of measurement.

Reading The Measurement

Step 1   Read the 0.025" graduation(s) on the beam (main scale). The graduations appear to the left of the "0" on the vernier

scale. (Each graduation equals 0.025".) Multiply 0.025 by this number.
Note: The example in Fig. 6-9A shows a beam reading of 0.075".

Step 2   Sight along the vernier scale to establish which vernier line (graduation) coincides with a line (graduation) on the beam.
          Note: In the example in Fig. 6-9B, the vernier scale reading is 0.100" + 0.005", or 0.105".

Step 3   Check the whole-inch number on the beam scale (for measurements of one inch and larger).

Step 4   Establish the overall vernier measurement. Add the whole-inch number, the 0.025" reading on the beam scale, and the one-thousandths indicated on the vernier scale.
          Note: In the example given in Fig. 6-9C, the overall reading is 4.287".

Step 5   Retake the measurement and recheck the reading for accuracy. Then, clean the instrument and safely store it.

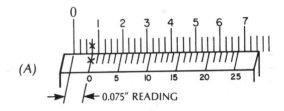

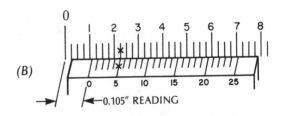

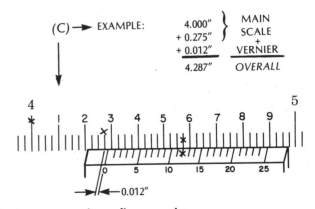

Fig. 6-9. Steps in reading a vernier.

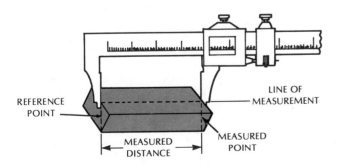

Fig. 6-8. Position of the vernier caliper for an accurate measurement.

MEASURING WITH THE VERNIER DEPTH GAGE

The vernier depth gage (Fig. 6-10) is essentially a depth gage. Locking screws, an adjusting nut, a rule, and a vernier scale are added. The rule in the inch-standard system is graduated in 0.025″.

Measurements are taken with the same care as for a regular depth gage with a line-graduated rule or a depth micrometer. The adjusting screw permits the blade to be brought against the work surface and locked at this position. However, taking a precise accurate measurement is difficult.

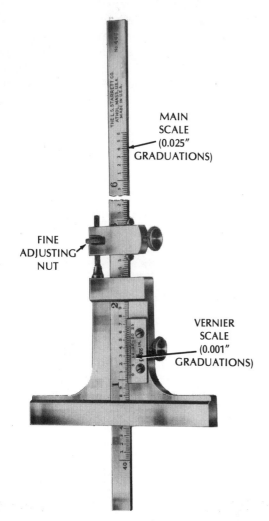

MAIN SCALE (0.025″ GRADUATIONS)

FINE ADJUSTING NUT

VERNIER SCALE (0.001″ GRADUATIONS)

Fig. 6-10. The main features of a vernier depth gage. (Courtesy of The L.S. Starrett Company)

MEASURING WITH THE VERNIER HEIGHT GAGE

The vernier height gage (Fig. 6-11) is widely used in layout work and for taking measurements to an

accuracy of 0.001″ or 0.02 mm. The three main parts are the base, the column (beam), and the slide arm. The main scale is on the column. The vernier scale is attached to the slide arm.

Measurements are taken in the same manner as with other vernier measuring instruments. Whole inch and 0.025″ (or centimeter and fractional millimeter) readings are obtained from the graduations on the beam column. The additional decimal values in multiples of 0.001″ (or 0.02 mm) are read on the vernier scale. These readings are taken at the two graduations that coincide. The vernier height gage measurement consists of the (1) inch, (2) 0.025″ graduations to the left of the zero index line on the vernier scale, and (3) the vernier scale reading in thousandths (0.001″).

There are a number of attachments used with the vernier height gage. A flat scriber (Fig. 6-12A) or an offset scriber (Fig. 6-12B) may be secured to the slide arm. This combination is used in layout work or for making linear measurements. Depth measurements may be taken with a depth-gage (Fig. 6-12C) or offset attachment. Many other special attachments are used for different measurements.

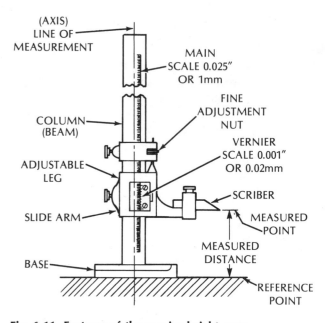

(AXIS) LINE OF MEASUREMENT

MAIN SCALE 0.025″ OR 1mm

COLUMN (BEAM)

FINE ADJUSTMENT NUT

VERNIER SCALE 0.001″ OR 0.02mm

ADJUSTABLE LEG

SLIDE ARM

SCRIBER

MEASURED POINT

MEASURED DISTANCE

BASE

REFERENCE POINT

Fig. 6-11. Features of the vernier height gage.

HOW TO CHECK THE VERNIER HEIGHT GAGE

The vernier height gage is usually used in combination with a surface plate. The surface plate and instrument should be checked for burrs and nicks. The vernier height gage should then be

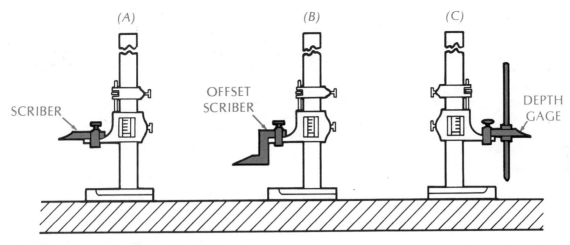

Fig. 6-12. Attachments for the vernier height gage.

carefully placed on the surface plate. Apply a slight force on the base. Test whether the instrument rests solidly or if there is any slight movement. A movement indicates that the instrument is not seating properly. It should be rechecked for burrs or other damage to the base.

Step 1   Use the flat extension arm. Securely clamp it to the slide arm of the instrument.

Step 2   Select a gage block of one inch or 25 mm size.

Step 3   Slide the extension arm over the gage block. Adjust the arm until the measured point is reached.

Step 4   Read the vernier measurement at this low end of the height gage. Compare the accuracy of the measurement with the dimension of the gage block.

Step 5   Repeat steps 3 and 4 with larger size gage blocks. The sizes chosen should cover the range of measurements for which the vernier height gage is to be used.
Note: The vernier height gage also may be checked to very close tolerances by clamping a dial indicator to the slide arm. Then, follow steps 2 through 5.

Step 6   Return the gage to the inspection section if these measurement tests indicate errors caused by inaccuracies in the instrument.

THE UNIVERSAL VERNIER PROTRACTOR

The vernier principle is also applied in making angular measurements. An instrument known as the *vernier bevel protractor* (Fig. 6-13) is used to measure angles to an accuracy of 5 minutes (5'), or 1/12th of one degree. The universal vernier protractor includes

a number of attachments. These make it possible to make a wide range of measurements. All of these measurements are within the same degree of accuracy as the standard vernier protractor.

The protractor main scale (dial) is graduated in whole degrees. On some instruments this scale is on the body of the instrument itself. In other cases the graduations are on the rotating turret. Depending on the manufacturer, the vernier scale is part of either the rotating turret or the body. Some main scales are arranged in four 90° quadrants (0° to 90° to 0° to 90° to 0°). Fig. 6-13 shows the main features of a vernier bevel protractor.

The vernier scale has 24 divisions; 12 are on each side of the zero (index) line. The 24 vernier scale divisions are numbered from 60 to 0 to 60, as illustrated in Figs. 6-14 A and B. Each vernier scale graduation represents 5'. The graduations on the vernier scale are numbered on *both* sides of the 0 index line. This arrangement makes it easy to read

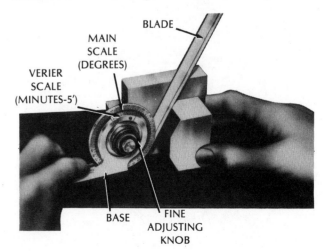

Fig. 6-13. Features of a vernier bevel protractor. (Courtesy of The L.S. Starrett Company)

the minute values. Other vernier scales are marked at every third graduation (60, 45, 30, 15, 0, 15, 30, 45, 60). Again, each line (graduation) on the vernier scale represents 5'.

In taking angular measurements it is important to read the vernier in the same direction from zero as the main scale reading. When the angle is an exact whole number of degrees, the index line (0) on the vernier scale coincides with a whole-degree graduation on the main scale. However, if the angle is more than an exact whole number of degrees, the fractional value (in 5' increments) is read on the vernier scale.

Fig. 6-14A shows a reading of exactly 17°. In Fig. 6-14B the index line (0) has moved past 12° on the main scale. The 50' line on the vernier scale coincides with a graduation (30°) on the main scale. Reading in the clockwise direction, the angle represented on the vernier protractor in Fig. 6-14B is 12° 50'.

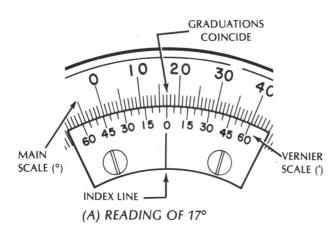

(A) READING OF 17°

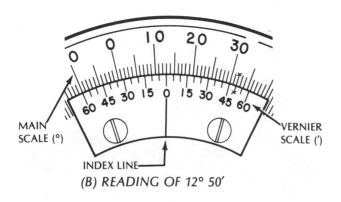

(B) READING OF 12° 50'

Fig. 6-14. Examples of 17° and 12° 50' readings.

┌─────────────────────────────────────┐
│ **HOW TO MEASURE WITH** │
│ **THE VERNIER PROTRACTOR** │
└─────────────────────────────────────┘

Step 1   Position the vernier protractor and the workpiece on a surface plate. Be sure all surfaces are clean.

Step 2   Adjust the blade of the vernier protractor with the sensitive adjusting screw. The protractor blade is set to the angle of the workpiece.

Step 3   Remove the vernier protractor. Read the number of whole degrees indicated on the protractor main scale. This value is located at the index line (0) on the vernier scale. (In Fig. 6-15, the main-scale graduation immediately to the left of the vernier-scale index line is 22°.)

Step 4   Continue reading in the same direction. Determine which graduation on the vernier scale coincides with a graduation on the protractor scale. (The vernier 35' graduation coincides with the main-scale 36° graduation in the example in Fig. 6-15.)

Step 5   Add the vernier scale reading (in minutes) to the whole degree reading on the main scale. This is the angle in degrees and minutes. (The whole and fractional degree reading in the example in Fig. 6-15 is 22° 35'. Example:

     Step 3   22°
     Step 4   35'
   Angle Reading   22°   35')

Step 6   Repeat steps 2 through 5 to check the accuracy of the measurement.

Step 7   Replace the vernier protractor in a protective case.

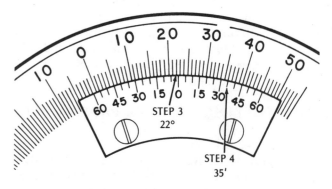

Fig. 6-15. Whole and fractional degree readings.

SOLID AND CYLINDRICAL SQUARES

Solid Square

The *solid square* is a precision layout and angle-measuring tool. Fig. 6-16 illustrates a common type of hardened steel solid square. The solid square may have either a flat or a knife edge. This square is widely used for checking the squareness of a part feature that is 90° from a reference surface. The square is brought into contact with the feature. Any variation from a true 90° may be detected by placing a white paper behind the square and observing whether any white is visible.

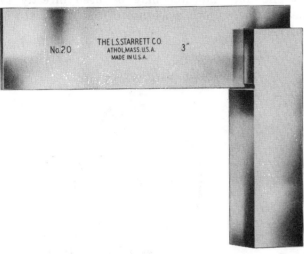

Fig. 6-16. A solid steel square. (Courtesy of The L.S. Starrett Company)

Cylindrical Square

The *cylindrical square*, shown in Fig. 6-17, is another tool that is used to make extremely accurate right-angle measurements. Markings on the cylindrical square show deviations in 0.0002″ steps from a true square condition.

The cylindrical square and the workpiece usually are placed on a surface plate. The part surface to be measured is moved into contact with the cylindrical square. The cylindrical square is rotated until all light between the part feature and the square is shut out. The deviation from squareness is read directly from the 0.0002″ graduations on the cylindrical square. Fig. 6-17 shows a common application of the cylindrical square.

TEST INDICATORS AND DIAL INDICATING INSTRUMENTS

Test indicators and *dial indicators* are *comparison instruments*. Unlike direct measuring instruments,

which incorporate a standard unit of length, test indicators and dial indicators first must be set to a reference surface. The unknown length of the part to be measured is then *compared* to a known length. Test indicators and dial indicators are positioned from a stand or adjustable holder to which they may be secured.

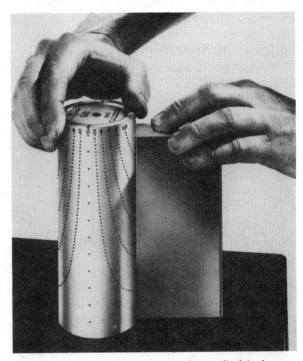

Fig. 6-17. Measuring squareness with a cylindrical square. (Courtesy of Brown & Sharp Manufacturing Co.)

Test Indicators

Two groups of indicators commonly are used in shops and laboratories. The first group is called *test indicators* (Fig. 6-18). These are used for trueing and aligning a workpiece or a fixture in which parts are positioned for machining. Once the workpiece or fixture is aligned, machining, assembling and other operations are performed. Test indicators also are used to test workpieces for roundness, parallelism, or for comparing a workpiece dimension against a measurement standard.

Fig. 6-18. A common dial test indicator and attachment. (Courtesy of The L.S. Starrett Company)

The test indicator is held securely in a stationary base (Fig. 6-18). The ball end of the test indicator is usually brought into contact with the workpiece surface. As the workpiece is moved, any deviation from the dimension to which the test indicator is set is indicated by movement of the indicator. The range of measurements is limited. Most test indicators employ a lever mechanism that multiplies any minute movement of the measuring (ball) end.

Dial Indicating Instruments

The *dial test indicator* (Fig. 6-19) contains a mechanism that multiplies the movement of a contact point. This movement is transmitted to an indicating hand. The amount of movement is read on a graduated dial face. Because the dial test indicator is more functional and more accurate than the test indicator, it has a wider range of applications.

The dial test indicator is a sensitive instrument. The internal mechanism consists of small gear trains or levers that multiply movement. These are actuated by the spindle movement. The multiplied motion is transmitted through a pinion to an indicating hand. The hand rotates across the face of a stationary graduated dial. The amount of spindle movement is read on the graduated dial.

The dial indicator serves two major functions.

● The dial indicator may be used to measure a length. This is the distance between a standard dimension to which the dial indicator is set and the length of the part being measured.

● The dial indicator may be used directly to measure how much a feature of the workpiece is *out-of-true*.

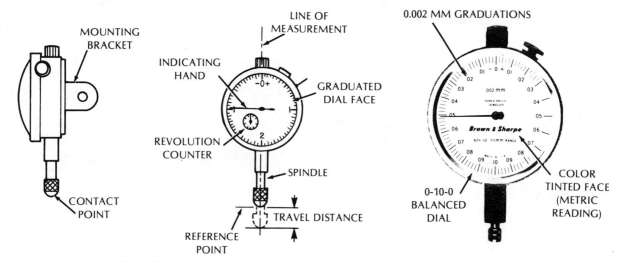

Fig. 6-19. Features of a dial indicator.

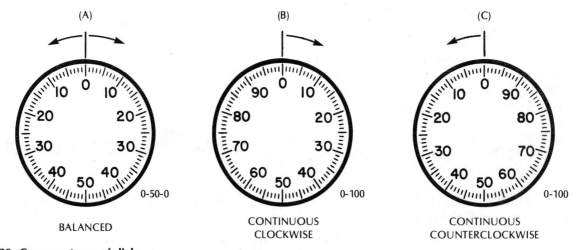

Fig. 6-20. Common types of dials.

Inch-standard dial indicators are commerically produced to measure to accuracies of 0.001″, 0.0005″, and 0.0001″. Metric dial indicators are accurate to 0.02 mm and 0.002 mm. The dial face graduations are marked to indicate the degree of accuracy.

Balanced And Continuous-Reading Dials—Some dials are graduated to show + and − variations from 0 to 0.025″ or 0 to 0.0005″. This type of dial is known as a *balanced dial*. Examples of the many types of balanced dials are: 0-50-0 (0-0.050″-0), 0-100-0 (0-0.100″-0) and 0-5-0 (0-0.0005″-0). A 0-50-0 balanced dial is shown in Fig. 6-20A.

Continuous dials are graduated from 0 to 0.100″. Readings beyond 0.100″ are made by multiplying the number of complete dial hand revolutions by 0.100″. The revolutions then are added to the reading of the indicating hand. Common continuous-clockwise and continuous-counterclockwise dials are shown in Figs. 6-20B and C.

Dial test indicators also are available for measuring dimensions up to ten inches. These long-range dial test indicators are equipped with revolution counters. The long-range indicator in Fig. 6-21 has a 0-2.000″ range, with a 0-0.100″ dial, a revolution counter (0-1.000″), and an inch counter (0-1.000″-2.000″).

Common Applications Of Dial Test Indicators—The dial test indicator is constructed so that the indicator mechanism may be used alone. It may be secured to a holding device and positioned to take a measurement. For example, to check the out-of-roundness of a part on a lathe, the indicator is clamped in a holder. The indicator holder, in turn, is mounted in a lathe

Fig. 6-22. Checking the concentricity (run-out) of a lathe spindle bore with a dial test indicator. (Courtesy of the Clausing Corp.)

tool holder. As the workpiece is slowly rotated, any roundness variation causes the contact end of the indicator to move up or down.

Another common application, checking the accuracy of a lathe spindle is shown in Fig. 6-22. A precision-machined test bar which is inserted in the taper of the spindle is used in combination with the dial test indicator.

Dial Indicating Gages—The dial test indicator often is combined with other gaging devices and measuring tools. A few examples are given in Fig. 6-23 to show the versatility of this instrument. When the dial test indicator is combined with a depth gage, it is called a *dial indicating depth gage* (Fig. 6-23A). When the dial indicator is adapted to testing holes for size and out-of-roundness (concentricity) or other surface irregularities, the instrument is called a *dial indicating hole gage*, or *dial bore gage* (Fig. 6-23B).

● The dial test indicator also is used for taking inside or outside measurements. Such instruments, called *dial indicating caliper gages* (Fig. 6-23C), have revolving counters. The counter permits measurement of dimensions up to three inches (75mm).

Dial indicating snap gages are another application of the dial test indicator. These gages are used for linear measurements as well as for measuring diameters.

Dial Indicating Micrometer—Inch-standard *dial indicating micrometers* (Fig. 6-24) can measure dimensional variations in ten-thousandths inch (0.0001″) and one-thousandths inch (0.001″). Metric dial indicating micrometers are used to measure to

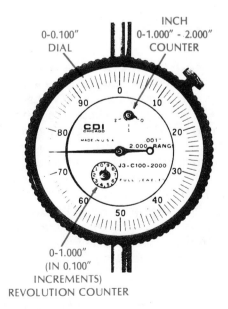

Fig. 6-21. A long-range indicator with revolutions counters.

Dial Indicating Depth Gage

Dial Bore Gages

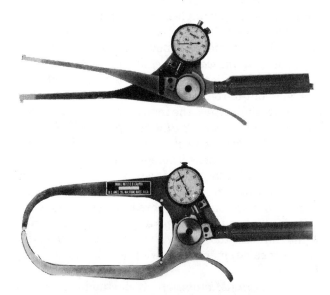

Dial Indicating Caliper Gages

Fig. 6-23. Dial indicating caliper gages. (Courtesy of Federal Products Corp., and the B.C. Ames Co.)

accuracies of 0.02 millimeters and 0.002 millimeters.

The lower anvil of the dial indicating micrometer is a sensitive contact point. It is connected directly to a dial indicator that is built into the instrument. When used as a micrometer, the thimble is set to the one one-thousandth (0.001″) setting nearest the required measurement and is locked in this position. As the dial indicating micrometer is brought into contact with the workpiece, it serves as a gage. Any variation from the required size appears on the dial indicator. The dial is graduated in increments of 0.0001″, with a range up to 0.001″.

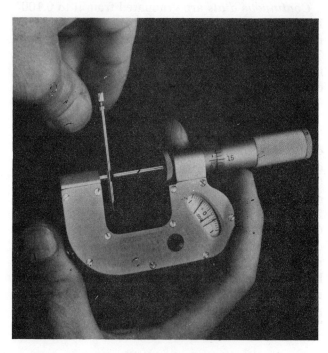

Fig. 6-24. A dial indicating micrometer (± 0.0001″). (Courtesy of Federal Products Corp.)

Universal Dial Indicator Sets—The term *universal dial indicator set* denotes that there are a number of attachments for holding and positioning the dial indicator. For instance, to check the height of a machined surface, the dial indicator head may be mounted on a *T-slot* base (Fig. 6-25). The indicator is set at a desired height. It is then moved along a surface plate and positioned over the part of the workpiece that is to be measured.

Other dial indicator applications include checking the runout, parallelism or face alignment of a workpiece. In these cases the dial indicator may be held in a tool post holder attachment on a lathe or other machine.

In many quality control and inspection laboratories an indicator and a test stand are used with the dial indicator. With these accessories it is possible to position the dial indicator to take measurements along any one of three axes. The dial indicator and test stand combination is used with a surface plate, which serves as a reference surface. Slip joints equipped with clamping screws permit as many as six movements along the three axes. The six possible movements used for positioning the indicator contact point on the workpiece surface are shown in Fig. 6-25.

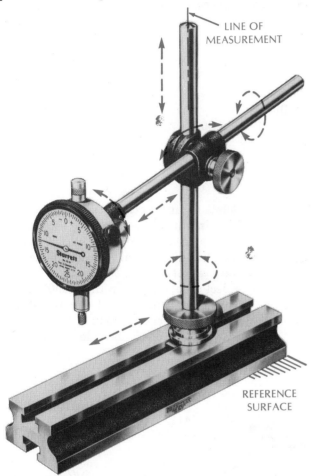

Fig. 6-25. Six positioning movements of a dial test indicator. (Courtesy of The L.S. Starrett Company)

PRECISION GAGE BLOCKS

Gage blocks (Fig. 6-26) are blocks of steel that are heat treated to a high hardness. Gage blocks are machined, ground and finished to precise limits of dimensional accuracy. The heat treating process makes it possible to retain accuracy and dimensional stability. The hardness produced by heat treating helps prevent wear and damage to the block surfaces.

Gage blocks are available in three general grades: (1) *master blocks*, which are accurate to within 0.000002″ or 0.00005 mm; (2) *inspection blocks*, which

are accurate to within 0.000005″ or 0.0001 mm; and (3) *working blocks*, which are accurate to within 0.000008″ or 0.0002 mm. Master and inspection blocks usually are used in temperature-controlled laboratories.

The *accuracy* of gage blocks relates to *flatness*, *parallelism* and *length*. Some sets include two additional *wear blocks* (0.050″). These are used as *end blocks*. They prevent wear on the other blocks.

Gage Block Sets And Applications

Practical gage block applications include:
- The checking of precision measuring instruments and gages
- The setting of other instruments, such as dial test indicators and height gages, to make comparison measurements
- Layout operations on machined surfaces where highly accurate linear measurements are required.

Angular surfaces may be checked and measured with angular gage blocks. Flat gage blocks may also be used in combination with other measuring instruments and setups.

Gage blocks are furnished in many combinations. Two common sets have 83 and 35 gage blocks each. The sizes of the gage blocks in the 83-piece inch-standard set are given in Chart 6-1. Note that in the *0.0001″ series* each block increases in size by an increment of 0.0001″.

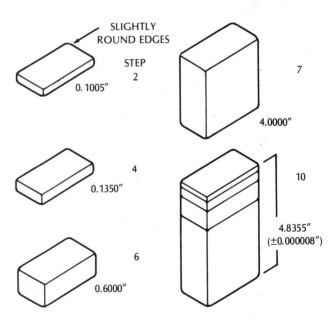

Fig. 6-26. Selection of gage blocks for a measurement of 4.8355″.

.0001 Series (9 Blocks)

.1001 .1002 .1003 .1004 .1005 .1006 .1007 .1008 .1009

.001 Series (49 Blocks)

.101	.102	.103	.104	.105	.106	.107	.108	.109	.110	.111	.112	.113
.114	.115	.116	.117	.118	.119	.120	.121	.122	.123	.124	.125	.126
.127	.128	.129	.130	.131	.132	.133	.134	.135	.136	.137	.138	.139
.140	.141	.142	.143	.144	.145	.146	.147	.148	.149			

.050 Series (19 Blocks)

| .050 | .100 | .150 | .200 | .250 | .300 | .350 | .400 | .450 | .500 | .550 | .600 |
| .650 | .700 | .750 | .800 | .850 | .900 | .950 | | | | | |

1.000 Series (4 Blocks)

1.000   2.000   3.000   4.000

Two .050 Wear Blocks

Chart 6-1 Series and sizes of gage blocks—83 piece set

The increment in the 49-block *0.001" series* is 0.001". In the *0.050" series*, there are 19 blocks varying in size from 0.050" to 0.950" in increments of 0.050". There are four blocks in the *1.000" series*, with sizes from 1.000" to 4.000".

Over one hundred thousand different measurements may be made with the 83-piece set by combining different gage blocks. The gage block surfaces are so precisely finished that when perfectly clean they may be carefully *wrung together* to become almost one gage block.

Gage blocks are assembled by overlapping one clean block on another. A gentle force is applied while sliding the blocks together. The gage blocks may be taken apart by sliding them in a similar manner. In all gage block applications it is important to use the least number of blocks possible.

**HOW TO SET UP
A GAGE BLOCK MEASUREMENT**

Example: Make a gage block setup for measuring 4.8355" ± 0.000008". (A set of *working gage blocks* will give this degree of accuracy.)

Selecting And Joining Gage Blocks

Step 1   Determine the last-place decimal value (number) in the required setup (.__ __ 5").

Step 2   Select a gage block in the 0.0001" series which ends in 5. In this case it is 0.1005".

Step 3   Select a gage block in the 0.001" series. The size must have the same second-and third-place decimal value (.__ 3 5) as the required dimension. A practical size for the example is 0.135"

Step 4   Select a gage block in the 0.050" series. This block must equal the remaining first-place decimal value of (.6__ __ __). The 0.600" block is selected for the example.

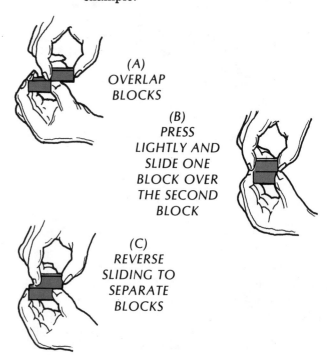

(A)
OVERLAP
BLOCKS

(B)
PRESS
LIGHTLY AND
SLIDE ONE
BLOCK OVER
THE SECOND
BLOCK

(C)
REVERSE
SLIDING TO
SEPARATE
BLOCKS

Fig. 6-27. Joining and separating gage blocks.

Step 5   Select a gage block in the 1.000″ series that is equal to the required whole number (4.___ ___ ___). The 4.000″ block is used for the example.

Step 6   Add the values of the gage block combination. Recheck it against the required measurement. The combination should contain the least number of blocks possible. The four gage blocks selected for the example measurement of 4.8355″ ± 0.000 008″ are shown combined in Fig. 6-26.

Step 7   Check each block to see that it is clean.

Step 8   Start with the largest blocks. Overlap each block as illustrated in Fig. 6-27A.

Step 9   Press the blocks together by hand.

Carefully slide each block into position. (Fig. 6-27B).

Step 10   Repeat the overlapping, pressing and sliding steps with each successive set of blocks.

Disassembly

Step 1   Reverse the assembly process (Fig. 6-27C). Start with the smallest block and slide it off the next block combination.

Step 2   Continue to slide each successive block off the combination.

Step 3   Rewipe the gage blocks. Insert each block in its proper position in the case.

SAFETY PRECAUTIONS IN USING PRECISION MEASURING AND LAYOUT INSTRUMENTS

- Check the workpiece to see that it is free of burrs and foreign particles. These impair the accuracy of all measurements.
- Clean all contact surfaces of vernier and dial indicating instruments, gage blocks and related accessories.
- Check each instrument for accuracy. For example, in the fully closed position, the inside and outside readings of the vernier caliper must be "0" (0.000″ or 0.00 mm).
- Bring the jaws or other instrument surfaces lightly into contact with the workpiece. Any force applied to the instrument will cause *springing* and inaccurate measurements.

- Take measurements as close to the instruments as possible. The longer the distance a measurement is taken from the base or frame of the instrument, the greater the probability of error.
- Check the alignment of the instrument, the condition of the reference surface, and the position of the workpiece before taking a measurement.
- Recheck each measurement to ensure that it is accurate.
- Thoroughly clean each precision tool. Instrument moving parts must be lubricated with a thin, protective film. All instruments should be carefully placed and stored in appropriate containers.

VERNIER AND DIAL INDICATING INSTRUMENTS AND GAGE BLOCK TERMS

Vernier instruments   Precision measuring instruments, layout measuring tools, and machine parts having a basic (primary) measuring scale and a vernier scale.

Vernier scale   A secondary scale attached to a movable leg or turret. The vernier scale usually contains one less graduation than the number on the primary scale of a fixed beam or head.

Primary (beam) scale   The graduations on the stationary frame or body of a measuring instrument. The graduations on the scale are in inch or metric units and a decimal part of the unit.

Coincide   The alignment of two scale graduations: A graduation on the vernier scale that aligns with one on the stationary part of the instrument.

Vernier reading   An additional decimal value of a measurement. The value is read on the vernier scale at the point where the vernier graduation coincides with a graduation on the primary scale.

Cylindrical square	A precision-finished cylinder which is graduated to show deviations in the squareness of a workpiece. The variation is read in steps of 0.0002″ from a perfectly square condition.
Alignment	The exact relationship between an instrument and a workpiece. A correct positioning of an instrument. Taking a measurement along the line of measurement between the measured points.
Comparison measurement	Setting a measuring instrument to a known length or measurement standard. The unknown length is compared with the instrument setting.
Dial indicator	An instrument which multiplies the contact point movement on a part to be measured. The linear movement may be read directly. The indicator dial may be graduated in decimal values of the standard inch or millimeter.
Test indicator	A measuring instrument in which movement is multiplied through a system of levers or gears. A minute variation in movement of the contact end is magnified and observable on a scale at the measurement end. An instrument for testing parallelism, alignment, trueing workpieces for machining, and making comparison measurements.
Gage blocks	Hardened, precision-ground and finished rectangular steel blocks used to lay out work and make measurements. Working to accuracies in ranges of 0.000008″ to 0.000002″ and equivalent metric values in decimal parts of millimeters.
Gage block measurements	The use of gage blocks in a combination that equals a required measurement. A measurement standard against which other instruments or parts may be set or checked.

SUMMARY

- Vernier scales attached to measuring tools increase the potential accuracy of the instruments.
- Vernier measuring tools are named according to the basic function of the tool. The vernier *height* gage, the vernier *bevel protractor*, and the vernier *caliper* are examples.
- The vernier scale on inch-standard instruments extends the accuracy range from 0.001″ to 0.0001″. Similarly, on metric standard instruments the basic 0.02 mm accuracy is increased to 0.002 mm.
- A vernier measurement includes the reading on the primary scale of the instrument plus the reading on the vernier scale.
- The vernier protractor applies the vernier principle to angular measurements. The vernier extends the accuracy of the angular reading to within 5 minutes (5′).
- Dial indicators are calibrated to show measurements over a wide range of sizes. The range of accuracy is within 0.0001″ and 0.001″. In the metric system, the accuracies are 0.002 mm and 0.02 mm.
- Universal dial indicator sets include a number of attachments. These permit a wide range of applications.
- Solid-steel and cylindrical squares are used to check squareness. The cylindrical square measures any deviation from a true square condition to within 0.0002″.
- Test indicators and dial indicators are designed with a multiplying system of levers or gears. Any minute movement is magnified and indicated on a dial or other graduated scale.
- Test indicators usually have a limited range of measurement. They are particularly useful in checking alignment, parallelism, or out-of-roundness.
- Dial indicators are used to measure linear movements. When set to a particular standard, it is possible to measure any variation on either the + or − side of the standard measurement.
- Gage block sets consist of a series of precise, rectangular blocks. These are wrung together to establish a specific dimension. The gage block then is used as a standard for comparison measurements.
- Working gage blocks are used for measurements within ± 0.000008″ (± 0.0002 mm). Master gage blocks, accurate to ± 0.000 002″ (± 0.00005 mm), are used in temperature-controlled laboratories.

---

1. Indicate the functions of (a) the beam containing the main scale of a vernier measuring instrument and (b) the vernier scale.
2. Use the inch standard 25-division vernier caliper to explain how to read a measurement of 4.377″.
3. State how the accuracy of a metric vernier height gage may be checked.
4. Give three major differences between a universal bevel protractor and a standard bevel protractor of a combination set.
5. State two distinguishing design features between a solid square and a cylindrical square.

6. Indicate the different purposes that are served by test indicators and dial indicating instruments in relation to vernier instruments.
7. Give four applications of dial test indicators.
8. List three examples of instruments that combine the measurement features of dial indicators with other measurement instrument functions.
9. Identify the three standard grades of gage blocks by name and degree of accuracy.
10. Explain how to build up a gage block measurement of 2.5255″ + 0.000 008″.
11. State two safety precautions to follow to ensure accurate, precision measurements.

PART TWO

BENCH WORK TECHNOLOGY AND PROCESSES

SECTION ONE

BENCH WORK AND BENCH HAND TOOLS

Work processes performed with hand-manipulated tools and instruments and portable power equipment are generally categorized as bench work or floor work. Included among these processes are layout, fitting, finishing, assembling, and various other noncutting, cutting, and forming operations. This section introduces bench and floor work and presents the principles of and correct practices for performing noncutting hand processes.

Unit 7

PREPARATION FOR BENCH WORK PROCESSES

Bench and floor work involve the use of hand tools and small portable equipment, layout tools, cutting and forming tools, and measuring instruments. The tool and instruments are all manipulated by hand. The accurate performing of each bench work process requires the development of hand skills and a working knowledge of related technology.

The term *bench work* refers to setting up, laying out and checking dimensions and measurements. Bench work also includes filing, sawing, threading and reaming operations, and assembling and fitting processes. Three typical bench-work processes are shown in Fig. 7-1.

A work area larger than a bench is sometimes required because of the size of the workpiece or unit. In such cases, similar operations involving identical tools and instruments are performed on the *floor* as *floor work*.

SCOPE AND REQUIREMENTS OF BENCH AND FLOOR WORK

Bench work requires the craftsperson and technician to:
- Read and communicate in the technical language of the trade
- Make necessary computations using technical manuals and handbooks
- Relate machine and other forming processes to bench finishing operations
- Plan a work procedure in which each operation is performed in proper sequence
- Check the accuracy at each step of an operation and make whatever adjustments are needed
- Use tools properly
- Observe safe practices, to prevent personal injury or damage to machines, tools and work.

BASIC BENCH WORK PROCESSES

Bench and floor work involve a wide variety of tools, accessories and instruments, and a great range of operations. The applications of these vary with each job. It should be noted that many of the same hand tools, measuring instruments, layout tools, and

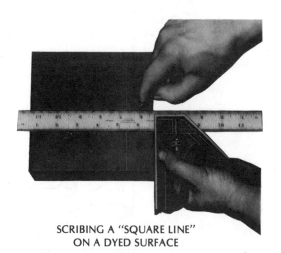

SCRIBING A "SQUARE LINE" ON A DYED SURFACE

HAND TAPPING

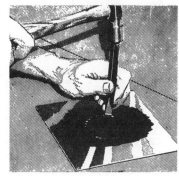

CHISELING

Fig. 7-1. Typical bench work processes. (Courtesy of The L.S. Starrett Company)

77

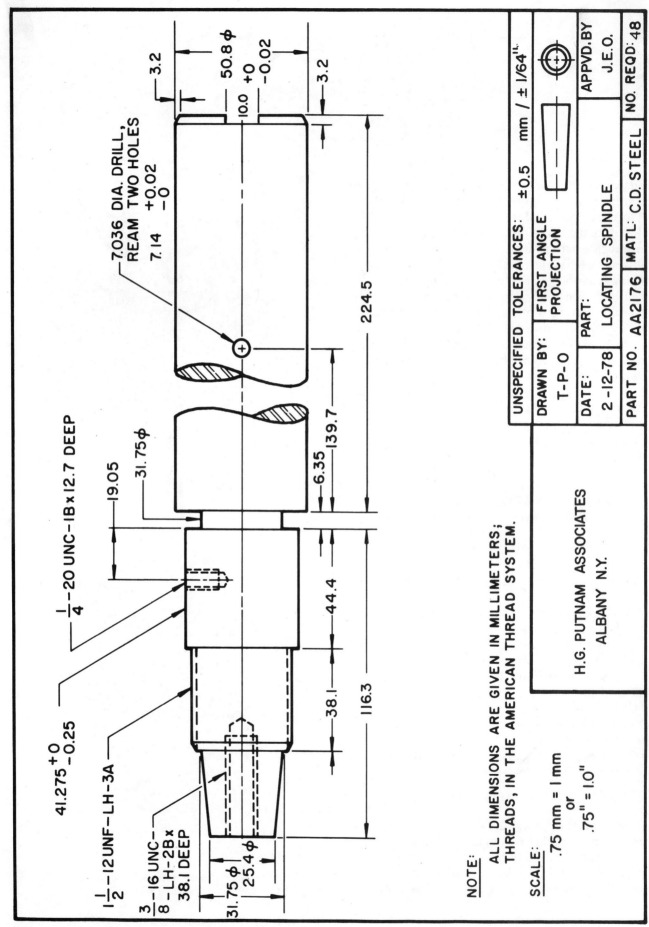

Fig. 7-2. Typical drawing from which a production plan may be developed.

3.2

50.8 φ

10.0 +0
      -0.02

3.2

7.036 DIA. DRILL,
REAM TWO HOLES
7.14 +0.02
      -0

224.5

139.7

31.75 φ

19.05

6.35

44.4

38.1

116.3

$1\frac{1}{2}$–12 UNF–LH–3A

41.275 +0
        -0.25

$\frac{1}{4}$–20 UNC–IB×12.7 DEEP

$\frac{3}{8}$–16 UNC–
8–LH–2B×
38.1 DEEP

31.75 φ

25.4 φ

NOTE:

ALL DIMENSIONS ARE GIVEN IN MILLIMETERS;
THREADS, IN THE AMERICAN THREAD SYSTEM.

SCALE:
.75 mm = 1 mm
or
.75" = 1.0"

H.G. PUTNAM ASSOCIATES
ALBANY N.Y.

UNSPECIFIED TOLERANCES:	±0.5 mm / ± 1/64".	
DRAWN BY: T-P-O	FIRST ANGLE PROJECTION	APPVD.BY J.E.O.
DATE: 2-12-78	PART: LOCATING SPINDLE	
PART NO. AA2176	MATL: C.D. STEEL	NO. REQD: 48

accessories are also used in machining operations.

Bench and floor work require the use of hand tools for:

- hammering
- holding
- turning and fastening
- punching and shearing
- sawing
- reaming
- threading
- filing and burring
- polishing and others.

Common hand tools and their proper use for performing these operations are covered in succeeding units. The remainder of this unit deals with:

- Specific information provided by work orders
- Work habits and attitudes
- Preparation of surfaces for accurately laying out a workpiece and performing bench operations
- Projection lines for transferring and locating lines and surfaces on drawings and in layout work
- Safety rules.

WORK ORDERS, PRODUCTION PLAN, AND THE BLUEPRINT OR SKETCH

The *work order* and/or *production plan* provide the following types of information about a particular part or unit: the name, function, quantity, work processes, schedule, etc. The work order and production plan are usually used with a mechanical drawing, freehand sketch, or blueprint.

The blueprint graphically describes parts and mechanisms. The blueprint description is complete and accurate. A part may be made of the same materials, machined to the same degree of accuracy and surface finish, and precision assembled anywhere in the world. Fig. 7-2 is an example of a typical working drawing from which a production plan and work order may be prepared. The production plan provides technical information about tool requirements, the machining sequence, tolerances, sizes, machine setups, cutting speeds and feeds, and other production conditions.

The blueprint and the work order provide all essential information. This information includes form, sizes, quantity, material, finish, and the part's relation to other fitted or mated parts. The same technical information is conveyed to the designer, draftsperson, engineer, mechanic, technician, and consumer. Each is then able to:

- Form a mental picture of the shape, features, material used, and the size of each feature
- Visualize the fabricating and manufacturing processes required to make the part
- Plan the sequence of steps for performing each process
- Translate the lines, symbols, views, sections, dimensions and notes found on drawings and sketches into occupational standards. These standards are then used in laying out and performing other bench and floor work processes and machining operations.

The blueprint and the work order or production plan are the starting point. The worker must be able to secure materials, produce the workpiece, and perform efficiently whatever bench and assembly operations are needed.

CONDITION OF THE WORKPIECE

Grease, dirt, and other foreign particles must be removed from a workpiece before layout processes are begun. Castings are usually machine tumbled to remove sand and rough projections. Forgings are ground to remove *fins* and rough edges. (Fins are excess material formed in the area where die sections meet.) Steel rods and bars that have protective coatings are immersed in a cleansing bath to remove the film, grease and dirt. In each case, the surface must be clean if a layout dye or other coating is to be applied.

Equally important is the need to continuously check the work to be sure that all burrs are removed. Burrs are a safety hazard because they can cause cuts and bruises. Burrs also can cause damage to parts in assembly, and inaccuracies in layout, machining, or finishing operations.

Burrs can be removed from hardened surfaces by grinding or stoning the edge or surface with a small abrasive sharpening stone. Burrs on internal surfaces such as holes or fillets are sometimes cut away with a hand scraper. Other edges are usually *broken* by filing them with a smooth-cut file.

PREPARING THE WORKPIECE

The sizes, shapes and other characteristics of materials drawn from supply sources should always be checked against the corresponding size, shape and other physical specifications in the work order. The material should be sufficiently oversize to permit the worker to perform the hand and machine processes necessary to produce a particular part.

One of the unique functions of all machine and metal trades is the production of parts of precise shape, size, and surface finish, accurately assembled. The material from which the part is made may be a forging, casting, rolled, or otherwise preformed shaped bar. It is often necessary when working with these materials to prepare the work surface. This permits layout lines to be drawn that are clearly visible and easy to follow. Such a precision layout using a vernier height gage is shown in Fig. 7-3.

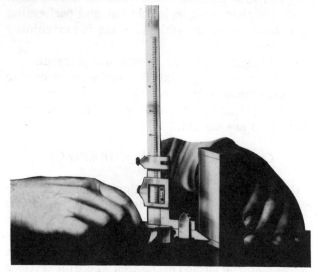

Fig. 7-3. Precision layout lines clearly visible on a coated work surface. (Courtesy of The L.S. Starrett Company)

The more practical and universally used surface coloring agents are layout dyes. These may be used with ferrous metals and with some nonferrous metals such as aluminum, brass, bronze, and copper. After the layout surface is *burred* and cleaned to remove any oil film, the dye solution is brushed or sprayed on.

Layout dye dries rapidly and provides a deep blue, lavender, or other colored surface. Finely scribed lines stand out sharply and clearly. Fig. 7-4 shows cleaned surface being sprayed with a layout dye.

The surfaces of rough castings and forgings may be painted with a thin white pigment that is mixed with alcohol. This white paint spreads easily and dries rapidly. It produces an excellent coating for layout operations. Ordinary white chalk is used sometimes. However, the chalk coating may wear or be brushed away before the operations are completed. This requires that the layout be repeated. For this reason, chalked coatings are not as desirable or practical as are dyes.

WORK HABITS AND ATTITUDES

The craftsperson works in an orderly manner. Attention is paid to the condition of all layout, cutting or other hand tools and instruments. Their placement and handling are important. The work place and the part or parts are all considered. The skill of the worker is affected by the safe handling and storage of all tools. Safe practices as applied to the individual bench tools and processes are treated in each unit at the time when each is described.

A number of hand tools have cutting edges or are heavy. These must be placed where they do not rest on or scrape against finished surfaces or precision measuring tools. Rough, unmachined surfaces must be positioned so that they are not in direct contact with finely finished surfaces or precision instruments. Separate places on the work bench should be planned for placing and working with different tools. The effects of burrs on accuracy and personal safety require their removal.

The skilled worker avoids scraping one surface against another. Such scraping produces scratched or scored surfaces. Instead, parts are gently picked *up* and carefully placed in position. Surfaces of different degrees of finish that are to be placed together should have a thin metal sheet or even paper placed between the surfaces to protect them. This is particularly true of parts that must be layed out before machining operations are performed.

PROJECTION LINES AND THE FEATURES OF A PART

The drawing the worker uses consists of one or more *views*. A *view* is a position from which a part is observed. Designers and drafting room, shop, and laboratory workers must represent the lines and surfaces in one view in correct relationship with the same lines and surfaces in another view. *Projection lines* are used to locate the exact positions of such lines and surfaces.

Projection lines are fine, unbroken lines (Fig. 7-5). They project from a line or surface in one view to the same line or surface in another view. As shown in Fig. 7-6, some projection lines may be drawn horizontally or perpendicularly from one view to another. Other projection lines in Fig. 7-6 are drawn at a 45° angle or as 90° arcs. The features in view I may be projected accurately to view III by using either angular or circular projection lines.

Fig. 7-4. Spraying layout dye on a cleaned surface. (Courtesy of The L.S. Starrett Company)

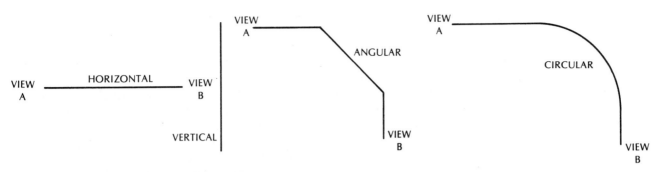

Fig. 7-5. Projection lines used on drawings and for laying out workpieces.

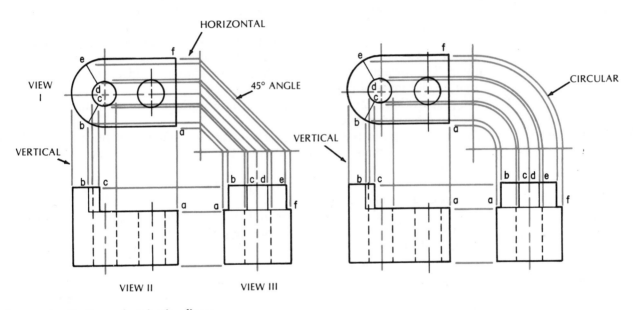

Fig. 7-6. Applications of projection lines

SAFE PRACTICES IN PREPARING WORK FOR BENCH OR ASSEMBLY PROCESSES

- Clean the work place, and remove grease and dirt from all tools.
- Remove abrasive and foreign particles and burrs from all workpieces.
- Place the cutting tools and rough hand tools side by side in one area of the work bench.
- Check the accuracy of all measuring tools and the condition of all cutting tool and work piece edges.

- Place a clean cloth or a soft-surfaced material under all precision measurement and layout tools.
- Pick *up* each tool to avoid sliding one tool on another. Sliding can cause scratching.
- Return each tool to its appropriate place on the work bench.
- Wipe off each tool and return it to its proper container or holder.

BENCH AND FLOOR WORK AND LAYOUT TERMS

Bench and floor work	A series of hand or portable machine processes performed at a bench or on the floor. Examples include laying out, cutting, forming, finishing, fitting, and assembling one or more parts.
Technical trade language	Terms, symbols, techniques, and devices used by craftspersons and technicians to communicate complete specifications about a part or mechanism. The language is universally understood. Blueprints, work orders, technical manuals, and handbooks are technical communications media.
Hand tools	Tools that are manipulated by hand. Common hand tools include those used for layout processes, holding, hammering, fastening, punching, shearing, sawing, filing, threading, burring, and polishing.
Work order	An information sheet used by a worker to produce a part or mechanism. It provides complete information about shape, size, quantity, material, and other specifications of a workpiece or assembly.
Production plan	The sequence of steps and the routing of a workpiece through a shop or laboratory.
Workpiece	A mass (material) having a particular shape (features), size, composition and surface finish. The workpiece may be cast, forged, preformed, welded, rolled, or machined from a solid mass of material.
Preparing the workpiece	The preparation of one or more surfaces of a workpiece by cleaning, burring, and removing foreign particles.
Surface treatment	The application of a coating on a work surface to make layout lines clearly visible. Surface treatment helps establish when the limits of a dimension and shape are reached.
Coloring agents	Whiting, multi-colored dyes, copper sulphate, and other solutions used in shops and laboratories to color a work surface.
Shop layout	Processes of marking the locations and shape of different features of a workpiece. Locations for positioning a workpiece as a guide for hand and machine operations. Marking the features of a required part. Lines that help the worker to establish when the limits of a shape or size are reached.
Projection lines	Fine, unbroken lines for projecting a line or a surface from one view to another. Layout lines for transferring a line or locating a point between views.
Work habits and attitudes	Qualities and attitudes developed by skilled craftspersons. Desirable performance standards on the job; concerns in maintaining an individually fair level of production. The safe handling of work, tools, and instruments to prevent damage and to avoid personal injury.

SUMMARY

- Machining processes are usually preceded by hand and other layout processes. These are followed by checking, fitting and assembly operations at the bench or on the floor.
- Bench and floor processes may be adapted to machine operations.
- Layout and measuring tools, and some of the hand cutting tools are also used on the lathe, milling machine, and other machine tools.

- The craftsperson must be able to:
  - Interpret specifications on drawings
  - Make calculations
  - Apply science and trade skill and technical knowledge to produce accurately and efficiently a required part or mechanism.
- Lines and surfaces are projected from one view to another. Projection lines are used in drafting and on shop layouts.

- The work order provides written directions about the requirements of a job. This information may be added to by a production plan. The plan gives the steps to produce a part, the routing of the part through several departments, degrees of accuracy, and other pertinent information.
- Work surfaces must be prepared to receive a coloring agent. Quick-drying paints and layout dyes are used to coat a work surface. Coated surfaces simplify layout operations and make all lines and dimensions clearly visible.

- Orderly work habits are the hallmark of the skilled craftsperson. The protection of tools and work pieces and their safe handling to prevent personal injury are essential practices.
- The condition of the workpiece affects the accuracy of layout, fitting, assembling, and other bench, floor, and machine operations. Burrs, grease, and other foreign particles must be removed.

UNIT 7 REVIEW AND SELF-TEST

1. List six common bench work processes.
2. Give four reasons why the worker depends on a blueprint and/or work order (production plan).
3. Explain briefly what is meant by (a) the condition and (b) the preparation of a workpiece.

4. a. Describe a *projection line*.
   b. State the purpose of projection lines on a drawing.
5. Set up a series of four conditions that constitute good work habits and attitudes.

Unit 8 NONCUTTING HAND TOOLS AND PROCESSES

The term *hand tools* applies to tools that are controlled by hand. Such tools are used to apply a force, to cut, or to form a surface. The term relates to hand operations that are performed on bench or floor work. Hand tools also are used in the setting up of machines and for some machine processes.

Bench hand tools fall into two groups of processes: *noncutting*, and *cutting and shaping*.

Noncutting Hand Processes:
- Hammering
- Clamping
- Positioning

Hand Cutting and Shaping Processes:
- Sawing
- Filing
- Chiseling
- Punching and Driving
- Threading
- Reaming

This unit deals with hand tools in the first group of bench work processes. Hand cutting and shaping tools are described in Section Two of this part.

HAMMERING TOOLS AND PROCESSES

Ball Peen Hammers

One of the everyday tools used in the machine and metal trades is the *ball peen* hammer. Two others that are not as common are the *straight peen* and the *cross peen* hammers. The three types are shown in Fig. 8-1. One end of each hammer is shaped as a

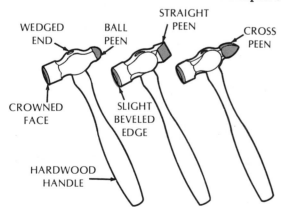

Fig. 8-1. Features and shapes of hammers.

cylindrical solid. The face is ground slightly crowned and the edge is beveled. This face is used to deliver a flat blow (force) and to shape and form metal. The ball-shaped face permits the operator to deliver a blow in a more confined area. A surface may be *peened* by shaping it through a series of cupped indentations.

Hammers come in many sizes depending on the work size and required force. Ball peen heads weighing just an ounce are used in fine instrument work and for small, precise layout operations. The weights range up to three pounds. The 12-ounce ball peen hammer is one of the most practical sizes for general machine work.

These types of hammers are usually forged to shape, hardened and tempered. The *eye* (hole) in the hammer head is tapered in two dimensions (depth and width) both ways. The handle is made of a close-grained hardwood. It is secured to the head with a metal wedge. This is driven in at the end of the handle. Safe handling requires that the handle be tight in the hammer head at all times. A large force may be delivered by a hammer at a desired point by firmly grasping the handle near the end and delivering a solid, sharp blow.

Soft-Face Hammers

There are many assembly and disassembly operations in which machined parts fit tightly together. These must be driven into place without damage to the part or any surface. The pounding faces of a soft-face hammer may be made of rawhide, plastic, wood, hard rubber, lead, brass, or other soft metal or composition (Fig. 8-2). In each instance, the soft-face hammer head must be softer than the workpiece. Some makes of soft-face hammers are designed so that the faces may be replaced when worn.

FASTENINGS AND FASTENER TOOLS

Screws, nuts, and bolts are called *fasteners*. Fasteners are turned with two major groups of hand tools: screwdrivers and wrenches. The construction and design of these tools vary depending on the

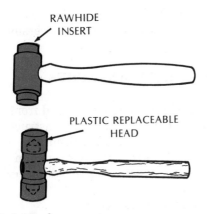

Fig. 8-2. Soft-face hammers.

application. For instance, small *jewelers' screwdrivers* are used for fine precision work. By contrast, some screw drivers for heavy-duty service have a square shank to which a wrench can be attached to gain added leverage (force) as shown in Fig. 8-3.

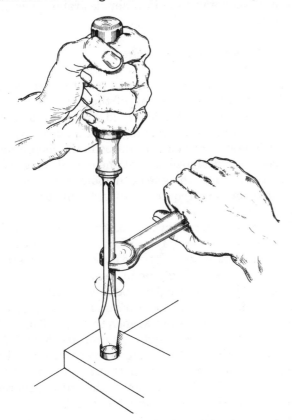

Fig. 8-3. A wrench attached to the square shank of a heavy-duty screwdriver provides added turning force.

Flat-And Phillips-Head Screwdrivers

A screwdriver (Fig. 8-4) usually combines four basic parts: handle, ferrule, shank, and blade. Other types omit the ferrule by bonding the handle to the

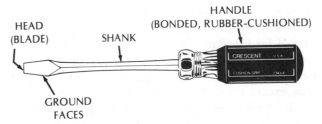

Fig. 8-4. The basic parts of a standard flat-head screwdriver. (Courtesy of The Cooper Group)

blade. The ferrule is a metal band, or bushing, that compresses the handle around the shank. In each case the screwdriver handle is designed to be grasped comfortably. It is also ribbed to improve gripping. This feature permits the screwdriver to be held securely for positioning the tip of the blade in the screw head and for turning. The handle may be made of wood, plastic or metal. Some handles are covered with rubber to provide better gripping and insulation.

A good screwdriver blade (shank) is made of a high-quality steel that has been forged to shape, hardened and tempered. The *flat-head* screwdriver point is ground to a slight wedge shape. The correct shape is shown in Fig. 8-5. Screwdriver sizes are designated by the length of the blade; typical sizes are 6", 8" and 10". The size of the tip is proportional to the blade length.

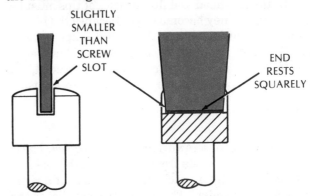

Fig. 8-5. Correct size and shape of a flat-head screwdriver head.

There are two basic shapes of tips: *flat head* and *Phillips head*. The flat-head shape fits screw heads that are slotted completely across the head diameter. The worker should select a tip size that is almost the width and length of the screw slot. The end of the blade should be flat so that it can be held firmly against the bottom of the slot.

The term *Phillips head* is a general one. It denotes a screwdriver tip that fits screws and bolts that have two crossed slots in the center of the head. The crossed slots extend across the screw head but stop short of the outer edge. This type of screwdriver is named for the manufacturer of the specific type of recessed screw which it fits, such as Phillips and Reed. These screwdrivers are similar to the flat-head type except for the shape of the shank and tip.

Care must be taken to select the size of screwdriver and tip that is designed for the particular screw. The size and shape of the *flutes* on the tip must fit the recessed groove of the screw head (Fig. 8-6). If the flutes are too small, they will contact only a fraction of the head area. Then, as force is applied, the screw slot will be damaged. If the tip is too large, only the top edge of the screw slot will be engaged by the flutes. As turning force is applied, a burred edge is produced on the screw head.

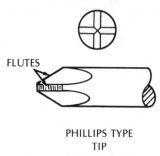

FLUTES

PHILLIPS TYPE
TIP

Fig. 8-6. Features of a Phillips screwdriver head.

Both the flat-head and fluted types of tips must be touched up as they become worn. Because the blades are made of hardened steel, the correct shape and size usually can be restored only by grinding. However, if only a small amount of reshaping is needed, an abrasive sharpening stone sometimes can be used.

Offset Screwdrivers

A third group of screwdrivers is called *offset*. These screwdrivers are made from a solid metal piece and usually have a tip on each end. One tip is at a right angle to the "handle." The other tip is parallel to the handle. Two common offset types are illustrated in Fig. 8-7. However, offset screwdrivers also come in Phillips and Reed types. Note that the offset screwdriver is one solid piece of metal.

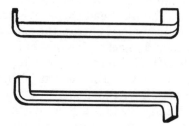

Fig. 8-7. Offset screwdrivers.

The offset screwdriver is used to hold and turn fasteners in spaces that are inaccessible to the straight-blade types.

HAND WRENCHES

The names of wrenches are derived from (1) their shape (such as *open end*), (2) their construction (such as *pipe* wrench), or (3) the requirements of the job (such as *adjustable* wrench). Four common groups of wrenches used in bench work are described in this unit. These are:

- Solid, nonadjustable wrenches
- Adjustable wrenches
- Socket-head wrenches
- Special spanner wrenches.

Each wrench should be made of a high-quality alloy steel which combines strength, toughness and durability. Wrenches also should be heat treated, to prevent wear and to gain other desirable physical properties.

Solid Nonadjustable Wrenches

The term *solid wrench* means that the wrench is made of one piece. The size of the jaw opening is permanent (nonadjustable). Solid wrenches come in sets with a range of jaw openings which accommodate the various sizes normally required in the shop. Solid wrenches are designed so that the overall length is proportional to (1) the size of the bolt or nut and (2) the force normally required for tightening or loosening operations.

The three general types (shapes) of solid nonadjustable wrenches are: (1) *open end*, (2) *socket*, and (3) *box*. Each of these wrench types is available in metric or inch sizes, and singly or in sets.

The size indicates the jaw opening or the distance across the flats on a bolt head or nut. For example, the jaws of a double-end solid wrench that are 20 and 25 mm fit over bolt heads and nuts that are 20 and 25 mm wide. A 3/8″ and 1/2″ hexagon box wrench may be used on hexagon-shaped parts that are 3/8″ and 1/2″ across the flats. Fig. 8-8 shows the three general shapes of open-end and box wrenches.

Solid Open-End Wrenches—Some wrenches are *open ended*. This means they are designed with either one open end or two open ends with different sizes. The open end permits the wrench jaws to be slid onto the sides of a bolt or nut from either the top or side, and from whatever position is most advantageous. Force is then applied to the end of the wrench.

Box Wrenches—The *box wrench* is another form of solid nonadjustable wrench. Box wrenches have

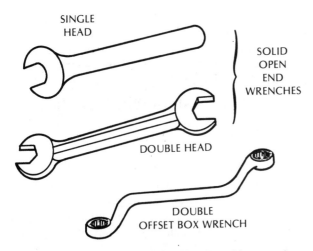

Fig. 8-8. General shapes of solid nonadjustable wrenches.

closed heads (ends). Each head accommodates a particular size of hexagon bolt or nut. The box wrench is used whenever space permits the head to be fitted *over* the bolt or nut.

Socket Wrenches—Socket wrenches (Fig. 8-9) are another type of solid wrench. Hollow sockets are usually fluted. Hexagon sockets fit over hexagon-head nuts and bolts. Square sockets are designed to fit over square-head nuts and bolts.

Socket wrenches provide flexibility. They may be used with a ratchet handle, which permits turning nuts and bolts in confined spaces. They also may be used with flexible adapters and extension bars, which make it possible to turn bolts and nuts that are inaccessible with regular wrenches. The parts of a typical socket-wrench set are shown in Fig. 8-9.

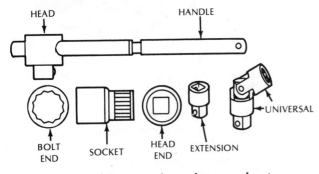

Fig. 8-9. Parts and features of a socket wrench set.

Socket and box wrenches should be used whenever possible in preference to open-end and adjustable wrenches. The proper size and shape of a socket or box wrench *nests*, or *houses*, the nut or bolt to prevent slippage. This permits a greater force to be applied, and makes it easier to manipulate in confined places.

Torque Wrenches—The proper functioning of many parts requires that the contact surfaces be secured together with a uniform (specified) force. The

tightening, or turning, force (torque) must be controlled to prevent overstressing and stripping of the threads or shearing off of the bolt or nut.

Tension or *torque wrenches* (Fig. 8-10) are used to measure and control turning force. A torque wrench is a force-measuring wrench. These wrenches have either a graduated dial or other type of scale which indicates the amount of the applied turning force in foot-or inch-pounds.

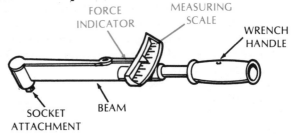

Fig. 8-10. A torque wrench.

Always be sure that the surfaces to be held together are free of burrs and that the threads are in proper condition. Otherwise, regardless of the torque wrench reading, the parts will not be seated properly.

Adjustable Wrenches

Adjustable wrenches are adaptable to a wide range of nut, bolt or workpiece sizes and shapes. Three common adjustable types are the adjustable open-end wrench, the monkey wrench, and the pipe wrench (Fig. 8-11). One adjustable wrench often can replace many solid (fixed-size) wrenches. The sizes of monkey and pipe wrenches are designated by the overall length—examples: a 10″ monkey wrench and a 12″ pipe wrench. The size of an adjustable open-end wrench indicates the maximum dimension of the jaw opening.

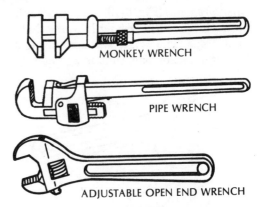

Fig. 8-11. Common types of adjustable wrenches.

Monkey, pipe and adjustable open-end wrenches require that the direction of pull (turning force) be *toward* the adjustable jaw. For personal safety and to avoid damage to the workpiece or part, these wrenches must be checked and adjusted repeatedly during the operation to maintain a close fit between the jaws and the work. The pipe wrench is self-adjusting to a limited extent. The jaws move together when a force is applied.

The pipe wrench is used for turning objects that are round or irregularly shaped. The two jaws have teeth. As pressure (force) is applied in the turning direction, the teeth grip the work and the turning force is transmitted to the part, causing it to turn. The teeth leave indentations and burrs which might cause the workpiece to be rejected.

Setscrew Wrenches

Hollow or socket-type setscrews and other bolts are turned by an offset wrench known as an *Allen* wrench. Fig. 8-12 shows one of these hexagon-shaped wrenches. Allen wrenches are made of alloy steel and are heat treated to withstand great forces. These wrenches are available individually or in sets, and in either millimeter or fractional-inch sizes. The size— for example, 4 mm or 1/8"—indicates the distance across the flats of the hexagon-shaped wrench stock.

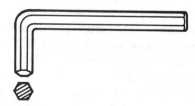

Fig. 8-12. An allen setscrew wrench.

Spanner Wrenches

Spanner wrenches (Fig. 8-13) are occasionally used in bench work. The *face spanner* has pins that fit into notches or holes. The *single end spanner* is used to turn a special nut or bolt that is notched.

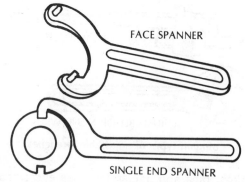

FACE SPANNER

SINGLE END SPANNER

Fig. 8-13. Single-end and face spanner wrenches.

Spanner wrenches are widely used on machine tools. The wrenches are used to turn the large nuts that draw and hold a chuck or other device securely to a tapered machine spindle. The reverse process unlocks the chuck so that it may be removed.

CLAMPING AND GRIPPING DEVICES

Bench Vises

The bench vise is a device that can be adjusted quickly to clamp work between its jaws. The vise is positioned on the bench at a height that permits the worker to comfortably perform bench work operations. A standard bench vise is shown in Fig. 8-14.

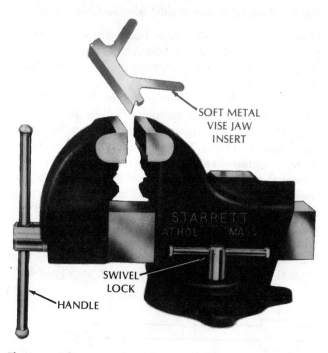

SOFT METAL VISE JAW INSERT

STARRETT ATHOL MASS

SWIVEL LOCK

HANDLE

Fig. 8-14. A bench vise with a soft-metal jaw insert. (Courtesy of The L.S. Starrett Company)

The bench vise top section can be turned horizontally on a simple swivel arrangement. A swivel lock secures or releases the vise top section from the base, which is attached to the bench. This swivel feature permits work of different sizes and shapes to be held in the position desired for performing bench operations.

The regular vise jaws are *serrated* (grooved) to provide a good gripping surface. Some workpiece surfaces, however, must be protected against the scratches, burrs and indentations caused by direct contact with the jaws. *Soft jaws* which fit over the regular vise jaws are used to prevent scoring of such workpiece surfaces. The soft jaws are made of materials that are softer than the workpiece.

Sometimes, other soft metal pieces are placed between the vise jaws and the surfaces of the work.

Pliers

Pliers are used for gripping, turning, and drawing wire and small parts and for cutting metal wires and pins. The two most commonly used pliers (Fig. 8-15) are *slip-joint* pliers and *needle-nose* pliers. Slip-joint pliers can be used for small- to medium-size work applications which require gripping action. The two-position joint of these pliers provides two ranges of jaw opening sizes. The handles are positioned comfortably for optimum gripping power for each opening size. *Needle-nose pliers* are most practical for holding small workpieces.

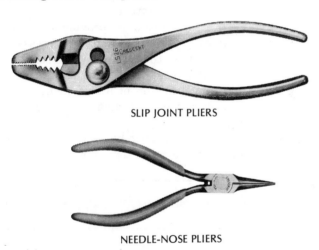

SLIP JOINT PLIERS

NEEDLE-NOSE PLIERS

Fig. 8-15. Two basic styles of pliers. (Courtesy of The Cooper Group)

Clamps

The "C" clamp (Fig. 8-16) usually is used for comparatively heavy-duty clamping. The "C" clamp also permits the clamping of parts that are wider than can be accommodated with parallel clamps. The "C"-shaped frame permits the clamping screw to fit into and be tightened in confined spaces.

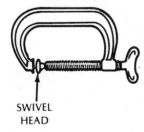

SWIVEL HEAD

Fig. 8-16. A common "C" clamp.

Parallel toolmaker's clamps have two jaws that provide parallel clamping surfaces. The distance between the two jaws is controlled by two screws (Fig. 8-17). The *adjusting screw*, which is nearest the jaws, is used to move the jaws together. The outside *tightening screw* provides leverage. When the jaws are parallel with the work faces, sufficient force may be exerted to securely hold the pieces. It is important that as much of the full surface of each parallel jaw be in contact with the work as is possible.

Fig. 8-17 shows a simple set of toolmaker's clamps. These holding devices generally are used for clamping workpieces with finished surfaces during layout, inspection, and basic machining operations. Because the clamping faces of each jaw are ground smooth, they will not mar a finished surface.

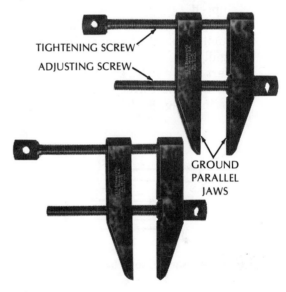

TIGHTENING SCREW

ADJUSTING SCREW

GROUND PARALLEL JAWS

Fig. 8-17. Toolmaker's parallel clamp set. (Courtesy of The L.S. Starrett Company)

INDUSTRIAL GOGGLES AND FACE SHIELDS

Although industrial goggles and face shields are not classified as tools, they are introduced at this point because their use is essential in preventing eye injuries. They are especially important in bench and floor work and in machining operations because of the ever-present danger of injuries from flying particles and other foreign matter.

The worker should obtain a personal pair of goggles or an appropriate shield. The shape and size should provide both comfort and unobstructed vision. The design also should provide protection during both heavy-duty bench and machine operations. The goggles should be ventilated to prevent *fogging*, yet should protect the eyes against grit and other small particles. Goggles which fit over vision-correcting eyeglasses also are available.

Face shields are preferred as a light weight and cooler eye protection device. The shields are practical in operations that do not involve dust, abrasives, and other fine particles.

HOW TO USE HAMMERS

General Hammering

Step 1   Select a hammer weight suitable for the size of the job. Inspect the handle to see that it is securely fastened to the head. Check the faces to be sure they are not chipped.

Step 2   Grasp the hammer near the end of the handle. Hold the hammer firmly but not rigidly.

Step 3   Start with a light, sharp blow to get the *feel* and the correct distance for hammering (Fig. 8-18).

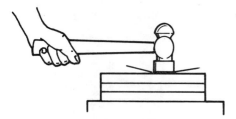

Fig. 8-18. Delivering a square, solid blow.

Step 4   Continue, striking heavier blows.

CAUTION: Watch the point of action and not the hammer head.

Riveting

Step 1   Strike a blow at a slight angle on the edge of the rivet head. This starts the riveting action.
NOTE: Sometimes a head is riveted by *peening*. Peening consists of a series of indentations around the periphery (edge) of a pin or other part. The indentations are produced by striking the head with the ball end of the hammer.

Step 2   Continue around the entire edge of the rivet head. Peen or flatten the edge as required. The shape of the formed rivet head is shown in Fig. 8-19.

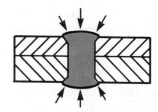

Fig. 8-19. Shape formed with ball or flat end of ball peen hammer.

Driving With Soft-Face Hammers

Step 1   Select a soft-face hammer that has softer head material than the part to be driven or assembled.

Step 2   Secure the part or mechanism in a vise or other holding fixture.
NOTE: The vise must provide adequate support to permit applying the necessary force to drive the part. Check to see that the driven part will clear the vise jaws.

Step 3   Apply a thin film of lubricant between the mating parts that are to be driven into position.

Step 4   Deliver sharp, firm blows to separate or to drive the parts together.

HOW TO USE SCREWDRIVERS

Step 1   Select the appropriate screwdriver type, tip, and size for the job.

Step 2   Inspect the tip to be sure the blade (flat-head type) or flutes (Phillips type) will enter the screw head. The tip or flutes must seat properly in the slot(s) in the screw head.

Step 3   Grasp the handle in one hand. Guide the blade or fluted tip into the screw slot(s) with the other hand (Fig. 8-20).

Fig. 8-20. Turning a Phillips head screw.

Step 4   Apply force toward the screw head. At

the same time, turn the screwdriver handle clockwise to tighten a regular right-hand screw (Fig. 8-21). Use a counterclockwise turning force to loosen such a screw.

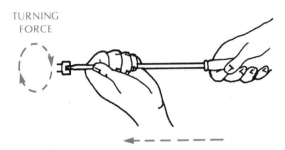

TURNING FORCE

Fig. 8-21. Force should be applied toward the screw as it is turned.

Step 5   Tighten all screws by applying uniform turning force. If uniform appearance is required, all screw slots should be aligned in the same relative position.

HOW TO USE WRENCHES

Open-End Wrenches

Step 1   Select an open-end wrench whose jaws fit the nut or bolt.

Step 2   Place the wrench in the position in which the greatest leverage can be applied.

Step 3   Pull the wrench handle clockwise to tighten a right-hand bolt or nut. Reverse the direction to loosen these types of bolts and nuts. Gradually apply the turning force. Use either one or both hands, depending on the job requirements.

CAUTION: When great turning force is needed, be sure that the wrench jaws do not spring out of shape. The fit between the wrench jaws and work should be checked repeatedly.

Step 4   Move (force) the wrench as far as space permits. Turn over the wrench so that the part can be turned an additional amount. Then, turn over and reposition the wrench on the part. Repeat these procedures until the part is properly tightened or has been removed.

Box And Socket Wrenches

Step 1   Select a correct size of box or socket wrench in preference to an open-end wrench, if possible.

Step 2   Determine what socket-wrench accessories, if any, are needed to provide the required leverage with the greatest efficiency.
NOTE: A solid or a ratchet handle may be used alone with the socket head. Other applications, however, might require the use of an adapter, universal joint or extension bar.

Step 3   Place the wrench in the position that will provide maximum leverage. Apply a small amount of lubricant to the threaded parts. Gradually apply force to the end of the wrench handle (Fig. 8-22). A clockwise force tightens and a counterclockwise force loosens a right-hand threaded part.

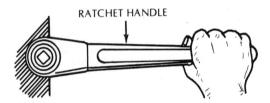

RATCHET HANDLE

Fig. 8-22. Applying a turning force with a ratchet handle.

CAUTION: The amount of force to use depends on the size and requirements of the job. Accurate judgment by the worker is important. Excessive force can *round over* the serrations in the wrench end or socket or damage the bolt head or nut, or break the part.

Torque Wrenches

Step 1   Use a torque wrench to uniformly tighten bolts and nuts. Pull the wrench handle until the amount of force read on the scale meets the job specifications.
NOTE: Parts should be assembled by first turning the bolts and nuts by hand. Then, use a conventional ratchet or socket until all bolts and nuts are uniformly seated. When two or more nuts are to be tightened, they should be turned alternately and uniformly.

Adjustable Open-End And Monkey Wrenches

Step 1   Select an adjustable open-end or monkey

wrench. The wrench size must be proportional to the size of the nut, bolt or part to be tightened or loosened.

Step 2   Adjust the wrench jaws to the approximate size of the part to be turned and place the wrench on the part. The adjustable jaw should be on the side *toward* which the turning force is to be applied.

Step 3   Adjust the jaws to fit the part.

Step 4   Apply force slowly in the direction of the movable jaw (Fig. 8-23). Continue to increase the force until the part begins to move.

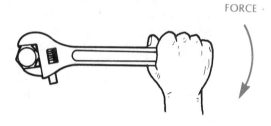

FORCE

Fig. 8-23. Correct hand and wrench position for applying turning force with an adjustable end wrench.

CAUTION: Repeatedly check the jaws and work to ensure that a close fit is maintained as increased force is applied.

Pipe Wrench

Step 1   Check the condition of the workpiece and the teeth of the pipe wrench jaws.

Step 2   Adjust the movable jaw until the wrench opening permits gripping the work in the *center* of the jaws.

Step 3   Apply force on the end of the handle *toward* the movable jaw. Put oil on any threads.

CAUTION: Keep checking the position of the wrench jaws in relation to the work. The wrench should have a *good bite* before full force is applied. This prevents slippage.

Socket-Head (Allen) Wrenches

Step 1   Remove dirt and other foreign particles from the screw head.

Step 2   Select the appropriate socket-head wrench that fits the screw head.

Step 3   Grip the socket-head wrench in one hand and apply appropriate force in the direction required for tightening or loosening. Judgment must be used in exerting force appropriate for the size and construction of the part.

Spanner Wrenches

Step 1   Select a hook spanner or a face (pin) spanner wrench that fits the particular part. The correct position and fit of a hook spanner is illustrated in Fig. 8-24.

Step 2   Apply the necessary force near the end of the handle. Reverse the hook spanner to turn a nut in the opposite direction.

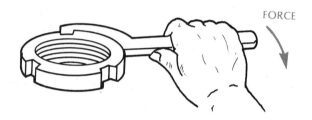

FORCE

Fig. 8-24. Position and fit of a hook spanner wrench.

HOW TO USE CLAMPING AND GRIPPING DEVICES

Bench Vise

Step 1   Clean the vise. Regularly oil the moving parts.

Step 2   Determine whether the workpiece may be clamped directly. Use soft jaws or other appropriately soft material to protect finished surfaces.

Step 3   Position the work between the vise jaws. *NOTE:* Assembly and other hand operations should be performed as close as possible to the vise jaws. Make sure the surfaces to be clamped are strong enough to withstand the compressing force of the vise jaws. Place a block below the work to prevent it from moving during hand operations.

Step 4   Bring the vise jaws against the work. Turn the vise handle to apply sufficient force to securely hold the workpiece in position.

Pliers—Noncutting Operations

Step 1   Select the proper type and size of pliers for the job.

Step 2   Position the workpiece between the nose of the pliers, then apply a gripping force to the handles (Fig. 8-25).

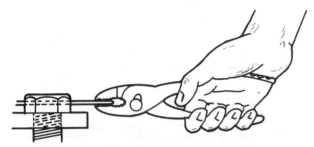

Fig. 8-25. Gripping and applying a force with pliers.

Step 3   Hold, pull, insert, bend or twist the part as required.

Pliers—Cutting a Part

Step 1   Insert the unhardened part between the cutting area of the jaws at the specified length.

Step 2   Shear the part by forcing the handles together.

"C" Clamps And Parallel Clamps

Step 1   Remove burrs and clean the workpieces and clamps. Apply a drop of oil to the moving parts of the clamps.

Step 2   Open the clamp so that the jaws are slightly wider than the thickness of the parts to be clamped. Use soft metal inserts between the clamp jaws and the work surfaces that must be protected.

Step 3   Position the clamp over the work so that bench or machine operations can be performed without interference. Clamps often are used in sets to hold parts more securely and accurately (Fig. 8-26).

Fig. 8-26. "C" clamps used in sets.

Step 4   Move the clamp jaws together. Turn the front adjusting screw of a parallel clamp

with the fingers to set the jaws to the width of the work. Bring the jaws parallel by turning the back (outside) tightening screw.

Step 5   Insert a pin or round metal object in the hole of the adjusting screw.

Step 6   Turn this screw as shown in Fig. 8-27.

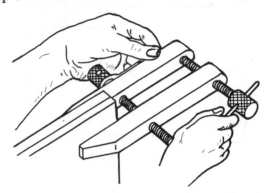

Fig. 8-27. The clamp jaws should be parallel for applying force.

This brings the jaws parallel. Once in position, apply an additional clamping force to the two jaws.

NOTE: When properly positioned and secured, the jaws should be parallel and tightly forced against the workpieces.

REPRESENTING PARTS BY A ONE-VIEW DRAWING

Many uniformly shaped flat or cylindrical parts are adequately described by a *one-view* drawing. The one-view drawing with dimensions and notes usually provides complete technical information. From it a part may be layed out, produced, and assembled. A one-view drawing saves drafting time and simplifies the reading of a blueprint.

A center line is used in one-view drawings to indicate a part that has symmetrical features. A diameter on a cylindrical part is dimensioned with the abbreviation DIA. This designation is used according to American Standards. In SI metrics, the symbol ⌀ replaces DIA. The two American standard diameters in Fig. 8-28A are indicated by DIA. The metric diameter sizes in Fig. 8-28B are given in mm, followed by the symbol ⌀.

DRAWING NOTES PROVIDE TECHNICAL INFORMATION

Notes, or written information, are included on drawings to supplement what is represented in graphic form, including dimensions. A note can save the drawing of a second or third view.

The note in Fig. 8-29 tells that the template is made of 3-mm thick, ground-flat stock. All other shape, feature, and size information is provided by the dimensioned one-view drawing.

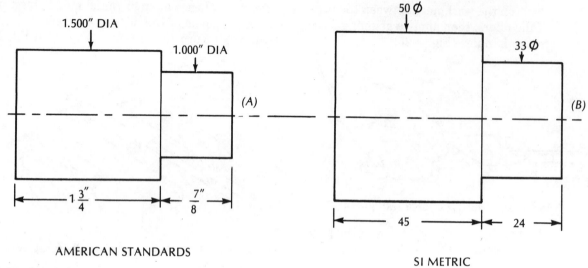

AMERICAN STANDARDS

SI METRIC

Fig. 8-28. A one-view representation of cylindrical parts.

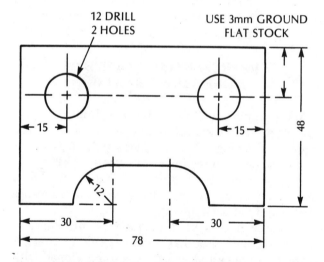

Fig. 8-29. Complete description provided by a one-view drawing and note.

SAFE PRACTICES FOR USING BENCH TOOLS

Hammering Tools

- Examine the hammer each time *before* it is used. The handle must be securely wedged in or bonded to the head. The flat (slightly crowned) face must not be chipped.
- Use a soft-face hammer if the work surface must not be dented by the hammer blow.

Fastening Tools

- Select a screwdriver with a tip which is only slightly smaller than the diameter of the screw head. The blade or flutes should fit snugly against the slot sides and should seat against the bottom of the slot.
- Firmly press the screwdriver tip into the screw slot or cross-slots while applying turning force. This helps prevent the tip from slipping out of the slot(s), which can cause burring of the screw head slot(s), and personal injury.
- Select the type, shape, and size of wrench opening most suitable for the application. Position the wrench jaws as close to the work as is possible. In this position, the wrench is less likely to slip.

- Turning force should be applied to an adjustable wrench only in the direction of the *movable* jaw.
- Apply turning force gradually, while carefully monitoring the position of the wrench and the work to detect signs of slipping. Excessive force can cause the jaws of any open-end wrench to spring.
- Protect the surfaces of a workpiece from indentions produced by pipe wrenches. Such cuts into a work surface can ruin the workpiece.

Clamping Devices

- Use soft jaws or other soft material when holding finished work surfaces in a bench vise. These prevent marring of the finished surface.
- Position the workpiece in a vise so that it does not overhang into an aisle or other area where a person might accidently brush against it.

NONCUTTING HAND TOOL TERMS

Ball peen, straight peen and cross peen hammers	Three common forms of hammer heads used in machine and metal trades. Each of these hammer heads has one end that is a cylindrical solid with a slightly crowned face. The opposite end is either shaped like a ball or formed into a triangular prism. The apex of the triangular section has a blunt rounded edge.
Soft-face hammer	A hammer whose end faces are softer than the workpiece. A hammer having crowned faces made of rawhide, plastic, or other relatively soft material.
Fasteners	Screws, nuts, bolts, keys, pins, and other devices that fasten or secure parts together.
Screwdriver head (flat and Phillips)	The end of the screwdriver blade that conforms to the shape and size of the screw slot(s) into which it is inserted.
Offset screwdriver	A single-piece metal screwdriver with two formed ends (tips). The ends may be on the same side or at right angles to the body.
Solid open-end wrenches (nonadjustable)	A one-piece steel wrench consisting of a body and one or two open ends. Each end is machined to fit a particular nut or bolt head size. The ends are open on the front side to permit sliding them into position on the nut or bolt.
Box wrenches	A solid-steel wrench with a hole machined into one or both ends. Each hole fits around a particular size and shape of bolt, nut or part.
Socket wrenches	A form of solid wrench. One end of each socket fits over a bolt or nut of a standard size and shape. The other end of each socket has a square hole into which fits the square

	shank of a ratchet or straight handle or an accessory such as an extension, adapter or universal joint.
Torque wrench	A calibrated inch- and foot-pounds measuring device which attaches to a socket. Simultaneously turns nuts and bolts and measures the turning force (torque).
Adjustable wrenches	A group of wrenches whose jaw openings can be adjusted to fit a number of different sizes and shapes of nuts and bolts or round or irregularly shaped parts. Three common types are 1) the monkey wrench and 2) the open-end adjustable wrench, for flat surfaces; and 3) the pipe wrench, for round (pipe) or irregularly shaped parts.
Socket head (Allen) wrench	A metal piece whose size and cross-section fit the hollow or socket type screw heads. One end is offset to form a handle for applying turning force.
Spanner wrench	A special form of wrench that fits around large, round, threaded collars. The collars have either slotted edges or holes in their faces. Single- and double-pin spanner wrenches fit these respective types of collar.
Pliers (slip-joint and needle-nose)	A gripping tool for turning, twisting, bending, shearing or manipulating wires, pins and similar small parts. A hand tool consisting of two parts that are joined together and move on a pivot. One end is shaped with serrated grooves to hold workpieces. Force is applied on the jaws by gripping the handles.
Clamps (C and parallel, or toolmaker's)	Work holding clamping devices of "C" or parallel forms.
Eye protective devices	Face shields and industrial goggles of nonshatterable materials and enclosures for protecting the eyes against flying particles or splattering materials.
One-view drawing	Representing and completely describing an object by a single view.
Drawing note	Written remarks on a drawing that supplement the graphic description and dimensions.

SUMMARY

- Bench tools are broadly grouped to serve such functions as:
  - Hammering
  - Clamping
  - Fastening
  - Forming
  - Sawing
  - Filing
  - Chiseling
  - Threading
  - Reaming
- The common fastening tools used in bench work and for other machining operations include: standard, Phillips head types, and offset screwdrivers.
- Wrenches are of the nonadjustable plain and socket types and the adjustable type. Other wrench types include setscrew wrenches, special spanner wrenches, and monkey and pipe wrenches.

- The ball peen hammer is used to apply a flat blow, to peen an edge, or to direct a blow to a limited area.
- Two common types of clamping devices are the bench vise and "C" and parallel clamps.
- Symmetrical flat and cylindrical parts are often fully described by a one-view drawing and supplemental notes.
- Safe work practices must be followed in bench work to:
  - prevent damage to finished work surfaces
  - avoid personal injury from sharp or burred edges or parts extending into aisle areas
  - protect tools and instruments from damage.

1. List four noncutting and four cutting and shaping bench processes.
2. Identify one type of hammer that (a) is used to drive a pin and (b) another that is used for assembling mating parts.
3. a. Describe a flat-and a Phillips-head screwdriver.
   b. State two advantages of the Phillips-head screwdriver over the flat-head type.
4. Recommend a hand wrench to use in each of the following applications:
   a. Tightening a hexagon nut in a confined recessed area
   b. Tightening large-size, square-head lag screws
   c. Assembling hexagon bolts in areas that are inaccessible to open end wrenches, where each bolt must be uniformly tightened
   d. Assembling and disassembling large diameter hexagon coupling nuts
   e. Turning hollow or socket type setscrews
   f. Tightening a shouldering collar that (1) is fitted with a single slot on the periphery and (2) another that is fitted with two slots on the face.
5. Describe how holding workpieces with a C-clamp differs from gripping with a set of parallel clamps.
6. State two factors that should guide the worker in selecting industrial goggles or a protective face shield.
7. List four general safety precautions to observe when selecting and using hammers, screwdrivers, and wrenches.
8. Tell when a one-view drawing is appropriate to fully represent a part on a drawing.
9. Cite four safe practices to observe when using bench tools.

SECTION TWO

CUTTING AND SHAPING HAND TOOLS

Bench and floor work processes that involve cutting and shaping with hand tools are: hand sawing; chiseling, punching, and driving; filing; scraping, burring, and polishing; internal and external threading; and reaming. The principles of and correct practices for performing these hand processes are presented in this section.

Unit 9

HAND SAWING PROCESSES AND TECHNOLOGY

Hand sawing of metals and other hard materials is an everyday shop process. The hand tool used for this purpose is the *hacksaw*. The correct use of the hacksaw enables the mechanic to efficiently and safely cut material and avoid breakage and unnecessary wear of the saw blade.

Two sets of related factors should be considered when selecting the correct hacksaw frame and blade:

• The shape, size, and kind (properties) of material to be cut

• The features of a particular saw blade. These include: *pitch*, *set* of the teeth, and length; and the metal of which the blade is made, and its heat treatment.

HACKSAW FRAMES

Hacksaws are named according to the shape of the handle, such as *pistol grip* and *straight handle*. Both of these types are available with either (1) a solid-steel (nonadjustable) frame that fits one length of blade or (2) an adjustable frame that accommodates various lengths of blades. The pistol-grip type with an adjustable frame is shown in Fig. 9-1. This particular adjustable frame accommodates hacksaw blade lengths ranging from 8″ to 10″ (200 mm to 250 mm).

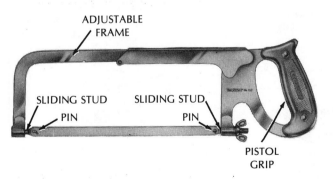

Fig. 9-1. An adjustable pistol-grip frame. (Courtesy of The L.S. Starrett Company)

The front and handle ends of the frame are equipped with *sliding studs*. These may be turned to and locked in any one of four different positions. They

permit turning the blade to 90°, 180°, 270° and the normal vertical position. This turning of the blade makes it possible to position the frame so that it clears the work during the cutting action.

There is a pin on each sliding stud. The frame is set to the blade length. The holes in the blade ends are then inserted over the stud pins. The blade is seated against the flat part of each stud. Tension is applied to hold the blade tightly in the frame. This is done by turning the threaded handle of the straight-handle type or the wing nut on the pistol-grip type. The selection of the hacksaw frame is a matter of personal choice. As with other hand tools, the frame and handle of the hacksaw should feel comfortable. The cutting operation should not become tiresome.

Another important consideration is the selection of the correct blade.

HACKSAW BLADES

Hacksaw blades are made from high-grade tool steel, tungsten alloy steel, tungsten high-speed steel, molybdenum steel, and molybdenum high-speed steel. The tungsten adds *ductile* strength to relieve some of the brittleness. The molybdenum adds to the wearing qualities.

All blades are hardened and tempered. There are three hardness classifications: *all hard*, *semiflexible*, and *flexible back*. Each term describes the heat treatment of a particular blade. *All hard* indicates that the *entire* blade is hardened. Although the all-hard blade retains a sharp cutting edge longer, it is relatively brittle and may be broken easily. By contrast, only the *teeth* on the flexible back blade are hardened. This blade can *flex* (bend slightly) during the cutting process without breaking.

The blade width is standardized at approximately 1/2″ or 12 mm. Various lengths of blades are available. The most common ones are 8″, 10″, and 12″ (200, 250 and 300 mm) long. *Length* refers to the center-to-center distance between the holes in the ends of the blade. Hand hacksaw blades are 0.025″ (0.6 mm) thick.

Two other important factors also enter into the selection of the blade:

- The *number* of teeth
- The *set* of the teeth.

Hacksaw blades are made with a specified number of teeth per inch. This is called the *pitch of the blade* (Fig. 9-2). A blade with 14 teeth per inch is a *14 pitch* blade. Blades are made with pitches that vary from coarse (14 pitch), to medium (18 or 20), to fine (24 to 32).

Fig. 9-2. The pitch of a hacksaw blade.

Table 9-1. MANUFACTURERS' RECOMMENDED PITCHES FOR HAND HACKSAWING

Material Types And Recommended Pitches	Cutting Conditions	
	CORRECT PITCH	INCORRECT PITCH
14 Pitch For large sections and mild materials	GOOD CHIP CLEARANCE	PITCH TOO FINE NO CHIP CLEARANCE RESULT: CLOGGED TEETH
18 Pitch For harder materials like tool steel, high-carbon, and high-speed steels		
24 Pitch For angle iron, brass, copper, iron pipe, and electrical conduit	AT LEAST TWO TEETH ON A SECTION	PITCH TOO COARSE ONE TOOTH ON SECTION RESULT: STRIPPED TEETH
32 Pitch For thin tubing and sheet metals		

The pitch to select depends upon the hardness, the shape, and the thickness of the part to be cut. Manufacturers' recommended hacksaw blade pitches for different materials are given in Table 9-1. Both the correct cutting action and the result of incorrect pitch are illustrated.

The general practice for cutting tubing, thin materials, or unusually shaped workpieces is to use a blade pitch which places two or more teeth in contact with the work at all times.

In addition to pitch, the hacksaw teeth are *set*. *Standard set* means that one tooth is offset to the right and the next to the left, etc. The set permits a groove *(kerf)* to be cut freely. The groove is slightly wider than the thickness of the blade. This prevents the blade from binding. Some fine-pitch blades have a *double alternate set,* or a *wave set.* The *wave set pattern* has alternate sets of teeth formed to the right and left of the back. Kerf, standard, and alternate (wave) sets are illustrated in Fig. 9-3.

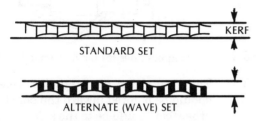

Fig. 9-3. Standard and alternate set patterns.

CUTTING SPEEDS AND FORCE (PRESSURE)

For hand hacksawing, blade manufacturers recommend a cutting speed of 40 to 50 strokes a minute. At this speed, the worker is able to control (relieve) the force on the teeth during the return stroke, and saw without tiring.

The amount of force to apply when cutting depends, again, on the skill and judgment of the operator. Enough force must be applied on the *forward* cutting stroke to permit the teeth to cut the material. The blade should not slip off or slide over the work. This action causes the teeth to become *glazed* and dulled. Broken blades often result from too much force on or twisting of the blade in the saw cut. When blades are subjected to too much speed and force, the heat

generated at the teeth is sufficient to draw the temper of the blade. The blade becomes soft and useless.

HOLDING IRREGULAR SHAPES AND THIN-WALLED MATERIALS

A general principle in hand hacksawing is to always engage two or more teeth. This minimum number tends to prevent *digging in, stripping* of the teeth, or *bending* of the part. Thin, flat materials should be held between blocks of wood. Tubing and other circular shapes should be *nested* in a simple form made of soft material. A wider area of clamping surface is thus provided.

On thin-gage tubing, a wooden plug may be inserted in the center. Sometimes, the hacksaw cut is made through both the wood support and the workpiece. This practice permits a greater number of teeth to be engaged. Breaking through or damaging the part is avoided.

Common practices for holding structural shapes are illustrated in Fig. 9-4A. Similarly, the three sketches in Fig. 9-4B show how thin-wall and irregular-section parts should be nested and held in a vise.

HOW TO USE THE HAND HACKSAW

Selecting The Correct Frame And Blade

Step 1   Examine the work order or blueprint.

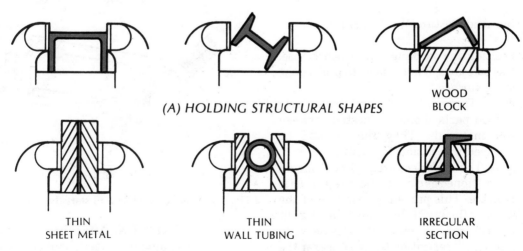

(A) HOLDING STRUCTURAL SHAPES

WOOD BLOCK

THIN SHEET METAL

THIN WALL TUBING

IRREGULAR SECTION

(B) HOLDING THIN WALL AND IRREGULAR PARTS

Fig. 9-4. Holding structural, thin wall, and irregular shapes in a vise.

Determine the nature and shape of the material to be cut.

Step 2 Select a hacksaw frame appropriate for the job.

Step 3 Select the length, pitch, set, and hardness of the hacksaw blade according to the job requirements.

Step 4 Place the sliding studs in the correct position for cutting. Then, insert the blade in the frame. The cutting teeth must point *away* from the operator and toward the front of the frame (Fig. 9-5).

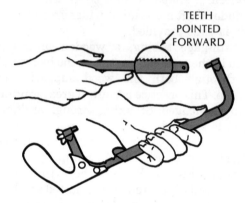

Fig. 9-5. Positioning a hacksaw blade in a frame.

Step 5 Adjust the tension of the blade in the frame. It should be tight enough to prevent the blade from *buckling* and *drifting* during use.

CAUTION: Excessive tightening may cause the frame to bend, the blade to break, or the stud pins to shear.

Work Position And Posture While Sawing

Step 1 Position the workpiece in a bench vise. Use soft jaws, if necessary, to protect the work surface.
Note: Thin-wall parts and irregularly shaped parts should be nested between soft materials. They then should be clamped in position for cutting.
Note: The workpiece area to be cut should be positioned as close to the vise jaws as is possible. This prevents springing of the work or flexing of the blade. If a long cut is required, it may be necessary to frequently reposition the part nearer the vise jaws.

Step 2 Grasp the front of the hacksaw frame with one hand. Hold the handle in the

other. The correct position of the hands is shown in Fig. 9-6.

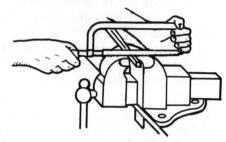

Fig. 9-6. Correct hand positions for holding a hand hacksaw.

Step 3 Take a position near the vise. One foot should be pointed toward the bench. The other foot should be to the side and back. This gives balance when sawing.

Step 4 Move the body on both the forward cutting and return strokes. Lean forward from the hips on the forward stroke. Return to the original position at the end of the return stroke.

Starting The Cut

Step 1 Break any sharp corner that may strip the saw teeth. This is done by filing a small groove, or indentation close to the layout line or a measured point.

Step 2 Place the front of the blade in the indentation, or guide it to the cutting line. To *guide the blade,* hold the handle in one hand. Place the side of the thumb (Fig. 9-7) against the flat portion of the hacksaw blade. Move the blade to the starting point of the cut.

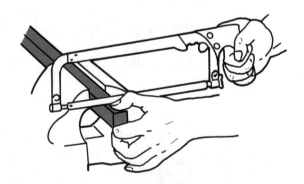

Fig. 9-7. Guiding the blade to start the cut.

CAUTION: After the cut is started, the thumb no longer should be used to guide the blade.

Step 3 Apply a slight, steady force. Push the saw

forward and at a slight angle across the surface of the work. The arrows in Fig. 9-8A show the forces that should be applied during the cutting stroke.

Note: The force depends on the thickness of the material and its composition. For example, a thick part made of machine steel requires a greater force than does a softer metal like brass or aluminum.

Step 4   Release the force during the return stroke. Raise the frame and blade. Draw the hacksaw straight back to the starting point. The dotted arrows in Fig. 9-8B indicate the releasing of the forces.

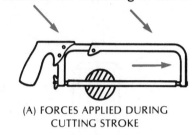

(A) FORCES APPLIED DURING CUTTING STROKE

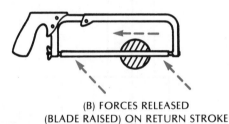

(B) FORCES RELEASED
(BLADE RAISED) ON RETURN STROKE

Fig. 9-8. Forces applied and released during hand hacksawing.

Step 5   Make each stroke as long as the blade, frame, and work area permit. Maintain long, steady strokes. Avoid the tendency to speed up.

Finishing The Cut

Step 1   Slow down the cutting action near the end of the cut. This is a safety precaution to prevent personal injury, damage to the work, or breaking of the blade. These result from abruptly breaking through the cut, losing balance, and having the

free end of the work tear away from the portion held in the vise.

Step 2   Clean the chips from the blade. Loosen the tension on the blade and frame. Remove the burrs from the workpiece.

REPRESENTING PARTS WITH TWO-VIEW DRAWINGS

A one-view drawing may not give full information about the shape, construction and size of every workpiece. The addition of other views often provides missing details. The term *two-view drawing* is descriptive of the number of views that are used to represent a part. The two views are selected by the designer or draftsperson. The names of the views may be different. They depend on how the object is viewed.

The T-stud in Fig. 9-9 may be represented by one of a number of different views. Five arrangements of views are illustrated in Fig. 9-10. In Fig. 9-10A the T-stud is viewed from a horizontal position. Three different sets of views are shown in Fig. 9-10B. In each, the T-stud is shown in a vertical position. Names are given to each of the two-view drawings. Normally, these names do not appear on the drawings. The drawings are completed by the addition of dimensions and tolerances. A note may be included on the drawing to indicate the kind of material and/or the quantity needed.

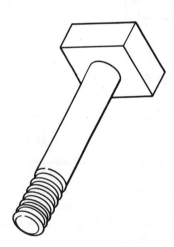

Fig. 9-9. A sketch of a T-stud.

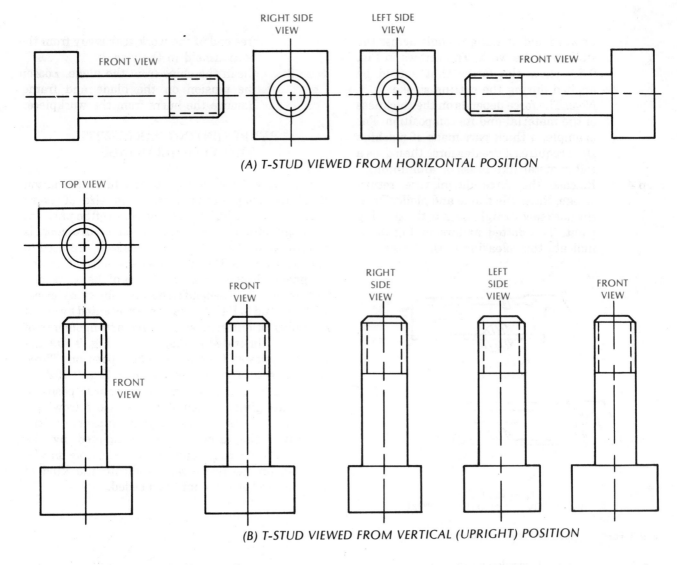

(A) T-STUD VIEWED FROM HORIZONTAL POSITION

(B) T-STUD VIEWED FROM VERTICAL (UPRIGHT) POSITION

Fig. 9-10. Two-view drawings of a T-stud.

SAFE PRACTICES IN HAND HACKSAWING

- Position the workpiece area where the cut is to be made as close to the vise as is possible. This practice prevents springing, saw breakage, and personal injury.
- Clamp the workpiece securely. However, remember that finished surfaces must be protected from damage caused by direct contact with the vise jaws.
- Nest thin-wall tubing and other easily damaged parts in suitable forms. Otherwise, they may be collapsed or distorted by the clamping force.
- Select the blade pitch and set that is most suitable for the material and the nature of the cutting operation. Stripped teeth often result when the pitch is too coarse and fewer than two teeth are in contact with the workpiece at one time.
- Apply force only on the *forward* (cutting) stroke. Relieve the force on the return stroke.
- Reduce the speed and the force at the end of a cut. This prevents the teeth from digging into the small remaining section of the material. It also prevents the cut portion from breaking away.
- Start a new blade in another place when a blade breaks during a cut. This helps prevent binding and blade breakage. Remember, the groove produced by a blade becomes narrower as the blade wears.

- Saw straight. If the cut *runs*, reposition the workpiece a quarter turn and start a new cut. This prevents cramping of the blade, and also helps guide the second cut in the correct line.
- Cut a small groove with a file in sharp corners where a saw cut is to be started. The groove permits accurate positioning of the saw, and also prevents stripping of the teeth.

- Clamp thin metal strips between two pieces of wood. Cutting through both the wood and the metal prevents the saw teeth from digging in and bending the metal. The work also can be cut more easily and accurately.

HAND HACKSAWING TERMS

Pistol-grip or straight-handle	Two general types of hand hacksaw frames.
Hacksaw blade	A thin blade of a high-grade steel and other special alloys. Teeth of a particular pitch and set are cut on one edge and hardened.
Pitch (range)	The number of teeth per inch or metric equivalent. The general range for hand hacksaw blades is 14 to 32.
Set	The alternate positioning of hacksaw blade teeth to cut a *kerf* (groove) wider than the blade body thickness.
Flexible blade	A hacksaw blade that is heat treated. The teeth are harder than the body, and the blade may be flexed.
All-hard blade	A heat-treated hacksaw blade whose body and teeth are uniformly hard.
Cutting speed	The number of strokes per minute recommended by hacksaw blade manufacturers. The speed at which maximum performance and control of the cutting process may be obtained.
"V" indentation (notching)	The filing of a slight groove at a sharp corner to prevent stripping the saw teeth during the start of the cut.

SUMMARY

- Hand hacksawing is a practical method of cutting materials of different degrees of hardness to a particular size or shape.
- The characteristics of both the workpiece and the hacksaw blade must be considered when selecting the correct blade.
- Hacksaw frames are adjustable to accommodate different lengths of hacksaw blades. The blade may be held securely in any one of four frame positions.
- Hacksaw blades may be all-hard, semiflexible, or flexible back. All blades are designated according to (1) the material of which they are made, (2) the pitch, (3) the set, (4) the length, and (5) the heat treatment.
- Blades are positioned correctly in the frame with the teeth pointed *toward* the front.

- The range of pitch for hand hacksaw blades is from 14 to 32. This range permits cutting solid, tubular, and thin sheet metals.
- Cutting speeds of from 40 to 50 strokes per minute are recommended. The speed should be reduced near the end of the cut.
- The cutting action is produced by *applying* a steady forward and downward force during the *forward* stroke.
- The force is *released* and the blade is *raised* slightly on the *return* stroke.
- Tool safety precautions are necessary to prevent personal injury and damage to tools and materials. Attention must be paid to:
  (1) Starting a cut on narrow-sectioned material
  (2) Positioning and holding the workpiece
  (3) Replacing a worn blade
  (4) Cutting action at the end of the cut.

UNIT 9 REVIEW AND SELF-TEST

1. Explain the design features of a hand hacksaw frame that permits the hacksaw blade to be positioned straight or at 90°, 180°, and 270°.
2. Identify four materials from which hacksaw blades are manufactured.
3. List the kind of information that must be supplied in ordering hand hacksaw blades.
4. State one principle to follow for hand hacksawing thin-sectioned materials and structural shapes.
5. Define the meaning of *cutting speed* as applied to hand hacksaw blades.
6. Indicate three possible causes of glazing and premature dulling of hand hacksaw blade teeth.
7. Tell how to start a cut at a sharp corner that may strip the saw teeth.
8. Give the reason for using a two-view drawing instead of a one-view drawing.
9. List a safe practice to follow in hand hacksawing when a blade breaks before the cut is completed.
10. List three cautions to observe in setting a hand hacksaw blade.

Unit 10

CHISELING, PUNCHING, AND DRIVING TOOLS AND PROCESSES

CHISELS

Many workpieces require the cutting away of metal or other hard material by a hand process called *chiseling*. Machine tools such as the lathe, milling machine, shaper and planer also remove metal, but with greater efficiency.

Cold Chisels are used for cutting metals and other materials on both bench and floor work. The three main parts of a cold chisel are: (1) the *cutting end* (edge), (2) the *body,* and (3) the *head* (Fig. 10-1).

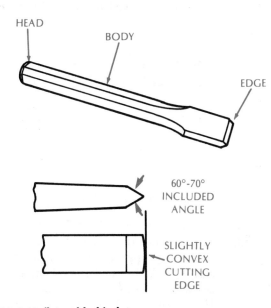

Fig. 10-1. A flat cold chisel.

Cold chisels may be driven (forced) with a hand hammer or a pneumatic (air) hammer. The body shape of the chisel may be hexagonal, octagonal, rectangular, square, or round. Cold chisels are usually forged to shape, hardened, and tempered. Cold chisels will cut materials that are softer than the hardness of the chisel itself.

Common Types of Cold Chisels

Cold chisels are named according to the shape of the cutting edge. The chisel that is most common is the *flat chisel.* This chisel is used to chip (remove)

material from a surface. The flat chisel also shears (cuts through) rivets and other parts. The cutting faces of this chisel are ground, in general practice, to an included angle of 60° to 70°. The cutting edge is slightly crowned. Fig. 10-1 shows the shape and cutting edge angles of a flat cold chisel.

Another common form is the *cape chisel* (Fig. 10-2A). This chisel is narrower in width. The cape

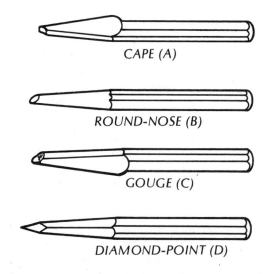

Fig. 10-2. Common types of cold chisels.

chisel is especially useful for cutting narrow grooves, slots, and square corners. The cape chisel is also used to cut shallow grooves in work surfaces that are wide. The flat chisel is then used to remove the material *(stock)* between the grooves.

The *round-nose chisel* (Fig. 10-2B) has a round cutting edge. It is designed for chiseling round and semi-circular grooves and inside round *filleted* corners. The *gouge chisel* (Fig. 10-2C) is used for chipping circular surfaces. The *square* or *diamond-point chisel* (Fig. 10-2D) is used for cutting square corners and grooves.

The term *chiseling* usually refers to two operations: (1) *chipping,* which is the process of removing or cutting away material, and (2) *shearing,* which refers to the cutting apart of metals or other materials. The solid vise jaw may be used as a blade for shearing. The chisel is considered the other shearing blade. A scissor-like action shears the material.

Holding, Positioning, and Striking Chisels

The holding, positioning, and striking of all types of chisels is important. Skillful cutting requires that the chisel be held steady in one hand, with the finger muscles relaxed. The cutting edge is positioned for the depth of cut. The chisel is again positioned for each successive rough cut and final finish cut. The positions of the hand and chisel in relation to the workpiece are shown in Fig. 10-3.

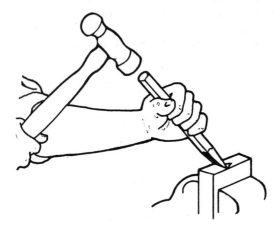

Fig. 10-3. The correct position of a flat cold chisel for chipping.

For the shearing operation, the cutting edge of the chisel is set at the cutting line. On mild ferrous metals a 1/16″ or 4 mm cut is considered a roughing cut; 1/32″ or 2 mm, a finish cut. The depth of the roughing cut may be increased to 3/32″ on softer nonferrous materials.

Quick, sharp hammer blows on the head of the chisel are most effective for cutting. For personal safety, it is important to follow the cutting edge, and not watch the head of the chisel. The depth of cut depends on the angle at which the chisel is held. The sharper the angle between the chisel and the workpiece, the deeper the cut. After each cut, the chisel is positioned for the next cut. The cutting process is then repeated. The chisel point may be lubricated with a light machine oil for easier driving and faster cutting.

CAUTION: When chiseling cast parts like cast iron, brass, or aluminum, the corners may break away below the desired layout line. The chiseling of cast parts is started from the outside edges. The cut is then finished near the center of the workpiece.

HOW TO USE HAND CHISELS

CAUTION: Put on a pair of safety goggles or a protective shield.

Straight Chiseling

Step 1   Position the part in a vise or other holding device. If the material is to be sheared, position the layout line at the level of the vise jaw or other hard backing surface.
NOTE: Put a wood or metal block under the work to prevent it from moving down.

Step 2   Select and inspect the head of the chisel.

CAUTION: A *mushroomed* head must be reground for safety. The corrected shape of a mushroomed head is illustrated in Fig. 10-4

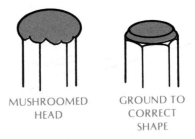

MUSHROOMED HEAD   GROUND TO CORRECT SHAPE

Fig. 10-4. Corrected shape of a mushroomed head.

Step 3   Check the angle (60° to 70°), crown, and sharpness of the cutting edge.

Step 4   Select a proper weight ball-peen hammer. Check the condition of the face and the secureness of the handle.

Step 5   Hold the chisel firmly in one hand. Position it against the workpiece. Steady the hand on the vise jaw, if possible. The correct chisel angle is illustrated in Fig. 10-5.

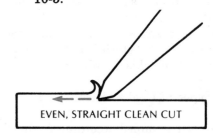

EVEN, STRAIGHT CLEAN CUT

Fig. 10-5. A chisel correctly positioned.

Step 6   Check the chisel angle position and the depth of cut.
NOTE: When the angle of the chisel in relation to the work is too steep, the chisel tends to *dig in*. This condition is shown in Fig. 10-6A. If the angle is too low, the chisel will climb away from the cut (Fig. 10-6B).

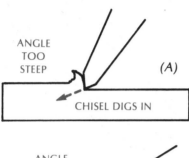

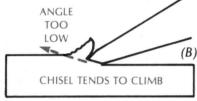

Fig. 10-6. Incorrect angles for chiseling.

Step 7   Start chiseling from one edge of the work, toward the *solid* jaw of the vise. The cutting edge of the chisel should be parallel to the layout line.
Step 8   Strike quick, sharp, hard blows. Watch the cutting edge.
Step. 9   Reset the chisel point at the next depth for the second cut. On *cast* parts the same depth of cut should be continued from the opposite edge.
NOTE: Repeatedly examine the chisel point. If it has dulled, regrind it. Fig. 10-7 illustrates a ragged cut produced by a dulled cutting edge.

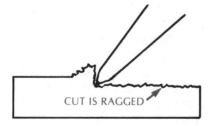

Fig. 10-7. The effect of a dulled chisel edge.

Step 10   Use a small quantity of a light machine oil to lubricate where the cutting action

is taking place.
Step 11   Reduce the depth of cut for the finish chipping operation.

Shearing a Strip

Step 1   Position the work in the vise at the desired height. Use the vise *solid* jaw as a cutting edge.
Step 2   Hold the chisel point at the cutting line. Position the chisel at an angle. One side of the chisel face should rest on the movable jaw. Start at the end of the workpiece.
Step 3   Strike a sharp blow. This produces a shear-like action that cuts the metal. The shearing action is shown in Fig. 10-8.

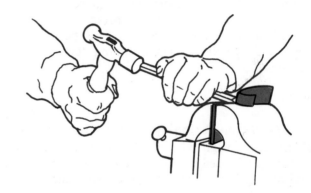

Fig. 10-8. Shearing a metal strip.

Step 4   Repeat the process. Each time, move the chisel a little less than its width. Position the chisel. Then, strike another quick, sharp blow.
NOTE: Sometimes, instead of shearing, the sides of an area to be removed are *nicked*. The part is then broken out by bending the scrap portion back and forward a few times.

Cutting a Hole

Step 1   Place the workpiece so that the hole to be cut is over a softer scrap material plate.
Step 2   Check the cutting edge and the condition of the head of a cape chisel.
Step 3   Position the cutting edge as close to the layout line as possible. Strike a quick, sharp blow to cut through the material.
Step 4   Reposition the chisel so that the cuts overlap. Continue until the excess part is completely cut and may be removed.

Step 5   Turn over the workpiece. Place it on a smooth, flat plate. Use a soft face hammer or mallet. Straighten (flatten) the material around the cutaway section. Finally, remove all burrs.

HAND PUNCHES

Hand punches are used in bench work for four general purposes:
- To make indentations
- To drive pins, bolts or parts
- To align parts
- To punch holes.

Punches are known by the shape and diameter of the point or shank end. Each type of punch is designed for a specific use. Punches are made in a range of sizes. In practice, where parts are to be driven in or removed, the diameter of the punch should be as large as possible. This permits the worker to deliver a solid, driving force with a hammer.

Solid Punches

The *solid punch* (Fig. 10-9A) has a short, tapered shank. It is designed to withstand heavy blows. The solid punch is used to *start* or *set* pins, bolts and parts and to punch holes in thin-gauge sheet metal.

Pin Punches

The *pin* punch (Fig. 10-9B) has a round shank. The shank is the same diameter for about half its length. It then tapers from this diameter to the body diameter. The pin punch is used to drive a pin, rivet, or other part through a hole. Because the long shank design cannot withstand heavy blows, the pin punch is used for *finish* driving.

Taper Punches

The *taper* punch (also referred to as a *drift* punch) has a gradual taper from the point end to the body. A slightly tapered punch, like the one shown in Fig. 10-9C, is useful for accurately aligning parts that require close-fitting pins or bolts. The taper punch is also used to finish drive pins, rivets and other small parts.

Hollow Punches

Hollow punches are used to cut holes in *shim* stock or other thin materials. Hollow punches (Fig. 10-10)

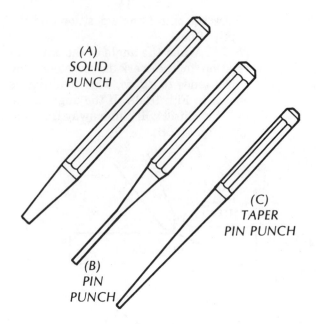

Fig. 10-9. Common shapes of hand punches.

Fig. 10-10. A hollow punch.

are made in sets from ¼″ (6 mm) to 4″ (100 mm) in diameter. Like all other punches, hollow punches are made of tool steel. They are hardened, tempered and ground. There is usually a tapered hole in the body. This hole permits the cut out blanks to be easily removed from the punch.

Prick and Center Punches

Two similar punches are used extensively for making cone-shaped indentations: the *prick punch* and the *center punch*. The name of each punch is derived from the operation each usually performs.

The *center punch* has a 90° included angle. This punch produces a centering location as a guide for positioning a drill. The shape of a center punch and its point are illustrated in Fig. 10-11A.

The *prick punch* (Fig. 10-11B) is smaller than the center punch because it is not subject to so heavy a striking force. The included angle of the point is 30°. Primarily, the prick punch is used to:

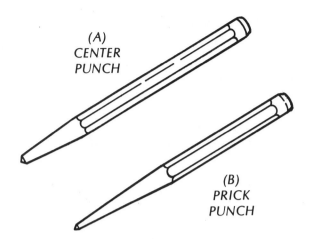

Fig. 10-11. Center layout punches.

- Form a small center hole from which circles may be layed out with dividers
- Make small, easily identifiable, cone-shaped depressions along scribed lines. (These prick-punched holes assist the worker during machining operations.)
- Mark the location of parts for proper identification and assembly.

DRIFTS

Drifts are used to drive parts and are usually round, soft-metal pieces. Drifts are slightly smaller than the diameter of a shaft, pin, stud, or other part that is to be driven. Brass rods are often used for drifts. The brass is softer than steel and other alloys. Such a drift withstands heavy blows and will not score or mark the part that is being driven.

Driving With Solid And Pin Punches

Step 1   Grip the workpiece. Position it so the pin may be driven out.
NOTE: Sometimes the part is positioned on a solid anvil. The pin is placed over one of the holes in the anvil. In other cases, the workpiece is placed on two parallel pieces of metal resting on another solid-metal plate.

Step 2   Select a solid punch. Its size must be a little smaller than the diameter of the pin. Fig. 10-12 shows a punch that is slightly smaller than the pin diameter.

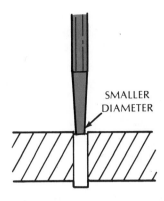

Fig. 10-12. A solid punch used to begin driving a pin.

CAUTION: Check the head of the punch. If it is mushroomed, *dress* (grind) it before using.

Step 3   Place the punch on the pin and center it. Hold the punch squarely with the fingers of one hand.

Step 4   Select a hammer suitable for the size of the job. Judge the amount of force needed to start and to drive the pin. Strike a solid blow.
NOTE: Check to see if the head of the pin or part is mushroomed. In such a case, the rounded pin head may need to be filed or ground away before attempting to drive the pin.

Step 5   Select a pin punch for finish driving, if necessary. The use of such a long pin punch to finish driving a pin or stud is illustrated in Fig. 10-13.

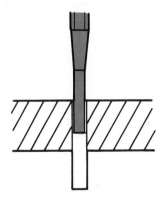

Fig. 10-13. A pin punch used to finish driving a pin through the part.

Step 6   Place the pin punch squarely on the driven pin. Continue to apply force to drive the pin out. The hammer blows should be lightened when the pin begins to move a slight distance each time it is hit.

Driving With A Drift

Step 1   Select a drift large enough for the size of the job and the amount of force to be used. The drift material should be *softer* than the part.

CAUTION: The face of the drift should be square. Check the head end. Remove any excess material caused by mushrooming.

Step 2   Hold the workpiece in a vise, other suitable holding fixture, or against a surface or part.

Step 3   Position the drift squarely over the gear, shaft, hardened pin, or other part that is to be removed.

Step 4   Use solid blows to start driving the drift. Reduce the force as the parts separate.

Aligning Parts

Step 1   Select the kind of taper pin punch that is appropriate for the job.
NOTE: The punch may be tapered to a point, slightly tapered, or of the same size as the pin or stud.

Step 2   Align the parts by sight. Insert the punch through the holes. Move the parts so they are aligned more closely.

Step 3   Tap the taper pin punch lightly. This action will move the parts into final alignment.
NOTE: The parts may need to be clamped to hold them in alignment. The taper pin punch is removed after aligning.

Punching Holes With A Hollow Punch

Step 1   Mark the center of the hole with a prick punch.

Step 2   Scribe a circle of the required diameter.

Step 3   Place the material so it rests flatly on a block. The hard wood or other flat material should be softer than the workpiece.

Step 4   Hold the hollow punch square with the workpiece. The diameter of the punch is

centered with the scribed circle (Fig. 10-14).

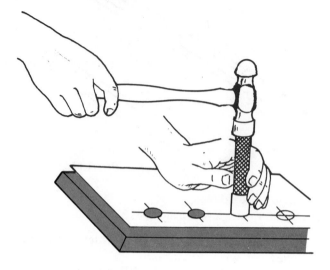

Fig. 10-14. Punching holes with a hollow punch.

Step 5   Strike a sharp blow. The force depends on (1) the thickness and softness of the material, (2) the firmness of the hammering block, and (3) the diameter of the hollow punch. The first blow should produce an impression.

Step 6   Raise the punch. Examine the impression. Be sure it is centered. If it is *off-center,* position the punch so it is concentric with the scribed circle. Then, strike a blow to get a correctly centered impression.

Step 7   Return the hollow punch to the correct impression. Continue to strike hard, sharp blows until the hole is punched.

Step 8   Place the sheet upside down on a flat metal plate. Use a soft faced hammer or mallet to flatten the metal part.

CAUTION: Remove all sharp edges and burrs. The edge (circumference) of the hole should be ground, filed or scraped.

Punching Holes With A Solid Punch

Step 1   Lay out the center of each hole. Scribe a circle of the required diameter.
NOTE: Small diameters usually are punched directly. The craftsperson centers the hole in relation to the center-punched center.

Step 2   Select a solid punch.

Step 3   Place the workpiece on a flat, solid block.

Step 4   Center the solid punch. Punch the hole (Fig. 10-15). Follow steps 4 through 8 under *Punching Holes With A Hollow Punch*.

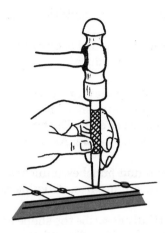

Fig. 10-15. Punching small holes with a solid punch.

SAFE PRACTICES FOR CHISELING, PUNCHING HOLES, AND DRIVING PARTS

- Wear safety goggles or a protective shield when chipping and driving parts. Flying chips and other particles may cause eye injury.
- Place a chipping guard or frame between the work area where a part is being chipped and other workers. The guard prevents flying chips from hitting other persons and avoids damage to property.
- Grind off any overhanging metal (mushroom) that may form on the head of the chisel or drift. Fragments of the mushroomed head may break away. Such pieces can fly off at high speed during hammering. The flying particles may cause personal injury or other damage.
- Prevent a workpiece from slipping during chiseling by placing a block under it.
- Check the condition of the hammer head to see that it is not chipped. The handle must be securely fastened to the head.

- Chisel *toward* the *solid* vise jaw.
- Limit the depth of the roughing cut to 1/16″ (2 mm). Allow 1/32″ (1 mm) for a finish cut. Heavier roughing cuts require excess force, which may cause the chisel to break or the cutting edge to dull quicker.
- Watch the chisel, punch, or drift point when striking sharp, heavy blows. This practice avoids hitting your hand and permits guiding the chisel point.
- Draw the cutting edge of the chisel, punch, or drift away from the workpiece about 1/16″ every few blows. This prevents hand fatigue and provides better control for cutting and driving.
- Avoid rubbing the edges of holes that have been punched in metal. Unless the solid or hollow punch is very sharp and the hammering plate is flat, the metal may be *torn* in cutting through. The edges of such holes may be ragged and *razor sharp*.

SHOP TERMS USED IN CHISELING, PUNCHING HOLES AND DRIVING

Cold chisel	A metal cutting tool used in bench work for removing material by chipping or by shearing.
Flat, round-nose, gouge, cape and diamond-point chisels	Five common types of cold chisels. Cold chisels are named according to the shape of the cutting edge, chip, surface, or groove that is produced.
Angle of cutting edge	The included angle between the two adjacent sides of the cutting point of a cold chisel.
Chiseling	A general term denoting two processes: chipping and shearing.
Mushroomed head	The *flowing over* of metal on a chisel head or punch or drift. The form of a chisel head produced by the continuous delivery of hard hammer blows. An unsafe condition of a chisel, punch or drift head.
Hand punches	Straight- or tapered-shank or hollow-center round punches. Punches used for four basic hand operations: to make indentations, align parts, drive parts, and blank holes.
Solid punch, pin punch, taper pin punch, center punch and hollow punch	Basic shapes of hand punches. The names partially describe the shape and function of each punch. Punches are obtainable in sets of various diameters.
Drift	A metal bar that is softer than the part to be driven. A driving bar shaped to correspond with a workpiece. The driving end is flat and at right angles to the body.
Punching	Producing a hole. Forcing a solid or hollow punch to cut through material.

SUMMARY

- Bench processes involve the removal or cutting of metal or other material. Cutting and driving tools are used for chipping, shearing, making of indentations, aligning, and assembling parts.
- The cold chisel is a hand cutting tool. The flat cold chisel is used for cutting and shearing. The cape, gouge, round-nose and diamond-point are other common cold chisel shapes. These chisels are used to cut flat, square, "V", and round grooves and surfaces.
- The depth of cut, when chiseling, depends on:
  - (1) The angle of the cutting point and its sharpness,
  - (2) The positioning of the chisel in relation to the workpiece,
  - (3) The hardness of the material, and
  - (4) The driving force.
- The recommended depth of cut for rough chipping of ferrous metals is 1/16". A 1/32" cut usually is taken when finish chipping.
- Shearing may be done by securely holding the workpiece and cutting it with a flat chisel. The layout line is positioned at the height of the vise solid jaw. The flat chisel is held at the cutting line. A sharp blow on the chisel causes a shearing, cutting action.
- Chisels, hand punches and drifts should be grasped by the fingers of one hand. To avoid fatigue, if possible, the hand should rest on a surface.
- Sharp, solid blows are needed in chiseling, punching, and driving. The operator must watch and guide the cutting edge, *not the chisel head*.
- Chipping and shearing operations should be *toward* the vise *solid* jaw.
- Six common types of hand punches are the solid punch, the pin or drift punch, the taper pin punch, the hollow punch, the prick punch, and the center punch.
- The prick punch and the center punch form indentations. These provide a guide for layout and drilling operations, respectively.
- The hollow punch is centered with a circle that is scribed to indicate its location. A hole is formed by driving the punch until the cutting edge pierces the workpiece.

- Small holes may be produced with the pin or tapered punch.
- Drifts are usually homemade pieces of a soft material. The drift conforms in size and shape to the part to be driven.
- Straight and taper pin punches are used to drive pins, studs, bolts, and other parts and to align holes in mating parts.
- Chipping operations are dangerous. Safety goggles or a protective shield and a screen are required to prevent injury.

UNIT 10 REVIEW AND SELF-TEST

1. a. Name three common hand chisels.
   b. Describe the main design features of the cutting end of each chisel.
   c. List an application of each chisel.
2. Tell what effect changing the angle of a chisel has on chiseling.
3. State three guidelines to follow when using a hand chisel for a shearing operation.
4. Distinguish between the design and use of a *solid hand punch* in comparison with a *hollow punch*.
5. Indicate two characteristics of drifts that differ from solid punches.
6. List the steps for aligning holes in two mating parts.
7. State three safety precautions for personal protection against injury during chiseling.

Unit 11

FILING: HAND FILE CHARACTERISTICS AND PROCESSES

The *file* is a cutting tool. Files are used extensively in bench and other hand operations to:
- Alter the shape of a part
- Reduce the size
- Remove burrs and edges
- Finish surfaces that have tool marks
- *Touch up* surfaces that require *fitting* in assembly.

The cutting edges of a file consist of a number of teeth. These are cut in parallel rows diagonally across the face. Since files are made of a high-grade tool steel and are hardened and tempered, the teeth are hard. A file therefore can cut softer materials like unhardened steels, cast iron, brass, and nonmetallic products.

Files are manufactured in a variety of shapes, sizes, and cuts. These characteristics accommodate many kinds of materials that require different surface finishes and processing.

CHARACTERISTICS OF FILES

Parts of a File

The common terms that are used to designate parts and features of a file are illustrated in Fig. 11-1. Note that the *length* is measured from the *point* to the *heel*.

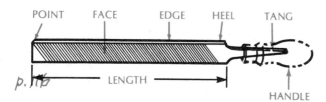

Fig. 11-1. The parts of a file.

Files are produced with different *cuts*. A file with a single series of parallel rows of teeth *(cuts)* is called a *single-cut* file. A *double-cut* file has two series of parallel rows of teeth. These cross each other at an angle. The angle of the *cut* on single-cut files varies from 65° to 85° with the edge (Fig. 11-2A). Double-cut files for general metal work have one set of cuts at a 40° to 45° angle from the edge (Fig. 11-2B). The

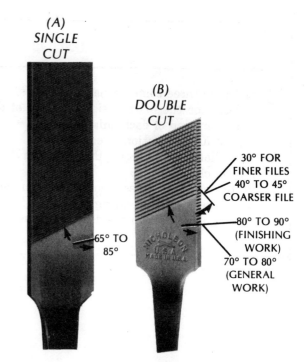

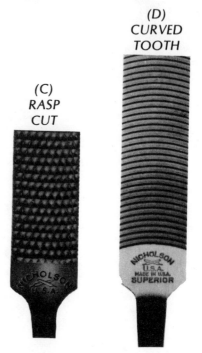

Fig. 11-2. Four different cuts of files.

second set (which cross) are at an angle of from 70° to 80°. The angles of the two cuts of finer double-cut files for finishing operations are 30° and 80° to 90°, respectively. Two other cuts of files are the *rasp cut* (Fig. 11-2C) and the *curved tooth* (Fig. 11-2D.)

The diagonal chisel-shaped cutting edge on single-cut files extends from one edge to the opposite edge. The alternate crossing of rows on double-cut files produces a series of sharp-pointed cutting edges in each row.

Coarseness of Cuts

The distance between each parallel row of cuts denotes the *coarseness*. The coarsest cut on files larger than 10″ is known as *rough cut*. The finest cut is *dead smooth*. In between there are a *coarse cut*, a *bastard cut*, a *second cut*, and a *smooth cut*. Each of these cuts is illustrated in Fig. 11-3.

The cuts are relative. The distance between the parallel rows and the depth of the cutting edge are proportional to the size of the file. A 16″ *rough-cut* file is coarser than a 10″ *rough-cut* file.

Numbers are used to indicate the cuts and relative coarsenesses of files that are smaller than 10″. There are ten different cuts: #00, #0, #1, #2, #3, #4, #5, #6, #7, and #8. The #00 cut is the coarsest, and #8 is the finest. Again, a #1 cut 8″ file is coarser than a #1 cut 4″ file.

Each cut of file is used for a particular hand filing process. The most commonly used files in machine shops are: the *bastard cut*, for rough filing; the *second cut*, for reducing a part to size; and the *smooth cut*, for finish filing.

File Names And Descriptions

The name of a file is associated with its cross section, particular use, or its general shape. The cross sections of eight common files are illustrated in Fig. 11-4. Brief descriptions of each follow.

Flat File—This file (Fig. 11-4A) has a rectangular cross section. The 12″ bastard-cut flat file is an all-around general utility file in the shop. Because it is double cut, it has pointed teeth for easier cutting. The edges taper toward the point. The faces of the body are *convex* and are thinner at the point.

Mill File—The mill file (Fig. 11-4B) is single cut. This is in contrast with most other files used in machine shops which are double cut. The chisel-shaped teeth produce a shearing action. While the filing process is slower than with a double-cut file, the mill file produces an excellent fine-grained surface. The 12″ mill files are popular for final finishing operations. Bastard mill files are used for filing soft metals such as brass and bronze. The second-cut mill file is used commonly for finish filing turned parts on a lathe.

Hand File—The edges of the hand file (Fig. 11-4C) are parallel. The faces are slightly convex. The second-cut hand file is adaptable for finish filing on lathe work and other flat surfaces.

Pillar File—This file (Fig. 11-4D) is rectangular in shape and narrower than the hand file. While larger sizes are used, the #00, #2, and #4 cuts of the 8″ pillar file provide a light, practical file.

Half-round File—This shape of file (Fig. 11-4E) is a practical combination of a flat- and curved-surface file. The bastard cut is excellent for rough filing; the second cut is used for finish filing.

Square File—The name indicates its cross section

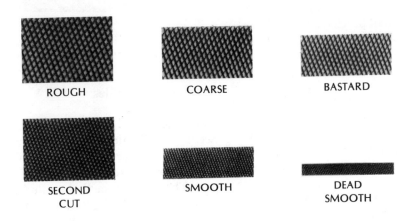

Fig. 11-3. Coarseness of regular files.

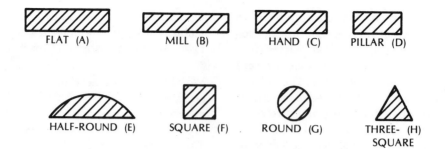

Fig. 11-4. Comparative shapes of eight common files.

and its application (Fig. 11-4F). It is used for filing square and rectangular holes and slots.

Round File—The round file (Fig. 11-4G), like the square file, is generally tapered toward the point. It is used to file circular and irregularly-shaped curved surfaces and fillets. The smaller sizes of this file taper sharply toward the point. They are known as *rat-tail files*.

Three-square File—This file (Fig. 11-4H) is triangular and has teeth cut on each of the three faces. The teeth on the double-cut file are very sharp. The shape makes it adaptable for finishing surfaces that are at an angle to each other.

PRECISION TYPES OF FILES

The files just described are generally classified as *machinists'* files. Three other groups are included from among the many designs of files. These are the *Swiss pattern, needle-handle,* and *special-purpose* files.

Swiss Pattern Files

The Swiss pattern series of files are used to accurately file and finish small precision parts. They are used commonly by tool and die makers, instrument makers, jewelers, and other highly-skilled craftspersons. The files have cross sections similar to machinists' files.

The major differences between Swiss pattern and machinists' (American pattern) files are:
- The points of Swiss pattern files are smaller
- The tapers of the body toward the point and at the tang are longer and sharper than the conventional American pattern
- The cuts of the Swiss pattern teeth are finer
- The specifications of Swiss pattern files are more exacting.

Six of the most common Swiss pattern files—usually called just *Swiss files*—are illustrated in Fig. 11-5. There are over 100 different shapes. The teeth are double cut on the faces. The teeth on the edges of the *knife* and *three-square* files are single cut. Some of these files are manufactured with *safe edges* (no teeth).

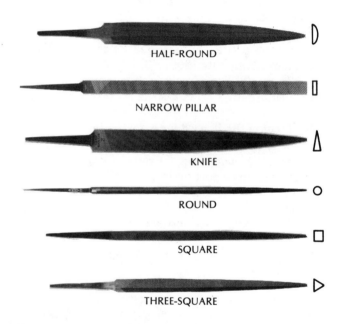

Fig. 11-5. Typical swiss pattern file shapes.

Needle-Handle Files

Another group of small precision files is known as *needle-handle* files. They are normally 4″ to 6″ long. They are used for fine, precision filing where a small surface area is to be formed or fitted with another part. The teeth of needle-handle files are formed over only a part of the body. The remainder of the body forms a small-diameter handle. The toothed portion tapers sharply from the body to almost a point.

These files also are designated by their shape, cut and length. The characteristics and common shapes of some of these files are illustrated in Fig. 11-6.

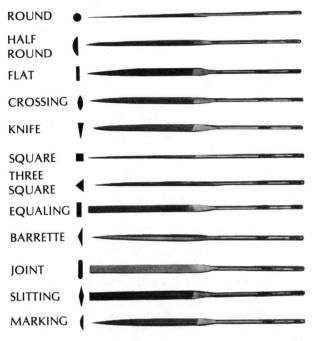

ROUND
HALF ROUND
FLAT
CROSSING
KNIFE
SQUARE
THREE SQUARE
EQUALING
BARRETTE
JOINT
SLITTING
MARKING

Fig. 11-6. A set of assorted needle round-handle files. (Courtesy of The Cooper Group)

Special-Purpose Files

Special-purpose files produce different kinds of cutting action and surface finishes on various materials. They are designed for greater efficiency than the general-purpose files. For example, although a mill file is widely used in lathe work, the *long-angle lathe file* has a number of advantages:

• The teeth, cut at a longer angle, provide an efficient shearing action. This contrasts with the comparative *tearing* and *scoring* action of general-purpose files
• Chatter marks are prevented
• Clogging of the teeth is reduced
• Materials are cut away faster
• A finer surface finish is produced.

Soft metals, plastics, stainless steels, and other materials each require special-purpose files. These permit maximum cutting efficiency, dimensional accuracy, and high-quality surface finishing. For very soft metals, a special single-cut coarse file with a short angle provides an efficient shearing action (Fig. 11-7).

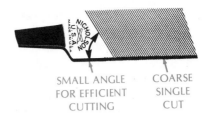

SMALL ANGLE FOR EFFICIENT CUTTING   COARSE SINGLE CUT

Fig. 11-7. A small-angle, single-cut coarse file for soft metals. (Courtesy of The Cooper Group)

FEATURES OF FILES

There are three features of files that are especially important in filing:
• A *safe edge*
• *Convexity*
• *Taper.*

Safe Edge

A *safe edge* is a file edge that has no teeth. The safe edge (Fig. 11-8) protects from any cutting action the surface against which the file is moved. The safe edge also permits filing to a sharper corner.

Files are manufactured with one safe edge (as on the hand file), two safe edges (pillar file), or none. The teeth are sometimes ground off the edges of a file to produce a safe edge.

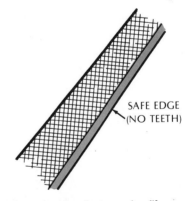

SAFE EDGE (NO TEETH)

Fig. 11-8. The safe edge feature of a file.

Convexity

Convexity of files means that the faces are slightly convex and bellied along their length, as shown in Fig. 11-9. The faces curve so that they are smaller at the point end of the file. Convexity is important for five reasons:
• A perfectly flat face would require a tremendous

downward force to make the teeth *bite* and a great forward force to produce a cutting action

• Control of a perfectly flat file would be difficult

• A flat surface could only be produced if every part of the file was perfectly straight

• File warpage, resulting from the hardening process, would produce an undesirable concave face

• Convexity makes it possible for the skilled worker to control the position and amount of cutting.

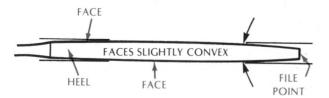

Fig. 11-9. The convexity feature of a file.

Taper

Most files taper slightly in width (Fig. 11-10). The taper on round, square, and three-square (triangular) files is more severe than that of other types. The cross-sectional area is reduced considerably from about the middle of the body to the point. A file is termed *blunt* if it has a uniform cross section along its entire length.

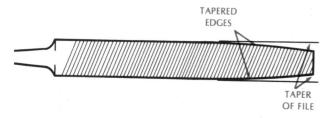

Fig. 11-10. The taper feature of a file.

THE CARE OF FILES

One common problem in hand filing is *pinning*. A *pin* is a small particle of the material being filed. The pin gets wedged (*pinned*) on the lead edge of the file teeth. The pins, as they move on the file across the face of the work, tear into and scratch the filed surface.

Pinning is caused by:

• Bearing too hard (applying too much force) on the file, particularly on new files

• Failure to regularly clean the file teeth.

Pinning may be avoided by:

• Brushing the workpiece with a *file brush* (Fig. 11-11) to remove the filings

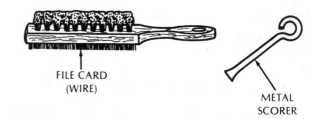

Fig. 11-11. Two accessories for file cleaning.

• Brushing the file teeth with a *wire file card*

• Pushing a *metal scorer* (Fig. 11-11) or a soft metal strip through the grooves of the teeth

• Rubbing ordinary chalk across the file teeth to *chalk* them.

FILE HANDLES

Part of the skill in using a file depends on the size, shape, and *feel* of the handle. A file may be controlled better and hand fatigue may be avoided if the handle is easy to grasp and hold. The handle should be selected for shape and size. The handle (1) must feel comfortable and (2) must give *balance* to the file.

Some handles are threaded. Others have a hole drilled in them. The hole guides the *tang* and prevents splitting. The drill size should be equal to the average thickness of the tang. The depth of the hole is the same as the length of the tang. The hole is tapered to match the tang shape by heating an old, worn file to a dull red. Then, using a pair of tongs, the heated tang is forced almost to its length into the handle and removed. The handle and file then are plunged into water to stop the burning action and to cool the file. The tang of the new file is inserted in the handle. The end of the handle then is tapped firmly on a hard surface to seat the tang (Fig. 11-12). With use, the handle may loosen. In such cases, the file is reset in the handle by retapping it.

Fig. 11-12. "Seating" a handle.

FILING PROCESSES

Cross Filing

Cross filing is often referred to as just plain *filing*. The file face is forced against the work and pushed forward on the cutting stroke. The force is relieved on the return stroke.

After a few strokes, when a complete surface area has been filed, the position of the file and the filing operation are changed. The strokes are *crossed* as shown in Fig. 11-13. This procedure continues until the part has been filed (*brought down*) to a layout line or dimension. As much of the file face as is possible should be used during each cutting stroke.

Fig. 11-13. The positions of a file in cross filing.

The surface of most materials may be filed directly. However, cast iron parts require a special procedure. The outer surface (*scale*) of a cast iron casting is often *file hard*. If a file is used directly on such a surface, the teeth will dull rapidly. This condition may be prevented either by rough grinding the surface or by chipping below the depth of the scale. Once the scale is removed, the cast iron is soft enough to be rough or finish filed.

Cross filing helps:
- Show unevenness in filing
- Keep a surface straight and flat
- Prevent fatigue of the hand and arm.

Drawfiling

Drawfiling produces a fine-grained surface finish. For drawfiling, a single-cut file is held with both hands, as illustrated in Fig. 11-14. The file is then forced on the workpiece as it is moved over the surface. A cutting action is produced by *drawing* the file alternately in each direction. Particular care must be taken to use a sharp file and to keep it clean.

Drawfiled surfaces are sometimes polished by using an abrasive cloth. A strip of abrasive cloth is held against the bottom face of the file. The abrasive is moved over the work surface the same as in drawfiling.

Fig. 11-14. The positions of the file, hands, and work in drawfiling.

Soft Metal Filing

The files and processes described thus far have dealt primarily with the filing of steels and other hard metals. These same files may not be suitable for soft materials because the chips can clog and wedge between the teeth.

The curved-cut file, mentioned earlier, has been especially designed for efficiently filing soft metals like brass, copper, aluminum, and lead. This file has chisel-edged teeth that produce an excellent finished surface. Clogging of the teeth is reduced by the shape of the grooves between the rows of teeth. The curved-cut file is also used for finish filing of parts turned on a lathe.

HOW TO CROSS FILE

Holding The File

Step 1   Position the work in a vise or other holding device so there will be no interference while filing. The height of the surface to be filed should be level with the worker's elbow.

Step 2   Select a file that will produce the fastest cutting action. This depends on the material, the size of the part, and the nature of the filing process.

Step 3   Check the tightness of the file handle. Clean the file teeth with a file brush, card, or metal scorer, as needed (Fig. 11-15).

Step 4   Hold the point end of the file in one hand, with the thumb and palm pushing on the

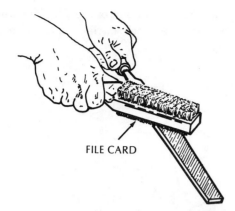

Fig. 11-15. Cleaning the file teeth with a file card.

Positioning the Body

Step 1 Stand close to the bench, with one foot pointed forward. Place the other foot from 8″ to 12″ to the side and about 16″ back.

Step 2 Lean the upper part of the body forward for most of the forward stroke. Bend the knees slightly. Apply force by pushing with the arms.

Step 3 Follow through with the arms to the end of the stroke.

Step 4 Bring the body back to the original position. Return the file to the start of the next stroke by moving it lightly across the work.
Note: Some surfaces are wider than the width of the file. In such cases, the file is moved to the next area to be filed during the return stroke. Each cut should overlap the preceding one.

Step 5 *Cross the stroke* during rough filing. This helps rest the arms and check the evenness of filing.

top face (A in Fig. 11-16). Curve the remaining fingers on the underface. Press the thumb, palm and fingers into the file point. The greatest downward force can be applied in this position.

Step 5 Grasp the file handle in the other hand. The palm should rest against the handle, with the thumb on top. The fingers are curled around the handle (B in Fig. 11-16).

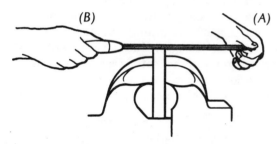

Fig. 11-16. The correct hand positions for heavy filing.

Note: Another common position of the hands is shown in Fig. 11-17.

Checking The Filed Surface

Step 1 Test the work surface for flatness and straightness (Fig. 11-18). Use a steel rule or the blade of a square. Check the work surface crosswise, lengthwise, and diagonally. Note any variation in accuracy.

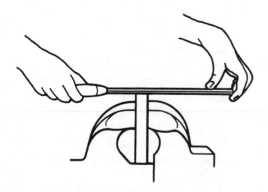

Fig. 11-17. The correct position of the hands for normal filing.

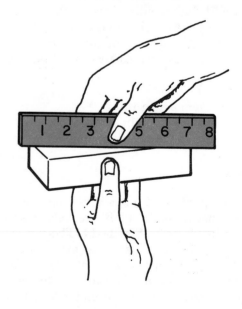

Fig. 11-18. Checking flatness with a steel rule.

Step 2   File the *high spots*. The convexity of the file makes it possible to pinpoint the filing area.

Step 3   Check the depth and shape of the filed surface either against the layout lines or by direct measurement.

Checking Squareness

Step 1   Remove burrs. Clean the workpiece and the square.

Step 2   Hold the solid base of the square against one face of the work. Gently slide the base down the face until the blade contacts the filed surface (Fig. 11-19).

Fig. 11-19. Checking a filed surface for squareness.

Step 3   Hold up the workpiece and square to the light. Note whether all the light is shut out between the blade and filed area. The presence of light indicates places where the surface was filed too much (*low*).

Step 4   Continue to file the high places. Recheck until the surface is square and filed to the correct size.

> **HOW TO DRAWFILE AND POLISH A FINISHED SURFACE**

Drawfiling

Step 1   Select a single-cut file appropriate to the size of the job and surface finish required. Mount and secure the workpiece.

Step 2   Hold one end of the file with one hand. Do the same on the other end. Fig. 11-20 shows how both hands should be positioned.

Note: Sometimes the handle is removed to get a better *balance on the file.*

Fig. 11-20. The hand positions and file movement for drawfiling.

Step 3   Press downward and forward on the forward stroke. Press downward and backward on the return stroke. The angle of the file may be changed in relation to the work. This may produce a better cutting action and grained surface.

Step 4   Clean the file regularly (Fig. 11-21). Brush the filings from the work. The filed surface should not be cleaned with the hand. Oil on the surface may cause the file to glide over the oily spots. This action causes uneven cutting.

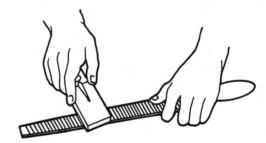

Fig. 11-21. Regular file cleaning with a file card.

Step 5   Continue drawfiling on the forward and return strokes. Check to see that the entire surface is *grained* (drawfiled).

Polishing

Step 1   Tear off a strip of abrasive cloth. It should be slightly wider than the file width.

Step 2   Curve one end of the strip around the file point. The abrasive particles should face the work surface. Draw the abrasive cloth along the file face and hold it taut (Fig. 11-22).

Step 3   Apply moderate pressure downward during the forward and return strokes. This force and motion causes the abrasive grains to cut as they move across the surface.

Note: Polishing may be done by holding the file and abrasive cloth in either the regular or drawfiling positions.

ABRASIVE CLOTH

Fig. 11-22. An abrasive cloth drawn taut against the file face.

Step 4   Examine the *graining* and polishing effect. Continue to polish until the required surface finish is reached.

HOW TO FILE A ROUNDED CORNER

Step 1   Clamp the workpiece so that the layout line is above the vise jaws.

Step 2   Begin by rough filing the top surface to within 1/64″ of the layout line.

Step 3   Continue to rough file at about a 45° angle. File to within 1/64″ of the layout line (Fig. 11-23A).

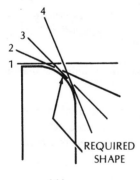

REQUIRED SHAPE

(A)
SEQUENCE OF CUTS

Step 4   File away the excess material on the corner to be rounded. File at a series of angles. The sequence of file cuts is shown in Fig. 11-23A.

Step 5   File away the crests formed by each previous angular cut with a finer cut file (Fig. 11-23B).

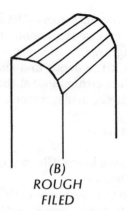

(B)
ROUGH FILED

Step 6   Finish filing by using a rounding, rocking motion. Continue filing until the surface conforms to the required curvature (Fig. 11-23C).

(C)
FINISHED FILED

Fig. 11-23. Steps in filing a rounded corner.

Note: After a few strokes, check the filed surface for *high* or *low* spots. Checking may be done against a layout line or with a template of the correct size and shape.

HOW TO FILE BURRS

Step 1   Select either a hand-smooth cut or a mill-smooth cut file.

Step 2   Mount the workpiece in a vise.

Note: When the workpiece is small, the skilled worker holds the part in one hand, as an easy way of removing burrs from all edges.

Step 3   Hold the file at a slight angle to the burred edge. Take light cuts to *break the*

Step 4   Move to the adjoining surface and repeat this filing process.

CAUTION: Burrs can be *razor sharp*. Scraping against them or rubbing with the fingers or hand can produce cuts or bruises.

Step 5   Continue the same steps to remove the burrs from all sides and surfaces of the workpiece.

SAFETY PRECAUTIONS IN FILING

- Avoid hitting a file directly, scraping files together, or placing one on another.
- Remove pins that cling to and filings that clog the file teeth. Removing pins prevents scratching or damaging a work surface.
- Use a file with a properly fitted, tight handle.
- Seat the tang in the handle. Tap the end of the wood handle on a solid surface.
- Avoid handling parts with burrs or fine, sharp projections. Remove burrs as soon as practical.
- Use the shape, cut and length of file that are most appropriate for the material and operation to be performed.
- The outer scale on a cast iron casting should be removed by grinding or chiseling before filing.
- Test the workpiece for squareness, straightness, and dimensional accuracy. Test often to avoid filing too far.
- Regulate the speed of filing to control the action and to prevent arm and hand fatigue.

HAND FILE AND FILING TERMS

Cut	A designation for rows of teeth that are formed on the faces of a file.
Single cut	A single series of parallel rows of file teeth. The teeth are cut diagonally across the file faces.
Double cut	Two series of parallel rows of teeth. The rows cross each other diagonally. The double cut produces sharp-pointed cutting edges as file teeth.
Coarseness of cut (files 10" and longer)	A relative measurement of distance between two parallel rows of file teeth. The range of coarseness is: *rough cut* (coarsest), *coarse cut, bastard cut, second cut, smooth cut,* and *dead smooth* (finest).
Coarseness of cut (files shorter than 10")	A range of cuts from coarse (#00) to fine (#8). Intermediate cuts are designated from #0 to #1 through #8.
File names	A name associated with a particular cross section (shape) or use. The most common names are: flat, mill, hand, pillar, half-round, round, square, and three-square (triangular).
Needle-handle files	A group of files 6" and smaller in length. Each file has a small, round handle as the body. The toothed portion tapers sharply to almost a point.
Swiss pattern files	A group of precision files generally applied in instrumentation and toolmaking. These files have longer tapers, finer cut teeth, and smaller point ends. Swiss pattern files meet more exacting specifications than do typical American pattern files.
Safe edge	A file edge without teeth.
Convexity	Slightly *bellied* faces that curve from the middle of the file to a smaller cross section at the point end.
Taper of files	The gradual reduction of the file width from the body to the point.
Pinning	The wedging of filings on the file face or in the grooves of file teeth.
Cross filing (filing)	Filing materials by diagonally crossing the file strokes.
Drawfiling	A process for graining a workpiece. Drawing a second-cut or mill file across a work surface, cutting on both the forward and return stroke.

Soft metal file and filing	A deeper cut and differently shaped groove which prevents filings from clogging the file teeth. Filing soft materials with a curved-cut file.
High spots	Areas of a surface that are higher than the specified dimension. File areas requiring the additional removal of material. Filing the raised spot to be even with the remaining surface.
Burrs	A *fin* or other fine, sharp, rough edge produced by a hand or machine process. An unsafe edge to handle. An edge that can impair the accuracy or the fitting of parts.

SUMMARY

- Filing is a cutting process. Filing alters the shape and size and the fit of a part. A particular surface finish may be produced by filing.
- *Swiss pattern* and *needle-handle* files are used for fine precision filing of small surfaces.
- *Safe edges* on files protect the surface against which the file edge rubs from being filed.
- Files have parallel rows of teeth. These are cut diagonally across each face. A *single cut* refers to one set of parallel rows. *Double cut* designates two sets that diagonally cross each other.
- The angles of single- and double-cut files vary for different materials and processes.
- The coarseness of a cut relates to the distance between the parallel rows of cuts. *Rough, coarse, bastard, second, smooth,* and *dead smooth* cuts are common degrees of coarseness.
- The coarseness of files that are less than 10″ long ranges from #00 (coarsest) to #8 (finest).
- Eight common shapes of files are: *flat, mill, hand, pillar, half-round, round, square,* and *triangular.*

- File faces are produced with *convexity.* This feature makes it possible to position and control the file to produce flat surfaces. Convexity makes it easier to file.
- The wedging of *pins* (filings) on the lead edge of file teeth causes a scratching action that can mark, or score, a surface. *Pinning* can be caused by too much force or by failure to clean the file teeth.
- A file should be fitted with a handle whose size and shape feels comfortable and balanced.
- *Cross filing* is the process of diagonally crossing the position of the file cuts.
- *Drawfiling* is the fine graining of a work surface. Drawfiling requires the *drawing* of a single-cut smooth file (using a slight force) across the work surface.
- Filed surfaces are checked for squareness and straightness by using a square and/or other straight-edged tool. *High spots* are sighted and then filed until the surface is accurate.
- Burrs and other raised indentations should be removed as a safety precaution. A *hand smooth-cut* or *mill smooth-cut* file is usually used for *burring.*

UNIT 11 REVIEW AND SELF-TEST

1. Identify (a) the types of cuts and (b) the angles of a general metal work file.
2. State how coarseness is designated for files under 10″ long.
3. a. List three files that are used to produce a flat surface.
   b. Give one application of each file.
   c. Describe at least one distinguishing design feature of each file.
4. Give reasons why *Swiss pattern files* are preferred by many precision mechanics in place of *American pattern files.*
5. Enumerate three advantages of using a *long-angle lathe file* to mill file lathe work.
6. Cite four reasons why files are produced slightly convex along their length.
7. State what corrective steps may be taken to avoid *pinning.*
8. Describe *drawfiling.*
9. List four safety precautions to follow in filing.

Unit 12

SCRAPING, BURRING, AND POLISHING: TECHNOLOGY AND PROCESSES

Scraping, burring, and polishing each involves a cutting action. Small amounts of material are removed.

Parts are *scraped* to produce an extremely accurate flat or circular bearing surface. A special surface finish effect also results from scraping.

Burring deals with the removal of rough, unsafe edges and other surface imperfections.

Polishing with coated abrasives produces smooth, fine-grained surfaces. Polishing does not necessarily improve the dimensional accuracy of a finished part.

SCRAPING TOOLS AND PROCESSES

Scraping is not as common a practice today as it was years ago. Modern grinding techniques and improved machining processes produce surface finishes of high quality and precise limits of accuracy. However, the scraping process is still performed when:

● The surface produced on a shaper, planer, milling or other machine has tool marks. These marks may not be suitable as a close-grained, smooth bearing surface.

● A surface slightly deviates from being flat and true. This can be corrected by hand scraping.

● A curved shaft on which another shaft, spindle, or circular part must ride is only a few thousandths of an inch away from a perfect fit and must be scraped into position.

● A surface is to be *frosted (flaked)*.

These last two terms refer to a repetitive design. This is produced by scraping in a particular pattern of strokes. Sometimes, these designs are circular and scroll shaped. In other cases, the scraping follows straight-line patterns. Strokes are alternated at right angles. There are many combinations.

Frosting is the surface finish design produced by hand scraping. A surface is frosted to:

● Improve the sliding quality between two mating surfaces and help eliminate *sticking*

● Improve appearance

● Provide *pockets* for lubricating fluid.

The tools used for scraping and producing surface finishes are simple. Flat and hook type scrapers are shown in Fig. 12-1. The cutting edge of the *flat scraper* is square with the faces and edges. However,

instead of being perfectly straight, the edge is slightly convex. This convex end permits the operator to control the cutting action and position the scraper on the *high spots,* and prevents scoring or scratching during the scraping process. The corners of the cutting edge often are rounded with a hand stone. This is a further precaution against scoring

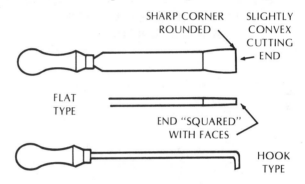

Fig. 12-1. Flat and hook type scrapers.

A second form of scraper is called a *bearing,* or *half-round, scraper* (Fig. 12-2). This specially shaped scraper has an offset cutting end that is spoon shaped. The cutting edges on the two sides extend along the lips of the spoon shape. They come to a sharp point. A half-round bearing scraper is suited for scraping and finishing round and curved surfaces that require precision fitting and/or a special surface finish.

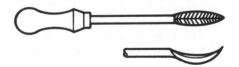

Fig. 12-2. A half-round (bearing) scraper.

The third form of scraper is the most common. It is called a *three-cornered scraper* (Fig. 12-3). Frequently, it is made from an old *three-square* file. The length is shortened, the teeth are ground off, and the sides are formed into a sharp point. The three-cornered scraper is a handy, practical cutting tool. It is used to remove a sharp edge from a hole, fillet, or other curved surface. The sharp edges may be the result of other hand or machining operations like drilling, reaming, or boring.

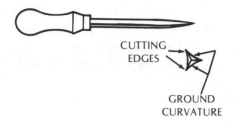

Fig. 12-3. A three-cornered (triangular) scraper.

Scrapers are also produced with carbide tips. Many scrapers are moved by hand. Others are held in hand power tools that produce the necessary movement for scraping.

HOW TO SHARPEN AND USE A HAND SCRAPER

Sharpening And Stoning A Scraper

Step 1   Grind the face(s) or end of the scraper.
Note:
● The flat scraper should be ground straight and square with a slightly convex end.
● The triangular scraper is ground with sharp cutting edges on each side. These taper to a point.
● The curved scraper is ground on the top lips to form a sharp point.

CAUTION: Keep the cutting edge of the scraper cooled during any grinding operation. Overheating causes the temper of the steel to be drawn. This ruins the sharp edge-holding capability of the scraper.

Step 2   Select a *clean* oilstone. This means there are no metal particles imbedded in the oilstone.

Step 3   Hold the flat scraper so that the cutting edge is at right angles to and flat with the face. Position the edge so that it forms a 45° angle to the direction of movement (Fig. 12-4).

Step 4   Apply a slight downward force while moving the scraper across the full face of the oilstone. Repeat this process until the grinder marks are evened out and there is a fine cutting edge.

Step 5   Turn the scraper 180°. Repeat steps 3 and 4.

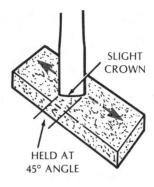

Fig. 12-4. Oilstoning a scraper end.

Step 6   Finish stoning both faces at the cutting end. The scraper is held flat and may be sharpened by moving it in a circular pattern (Fig. 12-5).
Note: Bearing (half-round) and triangular scrapers are finish sharpened after grinding. The ground cutting edges are positioned on the oilstone. As the edge of the scraper is moved over the face of the stone, it is *rocked*. This motion makes it possible to sharpen the entire cutting edge.

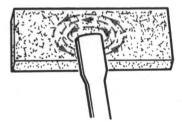

Fig. 12-5. Oilstoning flat scraper faces.

Flat Scraping

Step 1   Remove all burrs. Clean the work surface and the flat mating surface or surface plate with a solvent.

Step 2   Apply a thin coating of venetian red or prussian blue paint to the finished surface. Apply it with a cloth swab.

Step 3   Place the face of the workpiece gently on the coated flat surface. Slowly move it in a figure-eight pattern. This movement transfers the coloring to the *high spots*.

Step 4   Mount the workpiece in a vise or other holding device. Place the cutting edge of the scraper at an angle on a high spot. Apply a slight force while moving the scraper forward about one-half inch. This is the cutting stroke.

Note: The cutting action of scrapers is illustrated in Fig. 12-6. The cutting takes place on the forward stroke (Fig. 12-6A). Cutting is done on the backward stroke when a hook-type scraper is used for the flat scraper (Fig. 12-6B).

Note: Only two or three thousandths of an inch should be scraped at one time. The cutting edge should be oilstoned as soon as it loses its sharpness.

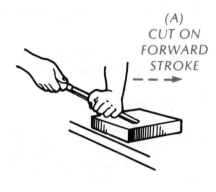

(A)
CUT ON FORWARD STROKE

(B)
CUT ON BACKWARD STROKE

Fig. 12-6. Correct hand positions and cutting angle for using a flat scraper.

Finish Scraping Flat Surfaces

Step 1   Repeat the process of scraping all the high spots. After each scraping process, return the workpiece to the surface plate, to locate any remaining high spots.

Step 2   Continue the scraping process until the surface is flat.

Finish Scraping Round Surfaces

Step 1   Place in the bearing or mating surface the circular part that is coated with a coloring agent. Rotate it part of a revolution to mark the high spots. Remove the part.

Step 2   Locate the cutting edge of the half-round (bearing) scraper on the high spot. Press down while turning the scraper. This produces the cutting action. The correct position and motion are illustrated in Fig. 12-7.

CUTTING MOTION

Fig. 12-7. Positioning of hands and half-round scraper for scraping a curved surface.

Step 3   Repeat until the high spots have been scraped.

Step 4   Continue the marking/scraping process until the part is well seated.

Breaking A Corner Or Removing A Burr

Step 1   Hold a sharp three-cornered scraper against the round or curved surface edge that is to be broken (burred).

Step 2   Apply a slight force on the handle while moving the scraper cutting edge around the surface that is being cut.

CAUTION: Position the hand against the workpiece wherever possible. This prevents the scraper from slipping, guides it, and provides a solid position from which to apply force.

BURRS AND BURRING

Burring with a hand file was treated briefly as one of the filing processes. Since a complete understanding of burrs is important to personal safety and to machine and tool safety, accuracy, and performance, and workpiece appearance, burring is covered in more detail in this unit.

A burr is the turned up or projecting edge of a material. A burr generally is produced by sawing, drilling, punching, shearing, and most cutting processes. An enlarged, razor-sharp burr is shown in Fig. 12-8. Burrs are also raised through careless handling, hammering, dropping and nicking. Damaged or spoiled work parts often result from failure to remove burrs before clamping or performing hand or machine operations.

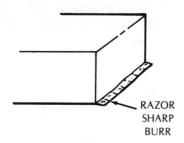

Fig. 12-8. An enlarged, razor-sharp burr.

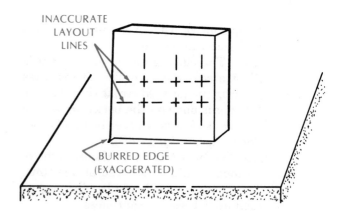

Fig. 12-9. A burr under the workpiece edge causes inaccurate layout.

Burred edges, nicks, dents, and ragged or unwanted sharp corners are usually removed by hand filing. The sharp edges and burrs on hardened steel pieces may be removed easily with a small oilstone. The burrs on curved surfaces may be *scraped* with the three-cornered scraper as described earlier. In summary, parts are burred to:
- Insure dimensional accuracy
- Provide for the exact fitting of mating parts
- Permit safe handling
- Avoid damage to tools and machines
- Improve the finish and appearance of the part.

Burring For Accuracy

Improper burring is a frequent cause of inaccurate layout work. It is impossible to accurately lay out a job if it rests on a burred edge or on a surface which is not clean. The raised surface area (mass) produced by a burr will not seat solidly against a finished surface. This added mass has the same effect as a shim of thin material placed between the surfaces.

Work that is being positioned for layout and machining should be free of burrs. The work must rest evenly on parallels, in a vise, on a table, or in any other holding device. A burr under one edge is sufficient to throw the job out of alignment, as illustrated in Fig. 12-9. The result is inaccurate, damaged, or spoiled work.

Measurements taken over burrs are another source of trouble. Burrs prevent holding the steel rule or other measuring tool sufficiently close to the measured point. This makes it difficult to obtain an exact reading. All measurements should be taken on surfaces that are free of burrs.

Burrs frequently cause problems when fitting parts together. Burrs may tear a surface against which they are moved. The projections also produce an *oversize* condition. In each instance, sharp corners,

turned over edges, and machine-produced burrs must be removed for parts to fit accurately.

Burring For Appearance

One of the hallmarks of a craftsperson is a properly burred workpiece. The rounding of corners and their removal adds to the finished appearance of a job. Industrial practices often require finishing in this manner. Many industrial blueprints include notations such as "BREAK EDGES" or "REMOVE ALL UNNECESSARY SHARP CORNERS."

Burring For Safety

To reemphasize, parts with burred edges, ridges, and sharp corners are not safe to handle. These razor-sharp edges cut, and even tear, metal against which they may be brushed or rubbed. Consequently, where possible, remove all unnecessary sharp corners and burrs to prevent personal injury.

There are instances where sharp corners are necessary, such as in mold making, die work, and for most other cutting tools. Sharp corners must be retained in such applications.

COATED ABRASIVES AND HAND FINISHING

Abrasives and their applications to machine cutting and surface finishing processes are covered in detail in the units on grinders and grinding operations. Only basic information on coated abrasives that relates to bench and floor work is presented at this point.

Coated abrasives are used in hand operations in machine shops. These abrasives produce a smooth or polished surface and remove very slight amounts of material. The edges of the abrasive grains and their ability to fracture and form new sharp edges as they wear make coated abrasives excellent for polishing.

Coated Abrasives

Coated abrasives are produced by bonding a particular type and quantity of abrasive grains (the *coating* and *bond*) on a suitable *flexible backing material*. The relationship of these materials is shown in Fig. 12-10.

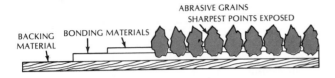

Fig. 12-10. Prime materials used in the manufacture of coated abrasives.

The abrasive grains are spaced evenly as either an *open* or *closed coating*. The grains are electrostatically imbedded in a bond in the backing material. The sharpest points of the grain are thus positioned and exposed.

An *open-coated abrasive cloth* means that only part of the backing surface is covered with abrasive grains (Fig. 12-11A). The open spaces between grains prevents the material from filling with chips or clogging.

By contrast, *all* of the backing surface is covered in *closed coating* (Fig. 12-11B). *Closed-coating abrasive* cloths, belts, and discs are used in heavy cutting operations that require a high rate of stock removal. The two basic open and closed coatings are shown in Fig. 12-11.

Forms of Coated Abrasives

Coated abrasives are produced in a number of forms. The most common forms are shown in Fig. 12-12. These include abrasive belts, grinding discs, and abrasive cloth. The abrasive cloth may be in sheet or roll form.

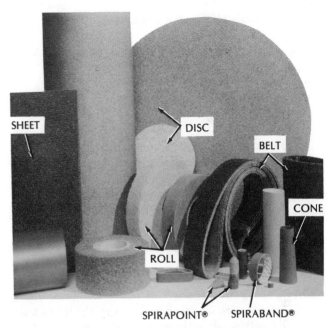

Fig. 12-12. Common forms of coated abrasives. (Courtesy of the Norton Company, Abrasives Marketing Group)

Abrasive belts provide an efficient cutting material and technique for polishing sheet or strip steel. *Grinding discs* are used for smoothening welds and for removing surface irregularities, projections, and burrs. *Abrasive cloth* is used primarily for hand polishing and surface finishing.

When selecting a coated abrasive, consideration must be given to two sets of factors:

● The type of material in the part and the required surface finish

● The qualities of the coated abrasive that are

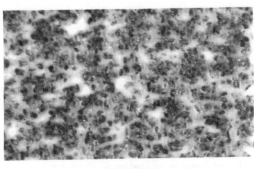

(A) OPEN

(B) CLOSED

Fig. 12-11. Two basic abrasives coatings. (Courtesy of the Norton Company, Abrasives Marketing Group)

needed to cut efficiently and to produce the desired finish.

Characteristics Of Coated Abrasives

The cutting efficiency of coated abrasives and surface finish depend on the:
- Abrasive material
- Bonding and backing materials
- Type of coating
- Grain size.

An *abrasive* is a hard, sharp-cutting, crystalline grain. Abrasives are applied in machine, metal, and other industries to grind, polish, buff, and lap metal surfaces to specific sizes and surface finishes. The abrasives may be *natural* or *artificial*.

All abrasives possess at least the following three properties:

(1) *Penetration hardness,* which is the cutting ability due to the hardness of the grain.

(2) *Fracture resistance,* which is the ability of the grain to fracture when the cutting edge has started to dull, and thereby produce a new sharp cutting edge.

(3) *Wear resistance* is the ability to resist rapid wear and to maintain its sharpness. Wear resistance is also affected by the chemical composition of the grains in relation to that of the work being ground.

Natural Abrasives—The natural abrasives that are still used in industry include *emery, garnet, flint, crocus* and *diamond.* Coated *garnet* and *flint* abrasive papers and cloths are used in the woodworking industry. *Crocus* cloth and *rouge* (a powder-size *flour*) are used for polishing and producing a very fine finish. *Crocus* is a red-iron oxide. Extremely small amounts of metal are removed with *rouge.*

Emery, a natural, hard abrasive, consists of *corundum* (aluminum oxide), iron oxide, and other impurities. Rock–size pieces are mined. These are broken into small particles, sifted through screens, and graded according to size. The grains then are bonded to a flexible backing. Fine, flour-size grains also are available in small quantities. These grains are used as an abrasive grit without a backing.

The coarse grades of emery are used for rapid cutting and removal of comparatively large amounts of material where surface finish is not of prime importance. Medium grades are used for removing surface variations caused by light tool marks or fine scratches. Fine grades of emery are used for polishing.

Hardness Of Selected Abrasives—Natural and artificial abrasives both are considerably harder than hardened steel. A few selected abrasives are listed in Table 12-1. The table shows *relative hardness values* in comparison with that of hardened steel.

Table 12-1. HARDNESS SCALES OF SELECTED MATERIALS

Material	Scales of Hardness	
	Moh's	Knoop
Crocus	6	
Steel (hardened)		740
Flint	6.9	
Topaz	8	1350
Garnet		1350
Cemented carbides		1400-1800
Corundum (aluminim oxide)	9	2000
Silicon carbide		2500
Boron carbide		2800
Diamond (natural or artificial)	10	greater than 7000

The older *Moh's Scale* has a range of minerals whose hardnesses are rated from 1 (talc) to 10 (diamond). Diamond is still the hardest known material.

The *Knoop hardness value* also is a numeral. This scale indicates hardness according to the depth of an impression made by a diamond tester under a fixed load.

According to the scales of hardness in Table 12-1, hardened steel has a value of 740. The cemented carbides register from 1400 to 1800. Diamond (either natural or artificial) is an excellent cutting material due to its extreme hardness properties (7000 and higher on the Knoop scale).

Artificial Abrasives—The aluminum oxides, silicon carbide, and boron carbide are synthetic (artificial) abrasives. As shown in Table 12-1, the hardness of each of these abrasives is higher than that of the natural abrasives. Artificial abrasives with ever-higher hardness qualities are being developed.

The important fact is that artificial abrasives are harder and have greater impact toughness than the natural abrasives. Artificial abrasives have these advantages over natural abrasives:

(1) Faster cutting characteristics

(2) Ability to retain their cutting properties longer

(3) Uniformity of quality

(4) Economical manufacture in a wider range of grain sizes.

○ *Silicon Carbide*. This artificial abrasive is made by fusing silica sand and coke in an electric furnace at high temperatures. The sand provides the silica; the coke, the carbon. Silicon carbide is used for coated abrasives, grinding wheels, and *abrasive sticks* and *stones*. Silicon carbide has a high grain fracture. This makes it desirable for grinding some of the softer metals like cast iron, bronze and copper. Nonmetallic materials that are ground with silicon carbide include marble, pottery and plastic.

○ *Aluminum Oxide*. This artificial abrasive is manufactured from aluminim oxide (bauxite ore). The ore is heated to extremely high temperatures, at which it is purified to crystalline form. The addition of titanium produces an abrasive that has added toughness qualities. With toughness, resistance to shock, and grain fracturing properties, aluminum oxide abrasives are suitable for grinding both soft and hard steels, wrought and malleable iron, and hard bronze.

Backing Materials

Paper, cloth, fibre, and combinations of these provide the *backing material* for coated abrasives. Cloth and fibre backings are the most durable. They are used for hand and machine cutting, buffing, and polishing operations.

The cloth backing may be lightweight and flexible (*jean*) or a heavier weight, stretch-resistant cloth (*drill*). *Drill backing* is used for belt or disc machine grinding.

Fibre, which is stronger and more durable than either cloth or paper, provides better wearing qualities for tough disc-grinding operations.

The commonest forms of cloth-backed coated abrasives for hand use are sheets and rolls. The sheets are 9″ x 11″ (225 mm x 275 mm). The rolls range from 1/2″ to 3″ (12 mm to 75 mm) wide. Other forms of coated abrasives for high-powered grinding machines include belts, discs, spiral points, and cones.

Bonding Materials And Coatings

Bonding Materials—The *bonding material* serves as an adhesive. This holds the grain particles to the backing and indirectly contributes to the cutting properties. Since coated abrasives may be used on dry or wet processes, a combination of bonds is employed.

Glues provide a bond for dry, light, hand grinding and polishing. *Synthetic resins* are practical for wet grinding with waterproof cloth. There also are modifications of glues, resins, and varnishes. The resin-bonded combinations, in particular, provide moisture- and heat-resistant properties. These are required for rough grinding and polishing applications and for toughness.

Grain Density (Coatings)—As stated earlier, there are two basic types of *grain density patterns*. The *open coating* provides definite amounts of open spaces in the backing surface. *Closed coating* means that the backing surface is completely covered with abrasive grains. The open coating is more adaptable for lighter hand grinding. The closed coating is more practical when a high rate of material is to be removed.

Grain Sizes

The abrasive grain sizes for coated abrasives range from extra coarse to extra fine. These and the intermediate degrees of coarseness are indicated by numbers. These range from 12 to 600. The general classifications and numbers are given in Table 12-2.

Table 12-2. COARSENESS DESIGNATIONS AND GRAIN SIZES

Relative Coarseness	Grain Sizes
Extra coarse	12, 16, 20, 24, 30, 36
Coarse	40, 50
Medium	60, 80, 100
Fine	120, 150, 180
Extra Fine	220, 240, 280, 320, 360
Flour	440, 500, 600

In the trade, the terms *grain sizes* and *grits* are used interchangeably. Usually, the coarser grains (grits) are used for heavy stock removal; the finer grains, for highly polished surface finishes. For hand polishing, it is good practice to start with as coarse a grain as is possible. The grain size must be consistent with the required surface finish. Reducing the grain size from coarser to finer during a series of final finishing operation speeds the cutting action. Fig. 12-13 shows enlarged examples of coarse-, medium- and fine-grain sizes for coated abrasives.

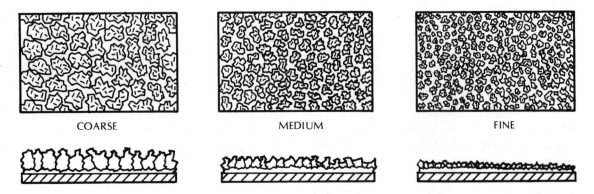

Fig. 12-13. Enlarged sections of coated abrasives showing grain sizes.

HOW TO HAND POLISH WITH COATED ABRASIVES

Selecting The Coated Abrasive

Step 1   Determine the abrasive grain size that is appropriate for the material to be polished and the required quality of surface finish. Select the coarsest grade possible. Move to finer grades as the process progresses.

Step 2   Tear a strip slightly wider than the width of the file or other backing object. The length should permit the abrasive cloth to be held tautly.
Note: The grain runs lengthwise on cloth backing sheets and rolls. It is easy to tear an even strip running with the grain. The direction of tear is illustrated in Fig. 12-14.

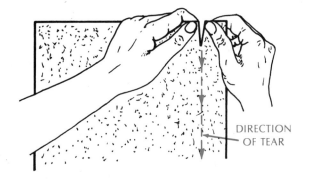

Fig. 12-14. Tearing an abrasive along the grain of the fabric.

Step 3   Grip the work securely in a vise or other holding device. Use soft jaws or pieces of soft stock to protect the finished surfaces.

Step 4   Check the surface of the backing block to see that it is free of burrs and is flat and smooth. A file, a hardwood block, or other suitable material, may serve as a backing.

Hand Polishing

Step 1   Fold the abrasive strip over the end of the file point or other backing block (Fig. 12-15). Hold it there with one hand. Bring the other end along the under face of the file or block. Pull it taut with the other hand.

Fig. 12-15. The correct method of holding an abrasive strip on a file face.

Step 2   Place the abrasive strip on the work surface. Apply a slight downward pressure. Move the abrasive strip in a straight line over the work surface. Take as long a stroke as possible. Cut on both the forward and return strokes. Two positions for holding a file and abrasive cloth are shown in Fig. 12-16.
Note: Avoid any rocking motion when working on a flat surface.

Step 3   Replace the abrasive strip. Change from a coarser grain to a finer one for fine finishing.
Note: A few drops of oil applied to a metallic surface helps the cutting action.

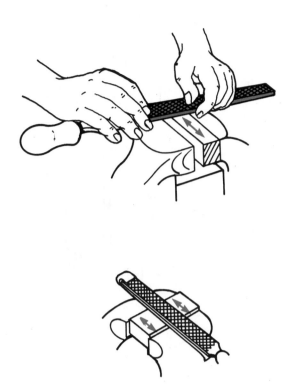

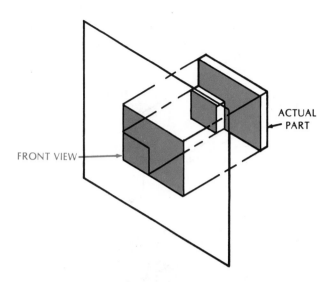

Fig. 12-17. Projecting the front view onto an imaginary screen.

Fig. 12-16. Polishing flat work surfaces.

The Top View

More details are obtained by viewing the object from a second position. This time the block is seen from the top. The edges that are visible from a position immediately above the object are shown in the imaginary plane in Fig. 12-18. This view is called the *Top View*.

REPRESENTING A PART WITH A THREE-VIEW DRAWING

One-view or two-view drawings are used to describe regular round and flat objects. These require simple machining operations. However, as the shape of an object changes, areas are often cut away. At the same time, the hand and machining processes become more complex. A drawing with a number of views is needed to describe the object accurately.

Another widely used shop drawing has three views. Design and drafting room persons use a combination of front, top, and side views to describe some objects. To repeat, the name of the view depends on how the object is viewed by the person making the drawing.

The Front View

The surface as seen by looking directly at the front of the object is called the *front view*. The object is first examined to determine which features can best be represented in this view. A rectangular block with a square projection (*pad*) at the left corner serves as an example. The shape of the front view of the block is projected onto an imaginary screen. This is shown in Fig. 12-17.

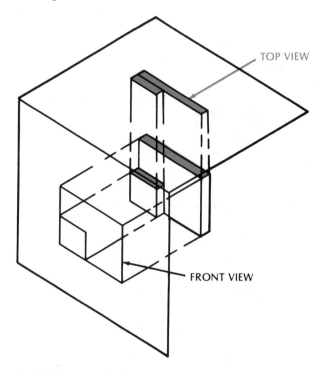

Fig. 12-18. Projecting the top view onto an imaginary screen.

The top view is rotated 90° (theoretically) to form a single plane with the front view. The relative

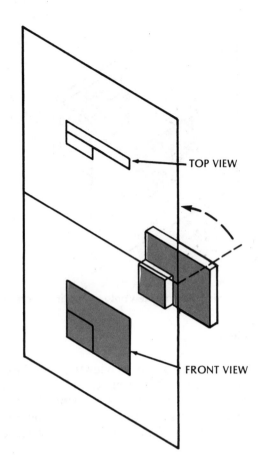

Fig. 12-19. Relative positions of the front and top views.

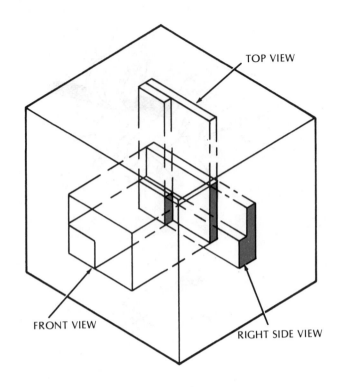

Fig. 12-20. Projecting the right side view.

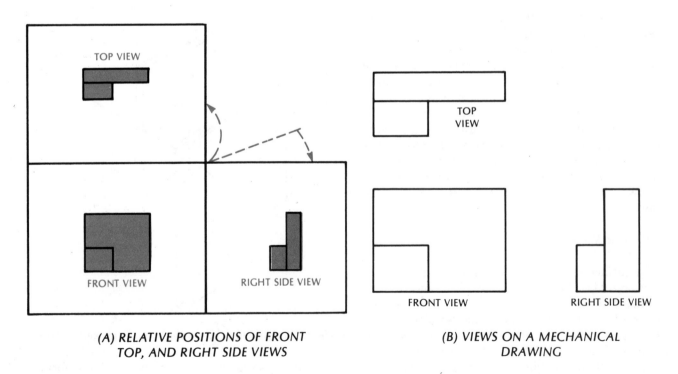

(A) RELATIVE POSITIONS OF FRONT TOP, AND RIGHT SIDE VIEWS

(B) VIEWS ON A MECHANICAL DRAWING

Fig. 12-21. Front, top, and right side views.

position of these two views are shown in Fig. 12-19.

The Side View

The third view of the block is drawn by looking at it from the right side. The shape and size are projected onto another imaginary screen (plane). Fig. 12-20 illustrates the *right side view*. This is often called simply a *right view*. If the object is viewed from the left side of the front view, the view is known as a *left side view,* or just a *left view*.

The Front, Top And Side Views

The relative position of the front, top, and right side views (Fig. 12-21A) as they appear on a mechanical drawing are shown in Fig. 12-21B. Although the views show the shape of the block, the drawing requires dimensions and notes. The drawing with the full shape and size description, dimensions, and notes is known as a *working drawing*. The craftsperson is able to produce the rectangular block from the working drawing (Fig. 12-22).

Other Views

The drawings of more complicated parts require the use of still other views and cutaway sections. Angular surfaces usually require an auxiliary view. The least number of views is used to adequately describe the object.

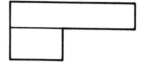

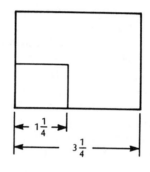
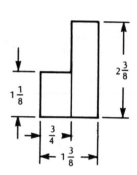

Fig. 12-22. An example of a working drawing.

SAFE PRACTICES IN SCRAPING, BURRING, AND POLISHING

- Hold the cutting face of a flat scraper flat with the workpiece but inclined at a cutting angle. This prevents gouging into or scratching the finished surface.
- Manipulate the three-cornered scraper using a properly secured handle. The cutting face should be moved along the contour of the edge that is to be removed.
- Steady the hand against the part when using a three-cornered scraper. This provides support in applying a slight force, guiding the scraper, and preventing the sharp point and edges from slipping away from the work to cause personal injury.
- Remove only small amounts of material either to precision finish parts or to produce a desired surface finish design.

- Remove sharp projecting edges and burrs which produce inaccuracies in layout, measurement errors, and improper fits. Burrs also cause personal injury and tool damage. Burrs should be removed carefully and as soon as practical.
- Select the abrasive grain size required to produce the desired quality of surface finish. Coarse grains cut deeply. They may produce surface scratches that are not removable within the dimensional limits of the finished part.
- Move a coated abrasive in one direction to polish a work surface. Crossing of strokes produces scratch marks. These may be objectionable.

SCRAPING, BURRING AND POLISHING TERMS

Scraping	A process of removing small quantities of material. Cutting into a surface with a flat, curved, or triangular scraper. A process of producing precision flat, curved, or circular mating surfaces.
Frosting (flaking)	Producing a repetitive design of scrolls, circles, or straight lines on a finished work surface.
Scraper	A hand cutting tool for precision finishing flat, round and other curved forms; fitting mating parts; and producing surface finish designs.
High spots	Surface markings produced by moving a workpiece to be fitted on a perfectly flat surface or mating part. Markings which reveal places on a surface that are high in relation to surrounding areas.
Breaking an edge	The removal of an unnecessary sharp edge by filing, scraping, or stoning.
Dimensional accuracy	A standard established for a dimension. A standard to ensure that parts will fit correctly and function properly.
Abrasive	A natural or synthetically produced hard material. The grains have properties of penetration hardness to cut softer materials; fracture resistance to expose new sharp cutting edges without losing the entire grain, and wear resistance to maintain sharpness.
Artificial abrasive	The synthetic production of materials having the properties of hard-cutting abrasive grains. Three broad categories of manufactured abrasives: aluminum oxide, silicon carbide and boron carbide.
Coated abrasive	A cutting product consisting of abrasive grains with special cutting properties, fixed grain quantities distributed over an area, a bonding agent, and a backing material.
Polishing	As used in this unit, the process of producing a smooth, fine-grained, finished surface. A surface finish for appearance or to improve the sliding qualities between two surfaces.

SUMMARY

- Scraping produces surfaces that are close-grained, smooth, and perfectly flat and true fitting.
- Frosted and flaked surface finishes are also scraped. Scroll or other straight line surface designs are formed by scraping minute amounts from the surface. Frosting improves the sliding quality and appearance, and provides lubrication pockets.
- The three common hand scrapers are the flat, bearing or half-round, and triangular. The first two produce precision-fit flat and curved surfaces. The triangular scraper is a practical tool for burring and for rounding over edges.
- Scraping requires a workpiece to be positioned on the mating part. High spots are located by a slight movement over a prussian blue or colored surface. The high spots are scraped. The process is continued until there is a perfect match.
- Scraping is done by locating the keen scraper cutting edge on the high spot. A little force is applied while moving the scraper over the high spot area.
- Coated abrasives are widely applied in hand finishing operations. These abrasives produce a smooth and highly polished surface.
- Abrasive grains, when properly selected, mounted, and held, produce excellent cutting edges.
- Coated abrasives are produced in forms such as belts, discs, abrasive cloths, and powders (flour).
- Cutting action and surface finish depend on the nature of the abrasive grains (size and material), bonding material (glues, resins and varnishes), coatings (open and closed), and backing material (paper, cloth and fibre).

- Abrasives may be natural (emery, garnet, flint, diamond) or artificial (aluminum oxide, silicon carbide, boron carbide).
- The polishing process should be started with the coarsest grain possible, consistent with the final surface finish required. Successively finer grains (grits) should be used until the final finish is reached.

- A burr is an unnecessary jagged edge produced by hand or machine processes.
- Burrs may cause inaccurate layouts, improper fitting of parts, damage to a workpiece and poor appearance. Burrs can scratch and mark tools and machines. Burrs are unsafe for the worker and produce safety hazards.

UNIT 12 REVIEW AND SELF-TEST

1. State why precision machined surfaces are often *frosted*.
2. a. Identify the design features of the cutting edge of a flat scraper.
   b. Give the reason for forming the cutting edge in this manner.
3. List the steps to follow in finish scraping round surfaces.
4. Cite a number of cutting processes and handling conditions that produce burrs.
5. Give three examples of how burrs affect layouts and dimensional measurements.
6. Contrast *open-coat* with *closed-coat* abrasive cloths.
7. Name three properties of all abrasives.
8. State three advantages of using artificial abrasives in place of natural abrasives.
9. Give two guidelines the skilled mechanic uses to finish a machined surface by hand polishing.
10. Explain why three-view dimensioned drawings are required to describe some parts.
11. Tell why the hand is steadied against a workpiece when using a three-cornered scraper.

Unit 13
THREADING (INTERNAL): CHARACTERISTICS, TOOLS, HAND TAPPING PROCESSES

Technical knowledge about the screw thread was recorded by the scientist Archimedes over 2,000 years ago. From that time until the Middle Ages, the Romans, Greeks, Arabs, and other civilizations designed, hand produced, and applied screws. Such screws were formed from a thread outline drawn around a cylinder. Such a layout is shown in Fig. 13-1. The *grooves* were then sawed and filed. Hardwood was used as the basic material.

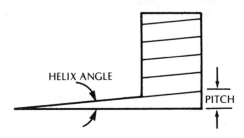

Fig. 13-1. Early layout of a screw thread.

When metal armor was introduced, its bolts and nuts also were hand made of metal. With the introduction of movable machines, like the printing press, came the need for components made of metal. The parts were activated (moved) by metal screws. Other parts that had to be fastened and held together securely also needed metal screws. Growing demands brought on more developmental work to produce screw threads faster and more accurately with machines.

HISTORICAL SCREW THREAD DESIGN AND MANUFACTURING DEVELOPMENTS

A number of Leonardo DaVinci's design sketches around the year 1500 provided technical information about and showed the features of a screw-cutting machine. Twenty years later, the Frenchman Besson was credited with developing the first foot-operated lathe spindle. This lathe was used for *chasing threads* with a *combing tool*.

The first breakthrough in the production of wood screws came in the mid-1700's. A mechanized factory was started by the Wyatt brothers of England. They mass produced 100,000 wood screws a week. Steam power and the industrial revolution brought about greater efforts toward the design and accurate

reproduction of *master screws*. These also are called *lead screws*. Lead screws were needed, in turn, to produce other forms and sizes of screw threads. During this period, after years of laboriously designing, forming, and testing, Henry Maudslay produced the first all-metal screw-cutting lathe in England. The lead screw was extremely accurate for its time. It had an error of 1/16″ in its length of seven feet. Through this accomplishment, this lead screw became a *master* for producing other lead screws. During most of the 1800's, screw threads were cut on lathes by single-point *chasing tools, thread-cutting dies,* and *thread-combing tools*.

Within the United States, a Franklin Institute committee in 1864 recommended a system of *screw thread standards*. The standards were largely developed by William Sellers. He had improved on the earlier work of Sir Joseph Whitworth of Great Britain. Sellers modified the sharp crest and root of a 60° V form by introducing a more practical *flat* crest and root. It should be noted that modifications of the screw thread standards established by Whitworth in 1841 became the basis of the British Standard Whitworth (BSW) thread.

By 1868, the Sellers Standard was adopted as the *United States Standard* (USS). This system was excellent for *coarse threads*. However, with industrial expansion, new machines and mechanical movements, and the later demands of the evolving automotive, aircraft, and instrumentation industries, a finer thread series was needed. The work of the Society of Automotive Engineers (SAE) on a fine-thread screw series was adopted in 1911.

Going back almost 25 years, another important cutting tool development took place in 1887. A. B. Landis introduced a *self-opening thread-cutting die head* in the United States. This further revolutionized thread cutting and the mass production of threaded parts.

FOUNDATIONS FOR THE AMERICAN NATIONAL AND UNIFIED SCREW THREAD SYSTEMS

The two systems, USS and SAE, provided standards for form, pitch and outside diameters. Further

140

standardization and improvements were needed. The National Screw Thread Commission of 1918, together with the SAE, the American Standards Association, the American Society of Mechanical Engineers, and other professional groups, developed the original *American National Screw Thread Standards.*

The new standards included different tolerances, clearances, and other improvements. They provided flexibility in thread forming and manufacturing. The American National Screw Thread Standard included the coarse-thread series of the USS and the fine-thread series corresponding to the SAE. By 1935, the American Standards Association (ASA) approved these standards. These later became the *National Coarse (NC)* and *National Fine (NF)* series.

Although efforts to standardize were carried on (notably by Great Britain, Canada, and the United States), threaded parts among these countries were not interchangeable. The British Whitworth thread had its 55° thread angle; the United States, the 60° standard. Finally, in December 1948, the systems were standardized into a *Unified Screw Thread System.* The *Unified Threads* are a compromise between the British and the American systems.

The design of the new thread is such that parts threaded to conform in shape to either the National Coarse or National Fine series are interchangeable. The British preference for the rounded crest and root is incorporated in the Unified Thread. This feature is important in mass production and the life of a screw thread.

THE INTERNATIONAL (ISO) METRIC THREAD SYSTEM

Paralleling developments in the *Unified* and *American National* systems were international movements. These were designed to modify the many European Metric systems into an internationally-accepted, single Metric system. The result was the *international metric thread form.* This form was established by the *International Organization for Standardization,* known as "ISO". This metric standard form is used in Europe and in many countries throughout the world.

The ISO form is similar to the Unified and the American National form, except for three basic differences:
● The depth of the thread is greater
● The sharp-V thread form is flattened at the crest to one eighth of the thread depth
● The root is flattened (with a rounded root preferred) to one eighth of the thread depth.

The ISO form is the foundation of the ISO metric screw thread series.

There are still many problems to be resolved among the Unified, American National, and the ISO Metric thread systems. Each system still includes a considerable number of *pitch combinations* in relation to diameters (pitch-diameter) in both the course-thread and fine-thread series.

The *Industrial Fasteners Institute (IFI),* after considerable worldwide research, has proposed a single *pitch-diameter series* of 25 metric threads. When adopted, this system may reduce and replace many of the 59 standard Unified and 66 standard ISO Metric pitch-diameter combinations. These combinations are now in the Unified and ISO Metric systems for the coarse-thread (UNC and NC) and fine-thread (UNF and NF) series.

INTRODUCTION TO SCREW THREADS

There are a number of approaches that might be taken in a study of threads and threading methods. One relates to design, the features of thread-cutting tools, and research. This is a laboratory approach. Another way is to treat mass production systems and tooling setups. Still another approach is to deal with the day-by-day need of skilled craftspersons and technicians to cut internal and external threads by hand and machine processes.

This last approach is the one adopted in this textbook. This particular unit, after the preceding brief historical treatment of important milestones, deals with:
● Functions and characteristics of screw threads
● Cutting tools and accessories for internal threading by hand
● Actual hand tapping processes.

Later units cover more advanced technology and precise machining methods, tools, thread fits, and thread forms. Thread formulas and other computations are treated as they relate to machine setups and processes.

Basic Thread Producing Methods

Eight basic methods are employed to produce threads. The threads meet high standards of form, fit, accuracy, and interchangeability within a thread system. The eight methods are:
● Hand cutting internal threads with taps and external threads with dies
● Machine tapping using formed cutters

• External and internal machine thread cutting with single-point cutting tools
• Machine threading using multiple-point cutting tools
• Machine milling with rotary cutters
• Machine grinding with formed abrasive wheels
• Machine rolling and forming
• Casting, using sand, die, permanent mold, shell casting, and other techniques.

One unique advantage of the casting method is that parts of some machines—like sewing and vending machines, typewriters, and toys—have internal threads cast in place.

The Nature Of Screw Threads

A screw thread is defined as a helical (spiral) ridge of uniform section (form and size). This ridge is formed on a cylindrical or cone-shaped surface. The basic features of a screw thread are illustrated in Fig. 13-2. Screw threads that are produced on the outer surface, like those of bolts and threaded studs, are called *straight external threads*. Screw threads that are cut on cone-shaped surfaces, like pipe threads, are *tapered external threads*. Threads that are cut parallel on inner surfaces, as are those of an adjusting nut or a hexagon nut fastener, are known as *straight* or *parallel internal threads*. External and internal parallel threads are referred to in the shop as just *outside threads* or *inside threads*.

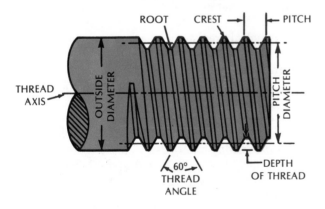

Fig. 13-2. Basic features of a screw thread.

Some threads may be advanced (moved) or tightened into a part by turning them clockwise. These are classified as *right-hand threads*. Threads that advance when turned counterclockwise are *left-hand threads*. Right- and left-hand threads each have special applications. Unless otherwise indicated on a drawing, the thread is assumed to be *right-hand*.

Functions Served By Screw Threads

Screw threads serve four prime functions:
• As a fastener to hold two or more parts securely in a fixed position, and to permit assembly and disassembly
• As a simple machine (in a physical science context). The screw thread transmits and increases force (power). A machine feed screw and a common screw jack (Fig. 13-3) are examples of this application.

Fig. 13-3. A screw thread increases force.

• As a mechanical device to produce motion, like a threaded spindle of a micrometer. Precise measurement standards may be established by this motion.
• To change a rotary motion to a straight-line motion. Fig. 13-4 shows a feed screw being turned. The screw raises or lowers a machine table.

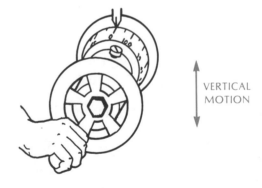

Fig. 13-4. A screw thread changes rotary to linear motion.

AMERICAN NATIONAL AND UNIFIED SCREW THREAD SYSTEMS

Each thread has a shape, or profile. This is called the *thread form*. Although the size of the thread form varies according to the dimensions of threaded part, the shape remains constant. The symbol U preceding the thread form indicates that it conforms to the

Unified System standards. It is accepted by Great Britain, Canada, and the United States. There are four basic *thread series* and one *special thread series* in both the Unified (Thread) System and the American National (Thread) System. The four basic thread series are:

(1) *Coarse-thread series.* This is designated as *UNC* in the Unified System. In the American National System, *NC* denotes *National Coarse.*

(2) *Fine-thread series.* These threads are marked *UNF* and *NF.*

(3) *Extra-fine-thread series. ÙNEF* and *NEF.*

(4) *Constant-pitch series,* as *8 UN, 12 UN,* and *16 UN* for the Unified System, and *8 N, 12 N* and *16 N* for the American National System.

There are a number of pitch series in the *constant-pitch* series. There is the 8 pitch, 12 pitch, and 16 pitch series, and others. *Constant-pitch* means there are the same number of threads per inch for all diameters that are included in the series. For example, the 12 N series always has 12 threads per inch regardless of the outside thread diameter.

The constant pitch series is used when the other three thread series fail to meet specific requirements.

There is a final designation for special conditions that require a nonstandard or *special thread series.* Some drawings show the code UNS, which indicates that the threads are *Unified Special,* or NS, which indicates *National Special* threads.

ISO METRIC THREADS

The ISO metric threads for all general threading and assembling purposes are identified with the *ISO Metric Coarse series.* The *ISO Metric Fine series* is employed in fine precision work. Table 13-1 provides an example of 12 common ISO metric thread sizes ranging from 2 to 24 millimeters. These cover about the same range as the fractional inch sizes in the National Coarse series up to 1″ outside diameter. There are additional standard sizes in the ISO Metric coarse series beyond this range.

The nine common thread sizes in the *ISO Metric Fine series* cover the equivalent range of the fractional inch sizes in the National Fine series up to one inch. Table 13-2 gives the *tap drill diameter* as equal to the outside thread diameter minus the pitch. To cite an example, the coarse series table shows that an ISO Metric threaded part with an 8 mm outside diameter has a pitch of 1.25 mm. This information would appear on an industrial print or drawing as M8-1.25 for the tapped hole it represents. The *tap drill* is 8 mm − 1.25 mm, or 6.75 mm. This value is rounded off to 6.8 mm for the tap drill size.

More complete tables cover the whole range of sizes and systems. These are contained in trade handbooks, Standards Association papers, manufacturers' technical and other printed manuals.

Table 13-1. SELECTED ISO METRIC THREAD COARSE SERIES (2-24 mm)

Outside Diameter (mm)	2	2.5	3	4	5	6	8	10	12	16	20	24
Pitch	0.4	0.45	0.5	0.7	0.8	1.0	1.25	1.5	1.75	2.0	2.5	3.0
Tap Drill Diameter	1.6	2.1	2.5	3.3	4.2	5.0	6.8	8.5	10.3	14.0	17.5	21.0

Table 13-2. SELECTED ISO METRIC THREAD FINE SERIES (2-24 mm)

Outside Diameter (mm)	8	10	12	14	16	18	20	22	24
Pitch	1.0	1.25	1.25	1.5	1.5	1.5	1.5	1.5	2.0
Tap Drill Diameter	7.0	8.8	10.8	12.5	14.5	16.5	18.5	20.5	22.0

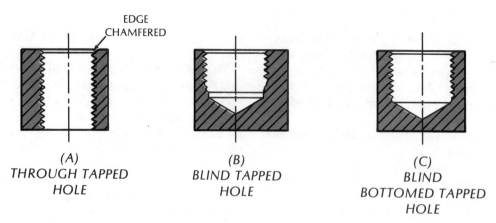

(A)
THROUGH TAPPED
HOLE

(B)
BLIND TAPPED
HOLE

(C)
BLIND
BOTTOMED TAPPED
HOLE

Fig. 13-5. General tapping requirements.

CUTTING INTERNAL THREADS WITH HAND TAPS

Internal threads may be formed with a cutting tool called a *tap*. Taps are used to meet three general tapping requirements. Sections of these appear in Fig. 13-5. Some holes are tapped through (Fig. 13-5A). Other holes are tapped part of the way (Fig. 13-5B); these are called blind holes. Other blind holes are *bottomed* with a thread (Fig. 13-5C).

Features Of Hand Taps

The process of cutting internal threads is referred to as *tapping*. The *tap* is a specially shaped, accurately threaded piece of tool steel or high-speed steel. The general features of a hand tap are represented in Fig. 13-6. A number of *flutes* (grooves) are milled into the body for the length of the threaded portion. These help form the cutting edges. The flutes also provide channels. Chips move out of the workpiece along the flutes during the cutting process.

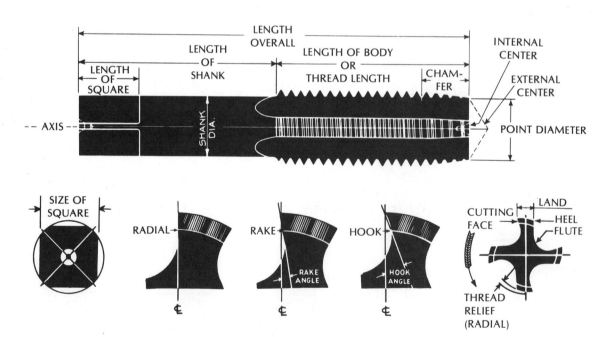

Fig. 13-6. The principal features of a hand tap. (Courtesy of TRW Greenfield Tap & Die Division)

Regular taps have straight flutes. Some taps are *spiral fluted* to produce a different cutting action. The shearing that takes place and the spiral shape push the chips ahead of the tap.

The *shank* is smaller in diameter than the tap drill size. This permits the last teeth on the tap to be employed in cutting a thread. At the same time, the tap may be turned until it completely moves through the workpiece. Another feature of the shank end is that it is square. This square shape provides a good bearing surface for a *tap wrench* and for turning the tap. There is also a *center hole* on the shank end. The center hole is used to align the tap at the line of measurement. Alignment is essential in starting the hand tap and in tapping on a machine.

After being machined almost to size and form, the threads are *relieved* to produce cutting edges. The taps are then hardened and tempered. The finer quality taps are finally ground to size and shape.

Tap Sets

Standard Sets—Taps that are larger than 1/4″ usually come in standard size *sets* for each thread size and pitch. A *set* consists of three taps: a *taper tap, plug tap,* and *bottoming tap* (Fig. 13-7).

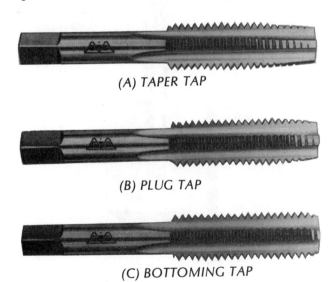

(A) TAPER TAP

(B) PLUG TAP

(C) BOTTOMING TAP

Fig. 13-7. A standard set of hand taps.

The *taper tap* is sometimes called a *starting tap* (Fig. 13-7A). It may be used to start the hand tapping process, particularly when threading a *blind hole,* or to tap a hole that is drilled completely through the part. Because the taper tap has a greater number of teeth engaged in cutting the thread, less force is required. Also, the tap is easier to align with the line

of measurement, and there is less likelihood of tap breakage. The taper tap has the first 8 to 10 threads ground at an angle. The taper begins with the tap drill diameter at the point end. It extends until the full outside diameter is reached.

The *plug tap* has the first three to five threads tapered (Fig. 13-7B). The plug tap is usually the second tap that is used for tapping threads. It may be used for tapping through a part or for tapping a given distance. Certain blind hole tapping is done with a plug tap if the last few threads need not be cut to the total depth.

The *bottoming tap* (Fig. 13-7C) is *backed off* (tapered) at the point end from one to one and one-half threads. This is the third tap that should be used when a hole is to be blind tapped and is to *bottom* at the end of the tap drill hole. The *bottoming tap* should be used after the *plug tap.* Extra care must be taken near the end of the thread to see that the tap is not forced (jammed) against the bottom of the hole. Since so much force is applied on the first few teeth that do most of the cutting, it is especially important to use the complete set of taps.

Machine-Screw Taps—Tap sizes that are smaller than 1/4″ are designated by whole numbers rather than a fraction. For example, $10-32$ means that the outside diameter (OD) of this machine screw tap is equal to the number of the tap (#10) x 0.013 plus 0.060″. For the 10-32 tap, the outside diameter, using the formula, equals $(10 \times 0.013) + 0.060″$, or 0.190″.

On smaller size taps, the amount of material to be removed in cutting the thread form is limited. The use of two or three taps in a set is not required.

Serial Set Taps—There are times when deep threads must be cut by hand in tough metals. The taps that meet these requirements have a different design than the standard taps. These taps are known as *serial taps.* A serial tap set is shown in Fig. 13-8. The outward appearances of these taps are similar to the taper, plug, and bottoming taps. They differ in the amount of material each tap cuts away.

The first two taps cut to a portion of the tooth depth. These taps are known as #1 and #2. They are identified by the number of grooves cut around the shank at the head end, one groove for #1 and two for #2. The #1 tap cuts a shallow groove (thread). The #2 tap cuts the thread deeper.

The #3 tap is the final *sizing tap.* It cuts to the required depth to correctly form the teeth.

Serial taps must be used in sets. The combination reduces the amount of force exerted on any one tap and helps avoid tap breakage.

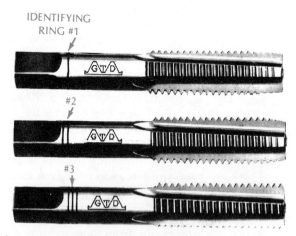

Fig. 13-8. A serial set of hand taps.

Other Types Of Hand And Machine Taps

Gun Taps—The *gun tap* derives its name from the action caused by the shape at the point end of the tap. The cutting point is cut at an angle. This design causes the chips to *shoot out* ahead of the tap. This cutting action is shown in Fig. 13-9. The *gun tap* is applied primarily in tapping through holes in *stringy* metals. The chips tend to *bunch up* and lodge in the flutes. Under these conditions, tapping requires that a greater force be exerted on the tap, increasing the probability of tap breakage.

Fig. 13-9. The cutting action of a plug gun tap.

The gun tap is a production tap. To withstand the greater forces required in machine tapping, there are fewer flutes on the gun tap and they are cut to a shallower depth.

Gun taps are furnished in three general forms. Because each performs a function that is different than that of a standard tap, the names vary. The gun tap names are: *plug gun, bottoming gun,* and *gun flute only.*

The *plug gun tap* is used for through tapping where the holes are open and the chips may shoot out.

The *bottoming gun tap* is used for bottoming tapping operations. The point end of the bottoming gun tap and the flutes have a special design. The flutes are cut deeper. These accommodate the metals to be tapped and the machine method of tapping. The design of the point angle causes the chips to break into fine pieces (Fig. 13-10). The chips are then removed easily.

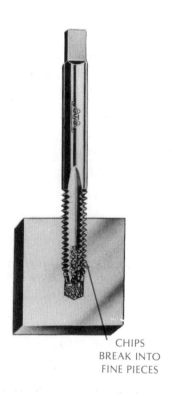

Fig. 13-10. The cutting action of a bottoming gun tap.

The *gun flute only tap* is used on soft and stringy metals for shallow through tapping. The design features of such a tap appear in Fig. 13-11. Due to the larger cross-sectional area, the added strength permits gun taps to be powered with higher-force turning devices than do regular machine taps.

Fig. 13-11. A gun flute only tap.

Spiral-Fluted Taps—Low-angle spiral-fluted taps are especially adapted for tapping soft and stringy metals. Such metals include copper, aluminum, die cast metals, magnesium, brass, and others. Instead of straight flutes, these taps have *spirals*. The spiral flutes provide a pathway along which the chips may travel out of the workpiece. The cutting action of a low-angle spiral-fluted tap is shown in Fig. 13-12.

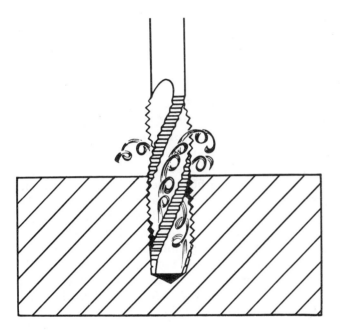

Fig. 13-12. The cutting action of a low-angle spiral-fluted tap.

High-angle spiral-fluted taps are excellent for tapping (1) tough alloy steels, (2) threads that are interrupted and then continue in a part, and (3) blind holes.

*Pipe Taps—*Three common forms of pipe taps are used to cut American National Form Pipe Threads. These include (1) *standard taper pipe threads,* (2) *straight pipe threads,* and (3) *Dryseal pipe threads.* The processes of producing pipe threads are similar to those used for tapping any other of the National Form threads. The *taper pipe tap,* shown in Fig. 13-13, forms *NPT* threads (National Pipe Taper). *Straight pipe taps* (Fig. 13-14) cut *NPS* threads. *Dryseal pipe taps* produce American standard *Dryseal pipe threads.*

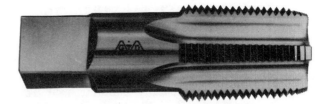

Fig. 13-13. A standard taper pipe tap (NPT).

Fig. 13-14. A straight pipe thread tap (NPS).

The shape of the standard taper pipe thread and the straight pipe thread are the same as the National Form. The thread angle is 60°. The crests and roots are flattened. The differences lie in the diameter designation and pitch. Tapered pipe threads taper 3/4" per foot.

Parts cut with tapered pipe threads may be drawn together to produce a rigid joint. When a pipe compound seal is used on the threads, the line is made gas or liquid leak proof.

Straight pipe threads are used in couplings, low-pressure systems, and other piping applications where there is very limited vibration.

Dryseal pipe threads are applied in systems requiring pressure-tight seals. This is accomplished by making the flats of the crests the same size or slightly smaller than the flats on the roots of the adjoining (mating) thread. Fig. 13-15A shows the initial contact between the crests and roots as a result of hand turning. The pressure-tight seal produced when the parts are further tightened by wrench is illustrated in Fig. 13-15B.

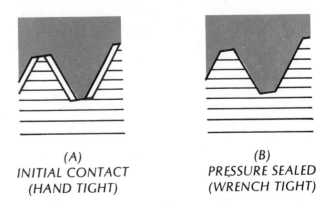

(A)
INITIAL CONTACT
(HAND TIGHT)

(B)
PRESSURE SEALED
(WRENCH TIGHT)

Fig. 13-15. How dryseal pipe threads produce a pressure-tight seal.

FLUTE DEPTH AND TAP STRENGTH

While referred to as hand taps, a number of taps also are used for machine tapping. The difference in some cases is in the material from which the tap is made. Hand taps are usually manufactured from high-carbon tool steel; machine taps, from high-speed steel. Tap sizes that are 1/2" or larger normally have four flutes. Smaller sizes may have two, three or four flutes. Machine screw taps usually have three flutes. The three flutes are cut deeper and provide extra space for chips in blind hole tapping or in tapping deep holes where the chips are stringy.

The two-fluted tap has even more flute space for chips in proportion to the amount of material to be removed. By cutting the flutes deeper, the tooth material between the grooved areas of the tap is reduced. This reduces the strength of the tap to resist shearing forces that may cause the tap to fracture and break.

In succession, the four-fluted tap is stronger than the three-fluted tap; the three-fluted tap has greater strength than the two-fluted tap. Where the grooves are cut shallower, as in the case of gun taps, the reverse is true. The tap strength is increased with the shallower-grooved gun taps.

TAP DRILL SIZES AND TABLES

A *tap drill* refers to a drill of a specific size. The tap drill produces a hole in which threads may be cut to a particular thread depth.

Tap Drill Sizes

The *tap drill size* may vary from the *root diameter* required to cut a theoretical 100% full thread to a larger diameter. The larger diameter leaves only enough material to cut a fraction of a full-depth thread. According to laboratory tests, a 50% depth thread has greater holding power than the strength of the bolt. In other words, the bolt *shears* before the threads of the tapped hole *strip*.

The material in a part, the length of thread engagement, and the application of the thread itself are considerations for establishing the percent of the full-depth thread that is required. Finer threads for precision instruments may be cut close to their full depth. By contrast, bolts, nuts, and deep-tapped holes in the NC, UNC, or ISO Metric coarse series may require a smaller percentage of the full-depth thread. The possibility of tap breakage is reduced when holes are threaded to a fraction of the full depth.

The accepted practice in shops is to use a tap drill that provides approximately 75% of the full thread. The tap drill size may be computed by formula. The easier and more accurate and practical way is to use a handbook or a tool manufacturer's *handy-reference table*.

When the tap drill size is to be computed, values may be substituted in the following formula:

$$\text{Tap Drill Size} \atop \begin{array}{c}(75\% \text{ of} \\ \text{Total Depth})\end{array} = \text{Outside} \atop \text{Diameter} - \left\{ 0.75 \times \frac{1.299}{\begin{array}{c}\text{Number of} \\ \text{Threads} \\ \text{Per Inch}\end{array}} \right\}$$

If a thread depth other than 75% is specified, substitute the required percent in place of the 0.75. The nearest standard size drill is selected as the tap drill.

Tap Drill Tables

The Appendix contains a series of thread tables. One is titled *Hole Sizes for Various Percents of Thread Height and Length of Engagement*. Two others relate to various thread dimensions in the UNC/NC and UNF/NF series.

Parts of two simplified tables with just a few thread sizes are used to illustrate tap drill size tables. The sizes all relate to a thread depth of approximately 75%. Refer first to Table 13-3, which covers UNC/NC threads. Note, as an example, that a 1/2-13 UNC/NC thread requires a 27/64" tap drill to produce a 75% thread. By contrast, Table 13-4 shows that a 10-32 UNF/NF thread requires a #21 tap drill (0.159") for a 75% depth thread.

Table 13-3. TAP DRILL SIZES (75% THREAD) UNC/NC SERIES

Thread Size and Threads per Inch	Major Outside Diameter (Inches)	Tap Drill Size	Decimal Equivalent of Tap Drill
		(75% Thread Depth)	
¼-20	0.2500	#7	0.2010
5/16-18	0.3125	F	0.2570
3/8-16	0.3750	5/16	0.3125
7/16-14	0.4375	U	0.3680
½-13	0.5000	27/64	0.4219
9/16-12	0.5625	31/64	0.4844

Table 13-4. TAP DRILL SIZES (75% THREAD) UNF/NF SERIES

Thread Size and Threads per Inch	Major Outside Diameter (Inches)	Tap Drill Size	Decimal Equivalent of Tap Drill
			(75% Thread Depth)
5-44	0.1250	#37	0.1040
6-40	0.1380	#33	0.1130
8-36	0.1640	#29	0.1360
10-32	0.1900	#21	0.1590
12-28	0.2160	#14	0.1820
¼-28	0.2500	#3	0.2130

TAP WRENCHES

Taps are turned by specially designed *tap wrenches*. Two general forms of wrenches are the *T-handle* and the *straight handle* (Fig. 13-16). Both of these have adjustable jaws. This adjustment permits the use of a single tap wrench handle to accommodate a number of taps with different sizes of square heads. However, the range of tap sizes is so great that tap wrenches are made in a number of sizes.

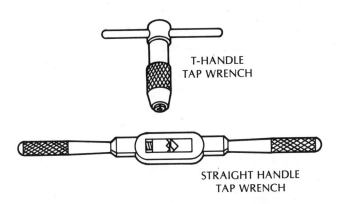

T-HANDLE TAP WRENCH

STRAIGHT HANDLE TAP WRENCH

Fig. 13-16. General forms of tap wrenches.

The size of the square head of a tap is proportional to the tap size. Care must be taken to select the appropriate size of tap wrench. Many taps are broken because excessive force is exerted through the leverage of a wrench that is too large. While designed principally for use as tap wrenches, they are also used with other square-shanked cutting tools for hand turning operations.

HOW TO TAP INTERNAL THREADS BY HAND

Selecting The Tap And Tap Wrench

Step 1   Examine the thread specifications on the work order, print or shop sketch.

Step 2   Determine the most practical tap to use (that is available). The correct tap depends on the material, length of thread, and percent of thread depth required. Select an appropriate size and type of tap wrench.

Step 3   Refer to a Table of Tap Drill Sizes for the thread size and form series. Determine the tap drill size that meets the job specifications.
Note: Shop prints normally carry a notation on the size of drill and depth, particularly for blind holes.

Step 4   Check the drilled hole for correct tap drill size and depth. Remove all chips. If possible, the drilled hole should be countersunk slightly. This helps in starting a thread and prevents throwing up a burr.

Starting The Tap

Step 1   Mount the workpiece securely in a vise. Check the hole to see that it is in a vertical position.

Step 2   Start with the taper tap. Tighten the shank end in the wrench jaws. Grasp the tap and tap wrench in one hand. Place the tap in the tap hole. Guide the tap with the other hand so that it is vertical. Fig. 13-17 shows the correct position of a tap at the start.

Step 3   Apply a slightly downward force while turning the tap clockwise at the same time. Continue to turn with one hand. Apply a constant force for two or three turns. This action should cause the tap teeth to start to form partial threads.

Checking For Squareness

Step 1   Remove the tap wrench carefully. Clean the work surface and check for burrs at the beginning of the thread. Check the tap for squareness (Fig. 13-18). Bring the blade of a square against the tap shank in two positions at 90° to each other.

Note: If the tap is not square, *back it off* by reversing the direction. Restart the tap by turning it clockwise. Apply a limited force in the direction from which the tap *leans*.

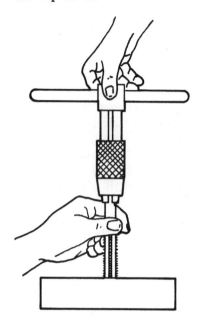

Fig. 13-17. Positioning a tap at the start.

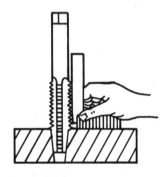

Fig. 13-18. Checking a tap for squareness.

CAUTION: Care and judgment are needed. The tap is brittle. Any sharp jarring or excess force at an angle may cause the tap to *snap* and break.

Step 2   Turn the tap two or three turns. Repeat the checking and adjusting steps until the tap is square with the workpiece.

Step 3   Apply a small quantity of an appropriate lubricant to the cutting edges of the tap. *Note:* Use turpentine or a mineral spirit for very hard materials. Steel, bronze and wrought iron require a lard oil. Soft

metals and cast iron may be hand tapped without a lubricant.

Tapping A Through Hole

Step 1   Turn the tap with both hands on the tap wrench handles. As the tap *takes hold*, the downward force may be released. *Note:* The two handle ends of small tap wrenches should be held and turned with a steady, gentle, even force (Fig. 13-19).

Fig. 13-19. Tapping with a T-handle tap wrench.

Larger straight-handle tap wrenches and taps are turned by grasping one end of the handle with one hand and the other end with the second hand. The correct position of the hands is shown in Fig. 13-20.

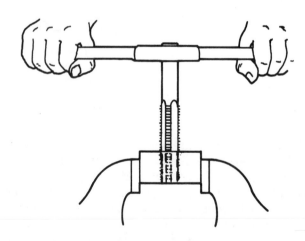

Fig. 13-20. Tapping with a straight tap wrench.

Step 2   *Back off* the tap after each series of two or three turns. This is done by reversing the direction of the tap. This action causes the chips to break. They then can move through the flutes and out of the work. Backing off also prevents tap breakage.

Step 3   Continue to turn the tap, repeating the steps for cutting, lubricating and backing off. The starting taper tap may either be moved through the workpiece or removed by reversing the direction.

Note: It is good practice to follow the taper tap with a plug tap. The plug tap then serves as a finishing or *sizing tap.*

Tapping A Blind Hole

Step 1   Determine in advance how much of the tap should be above the work surface when the work is tapped to the required depth. Start to tap.

Note: The taper tap may be used to start the tapping of deep blind holes. The plug tap is used for tapping shallow blind holes.

Step 2   Remove the tap and the chips in the blind hole.

Note: It may be necessary to remove the part from the vise and use an air line to clear the chips.

CAUTION: Goggles must be worn during the chip-clearing process. The workpiece should be turned away from the operator and covered with a wiping cloth. This helps to prevent the chips from flying in the area.

Step 3   Reposition the work in the vise. Continue to tap. Check the depth occasionally. A steel rule placed on the work surface and next to the tap body may be used to roughly check the depth.

Note: It may be necessary to remove the chips a second time from the blind hole.

Step 4   Repeat the tapping steps with a bottoming tap if the threads must be bottomed.

CAUTION: The first few teeth of the bottoming tap cut to the full thread depth. This requires considerable force. The tap must be turned slowly near the end of the hole and stopped as soon as the tap bottoms.

Step 5   Remove the tap when the thread is bottomed. File away any burrs raised on the surface by tapping. If possible, take a countersink and turn it by hand in the tapped hole. This removes any fine rough edge at the top of the thread. Finally, thread the plug tap through the tapped hole to remove any remaining particles.

Step 6   Clean the workpiece. Check the accuracy of the thread with a thread plug gage, the mating threaded part, or a bolt or screw.

TAP EXTRACTORS

The name *tap extractor* indicates that this tool is used to extract parts of taps that have broken off in a hole. The extractor has prongs. (See Fig. 13-21) These fit into the flute spaces of the tap. The prongs are pushed down into the flutes. The steel bushing on the extractor is moved down as far as possible to the work. In this position, the bushing provides support for the prongs. The extractor is turned *counterclockwise* with a tap wrench to remove a broken *right-hand* tap. After the broken tap is removed, the tapped hole should be checked to see that no broken pieces remain in the hole.

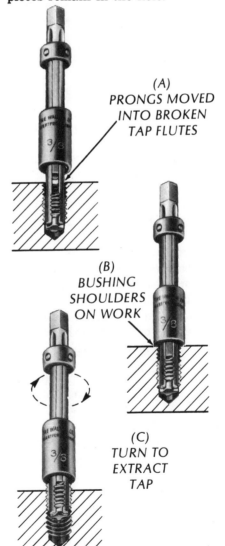

(A)
PRONGS MOVED INTO BROKEN TAP FLUTES

(B)
BUSHING SHOULDERS ON WORK

(C)
TURN TO EXTRACT TAP

Fig. 13-21. The application of a tap extractor.

Step 1   Remove any fractured tap particles near the top surface with a pair of pliers. **CAUTION:** Taps normally break into *sharp* fragments. These should be removed only with other hand tools.

Step 2   Select a tap extractor appropriate for the tap size. Slide the extractor prongs down into the tap flutes as far as possible (Fig. 13-21A).

Step 3   Move the steel bushing over the prongs and down to the work surface (Fig. 13-21B).

Step 4   Turn the extractor with a tap wrench in the direction opposite that used during the tapping process (Fig. 13-21C).

Step 5   Continue turning the extractor a few turns. Then reverse the turning to dislodge any chips. These may cause the tap to bind. Repeat this process until the tap is removed.

Step 6   Clean the part. Recheck the hole for remaining chips and broken tap particles. Check to see that the tap drill size is correct.

Step 7   Retap the hole.

SAFE PRACTICES WITH THREADING TOOLS AND PROCESSES

- Check the tap drill and hole size before the tapping process is started. Undersized tap drills require excessive force.
- Use a standard taper tap, wherever practical, as a starting tap.
- Use the bottoming tap as a finishing tap for blind tapped holes.
- Reverse a standard hand tap during the cutting process. This breaks the chips, makes it easier to move them out of the tapped hole, and reduces the force exerted on the tap.
- Select a size and type of tap wrench that is in proportion to the tap size and related to the nature of the tapping process. Excessive force applied by using oversized tap wrenches is a common cause of tap breakage.
- Exercise care with bottoming taps, particularly in forming the last few threads before bottoming. Avoid excessive force or jamming the point of the tap against the bottom edge of the drilled hole.

- Back out a tap that is started incorrectly. Restart it in a vertical position, square with the work surface. Any uneven force applied to one side of a tap handle may be great enough to break the tap.
- Use caution when using an air blast. Chips should only be blown from a workpiece when (1) each operator in a work area uses an eye protection device, and (2) the workpiece is shielded to prevent particles from flying within a work area.
- Handle broken taps carefully. Taps fracture into *sharp, jagged pieces.* Workpieces with parts of broken taps should be placed so that no one brushes up against them. Use only a hand tool, like a pair of pliers, to grasp or try to move a broken tap.
- Chamfer the ends of a tapped hole with a countersink. A triangular scraper may also be used to remove any burrs produced by tapping.
- Select cutting fluids (lubricants) appropriate for the material being tapped. The cutting fluids reduce the force required for tapping, help the tap cut easier, and produce a smoother finished thread.

SCREW THREAD AND HAND TAPPING TERMS

Internal or external screw thread	A helical ridge of uniform section. A ridge formed in a straight or tapered cylindrical hole (internal) or outside diameter.
USS, SAE	Early coarse and fine thread series standards. Beginning standards that are presently incorporated in the American National Form Coarse Series (NC) and Fine Series (NF).
Unified thread form	Thread form standards adopted by Great Britain, Canada and the United States. Standards that ensure interchangeability of threaded parts conforming to the national standards of these countries. A 60° included angle thread form with crest and rounded root. A thread form that incorporates features of the British Standard Whitworth and the American National Form.
National Coarse and National Fine	Two basic thread series built upon the American National Form. Coarse Series (NC) threads are applied in general and rough applications. National Fine Series (NF) threads are used on parts and instruments requiring finer, more precise threads.
International (ISO) Metric Thread System	A Metric thread system accepted as an international standard. A thread form having a 60° thread angle. The crests and roots are flattened at 1/8th of the depth. A rounded root is recommended. Drawings are dimensioned with Metric thread specifications.
Pitch-Diameter Series	A thread system proposed by the Industrial Fasteners Institute. The system is intended to reduce and replace current Unified and ISO Metric pitch-diameter combinations.
Constant-Pitch Series	A threaded series in which the pitch is the same for all diameters that are included in the series. *Examples:* 8 pitch, 24 pitch, 32 pitch.
ISO Metric thread designation	Specifications of a metric thread giving the outside diameter and pitch in millimeters. *Example:* M20-2.5, an outside diameter of 20 millimeters and a pitch of 2.5 millimeters.
Flutes	Channels running parallel to a tap axis and cut below the thread depth. Grooves which form the teeth of a tap. Channels into which chips flow and are removed during the cutting process.
Relieved	The reduction in size of the mass behind a special shape. This produces a cutting edge and permits a cutting tool to be turned with minimum resistance (*drag* or friction).
Bottoming	A common threading term used to indicate that as close to a full thread as possible must be produced at the bottom of a hole.
Serial set taps	A set of three progressively larger taps (#1, #2, and #3). A set where the #3 tap conforms to the correct thread specifications. A set of three taps, each of which must be used in succession to produce a full thread form.
Spiral-fluted taps	Specially designed taps for soft, stringy metals (low-angle) or tough alloy steels (high-angle).
Tap drill	A standard drill corresponding in size to the diameter to which a hole must be drilled. The tap drill permits cutting a thread to a specified depth.
Tap extractor	A pronged tool inserted in the flutes of a broken tap and turned to extract the tap.

SUMMARY

- The early forms of screw threads were layed out around a cylindrical surface. The threads were sawed and filed to shape.
- Screw thread standards established by William Sellers represented modifications of the work of Whitworth and the British Standard System.
- There are eight basic methods of producing threads: hand cutting, machine cutting with formed cutters, machine cutting with single-point and multiple-point cutting tools, milling, grinding, rolling and forming, and casting.
- Screw threads serve: as fasteners, to change force, to produce motion, and as a design feature on measuring instruments.
- Screw threads are right- or left-hand, and straight or tapered.
- Five common thread series in the Unified and American National Thread systems are:

  ○ Coarse: NC/UNC
  ○ Fine: NF/UNF
  ○ Extra Fine: NEF/UNEF
  ○ Constant Pitch: N/UN
  ○ Special: NS/UNS

- The American National Thread Form for the NC series evolved from the earlier Sellers Standards and the USS coarse-thread series. The NF series developed from the former SAE fine-thread series.
- The Unified Thread Form has a 60° included angle. The thread has a flat crest (1/8 P) and root (1/6 P), preferably rounded.
- The ISO Metric thread system has a 60° included angle with a flat crest and root. Each of these is equal to 1/8 P, with a preferred rounded root. Dimensions are in metric units of measurement.
- ISO Metric coarse and fine-thread series threads are identified on drawings by the letter M followed by the outside diameter and pitch size. For example: M12-1.75.
- Standard hand tap sets include taper (for starting), plug (for through and deep-hole tapping), and bottoming taps.

- A cutting fluid, appropriate for the material being tapped, helps the cutting action, improves the quality of the finished threads, and reduces the force required for tapping.
- The tap drill produces a hole size with enough remaining material to form a thread of a specified depth.
- Machine screw taps under 1/4″ are designated by number. Usually a single tap is used to cut the thread to a required percent depth.
- Serial set taps are designated for cutting tough materials. Taps #1, #2, and #3 each cut a portion of the full thread depth. The #3 tap is the final, finishing tap.
- The plug gun, bottoming gun, and gun flute only taps are production taps. These are used for stringy metals where chips adhere to straight-fluted taps.
- The low-angle spiral-fluted tap is designed for tapping soft metals. The high-angle taps are for tough alloy steels and blind holes.
- Straight (NPS), standard taper (NPT), and American Dryseal are three common forms of pipe threads.
- There are usually fewer flutes (2 and 3) on small-sized taps. These are cut deeper for greater chip clearance. These taps may be sheared and fractured easily.
- Tap drill tables provide tap drill sizes based on length of thread, nature of material, and percent of full depth required. The formula for computing tap drill sizes for the National Form is:

$$\text{Outside Diameter} - \left\{ \frac{1.299}{\begin{array}{c}\text{Number}\\\text{Threads}\\\text{per Inch}\end{array}} \times \begin{array}{c}\text{\% Thread}\\\text{Depth}\\\text{Required}\end{array} \right\}$$

- A tap wrench should be selected in terms of the tap size and the location of the tapped hole.
- Taps are brittle because of their construction and hardness. Extra care must be taken to see that the tap drill size is correct, the tap is started squarely, the workpiece is secured so that it will not move, a proper cutting fluid is used, the chips are broken regularly and are removed, and no undue force is applied on the tap.

UNIT 13 REVIEW AND SELF-TEST

1. Describe the significant design changes of the early Sellers Standard for coarse screw threads.
2. Tell why it was necessary to adapt the SAE Screw Thread System in 1911.
3. State what compromises were made in the 55° British Whitworth thread form and the United States 60° standard thread form to standardize them into the Unified Screw Thread System.
4. Identify how the Industrial Fasteners Institute (IFI) proposes to reduce and replace the 125 Unified and ISO Metric pitch-diameter combinations for the coarse- and fine-thread series.
5. List four basic hand and machine (lathe cutting) methods of producing threads.
6. Differentiate between the pitch to diameter relationship of threads in the UNC series and the *constant-pitch series* (UN) for the Unified System.
7. Show how information about ISO Metric threads with a 25 mm outside diameter and a pitch of 4 mm is represented on an industrial drawing.
8. Identify when each of the following taps is used: (a) *taper,* (b) *plug,* and (c) *bottoming.*
9. Tell when (a) *low-angle spiral-fluted* and (b) *high-angle spiral -fluted* taps are used.
10. Give one design feature of an American Standard Dryseal Pipe Thread.
11. State why general-purpose threading to 75% of full depth is an accepted practice.
12. Explain how part of a tap that breaks in a hole may be removed safely.
13. List four safety practices to follow to avoid personal injury or tap breakage when threading.

Unit 14
THREADING (EXTERNAL): DRAWINGS, DIES, AND HAND PROCESSES

Historical and basic technical information about standards for thread forms and thread series, hand cutting tools, and tapping processes were covered for internal threads in Unit 13. This unit provides additional details that relate principally to external threads. Techniques are described for representing and specifying screw threads on shop drawings, prints and sketches.

Today, it is still necessary and practical to cut threads by hand. While such threads may not be as precise as those produced by machining processes, hand cut threads do meet general requirements. The tools and processes used in bench work for cutting external threads and for measuring pitch are also dealt with in this unit. More advanced technology and processes are included in later units. These deal with different forms and precision methods of cutting internal and external threads.

SCREW THREAD SPECIFICATIONS AND DRAWING REPRESENTATION

Drawings Of Screw Threads

The designer's concept of an exact part is communicated to the craftsperson through the medium of a drawing. The craftsperson must be able to accurately interpret design features, dimensions, and the relationship of one part with another. Each line, view, dimension and note on a drawing has a particular meaning for the worker.

To review: Threads may be external or internal. A thread may extend entirely through or over a given length. The thread may also be bottomed at the depth of a drilled hole, or it may go only a specified distance. Similarly, external threads may be cut to a shoulder or threaded part way.

All of these conditions, and others, are usually shown graphically (represented) on drawings and sketches in one of three ways:

- Pictorial representation
- Schematic representation
- Simplified representation

Fig. 14-1 illustrates how internal and external threads are represented pictorially, schematically, and through simplified techniques.

Dimensioning Screw Threads On Drawings

To provide all essential details, a drawing must include full specifications for cutting and measuring the threads. Technical information about thread sizes, depth or length, class of threads (*fits*), and surface finish appear near the threaded portion. Sometimes the thread length is specified by a dimension on the part. Other instructions may be stated under *Notes*.

Dimensions for internal and external threads are added to the part drawing. A simplified drawing of threads with a dimensioning code and dimensions is provided by Fig. 14-2. Each drawing includes a series of encircled numerals. The type of dimension represented by a particular numeral is explained by a code (Fig. 14-2A). The specifications of the external thread are given on a drawing as shown in Fig. 14-2B. The details of the blind threaded hole (internal thread) appear on a drawing as illustrated in Fig. 14-2C.

MEASURING SCREW THREAD PITCH

Two simple screw thread measurement techniques may be used to establish the pitch of a particular threaded part. The pitch is the distance at the same point on a thread form between two successive teeth. The first method requires the use of a *screw pitch gage;* the second, a steel rule.

Screw Pitch Gage

A *screw pitch gage* (Fig. 14–3) has a series of thin blades. Each blade has a number of teeth. The teeth match the form and size of a particular pitch. Each blade is marked for easy reading.

Pitch is checked by selecting one of the thread gage blades. The blade teeth are placed in the threaded grooves of the actual part, as shown in Fig. 14-3. The blade is *sighted* to see whether the teeth match the teeth profile of the workpiece. Different blades may

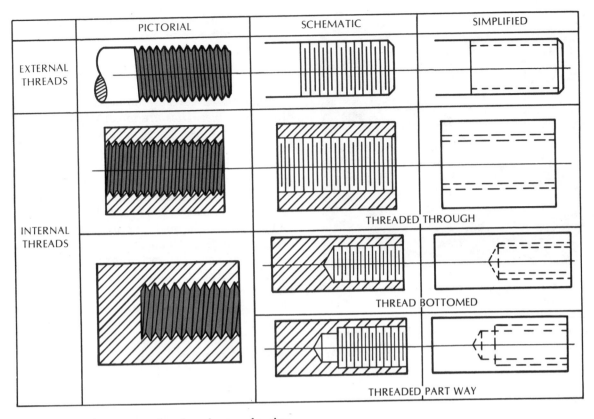

Fig. 14-1. Techniques for representing threads on a drawing.

	PICTORIAL	SCHEMATIC	SIMPLIFIED
EXTERNAL THREADS			
INTERNAL THREADS		THREADED THROUGH	THREADED THROUGH
		THREAD BOTTOMED	THREAD BOTTOMED
		THREADED PART WAY	THREADED PART WAY

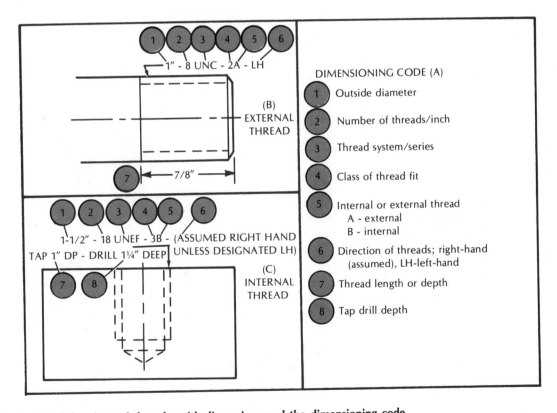

Fig. 14-2. Simplified drawings of threads, with dimensions and the dimensioning code.

1" - 8 UNC - 2A - LH

(B) EXTERNAL THREAD

7/8"

1-1/2" - 18 UNEF - 3B - (ASSUMED RIGHT HAND UNLESS DESIGNATED LH)

TAP 1" DP - DRILL 1¼" DEEP

(C) INTERNAL THREAD

DIMENSIONING CODE (A)

1 Outside diameter

2 Number of threads/inch

3 Thread system/series

4 Class of thread fit

5 Internal or external thread
   A - external
   B - internal

6 Direction of threads; right-hand (assumed), LH-left-hand

7 Thread length or depth

8 Tap drill depth

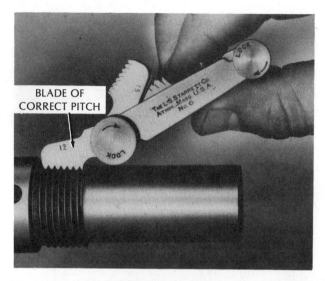

Fig. 14-3. A screw pitch gage used to determine thread pitch. (Courtesy of The L.S. Starrett Company)

need to be tried until one is found that conforms exactly to the workpiece teeth. This is the required pitch.

The Steel Rule

The screw pitch may also be measured by placing a steel rule lengthwise on a threaded part (Fig. 14-4). An inch graduation is usually placed on the crest of the last thread. The number of crests are counted to the next inch graduation on the rule. This number of threads represents the pitch. This technique is shown in Fig. 14-4.

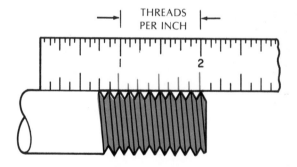

Fig. 14-4. Using a steel rule to measure the number of threads per inch.

If the threaded portion is less than one inch, the number of threads in the given distance may be multiplied by the ratio that the fractional part bears to one inch. The pitch of a metric thread is expressed in millimeters. As stated earlier, the pitch represents

the distance from a specific point on one tooth to the corresponding point on the next tooth.

EXTERNAL THREAD CUTTING HAND TOOLS

External threads are hand cut with cutting tools known as *dies*. *Dies* are made in a variety of sizes, shapes and types. These depend on the thread form and size and the material to be cut. Some dies are solid and have a fixed size. Other dies are adjustable. Dies are made of tool steel or high-speed steel, and other alloys. Dies are hardened and tempered.

The adjustable type is *split*. The *split die* permits adjustments to be made to produce a *smaller* or *larger size thread*. Sometimes the size adjustment is made with an adjusting screw in the die. In other instances, screws on the *die holder* provide for adjustments. Once adjusted, the position is secured. Another common type of adjustable die contains two *die halves* that are held in a *cap*. The three forms of hand threading dies are illustrated when each one is described in more detail.

Solid Dies

This form is rarely used in the shop. The solid square- or hexagon-shaped die has a fixed size (Fig. 14-5). It is applied principally to *chase* threads that have been poorly formed or have been damaged. Such

Fig. 14-5. A solid hexagon die. (Courtesy of TRW Greenfield Tap & Die Division)

threads require that a sizing die be run over them to remove any burrs or nicks or *crossed threads*. The damaged threads are reformed to permit a bolt or other threaded area to turn in a mating part. The hexagon shape may be turned with a socket, ratchet, or other adjustable wrench.

Adjustable Split Dies

The adjustable split die is available in three common types.

The type in Fig. 14-6 must be adjusted each time the die is changed in the die holder. The adjustment is made by turning the adjusting screws in the *die holder* (sometimes called a *die stock*). The split die is placed in the die holder and tried on a correctly sized threaded part. The screws are adjusted until there is a slight drag between the die and the threaded part. In other cases, the die is adjusted by a *cut-and-try* method. The die is *opened*. A few full threads are cut. The workpiece is *tried* on the mating part. The die is adjusted until the correct size is reached. The die is then locked at this setting.

The second type of split-fixed adjustable die has a screw within the die head (Fig. 14-7). This screw adjustment opens or closes the die thread teeth. This provides a range of adjustments for cutting *oversized* or *undersized* threads. The die is mounted in a die holder. The die is held at the set position with the adjoining screws in the die holder.

The third basic type has three essential parts: a cap, a guide, and the *die halves* (which are the threading die). All of the parts and the complete assembly are shown in Fig. 14-8. The sides of the two die halves are cut at an angle. The die halves are held securely against the machined, tapered surfaces in the cap. The threads on the die halves are tapered at the front end. The taper permits easier starting and relieves excessive pressure on the first few cutting teeth. The cap has two adjusting screws. These move and position the die halves. The cap also has a lock screw recess by which the die is secured in the holder.

The *guide* serves two purposes:

● It forces the tapered sides of the die halves

against similarly tapered sides in the cap. With this force the two halves are held tightly in position so that the size cannot change.

● It pilots the die (halves) so that it is centered over the workpiece. Once the die halves are set to the required size, they are locked in position with the cap. The assembled die is then locked in the die holder.

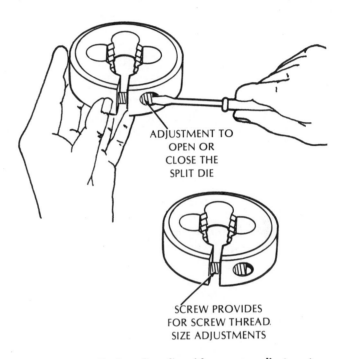

ADJUSTMENT TO OPEN OR CLOSE THE SPLIT DIE

SCREW PROVIDES FOR SCREW THREAD SIZE ADJUSTMENTS

Fig. 14-7. A split threading die with a screw adjustment.

The cutting teeth of the die are formed by removing metal from portions of the thread form. Instead of flutes, the area in back of the threaded section is removed. This provides both a cutting edge and ample room for chips to clear the die. Since dies have more body than do taps, they are considerably stronger, and the chip channels may be cut deeper. The teeth are *relieved* in back of the cutting edge to make cutting possible and to avoid binding.

Pipe Thread Dies

Pipe threads are tapered in order to make tight joints in air and liquid lines. Pipes are measured (sized) according to their *inside* diameter. Because of this measurement, pipe threads are larger in diameter than regular screw threads. The taps and threading dies are, correspondingly, larger for a specific diameter than regular taps and dies.

A 1/2″ standard taper pipe thread (NPT) requires a tap drill of 23/32″. This size drill permits the hole to be tapered with a thread taper of 3/4″ per foot. The

ADJUSTING SCREWS

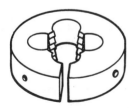

Fig. 14-6. The split die thread size is controlled by three adjusting screws.

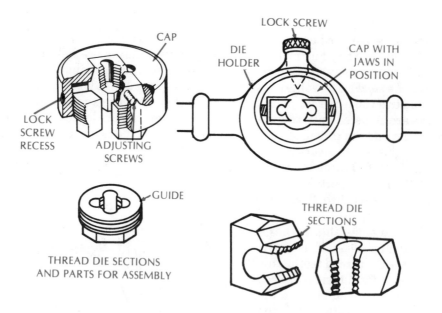

Fig. 14-8. Features and assembly of the dies and the guide.

actual outside diameter of the 1/2″ threaded pipe is 0.840″. Other common National Taper Pipe Thread sizes are given in Table 14-1.

Table 14-1. NATIONAL STANDARD TAPER (NPT) PIPE THREADS

Nominal Pipe Size	Threads per Inch	Outside Diameter	Tap Drill Size
⅛	27	0.405	(R) ¹¹/₃₂
¼	18	0.504	⁷/₁₆
⅜	18	0.675	³⁷/₆₄
½	14	0.840	²³/₃₂
¾	14	1.050	⁵⁹/₆₄
1	11½	1.315	1⁵/₃₂
1¼	11½	1.660	1½
1½	11½	1.990	1⁴⁷/₆₄
2	11½	2.375	2⁷/₃₂

Dies for pipe threads may be solid or adjustable. On the adjustable type the dies vary in thickness to correspond to the length of pipe thread that should be cut for a specific diameter. The die halves are held in an *adjustable stock* that serves as the die holder. There are indicating marks on both the die and the stock. When the die halves are aligned with these marks, the die is set to cut a standard pipe thread. The adjustable stock is provided with a positioning cap. This cap centers the work with the die halves.

Where single pipes are to be threaded by hand, a pipe plug or the mating part may be used as a gage. A pipe thread is cut to near its full depth, tried for size and fit, and then cut to the correct depth. Since burrs are raised easily in pipe threading, care must be taken to remove the sharp burrs. The work surfaces must also be cleaned before the parts are assembled.

Straight, taper, and Dryseal pipe thread standards are found in Handbooks. The tapers present an added problem of fitting the parts carefully. The same attention must be paid to start taps and dies squarely. An appropriate cutting lubricant must be used. The dies must be backed off to break the chips. The chips must be removed. The workpiece must be burred and cleaned.

HOW TO CUT THREADS WITH A HAND THREADING DIE

Selecting The Threading Die

Step 1   Select the adjustable threading die that meets the job requirements for thread form and size. Select an appropriate type and size of die holder.

Step 2   Check the outside diameter of the workpiece. Chamfer the sharp edge to the depth of one thread by turning or hand filing. This is necessary to start the die squarely so that the thread will be cut straight.

Adjusting The Threading Die

Step 1   Test the adjustable threading die. Thread (turn) the mating part into the die by hand. Adjust the screw on the screw-adjusting type until a slight force is needed to turn the die.

Note: On split dies, the die may be positioned in the holder. The adjusting screws are then tightened to bring the die to size and to lock it in this fixed position (Fig. 14-9).

Note: The lead side (tapered threads) of the die should be placed opposite the shoulder of the die stock.

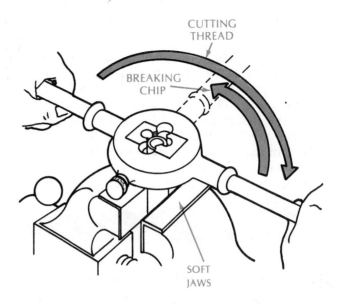

Fig. 14-10. Cutting a right-hand thread, and reversing the stroke to break the chip.

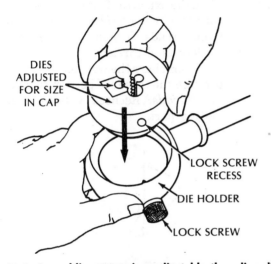

Fig. 14-9. Assembling a two-jaw adjustable threading die.

Step 2   Tighten the lock screw in the die holder.

Starting The Thread

Step 1   Mount the work securely in a vise. Use protecting soft jaws where required. Place the die on the work. The tapered starting cutting area is on the bottom facing the workpiece.

Step 2   Apply a cutting lubricant to the work surface and the die cutting edges. Press down firmly. At the same time, move the die on the top surface of the work.

Step 3   Continue to apply a downward and circular force as shown in Fig. 14-10. Make two or three revolutions to start the first few threads. Back off the die. Check the threads for squareness and to see that they are started correctly.

Step 4   Continue to turn the die until a few full threads are produced. Remove the die.

Clean the part. Check the size against the mating part, a ring gage, or other measuring tool. If necessary, adjust the die jaws.

Note: A finer finished thread is produced when the die is opened as much as possible. The thread is cut to length. This step is then followed by readjusting the die and taking a fine cut to produce the required fit.

Note: Like tapping, the direction should be reversed every few turns. This action breaks the chips and clears them away from the work surface and die.

Note: A cutting lubricant should be applied throughout the cutting process.

Step 5   Thread to the required length. Remove the die. File off the burrs. Clean the threads. Retest the threaded part for size by turning it into the mating part. If required, readjust the die and take a finishing cut over the threaded part.

Note: When a thread is to be cut to a shoulder or close to a head, the die is turned 180° (*inverted*). The last few threads are cut with the die in this position.

CAUTION: Chips are usually blown out of dies and from the workpiece by an air blast. Eye protection devices must be used during this process and safety precautions must be followed in directing the air blast.

SCREW EXTRACTOR

Bolts, screws, studs, and other threaded parts may be sheared off, leaving a portion in the tapped hole. This portion may be removed by using a tool called a *screw extractor*. This tool is a tapered spiral metal form. The features of a right-hand thread screw extractor are illustrated in Fig. 14-11. The spiral is formed in the reverse direction from the threaded part that is to be removed. In other words, a left-hand spiral is needed for a right-hand thread.

Fig. 14-11. Right-hand thread screw extractor.

A hole is first drilled into the broken portion. The diameter of the hole must be smaller than the root diameter of the threads. The drill size to use is generally given on the shank of the tap extractor. As the point of the spiral taper is inserted and turned counterclockwise (for right-hand threads), the extractor feeds into the hole. The spiral surfaces are forced against the sides of the hole in the broken part. The extractor and part become bound as one piece. The threaded portion then may be removed by turning the extractor.

HOW TO EXTRACT A BROKEN THREADED PART

Drilling A Pilot Hole

Step 1   Select a screw extractor which is appropriate for the size of the stud or broken part that is to be removed.

Step 2   Determine the size of the drill and the depth to which the pilot hole must be drilled. The diameter must be smaller than the root diameter of the thread. Drill the pilot hole.

Extracting A Broken Part

Step 1   Insert the screw extractor in the drilled hole. Turn gently and apply a downward force to *set* the extractor. Fig. 14-12 shows the position of the extractor in the pilot hole of a broken part.
Note: Apply a penetrating oil or lubricant around the threads only. Keep the fluid away from the extractor surfaces.

Step 2   Turn the extractor firmly and slowly in a counterclockwise direction for right-hand threads. Turn in a reverse direction for left-hand threads.

Step 3   Clean and burr the workpiece. Run a tap through the threaded portion to *touch up the threads*. Recheck with a gage or mating part.

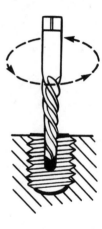

Fig. 14-12. A screw extractor turned into the pilot hole previously drilled in the broken part.

SAFE PRACTICES IN CUTTING EXTERNAL THREADS BY HAND

- Analyze and correct the causes of torn threads. Torn threads indicate any one or all of the following conditions:
  - The jaws may be set too deep
  - The threads are being cut at an angle
  - The diameter of the workpiece is too large
  - The cutting teeth are dull
  - A proper lubricant (if required) is not being used.
- Remove burrs produced by threading. The beginning thread should be chamfered to start the thread squarely and easily.

- Position the workpiece as close to the vise jaws as possible. Use soft jaws where necessary.
- Exercise care in threading long sections. Turn the die holder from a position in which the hands do not pass across the top of the threaded part.
- Check adjustable dies as soon as practical to avoid threading undersize. Open the die and cut a full thread. Then, adjust the die until the required fit is produced.
- Check a threaded hole against a gage or mating part for surface finish and class of thread (fit).

SCREW THREAD DRAWING AND HAND THREADING TERMS

Thread representation
Graphic techniques of drawing threads in order to describe them accurately and provide full thread specifications. Pictorial, schematic, and simplified techniques of drawing and sketching a thread. A drawing with all features and dimensions.

Standard dimensioning of a screw thread
A set of symbols, forms and technical data. A universally accepted dimensioning system which provides the full specifications of a screw thread.

Screw pitch measurement
Establishing the number of threads per inch in the Unified and National Form thread systems. Measuring the distance between teeth in millimeters in the ISO Metric system. Gaging the pitch of a screw thread with a screw pitch gage, a steel rule, or by other measuring techniques.

Dies (screw thread)
A screw thread forming tool. The cutting edges conform to a required screw thread. Solid, split adjustable, or two-part adjustable screw thread cutting tools.

Adjustable die (thread size adjusting)
A thread cutting tool which provides for variations in thread size. Die segments may be expanded or reduced with adjusting screws.

Adjustable die guide
A threaded part that screws into the cap against the two die sections (halves) to lock them in position. The hole in the guide centers the work surface in relation to the die. The guides help to start the threads squarely.

Nominal pipe size
The *inside* diameter of a standard pipe.

Screw extractor
A hardened, spiral-fluted tool. When turned into a drilled hole in the direction in which the screw thread can be backed off, the extractor turns the broken portion out of the mating part.

SUMMARY

- Screw threads are drawn pictorially, schematically, or by simplified representation techniques to represent particular features. Dimensions provide further specifications about size, degree of accuracy (class of thread fit), depth or length, etc.
- The pitch of a screw thread in the Unified, American National, and ISO Metric thread systems may be measured directly with a steel rule or gaged with a screw pitch gage.
- Three common forms of dies for regular straight threads and pipe threads include the solid die, split die, and adjustable split die.
- Split dies and adjustable split dies are adjusted to the specific size of the workpiece. They are then locked in this position. Adjusting screws in the die head and the die holder provide for a variation in pitch diameter.
- The pitch diameter of a thread controls the fit of the thread. The die head is locked in a fixed position (pitch diameter) by a lock screw in the die holder.
- The die is positioned over the workpiece and centered by a guide. This aligns the die in relation to the work surface.
- Pipe thread diameters are expressed in terms of the *inside* diameter of the pipe. The tap drill size permits threads to be cut at full thread depth and at a taper of 3/4″ per foot.
- The threading die halves of a pipe thread die are aligned with an index line on the die stock. The die halves are adjusted to produce a thread that meets specific job requirements. Once set, the threading dies may be used to thread a number of pieces.

UNIT 14 REVIEW AND SELF-TEST

1. Explain the statement: *A drawing provides full specifications for cutting and measuring a required thread.*
2. Give three reasons why adjustable split dies are preferred over solid dies.
3. State three major differences between pipe thread dies and regular screw thread dies.
4. List the steps required to thread a workpiece on the bench using an appropriate size adjustable threading die.
5. Identify how a broken threaded part may be removed.
6. State three conditions that result in cutting torn external threads when threading by hand.

Unit 15

REAMING: HAND REAMERS AND PROCESSES

This unit is limited to hand reamers and hand reaming processes. *Solid (fixed), adjustable,* and *expansion* types of reamers are covered. Consideration is given to *straight, spiral* and *taper* forms. The features of hand reamers are illustrated to help explain the shearing action. A few commonly used cutting fluids are considered. The fluids are related to cutting action and surface finish. General applications are followed by step-by-step procedures on how to safely use each type of hand reamer.

Characteristics of machine reamers, their cutting action, speeds and feeds, cutting fluids, reaming processes, and other machining technology are treated in later units on machine reaming.

COMPARISON OF DRILLING AND REAMING PROCESSES

Many machined holes must be held to close tolerances for roundness, straightness, quality of surface finish, and dimensional accuracy. Drilling and other machining processes produce holes that have comparatively rough surfaces. These holes are larger in diameter than the drill size. They may not be perfectly straight and round. A drilled hole may vary up to +0.005″ in diameter for drill sizes up to 1/2″. The variation may be as high as 0.020″ for 1″ and larger sizes of drills. When the drill is guided by a drill bushing in a fixture, greater accuracy is possible.

Whereas a drill is used to produce a hole in a solid mass, the *reamer* is a *finishing tool*. The reamer *reams a hole precisely to size*. Parts may be *hand reamed* or *machine reamed*. Only a limited amount of material is removed as a finishing cut with a hand reamer. By contrast, a greater amount may be left for machine reaming.

HAND REAMING

Reaming is the process of shearing away small amounts of material from a previously prepared hole. The process produces a hole that is:

● Perfectly round
● Held to a specified diameter within

+0.001″/−0.000″ (or to a closer tolerance)
● Smooth and has a high-quality surface finish
● Straight.

Reaming is performed by a multiple-fluted cutting tool known as a *reamer*. The cutting action takes place by rotating and feeding the cutting edges on each flute of the reamer against the periphery (surface) of a hole. The cutting on many hand reamers is done along the *tapered starting end*. The amount of material removed in hand reaming ranges from 0.001″ to 0.003″, and occasionally up to 0.005″.

FEATURES OF HAND REAMERS

The main features of the cutting end of a spiral-fluted hand reamer are identified in Fig. 15-1. The hand reamer, whether straight or spiral (helical), has a number of flutes. These are cut into and along the body. The flutes are machined so that the front face is *radial*. This means that if the edge of a steel rule were placed on the front face of the reamer, it would pass through the center.

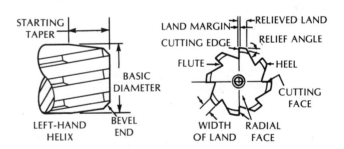

Fig. 15-1. The main features of the cutting end of a spiral-fluted hand reamer.

Each flute has a *land*. Each *land* has a small *margin*. This is a circular width behind the cutting edge of the reamer. The margin extends from the cutting face to the *relieved area*. Beyond the margin, the land is *relieved*. This prevents the reamer from binding in the hole.

The shearing, cutting action is produced by the *starting taper*. Beginning at the point of the reamer, the flutes are ground at a slight taper. This taper merges into the outside surface *(periphery)* of the

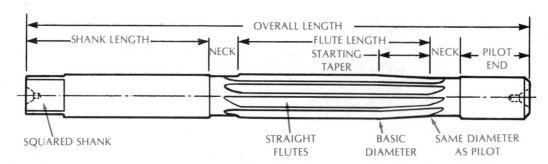

Fig. 15-2. The unique features of a straight-fluted hand reamer.

reamer. The taper usually extends back for a distance slightly longer than the diameter of the reamer.

Additional features of a hand reamer are illustrated in Fig. 15-2, which shows the straight-fluted hand reamer.

Whenever a reamer becomes dull, the starting taper or other cutting edge on each flute must be accurately ground on a tool and cutter grinder.

The *shank* of each reamer is smaller than its outside diameter. In reaming it is desirable to turn the reamer clear through the workpiece whenever possible. This avoids *scoring the inside finished surface*. The end of the shank is square to permit turning a reamer with a tap wrench. Both ends of the reamer have center holes for two purposes:

● To provide a bearing surface for grinding and sharpening. The bearing surface is concentric with the periphery of the reamer.

Fig. 15-3. A multiple-diameter, straight-fluted hand reamer with a pilot. (Courtesy of the Acme-Cleveland Corp.)

● To position the reamer on the machine (where the hole may have been drilled or rough reamed), with a machine center. This helps to align the reamer with the line of measurement of the workpiece.

Hand reamers are usually made of carbon steel and high-speed steel. Reamers are manufactured in standard dimensional sizes. Special multiple diameters are designed for applications which require the simultaneous reaming of more than one diameter of a hole. Some hand reamers, like the one shown in Fig. 15-3, have a *pilot*. The *pilot* is undersized. The pilot

helps to accurately position and guide the reamer.

BASIC TYPES OF HAND REAMERS

Solid Hand Reamers (Straight Holes)

Solid hand reamers may have flutes that are *straight* (parallel) or *helical* (at an angle). A straight-fluted reamer is illustrated in Fig. 15-4A. A spiral-fluted reamer appears in Fig. 15-4B. These reamers produce straight holes of a specified diameter, good surface finish, and to close tolerances. Straight and helical reamers may be purchased in inch standard and metric standard sets. The reamer sizes are in increments of 1/64" and 1/32" and metric equivalents, respectively. Reamers are also furnished in special sizes.

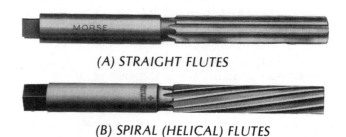

(A) STRAIGHT FLUTES

(B) SPIRAL (HELICAL) FLUTES

Fig. 15-4. Solid (fixed-size) hand reamers. (Courtesy of Morse Cutting Tools Division, and the Acme-Cleveland Corp.)

The outside diameter of an old reamer may be ground to within 0.003" to 0.005" of a standard size. This makes it a *roughing reamer*. A roughing reamer is used to correct inaccuracies caused by the cutting action of a drill. A *finishing reamer* of the exact size is then used as a *second reamer*. The finishing reamer produces the accurately reamed hole.

Spiral-fluted reamers are machined with a *left-hand helix*. This design permits the cutting edge to cut as the reamer is turned and fed into the work. If the helix were right-hand, the reamer would thread into the work, score the surface, and produce an inaccurate hole.

Solid Hand Reamers (Tapered Holes)

The flutes of reamers may also be tapered for reaming tapered holes. There are three basic types of *taper reamers.*
- *Pin*
- *Socket*
- *Burring.*

Taper Socket Reamer—Taper socket reamers also produce tapered holes. However, they are designed to receive tapered parts. These conform to Morse, Brown and Sharpe, or metric standard tapered socket, tool shank and other spindle standards. Since these reamers must remove a considerable amount of material, they are produced in a set of two reamers. One reamer is for *roughing;* the other, *finishing.* A roughing and finishing set is illustrated in Fig. 15-5 (A and B).

(A) HAND ROUGHING MORSE TAPER HOLE REAMER (STRAIGHT FLUTE)

) HAND FINISHING BROWN & SHARPE TAPER REAMER (STRAIGHT FLUTES)

) HAND FINISHING BROWN & SHARPE TAPER REAMER (SPIRAL FLUTES)

Fig. 15-5. Taper hole hand reamers. (Courtesy of Morse Cutting Tools Division)

Fig. 15-5C is a second type of finishing reamer. This one has spiral flutes. A series of shallow grooves is cut across each flute of a roughing reamer and in a staggered pattern. The grooves relieve some of the force required to turn the reamer because they reduce the area over which the cutting takes place.

Burring Reamer—The *burring reamer* is a third form of reamer. It is not considered a finishing tool in terms of producing a precision surface finish or accurate diameter. The burring reamer has a steep taper. This reamer is used to burr tubing and pipes and to enlarge the diameter of a hole in thin metal. The burring reamer has straight or helical flutes. The reamer is turned with a hand brace instead of a tap wrench. A spiral-fluted hand burring reamer is shown in Fig. 15-6.

Fig. 15-6. Spiral-fluted hand burring reamer. (Courtesy of the Acme-Cleveland Corp.)

Expansion Hand Reamers

A standard straight and a spiral-fluted expansion hand reamer are illustrated in Fig. 15-7. The spiral-fluted reamer has a pilot end. The body of the *expansion reamer* is bored with a tapered hole. This hole passes through a portion of its length. A section of each flute is slotted lengthwise into the tapered hole. The hole is threaded at the end to receive a tapered plug. As this plug is turned into the tapered portion of the body, the flutes are expanded.

TAPERED PLUG

(A) STRAIGHT FLUTED

(B) SPIRAL FLUTED

Fig. 15-7. Expansion hand reamers with pilot ends. (Courtesy of Morse Cutting Tools Division)

The limit of expansion varies from 0.006″ for 1/4″ diameter reamers, to 0.012″ for 1″ to 1 1/2″ diameter, to 0.015″ for 1 9/16″ to 2″ reamers. The pilot on the reamer is undersize. It serves to align the cutting edges concentric with the hole. Expansion hand reamers may be straight fluted or left-hand helical fluted. This form is recommended for holes that are partly cut away so that the surface is *interrupted*.

The expansion reamer may be increased in size only to produce *oversized holes*. These require the removal of a small amount of material over a nominal size. The cutting action takes place on the starting taper of each flute. After use, the adjusting screw should be loosened. Loosening permits the flutes to return to their normal position. Otherwise, there is a possibility that the flutes may become permanently set, losing the adjustable feature.

The helical-fluted expansion reamer operates on the same principle of adjustment as the straight-fluted type. The helical-flute is especially adapted, as stated earlier, to reaming holes that are grooved or in which the surfaces are interrupted.

One great advantage of expansion reamers is longer *tool life*. The expansion feature permits the reamer to continue to produce holes of a standard size after the reamer has been sharpened many times.

Adjustable Hand Reamers

Adjustable hand reamers are practical for reaming operations that require a diameter that is larger or smaller than a basic size. A complete set of adjustable hand reamers makes it possible to produce holes over a range of three inches, starting at 1/4″ diameter. Each reamer may be expanded to produce intermediate diameters. These overlap the next size of adjustable reamer.

The *blades* of adjustable hand reamers may either be ground or replaced. The inserted blades provide the long-life of the reamer. The ease of adjusting over a wide dimensional range and the blade replacement feature make this reamer a practical, efficient tool.

The replaceable blades fit into slots that have tapered bottoms. The blade holder may be positioned along the threaded portion of the body. The cutting blades are adjusted to a required size and are held in place with two adjusting nuts. An adjustable hand reamer with replaceable blades is shown in Fig. 15-8.

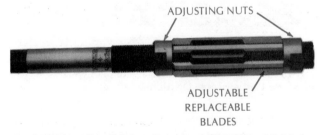

ADJUSTING NUTS

ADJUSTABLE
REPLACEABLE
BLADES

Fig. 15-8. An adjustable hand reamer with replaceable blades. (Courtesy of Morse Cutting Tools Division)

CHATTER MARKS RESULTING FROM REAMING

Chatter marks describe uneven, wavelike surfaces. These have undesirable high and low spots. Chatter marks which result from reaming may be caused by any one or combination of the following conditions:

- The speed at which the reamer is being turned is too fast
- The distance the reamer is advanced each revolution (feed) is too little

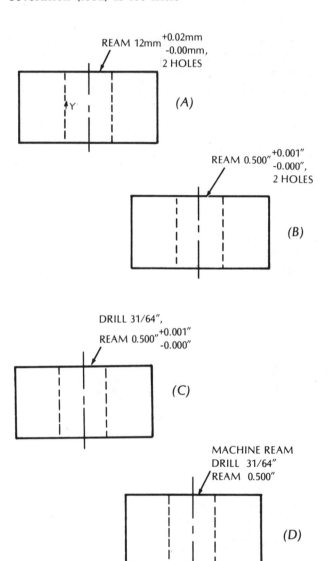

REAM 12mm $^{+0.02mm}_{-0.00mm}$, 2 HOLES

Y

(A)

REAM 0.500″ $^{+0.001″}_{-0.000″}$, 2 HOLES

(B)

DRILL 31/64″,
REAM 0.500″ $^{+0.001″}_{-0.000″}$

(C)

MACHINE REAM
DRILL 31/64″
REAM 0.500″

(D)

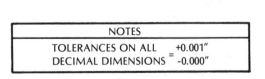

NOTES		
TOLERANCES ON ALL DECIMAL DIMENSIONS	=	+0.001″ -0.000″

Fig. 15-9. Techniques of representing and dimensioning reamed holes.

• The workpiece is not held securely; there is too much spring to the tool or the work

• The reamer is unevenly forced into the workpiece

• Chips are not being removed and are clogging the flutes

• The drilled or rough reamed hole is incorrect in terms of size, surface condition, roundness, or straightness.

The remedy in each case is to correct the condition, making whatever changes are necessary in speed, feed, proper preparation of the hole to be reamed, etc. Chatter marks may also be avoided by using a left-hand spiral reamer that may be fed evenly. Some reamers are *increment cut*. The cutting edges are unevenly spaced. This permits overlapping during the cutting process, which prevents a ragged surface from being formed.

REPRESENTING AND DIMENSIONING REAMED HOLES

Drawings and sketches of parts that require holes to be reamed are dimensioned in several ways. Four common techniques are illustrated in Fig. 15-9. In (A) and (B) the metric and inch standard dimensions indicate only the dimension of the reamed hole. Normally, with this method of dimensioning the holes are considered to be hand reamed. The holes would either be drilled to within 0.003″ to 0.005″ (0.08 mm to 0.12 mm) or rough reamed to within 0.001″ to 0.003″ (0.02 mm to 0.08 mm) of the finished size.

The holes represented in Fig. 15-9 C and D are machine reamed. The accuracy is indicated along with the dimension in Fig. 15-9C. The accuracy of the reamed hole in Fig. 15-9D is covered by the tolerance in the note. The tolerance applies generally to all dimensions on the drawing. Note in both (C) and (D) that a larger amount of material is to be left for machine reaming.

CUTTING FLUIDS

The use of a cutting fluid on certain materials makes the hand reaming process easier. Less force is required and a higher quality of surface finish is possible. Most metals except cast iron are reamed using a cutting fluid. Generally, for nonferrous metals like aluminum, copper and brass, a *soluble oil or lard oil compound* is recommended. Cast iron is *reamed dry. Soluble oils and sulphurized oils* may be used on ordinary and hard steel and stainless steel.

Recommended cutting fluids for various ferrous and nonferrous metals are listed in Appendix Table

A-28. While recommended primarily for machine reaming, the cutting fluids may also be used for hand reaming.

HOW TO USE HAND REAMERS

Solid Hand Reamers—Straight Hole Reaming

Step 1   Measure the outside diameter of the machined part. Check the drawing. Establish what tolerance is allowed between the part size and the reamed hole.

Step 2   Select the correct size of solid hand reamer. Check the outside diameter. (The outside diameter of the reamer in Fig. 15-10 is being measured with a micrometer.)

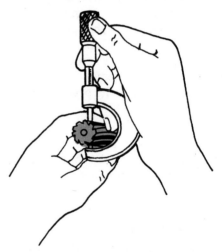

Fig. 15-10. Checking the outside diameter of a solid hand reamer.

Step 3   Remove burrs from the holes and other surfaces. Check the condition and size of the hole. The allowance for hand reaming should range from 0.002″ to 0.005″ (0.05 to 0.12 mm) maximum.
Note: If the hole is *smaller* than the hand reaming allowance, it may be necessary to rough ream with a roughing reamer.

Step 4   Examine the areas of the workpiece that are to be clamped. Be sure they can withstand the force of the vise jaws without bending or cracking.
Note: Sometimes it is more practical to hold the workpiece in a vise so that the hole is reamed in the horizontal position.

In other cases the reamer may be positioned vertically in the vise. The workpiece is then turned on the reamer.

Step 5   Select a suitable tap wrench. Start the tapered cutting edge in the hole to center the reamer.

Step 6   Turn the reamer clockwise slowly while feeding the reamer into the workpiece.
Note: If the reamer is not centered and is turned too fast, the cutting edges may produce a corrugated chatter-marked surface.

Step 7   Apply a cutting fluid where required between the cutting edges and the inside surface of the hole.

Step 8   Continue to turn the reamer clockwise. Feed it into and through the workpiece.
Note: Where very close tolerances of +0.0005″ (+0.01 mm) must be held, the reamer is sometimes removed when it cuts to its full diameter. The size of the reamed portion is checked. If necessary, the reamer is changed.

Step 9   Remove the chips from the hole and reamer flutes. Examine the cutting edges and lands. Remove any burrs. Replace the reamer in its proper storage container.

CAUTION: The area around a reamed hole may be very sharp. If possible, burr as soon as practical.

Step 10   Inspect the workpiece for quality of surface finish. Insert the mating part into the reamed hole. Check the parts against the job specifications.

Solid Hand Reamers—Taper Reaming (Taper Socket)

Step 1   Check the drawing or measure the tapered mating part to establish the required size of taper reamer.

Step 2   Check the diameter of the hole to be reamed.
Note: There should be an adequate amount of material left to permit reaming at the small end of the taper.

Step 3   Determine how far the reamer must enter the hole to produce the required diameter at the large end of the taper.

Step 4   Insert the roughing taper reamer in the hole. Turn it clockwise while applying a

slight downward force. Use a steady motion.
Note: Taper reamers tend to feed into the work. No undue force should be applied.

Step 5   Apply a cutting fluid if the material being reamed requires it.

Step 6   Remove the taper reamer frequently. Clear away the chips.

CAUTION: Continue to turn the reamer in a clockwise direction when removing it.

CAUTION: Use a wiping cloth, or cover the workpiece and direct the air stream downward and away from other workers.

Step 7   Check the tapered hole against the sleeve, socket, or other mating part. Leave enough material for the finishing reamer.

Step 8   Change from a roughing to a finishing reamer. Continue to ream to the required depth.

Step 9   Remove the burrs at the large and small diameters of the tapered hole.

Step 10   Wipe the reamer clean. Check and remove any burrs. Oil the cutting edges of the reamer. Store the reamer in a suitable container.

Solid Hand Reamers—Taper Reaming (Taper Pin)

Step 1   Check the mating parts to see that the holes are aligned. The parts should be held securely in the position in which they are to be joined with a taper pin.

Step 2   Check the reamer condition and size. Start the reamer and clean the chips periodically. (The taper pin reaming of two mating parts is illustrated in Fig. 15-11).

CAUTION: Taper reamers are hard and fragile. If the reamer draws into the work, release all force. Continue to turn the reamer slowly in the cutting direction while withdrawing it. Then, carefully start the reaming process again.

Step 3   Clean the reamer and the part. Insert the taper pin. Check its depth against the job requirements.

Step 4   Repeat the reaming process until the taper pin fits accurately.

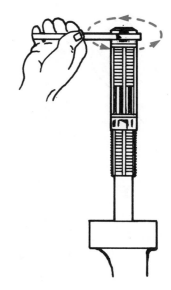

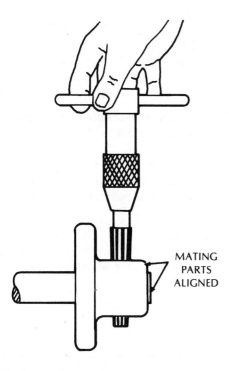

Fig. 15-11. Taper pin reaming.

Fig. 15-12. Setting the blades of an adjustable hand reamer to a required diameter.

Adjustable Blade Hand Reamers—Adjusting For Size

Step 1   Measure the diameter of the reamer with a gage or micrometer.

Step 2   Mount and secure the reamer in a vise.

Step 3   Adjust the reamer diameter. Turn the lower adjusting nut a quarter turn. Tighten the upper adjusting nut the same amount. (The position of the reamer and the turning of the top adjusting nut are shown in Fig. 15-12.)
Note: The blades are moved back from the reamer point to enlarge the diameter and vise versa.

Step 4   Continue the adjusting process until the correct reamer size is reached.

CAUTION: A quarter turn of the adjusting nut is recommended to prevent the blades from riding out of the slots and to prevent chips from entering under the blades.

Adjustable Blade Hand Reamers—Checking The Workpiece And Reaming

Step 1   Check the hole size. Mount the work (or the reamer if the workpiece is to be turned on the reamer).

Step 2   Start reaming. Continue part way until the reamer starts to cut to its full diameter.

Step 3   Continue to turn the reamer while withdrawing it. Clean the workpiece. Be careful of the sharp burred edge.

Step 4   Check the partially reamed hole for size. Adjust the reamer, if necessary.

Step 5   Continue to ream through. Remove burrs. Clean the workpiece. Test for size and surface finish.
Note: It may be necessary to readjust the blades and to take another very light finishing cut.

Step 6   Loosen the top adjusting nut one-quarter turn. Wipe the reamer clean. Examine and oilstone any burrs. Oil the reamer. Replace it in a container.

Expansion Hand Reamers—Setting The Hand Reamers

Step 1   Check the cutting edges, lands and pilot. See that they are free of burrs.

Step 2   Check the diameter of the predrilled or rough-reamed hole. There must be enough material left for reaming.

Step 3   Measure the outside diameter of the reamer at several places. Turn in the tapered screw to enlarge the reamer diameter. (Fig. 15-13 shows how to expand the flutes to a required diameter.)

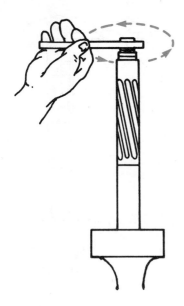

Fig. 15-13. Expanding the flutes of an expansion hand reamer to a required diameter.

Back out the screw to reduce the diameter.

CAUTION: The reamer should be expanded only within its design limits. Otherwise, the flutes may be damaged.

Step 4   Mount the workpiece, position the reamer, and start and finish the reaming process.
Note: Check the reamed diameter as soon as the reamer has cut to the full depth. Adjust the reamer if necessary.

CAUTION: Back off the tapered screw after reaming. This protects the flutes and makes it possible for the reamer to adjust to its normal undersize position.

SAFE PRACTICES IN THE CARE AND USE OF HAND REAMERS AND REAMING PROCESSES

- Examine the cutting edges and lands of reamers for nicks, burrs, or other surface irregularities. These produce surface scratches and imperfections. Burrs are removed by hand with an oilstone.
- Clean and oil the surfaces of reamers before storing. Oiling prevents rust spots from forming. Rust produces scratches in the finished surface.
- Store reamers in special holders or in separate containers so that they will not come in contact with other reamers (Fig. 15-14).
- Turn reamers in a right-hand (clockwise) direction only. Where possible, the reamer should be moved completely through the workpiece while being turned. Otherwise, keep turning the reamer clockwise while pulling outward on the tap wrench.
- Leave not more than 0.005″ or 0.1 mm of material for hand reaming. Holes that must be held to tolerances of +0.0005″/−0.0000″ should be rough reamed to within 0.001″ to 0.003″, then finish reamed.
- Examine each drilled hole before reaming it. The hole must not be tapered or *bell-mouthed*, out-of-round, or have a rough-finished surface. Any of these conditions may cause tool breakage and an inaccurately reamed hole.
- Use a left-hand spiral-fluted reamer for reaming holes in which there is an intermittent cut.

- Hold the workpiece reasonably rigid. This permits the reamer to be turned by an equal force applied to both handles of the tap wrench.
- Rotate the reamer clockwise and slowly. Allow it to align itself in the hole. The feed should be continuous and deeper per revolution than that used for drilling. The feed should approach one fourth the diameter of the reamer per revolution.
- Grind the cutting edges of all reamers concentric on a tool and cutter grinder. Each tooth cutting edge must cut uniformly.
- Start the reamer on an even surface. A reamer tends to feed toward the point of least resistance.
- Back off the adjusting screw on an expansion reamer. Backing off prevents a permanent set in the flutes (cutting edges).

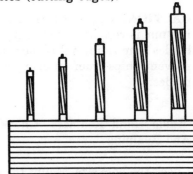

Fig. 15-14. Correct storage of reamers.

HAND REAMER AND HAND REAMING TERMS

Reaming	A shearing/cutting process for producing precision holes. Cutting holes that are dimensionally accurate, round, straight, and have a high-quality surface finish.
Hand reamer	A multiple-fluted cutting tool that is turned and fed by hand. A cutting tool for removing small amounts of material to finish a hole to a required dimension.
Starting taper	The cutting edges of straight hole reamers. A tapered end that begins with a diameter which is smaller than a basic size. The taper extends along the lands until it merges with the outside (basic) diameter of the reamer.
Periphery (reamer)	The external cylindrical surface of a reamer.
Solid hand reamer	A nonadjustable straight or helical-fluted hand reamer of a fixed diameter.
Taper pin reamer	A straight or helical-fluted reamer whose cutting edges are tapered. The taper conforms to the standard dimensions of tapered pins.
Taper socket reamer	A straight or helical-fluted reamer having tapered flutes. The taper of the flutes meets a specific standard. The standard relates to sockets, tapered spindles, or other tapered mating parts.
Expansion hand reamer	A hand reamer having flutes. These may be expanded by a tapered screw to enlarge the reamer size slightly.
Adjustable blade hand reamer	A hand reamer that has a series of blades. These are adjustable above or below a basic size. A reamer whose blades are bottomed against a tapered nesting part. The reamer size is set by turning two adjusting nuts.
Burring reamer	A straight or spiral, sharply tapered reamer. A reamer for burring tubes and pipes. A reamer used to roughly enlarge holes in thin-gage metals.
Chatter (reaming)	An imperfect reamer cutting action which produces inaccurate high and low surface areas.
Ream $+0.0005''$ $-0.0000''$	A standard technique of dimensioning a drawing to indicate maximum and minimum dimension limits. Dimensioning the required limits of accuracy of a reamed hole.
Left-hand helix (reamer)	A reamer with flutes cut at an angle to the body and counterclockwise. The left-hand direction of the flutes. This is the reverse of the cutting direction.

SUMMARY

- Parts are reamed to produce holes that are dimensionally accurate, round, straight, and have good surface finish.
- The condition and size of a predrilled hole affects the degree of accuracy and finish of a reamed hole.
- Reaming is a *finishing* process. The amount of material left for finish hand reaming should not exceed 0.005″.
- Three common types of hand reamers are: solid, adjustable, and expansion. These are available in straight, spiral (helical), and tapered forms.
- The cutting action of hand reamers is produced by the cutting edges of the flutes and blades. The starting taper merges with the outside diameter of the reamer.

- Reamer flutes are machined straight or with a left-hand helix. This permits the reamer to be fed into the work rather than be drawn in.
- Reaming requires an equal force to be exerted. A reamer is turned at a slower speed but at a faster feed than used for drilling.
- Taper pin reamers accommodate the full range of standard taper pins. Reamer sets provide for size overlapping.
- Taper socket reamers are designed for standard Morse, Brown and Sharpe, and other tapers. Taper reamers are used for sockets, tapered shafts, shanks, and other holders.
- The burring reamer is a nonprecision reamer. It is used for burring the inside diameter of tubes and

pipes or for enlarging a hole in thin metal.

- The expansion hand reamer may be expanded only *above* a basic dimension. An undersized pilot end positions the blades concentrically with a hole. The pilot guides the reamer so that it cuts evenly.
- Helical-fluted reamers are recommended for reaming holes that are interrupted.
- Adjustable hand reamers may be adjusted for reaming holes that are larger or smaller in diameter than a basic size. The blades may either be sharpened or replaced. The range of adjustment permits setting the reamers in a set to all intermediate diameters between 1/4″ and 3″.
- Chatter is an undesirable wavylike unevenness of a surface. Chatter marks result from excessive speed, too slow a feed, poor mounting, an improperly prepared hole, and various other causes.
- Reamed holes are represented on drawings with dimensional information about hole sizes and tolerances. In some cases, cutting tool sizes, processes, and quality of surface finish are given.
- Cutting fluids, when required for a particular material, improve the smoothness of the cutting action, reduce the force required for reaming, and increase the accuracy of the reamed hole.
- Safe practices must be followed when checking for and removing burrs and turning, feeding, cleaning, and storing reamers. These practices help protect the reamers, improve the quality and dimensional accuracy of reamed holes, and prevent personal injury.

UNIT 15 REVIEW AND SELF-TEST

1. Compare the variation from the nominal hole size for 12.5 mm and 25 mm diameter holes that are produced by (a) drilling and (b) reaming.
2. Describe each of the following design features of a hand reamer:
   a. *radial flute face*
   b. *margin*
   c. *starting taper*.
3. State why it is common practice to turn a straight hand reamer through a workpiece.
4. Tell what function is served by the series of shallow grooves cut into each flute in a roughing taper socket reamer.
5. Indicate the limit of expansion on the following diameter ranges of expansion hand reamers:
   a. up to 6 mm
   b. 1″ to 1½″
   c. 39 to 50 mm.
6. Cite three advantages of *adjustable blade reamers* over *expansion hand reamers*.
7. Give three common corrective steps that are taken to eliminate chatter during hand reaming.
8. Describe the meaning of a note on a drawing that specifices a *tolerance*.
9. List four cautions that must be observed to avoid personal injury and/or damage to hand reamers.

PART THREE

BASIC MACHINE TOOLS

SECTION ONE
METAL-CUTTING SAWS: TECHNOLOGY AND PROCESSES

The cutting apart of metals and other hard materials is referred to as *metal sawing* or *cutting-off*. There are three general designs of metal-sawing machines:

- *The power hacksaw,* which uses a saw-toothed blade
- *The horizontal (cutoff sawing) band machine,* which uses a saw band. (The vertical band machine, which is more universal and advanced than the horizontal type, is covered in a later section. The vertical type has many accessories and is adaptable to contour cutting, three-dimensional cutting, and rough and finish filing of regular- and irregular-shaped surfaces.)
- *The cold saw,* which uses a circular metal-cutting saw. (This saw is designed for heavy-duty operations, such as cutting off large-size stock. Because it has limited application in average machine shops and toolrooms, the cold saw is not covered here.)

This section describes the types, features, and operation of power hacksaws and horizontal band machines. Basic cutoff operations performed with these machines are presented in terms of everyday applications. Because the craftsperson must select the correct cutting speed, feeds, cutting fluids, machine adjustments, and cutting blades or saw bands, these are treated in depth. Also presented are step-by-step procedures for mounting workpieces, setting up the machines, and performing the other processes essential to accurate and efficient cutting-off.

Unit 16

POWER HACKSAWS: FUNCTIONS, TYPES, AND PROCESSES

Power hacksaws are used to cut to length sections of rods, bars, tubing and pipe, castings, forgings, and other parts. The hardness range of these materials is from comparatively soft nonferrous metals (like aluminum, brass, and bronze) to mild steels, tool steels, and harder alloys. The sections of stock may be cut square (straight) with the work axis or at an angle up to 45°.

The power hacksawing process is similar to hand hacksawing. The cutting action takes place by applying force on the teeth of a saw blade as it moves across the work surface. A series of chips are cut by the shearing action of the teeth. The cutting action continues until the section is cut off. Materials that are smaller than 1/2" are usually cut by hand; larger sizes, by machine.

TYPES OF POWER HACKSAWS

Power hacksaws may be classified as *utility*, *heavy-duty*, and *heavy-duty production machines*. The first two types may be either *dry-cutting* or *wet-cutting*.

Wet-cutting machines normally operate at higher speeds. Thus, they may be used on many different materials. The base of the wet-cutting machine is enclosed. This area serves as a reservoir for the coolant and houses the circulating pump. A two-speed, two-feed, wet-cutting power hacksaw is illustrated in Fig. 16-1. This machine operates at 85 and 170 surface feet per minute (sfpm).

Fig. 16-2 shows a heavy-duty power hacksaw. It has six *cutting speeds*: 40, 51, 64, 80, 102, and 128 strokes per minute. The *cutting speeds* are 36, 46, 54, 72, 92, and 108 sfpm. This particular model may be equipped with a roller conveyor unit for feeding the stock to length.

One of the most common utility power hacksaw sizes has a rated capacity of 6" x 6" (150 mm x 150 mm). In other words, the maximum size of workpiece that may be held in the vise and cut is 6" x 6", or 150 mm x 150 mm. The rectangular capacity of the larger power hacksaw shown in Fig. 16-2 for *right angle* (regular) *cuts* is 11" x 7 7/8" or 280 mm x 200 mm. The maximum width of stock that may be

Fig. 16-1. Utility type two-speed wet-cutting power hacksaw. (Courtesy of Kasto-Racine, Inc.)

177

Fig. 16-2. Heavy-duty type six-speed/feed wet cutting power hacksaw. (Courtesy of Kasto-Racine, Inc.)

Table 16-1. RECOMMENDED PITCHES, CUTTING SPEEDS, AND FEEDS FOR POWER HACKSAWING FERROUS AND NONFERROUS METALS

Material		Pitch (Teeth per Inch)	Cutting Speed (Feet per Minute)	Feed (Force in Pounds)
Iron	cast	6 to 10	120	125
	malleable	6 to 10	90	125
Steel	carbon tool	6 to 10	75	125
	machine	6 to 10	120	125
Aluminum	alloy	4 to 6	150	60
Brass	free machining	6 to 10	150	60

accommodated at a 45° angle setting is 6 5/8" or 168 mm.

Power hacksaws have a *reciprocating stroke*. On the cutting stroke the blade is forced into the material. Each tooth takes a *cut*. At the end of this stroke the blade is raised automatically. The teeth are not in contact with the workpiece as the blade is returned to its starting position. This cycle is repeated until the workpiece is cut and the machine stops.

POWER HACKSAW BLADES

Power hacksawing requires a greater force than hand hacksawing. The cutting action takes place over a larger surface area. The machine saw blades are thicker, wider, and coarser. They are usually made of high-speed steels, high-speed molybdenum steels, or other tough alloys. Cutting fluids are used.

Tables of machinability ratings of metals indicate the range of ease or difficulty in cutting a given metal. The tables provide information for establishing whether a cutting process should be carried on *wet* or *dry* and what kind of a cutting fluid should be applied.

Other tables are used to determine the correct pitch, cutting speeds, and feeds for power hacksawing. The particular material to be cut and the size and shape of its sections must be considered. Table A-6 in the Appendix lists the recommended pitches, cutting speeds and feeds for selected ferrous and nonferrous metals. A portion of this table is illustrated here in Table 16-1. The pitch recommended represents an average for material sizes of two inches or smaller. Coarser pitches may be used for thicker or larger sizes. Finer pitches are used for thinner sections.

Selection Of Power Hacksaw Blades

Some of the principles governing the selection of hand hacksaw blades apply equally as well to power sawing.

● There must be at least two teeth in contact with the work surface at all times
● The greater the cross-sectional area, the coarser the required pitch. (The coarser pitch provides greater chip clearance.)
● Easily machined and soft materials require a coarse pitch and large chip clearance
● Hard materials and small cross-sectional areas require finer pitches.

While some power hacksaw blades are made of carbon alloy steel, high-speed steel and other alloy steel blades are more durable. High-speed tungsten, high-speed molybdenum, and molybdenum steels are excellent cutting blade metals.

Standard blade lengths range from 12" to 14", 16", 18", and 24". The length depends on the machine size and the nature and size of the material to be cut. The general range of pitch is from 4 to 14 teeth per inch. General metric size blade lengths are 300 mm, 350 mm, 400 mm, 450 mm, and 600 mm. The common metric pitch range is from 6 mm to 2 mm.

Set Patterns For Saw Blade Teeth

The teeth on power saw blades are usually *raker set* or *wave set*. The *raker set* pattern consists of a repeat design for every three teeth. Fig. 16-3A shows the raker set. One tooth is *unset*. This is followed by two teeth. One of these is offset to the right. The other tooth is offset to the left. This pattern continues for the length of the blade. The raker set blade is recommended for heavy work on bar stock, forgings, die blocks, and parts that have a constant cross section.

A second common *set pattern* is called *wave set* (Fig. 16-3B). Blades with *wave set teeth* are used for cutting materials with cross-sectional areas that change. Wave set teeth are used where there is a considerable range of material sizes to be cut. Structural forms and pipes are examples of parts that have changing area sizes. *Wave set teeth* are offset in groups. These alternate from right to left to form a wave pattern.

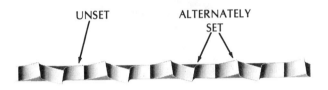

(A) RAKER SET

(B) WAVE SET

SETS OF TEETH WITH ALTERNATE OFFSET

Fig. 16-3. Raker and wave set patterns for saw blade teeth. (Courtesy of The Cooper Group)

A third set pattern, which has limited use in metal working, is called *straight set*. In this older pattern the teeth are offset alternately to the right and left of the blade.

POWER DRIVES AND MACHINE OPERATION

Most utility power hacksaws, whether wet-or dry-cutting, are equipped with an oil hydraulic system. This provides a smooth uniform control for the speed and cutting force. The hydraulic system actuates the saw frame. The system controls the downward feed (force) during the cutting stroke. It also automatically raises the frame and blade on the return stroke. At the end of the cut, the blade and frame are brought up to the highest position to clear the work.

Usually, dry-cutting power hacksaws have two cutting speeds: 70 and 100 strokes per minute. Where a greater range of materials and sizes is to be cut, wet-cutting machines are made with a four-speed drive motor. Four-speed machines may cut at 45, 70, 100, or 140 strokes per minute. The general range for six-speed machines is 45, 60, 80, 85, 110, and 150 strokes per minute. These machines are provided with a speed/feed chart for different materials. The cutting speeds are obtained by positioning the change speed lever at the required speed setting.

MAJOR MACHINE PARTS

The power hacksaw is a comparatively simple machine. A utility type wet-cutting power saw is shown in Fig. 16-4 to illustrate the major parts of a power hacksaw. There is a table mounted on a base.

Fig. 16-4. Utility type four-speed (6″ × 6″) wet-cutting power hacksaw. (Courtesy of Kasto-Racine, Inc.)

The saw frame is pivoted on the table. The frame is driven by an offset cam arrangement. The power drive is provided by an electric motor that is connected to a gear change box. As stated previously, an hydraulic system controls the cutting force, the return stroke, and the final positioning of the frame and blade at the end of the cutting cycle.

The work is held in a vise. The vise has a *fixed jaw*

and a *movable jaw.* Some vises are designed for making square cuts. Other vises have *elongated slots* in the base. These permit the fixed or solid jaw to be positioned at right angles to the blade. The fixed jaw may also be positioned at any angle up to 45°. The movable jaw may be swiveled. Force is applied by the movable jaw to hold the workpiece securely against the face of the solid jaw.

An *adjustable work stop* is attached to the table. The stop is set at a particular linear dimension and locked in place. The workpiece then is positioned against the stop. Additional pieces of the same length may be cut without further measurement. Long, overhanging bars of stock are usually supported by a floor stand that is adjusted to the machine table height (Fig. 16-5).

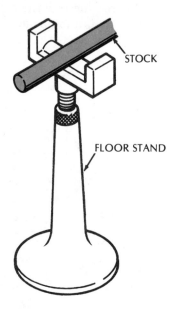

Fig. 16-5. Supporting the end of a long bar on a floor stand.

HEAVY-DUTY AND PRODUCTION HACKSAWS

Heavy-duty and *production power hacksaws* are built heavier than the utility type. These machines have additional accessories to those used on utility hacksaws. On most heavy duty hacksaws the work is positioned, clamped in the vise, and the cut is started by the worker.

An automatic heavy-duty power hacksaw is shown in Figure 16-6. Production power hacksaws have hydraulically-operated work tables (*carriage*). When the movable carriage is fully loaded and the cutoff length is set, the cutting-off process is a continuous one. The carriage automatically feeds to the required

length. The cut is started and completed and the saw frame is brought back to its full upright position. When all the workpieces are cut, the machine stops automatically.

Fig. 16-6. Automatic heavy-duty power hacksaw. (Courtesy of Kasto-Racine, Inc.)

Square and round bars, structural, and other shapes may be *stacked* for production cutting. *Stacking* refers to the grouping of many pieces together. They are clamped at one time, positioned, and cut to the required length. The operation of the vise jaws, the resetting of the stacked bars to the required length, the raising of the frame, the application of the cutting force, and the circulating of the cutting fluid are all controlled hydraulically.

On utility, heavy duty and production hacksaws of the wet-cutting type, the cutting fluid is recycled through screens into a reservoir. The larger chip particles settle in a container. The chips must be disposed of regularly. The fine chip particles settle as sediment at the bottom and must be scraped out. For hygienic reasons and to retain the qualities of the cutting fluid, it is necessary to completely clean and wash out the reservoir and fluid circulating parts. The cutting fluid must also be brought back to a specific strength.

> ### HOW TO OPERATE A UTILITY TYPE POWER HACKSAW

Cutting Off A Square Section

Step 1   Determine the pitch of the blade. Select a cutting speed according to the shape, size, and kind of material to be cut.

Step 2 Set the speed-change lever at the required speed.

Step 3 Check the direction of the teeth on the blade.
Note: The blade may be changed by placing the new blade on the pins in the frame. The teeth are pointed so that they cut on the *draw* stroke. Turn the clamping screw until the blade is held tautly in the frame.

Step 4 Check the position of the stationary vise jaw for squareness.

Step 5 Position the adjustable work stop at the required cutoff distance from the saw blade.
Note: Sometimes the length is marked out directly on the workpiece.

Step 6 Move the workpiece into the vise until it touches the work stop.
Note: The end of the part is checked first to see that the surface is square and flat. It may be necessary to *square the end* by cutting off a thin wafer-like section.

Step 7 Support the end of any long bar on a floor stand (Fig. 16-5).

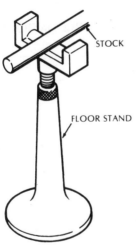

STOCK

FLOOR STAND

CAUTION: Place either a *warning flag*, cloth, or a protecting screen around the end of the bar. Care must also be taken that the length being cut does not project into an aisle or walking area.

Step 8 Position the cutting fluid nozzle. The fluid must be directed to the cutting area of the blade and workpiece.

Step 9 Recheck the vise to see that the work is held securely. The blade must be taut. The speed and blade pressure also require checking.

Step 10 Turn the power switch to *ON*. Position the frame with the handle. Bring the blade down almost to the work.

Step 11 Move the clutch handle to the engaged position. This automatically starts the cutting action.

Step 12 Remove burrs from the cutoff section.

CAUTION: Although the cutting fluid serves as a coolant, the part nevertheless may be too hot to handle safely. Allow it to cool before filing the burrs.

Cutting Off An Angular Section

Step 1 Select the proper power hacksaw blade.

Step 2 Set the machine at the recommended speed and feed.

Step 3 Loosen the *stationary* vise jaw. Position it at the required angle. (A stationary vise jaw positioned at a 45° angle is shown in Fig. 16-7.)

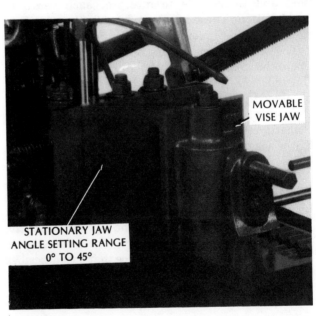

MOVABLE VISE JAW

STATIONARY JAW ANGLE SETTING RANGE 0° TO 45°

Fig. 16-7. Vise jaws permit angle settings to 45°. (Courtesy of Kasto-Racine, Inc.)

Note: The vise base is graduated in degrees. Tighten the stationary jaw at the correct angle.

Step 4 Mount the workpiece in the vise. The movable jaw is moved to the required angle.

Step 5 Measure the cut-off length of the workpiece.

Note: The end of a steel rule is usually placed against the teeth of the blade. The cutoff length is represented by the measurement to the end of the workpiece.
CAUTION: *Before* making this measurement, be sure the machine power switch is in the *OFF* position and the blade is *not* moving.

Step 6   Tighten the movable jaw. Check to see that the workpiece is held securely at the correct angle.

Step 7   Proceed with the cut by following Steps 7-12 under *Cutting Off a Square Section.*

SELECTION AND IDENTIFICATION OF VIEWS

One-, two-, and three-view drawings have been described up to this point. In each instance, the main purpose of a view is to provide information. The number of views must be adequate for the craftsperson to accurately conceive the shape and size of the part and to be able to interpret each detail of construction. These must conform to the specifications of the designer.

The selection and arrangement of views depends upon the complexity of the part. The details for construction, inspection and assembly must be obtained from the drawing. The front, top, and either right-side or left-side views are commonly used. Sometimes, a *bottom view* or a *back view* is added. These views provide other essential shape and size data.

The name and location of each view is illustrated in Fig. 16-8. The abbreviations of the views generally used on drawings are given in Table 16-2. As stated before, the front view may have no relationship to the actual front surface of the part. The front view is selected on the basis of the surface that best describes the general shape of the part.

Note in Fig. 16-8 that three positions for the back or rear view (B.V.) are shown. This view may be projected from a number of combinations of front view and right side, left side, or top views. The angular surface of the part is projected in its true shape and size in an auxiliary view (AUX. V.). The selection of the least number of views depends on the complexity of the part.

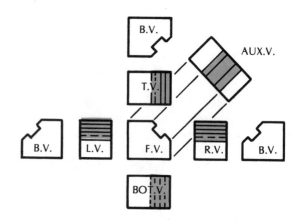

Fig. 16-8. Positions and identifications of views.

Table 16-2. ABBREVIATIONS FOR VIEWS

View	Abbreviation
Front View	(F. V.)
Right Side or Right View	(R. V.)
Left Side or Left View	(L. V.)
Top View	(T. V.)
Bottom View	(Bot. V.)
Back or Rear View	(B. V.)
Auxiliary View	(Aux. V.)

SAFE PRACTICES IN POWER HACKSAWING

- Position the cutting teeth and the blade to cut on the draw stroke.
- Tighten the blade tension until it is adequate to hold the blade taut during the cutting operation.
- Check the blade pins regularly to see they are not being sheared.
- Check the workpiece to be sure it is tightened securely before starting the cut.

- Make sure the blade is moved away from the work before starting the power hacksaw.
- Start a new blade (after a cut has been started) in a new location. Otherwise the teeth may bind in the old kerf and break the blade.
- Direct the flow of the cutting fluid, when required, over the cutting area. The flow must be as close to the cutting saw teeth as possible.

- Support the ends of long pieces that project from the power hacksaw, using a roller stand.
- Place a protecting screen or a danger flag at the end of parts that extend any distance from the saw frame.
- Cool the cutoff section before handling. Cooling helps to avoid burns and cuts resulting from hot, burred pieces.
- Remove cutting fluids and clean the reservoir regularly.

POWER HACKSAW AND SAWING TERMS

Cutting-off (metal sawing)	The sawing apart by the shearing action of a series of cutting teeth. The removal of a section of metal or other material by a metal sawing process.
Utility power hacksaw	An all-purpose wet-or dry-cutting machine for cutting off square or angular sections of workpieces. A sawing machine. Workpieces are cut off by a hacksaw blade that moves in a reciprocating motion.
Draw stroke	A stroke of a power hacksaw. A downward force is applied on the cutting teeth of the blade as it is drawn through the workpiece. The cutting of a saw kerf by the shearing action of the saw teeth on the cutting stroke.
Machinability	Properties of a material which relate to the ease or difficulty with which it may be cut or formed. Qualities of a material that influence the selection of a blade, its pitch, machining speed and feed, and an appropriate cutting fluid.
Raker set	A pattern of setting teeth to cut a kerf larger than the width of the hacksaw blade. A pattern of three teeth. An unset tooth is followed by one tooth offset to the right. Another tooth is offset to the left of the blade body.
Cutting speeds (power hacksaw)	The number of full strokes the saw frame makes per minute.
Four-speed (power hacksaw)	A range of four speeds. A speed range which permits the hacksaw blade to make 35, 70, 100, or 140 strokes per minute.
Elongated slots	Two parallel slots in the vise base. Long slots that permit the stationary jaw to be positioned square or at an angle to the cutting blade.
Production power hacksaw	A power hacksaw with production capabilities. Multiple pieces of stock are bundled and clamped in position. The cutting fluid is controlled, the parts are cut off, the hacksaw is stopped automatically

---

SUMMARY

- Two common power metal cutting saws are the power hacksaw and the horizontal band (cutoff sawing) machine.
- Mild to hard ferrous metals and soft to hard nonferrous metals and other materials may be cut off by power hacksawing. The section may be square, round, structural, or another preformed shape.
- Power hacksawing is the process of shearing a narrow width (kerf). The cutting teeth on the blade are forced to cut into the workpiece as the blade is drawn through.
- Dry-cutting power hacksaws operate at slower speeds. These hacksaws cut materials that do not require a cutting fluid or where there is a limited amount of heat generated. Further, any small amount of heat does not affect the hardness or efficiency of the saw teeth.
- Wet-cutting power hacksaws of the utility and heavy-duty types are used when a great deal of heat is generated. A cutting fluid serves as a coolant to control the heat. The lubricating qualities of the cutting fluid protect the blade surfaces from being scored. Cutting efficiency is improved.
- The pitch selection of power hacksaw blades is governed by the same principles that apply to hand hacksawing.
- Common lengths of power hacksaw blades range from 12″ to 24″, and pitches range from 4 to 14. Metric lengths range from 300 mm to 600 mm, with

pitches from 6 mm to 2 mm.

- A raker set tooth pattern consists of sets of three teeth. An unset tooth is followed by two other teeth that are alternately offset.
- The hydraulic system of a power hacksaw:
  - actuates the saw frame
  - controls the down feed during cutting
  - raises the frame and blade on the return stroke
  - maintains a constant flow of cutting fluid
  - automatically raises the frame to clear the work at the end of the cut
  - stops all motion.
- Two-speed power hacksaws generally operate at 70 and 100 strokes per minute. Four-speed machines have a range of 35, 70, 100, and 140 strokes per minute.
- Work may be held in a vise and positioned square or at an angle to the cutting blade. The stationary vise jaw may be set to a required angle.

- Quantities of regular or irregular shaped bars or extruded pieces may be nested as a solid mass and cut together.
- Production power hacksaws are designed to automatically load, position, clamp, and cut off sections, and then stop.
- Accurate shape, size, and construction details of a part are described on a drawing. The number and arrangement of views may include various combinations of front, right-side, left-side, top, bottom, back (rear), or auxiliary views.
- The cutting fluid and reservoir must be kept clean. The fluid must be tested regularly. Testing and correcting ensures that the evaporation of liquid does not change the efficiency of the cutting fluid.
- Personal and machine tool safety-related precautions must be observed. These cover all steps in handling the workpiece, cutting tools, and the power hacksaw.

UNIT 16 REVIEW AND SELF-TEST

1. Describe one cutting cycle of a common 6″ × 6″ (150 mm × 150 mm) utility power hack saw.
2. Cite three guidelines that especially apply to selecting power hacksaw blades.
3. Give a general application of the following set patterns for power hacksaw blade teeth:
   a. *raker set*
   b. *wave set*
   c. *straight set*.
4. Explain two design features of six-speed power hacksaws that make this machine more versatile than a two-speed machine.
5. Describe how stacked bars are automatically cut on a production power hacksaw.
6. Tell when and why an auxiliary view is used on a drawing.
7. State three safe practices that relate particularly to power hacksawing.

Unit 17

THE HORIZONTAL BAND MACHINE: CUTOFF SAWING PROCESSES

The *horizontal band machine (cutoff sawing)* is a second type of power saw. This machine is often referred to as a *metal-cutting band saw,* a *cutoff band saw,* or a *cutoff band machine.* The term *band* indicates that the cutting-off process requires a closed saw blade.

There are also *vertical* band machines with saw bands and file bands. These machines are used for intricate and precise sawing and filing processes.

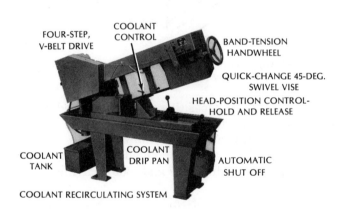

FOUR-STEP, V-BELT DRIVE
COOLANT CONTROL
BAND-TENSION HANDWHEEL
QUICK-CHANGE 45-DEG. SWIVEL VISE
HEAD-POSITION CONTROL-HOLD AND RELEASE
COOLANT TANK
COOLANT DRIP PAN
AUTOMATIC SHUT OFF
COOLANT RECIRCULATING SYSTEM

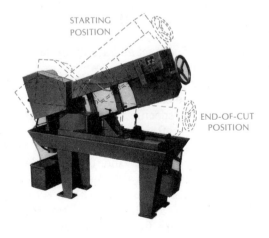

STARTING POSITION
END-OF-CUT POSITION

Fig. 17-1. Utility type horizontal band machine. (Courtesy of Do All Company)

Fig. 17-2. Automatic horizontal production band machine. (Courtesy of Do All Company)

BASIC TYPES AND FEATURES OF HORIZONTAL BAND MACHINES

Metal-cutting band machines of the cutoff sawing type may be either *dry-cutting* or *wet-cutting.* They may also be of a *utility* or *heavy duty* design. A utility type is illustrated in Fig. 17-1. A heavy duty machine for production sawing is shown in Fig. 17-2. Parts of these cutoff machines perform the same functions as power hacksaws. The bases, vises, and tables of power hacksaw and band types have similar features and functions. However, the cutting mechanisms differ considerably.

The saw frame of the band machine has two wheels. These hold and drive a continuous (closed) saw blade. A *tension control* adjusts the blade to track properly at the correct tension. The teeth are thus forced through the work. Blade guide inserts are provided for positioning the blade vertically at the cutting area. The inserts guide the blade to cut

185

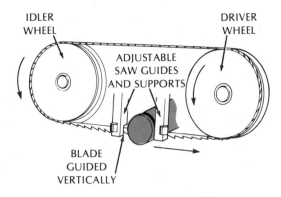

Fig. 17-3. Simple design features of a band machine.

squarely. The saw frame is hinged. This permits raising the saw band (blade) to clear the work, and lowering it to take a cut. A *pneumatic* system controls the circulation of the cutting fluid and many of the machine mechanisms.

OPERATING PRINCIPLES

The simple design features of a band machine are shown in Fig. 17-3. The continuous saw band revolves around the *driver* and *idler wheels*. Attached to the frame are two adjustable *band (saw) supports with guide inserts*. These serve two functions:
• Guide the blade in a vertical position so it does not bend from the work
• Support the blade so that a cutting force can be applied.

The metal-cutting action takes place by continuously feeding the cutting teeth into the work. This is possible because the blade revolves in one direction only. This is in contrast with the power hacksaw where the cutting is done during only part of the complete stroke (cycle).

The metal-cutting band machine may be operated at higher cutting speeds than the reciprocating power hacksaw.

DRY CUTTING AND WET CUTTING

Materials are *dry-cut* in both power hacksawing and band machine sawing when:
• A slow cutting speed is used and the frictional heat that is generated during the cutting process is minimal
• The metal is comparatively soft and easy to cut
• A material, like gray and malleable cast iron, produces loose graphite that acts as a lubricant
• The cutting action on hard materials (using a cutting fluid) has a tendency to produce a work-hardened surface.

Wet-cutting is recommended to:
• Dissipate the heat generated over the small surface area of each saw tooth
• Remove heat to prevent softening the cutting edges of the saw teeth
• Reduce the friction between the chips and the saw teeth

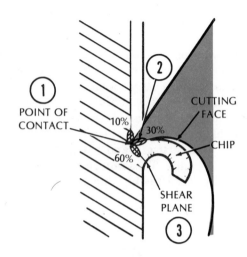

Fig. 17-4. The three main places where heat is generated by the cutting action of saw teeth.

• Prevent the depositing of metal (caused by the cutting action) at or near the edges of the teeth
• Clear chips away from the workpiece
• Keep the sides of the saw blade from being scored
• Increase productivity and tool wear life.

CUTTING ACTION AND CUTTING FLUIDS

The action of the saw teeth cutting through a workpiece produces a tremendous amount of heat. The heat is generated in three main places (Fig. 17-4):
• At the point of contact and along the cutting face
• On the cutting edge
• At the *shear plane* in the forming of the chip.

The heat must be carried away, and the temperature held within a specific range. Excess heat can cause the teeth to soften and dull.

The chart in Fig. 17-5 shows the effect of heat on the hardness of three major kinds of saw blades. Note that the teeth of the carbon alloy saw blade soften at a comparatively lower temperature than do those of the high-speed steel saw blade. As the saw teeth dull and wear, the cutting efficiency is reduced drastically. Over this same temperature range, tungsten carbide saw blade teeth maintain their cutting qualities.

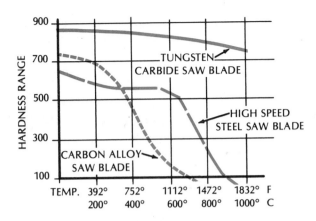

Fig. 17-5. The effect of heat on the hardness of saw band teeth.

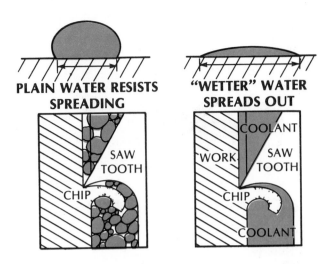

Fig. 17-6. Chemical additives improve the heat dissipation properties of cutting fluids.

Cutting fluids are used to *dissipate the heat* produced by sawing and cutting. The fluids also *wash chips away* from the cut and *increase the saw life*.

There are three basic categories of cutting fluids that are widely applied in wet-cutting. These serve in different degrees as either a *coolant*, a *lubricant*, or both. In each instance, the cutting fluids increase the cutting efficiency. *Straight oils, soluble oils* and *synthetic (chemical/water) cutting fluids* are three common groups. These are considered in terms of cutoff sawing operations.

Straight Oils

Straight oils are mineral oils. They are used on very tough materials that must be cut at slow speeds. The straight oil has *lubricity*. This means it has high lubricating properties in contrast with its heat removing capability. Straight oils may have sulphur or other chemical compounds added (*additives*).

Soluble Oils

Soluble oils are used in the cutting-off of a wide variety of materials. Since band machine sawing is done at high speeds, a great deal of heat is generated. It is necessary to remove the heat rapidly to maintain efficient cutting and to protect the life of the saw teeth and saw band.

Soluble oils are mineral oils. These are *emulsified* into fine particles or *globules*. The soluble oils mix readily with water. The resulting mixture combines the properties of the straight oils for lubricity and the high-cooling, or heat dissipating, rate of water.

The more concentrated the mixture, like one part of soluble oil to three parts of water (1:3), the greater the lubricity. The more dilute the mixture (like 1:7), the greater the heat removal capability. In any water solution there is natural evaporation and other losses

due to heat. This requires that the mixture be tested regularly to correct the water and soluble oil proportions.

Synthetic (Chemical-Water) Cutting Fluids

The *synthetic cutting fluids* do not contain mineral oils. These fluids meet the needs for a high cooling capacity. Synthetic cutting fluids are applied at the higher cutting speeds where a high rate of heat removal is essential. Chemicals are added to water to produce *wetter water*. Water tends to resist spreading because of its surface tension. This surface tension may be reduced by chemicals as shown in Fig. 17-6. The cutting fluid then has better characteristics of spreading and flowing easily and quickly. This combination of properties provides for a high rate of heat removal. Another desirable quality is that many synthetic cutting fluids are transparent. This makes the cutting area and action more visible.

KINDS AND FORMS OF SAW BLADES (BANDS)

There are three general kinds of cutoff saw blades:
● Carbon alloy
● High-speed steel
● Tungsten carbide.
Carbon alloy saw blades are generally used in the tool room, maintenance shop, and light manufacturing where accuracy is required. The *high-speed steel blades* are used in heavy-duty and full-time production work. *Tungsten carbide blades* are suited for heavy production and for rough cutting through tough materials.

Carbon alloy and high-speed steel saw bands have three forms of saw blade teeth:

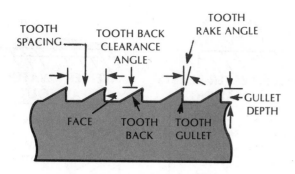

Fig. 17-7. General features of saw tooth forms.

- Precision
- Buttress
- Claw.

Tungsten carbide saw bands have a special tooth form.

The general design features of saw tooth forms are illustrated in Fig. 17-7. Each form is described.

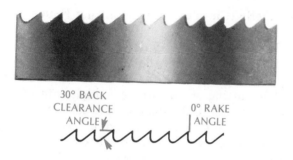

Fig. 17-8. Precision tooth form.

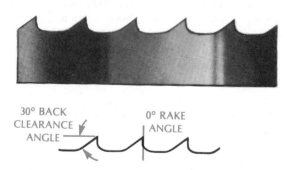

Fig. 17-9. Buttress tooth form.

Precision Form

The precision form is the most widely used tooth form. The tooth has a rake angle of 0°, a back clearance angle of 30°, a deep *gullet,* and a radius at the bottom. Each of these features is shown in Fig. 17-8. The precision form produces accurate cuts and a

fine, finished surface. The clearance angle and gullet provide ample chip capacity for most cutoff sawing operations.

Buttress Form

The *buttress* form (Fig. 17-9) is also known as *skip tooth.* The 0° rake angle and the 30° back clearance angle are similar to the precision form. However, the teeth are spaced wider apart. This spacing provides greater chip clearance. Buttress form teeth cut smoothly and accurately. This tooth form is recommended for thick work sections, deep cuts, and soft material.

Claw Tooth Form

The *claw tooth* (Fig. 17-10), is also called a *hook tooth.* The tooth has a *positive rake angle.* The clearance angle is smaller than that of the precision or buttress forms. The *gullet* is especially designed for *stress proof.* The claw tooth makes it possible to cut at a faster rate and at reduced feed pressures. These features provide longer tool life.

Tungsten Carbide Form

This tooth form has a positive rake angle and a smaller clearance angle than any of the other tooth forms. Fig. 17-11 illustrates these characteristics.

Fig. 17-10. Claw tooth form.

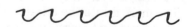

Fig. 17-11. Tungsten carbide tooth form.

The tungsten carbide teeth are fused into a fatigue-resistant blade. This combination is necessary on heavy, tough cutoff sawing operations.

CUTOFF SAWING REQUIREMENTS AND RECOMMENDATIONS

More cut-off sawing operations are required in the manufacture of parts than any other single machining process. The craftsperson must make a number of decisions that center around the saw band alone. The following are some of the major considerations:

- Kind of saw blade
- Form of tooth
- Cutting fluid, if required, and its rate of flow
- Pitch of the teeth
- Velocity of the saw band in feet per minute.

The decisions must be based on certain information:

- The size and shape of the workpiece
- The properties of the material to be cut
- The required quality of the finish of the sawed surface
- The quantity to be cut off
- The overall cross-sectional area if multiple pieces are cut at the one setting.

Band machine and saw band manufacturers provide tables. These aid the worker in selecting the correct saw band for maximum cutting efficiency. Such information is further simplified and combined in a *job selector*. A section of a job selector is illustrated in Fig. 17-12.

Fig. 17-12. Section of a band machine job selector. (Courtesy of Do All Company)

STACKING FOR QUANTITY SAWING

Stack sawing is a technique of placing and holding together a number of rectangular, square, round, or regular-shaped bars of stock. These are all cut at the one setting. Fig. 17-13A shows some simple stacking patterns for regular-shaped solid bars. The actual stacking and cutting of multiple pieces of angle iron is pictured in Fig. 17-13B.

Stacking results in greater productivity. *Setup* and *shutdown time* are reduced considerably from what they would be if each piece was cut individually. Stacking increases the number of pieces that are cut off during each sawing cycle.

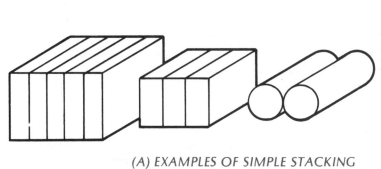

(A) EXAMPLES OF SIMPLE STACKING FOR MULTIPLE CUTTING

(B) CUTTING OFF MULTIPLE PIECES OF ANGLE IRON

Fig. 17-13. Stacking for quantity sawing. (Photo courtesy of Do All Company)

HOW TO SELECT THE SAW BLADE AND DETERMINE CUTOFF SAWING REQUIREMENTS

Sample Job Requirement: Select the correct saw blade for cutting off a quantity of workpieces. The required lengths are to be cut from a 2″ diameter bar of low carbon steel in the 1015 to 1030 range. A band machine with a welded and dressed blade is to be used.

Step 1   Turn the *job selector* to the kind of material to be cut off (Fig. 17-14).

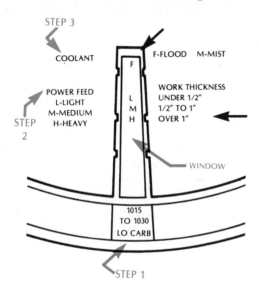

Fig. 17-14. Reading coolant and feed requirements on the job selector.

Note: The band machine settings and operating conditions are read in the *window* above *1015-1030 lo carbon.* This setting is shown in Fig. 17-14. The feed, band pitch, kind of saw blade, and band velocity are given in the left column opposite the work *thickness.* In this case, over *1".*

Step 2   Read the setting for the *power feed* (Fig. 17-14). Set the machine for a *heavy power feed* (H).

Step 3   Read the *rate of coolant flow* (Figure 17-14). The selector shows *flood* (F).

Step 4   Refer to a machinability table on cutting fluids for ferrous metals. (Appendix A-28 is an example.) Select the recommended fluid in the low-carbon steel column (machinability group II) under sawing

processes. Either a sulphurized oil (Sul), a mineral-lard oil (ML), or a soluble or emulsified oil or compound (Em) is recommended.

Note: Some job selectors provide this information.

Step 5   Read on the job selector the recommended kind of saw blade and pitch (Fig. 17-15).

Note: The "6P" indicates: *6 pitch, precision form, carbon alloy blade.*

Step 6   Select and mount the appropriate saw band.

Step 7   Read on the job selector the recommended *band velocity.* The setting recommended in Fig. 17-15 is *155 feet per minute.* Set the band velocity at this speed.

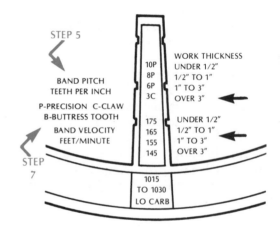

Fig. 17-15. Reading pitch, type of blade, and velocity on the job selector.

HOW TO OPERATE A HORIZONTAL CUTOFF BAND MACHINE

The operating steps which follow assume that:

• The saw band has been properly mounted over the idler and driver wheels and between the saw guides

• The tension handwheel has been adjusted to provide the correct band tension

• The wire chip cleaning brush is in contact with the teeth to remove the chips

• The wheel guards are in a locked position.

Checking The Saw Band

Step 1   Refer to the job selector. Check the type of saw band, tooth form, pitch, band speed and feed, and coolant that are recommended.

Step 2   Set the machine to the correct speed and feed rate.

Adjusting The Saw Arms And Guides

Step 1   Position the two saw arms (Fig. 17-16). The saw guides are placed as close to the workpiece as possible. Tighten the hand knobs.

Fig. 17-16. Band machine support arm setup for regular cutting off. (Courtesy of Do All Company)

Step 2   Test to see that the guides are secure. The saw band must be placed between the saw guide inserts. The teeth are pointed downward and toward the solid jaw of the vise.

Mounting The Workpiece

Step 1   Position the stationary vise jaw either square or at the required angle. (The structural section in Fig. 17-17, for example, shows the position of the I-beam for a square cut.)

Step 2   Move the workpiece through the vise and locate it for the cut.
Note: There are three common ways of setting the workpiece at the required cutoff length. (1) The length may be layed out and marked on the stock. (2) The adjustable stop may be set for the required length. (3) The distance may be measured directly from the end of the workpiece to the saw band.

Fig. 17-17. Making a square cut in a structural section. (Courtesy of Do All Company)

Cutting Off The Workpiece

Step 1   Position the coolant nozzle. Direct the coolant through the saw guides and workpiece to reach the teeth as soon as possible.

Step 2   Push the starter control button. Regulate the coolant flow.
Note: The saw band feeds to and through the work automatically. The machine stops when the section has been cut off.

CAUTION: Watch the blade for *walk off.* This indicates the blade is getting dull and requires replacing.

Step 3   Burr the cutoff section. Check its length.
Note: Multiple pieces should be set to length against the adjustable stop.

Step 4   Clean the machine. Return the remaining stock to storage.

SAFE PRACTICES IN BAND MACHINING (CUTOFF SAWING)

- Position the saw band vertically in the saw guides. The distance between the saw arms should be slightly longer than the width of the part to be cut.
- Check the blade tension. It must be adequate to permit proper tracking on the driver and idler wheels and to transmit sufficient force to cutoff the workpiece.
- Determine the speed, feed, kind, and form of saw blade teeth, and the cutting fluid. These are obtained from the machine and blade manufacturers' specifications.
- Lock the frame guards in the locked position *before* any control switches are turned *ON*.

- Direct the cutting fluid to provide maximum lubrication between the blade, the guides, and the workpiece. The positioning must also provide for dissipating the heat generated at the cutting edges.
- Clean the chip tray, reservoir, and cutting fluid system regularly.
- Replace water and other compounds regularly to soluble oil and synthetic cutting fluids when the solution becomes weak. The required properties must be maintained for cooling and lubricating.
- Position the power controls in the OFF position when making blade or work adjustments.

BAND MACHINING (CUTOFF SAWING) TERMS

Band machine (cutoff sawing)	A rotating machine with driver and idler wheels. The wheels drive and force a metal-cutting sawtoothed band to shear and cut off materials.
Blade supports	Two adjustable supports on the band machine frame. Arms which are positioned to provide support for the saw band.
Precision form	A saw tooth form having a 0° rake angle, a 30° back clearance angle, a deep gullet, and a bottom radius. An accurate-cutting, fine-finished-surface-producing saw tooth form.
Buttress (skip tooth) form	A saw tooth having a 0° rake and 30° back clearance angle. The teeth are spaced farther apart than are those of the precision form.
Claw (hook tooth) form	A saw tooth form that has a positive rake angle and a smaller clearance angle and wider-spaced teeth than the precision form.
Tungsten carbide form	Tungsten carbide fused on a fatigue-resistant blade back. A specially shaped, positive-rake-angle tooth form. A tooth form that is particularly adapted for heavy sawing operations through tough materials.
Job selector	An information dial guide on band machines. Manufacturers' recommendations on saw blade characteristics, speeds, feeds, cutting fluids and flow rate, and other machine operating conditions.
Blade guides	Inserts in the blade supports that nest the saw blade (above the teeth). Guiding the saw blade in a vertical direction through the workpiece.
Wet-cutting	The use of a cutting fluid to dissipate heat, flow chips away, reduce friction, and improve cutting efficiency.
Cutting fluids	Fluids that have two basic groups of properties: (1) rapid heat removal, and (2) lubricity. Lubricity helps produce a finer cutting action and reduce friction.
Straight oils	Mineral oils used on tough materials at low cutting speeds where a high degree of lubrication is required.
Soluble oils	A mixture of an emulsified mineral oil and a water soluble oil.
Synthetic cutting fluids	A mixture of chemical additives that improve the wetting and cooling properties of water.

SUMMARY

- The horizontal band machine is one of the newer developments in power metal-cutoff sawing.
- Utility and heavy-duty production types of horizontal band machines are available with stationary or angle-setting vises.
- The accuracy of the cut depends in part on the close positioning of the band supports in relation to the overall width of the material to be cut. The careful adjustment of the guides that steady the saw band also influences the accuracy.
- The saw band must track evenly on the idler and driver wheels. The tension on the saw band must be adequate to drive the saw teeth to cut uniformly.
- Horizontal band machines may be operated dry at slow speeds. Dry-cutting is used when a limited amount of heat is generated or the material does not require a cutting fluid.
- Wet-cutting is employed when (1) there is considerable heat to be dissipated, (2) friction is to be reduced to a minimum, (3) chip deposits are to be prevented from forming on the teeth, and (4) the chips are to be carried away.
- The cutting action of the band machine is produced by the continuous revolution and feeding of a saw band.
- Slow-speed cutoff sawing operations require a high degree of lubricity. A straight mineral oil is satisfactory.

- Soluble oils are used when a combination of a coolant and a lubricant is required.
- Synthetic cutting fluids with a chemical additive make a water base a more efficient cooling agent. These fluids are recommended for cutoff sawing operations that require a high level of heat removal.
- Carbon alloy, high-speed steel, and tungsten carbide are three basic kinds of saw blades. They are used for general precision sawing, production, and heavy-duty cutting through tough materials, respectively.
- The precision and buttress tooth forms have a 0° rake and a 30° back clearance angle. They differ in the distance between teeth, the depth of the gullet and bottom form, the quality of the finished surface produced, and the cutting time.
- A job selector chart provides saw blade and machining information. Recommendations include: the kind of saw blade and tooth form, cutting speeds, feeds, cutting fluids, and other operating conditions. These are read easily on the chart. The technical information is related to the type of material to be cut off, its size, and the surface finish required.

UNIT 17 REVIEW AND SELF-TEST

---

1. List three distinct design features of the band machine and blade.
2. Explain why the horizontal band machine cuts faster than a reciprocating power hacksaw.
3. State three conditions under which dry band machine sawing is recommended over wet-cutting.
4. Identify one saw blade material of which the cutting quality of the saw blade teeth is maintained longer (over the same temperature range) than that of a high-speed steel blade.
5. State why synthetic cutting fluids are used in cutting-off processes which require a high heat-removal rate.
6. Distinguish between (a) the *precision form* and (b) the *tungsten carbide form of saw band teeth*.
7. Indicate the kind of information that is provided by a *job selector* on a horizontal cutoff band machine.
8. List the preliminary steps that must be taken before mounting and cutting off parts on a horizontal cutoff band machine.
9. Identify three safe practices to observe when band machining (cutoff sawing).

SECTION TWO

BENCH AND FLOOR GRINDERS: TECHNOLOGY, PROCESSES, AND PRACTICES

This section presents the principles of and procedures for correctly using general-purpose bench and pedestal grinders to shape and size small tools and workpieces made of hardened steel. Included are: abrasives and grinding wheel technologies related to bench and floor grinding; procedures for accurately and safely mounting, truing, and dressing grinding wheels; and step-by-step procedures for grinding representative tools.

Unit 18

HAND GRINDING: MACHINES, ACCESSORIES, and BASIC PROCESSES

Cutting, marking, forming, punching, and other types of small tools are made of hardened steel. Through constant use, the cutting edges, heads, and other contact surfaces may become damaged or dull. These tools may be sharpened and returned to good working condition by *hand grinding* and *stoning*.

GRINDING AND STONING PROCESSES

Grinding is the process of cutting away particles of a material. Grinding reduces a part or tool to a particular size and shape. The cutting is done by countless numbers of abrasive grains that are bonded together. As each cutting edge is forced into the surface of a workpiece, it cuts away a small chip. By rotating a grinding wheel, a continuous series of cutting edges are placed in contact with the work or tool. These abrasive edges grind away the surface.

Stoning is another hand grinding process. In this instance, an abrasive *oilstone* is used. One purpose of *stoning* is similar to burring with a file. However, with hardened materials an abrasive stone is required to do the cutting. Stoning also is used to produce a fine cutting edge. This results in a smoother cutting action, a finer surface finish, and longer tool life.

BENCH AND FLOOR (PEDESTAL) GRINDERS

The grinding of the cutting edges of tool bits, drill points, chisels; the forming of blades and other tools; and various other metal-removing processes all may be performed on a simple grinding machine. This machine is called a *bench grinder* if it is mounted on a bench as shown in Fig. 18-1. The terms *floor or pedestal grinder* are used interchangeably when the grinding head is mounted on a pedestal for floor work (Fig. 18-2). Pedestal grinder sizes range from those used for small tool grinding operations to larger ones used for heavy-duty rough snagging processes on large forgings and castings.

This unit deals with the general-purpose bench grinder and a pedestal grinder of the same size. The principles of operation are identical. Basic technology related to abrasives and grinding wheels is

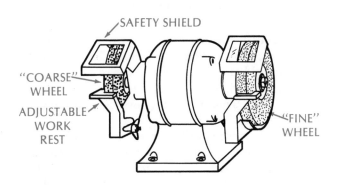

Fig. 18-1. A bench grinder.

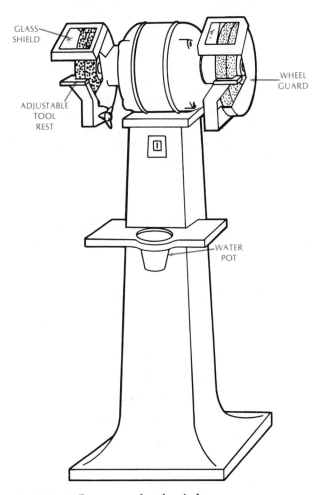

Fig. 18-2. A floor, or pedestal, grinder.

195

presented to the extent it is applied to bench and floor grinding.

Safe practices to follow in mounting, truing, and dressing a grinding wheel are described. Sample techniques for tool grinding, using a few representative tools, are covered.

CONSTRUCTION AND FEATURES OF GRINDERS

Grinding operations on the bench or floor grinder are referred to as *offhand grinding*. This means the object is held and guided across the revolving face of the grinding wheel by hand. The tool or part is held by one hand. The force and movement are directed with the other hand.

Offhand grinding differs from more precise (precision grinding) operations. In precision grinding, a part is positioned in a fixture or other machine accessory. The relationship between the wheel face and the work part is precisely controlled.

Regardless of the type, bench and pedestal (floor) grinders have a direct or belted power drive to a *spindle*. The spindle is flanged and threaded at both ends to receive two grinding wheels. One grinding wheel is *coarse*, for rough cutting operations. The wheel on the opposite end is *fine*, for finish grinding.

The grinding wheel is held between two *flanged collars*. There is a simple *adjustable tool rest* for each wheel. These rests provide a solid surface upon which the part may be placed and steadied while being ground. The tool rest may be positioned horizontally or at an angle to the wheel face.

Wheel guards are an important part of the grinder. The guards help protect the worker against flying abrasives and ground-off particles. Fig. 18-3 shows the principal design features of the wheel guard and spindle end. Safety glass shields permit the work

surface to be observed and provide additional protection against flying abrasives and other particles. However, because of the hazards of machine grinding it is also necessary for the worker to use personal eye protection devices.

GRINDING WHEELS

Mounting On A Spindle

Grinding wheels are mounted directly on the spindles of bench and floor grinders. Fig. 18-4 shows such a mounting. One end of the spindle has a right-hand thread; the other, a left-hand. The thread direction safeguards the wheel from loosening during starting, stopping, and while there is cutting action. The force exerted by the motion and cutting of the wheel is in the direction in which the clamping nuts tighten. Before mounting it, the grinding wheel should be tapped lightly with a soft material to *ring test* it to check for cracks.

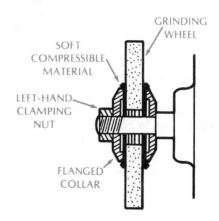

Fig. 18-4. A grinding wheel assembly.

Grinding wheels have a soft metal core that is bored accurately to the diameter of the spindle. The wheel slides on the shaft with a *snug fit*. The wheel should never be forced on the spindle or tightened with too great a force against the flanged collars. A soft compressible material (washer) is placed between the sides of the wheel and the flanged collars. This helps avoid setting up strains in the wheel. The clamping nut is drawn against the flanged collar only tight enough to prevent the grinding wheel from turning on the spindle during cutting.

Truing And Balancing Grinding Wheels

The wheel is tested for *trueness* and *balance* before

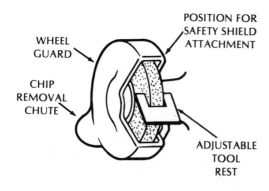

Fig. 18-3. A wheel guard and attachments.

grinding. *Trueness* refers to the concentricity of the outside face and sides of the grinding wheel as it revolves at operating speed. *Balance* relates to whether the *mass* of the grinding wheel is evenly distributed about its axis of rotation. Centrifugal force is produced if the wheel is out-of-balance. Excessive force may cause the wheel to fracture and fly apart. A fractured wheel is a safety hazard and should be replaced. Grinding wheel manufacturers true and balance the wheels so that they are ready to be mounted and used directly.

Truing is a process of shaping the grinding wheel. The wheel may be *dressed* so it has a straight, angular, or special contoured face. In each case, (after truing) the grinding *face* (periphery of the grinding wheel) must run concentric with the spindle *axis*.

DRESSING AND HAND DRESSERS

Dressing

During the grinding process, some of the sharp-edged abrasive grains become dull *(glazed)*. A loaded, glazed wheel is shown in Fig. 18-5. The grains may not fracture as they should to provide new sharp cutting edges. At the same time, metal particles from

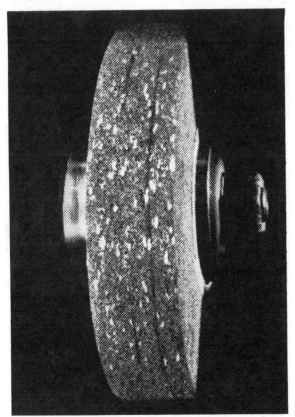

Fig. 18-5. A glazed (loaded) wheel before dressing. (Courtesy of The Desmond-Stephan Manufacturing Co.)

the workpiece or tool may become imbedded in the wheel. Such a wheel is said to be *loaded*. Under these conditions, the wheel must be *dressed*.

Dressing is the process of reconditioning a grinding wheel. This is necessary for efficient grinding and to produce a good quality of surface finish. Fig. 18-6 illustrates an open, dressed wheel. The grains are sharp. Fast cutting takes place.

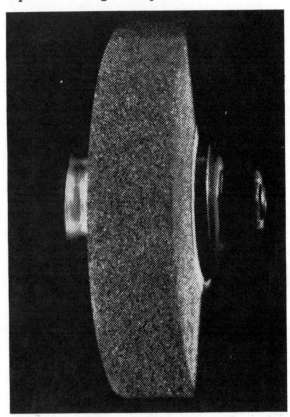

Fig. 18-6. An open, dressed wheel (sharp grains, fast cutting). (Courtesy of The Desmond—Stephan Manufacturing Co.)

Hand dressers are used for both dressing and truing grinding wheels. The two common hand types to be covered at this time are the *abrasive stick* and the *mechanical dresser*.

Dressers

Abrasive Stick—This type of dresser is used to remove particles which produce a glazed surface and to sharpen the cutting face. The abrasive stick, shown in Fig. 18-7, is held by hand. It is brought carefully into contact with the grinding wheel. Usually, a few passes across the face are sufficient. This action corrects the glazed condition and produces sharp-cutting abrasive grains.

Fig. 18-7. An abrasive dressing stick. (Courtesy of Do All Company)

Mechanical Dressers—Mechanical dressers are usually used for heavier dressing and truing operations than is the abrasive stick. *Mechanical dressers* are designed with many different forms of metal wheel discs. These are mounted on a shaft at one end of a hand holder. The discs have star or wavy patterns. They are made of hardened steel and are replaceable. There is a circular disc separator between each pair of star discs, and a great deal of *side play* between the discs.

As the disc faces are brought into contact with the face of the grinding wheel, they turn at considerable velocity in comparison with the speed of the grinding wheel. The action between the dresser discs and the grinding wheel causes the abrasive grains to fracture. New sharp-cutting edges are exposed. At the same time, the imbedded particles are *torn out*. The

wheel is also trued during this process.

A diamond dresser, guided under more controlled conditions, is used to true a grinding wheel to a finer degree of accuracy.

Mechanical dressers are practical where speed in reconditioning a grinding wheel is more essential than accuracy. Four common forms are illustrated in Fig. 18-8. A general-purpose dresser is shown in Fig. 18-8A. This type has a cast iron handle and star-shaped teeth on a hardened carbon steel cutter. This dresser is used for fast dressing of grinding wheels with diameters ranging from 6″ to 36″ (150 mm to 900 mm).

The dresser in Fig. 18-8B has corrugated cutters. These have a shearing action and produce a finer surface finish than the other dressers in Fig. 18-8.

The six-bearing-hole hexagon (hex) dresser in Fig. 18-8C is ruggedly designed to withstand hard use. The hex dresser is suited for high-speed wheels that operate within the range of 9,500 to 12,000 surface feet per minute (sfpm). The bearings in the side washers are hardened steel. As one set of bearings wears, the side washers are turned one sixth of a turn to employ the next set of bearings.

The heavy-duty dresser (Fig. 18-8D) has dust protected ball bearings for the cutter spindle. This dresser is practical on large, coarse grinding wheels, particularly those that are bakelite and rubber bonded.

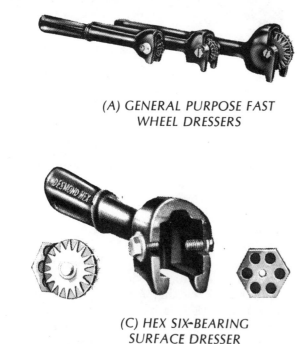

(A) GENERAL PURPOSE FAST WHEEL DRESSERS

(C) HEX SIX-BEARING SURFACE DRESSER

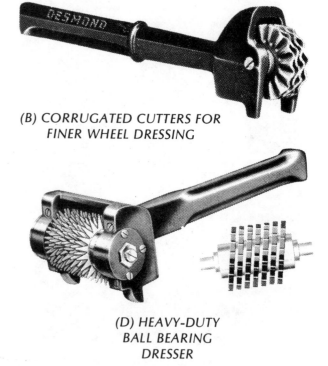

(B) CORRUGATED CUTTERS FOR FINER WHEEL DRESSING

(D) HEAVY-DUTY BALL BEARING DRESSER

Fig. 18-8. Four general types of mechanical wheel dressers. (Courtesy of The Desmond-Stephan Manufacturing Co.)

Wheel Conditions Requiring Dressing

The craftsperson must make decisions as to when a grinding wheel must be dressed. Dressing is needed under the following conditions:

- An uneven *chatter surface* is produced
- The workpiece is heated excessively during grinding
- Colored *burn marks* appear on the work surface
- There are *load lines* on the workpiece, indicating that the wheel is loaded with metal particles
- There is considerable vibration when the wheel is in contact with the workpiece.

HOW TO USE AND DRESS A GRINDING WHEEL (BENCH AND FLOOR GRINDERS)

Using A Mechanical Dresser

Step 1   Adjust the work rest (tool rest). The front face may be used as a guide for the mechanical dresser. The distance between the work rest and the wheel should permit the discs to almost touch the face of the grinding wheel.

Step 2   Remove the dresser. Check to see that the work rest and wheel guards are secure. The safety shield should be positioned to provide good visibility. Safety goggles should be worn.

CAUTION: Check to see that the wheel is tight on the spindle before truing.

Step 3   Start the grinder.

CAUTION: In all grinding operations, stand to one side *out of the path* of the grinding wheel. This is the operator's position at start-up and during the grinding operation.

Step 4   Place the *guiding lugs* of the dresser against the front edge of the work rest.

CAUTION: There should be no contact at this time between the cutter and the revolving grinding wheel.

Step 5   Hold the dresser down on the work rest and back against the edge with one hand. Move the dresser across the wheel face, guiding it with the other hand. Fig. 18-9 shows the position of the hands and dresser.

Fig. 18-9. Hand dressing a grinding wheel with a mechanical dresser. (Courtesy of The Desmond-Stephan Manufacturing Co.)

Step 6   Tilt the dresser upward a small amount. Move it across the wheel face.

CAUTION: The dresser should be raised slightly after each pass. Otherwise, there may be a tendency to *hog in*. Excessive amounts are cut out in hogging, producing undesirable results.

Step 7   Continue to raise the dresser for each pass across the face of the wheel. After a few passes, the wheel should run true, the face should be straight, and the glazed condition and dull grains eliminated. These corrected conditions produce an efficient cutting action.

CAUTION: There are a great number of abrasive and metal particles that are released at high speeds during the dressing process. Care must be taken to avoid inhaling these particles.

Step 8   Stop the machine. Reset the work rest to within about 1/16" of the wheel face.

HOW TO GRIND A FLAT SURFACE

Example: Grinding a flat point end and sides on a tool.

Grinding The Point

Step 1   Check the grinding wheel for cutting

Step 2   ability and trueness. See that the machine guards, safety shields, and the work rest are in position and are secure. Start the grinder. Hold the flat face of the tool blade down against the work rest with one hand. The tool is held at right angles to the wheel face. With the other hand, ease the edge of the tool into contact with the grinding wheel face.

CAUTION: The tendency with narrow width tools is to jam them into the wheel face. This may cause abrasive grains to be dug out or the edge of the wheel to chip. Either condition is unsafe.

Step 3   Move the tool edge across the whole face of the wheel (Fig. 18-10). Use one hand to keep a steady downward force. With the other hand, move the tool edge across the wheel face. Continue the grinding process until the tool face is ground square. First, rough grind with the coarse wheel. Then, finish grind.

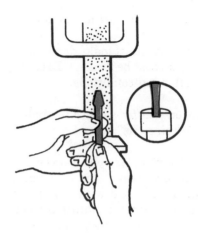

Fig. 18-11. Grinding a flat side.

Step 3   Turn the tool 180° and repeat step 2 on the second face. Continue to grind. Check the two faces for parallelism and thickness.

Step 4   Stone the edges to remove any burrs (Fig. 18-12). Clean the machine.

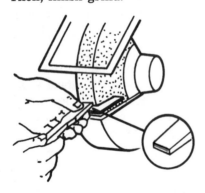

Fig. 18-10. Grinding a flat end.

CAUTION: The hardened tip becomes heated. It must be frequently dipped in water to prevent drawing the temper and softening the part.

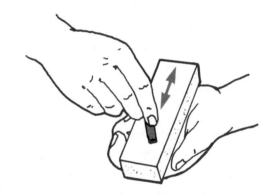

Fig. 18-12. Stoning an edge.

HOW TO GRIND AN ANGULAR SURFACE
AND A ROUND POINT

Grinding The Sides And Flat Faces

Step 1   Repeat steps 2 and 3 to grind the sides of the tool (or part) flat.

Step 2   Hold one face of the tool against the wheel at an angle *(tangentially)*. Move it across the face of the wheel (Fig. 18-11). The ground surface should be parallel with the tool axis.

Grinding An Angular Surface

Step 1   Set the work rest at or near the required angle. Move the work rest to within 1/8″ of the grinding wheel.

Step 2   Hold the tool or workpiece firmly on the work rest with one hand.

Step 3   Bring the tool into contact with the revolving grinding wheel face. At the same time, guide the tool across the entire wheel face with the other hand.

Step 4   Continue grinding until the required width of the angular surface is reached. *Note:* A slightly different procedure may be followed in grinding the convex angular face on a cold chisel. The angular and convex surface is produced by holding the flat surface of the chisel against the work rest at an angle (Fig. 18-13A). At the same time, the chisel point is swung through a slight arc (Figs. 18-13B and C). This cutting action produces the convex cutting face.

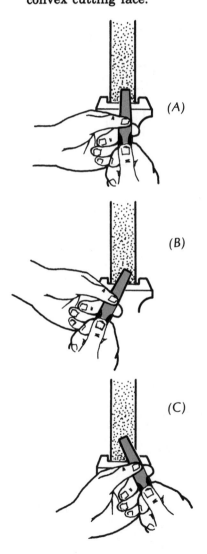

Fig. 18-13. Grinding an angular convex cutting edge.

Grinding A Point

Step 1   Hold the tool against the work rest at the angle to which the point is to be ground. *Note:* For a center punch, the included angle is 90°. Prick punches are ground at 30°.

Step 2   Bring the tool point carefully against the revolving grinding wheel. At the same time, turn the tool with the fingers of the other hand.
Note: While the tool or part is being rotated and ground, it is also moved across the face of the grinding wheel.

Step 3   Check the ground point angle against a gage or protractor. Change the angle position, if necessary. Then continue to grind to the required angle.

CAUTION: Quench the tool point frequently in water so that the tool retains its hardness. Nonhardened workpieces should also be quenched in water. This keeps the temperature within the range in which the part may be handled safely.

HOW TO GRIND A MUSHROOMED EDGE

Step 1   Steady the tool or part against the work rest. Hold it at an angle with one hand.

Step 2   Bring the mushroomed edges into contact with the grinding wheel. This is done with the other hand.

Step 3   Rotate the part continuously as it is being ground. Continue to grind around the edges of the tool. Grind at a slight angle until all mushroomed edges are ground away.

REPRESENTING ANGULAR SURFACES ON A DRAWING

Regular views are used to represent parts whose surfaces may be projected in their true shapes and sizes. The projections are made on either the horizontal or vertical plane or a combination of the two.

Auxiliary Views and Angular Features

When a part has an angular (slant) surface, the horizontal or vertical projections may not accurately show the design features. The true shape and size cannot be projected in the conventional way. In such cases, an *auxiliary view* is used. This view accurately represents the angular surface. An auxiliary view refers to a view of the angular features. The auxiliary view is in addition to other regular views. The auxiliary view shows the true shape and size of the features of an inclined (angular) surface.

An auxiliary view is *parallel* to the angular surface. The auxiliary view may be a full view. In other instances, a partial auxiliary view may show only the angular surface features. The features are projected on an imaginary parallel plane. Surfaces, circles and other features that are distorted in a regular plane (Fig. 18-14) are shown in their true shape and size on the auxiliary view.

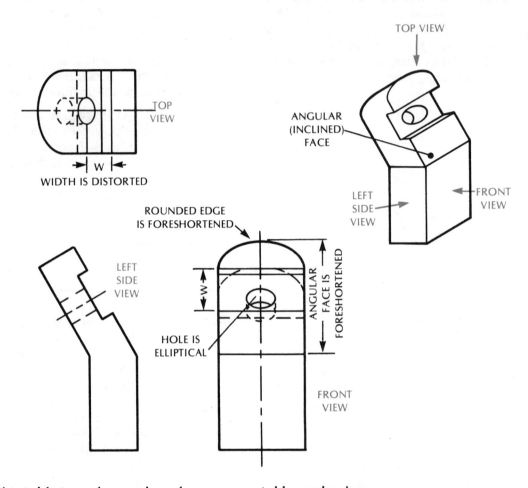

Fig. 18-14. Distorted features of an angular surface as represented by regular views.

Fig. 18-14 includes a pictorial sketch of a part with an inclined surface. The slot, rounded edge, and hole are distorted in the regular front, left, and top views in Fig. 18-14.

By contrast, a partial auxiliary view is used in Fig. 18-15. Lines 1 through 7 are projected at right angles to the angular surface. The circular shape, hole, and slot are shown in their true shape and size in the auxiliary view. When properly dimensioned, this view provides full and accurate information about the details of the angular surface.

Positions And Names Of Auxiliary Views

An auxiliary view may be an auxiliary front, top, left, right, or back view. Complex parts may require more than one auxiliary view. If one auxiliary view is projected from a *primary auxiliary view,* it is called a *secondary auxiliary view.*

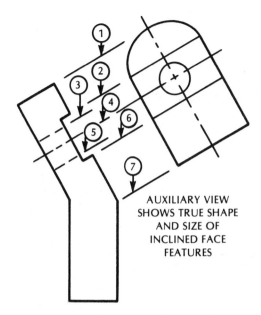

AUXILIARY VIEW SHOWS TRUE SHAPE AND SIZE OF INCLINED FACE FEATURES

Fig. 18-15. A partial auxiliary view of an angular face.

SAFE PRACTICES IN HAND GRINDING

- Secure the collars against the sides of the grinding wheel by applying a *slight* tightening force to the spindle nuts. Apply only enough force to prevent the wheels from turning on the spindle during the grinding process. Excessive force can strain and fracture the wheels, creating an unsafe condition.
- Use a soft, compressible material between the faces of each grooved flange (collar) and the sides of the grinding wheel.
- Store grinding wheels carefully in a special rack. Damage may result if the wheel is laid flat or placed where it can be hit.
- Correct out-of-true grinding wheels. These produce irregular ground surfaces. At the same time, centrifugal forces are set up. These forces can fracture the wheel, causing personal injury and machine damage.

- Dress glazed and dull grinding wheels. These conditions make the wheel inefficient and produce surface imperfections in the workpiece. After a grinding wheel is trued or dressed, the tool rest must be readjusted to within 1/16″ of the wheel face.
- Provide a small space between the tool rest, the dresser, and the grinding wheel. The tool rest supports and guides the dresser and permits it to be positioned at a cutting angle.
- Keep machine guards and shields in position. Put on safety glasses *before* the grinder is started. Keep them on throughout the grinding process.
- Maintain grinding wheel speeds *below* the maximum revolutions per minute specified by the manufacturer. This number is marked on the grinding wheel.
- Perform grinding operations from a position that is *out of direct line* with the revolving wheel.

BENCH AND FLOOR (PEDESTAL) GRINDER TERMS

Offhand grinding	The shaping and reducing to size of a workpiece or tool on a bench or floor grinder. Positioning, holding, and guiding a workpiece or tool against a revolving grinding wheel. Producing a cutting edge, a correct working surface, or forming a hardened or other part to a specific size and shape by hand grinding.
Stoning	A hand finishing process for removing burrs from a hardened metal surface. Finish grinding a fine cutting edge. Forcing a metal part against an abrasive sharpening stone to remove a limited quantity of metal.
Bench and floor (pedestal) grinder	A simple, basic grinding machine mounted on a bench or a pedestal for floor grinding. A flanged, threaded spindle with separate collars against which grinding wheels are secured. The spindle may be belt driven or have a direct motor drive.
Bench and floor (pedestal) grinding	The process of cutting away material with grinding wheels. The grinding of cutting, fastening, forming, and other small tools (like tool bits, drills, chisels and punches). Sharpening or forming a part to a desired shape or size. Grinding by hand with the aid of a work rest on a bench or floor grinder.
Trueness	The *concentricity* of the outer surface (periphery) of a grinding wheel in relation to its axis.
Truing	The process of cutting away the outer surface (periphery) of a grinding wheel. Producing a grinding wheel that runs concentrically and true.
Loaded wheel	A condition in which a great number of particles of the material being ground become lodged in the grinding wheel. A grinding wheel condition that (1) limits the quality of the grinding process, (2) produces additional heat, and (3) reduces the cutting efficiency.
Mechanical dresser (star dresser)	A series of hardened star- or corrugated-shaped discs with a circular separator between each pair. Discs that rotate freely in a holder when brought into contact with a grinding wheel face. A common hand dressing tool used to true and recondition a grinding wheel. Mechanically producing sharp cutting faces on abrasive grains.

SUMMARY

- Hardened cutting, forming, fastening, layout, and other tools and parts require grinding. Grinding reduces a tool to a correct form, size, and efficient cutting or operating condition.
- Bench and floor grinders are made to accommodate a wide variety of grinding wheel sizes. The grinding processes range from *finish grinding* small surfaces, to *rough snagging* on welded, forged, cast, and other manufactured parts.
- Bench and floor (pedestal) grinding is a cutting process. The cutting tool consists of a series of sharp-cutting abrasive crystals (grains). Each grain removes chips of material.
- Workpieces and tools are ground by holding the part against a work rest. The surface may be guided straight across the face of the grinding wheel. The part may also be held at an angle or turned. A desired angle, point, flat, or convex surface is produced by manipulating the tool or part by hand.
- Bench and floor (pedestal) grinders usually have one coarse and one fine wheel. The coarse wheel is used to remove metal as quickly as possible for rough grinding. The fine wheel produces a fine-grained surface finish and a good-quality cutting edge.
- Grinder spindles are threaded right- and left-hand. These prevent the clamping nut and grinding wheel from loosening during the grinding process.
- Grinding wheels must be true and in balance to grind properly and produce quality finished surfaces.
- Care must be taken in mounting and securing a grinding wheel. The bore should be the exact size of the spindle. *Blotters* should be placed between the

faces of the flanges and the wheels. The clamping nut should be tightened carefully. Apply only enough force to prevent the wheel from turning on the spindle during the grinding process.

- Wheel dressers of the abrasive stick or mechanical type are used to dress loaded wheels. Dressers are used to true the face and produce new sharp-cutting faces on the abrasive grains.

- Chatter marks and vibration are signs that a wheel must be *trued*. Excessive heating, burn marks, and load lines indicate that a wheel must be *dressed*.

- Bench and floor grinders are practical for hand grinding angular, flat, and conical tool points and edges.

- Flying abrasive and metal particles and the possible fracturing of an abrasive wheel require that grinders be adequately guarded. Safety shields must be in place. Protective goggles should be worn at all times.

UNIT 18 REVIEW AND SELF-TEST

1. Compare the *stoning* of burrs with the removal of burrs by filing.
2. Describe the mounting of grinding wheels on bench and floor grinders.
3. State why a six-bearing hole hexagon hand dresser is used for dressing large, heavy-duty grinding wheels.
4. Indicate three unsafe conditions that require the worker to dress a grinding wheel.
5. List the main steps for offhand grinding the flat face on a tool.
6. Tell why a *partial auxiliary view* is generally used in addition to other regular views of parts with angular features.
7. Give three safe practices to follow before starting a bench or floor pedestal grinder.

PART FOUR

DRILLING MACHINES: TECHNOLOGY AND PROCESSES

SECTION ONE

DRILLING MACHINES AND ACCESSORIES

This section describes the design features and functioning of conventional and unconventional drilling machines. Conventional types use "conventional" hole-making processes and tools such as drills and reamers. Unconventional types employ "unconventional" hole-making processes that involve newer technologies such as ultrasonics, laser beam, and electrochemical cutting "tools". Also described in this section are the work-holding devices and accessories commonly used in hole-making processes.

UNIT 19
NONPRODUCTION AND PRODUCTION DRILLING MACHINES AND PROCESSES

Holes are an essential feature in design and manufacture. Holes are produced in simple parts as well as in complex machines. The term *hole making* relates to machining processes that:

• Produce a hole of any size or shape where none previously existed

• Alter the size of a hole, as in reaming or boring

• Enlarge part of a regular hole, as in countersinking (at an angle) or counterboring (to produce a recessed surface), or an irregular-shaped hole

• Tap and form internal threads

• Grind and produce a hole to precise dimensions.

These are general processes performed on both nonproduction and production drilling machines. The work must be positioned and held securely for all processes. Various types of cutting tools and drilling machines are required. This unit deals with the basic work-holding devices, machine accessories, cutting tools, and *conventional* drilling machines. Brief descriptions are also given for newer hole-producing methods and *unconventional drilling machines*.

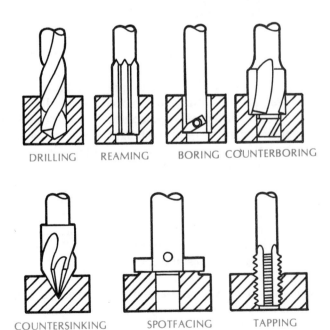

DRILLING   REAMING   BORING   COUNTERBORING

COUNTERSINKING   SPOTFACING   TAPPING (THREADING)

Fig. 19-1. Basic drilling machine processes.

CONVENTIONAL DRILLING MACHINES

To review, *hole making* refers to many different processes. Some of these are *drilling, reaming, counterboring, countersinking, spotfacing,* and *tapping* (Fig. 19-1). These processes are performed on a group of machines that are identified as *drilling machines*. Some drilling machines are light and simple. The operations are controlled by the operator, who feeds a cutting tool into the work. Drilling machines are usually of the bench or floor type. Other drilling machines are larger, for heavy-duty operations. In mass manufacturing there are multiple-spindle drilling machines. These are used for the rapid production of many holes.

Some work may be brought conveniently to the machine and positioned for drilling and other operations. Large, heavy, cumbersome parts that are cast, forged, or produced by other methods are secured on the machine table. The hole making operations are performed by positioning the movable tool head and the cutting tool at a desired location. The *radial drill press* is used for such operations.

Most holes are machined *straight,* at a right angle to a surface. Other holes are drilled at various angles. Some require *deep-hole drilling*, in which the workpiece is rotated instead of the drill. Drilling machine tables may be stationary or adjustable for angle operations. Hole making in each instance is *conventional*.

Six different groups of conventional machines are described in this unit:

• Bench and pedestal (floor) sensitive and heavy-duty drilling machines

• Multiple-spindle types, for production

• The deep-hole drilling machine

• The radial drill press, for operations requiring a long overhanging arm and a machine head that may be swung through a large radius

• The multi-operation turret drilling machine

• A *numerically-controlled unit,* the operations of which are programmed.

Sensitive Drill Press (Bench And Floor Types)

The simplest drilling machine design and the

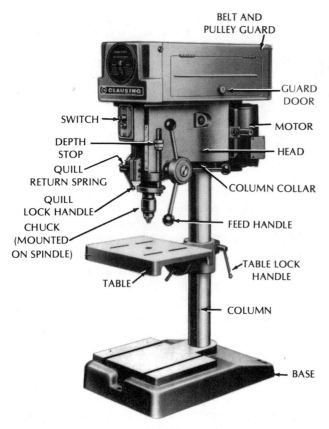

Fig. 19-2. Features of a sensitive drill press (bench model).

easiest to operate is a *sensitive drill press* (Fig. 19-2). When the length of the drill press column permits the machine to be mounted on a bench, it is referred to as a sensitive *bench* drill press. The same machine is also produced in a floor model. This model is referred to either as a sensitive *pedestal* or sensitive *floor* drill press.

The machine size indicates the maximum work diameter that can be accommodated. For instance, a 24″ drill press will accommodate a workpiece with up to a 24″ diameter (12″ radius). The speed range of the sensitive drill press permits drilling holes from 1/2″ down through the smaller fractional, number, letter and metric drill sizes. The smaller the drill size, the higher the speed required. Bench and floor drill presses are widely used for drilling, reaming, boring, countersinking, counterboring, spotfacing, and other operations.

Major Components Of The Sensitive Drill Press

Regardless of the manufacturer, there are four major components of the sensitive drill press:
- Base
- Work Table
- Head

- Drive Mechanism.

Machine Base—The *base* of most drilling machines has a stationary upright column. Its axis is at a right angle to the base. The upright column and base are rigid. The workpiece may rest on the machined surface. Long parts may be accurately positioned and held on the machined base.

Fig. 19-3. An adjustable drill press table with elongated slots.

Work Table—Mounted on the upright column is a *work table*. This may be moved vertically to accommodate different work lengths. The work table is then locked at the required height. The work table may be held *square* in relation to the spindle, or it may be adjustable for angle work. A common type of adjustable table is shown in Fig. 19-3. The tables have either elongated slots or grooved slots. The grooved slots provide a surface against which *T-head* bolts or nuts may be tightened. The slots permit locating the work in relation to the spindle. The work table may be round, square, or rectangular.

Drill Press Head—The *head* includes a sleeve through which a spindle moves vertically. The spindle vertical movement is controlled by a feed lever that is hand operated. Where there is a power feed, the spindle may be moved up and down by both hand and power.

The end of the spindle has a *taper bore*. Drilling attachments, chucks, and cutting tools have a corresponding *taper shank and tang*. There are *adapters* to fit the taper bore. The adapters accommodate tool shanks or other parts that have a smaller or larger taper.

Drill chucks are a practical and widely used device for holding straight-shank tools and pieces of round stock. The chuck jaws are adjustable over a wide range of sizes. For example, one drill chuck may hold all diameters from a #60 drill to 3/8" (or 10 mm). The cutaway section of a drill chuck in Fig. 19-4 shows the internal design and operation. A typical adapter is also included. The shank (long taper) fits the spindle. The short taper is fitted to the taper bore of the drill chuck.

Drive Mechanism—The *spindle* may be turned at different speeds (revolutions per minute) by a *V-belt drive mechanism*. Such a drive between a drive motor and the spindle is shown in Fig. 19-5. The required operating speed depends on the material, the nature and size of the operation, the type of cutting tool, and the desired surface finish. Speed adjustments are made by changing the *step-cone pulley* over which the V-belt rides. The range of bench and floor drill press speeds is from 30 to 5,000 revolutions per minute.

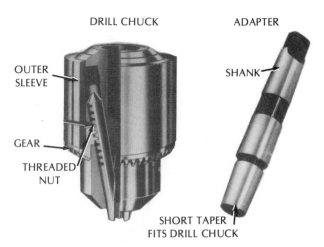

Fig. 19-4. A drill chuck with a spindle adapter.

The Standard And Heavy-Duty Upright Drill Press

The *standard upright* drill press is of heavier construction than the sensitive type. In the shop the standard upright drill press (drilling machine) is commonly called just a "drill press." The column may be round, cast, or fabricated in a rectangular form. Speed variations on the lighter upright drill presses are obtained by a variable speed drive like the one

Fig. 19-5. A five-step grooved pulley arrangement for a V-belt drive.
(Courtesy of the Clausing Corp.)

Fig. 19-6. A quick-setting variable speed drive. (Courtesy of the Clausing Corp.)

illustrated in Fig. 19-6. The operator sets the variable speed control at the desired speed.

Heavy-duty drilling machines use a *gear drive system* (Fig. 19-7). The selected speed is obtained by shifting gear selector handles while the machine is stopped. Different combinations of gears are thus engaged. The range of speeds is increased further by using a two-speed motor. This provides speed versatility for a *low-speed range* and a *high-speed range*.

The work table may be round or rectangular. Since it is heavy, the work table is raised or lowered by a *table adjusting crank*. This crank engages a thread and gear mechanism. A coolant pump and system are included to circulate cutting fluids. A coolant is essential for heavy-duty operations in which a great deal of heat is generated.

The floor-type standard drill press has both a manual (hand) feed and an *automatic mechanical feed*. A variable feed range from 0.002″ to 0.025″ is general. The *feed control lever* is set at the required feed per revolution. The cutting tool automatically advances the feed distance at each complete turn.

The Gang Drill

The *gang drill* is one type of multiple-spindle

drilling machine (Fig. 19-8). *Gang* refers to a number of spindles that are lined up in a row. The spindles may be driven individually or in the gang.

The number of spindles is determined by the number of operations to be performed. Holes in a workpiece may be drilled, reamed, bored, tapped, etc. The gang drill is often used for a series of drilling operations. Each hole may be of a different diameter. The combinations are endless.

One or more operators may perform the operations at one gang drill. The work is started at a beginning station. It is moved to the second spindle *(station)* for the next operation. The work is moved from spindle to spindle until all the operations are completed. Each cutting tool is usually positioned in relation to a specific dimension. *Drill jigs* or special work-holding *fixtures* are used for this purpose. On especially large drilling machines, the table on which the work is held may be positioned by hydraulic feed mechanisms that are numerically controlled.

Gang drills have an important advantage over other drilling machines. The work may be moved easily from spindle to spindle.

The sizes of cutting tools, the nature of the operations, the range of speeds and feeds required, and the material to be cut are major considerations. These determine the machine size, motor drives,

GEAR DRIVE
CONTROL
LEVERS

Fig. 19-7. A heavy-duty floor type drilling machine.
(Courtesy of Do All Company)

speeds, and whether the feed will be manual, mechanical, or hydraulic.

The Multiple-Spindle Drilling Machine

The *multiple-spindle* (or *multi-spindle*) *drilling machine* is a mass-production hole-making machine (Fig. 19-9). It consists of a number of spindles. These may be positioned at desired locations. Each spindle has a universal joint. The driving section may be expanded or brought closer together to change a spindle to a new location.

It is important that a *guideplate* (mounted with the drill head above the work) or a drill jig be used in multiple-spindle drilling operations. The drill lengths are also staggered to distribute the starting load.

Way-Drilling Machine

The *way-drilling machine* is an adaptation of the multiple-spindle drilling machine. The difference is that the spindles of the way-drilling machine operate from more than one direction. The production of automotive engine blocks is an example. A series of holes is produced simultaneously on several sides.

The Deep-Hole Drilling Machine

One big problem in drilling deep holes is to produce a hole that continues to follow a required axis. The longer the hole, the greater the tendency for the drill to be *deflected. True, straight holes* may be drilled by holding the tool stationary and rotating the work. Using this method, the cutting tool follows the axis of rotation. A straighter and more precisely drilled hole is produced.

The drill is supported and guided by a *guide bushing*. It is necessary in deep-hole drilling to keep a coolant on the drill point. The chips must be removed continuously. The drill should be withdrawn after the drill has advanced to a depth about equal to its diameter. On the larger deep-hole drilling machines, the drill is drawn out of the workpiece by an hydraulically operated head.

Special drills are available that permit a cutting fluid to flow through the drill body to the cutting edges. The fluid cools the cutting edges, drill, and workpiece. At the same time, chips are forced out of the drilled hole.

The Radial Drill

The *radial drill* name is derived from its design. A radial drill, with the major components identified, appears in Fig. 19-10. In place of a stationary tool head consisting of a spindle and drive mechanism, the tool head on the radial drill is *movable*. The tool head moves along a heavy arm that extends from a large-diameter column. The arm and tool head may be swung through a *radius* of a few feet on the small machines, to ten and more feet on extremely large machines and special applications of the radial drill. The arm may be swiveled around a large segment of the column.

The heavy workpiece is positioned and strapped to the base. The tool head may be moved with a handwheel. It may also be motor driven rapidly to near the desired hole location. The cutting tool is usually brought by hand to the exact point where an operation is to be performed.

The spindle is moved to the work until the cutting

Fig. 19-8. An eight-head (station) gang drilling machine. (Courtesy of the Clausing Corp.)

tool is almost in contact with the work. The automatic feed is engaged. Some machines have automatic *trips* to regulate the depth. A predetermined depth may be set. As the spindle and cutting tool reach the required depth, the feeding stops.

The radial drill may produce *straight holes* or angular holes. The holes may be located from a direct layout on the workpiece or with a *template* or a drill jig.

The Turret Drilling Machine

The *turret drilling machine* has a number of drill head units. The *turret* of the drilling machine may have six, eight or ten separate positions (stations). Fig. 19-11 provides an example of a six-station turret

drilling machine. A different cutting tool or size may be inserted in each *head*. For example, a hexagonal turret head may be set up with a combination of center drill, pilot drill, larger-sized drill, spotfacer, counterbore and reamer.

This arrangement represents the sequence in which the hole is started with the center drill, drilled with the pilot drill at a high speed, followed with the reamer drill, spotfaced, counterbored to a particular depth, and, finally, reamed.

The turret drilling machine eliminates tool-changing time. Also, a single machine may be used for all processes. The smaller-sized machines may be hand fed. The larger machines, on which some other operations like face milling are performed, may be manually operated. Other turret drilling machines may be *tape operated* by a numerical control unit.

The Numerically-Controlled Drilling Machine

A turret drilling machine or any other production machine may be *tooled up* to perform a combination of hole making and additional machining operations. These operations may be performed automatically according to a sequence of directions that are *programmed*. In *programming,* most of the mental and physical processes the craftsperson follows in setting up and machining a part must be recorded. Such information as the type and sequence of each operation, the dimensional position of each operation, the tool number, the depth of cut, the cutting

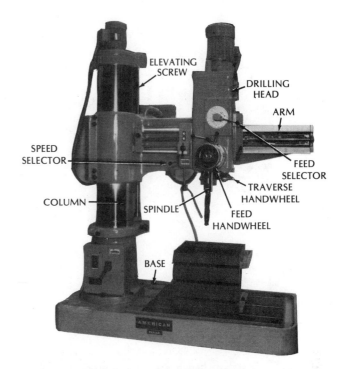

Fig. 19-10. The features of a radial drill.

speeds, the rate of feed, the speed of the spindle, all must be coded in a series of letters and numbers. The letters and numbers, when transferred to a tape, provide directions for the numerical control system. The tape has a different combination of perforations for each letter and number that has been punched into it.

The drilling machine or other machine tool is controlled by the numerical control system. As the tape moves, the workpiece and cutting tool are positioned in proper sequence. The operations are performed to the required depth. The degree of dimensional accuracy is ± 0.001′ (± 0.02 mm). After all operations are performed, the workpieces are changed by the operator. The table, spindle, and other movements, and the various feeds and speeds are all numerically controlled. The system uses electronic, hydraulic, pneumatic, and other automatic mechanical devices. Each device is activated by the tape and the numerical control unit.

It requires a craftsperson to initially set up the tools and the work-holding fixtures and devices. During the processes, the cutting qualities of each tool, the surface finish, and the accuracy of each operation must be observed by the operator. The setup person, programmer and operator must know the technology and drilling processes. This knowledge is essential to program effectively and to actually operate a numerically-controlled machine tool.

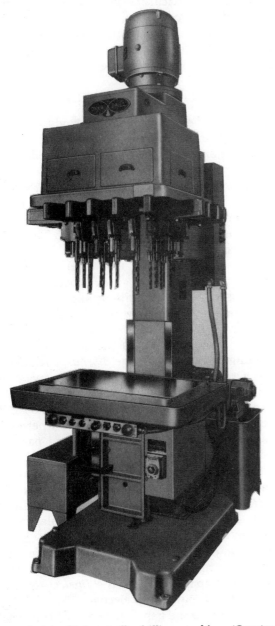

Fig. 19-9. A multiple-spindle drilling machine. (Courtesy of South Bend Lathe, Inc.)

Fig. 19-11. A six-station, bench model turrot drilling machine. (Courtesy of Burgmaster Division, Houdaille Industries, Inc.)

Special Hole-Producing Machines

The *drilling head* is a versatile unit. One head may be used, or any number of heads may be combined. They may be positioned vertically, horizontally, or at a simple or compound angle. A single operation like drilling may be performed. Other operations may be combined to include drilling, reaming, boring, etc.

The workpiece may be positioned and held in a drill fixture. The part may also be held in a drill jig with a *drill bushing* to guide the drill. The table may be round, rectangular, or rotary. The combination of machine and work-holding devices and cutting tools depends on the special requirements of the job. *Special hole-making machines* incorporate many of these quantity production requirements and design features.

UNCONVENTIONAL DRILLING MACHINES

The six basic groups of *conventional* drilling machines are contrasted with some of the newer hole-producing processes. These processes require different machine design features, cutting tools, accessories, and cutting fluids. The processes use ultrasonic, laser beam, chemical, electrical discharge, and other forms of energy, techniques and media. This group of hole-making machines and processes is called *unconventional*.

Unconventional machine tools produce holes with-

out using regular drills, reamers, and other such cutting tools. Harder metals and ceramic materials may be penetrated. Irregular-shaped holes may be produced with comparatively inexpensive *unconventional tools*. Some of these tools are consumed in the process. Problems of heat effect and warpage of the work part are eliminated.

Unconventional hole-producing machines are an answer to the machining of intricate and irregular-shaped holes. Unconventional machines may also produce straight, cylindrical holes in extremely hard and tough metals and other materials.

A few unconventional machines are described. These require electrical energy, and chemical, sound, velocity, and laser beam phenomena. The descriptions show some of the unique features and differences between unconventional and conventional drilling machines and processes.

Electrical Discharge Machine (EDM)

The *electrical discharge machine* and process was industrially accepted and marketed beginning in 1946. Holes are produced on an electrical discharge machine without using a movable tool and without tool contact with the workpiece. *Electrical discharge machining* (EDM) uses electrical energy to remove metal of any hardness. Holes that have an irregular and intricate contour may be produced. There is no friction-produced heat, so there is no warpage. Tolerances of ± 0.002″ (± 0.05 mm) are practical.

The *cutting tool,* called an *electrode,* is made of either brass, copper, graphite or alloys of copper or silver. The tool is hollow to permit pumping a *diaelectric* through it.

) An oil *diaelectric* covers the workpiece during machining. It helps flush the *discharge* (chips) and acts as a coolant and as the electrical conductor. The principle of EDM drilling is illustrated in Fig. 19-12.

A good, quality surface finish is produced. EDM is especially valuable in producing holes where other processes may distort the material and part. Machining a cluster of holes having a thin wall thickness between the holes, like 0.005″, is another example. The holes may be *through holes* or *blind holes.* Materials such as carbides and stainless steels, as well as softer and nonferrous metals, may be worked by EDM.

Electrical energy is required in EDM. The tool is positioned so there is a gap between it and the work surface. The gap is filled with the *diaelectric fluid.* A high-frequency pulsating electric current creates sparks that jump the gap. The bombardment vaporizes the material under the tool, producing the desired hole. The tool is consumed during the process.

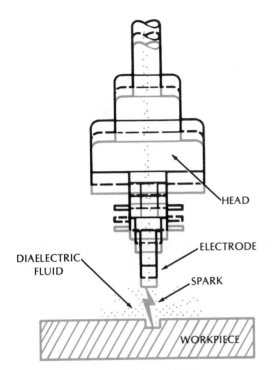

HEAD

ELECTRODE

DIAELECTRIC
FLUID

SPARK

WORKPIECE

Fig. 19-12. The principle of EDM drilling.

There are variations of EDM in which there is a constant reversal of polarity. This reduces wear on the tool and produces a higher removal rate. However, because the surface finish is rougher, this technique is used for roughing operations.

Electrochemical Machining (ECM)

There are many similarities between *electrochemical machining* (ECM) and EDM. The workpiece may be of any metal capable of conducting electricity. A tool made of material similar to that used for EDM is required. The tool acts as the *cathode*. The workpiece serves as the *anode*. An electrolyte is passed through the gap between the tool and the work at a high rate of flow (velocity). The metal is dissolved and removed by using direct current (DC). A low voltage (30 volts) and high current (2,000 amperes per square inch) are required.

ECM is a faster metal-removing process than EDM. Additional operations may also be performed by ECM, such as etching, deburring and face milling.

Ultrasonic Machining (USM)

Ultrasonic machining (USM) operates on the principle of utilizing high-frequency, low-amplitude vibrations. (The frequency is above the audible range.) The ultrasonic vibrations produce holes and other cavities.

USM requires a tool of a ductile, tough material similar to that used in EDM. A water abrasive mixture flows between the tool and the work. Cutting action is produced by the impact of the abrasive particles.

Holes up to two inches and deeper may be ultrasonically machined. Such holes are machined burr-free and to tolerances of ± 0.001″ (± 0.02 mm). EDM hole-making processes may be applied to hardened steel, carbides, diamonds and other gems, ceramic and other exceedingly hard materials.

Laser-Beam Machining (LBM)

Laser-beam machining is used to machine precise, accurate holes of 0.005″ to 0.010″ (0.1 to 0.25 mm) diameter in refractory metals, ceramics and thin materials. Such holes are produced without warpage or cracking by laser-beam machining. The hole-making process is accomplished by melting and vaporizing the work with a laser beam of intense single-directional light.

Abrasive Jet (AJM) And Plasma Arc Machining (PAM)

Abrasive jet machining (AJM) uses an extremely high-speed stream of abrasives to do the cutting. These are transported to the work surface in a gas. The process is used to cut intricate shapes in thin, hard metals and other materials in which heat must be avoided.

Plasma-Arc Machining uses a high-velocity jet of a high-temperature (10,000°F) ionized gas. The jet cuts through metals up to six inches thick and is useful in hole making. The surfaces produced are smooth.

This group of *unconventional* machines and techniques of hole making make it possible to:

● Produce intricate shapes that cannot be economically machined on other machines
● Penetrate hard, difficult materials to machine
● Machine holes to high degrees of dimensional accuracy and quality of surface finish.

WORK-HOLDING DEVICES

In a conventional hole-drilling operation the hole may be layed out. The hole location is center punched. The workpiece is then secured on a drill press table and the hole is drilled. The *accessory* in which the work is positioned and held may be a *work-holding or clamping device.* Some of the common methods of positioning and holding a workpiece include:

● Clamping directly in a vise. The size of the part

and the nature of the operation determine how the vise should be held. The vise may be held by hand when there is just a slight twisting force, as when drills smaller than 1/4″ or 6 mm are used. Heavier drilling requires the use of a *stop*. The workpiece is clamped when a large twisting force is to be exerted.

● Mounting on parallels or *V-blocks*. The part is then clamped in position with *hold downs, straps, clamps* and *T-bolts*.

● *Nesting* and holding in a specially designed form. The drill or cutting tool may be *guided* for location by a *drill bushing*. The drill bushing also helps produce a concentric hole. Another holding device is called a *drill fixture* or *fixture*.

Work-Holding Vises

Workpieces of regular shape (like square, rectangular and round) may be held in a standard *drill press vise* (Fig. 19-13). After the hole location is layed out, the work is placed on parallels. The work then is securely tightened in the vise and positioned on the drill press table for the hole-making operation. The

Fig. 19-14. A universal (three-way) vise. (Courtesy of Universal Vise & Tool Company)

Fig. 19-13. A multi-purpose drill press vise. (Courtesy of Do All Company)

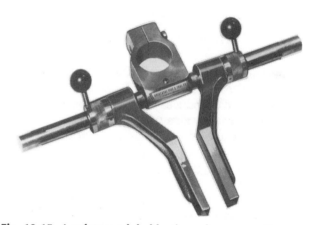

Fig. 19-15. A safety work holder for column-mounting on a drill press. (Courtesy of Universal Vise & Tool Company)

drill press vise should be clamped to the work table for heavy drilling operations.

An *angle vise* is used for some angular machine operations. The vise holds the work at the required angle, eliminating the need for tilting the work table. Compound angles require that the work be positioned and held in a *universal machine vise* like the one shown in Fig. 19-14. This vise permits three separate angular adjustments: 360° in a horizontal plane, within 90° from a horizontal to a vertical position, and through 90° in the third plane.

Although there are other vise designs, the standard, angle, and universal vises are the three basic types.

The *safety work holder* (Fig. 19-15) is another holding device. It fits around the machine column and has two easily adjusted arms. These hold the work in position on the work table.

Strap Clamps And T-Bolts

Irregular–shaped workpieces are sometimes held directly on the machine table. A number of differently shaped *clamps* (straps) are used for this purpose. Each clamp has an elongated slot. A T-bolt fits in the slot or in a machined groove in the table and passes through the clamp. The opposite end of the clamp is raised by *blocking*. It is then positioned to apply a maximum force on the workpiece.

The flat surface of the clamp should be parallel to the table. This permits the force applied with the T-bolt to hold the workpiece securely. Also, the workpiece is not forced out of the position required for the machining operation. Examples of square head, cut-away head, and tapped T-head bolts are shown in Fig. 19-16.

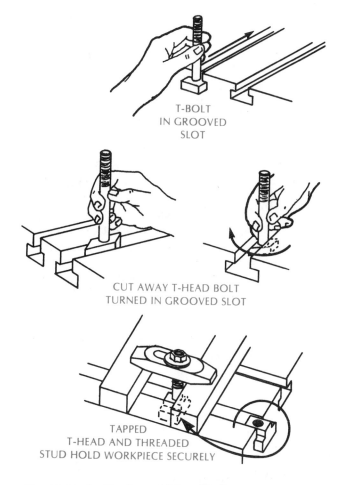

T-BOLT
IN GROOVED
SLOT

CUT AWAY T-HEAD BOLT
TURNED IN GROOVED SLOT

TAPPED
T-HEAD AND THREADED
STUD HOLD WORKPIECE SECURELY

Fig. 19-16. Applications of T-head bolts.

The judgment of the craftsperson is important. The force exerted in the clamping process should not cause the part to be bent or fractured. Applying excess force causes such conditions.

The *blocking* to raise the strap to a correct height may be done in several ways. A *step block* with a number of steps of different heights is produced for this purpose. Sometimes, sets of *blocks* are used. These may be added together to reach a particular height. An *adjustable jack* permits even greater variation in height and makes adjustment easier.

The four common shapes of strap clamps, each of which has a number of variations, are:

- Straight strap clamp
- Finger strap clamp
- Bent strap clamp
- U strap clamp

These are commerically produced from high-quality steel. The clamps are forged to shape and are heat treated. They are available in many sizes to accommodate a great range of work sizes. Although described here in terms of drilling machines and processes, these clamps are widely applied through-

out industry. The clamps are used for bench, machine and assembly processes wherever work parts must be held securely. The terms *strap, clamp,* and *strap clamp* are used interchangeably.

Straight Strap Clamp—The straight strap clamp is a rectangular bar with an elongated slot running lengthwise (Fig. 19-17). The two ends are flat. One end rests on the workpiece; the other, on the blocking. The necessary force is produced by tightening the T-bolt washer and nut assembly. Some straight strap clamps have an adjusting screw on one end.

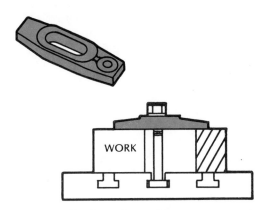

WORK

Fig. 19-17. Application of a straight strap, a T-bolt, and a packing block.

Finger Strap Clamp—It is not practical to have a clamp located on the top surface of all workpieces. In this position the clamp interferes with work processes. Some such workpieces have slots or holes in the sides. The *finger clamp* has either a flat or round *finger* (Fig. 19-18). The finger may be inserted into a slot or hole in the workpiece side, for side clamping.

Fig. 19-18. A finger strap, for side clamping.

Fig. 19-19. A bent-tail strap (gooseneck clamp).

Bent Strap Clamp—One common type of bent strap clamp is called a *bent-tail* or *gooseneck* clamp (Fig. 19-19). One end is offset for convenience in clamping on a surface and to lower the height that the T-bolt and nut projects. Sometimes the top and side of the offset help *nest* against the workpiece.

Another form of bent strap clamp is curved. One end of the arc rests on the work surface. The other end is clamped on a protective shim on the table or a block.

U-Strap Clamp—The *U-strap clamp* provides greater movement so that the clamp may be brought close to the workpart. The clamp has an open end (Fig. 19-20). This permits the clamp to be positioned by simply loosening the T-bolt. The clamp may then be secured in a desired position.

Fig. 19-20. A U-strap with a finger.

The Angle Plate

The *angle plate* that is used on a drilling machine usually has a number of holes or elongated slots. These permit strapping the angle plate to the table and securing the workpiece to the right-angle face. Fig. 19-21 shows a typical angle plate setup. Angle plates are machined precisely. After a workpiece is layed out, it is accurately positioned and strapped to the angle plate. The underside of the part is *jacked up*. This is a necessary precaution to ensure that the force of the machining operation does not cause the workpiece to shift position.

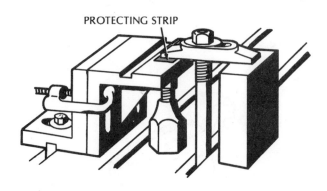

PROTECTING STRIP

Fig. 19-21. An angle plate setup.

Drill Jigs And Fixtures

These were described earlier as devices for positioning and holding the workpiece. Drill jigs and fixtures are widely used with workpieces of irregular shape when a quantity must be machined accurately.

A *drill fixture* is a specially designed holding device. The fixture contains a *nesting* area into which the workpiece fits. This arrangement positions and holds the workpiece. The relationship is fixed between the workpiece, the spindle of the machine, and the cutting tool.

A *drill jig* is an extension of the drill fixture. It serves the same functions and, in addition, has *guide bushings*. These bushings guide cutting tools such as drills and reamers to a precise dimensional location. The guide bushings help produce accurate concentric holes. Fig. 19-22 shows a simple drill jig for accurately drilling a hole in a cylindrical part.

The drill jig and drill fixture provide fast methods of obtaining exact hole locations. The parts do not require a layout or center drilling. Also, *runout* of the drill point is avoided.

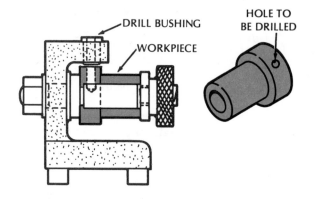

HOLE TO BE DRILLED

DRILL BUSHING

WORKPIECE

Fig. 19-22. A simple drill jig.

Accessories

The term *accessories* refers to the tools and the mechanical parts and machines that are used in conjunction with a machine tool. These aid in setting up and holding a workpiece so that a particular process may be performed. Four of the accessories that are widely used in hole-producing operations are: *parallels, step blocks, shims* and *packing blocks,* and *universal swivel blocks.* The clamps (straps), bolts, and V-blocks, which are also accessories, were described earlier.

Parallels—Parallels, or *parallel bars,* are square or rectangular bars. They are made of steel or granite

PARTS FOR ADJUSTABLE
JACK COMBINATIONS

SPACER
TUBES

TUBE
COUPLINGS

SWIVEL BASES

GOOSENECK
STRAPS

STRAIGHT
STRAPS

ASSEMBLED
JACK AND STRAIGHT
"U-" STRAP

Fig. 19-24. Parts for adjustable jack combinations. (Courtesy of Universal Vise and Tool Company)

(Fig. 19-23). Precision steel parallels are machined, heat treated, and ground accurately for parallelism and size. The bars are mostly used in pairs.

The workpiece is seated against the parallel faces for alignment purposes. Parallels are also used to raise the workpiece in a vise or other holding device, for convenience in performing a required operation. The clearance between the workpiece and the table or vise is important. This is the case in drilling, reaming, tapping, and other operations in which the point and the cutting edges of the tool must cut *through* the workpiece.

Jack Screw—The jack screw, or *jack*, is a simple, easily adjusted support device. The jack is designed to provide for a wide range of adjustment. Thus, it is possible to make up whatever length is required between blocks, a machine table, and a work surface. All jacks consist of a base, an elevating screw, and a swivel head. Jacks are placed under a workpiece and brought into contact with the surface that is to be supported. A jack may also be used under a clamp to replace a number of blocks.

Combination sets of jacks and straps, like those in Fig. 19-24, can be used for a wide range of clamping applications. The jack is brought to the required height. A T-bolt, which extends through the U-strap, is used to apply force on the workpiece.

Step Blocks and Packing Blocks—The name *step block* is derived from the fact that each block consists of a series of steps. These permit the end of a clamp to be located at whatever height (step) is needed for the job. The step also provides a surface against which a clamping force may be applied.

Fig. 19-23. Parallel bar sets.

Packing blocks are individual blocks of different sizes. The blocks may be used alone or in any combination. When a desired height is reached, a force may be applied correctly on the workpiece. Examples of step blocks and packing blocks are shown in Fig. 19-25.

Shims and Protecting Strips—Shims are thin metal strips. They are used to compensate for surface variations in measurement. Since it is necessary to have a good bearing surface so that the workpiece will not move during a cutting process, *shims* may be inserted between the workpiece and a parallel. Shims *level* an uneven surface.

Protecting strips are thin metal pieces. These are

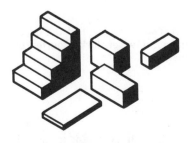

Fig. 19-25. Step blocks and packing blocks.

221

placed under any metal part, like a clamp. Protecting strips are used to prevent damage to a finished work surface or machine table when a force is applied in a workholding setup.

Universal Swivel Block—Many surfaces that are tapered or of irregular shape must be held securely. A *universal swivel block* is used with a vise for such purposes (Fig. 19-26). The universal swivel block consists of two major faces. One face on the body is flat. This face rests against the stationary jaw of the vise.

Fig. 19-26. An irregular workpiece held with a universal swivel block. (Courtesy of Universal Vise & Tool Company)

The other face has two *buttons*. This face is adjustable. It may be positioned against an irregular or tapered surface. The universal motion permits the swivel block to accommodate the shape of the workpiece. The universal swivel block with the univeral motion is shown in Fig. 19-27.

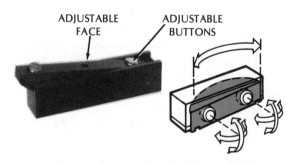

Fig. 19-27. Universal movement of the adjustable face and buttons of a universal swivel block. (Courtesy of Universal Vise & Tool Company)

V-Blocks

V-blocks are widely used to position and hold cylindrical parts for layout, machining and assembly operations. A *V-block* is a rectangular metal solid. It has V-shaped angular surfaces on at least two opposite sides (Fig. 19-28).

Round workpieces of different diameters may be centered and nested in V-blocks. A *U-shaped* clamp is provided to hold the workpiece in position. Some V-blocks are designed with a hole in the center of each block. This makes it possible to drill through holes without damage to the V-grooves and block.

V-blocks are made in pairs. Many kinds of V-blocks are used with machine and bench vises and other work-holding devices for round parts. Their function in each case is to position and hold a workpiece.

When a hole is to be drilled through the axis of a cylindrical workpiece, a centerline is usually scribed on the end. The workpiece is mounted in V-blocks. The centerline (axis) is then aligned with a steel or other square head. The square is held against the surface of the machine table or other layout plate. The alignment for the workpiece is shown in Fig. 19-29.

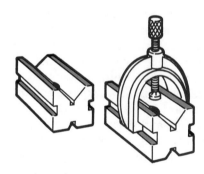

Fig. 19-28. A V-block set and a U-clamp.

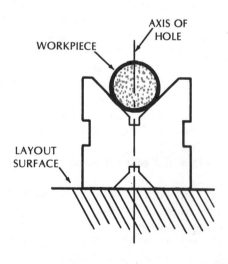

Fig. 19-29. Aligning the centerline of a hole to be drilled.

SAFETY PRECAUTIONS IN DRILLING MACHINE OPERATION AND MAINTENANCE

- Check the table surface, work holding device, and the workpiece. Each must be free of chips and foreign particles. Cleaning should be done with a brush or a wiping cloth with the machine turned *off*.
- Lock a movable table to the column prior to performing any hole-producing operation.
- Check the table surface for burrs and nicks. Only the required work-holding devices and measuring tools should be placed on the table.
- Place some protective material under any cutting tool that is to be removed with a drift from the spindle socket.
- Fasten each workpiece securely to withstand all cutting tool forces. This is necessary for personal safety and to avoid damage to the machine, work, cutting tool and other accessories.
- Place a protective sheet of soft metal between any rough surface of the workpiece and parallels or other machined surface.
- Tighten T-bolts to hold the workpiece in position and to prevent movement. No extra force should be applied. Excessive force may cause springing of the work or clamp, or damage to the machine table.

- Raise the bottom surface of the workpiece. The cutting tool must clear the work-holding device at all locations where through holes are to be drilled. On mechanically-fed drilling machines, set the *trip mechanism*. The down-feed must be stopped before the cutting tool cuts into the base, table, or accessories.
- Stop the machine when a feed change requires the shifting of a gear lever.
- Check to see that covers and guards are in place and secured. All extra tools and parts should be removed *before* starting the machine.
- Check personal clothing. Loose clothing may be caught by a revolving spindle, chuck or cutting tool.
- Use eye-protection goggles or a protective shield.
- Direct carefully any air blast used for cleaning the work, machine, or tools. The blast should be downward and away from any other operators. Also, cover the work and the area around it. This step restricts chips and other particles from flying.

DRILLING MACHINE TECHNOLOGY TERMS

Hole making	A series of machine processes for producing, altering, and enlarging a hole in part or total. Producing regular through and blind holes, threaded holes, and irregularly shaped holes.
Conventional drilling machine	A machine that produces holes by using standard cutting tools. Cutting action produced by revolving and feeding a standard tool into a workpiece.
Basic types of drilling machines	A classification given to upright bench and floor drill presses. Drill presses of the sensitive and heavy-duty types, and multiple-spindle, radial, and turret drilling machines, and adaptations of these machine tools.
Multiple-spindle drilling machine	A drilling machine with many spindles that may be positioned to produce a number of holes at one time. A mass-production hole-producing machine.
Deep-hole drilling	A process of rotating the workpiece and feeding a guided stationary drill. A hole drilled to a depth which exceeds approximately six times the drill diameter. A process of producing a deep hole concentric with the axis of the workpiece.
Radial drill	A machine tool for producing holes. A movable tool head extending radially from a machine column. A drilling machine with a head that may be positioned within the radius of the radial arm.
Turret drilling machine	A series of cutting tool stations on a drilling machine. The automatic positioning of the cutting tools for a preset sequence of operations. A six-, eight- or ten-sided turret that has a cutting tool positioned in each turret station.

Numerically-controlled machine tool	A system of machining in which each operation is controlled by a perforated tape. A tape that has a definite code of letters and numerals to indicate the nature, sequence and timing of machining operations. A machine tool that, once set up, is fed control information by tape. A machine tool in which all movements are controlled by programmed information. All machine components are operated by mechanical, electrical, hydraulic, pneumatic, or electronic devices.
Unconventional drilling machines	The making of irregular- and intricate-shaped holes using nonconventional tools. Properties of sound, light, heat, electricity, chemical reactions, and velocity (abrasive jet machining) are used to remove material.
Work-holding device Accessories	A vise, clamp, fixture or drill jig that is used to position and hold a workpiece. Tools, parts and mechanical devices that may be added to a machine or a setup. Parts needed to position, hold, or safely perform certain operations or to extend the range of operations.

SUMMARY

- Conventional drilling machines produce and re-form holes to various sizes and shapes.
- Standard drills, reamers, boring tools, spotfacers, counterbores, countersinks, machine taps and other cutting tools are used in conventional hole-producing methods.
- Drilling machines may be grouped according to size, number of spindles, and other design features. The groups include upright bench and pedestal drill presses of the sensitive and heavy-duty types, and multiple-spindle, deep-hole, radial, and turret drilling machines.
- The nature, sequence and processes performed on drilling machines may be programmed and numerically controlled. The machine must, however, be set up first. The cutting action must be checked regularly by the operator.
- The positioning and feeding of a cutting tool on a sensitive drill depends on the judgment and feel of the operator.
- Heavy-duty drilling machines have variable speed controls, a great number of feeds, and are equipped for both manual and automatic feeds. Coolant systems are required to lubricate and dissipate the heat developed during cutting processes.
- The gang drill, multiple-spindle, and way-drilling machines are production types of drilling machines. Many holes of the same or different sizes may be produced.
- When the depth of a hole exceeds five or six times the drill diameter, a deep-hole drilling technique is used. The work is rotated. The drill remains stationary. Guided into the workpiece, the drill produces a straight, accurate hole.

- Heavy workpieces that cannot be positioned and secured on a standard drilling machine may be secured on the work table of a radial drill. A movable head, riding on an extended arm, may be positioned to perform drilling and other operations.
- A series of machining processes may be performed on a turret head. Each operation is positioned in sequence over a workpiece.
- Intricate hole shapes may be produced accurately in thin or extremely hard metals and other materials that are affected by heat. Unconventional hole-producing processes are being widely applied in these cases.
- Unconventional drilling machines require specially designed tools. The cutting or disintegrating action is produced through:
  o Electrical discharge machining (EDM)
  o Electrochemical machining (ECM)
  o Ultrasonic machining (USM)
  o Laser-beam machining (LBM).
- Other nonconventional hole-forming processes include the shooting of abrasives at high velocity to do the cutting, as in abrasive jet machining (AJM). Plasma arc machining (PAM) requires a high velocity gas at a temperature in the 10,000°F range.
- The standard design features of the bench drill press are common to all other drilling machines.
- Regular, angle, and compound drill press vises hold and position workpieces. These vises are used for straight, single-angle, and compound-angle machining operations.
- The four most widely used strap clamp shapes are the straight, finger, bent-tail (gooseneck and arc styles), and the U-clamp. T-bolts are used as fasteners for holding parts to the machine table.

- Parallels and V-blocks provide precision surfaces. Workpieces may be positioned and held on these surfaces.
- Jacks, step blocks, regular blocks, and universal swivel blocks are common accessories. They are used for positioning and securing workpieces.
- Safety precautions must be observed in tool and machine setting, cutting, and cleaning and burring of the workpiece. The operator must pay particular attention to loose clothing or wiping cloths. These should be secured to prevent contact with an operating machine part or tool.

UNIT 19 REVIEW AND SELF-TEST

1. State three main differences between the operation of (a) a *sensitive* type *bench drill press* and (b) a *multiple-spindle drilling machine.*

2. Distinguish between *gang drilling* and *turret drilling machines.*

3. Indicate the kinds of specifications the craftsperson must code in a series of letters and numbers to transfer to a tape for a numerically-controlled drilling machine.

4. Identify (a) three nonconventional drilling machines, (b) the forms of energy used by each, and (c) the type of hole-producing tool or processes required for each.

5. Give three examples of the unique hole-producing functions of nonconventional drilling machines.

6. Cite the advantages of using a safety workholder attachment on a drill press to position and hold a workpiece.

7. Describe briefly the function served by each of the following work-holding devices or accessories as related to drilling machines:
   a. *step blocks*
   b. *adjustable jack*
   c. *gooseneck clamp*
   d. *drill fixture*
   e. *universal swivel blocks.*

8. List three safety precautions to follow in setting up for drilling, to avoid personal injury.

SECTION TWO

CUTTING TOOLS AND DRILLING MACHINE OPERATIONS

This section presents the technologies involved in and procedures for hole-making processes performed on drilling machines—drilling, reaming, threading, countersinking, counterboring, spotfacing, and boring.

Unit 20

DRILLING: TECHNOLOGY AND PROCESSES

The tool most commonly used with the drill press is the *twist drill*. This name usually is shortened to just "drill". In principle, a drill has:

• An angular point. This is where the cutting action takes place.

• A series of flutes cut along the body. The flutes permit forming of the cutting edges and the removal of chips.

• A channel through which a cutting fluid may reach the drill point.

Drilling processes are performed on metals and other materials. These vary in hardness. Holes are drilled to different depths and degrees of accuracy. Some drills are started in center-punched locations. Other drills are positioned and guided by fixtures and drill jigs.

The actual process of drilling is largely influenced by a thorough understanding of twist drills. The operator must know about the materials used in drills, the parts and functions, the grinding and checking of the drill point, relationships of drill sizes in the inch and metric systems, and speeds and feeds. This unit deals with the background technology related to twist drills, drill grinding, and drilling processes.

DESIGN FEATURES OF TWIST DRILLS

Materials

There are four common materials used in the manufacture of drills: carbon steel, high-speed steel, cobalt high-speed steel, and tungsten carbide. Although certain of the design features are similar, each material is adapted to special drilling requirements.

Carbon Steel Drills—Drills of carbon steel are used in comparatively slow-speed drilling operations. The limited amount of heat that is generated by the cutting action must be dissipated quickly. The hardness and cutting ability of a carbon steel drill is lost once the drill point is overheated and the *temper is drawn. Temper* refers to the *degree of hardness* and the *relieving of internal stresses* within a tool. Overheating causes the drill to soften and lose its cutting edges. A carbon steel drill may be identified during drill grinding by the *white sparks* in a *spangled pattern.*

High-Speed Steel Drills—High-speed drills have largely replaced carbon steel drills. As the name implies, a high-speed steel drill retains its cutting edges at high cutting speeds and at temperatures up to 1000°F. The cutting speeds are twice as fast as for carbon steel drills.

Fig. 20-1. A general-purpose high-speed steel (HSS) drill. (Courtesy of Do All Company)

High-speed steel drills (Fig. 20-1) are used in production. Such drills are versatile cutting tools for ferrous and nonferrous metals and other hard materials. The shank end of a high-speed steel drill is stamped *HS* or *HSS*. This mark identifies the metal used in the drill. When being ground, a high-speed steel drill gives off a dull red spark.

Cobalt High-Speed Steel Drills—The addition of cobalt to high-speed steel produces properties which are necessary for tough drilling operations. Cobalt high speed drills are used for drilling tough castings and forgings, armor plate, and work-hardened stainless steel, silicon chrome, and other hard steels. Cobalt high-speed steel drills are run at high cutting speeds. A steady, uninterrupted feed is used. Such drills may be operated at higher temperatures than high-speed steel drills. Fig. 20-2 illustrates a heavy duty cobalt high-speed steel drill.

Fig. 20-2. A heavy-duty cobalt high-speed steel drill.

Tungsten-Carbide Tipped Drills—The *carbide-tipped* and the solid carbide drill are used for drilling abrasive materials such as glass-bonded plastics; brass, aluminum, bronze, and other nonferrous metals, and hard-scaled castings.

A *carbide-tipped* drill is produced by altering the

spiral flute of a high-speed steel body. A tungsten carbide cutting tip is inserted and the clearance angle is changed. A common two-flute carbide-tipped drill is shown in Fig. 20-3. Carbide-tipped drills must be sharpened on special grinding wheels. A silicon carbide or diamond impregnated grinding wheel is used.

Fig. 20-3. A two-flute carbide-tipped drill.

Types Of Twist Drills

Two-Flute Twist Drill—The most commonly used twist drill for all-purpose and production drilling is the two-flute high-speed steel drill (Fig. 20-4). As the name indicates, there are two flutes. These may be either straight or spiral. The depth of the flutes varies. The depth depends on the degree of strength and rigidity required.

In addition to the standard form, *heavy-duty two-flute drills* are manufactured. These are used in drilling operations which require a maximum-strength drill.

Fig. 20-4. A two-flute general-purpose drill.

Three-and Four-Flute Drills—Three-and four-flute drills are used where holes are to be enlarged and close tolerance control is necessary. These drills are referred to as *core drills*. The enlarging is done on holes that may be formed by casting, drilling, punching, and other production methods.

Sub-Land Drills—A sub-land drill (Fig. 20-5) is an adaptation of a two-flute drill. The cutting end is reduced in diameter for a distance along the body. The cutting edges between the first and second diameters may be ground to produce a square, round, or angular connecting area.

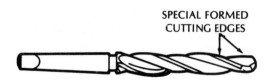

Fig. 20-5. A sub-land drill.

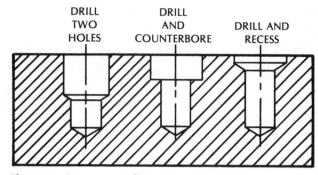

Fig. 20-6. Common applications of sub-land drills.

Three common applications are illustrated in Fig. 20-6. The advantage of using a sub-land drill is that it is possible to drill two or more concentric holes of different diameters. Only one drill and setup are required.

Oil-Hole (Coolant-Feeding) Drills—Many machining processes require a coolant and a lubricant. The use of a cutting fluid increases cutting tool life, production, dimensional accuracy, and the quality of surface finish. An adequate supply of cutting fluid may be provided from a coolant nozzle. The coolant may be *flushed* into the drilled hole *under a slight pressure*.

However, for deep-hole and heavy-duty drilling it may be necessary to use a *coolant-feeding drill*. A phantom view of this type of drill is shown in Fig. 20-7. A coolant-feeding drill has a hole drilled in the core or through the clearance area. The hole extends to the point of the drill. The cutting fluid is forced under pressure through the drill. The fluid thus reaches the area where the cutting action takes place. The cutting fluid keeps the cutting lips flooded and cooled. The cutting fluid also helps move the drill chips from the workpiece.

Functions Of Twist Drill Parts

There are two main areas of a twist drill: a *shank* and a *body*. The *shank* provides a gripping surface. A force may be exerted through the shank to turn and feed the drill. The *body* is designed to provide efficient cutting action, chip removal, and heat control. The body permits the cutting fluid to reach the cutting edges. The principal design features and

Fig. 20-7. A phantom view of a coolant-feeding drill. (Courtesy of Do All Company)

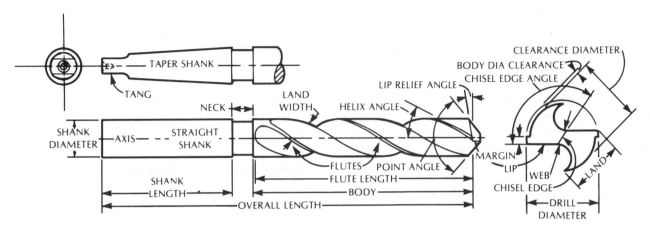

Fig. 20-8. The main design features of a twist drill.

part names are given in Fig. 20-8.

The Shank—Twist drill shanks are either straight or tapered. The standard *straight shank drill* usually fits into and is driven by a *drill chuck*. The straight shank is found mainly on *letter size, number size, metric,* and *fractional size* drills. These are 1/2" (12 mm) or smaller in diameter. The straight shank of a larger size drill is sometimes reduced in diameter to fit a regular drill chuck.

The end of a *taper-shank drill* is machined flat for a short distance. This portion forms a *tang*. The tapered area *(taper shank)* fits into the tapered drill press spindle. The taper shank permits the drill to be centrally aligned. The *tang* serves three functions:

● Together with the holding power of the taper shank, the tang helps transmit the spindle force to the drill.

● The tang prevents the drill from turning in the spindle

● The tang provides a way of removing a drill.

The Body—The drill body has a number of design features which relate to the cutting action. The body includes the *flutes* and the *point* of the drill. The flutes are formed and serve functions similar to the flutes of other cutting tools. Flutes are specially formed spiral grooves. They are cut around the body. The body extends from the drill point to the shank. The flutes serve three major purposes:

● The flutes cause the chips to curl

● The flutes provide a channel through which chips are fed out of the work

● The flutes provide a passage through which cutting fluid can reach the cutting edges.

The Web—The area between the two flutes is called the *web*. The cross-sectional area of the drill web increases from the point to the shank. The *web size* of a standard twist drill depends on the number of flutes and the drill size.

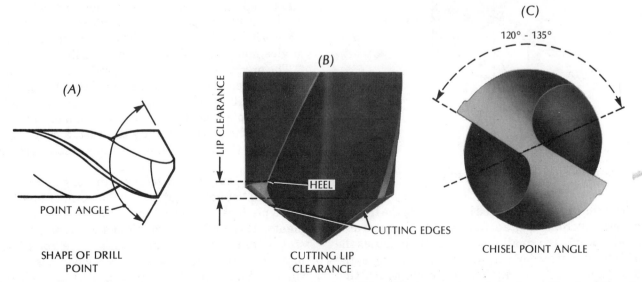

Fig. 20-9. The shape and angles of a drill point for general-purpose drilling. (Photos courtesy of the Acme-Cleveland Corp.)

The Land—The outer surface between the flutes is the *land* of the drill. If the land were the same size as the outside diameter, there would be a great amount of friction in drilling. The friction is reduced by leaving only a small *margin* of the drill diameter. The remaining portion of the land is *relieved* (the diameter is reduced).

The Point—The cutting action of a twist drill results from the design of the *point*. The point includes the cone-shaped area formed with the flutes (Fig. 20-9A). The angle of the cone depends on the material to be drilled, the drill size, and the nature of the operation. *Cutting edges,* or *lips,* are formed by grinding a *lip clearance* on the angular surface of each flute (Fig. 20-9B). In other words, the *lip* is ground at a clearance angle. The area behind the cutting edge slopes away to produce a chisel-edged shape. This is the edge that provides the shearing, cutting action. Since the clearance angles are on opposite sides of a two-flute drill, a flat surface is formed across the drill axis *(dead center).* A *chisel point angle* of 120° to 135° (Fig. 20-9C) is used for general drilling.

Although there are many adaptations of drills, the functions and design features just covered are basic.

Drill Point Shapes

Cutting conditions vary for different materials. To compensate for these, it is necessary to change the shape of the cutting edge and the cutting angle. The *standard drill angle* for all-round drilling operations is 118°. There is a 59° angle on each side of the drill axis. The angles are shown in Fig. 20-10.

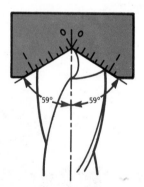

Fig. 20-10. A drill point correctly ground at 118°.

The *cutting edge* is formed by relieving the flute. A general *clearance angle* of from 8° to 12° is illustrated in Fig. 20-11. Too small a clearance angle causes the drill to rub instead of cut. At too steep an angle, the drill *tears into* the work. This may cause the cutting edges to dull rapidly or parts of the drill to break

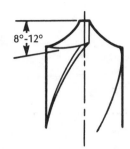

Fig. 20-11. The clearance angle range for general-purpose drilling.

Table 20-1. RECOMMENDED LIP CLEARANCE ANGLES

Inch Standard Drill Ranges		Lip Clearance Angle
Drill Sizes	Drill Diameters	
#80 to #61	0.0135 to 0.0390	24°
#60 to #41	0.0400 to 0.0960	21°
#40 to #31	0.0980 to 0.1200	18°
⅛ to ¼	0.1250 to 0.2500	16°
F to ¹¹/₃₂	0.2570 to 0.3438	14°
S to ½	0.3480 to 0.5000	12°
³³/₆₄ to ¾	0.5156 to 0.7500	10°
⁴⁹/₆₄ and up	0.7656 to	8°

away. After extensive research, drill manufacturers have prepared tables of recommended lip clearance angles. Table 20-1 contains such information for drill sizes #80 to 1″ and larger. The same lip clearance angles apply to equivalent metric drill sizes.

The precision to which a drill produces an accurate size hole depends on the grinding of the cutting lips. They must be equal in length and at the same angle. The three photos in Fig. 20-12 show common errors in grinding the cutting lips. By contrast, a continuous steel chip (Fig. 20-13) indicates a correctly ground drill.

Thus far, only the *chisel point* in the general-purpose drill has been treated. The *flat* point of this drill helps produce a hole that meets most work requirements. These relate to conditions of *roundness* and *size* within *rough tolerance limits.* The *roundness* may be more accurately produced when the chisel shape point is guided by a *drill bushing* in a drill jig.

Another shape of drill point is called the *spiral*

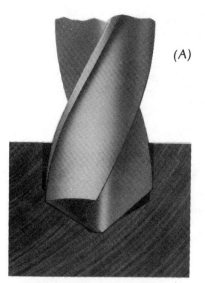

(A)

LIPS OF EQUAL LENGTH BUT
AT UNEQUAL ANGLES

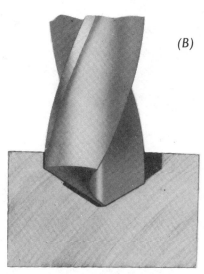

(B)

LIPS OF UNEQUAL LENGTH BUT
AT EQUAL ANGLES

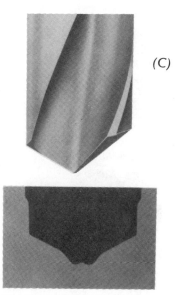

(C)

LIPS OF UNEQUAL LENGTH AND
AT UNEQUAL ANGLES

Fig. 20-12. Common errors in grinding lips. (Courtesy of the Acme-Cleveland Corp.)

point. The spiral point, in contrast to the chisel point, has a sharply pointed end. With this shape the drill may be started accurately without center punching, center drilling, or using a drill jig. The hole produced with the spiral point is concentric with the axis and is round. The photos in Fig. 20-14 provide a comparison of *roundness* produced by a spiral and a chisel point drill.

Spiral points may be ground on standard drills. A special *spiral-point drill-grinding machine* is used for this purpose.

The self-centering feature of the spiral-point drill eliminates the need for a drill-positioning device. The spiral-point drill is particularly adaptable to numerically-controlled drilling operations. Features of drills ground to spiral and chisel points are shown in Fig. 20-15. Spiral-point drills have longer tool life than drills ground with a chisel point. Test records of spiral-point drills show that a greater number of workpieces may be produced between each drill point grinding. This production is in contrast with drill points ground in the conventional way.

Fig. 20-13. A continuous steel chip indicates a correctly ground drill point. (Courtesy of the Acme-Cleveland Corp.)

CHISEL POINT   SPIRAL POINT

Fig. 20-14. Comparison of the roundness produced by two different cutting points.

231

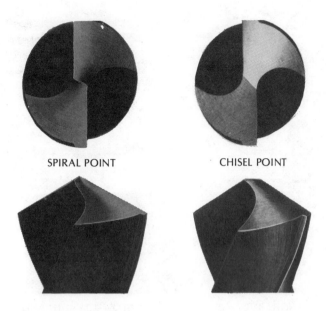

SPIRAL POINT   CHISEL POINT

Fig. 20-15. The features of drills ground to spiral and chisel points.

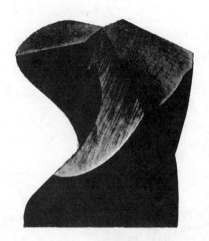

Fig. 20-16. A split-point drill grind.

The split point reduces tool wear. The grinding does not disturb the cutting lips. There is also less tool breakage when drilling into hard outer surfaces *(scale)* of castings and forgings.

THE DRILL POINT GAGE

The accuracy of the *drill point angle* and the *dead center* may be measured with a *drill point gage* (Fig. 20-17). A common drill point gage has a graduated blade shaped to the required drill angle. This blade is attached to a steel rule as shown in Fig 20-17.

The body of the drill is held against the edge of the steel rule. The point end is brought up to the blade. Any variation between the cutting edge and the correct angle may be sighted. The location of *dead center* is read on the graduated blade. The drill is then turned 180°. A fractional dimension reading is again taken. The drill point angle is correctly ground when there is no light between the cutting edge and the angular blade. The lips are correct when the

Split-Point Drill

Another type of point, the *split point,* reduces the web to a minimum shape. The point forms two additional cutting edges. The photo in Fig. 20-16 shows that the clearance angle of the added cutting edges is formed by grinding. Part of the web section is ground parallel with the drill axis. This design permits:

- Easier and faster cutting
- Easier access for the chips to clear the cutting edges and workpiece
- Less force to be used in the drilling process
- Cutting with less power
- Less heat to be generated from cutting.

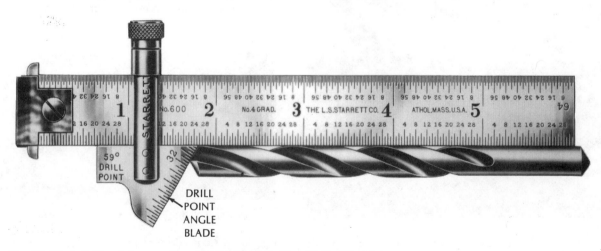

Fig. 20-17. A drill point gage. (Courtesy of The L.S. Starrett Company)

dimensional reading to the chisel point of each lip is the same.

TAPER DRILL SHANK ADAPTERS

Adapters are used to accommodate the difference between a taper-shank drill and the drill press spindle taper. Two common types of adapters are called a *drill sleeve* and a *taper socket*.

Drill Sleeve

When the taper shank of a drill is smaller than the spindle taper, it may be brought up to size with a drill sleeve (Fig. 20-18).

Fig. 20-18. A drill sleeve. (Courtesy of the Acme-Cleveland Corp.)

The standard taper that is used on taper-shank drills is known as a *Morse taper*. The same standard is used on many other cutting tools like reamers, boring tools, and counterbores. The Morse tapers are numbered from 0 through 7. The number 2, 3 and 4 tapers are used on drills which range from 3/8" to 1½" diameter (10 mm to 38 mm). For example, a drill sleeve is required for a 1/2" drill that has a #2 Morse-taper shank when it is to fit a spindle with a #3 or a #4 Morse taper.

Fitted (Taper) Socket

The *taper socket* is also known as a *fitted socket* (Fig. 20-19). It has an external Morse-taper shank. The body has an internal taper. There is an elongated slot at the small end of the taper. This slot is where the tang of the drill or cutting tool fits. It also serves as a way of removing a tool.

Fig. 20-19. A fitted (taper) socket. (Courtesy of the Acme-Cleveland Corp.)

A fitted (taper) socket often provides a method of extending the length of a cutting tool. Different combinations of taper sizes are available. For example, a fitted (taper) socket with a #3 Morse taper

shank may have a #2 taper in the body. The #3 taper fits a #3 taper spindle. The #2 taper accommodates a drill or reamer that has the same size taper shank.

THE DRILL DRIFT

The *drill drift* is a flat, metal, tapered bar (Fig. 20-20). It has a round edge and a flat edge. The round edge fits the round portion (end) of the elongated slot of the drill sleeve or the fitted socket. The flat edge rests and drives against the tang end of the cutting tool. The joined tapered surfaces are *freed* by a sharp, slight tap on the drift. The position of a drill drift in relation to the drill tang and elongated slot is shown in the cutaway portion of Fig. 20-20.

The *safety drill drift* is another type of drift. A floating handle is attached to the drift. The body of a cutting tool may be held in one hand. The other hand is used to quickly slide the handle so it delivers a sharp tap. The force disengages the tapers. It is good practice to protect the surface under the drill so that it will not be hit by the cutting tool when it is freed.

Fig. 20-20. A drill drift positioned to remove a drill. (Courtesy of the Acme-Cleveland Corp.)

DRILL POINT GRINDING

Drill points may be ground *off-hand* or on a *drill point grinding machine*. Another method is illustrated in Fig 20-21. A drill point attachment is fitted to a bench or floor grinder. Regardless of the method, the drill must be positioned, held, and fed into the

Fig. 20-21. A drill point attachment for a bench grinder. (Courtesy of the Clausing Corp.)

grinding wheel. This action produces a cutting edge and clearance in correct relation to the flute.

There are two main differences between off-hand grinding and grinding with a drill point machine or attachment. First, the skill and judgment of the operator controls the accuracy of the drill point when grinding by hand. Secondly, the face of a straight grinding wheel is used in hand grinding. The cutting is done in machine grinding on the side face of a *cup wheel*.

In drill point grinding, particular attention must be paid to the following points:

• *Clearance angle.* Too great a clearance angle produces a cutting edge that wears rapidly and may chip. Too small a clearance angle produces excess heat. The drill rubs instead of cutting.

• *Heat generated.* The cutting edges must be kept cool when being ground. This is especially important in grinding carbon steel drills. Otherwise, the temper may be drawn.

• *Lips.* The *lips* must be ground evenly so they are not off-center. This means the lips are ground to an equal angle and length.

• *Web.* The web thickness increases when a drill is ground a number of times. The web must be thinned to the same size as the original drill. A thick web

increases the feed pressure, thereby reducing the cutting efficiency.

DESIGNATION OF DRILL SIZES

There are three standard designations of drill sizes in the inch-standard system and one in the metric system. In the first instance, a drill may be given a *letter size, wire number size,* or a *fractional inch size.* Each drill has its equivalent decimal value. The metric sizes are given in terms of millimeters. The common sizes of straight-shank drills in the metric system, for example, range from 0.20 mm to 16.00 mm.

Table 20-2. SELECTED DRILL SIZES AND DECIMAL EQUIVALENTS

Number (Wire)	Letter	Fractional (inch)	Metric (mm)	Decimal Equivalent
1				0.2280
			5.8	0.2283
			5.9	0.2323
	A			0.2340
		¹⁵/₆₄		0.2344
			6.0	0.2362
	B			0.2380
			6.1	0.2402
	C			0.2420

Number Size Drills (Wire Gage Drills)

The sizes of this series of drills range from #80 (0.0135″) to #1 (0.228″). The most widely used set ranges from #1 to #60 (0.040″).

Letter Size Drills

A second series of drill sizes is designated by letters, from "A" (0.234″) to "Z" (0.413″). Letter size drills extend the range from the last diameter in the number size drill, which is number "1" (0.228″).

Fractional Size Drills

A third series of drill sizes begins at 1/64″ diameter. Each successive drill is 1/64″ larger. The general range of diameters extends to 3″. When a series of fractional size drills is part of a set, for example from 1/64″ to 1/2″, it is sometimes called a *jobbers set*.

Part of a drill size table with selected number, letter, fractional, and metric sizes is represented in Table 20-2. Note that there is a variation in diameter among all four drill size designations. These provide an adequate range for most hole drilling from a #80 drill (0.0135″) to 3/4″ or 19.0 mm (0.748″).

Drill Size Measurement

Drill sizes may be measured with a micrometer. A micrometer measurement is taken across the margins of a two-flute drill. This gives an accurate measurement.

Another quick, common practice in the shop is to use a *drill gage* in the same series as the drill. A drill gage is a flat plate with the same number of holes as there are drills in the series. A number (wire gage) size drill gage is shown in Fig. 20-22. A letter size drill is gaged with a *letter size drill gage*. There are *number size, letter size, fractional size,* and *metric size drill gages*. The decimal equivalent appears under each drill size designation on each drill gage. Fig. 20-23 illustrates how the gage and drill are held when checking a drill size.

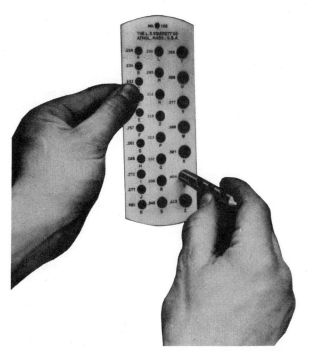

Fig. 20-23. Drill size measurement with a drill gage. (Courtesy of The L.S. Starrett Company)

DRILLING PROBLEMS: CAUSES AND CORRECTION

A number of problems can arise in drilling if correct practices are not followed. These affect dimensional accuracy, surface finish, and drilling efficiency. Tool life is increased and the accuracy and quality of drilling are improved when correct practices are followed.

Table 20-3 lists a number of common problems that relate to drills and drilling procedures. Possible causes of each condition are given. These are followed by a series of check points for correcting each problem. The craftsperson must be able to recognize a problem, establish the cause, and take corrective steps.

CUTTING SPEEDS FOR DRILLING

Every cutting tool performs most efficiently and with greater tool life when it is cutting at the correct cutting speed. The *cutting speed* is expressed as the *surface feet the circumference of a drill travels in one minute*. This value is stated as *surface feet per minute (sfpm)*.

Calculating Cutting Speed

The *surface feet per minute (cutting speed, or circumferential speed)* may be calculated with the

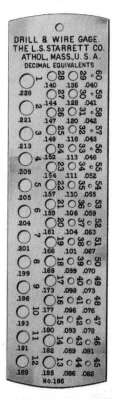

Fig. 20-22. A number (wire gage) size drill gage. (Courtesy of The L.S. Starrett Company)

DRILLING MACHINES

Table 20-3. DRILLING PROBLEMS: CAUSES AND CORRECTIONS

Problem	Possible Cause	Correction
Drill breaking	• Dull drill point • Insufficient lip clearance • Spring in the workpiece • Feed too great in relation to cutting speed	• Regrind drill point • Check angle of lip clearance • Reclamp workpiece securely • Decrease feed or step up the cutting speed
Drill breaking (nonferrous and other soft materials)	• Flutes clogged with chips • Cutting point digging into the material	• Use appropriate drill for the material • Remove the drill a number of times so the chips clear the hole • Increase the cutting speed • Grind a zero rake angle
Rapid wearing of cutting edge corners	• Cutting speed too fast • Work hardness; hard spots, casting scale, or sand particles	• Reduce cutting speed • Sand blast or tumble castings, or heat treat • Change method of lubricating and check for correct cutting fluid
Tang breaking	• Taper shank improperly fitted in sleeve, socket, or spindle	• Check condition of adapter or spindle • Clean taper hole and remove nicks or burrs by reaming carefully
Margin chipping (using drill jig)	• Jig bushing too large	• Replace drill bushing with same size as drill
Cutting edge and lip chipping	• Lip clearance too great • Rate of feed too high	• Regrind to proper lip clearance • Reduce the feed rate
Cracking or chipping of high-speed steel drill	• Excessive feed • Improper quenching during grinding or drilling	• Reduce the feed rate • Quench gradually with warmer coolant when grinding
Change in cutting action	• Drill dulling; cutting edge damaged • Variations in the hardness of the workpiece • Change in the amount of cutting fluid reaching the drill point	• Regrind the drill point correctly • Anneal the workpiece • Increase the direction or rate of flow of the cutting fluid
Oversize drilled hole	• Improper grinding of point with equal angles and length of cutting edges • Excessive play in spindle • Slightly bent drill	• Check reground cutting angle and length of cutting edges • Adjust or correct amount of free movement in spindle • Straighten drill carefully
Cutting on one lip	• Angle of cutting lips varies • Unequal length of cutting lips	• Regrind and check accuracy of cutting angle and length of cutting edges
Poor quality of surface finish (rough drilled hole)	• Improperly ground drill point • Excessive drill feed • Wrong cutting fluid or limited rate of flow • Workpiece not correctly secured	• Regrind and check accuracy of drill point • Decrease drill feed • Change cutting fluid for the kind of material • Reposition and secure the workpiece securely
Drill splitting (along the center)	• Excessive feed • Insufficient lip clearance	• Decrease drill feed • Increase the lip clearance to the correct angle

simple formula:

$$RPM = \frac{Cutting\ Speed \times 12}{Drill\ Circumference}.$$

Values are *rounded off* to produce a more simplified formula that is used in the shop:

$$RPM = \frac{CS \times 4}{d}$$

in which:

CS = Cutting speed
RPM = Revolutions per minute
d = Drill diameter.

Manufacturers' tables contain recommended cutting speeds for different types of cutting tools and materials. Table 20-4 shows part of one such table

Table 20-4. PARTIAL TABLE OF CUTTING SPEEDS

	Cutting Speeds for High-Speed Steel Drills	
	Material	Cutting Speed Range (Surface Feet Per Minute, sfpm)
Steel	low-carbon	80 to 150
	medium-carbon	60 to 100
	high-carbon	50 to 60
	tool and die	40 to 80
	alloy	50 to 70

that lists the high-speed steel drill speeds for various steels. These values are reduced by 40 to 50 percent when a carbon steel drill is used instead of a high speed drill.

The craftsperson must determine the spindle *revolutions per minute* (RPM) from the information provided in a cutting speed table. The RPM must produce the required cutting speed for a particular drilling operation. As an example, assume that a series of holes is to be drilled in a low-carbon steel. The cutting speed of 150 sfpm is to be used with a 1/2" drill. These known values are inserted in the simplified RPM formula as follows:

$$RPM = \frac{150 \times 4}{1/2} = 1,200.$$

The drill must turn *1,200 rpm* to produce a cutting speed of 150 sfpm.

Most drill presses with a step-cone drive have a plate indicating the speed (rpm) at each step. The V-belt is positioned on the step that will produce the speed closest to the required rpm. The *smallest* step on the motor produces the *slowest* spindle speed.

On a *variable speed drive* the rpm is adjusted to the desired speed (rpm).

Factors Affecting Cutting Speed

The cutting speed that the worker finally selects is influenced by seven different factors:
- The material of the workpiece
- The material of which the cutting tool is made
- The size of the hole
- The quality of surface finish that is required

- The nature and flow rate of the cutting fluid
- The type and condition of the drilling machine
- The manner in which the workpiece is mounted and held.

Care must be taken in using cutting speed tables. The table must show the cutting speeds for the material in the cutting tool. In other words, cutting speeds are higher for high-speed drills than for carbon steel drills. The cutting speeds of carbide-and other alloy-tipped drills are even higher than those for either high-speed or carbon steel drills.

Table 20-5. FEEDS FOR SELECTED DRILL SIZES

Drill Size (inches)	Feed (inches per revolution)
Small #sizes to 1/8"	0.001 to 0.002
1/8 to 1/4	0.002 to 0.004
1/4 to 1/2	0.004 to 0.007
1/2 to 1	0.007 to 0.015
Larger than one inch	0.015 to 0.025

CUTTING FEEDS FOR DRILLING OPERATIONS

Cutting feed indicates the distance a tool cuts into the workpiece for each revolution of a cutting tool. Tables of cutting feeds are provided by cutting tool manufacturers. The tables provide a recommended range of feeds according to the size of the tool. Table 20-5 gives feeds for selected drill sizes. These range from the smallest number size drill, to drills of 1" diameter and larger.

The *feed indicator* on machines that are so equipped is set within the recommended feed range. The feed usually depends on the required surface finish. When feeding by hand, the operator must determine the correct feed. To repeat, a well-formed spiral chip is an indicator that the drill is feeding properly. Excessive force in feeding produces additional heat and may cause the cutting edge to break away. Sometimes the drill fractures. Insufficient force (feed) may cause the drill to scrape instead of cut. This action dulls the cutting edge.

A cutting fluid, either brushed or flowed over the workpiece and drill, improves the cutting action. A finer surface finish is also produced.

HOW TO GRIND A DRILL

Offhand Grinding

Step 1   Check the trueness of the coarse and the fine bench grinder wheels. The work rest should be horizontal and within 1/16″ of the wheel face.

Step 2   Examine the point of the drill. Be sure it is not *burned*. Check the condition of the margins and the dead center.
Note: The drill point may need to be ground back beyond the point of damage.

Step 3   Hold the drill on the work rest with one hand. The drill should be positioned at a 59° angle to the wheel face (Fig. 20-24). The forefinger of one hand serves as a positioner. It also applies a slight force to hold one cutting edge steady against the coarse face wheel.

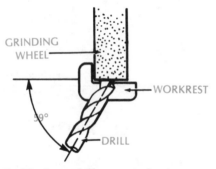

Fig. 20-24. Positioning a drill on a workrest.

Step 4   Hold the shank end of the drill in the other hand (Fig. 20-25). Move the shank end down. At the same time, apply force against the cutting face with the other hand.

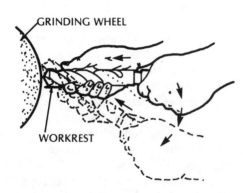

Fig. 20-25. Correct hand action in drill grinding.

Note: There is a natural tendency to move the drill in an arc. This is the pattern to which the drill is to be ground
CAUTION: Quench the drill point frequently. Cooling avoids *burning* the cutting edge. Cooling also makes it possible to handle the drill safely.
CAUTION: Avoid overheating the point and rapid quenching. This may cause microscopic cracks along the cutting edges. Such cracks cause the lips to break down rapidly.

Step 5   Check the 59° angle against a drill gage.

Step 6   Repeat steps 3 and 4 until the cutting edge and lip clearance are correct. A sharp cutting edge is required.
Note: It is good shop practice to alternate the grinding process between the opposite cutting edges.

Step 7   Check the drill point for (a) the correct cutting angle of 59°, (b) the length of the cutting edges, and (c) the required clearance angle (usually 8° to 12°).
Note: Finish grind the cutting edges on the fine grinding wheel. This produces a finer cutting edge and increases tool wear life.

Thinning The Web

Step 1   Hold the cutting edge of the drill parallel to the side of a straight grinding wheel. The wheel must have a square corner with a very slight radius.

Step 2   Grind the front end of each flute. The web thickness must be ground equal to that of a new drill of the same size.

Using A Drill Grinding Attachment

Step 1   Clamp the drill in the V-block. The point must overhang.

Step 2   Set the end guide *(tailstock center)* so that the tool point may be brought into contact with the front face of a cup wheel.

Step 3   Move the *setting finger* on the grinding attachment to contact and position the flute to be ground. The contact point should be close to the wheel.

Step 4   Set the base angle of the attachment at 59°. The body scale is set to the clearance angle of from 8° to 12°.

Step 5   Feed the drill until contact is made with the wheel face (Fig. 20-26). The move-

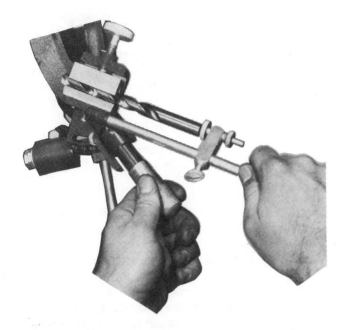

Fig. 20-26. Feeding a drill with a drill grinding attachment. (Courtesy of the Clausing Corp.)

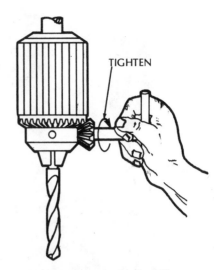

Fig. 20-27. Chucking a straight-shank drill.

ment of one lever produces the 59° angle. The other lever is moved to produce the clearance angle.

Step 6   Turn the drill 180°. Align the second flute. Grind to the same depth as the first cutting edge.

Note: The same setup and grinding principles are applied when a grinding fixture is used. The tool is fed with a vertical movement of the cutter grinder head.

HOW TO DRILL HOLES

Flat Surfaces Held In A Drill Vise

Step 1   Select the correct size drill. Check the shank to see that it is free of burrs. Check the drill point for the accuracy of the cutting and clearance angles.

Step 2   Determine the required speed and the feed if a power feed is to be used. Set the speed selector and the rate of feed.

Step 3   Move the workpiece so that the center of the required hole aligns with the axis of the spindle and drill.

Note: A center drill is commonly used. Center drilling helps start the drill in the exact location.

Step 4   Clamp the work-holding device securely to the table. On smaller size drills the vise may be held by hand.

CAUTION: The vise should be held by hand *only* when there is no possibility that the vise may turn. Another precaution is to place a *stop* in a table slot. The vise is brought against the stop. This prevents the twisting action of the drilling operation from turning the holding device.

Step 5   Chuck a straight-shank drill or center drill (Fig. 20-27).

Step 6   Start the machine. Bring the drill point to the work. Apply a slight force to start the drill.

CAUTION: If an angle plate setup is required, the overhanging leg should be supported by a jack (Fig. 20-28).

Step 7   Set the coolant nozzle if considerable heat might be generated. The cutting fluid should flow freely to the point of the drill.

Note: In general practice, where there is limited heat a cutting fluid is brushed on (Fig. 20-29).

Step 8   Continue to apply a force to feed the drill uniformly through the workpiece.

Note: On deep holes (depth in excess of 5 to 6 times the diameter of the drill), feed the drill until it is the equivalent of one diameter. Then remove the drill. This helps clear the chips and permits the cutting fluid to flow freely into the holes. The power feed is set so that it *trips off*

Fig. 20-28. An angle plate setup for drilling. (Courtesy of the Clausing Corp.)

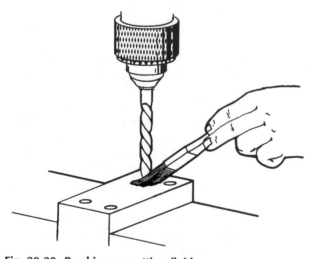

Fig. 20-29. Brushing on cutting fluid.

Drilling A Hole In A Cylindrical Surface

Step 1 Position the workpiece in a set of V-blocks. *Square* the centerline on the end surface with the machine table. This also aligns the hole in relation to any other hole or surface.

Step 2 Select a center drill. The pilot and outside diameter should be in proportion to the size of the hole to be drilled. Chuck the center drill.

Step 3 Set the center punched indentation central with the point of the center drill. Clamp the workpiece and setup if the cutting force requires it.

Step 4 Determine the speed and feed. Set these controls. Start the machine.

Step 5 Bring the center drill carefully to the workpiece. Start drilling. Feed slowly until the point end and the angle portion provide a large enough area in which to center a drill.
Note: A cutting oil is usually applied with a brush to the drill point. This produces a finer cutting action. The machinability of the material and the tool life and productivity also are increased.

Step 6 Replace the center drill with the required drill.
Note: A pilot hole is often drilled first. This is followed by a larger diameter drill, for more efficient drilling (Fig. 20-30).

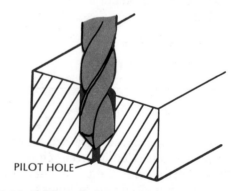

PILOT HOLE

Fig. 20-30. Drilling of a larger pilot hole for large-size drills.

Step 7 Change the cutting speed if necessary. Start the drill and feed it by hand. Engage the power feed for large-diameter holes.
Note: The lips of a drill are ground to a zero rake angle for drilling soft materials such as aluminum and brass.

after the drill clears the workpiece on a through hole. The depth of a blind hole may be read directly on a graduated scale on the drill head.

Step 9 Stop the machine. Remove the chips with a brush. The burrs may be cut away with a triangular scraper or a countersink.

Step 8   Drill and burr, following the same steps used in drilling holes in flat surfaces.

Step 9   Stop the machine. Clean the workpiece, machine surfaces and cutting tools.

HOW TO REMOVE A DRILL

Removing A Taper Shank Drill

Step 1   Move the spindle down so that the drill moves only a short distance when the tapered surfaces are freed.

Step 2   Place a protective block of soft material under the drill. Small-size drills may be held by hand.

CAUTION: The drill must be checked carefully to be sure it is cool enough to be handled. A wiping cloth may be placed around the drill for handling large sizes.

Step 3   Slide the drift handle quickly to deliver a sharp, light blow on the drift. The force causes the taper-shank surfaces to become free.

Note. A slightly different procedure is used when the drift must be tapped firmly with a hammer. In such instances, the drift is held in one hand. The end of the drift is tapped firmly. The drill is released against a protecting block.

Removing A Straight-Shank Drill

Step 1   Insert the pin of the chuck wrench into the chuck body. The wrench teeth must mesh properly with those in the *releasing collar.*

Step 2   Turn the wrench handle *counterclockwise,* to release the drill.

Step 3   Examine the shank of the drill. Remove any burrs or cut edges. These are produced if the drill turns in the chuck during the drilling process.

SAFETY PRECAUTIONS IN DRILLING

- Use the kind of twist drill (material) that is suited to the hardness of the workpiece and the cutting speed.
- Avoid softening a carbon steel drill point by attempting to drill hardened parts. Keep the cutting speed within the recommended sfpm range. Use a coolant where possible.
- Check the straight and taper shanks and tangs of drills, sleeves and drill sockets. They must be free of burrs and nicks. The tapers and tangs are checked for correct taper fit.
- Drive a taper shank drill to hold it in a drill socket or sleeve. Tap the end of the socket firmly with a soft-face hammer.
- *Set* the tapers on drills, adapters, chucks, etc., by a firm tap with a soft-face hammer.
- Check the round and flat edges of a drill drift. Remove any burrs or nicks.
- Engage feed gears on a power-fed, heavy-duty drilling machine only *after* stopping the machine. The gear lever then can be positioned at the proper feed. The expert craftsperson often shifts gears by shutting off the power. The gears are then carefully engaged as they barely turn to a stop.
- Check the workpiece to see that it is adequately supported to prevent springing during drilling.
- Grind the cutting edges of the drill to the same length and angle. Otherwise, oversized, roughly drilled, and uneven holes will be produced.
- Move (flow) chips out of the drilled hole to prevent clogging, the generation of excess heat, and drill breakage.
- Quench carbon steel drills regularly during grinding. This prevents softening (drawing the temper). A *burned drill* must be reground beyond the softened portion.
- Avoid excessive speed, feed, heat, and force. These can cause rapid wearing of the drill point and drill margin or drill breakage.
- Check the lip clearance angle. Too great a lip clearance angle causes chipping of the cutting edge. Too small a clearance angle and excessive speed causes drill splitting.
- Wear a protective shield to prevent eye injury and to keep hair away from a revolving spindle and cutting tools.
- Avoid wearing loose clothing near rotating parts. Remove all wiping cloths from the drilling area *before* starting the machine.

DRILLING MACHINE TERMS: TECHNOLOGY AND PROCESSES

Carbon steel drill	Generally, a two-flute twist drill is made of a tool steel and is hardened and tempered. Used in drilling operations where a limited amount of heat is generated.
High-speed steel and cobalt high-speed steel drills	A twist drill capable of cutting at high speed without changing the temper, softening the cutting edges, or reducing the cutting efficiency. The addition of cobalt to high-speed steel produces a cutting material that is capable of drilling through tough metals and other compositions.
Carbide tipped drill	Inserts of tungsten carbide on the drill point and point end of a regularly formed twist drill. Cutting edge and body inserts of tungsten carbide. The inserts make it possible to drill extremely hard surfaces and tough materials.
Twist drill	A cutting tool with two (the most common), three, or four flutes. A shank permits gripping to apply a cutting force. The flutes provide a channel for removing chips and forming the cutting edges. An angular pointed cutting tool with cutting edges extending from a center to the outside diameter.
Drill point angle	The included angle of the drill point. Equal angles on each side of a drill axis (centerline). A 118° included angle of a drill point for general, everyday applications.
Drill clearance angle	The angle at which the cutting edge of a drill is ground away in relation to a plane. The plane is at right angles to the drill axis.
Drill sleeve	An adapter. The outside taper fits the drill press spindle. The inside taper accommodates the taper shank of a drill.
Taper (fitted) socket	An adapter for taper shank drill sizes that are larger or smaller than the spindle taper.
Letter, number, fractional, and metric drill sizes.	Separate systems of designating drill sizes. A combination of systems having different drill sizes among those in each series. Four series of small drill sizes that permit drilling a wide range of holes.
Drill point gage	A flat, angular, graduated blade that may be attached to a steel rule. A gage for comparing the drill point angle with a standard. A gage for measuring the cutting lips of a drill and for checking the angle.
Drill point grinding	The positioning and grinding of the cutting edges of a drill at a correct drill point angle. Offhand grinding, or grinding with a drill point attachment, or a drill point grinding machine.
Cutting speed (drilling)	The number of surface feet the circumference of a drill travels in one minute (sfpm). Recommendations of sfpm for a particular type of cutting tool and process.
Cutting feed	The distance a drill advances for each revolution. The feeds recommended by manufacturers for drills of different sizes and materials.
V-block (drilling)	A layout, positioning, and holding device. A squared metal block with V-slotted sides. These *nest* cylindrical workpieces.

SUMMARY

- Drills are manufactured from carbon steel, high-speed steel (HSS), and cobalt high-speed steel. The HSS bodies may also be tipped with tungsten carbide cutting edge inserts.
- The two-flute drill is an all-purpose type. The design features include:
  - An angular point on which the cutting edges are formed
  - Flutes which are connected with a web

- Lands which are relieved to leave a thin margin
- A channel to remove chips and to feed cutting fluid
- A shank.
- Special sub-land (multidiameter) drills produce two or more holes of different diameters. The connecting shoulder is also cut during the one operation.
- Oil-hole drills feed a cutting fluid through the drill

shank and body to the cutting edges.

- The holding power of two mating tapers and a fitted tang is adequate to prevent a drill socket, sleeve, or other taper shank tool from turning during a drilling process.

- The most widely used drill point angle is 118°, with an 8° to 12° clearance angle. The cutting point angle and the length of the cutting edge may be checked with a drill point gage.

- The spiral-point drill is self-centering. It produces a concentric, dimensionally accurate hole.

- Drill points may be ground offhand. Other methods include the drill point attachment and a drill point machine.

- The drill point must be quenched when grinding, to dissipate any great amount of heat.

- The four main drill size designations are:
  ○ Number sizes (#1 to #80; 0.228″ to 0.0135″)
  ○ Letter sizes (A to Z; 0.234″ to 0.413″)
  ○ Fractional sizes (1/64″ to a number of inches). The increments are 1/64″ and larger
  ○ Metric sizes, in millimeters.

- Drill sizes are measured most accurately with a micrometer. The size may also be checked with a drill gage.

- Drill breakage, rapid dulling and wearing, oversized holes, rough hole surfaces, and broken tangs result from one or more of the following causes:
  ○ Improper fitting of tapers and tang
  ○ Too great or too small a clearance angle, cutting angle, rate of feed or speed
  ○ Unequal cutting angles and/or length of cutting edge
  ○ Too thick a web
  ○ A burned drill point.

- The rpm required for a recommended cutting speed (sfpm) may be found by the formula:
$$RPM = \frac{CS \times 4}{d}.$$

- The sfpm and recommended feed for each particular type of drill are provided in manufacturers' tables.

- Cutting speed depends on the material in the workpiece and cutting tool, the hole size, the required surface finish, the use of a cutting fluid or coolant, the condition of the machine, and how the work is secured.

- The drill sleeve and taper socket are two common adapters for taper shank tools. The Morse taper series is used on drilling machine spindles and taper shank drills, reamers, and other tools.

- The flat drift and the safety drill drift fit in the elongated tang slot. The rounded edge fits against the round portion. The flat edge drives against the tang of the tool to free the taper surfaces.

- V-blocks are a practical positioning and holding device for round workpieces.

- Personal safety requires the use of a protective shield. A check is required to see there is no loose clothing near a revolving machine part.

- Tools and workpieces must be protected from the effects of excessive heat and force, improper fitting of cutting tools in holders, and improper grinding and machine processes. The work must be held securely, either in a vise or by clamping to the table.

UNIT 20 REVIEW AND SELF-TEST

1. a. Cite the effect the addition of *cobalt* has on the cutting ability of high-speed drills.
   b. Identify the kinds of materials that may be efficiently drilled with cobalt high-speed steel drills.
2. Explain the principal design feature of a coolant-feeding drill.
3. Tell what purpose a drill bushing in a drill jig serves in drilling.
4. Give three advantages of using a *spiral-point drill* over a *chisel-point drill*.
5. Differentiate between two common types of adapters for drills.

6. Indicate three checkpoints to observe in grinding drill points.
7. Name four common drill size gages.
8. List five factors a drilling machine operator must consider when selecting the cutting speed for a particular drilling operation.
9. Give two cautions to observe in grinding a drill offhand.
10. List the steps to follow in setting up a cylindrical part and drilling a hole at a right-angle to the axis.
11. Provide a list of four safety precautions to follow in drilling, in addition to those listed in Test Item 9.

Unit 21
MACHINE REAMING: TECHNOLOGY AND PROCESSES

Holes may be produced to meet accurate dimensional tolerances, standards of roundness, and quality surface finishes by *reaming*. Unit 15 described reamers and the reaming process in terms of hand reaming. Industry also requires that reaming be done by machine. The drilling machine, lathe, milling machine, jig borer, and other machine tools are used for machine reaming.

There are common design and cutting principles that apply to both hand and machine reamers and reaming processes. However, the applications are so varied that many different reamer materials, types, shapes, and cutting angles are required. This unit provides additional information about the functions of machine reamers. General design features, a few machine setups, and actual reaming using a drill press are covered in detail.

To review, *reaming* serves four main purposes:
- To enlarge a hole to a precise dimension
- To form a straight (concentric) hole when the hole first is *bored* concentric and then reamed to size
- To produce a smooth hole with a high-quality surface finish
- To correct minute out-of-round conditions of a drilled hole.

UNDERSIZE HOLE ALLOWANCE FOR REAMING

A sufficient amount of material must be left in a predrilled or machined hole to permit accurate reaming. The amounts and conditions for hand reaming to close tolerances were covered as a bench work process in Unit 15. The amount stated for hand reaming ranges from 0.002″ to 0.005″, or 0.05 mm to 0.12 mm.

By contrast, larger amounts of material must be left for *machine reaming*. A common practice is to drill a hole 1/64″ smaller (*undersize*). This operation is followed in *one-step reaming* with a fluted reamer. The reamer is able to correct minor surface and shape irregularities of the hole.

Table 21-1 gives maximum material allowances for five hole sizes. The sizes range from 1/4″ to 3″ and metric equivalents from 6 mm to 72 mm.

Table 21-1. MATERIAL ALLOWANCES FOR REAMING (GENERAL)

Inch Standard		Metric Standard	
Reamer Diameter	Maximum Allowance	Reamer Diameter	Maximum Allowance
¼	0.010	6 mm	0.2 mm
½	0.016	12 mm	0.4 mm
1	0.020	24 mm	0.5 mm
2	0.032	48 mm	0.8 mm
3	0.047	72 mm	1.2 mm

Holes that require machining to finer dimensional tolerances and a higher quality surface finish are often produced by a *two-step reaming process*. A hole is first rough reamed undersize to within 0.002″ to 0.005″. A second fluted machine reamer is used for finish reaming.

FEATURES OF MACHINE REAMERS

Many terms which designate a particular part or function of a reamer apply to both hand and machine reamers. However, since there are many types of machine reamers with cutting actions that differ from hand reamers, these features are described.

The shank of a machine reamer is either *straight* or *tapered*. The tapered shank has a *tang*. The flutes may be *straight* or *helical* with either a *right-hand* or *left-hand helix*. The body and shank may be either one piece, or the fluted body may be a separate cutting tool fitted to an arbor.

The cutting action is performed by the cutting edges. The cutting end features of a machine reamer are labeled in Fig. 21-1. Each cutting edge is formed by a *radial rake angle*, a *chamfer angle* from 40° to 50°, and a *chamfer relief angle*. A machine reamer cuts to the outside diameter (*margin*).

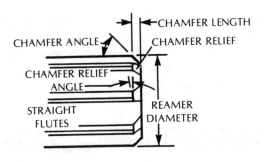

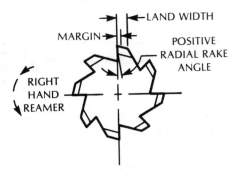

Fig. 21-1. The cutting end features of a machine reamer.

TYPES OF MACHINE REAMERS

The following eight common types of machine reamers are described and illustrated:
- General *jobber's reamer*
- *Shell reamer*
- *Fluted chucking (machine) reamer*
- *Rose chucking reamer*
- *Expansion chucking reamer*
- *Adjustable chucking reamer*
- *Step reamer*
- *Taper* (machine) *reamer.*

Each of these types is available with straight or taper shank and straight or spiral flutes.

Jobber's Reamer

The *jobber's (chucking) reamer* is a one-piece machine reamer. Jobber's reamers are general-purpose reamers. A straight-flute jobber's reamer with a taper shank is illustrated in Fig. 21-2.

Shell Reamer

The *shell reamer* consists of two parts. An *arbor* and *shank* form one part. The second part is a reamer shell that includes the flutes and cutting teeth. The bore of the shell has a slight taper which fits the arbor

taper. A pin (*lug*) in the arbor fits the two slots in the shell. Fig. 21-3 illustrates straight-and spiral-flute shell reamers and an arbor.

A shell reamer is used as a finishing reamer for dimensional accuracy. The shell is replaceable, making this reamer an economical type. Also, the same arbor may be used with a number of different sized shell reamers.

Fig. 21-2. A straight-flute jobbers reamer. (Courtesy of Morse Cutting Tools Division)

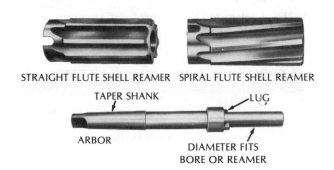

Fig. 21-3. Shell reamers and an arbor. (Courtesy of Morse Cutting Tools Division)

Fluted Chucking Reamer

The *fluted chucking reamer* is designed for cutting on the end and along the teeth. The teeth have a slight chamfer ground on the ends, for end cutting. A clearance angle is also ground along the entire length of each tooth. The fluted reamer is used for finish and difficult reaming operations. Straight-and taper-shank fluted chucking reamers are shown in Fig. 21-4.

A circular margin of 0.005″ to 0.020″ (0.12 mm to 0.5 mm) runs the length of the flute. The lands are *backed off* to provide *body clearance*. The cutting end has a 45° bevel. Helical flutes are especially adapted to produce smooth, accurate holes and a free-cutting action. A helical-fluted reamer must be used when there is an interruption in the roundness of a hole, such as a keyway or a cutaway section.

Rose Chucking Reamer

A *rose chucking reamer* (Fig. 21-5) is a coarse-toothed reamer in comparison with a fluted machine reamer. The teeth are ground at a 45° *end-cutting angle*. The beveled teeth do the cutting. The rose chucking reamer is used where a considerable amount of material is to be removed. Under these conditions, the rose reamer produces a rough, reamed hole.

Fig. 21-4. Straight-and taper-shank fluted chucking reamers. (Courtesy of Morse Cutting Tools Division)

Fig. 21-5. A rose chucking reamer with right-hand spiral flutes. (Courtesy of Morse Cutting Tools Division)

These reamers are used to rough ream to within 0.003″ to 0.010″ (0.08 mm to 0.3 mm) undersize. The teeth have a *back taper of 0.001″* per inch running the entire length of each flute. The *back taper* provides hole clearance so that the reamer does not bind in deep-hole reaming. The outside diameter is ground concentric. There is no land.

Expansion Chucking Reamer

As the name suggests, the diameter of an *expansion chucking reamer* (Fig. 21-6) can be increased slightly. An adjustment feature on the end permits enlarging the reamer to produce a hole that is larger than a standard size. Compensation can also be made for wear. This feature increases the tool life.

Fig. 21-6. An expansion chucking reamer. (Courtesy of Morse Cutting Tools Division)

Adjustable Chucking Reamer

Adjustable chucking reamers are produced with

high-speed steel (HSS) blades (Fig. 21-7A) or carbide-tipped, high-speed steel blades (Fig. 21-7B). The blades can be adjusted to increase the range of the reamer up to 1/32″ (0.8 mm) in diameter. The blades are replaceable.

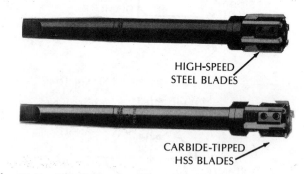

HIGH-SPEED STEEL BLADES

CARBIDE-TIPPED HSS BLADES

Fig. 21-7. Adjustable chucking reamers. (Courtesy of the Acme-Cleveland Corp.)

The carbide-tipped chucking reamer is especially adaptable to the reaming of ferrous and nonferrous castings and other materials. These materials require the cutting tool to have *high abrasion-resistant qualities*. The cutting tool must withstand high cutting speeds and temperatures.

Step Reamers

Step reamers produce two or more concentric diameters in one operation. Fig. 21-8A is a line drawing of a two-step reamer. A common combination reamer is ground with one portion of the fluted area at a smaller diameter, for rough reaming. This is followed by the standard diameter for final precision reaming. Three examples of step reaming combinations are illustrated in Fig. 21-8B.

(A) TWO-STEP REAMER

TWO REAMED DIAMETERS

ANGULAR SHOULDER   ROUND SHOULDER   ROUGH AND FINISH REAMED

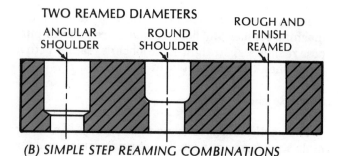

(B) SIMPLE STEP REAMING COMBINATIONS

Fig. 21-8. A two-step reamer and its applications.

Taper Reamers

Machine taper reamers are of similar design to hand taper reamers. The tapers conform to standard Morse, Brown and Sharpe, SI Metric system, and other special tapers. A standard Morse taper machine reamer is shown in Fig. 21-9.

TAPER SHANK

HELICAL FLUTES

Fig. 21-9. A Morse taper machine reamer. (Courtesy of Morse Cutting Tools Division)

MACHINE REAMING SPEEDS AND FEEDS

Machine Reaming Speeds

The speed at which a hole is reamed depends on the following conditions:
- The required dimensional accuracy
- The material to be reamed
- The kind and flow of the cutting fluid on the cutting edges and workpiece
- The desired quality of surface finish
- The type and condition of the machine spindle
- The size of the hole to be reamed.

The cutting speed for general machine reaming operations is approximately 60% of the cutting speed required for drilling the same material. A table may be used to determine the correct cutting speed. Table 21-2 gives some examples of cutting speeds for HSS drills in surface feet per minute (sfpm) and meters per minute (m/min). The values are converted to cutting speeds for reaming by multiplying by 0.6 (60%).

Machine Reamer Feeds

Like hand reamers, machine reamers should be fed two or three times deeper per revolution than in drilling. The amount of feed depends on the material and required quality of surface finish. As a general rule, the highest rate of feed should be used to produce the desired finish. Operator judgment is important. A starting feed of from 0.0015″ to 0.004″ (0.04 mm to 0.1 mm) per flute per revolution may be increased by observing the cutting action, dimensional accuracy, and surface finish.

Table 21-2. SELECTED CUTTING SPEEDS FOR HIGH-SPEED STEEL (HSS) DRILLS

Material		Cutting Speed	
		(sfpm)	(m/min)*
Steel	Low carbon (.05-30%)	80-100	24-30
Cast iron	Soft gray	100-150	30-45
Nonferrous	Aluminum and alloys	200-300	60-90
Plastics	Bakelite, et al	100-150	30-45

*Meters per minute

REAMER ALIGNMENT AND CHATTER

Reaming accuracy depends on *alignment*. The axis of the hole to be reamed, the reamer, and the machine spindle must be aligned. When a reamer bushing is used on a jig, this, too, must be aligned with the reamer axis. Any *misalignment* produces excessive wear on the bushing and reamer. The hole is also reamed inaccurately. A *floating reamer holder* reduces misalignment.

Chatter always affects dimensional accuracy and the quality of the machined hole. As mentioned earlier, chatter is caused by:
- Excessive cutting speed
- Too light a feed
- Improper grinding with excessive clearance
- An incorrect setup in which the workpiece is not held securely
- *Play*, or looseness, in the spindle or floating holder.

When chatter is produced, the operation must be stopped and the cause corrected.

CUTTING FLUIDS FOR REAMING

The reaming of most materials requires the use of a cutting fluid. As in other machining operations, the cutting fluid is a coolant. Where necessary, the cutting fluid also serves as a lubricant. A cutting fluid helps produce a finer cutting action and surface finish, and prolongs tool life.

Single workpieces of gray cast iron are reamed dry. A stream of compressed air is used as a coolant in production reaming of cast iron. Mineral-lard oils and sulphurized oils are recommended in general reaming practices. Tables of cutting fluids and manufacturers' data are referred to in selecting an appropriate cutting fluid. This is particularly important when the cutting fluid is flowed on the work through a coolant system.

VARIABLES IN REAMING

The cutting action of reamers depends on a number of variables. The judgment of the craftsperson is based on considering these factors:
- Speed and feed
- Material of the workpiece
- Condition of the machine
- The rigidity of the setup
- The kind of reamer used
- The rake of the cutting edge
- The clearance angle
- The required degrees of dimensional accuracy and surface finish
- The amount of stock to be removed
- The proper cutting fluid, where required, and its application
- The accuracy and type of reamer guide bushings, when used
- The condition of the drilled hole.

COMMON CAUSES OF REAMER BREAKAGE OR EXCESSIVE WEAR

Excessive wear and breakage of reamers may be caused by any one or a combination of the following incorrect practices or conditions:
- The taper socket or machine spindle may have burrs or other foreign particles on the surfaces
- The setup has one or more parts out of alignment. (This causes excessive wear on the lands of the reamer and produces a *bell-mouthed hole*.)
- The cutting speeds may be too fast or too slow
- The composition of the cutting fluid is not suited to the kind of material being reamed. (Also, lubrication may be lacking between a guide bushing and the reamer.)
- The stock allowance may be too much or insufficient for reaming
- The drilled hole may be too rough, tapered, or bell-mouthed. (These cause a wedging action.)
- Faulty grinding, such as:

  ○ Improperly ground clearance of the cutting edges
  ○ Unbalanced grinding of the cutting edges.

(Only a few teeth cut. These teeth take the whole cutting load.)
  ○ Grinding a wrong end-cutting angle
  ○ Producing *saw-toothed* cutting edges, which result from too coarse a grind
  ○ Grinding the cutting edges so fast and with such a heavy cut that grinding cracks are produced
- A reamer bushing may be oversize or undersize
- An improperly designed reamer may be used
- Spindle play, lack of rigidity in the setup, and chattering
- Incorrect rotational direction of the workpiece or reamer
- Forcing the reamer when entering the workpiece or when bottoming
- Failure to cut below scale and hard spots in cast and forged materials
- Nonremoval of abrasive particles before reaming.

REPRESENTING AND DIMENSIONING DRILLED AND REAMED HOLES

Center-to-center distances between holes, their location in relation to a special surface, hole size, and allowable tolerance are all provided on an industrial blueprint. As stated before, drawings are the common language among the designer, builder, and operator. The workpiece is represented on a drawing with lines, views and dimensions.

Standard Hole Dimensioning Practices

Drilled and reamed holes are usually located by centerlines (Fig. 21-10). These establish the hole location in relation to other holes and surfaces. The hole itself is shown as a circle. The dimension

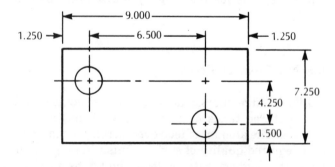

Fig. 21-10. Centerlines used to locate holes.

of the hole and the required degree of accuracy may be given as DRILL ½ $^{+0.005}_{-0.000}$". If the hole

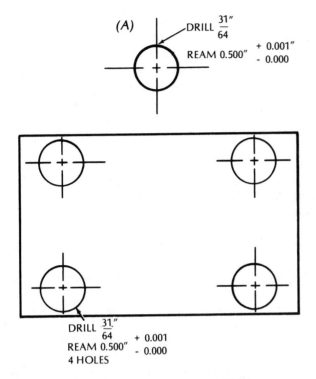

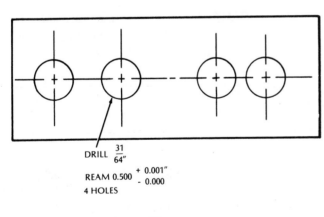

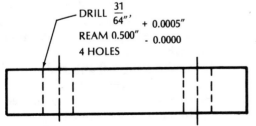

Fig. 21-11. Dimensioning a drilled and reamed hole.

(B) DIMENSIONING MULTIPLE DRILLED AND REAMED HOLES

Fig. 21-12. Representing and dimensioning reamed holes.

is to be machine reamed to a closer tolerance, the dimension would read as shown in Fig. 21-11. Just one hole may be dimensioned for a number of holes of the same size. The example in Fig. 21-11A shows that a 31/64″ hole should be reamed to 0.500″ with a tolerance of $^{+0.001″}_{-0.000}$. The workpiece in Fig. 21-11B contains four holes with these same dimensions.

A hole that is represented and dimensioned in a one-view drawing appears in Figure 21-12. Two or more holes that are to be aligned may carry a note to "REAM IN ASSEMBLY" (Fig. 21-13).

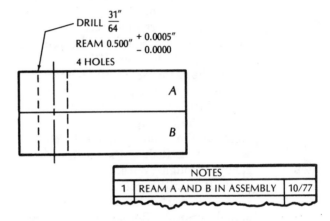

Fig. 21-13. A notation to ream in assembly.

Holes In Round Pieces

Many round pieces are represented in a single view. Dimensions of round surfaces that are concentric with the axis are indicated by the sign ⌀ (Fig. 21-14). An additional dimension depth is shown in Figs. 21-14 and 21-15.

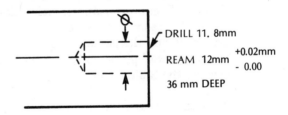

Fig. 21-14. Use of symbol and other dimensional information.

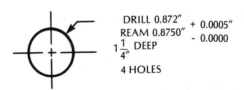

Fig. 21-15. Notes for reaming holes.

> ## HOW TO MACHINE REAM HOLES

Reaming Through Holes

Step 1   Align the hole axis and spindle axis. Center drill. Drill the hole with the reamer drill.

Step 2   Measure the reamer diameter. Check the taper shank, spindle, and adapter (if used) for burrs or nicks.

Step 3   Determine the reamer speed and feed. Set the spindle speed at the required sfpm and feed per revolution.

Step 4   Guide the reamer carefully to start it in the drilled hole. Apply a cutting fluid with a brush or spout container so that it reaches the cutting lips.
Note: If the reamer starts to chatter, withdraw it. Reduce the surface speed.

Step 5   Ream through. Withdraw the reamer while it is still revolving.

Step 6   Clean the workpiece. Check the dimensional accuracy. Reream if necessary.

Step 7   Remove burrs from the workpiece and reamer.

CAUTION: Replace the machine reamer in an appropriate holder.

Reaming Two Aligned Holes

Step 1   Clamp the workpieces so that the axis of the holes to be reamed are aligned.
NOTE: Whenever possible, holes that are to be reamed together should be drilled during the same setup.

Step 2   Strap the workpieces on parallels in a vise, on the table, or in another holding device.

Step 3   Align the axis. Drill the reamer-size hole. Follow with a finish-size reamer.

Step 4   Check the dimensional accuracy. Reream if necessary.

Producing A Duplicate Reamed Hole

Step 1   Align the second workpiece in relation to a hole that is already reamed. Clamp securely.

Step 2   Use a drill the size of the reamed hole. Align the drill with the axis of the previously reamed hole.
NOTE: A transfer punch is sometimes used to locate the position of a reamed hole in another workpiece.

Step 3   *Spot* the center by drilling for a short distance to produce a cone-shaped center. Remove the *spotting drill.*

Step 4   Use the reamer (undersize) drill and drill the hole.

Step 5   Select an appropriate type of machine reamer. Machine ream.

Step 6   Test for dimensional accuracy.

Step 7   Insert a *pin* (*plug*) that fits snugly in the reamed hole.

Step 8   Proceed to spot, drill, and ream the second hole.

Step 9   Insert a pin or plug in the second reamed hole if additional holes are to be reamed. This keeps the holes in the correct position.

Step 10   Continue to spot, drill, and ream any remaining holes. The two locating pins are enough to maintain the hole locations accurately.
Note: Continue to check the dimensional accuracy and the quality of surface being produced.

Step 11   Clean the workpieces, tools and machine. Carefully remove the reamer. Check and stone off any burrs.

CAUTION: A sharp fine burr may be produced in machine reaming. Avoid rubbing over a reamed hole.

> ## HOW TO HAND REAM ON A DRILL PRESS

Note: It is assumed that the workpiece has been aligned and the reamer hole has been drilled to within 0.002″ to 0.005″ (0.05 mm to 0.12 mm) of the required size.

Step 1   Mount a center point in a drill chuck.

Step 2   Select an appropriate size tap wrench for the size reamer to be used.

Step 3   Bring the square shank end of the hand reamer to the center point.

Step 4   Hold the cutting edge of the reamer close to the hole to be reamed.

Step 5   Bring the spindle down. Align the center point with the centered end of the reamer in order to guide it.

Step 6   Turn the reamer so the cutting edges touch the workpiece.
Note: An equal force must be applied to both handles. Brush a small quantity of cutting fluid on the reamer and cutting area.

Step 7   Continue to turn and feed the reamer. Remove the reamer when the hole has been reamed a short distance.

CAUTION: The reamer is turned *clockwise* for both reaming and withdrawing.

Step 8   Check and correct any condition causing chatter, improper cutting, an inaccurately reamed hole, a poor quality of surface finish, or misalignment.

Step 9   Ream through (or to the required depth).

Step 10   Deburr the reamed hole. Check for dimensional accuracy. Reream if an undersized hole is produced.
Note: It is good practice to countersink the edge of a hole before reaming, whenever possible.

Step 11   Clean the workpiece, machine, and tools. Examine and remove any burrs. Carefully store all tools.

SAFE PRACTICES IN MACHINE REAMING

- Examine the shank, tang, and margins of the reamer for nicks and burrs. These impair accuracy and produce scratches in the work surface. Remove burrs with an oilstone.
- Store reamers in separate storage sections. This prevents reamers from rubbing against or touching any tool, instrument, or hardened part.
- Set a taper shank reamer in a sleeve, socket, or taper spindle by tapping firmly and gently with a soft-face hammer.

- Turn the reamer in a clockwise direction only. This prevents damage to the margins, cutting edges, and work surface.
- Align the axis of the reamer, workpiece and spindle to avoid *drifting* of the reamer and assure reaming to specifications.
- Strap workpieces securely. Use adequate blocking and protecting strips.
- Stop the operation whenever chatter occurs. Diagnose the cause. Take steps to remedy the condition.
- Withdraw the machine reamer while it is still turning.

MACHINE REAMER AND REAMING TERMS

Undersize hole allowance (Stock removal allowance)	Amount of material left for finish reaming. Manufacturers' recommendations of how much material to leave. The material left to produce a dimensionally accurate hole that meets specific surface finish requirements.
General reaming allowance	A common shop practice of allowing 1/64″ for machine reaming. An allowance that will produce a dimensionally accurate hole with a fair surface finish.
Two-step reaming	Producing two concentric reamed holes simultaneously with one reamer. Using a single two-step reamer to ream two different size holes and the shoulder in one operation.
Chamfer angle	An angle ground uniformly on each flute to produce a cutting edge.
Reamer body	That portion of a reamer consisting of flutes, the area between them (known as *lands*), and the margin.
Relief, or body clearance	A relief machined in back of the margin and sloping toward the flutes.
Rake angle	The angle of a cutting face in relation to the center line of the reamer. A line extending from the front edge of a margin through the center.

Cutting lip clearance	The relief for the cutting lip of each tooth.
Shank (straight and taper)	An area of the reamer by which it is held and positioned centrally with the spindle. The reamer area through which force is applied.
Machine (chucking) reamers	Two types of machine reamers: rose reamers and fluted reamers.
Rose reamer	A reamer with teeth beveled on the end and provided with a clearance. A machine reamer on which the cutting takes place on the beveled end-cutting teeth. A reamer with lands that are not backed off but, instead, taper slightly (0.001″ per inch) along the full length of the land.
Fluted reamer	A machine reamer with narrow lands backed off the entire length. A rough finishing reamer with a greater number of cutting teeth than a rose reamer. Cutting is done with cutting teeth that are rounded or beveled, and relieved.
Shell reamer	A replaceable reamer section having a slight taper bore. A shell cutter with a taper bore that aligns on a tapered arbor. Slots in the shell and the tapers transmit the cutting force.
Cutting speed (reaming)	The sfpm recommended for a particular size reamer to cut to a desired degree of accuracy.
Reamer alignment	A condition in which the drilling machine spindle is positioned with the axis of the reamer and hole to be reamed.
Representation of reamed holes	Techniques of representing a workpiece. A dimensioned drawing or sketch that supplies full specifications for reaming holes.
Chatter	An imperfect, ridged, wavy surface. An imperfect surface produced by incorrect alignment, reaming, work holding practices, or incorrect speed.

SUMMARY

- Hand reaming produces finely finished hole surfaces. These are held to close dimensional tolerances.
- Machine reamed holes meet general standards for producing holes. A two-step roughing and finish reaming process with spiral-fluted reamers is used to obtain a higher degree of finish and dimensional accuracy than with other machine reamers.
- Precision hand and machine reaming allowances range from 0.002″ to 0.005″ (0.05 mm to 0.12 mm). The general rule for rough machine reaming is to allow 0.010″ for ¼″ diameter reamers to 0.020″ (1″) to 0.032″ (3″). Similar amounts in fractional millimeters are allowed on metric size holes.
- The fluted chucking reamer cuts at the front cutting lips and along the cutting edge of each margin. The lands are backed off, leaving a narrow margin.
- Machine reamers have either a straight or taper shank. The teeth may be straight or have a right-hand or left-hand helix.
- The cutting action of machine reamers is produced by cutting teeth on the front end. The teeth are beveled at a 40° to 50° angle. The teeth are relieved to form a cutting lip.
- A jobber's reamer is a general-purpose reamer.

- A shell reamer includes a replaceable reamer shell with a slight taper bore that fits an arbor. The shell reamer is a finishing reamer for dimensionally accurate holes.
- Machine reaming speeds are slower than those used for drilling. The feeds are considerably greater.
- Manufacturers' recommended cutting speeds, in sfpm, should be followed for the reamer size and the kind of material to be reamed.
- The rose reamer cuts on the beveled-angle cutting teeth only. This reamer is adaptable for rough reaming undersized holes. The teeth taper slightly along their length to prevent rubbing. The outside diameter of the teeth is concentric and not relieved.
- Expansion chucking reamers have the added feature of being adjustable. With them, holes may be reamed slightly larger or smaller than a standard size.
- Step reamers produce multiple concentric reamed holes.
- Adjustable blade reamers permit adjustments up to 1/32″ (0.8 mm). The blades may be reground and replaced.
- Reamed holes are dimensioned on sketches with drill and reamer size, depth, and the number of

holes of a similar diameter. The surface finish may also be included.

- Chatter may be overcome by:
  - Reducing cutting speed
  - Increasing the feed
  - Reducing the clearance on the cutting lips
  - Correct alignment
  - Securing the workpiece properly
  - Removing spindle play.
- A coolant and a lubricant, where applicable, should be used in machine reaming.
- Spiral-fluted reamers produce a finer shearing action. They are most practical where an interrupted section in the workpiece is to be reamed.
- A reamed hole should be checked at the beginning and end of the reaming process for dimensional accuracy and quality of surface.
- A reamer that is being removed from a hole must be turned in the same direction as that used for cutting.
- Reamer burrs may be removed with a countersink or triangular scraper.
- Clean and oilstone any burrs or nicks that may form on the margins, cutting lips, and shank.
- Safety precautions must be observed. Wear a shield to keep hair away from moving parts. Wear an eye-protection device to protect the eyes from chips and other particles. Loose clothing is dangerous and should be secured.

UNIT 21 REVIEW AND SELF-TEST

1. Tell how the cutting edges of a machine reamer are formed.
2. Distinguish between general applications of *fluted chucking reamers* and *rose chucking reamers*.
3. Make a rule-of-thumb statement about the cutting speeds for reaming compared with those for drilling the same material and hole size.
4. State three considerations and/or checks by the machine operator before increasing the cutting feed for a machine reaming process.
5. List five incorrect practices requiring corrective operator action to avoid excessive reamer wear and breakage.
6. Explain the function of a *floating reamer holder*.
7. Indicate how mating holes that are to be reamed are specified on a drawing.
8. Set up the series of steps to be followed in producing duplicate reamed holes.
9. Tell how to safely withdraw a machine reamer to prevent scoring a finish reamed hole.

Unit 22 MACHINE THREADING ON A DRILL PRESS

Hand and machine methods are used to cut and to form screw threads. Some threads are internal; others are external. The design features of taps and dies and their uses in hand threading were covered as bench work processes in Units 13 and 14. The production of a great number of duplicate threaded parts requires that the threads be *machined*.

This unit deals with the thread-cutting tools and accessories that are generally used on drilling machines. The thread-cutting tools either tap internal threads or cut external threads. American Unified and Metric systems of dimensioning screw threads are illustrated. Common tapping troubles and their causes are diagnosed. Recommended corrective measures are presented.

EXTERNAL MACHINE THREADING ON THE DRILL PRESS

Releasing-Type Die Holder

The *solid adjustable spring die* is one type of external thread-cutting tool. The die is held in a releasing die holder. A solid adjustable spring die and one that is mounted in a releasing holder are shown in Fig. 22-1. The holder is usually mounted in a chuck. The dies are adjusted for thread depth by adjusting the cap against the tapered portion of the spring die. As the die turns, it threads to the predetermined depth. When this point is reached, the die releases and remains stationary. The die holder continues to revolve with the drill press spindle.

The die is removed by reversing the spindle direction. This action causes the die holder to again engage the die and *back off* from the thread. Only a *slight* force should be applied when backing off, to avoid damage to the first thread.

Retractable Dies (Chasers)

Threads may also be produced with *die heads*. Die heads include die sections (called *chasers*), a holder (*carrier*), and a shank. The die heads may be *stationary* (hand-operated) or *revolving*. In both cases, the chasers are *closed* to cut a thread. The chasers must then be *retracted* (opened) after the thread is cut.

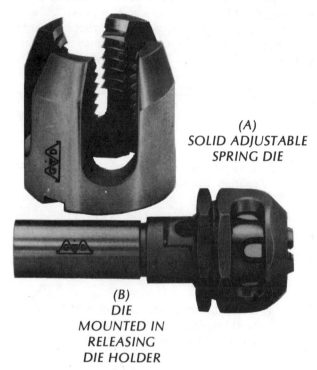

(A)
SOLID ADJUSTABLE SPRING DIE

(B)
DIE MOUNTED IN RELEASING DIE HOLDER

Fig. 22-1. An adjustable spring die and holder setup. (Courtesy of TRW Greenfield Tap & Die Division)

Stationary die heads are not used on drill presses. They instead are adapted to automatic and hand screw machines, turret lathes, and other turning machines on which the die head does not rotate. The model shown in Fig. 22-2 is used for cutting right-or left-hand threads. Chasers and carriers are changed for the *hand of thread* and size. The carriers are rigidly supported in the die head. They are easily adjusted for size. This die head has a front end *trip* that ensures machining a thread to an accurate length and close to a shoulder.

Revolving Die Head

A *revolving die head* is used for external threading on drill presses and other machines (Fig. 22-3). The head is designed to rotate about the axis of the workpiece. The chasers in the revolving die head are actuated by the machine and close automatically. The chasers are replaceable and come in sets for a required diameter and pitch. The chasers are adjustable to accommodate different depth and tolerance requirements.

Revolving die heads are designed to be opened manually or automatically. They are available for fine and coarse threading on all kinds of materials. The die head may revolve or remain stationary.

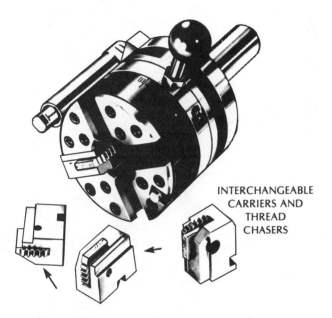

INTERCHANGEABLE CARRIERS AND THREAD CHASERS

Fig. 22-2. A self-opening insert chaser (stationary type) die head. (Courtesy of Brown & Sharpe Manufacturing Co.)

MACHINE TAPPING ATTACHMENTS

The skilled craftsperson often uses power to cut a thread with a tap on the drill press. The machine must be equipped with a spindle-reversing mechanism for such an application.

The tap is secured in the drill chuck and the hole is tapped. The spindle is then reversed. The tap is *backed out* and removed. Care must be taken to apply only a limited force and to see that the tap starts to cut immediately. Further, the tap must thread completely out of the workpiece without rubbing on the first threads.

Tapping is also done with other machine attachments and accessories. These provide for quality production and reduce tap breakage. An all-purpose drill press tapping attachment is shown in Fig. 22-4.

Reversing Spindle Tap Driver

Reversing spindle tap drivers are used on drilling machines that have a reversing spindle. The *tap driver* holds the tap securely while it is turned to cut the thread and while the tap is removed. The spindle direction is reversed to remove the tap.

Nonreversing Spindle Tap Driver

Nonreversing spindle tap drivers are adapted to drilling machines where the spindle rotates in one direction only. The tapping attachment may be mounted in the drill chuck. The attachment may also

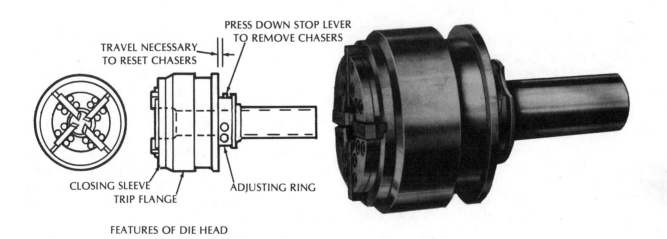

PRESS DOWN STOP LEVER TO REMOVE CHASERS

TRAVEL NECESSARY TO RESET CHASERS

CLOSING SLEEVE
TRIP FLANGE

ADJUSTING RING

FEATURES OF DIE HEAD

Fig. 22-3. A rotary self-opening die head. (Courtesy of TRW Greenfield Tap & Die Division)

Fig. 22-4. An all-purpose drill press tapping attachment. (Courtesy of Buck Supreme Inc.)

be held directly in the tapered spindle. The tap is *chucked* in the attachment and fed to the workpiece. The downward force causes the *forward clutch* to engage and drive the tap. Raising the spindle lever releases the torque on the tap. The reversing mechanism in the attachment is engaged. The tap is backed out. Fig. 22-5 shows a nonreversing drill press tapping attachment.

Torque-Driven Tapping Attachment

Friction clutch mechanisms are provided for closely controlling the amount of torque (force) that may be applied. *Torque-setting devices* are set according to the size and strength of the tap. When the torque setting is exceeded, the mechanism releases the tap. This prevents tap breakage. Accumulation of chips in the tap flutes, forcing of the tap at the bottom of a blind-hole tapping operation, and improper use of cutting fluids often require that excessive force be applied to a tap. These conditions and the resultant force may fracture the tap. This is prevented by using a torque-driven tapping attachment such as that shown in Fig. 22-6.

Micro And Heavy-Duty Tapping Attachments

Micro-tapping attachments (Fig. 22-7) are designed

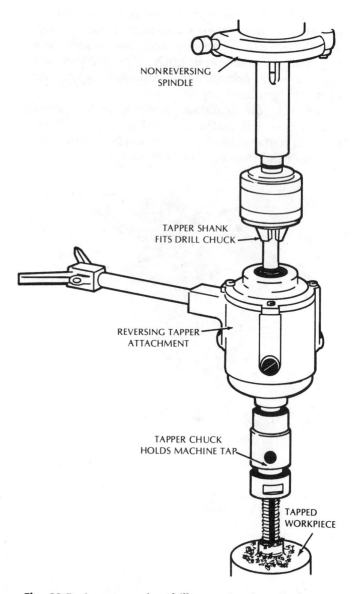

Fig. 22-5. A nonreversing drill press tapping attachment. (Courtesy of Buck Supreme Inc.)

for small taps in the #00 to #10 range. *Heavy-duty, self-contained tapping attachments*, by contrast, have *speed reducers* (Fig. 22-8). These are used for tapping operations that normally require a heavy-duty drill press. The speed-reduction feature increases the torque driving capability of the attachment. The *floating driver* feature compensates for any slight misalignment.

An all-purpose drill press tapping attachment works equally well with mild steels, tool steels, and other nonferrous metals. Lubrication systems are provided to concentrate the cutting fluid within the cutting area.

Fig. 22-6. A torque-driven tapping attachment. (Courtesy of Do All Company)

Fig. 22-7. A micro-tapping attachment. (Courtesy of Do All Company)

Fig. 22-8. A heavy-duty, self-contained tapping attachment. (Courtesy of Do All Company)

FACTORS AFFECTING MACHINE THREADING EFFICIENCY

Cutting Speed Considerations

There are many variables which affect the cutting speed required for machine tapping and other thread cutting processes. Each of the following factors must be considered before the craftsperson makes a judgment about cutting speed:
- The kind of material to be tapped
- The hardness and heat treatment of the workpiece
- The pitch of the thread
- The length of the threaded section
- The chamfer in the hole or the beveled edge of the workpiece
- The required percent of full thread depth
- The quality and fit (class) of the thread
- The quantity and type of cutting fluid

257

Table 22-1. RECOMMENDED CUTTING FLUIDS FOR TAPPING SELECTED METALS

	Metal	Recommended Cutting Fluid for Machine Tapping
Ferrous	Plain carbon and alloy steels Malleable iron Monel metal	Sulfur-base oil with active sulfur
	Tool steel High-speed steel Stainless and alloy steels (heat treated to a higher degree of hardness)	Chlorinated sulfur-base oil
Nonferrous	Aluminum, brass, copper, manganese bronze, naval brass, phosphor bronze, Tobin bronze	Mineral oil with a lard base
	Aluminum and zinc die casting metal	Lard oil diluted with up to 50% kerosene

• The designs of the drilling machine and the machine tapping attachment.

The recommended cutting speeds for general machine tapping are given in Appendix Table A-20. These cutting speeds must be adjusted to compensate for the preceding factors. In addition, the following conditions must be considered:

• As the length of a tapped hole increases, the cutting speed must be *decreased*. The chips tend to accumulate. This prevents the full flow of the cutting fluid, and additional heat is produced on the cutting edges.

• Taps with long chamfers (taper taps) may be run faster than plug taps in holes that are tapped for a short distance. Plug taps have just a few chamfered threads.

• Plug taps may be run faster than taper taps in holes that are to be tapped to a great depth.

• A 75% depth thread may be cut at a faster speed than a full-depth thread.

• The cutting speed for coarse thread series taps that are larger than 1/2" is slower than fine thread series taps of the same diameter.

• The cutting speed for taper thread (pipe) taps is from one half to three quarters of the speed used for straight thread tapping.

• Greater tapping speeds are employed when tapping is automatically controlled (versus manual operation).

General Cutting Fluid Recommendations

Cutting fluids serve the same functions in machine tapping as they do in reaming, countersinking, counterboring, and other machine processes. Friction is reduced, tap life is increased, surface finish is improved, and thread dimensions may be better controlled when the correct cutting fluid is used. Table 22-1 lists a number of metals and recommended cutting fluids.

Plastic and cast iron parts are machine tapped dry or with compressed air. The air jet removes the chips and cools the tap.

ANALYSIS OF COMMON TAPPING PROBLEMS

Certain precautions must be taken to produce quality threads and to prevent damage to a tap or workpiece. For example, the drill press spindle must be free to slide. Although no feed is necessary for tapping, a slight force is required at the start. The tap must *bite* into the workpiece. Otherwise the top of the hole will be reamed at a taper. Once started, the tap feeds itself.

It is a good machining technique to occasionally turn the tap backward to *break the chip*. On soft materials, the tap should be backed out several times to break and remove the chips.

Table 22-2 analyzes the causes of eight common tapping problems and suggests corrective measures that may be taken for each.

Table 22-2. GENERAL TAPPING PROBLEMS: CAUSES AND RECOMMENDED CORRECTIVE STEPS

Problem	Probable Causes	Corrective Measures
Tap breaking	• Tap drill size too small	• Use correct size tap drill
	• Dull tap	• Sharpen cutting edges
	• Misalignment	• Check holder and tap. Align concentric with axis of tapped hole.
	• Excessive force in bottoming	• Drill tap hole deeper if possible • Reverse the tap direction earlier • Correct the torque (driving force) or replace the machine tapper
	• Tapping too deep	• Use spiral point or serial taps
Teeth chipping	• Chips loading on cutting edges	• Back the tap to break the chips • Check the cutting fluid
	• Jamming the tap at the bottom in blind hole tapping	• Correct the reversing stop or reverse sooner • Drill a deeper hole
	• Chips packed in the blind hole	• Remove and clean the tap during tapping
	• Work hardening	• Check prior hole-producing processes
Excessive tap wear	• Sand and abrasive particles	• Machine tumble or wire brush to remove foreign matter
	• Incorrect or inadequate quantity of cutting fluid	• Consult chart for appropriate cutting fluid • Position the lubricant nozzle so that the cutting fluid reaches the cutting edges
	• Worn and dull tap	• Sharpen a dull tap • Replace a worn tap
Undersized threads	• Enlargement and shrinking of work-piece during tapping	• Use an oversized tap • Sharpen the cutting faces so that the tap cuts freely
Oversized threads	• Loading	• Check both the quality of the cutting fluid and the quantity that reaches the cutting edges
	• Misalignment	• Align the tap and work axes before starting to tap
	• Worn tapping attachment or worn machine spindle	• Adjust the *play* in the spindle or tapping attachment.
Bell-mouthed hole	• Excessive floating of machine spindle or tapping attachment	• Position the machine spindle and tapping attachment accurately
	• Misalignment	• Align the axes of the tap and workpiece
Torn, rough threads	• Incorrect chamfer or cutting angle	• Grind the starting teeth and cutting angle properly on a tool and cutter grinder
	• Dull tap teeth	• Grind the cutting faces of the tap
	• Loading	• Back the tap and remove the chips
	• Inadequate cutting fluid	• Consult the manufacturer's recommendations for the appropriate cutting fluid
Wavy threads	• Misalignment	• Align the axes of the spindle, tap, tap holder, and workpiece
	• Incorrect thread relief or chamfer of teeth	• Grind the cutting edges and chamfered teeth concentrically

FEATURES OF MACHINE TAPS

The terms and general features of taps were covered under hand taps and tapping processes in Units 13 and 14. Machine taps are similar except for two major differences. First, the shank end of a machine tap is shaped to fit the tapping attachment holder. Instead of the square head that provides a gripping surface for hand taps, the whole shank of a machine tap is round. The shank with a groove cut lengthwise fits a particular size and shape of chuck.

Secondly, machine tapping requires the use of a high-speed steel (HSS) tap. The HSS tap withstands higher operating temperatures and speeds than a carbon steel hand tap.

DUAL SYSTEM OF REPRESENTING SCREW THREADS

Screw thread standards are still being designed and approved by American, European, and other engineering and standards-setting bodies. The two basic systems of screw threads include the *Unified Inch System* of the United States, Britain and Canada, and the *SI Metrics System*. Ultimately, the International Organization for Standardization of the Metric Thread Series (ISO) may produce a single, composite system. Until that time, the craftsperson must be able to interpret drawings on which thread dimensions are designated in *American Unified thread sizes* or *SI Metric* sizes, or both.

There are no precise metric equivalents for many American Unified thread sizes. Engineering tables provide data on the present American Unified thread sizes, the outside diameter and pitch of SI Metric sizes, and the best SI Metric equivalent.

Drawings may be *dual dimensioned* to designate the thread size in both the SI Metric and the American Unified thread series. The dual dimensioning of a 1/2"-20 American Unified Fine thread is illustrated in Fig. 22-9. The outside diameter is .500". There are 20 threads per inch. Although the thread form between the two systems is different, the *best metric equivalent* that permits interchangeability is M12.7-1.27. The *M12.7* gives the equivalent outside diameter. The *1.27* (mm) pitch corresponds with the 20 threads per inch (UNF).

CHECKING AND MEASURING A THREADED PART

Internal and external threads may be measured in various ways. When a thread is machine tapped or

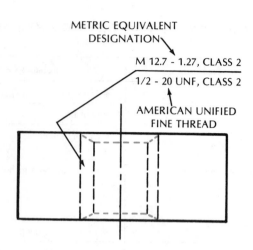

Fig. 22-9. Dual dimensioning of a tapped hole.

cut with a die on a drilling machine, its size may be checked with a bolt, nut, or the mating part. Threads that are held to closer tolerances may be gaged with *thread plug* or *ring gages*. The pitch diameter of a thread may be measured with a thread micrometer or by using three wires and a standard micrometer.

The threaded parts must be cleaned thoroughly. Burrs must be removed before any instrument or gage is used.

HOW TO MACHINE TAP (DRILL PRESS)

Step 1   Select the size and type of machine tap to use. These depend on the accuracy of the required thread and the kind of material to be tapped.

Step 2   Check the drawing specifications for the percent of full depth of thread required. Check the size of the tap drill hole.

Step 3   Select a tapping attachment that accommodates the tap size and is appropriate to the design of the drill press.
Note: A reversing tap driver may be used on a drill press with a reversing spindle. A nonreversing tapping attachment is practical for any drill press. Heavy-duty threading may be done with a speed-reducing unit.

Step 4   Mount the tapping attachment in a chuck or in the tapered spindle. Adjust the torque according to the size of tap and the material to be tapped.

Step 5   Set the spindle speed. Position the cutting fluid nozzle.

Step 6   Align the axis of the tapping attachment and tap with the center line of the workpiece. Check to see that the work is held securely.

Step 7   Start the drill press. Bring the tap carefully to the chamfered edge of the drilled hole. Apply a slight force so that the tapered threads of the tap begin to engage.

Step 8   Continue to cut the thread. Continuously apply a cutting fluid.
CAUTION: Back the tap out of the hole. Brush away the chips in deep-hole or blind-hole threading. Chips tend to accumulate.

Step 9   Reverse the spindle direction after the thread is cut when using a reversing tap driver.
Note: To engage the reversing mechanism of a nonreversing attachment, move the spindle handle upward.
CAUTION: Avoid using any force when withdrawing a tap. This is particularly important on the last thread.

Step 10   Clean the workpiece and tap. Remove burrs. Check the threads with a bolt, stud, or other thread gage.

CAUTION: The workpiece and tap should be covered if air pressure is used. The air blast should be directed away from the operator.

> ### HOW TO CUT AN EXTERNAL THREAD ON A DRILL PRESS

Step 1   Select a solid adjustable spring die and holder or a retractable die and head. The die and holder should be appropriate to the size of thread to be cut.

Step 2   Set the releasing die holder or the chasers to the required depth. Then release.
Note: A bolt, stud, or other threaded part is often used in the shop for setting an adjustable solid die or chasers.

Step 3   Mount the threading attachment in a chuck or drill press spindle.

Step 4   Check to see that the workpiece is held securely. The axes of the workpiece, die, and holder must be aligned. Position the cutting fluid nozzle.

Step 5   Set the spindle feed.

Step 6   Bring the die into contact with the chamfered surface of the workpiece.
Note: When only one part is to be threaded, the cutting fluid is often brushed on.

Step 7   Reverse the cutting process. This breaks the chips and prevents the formation of an excessive amount of chips. These tend to clog the die.

Step 8   Continue to cut the thread. Reverse the spindle direction if a releasing die holder is used.
CAUTION: Run the die completely off the workpiece to clear the first few threads and to avoid damaging them.

Step 9   Clean the workpiece, holder and cutting surface. Remove burrs.

Step 10   Check the quality of threads and size. Use a nut, mating part, or thread ring gage to measure the finished threads.

> ### SAFE PRACTICES IN MACHINE THREADING AND TAPPING

- Align the axes of a thread-cutting attachment, tap or die, and the workpiece. Alignment helps prevent damage to the tap or die and produces an accurately threaded part.
- Direct an adequate supply of cutting fluid to reach the cutting edges of the threading tool. A cutting fluid improves tool life, inceases cutting efficiency, and helps produce a quality surface finish.
- Avoid applying excessive force or jamming of a tap. These cause a tap to fracture.

- Free the drill press spindle so the tap, when once started, works freely into or out of the workpiece.
- Remove a broken tap safely with a tap extractor. Avoid brushing the hand over the broken tap.
- Clean and remove burrs before measuring or assembling a workpiece.
- Observe general safety precautions related to rotating machines and the safe handling of cutting tools.

MACHINE (DRILL PRESS) THREADING TERMS

Releasing type (external threading) holder	A chucking device with a tapered nose in which a solid adjustable spring die is held. A die holder which permits a die to release at a preset depth and torque.
Retractable die (chasers)	A series of chasers mounted in a die head. The die head may be *closed for threading* or *opened after a thread is cut*.
Revolving die head	A die head with replaceable thread chasers. A die head that rotates about the axis of a workpiece.
Spindle reversing mechanism (drilling machine)	A design feature of a drilling machine. A device for reversing the direction of a spindle to *back out* a machine tap or die.
Reversing tap driver	A tapping device used on a reversable spindle drill press. A mechanism for both driving and removing a machine tap.
Nonreversing tap driver	A machine tap holding and driving device that has an internal reversing clutch. Reversing takes place when a downward or upward force is exerted on the spindle.
Torque setting	A design feature of machine threading attachments. Setting the force needed to turn a tap or die during machine threading.
Threading efficiency factors	A series of variable conditions that the craftsperson must consider when producing threaded holes and parts. These must meet standards which conform to quality and quantity specifications.
Cutting speed considerations	Background information that must be weighed when selecting the most appropriate cutting speed in machine tapping processes.
Dual representation of screw threads	Design, drafting, and engineering practices used in industry to represent and dimension screw threads. Dual representation gives the best equivalent in both the SI Metric and Unified Thread Series for a particular size thread.
Thread plug or ring gage	Threads ground precisely on a tool for measuring and gaging purposes. Gaging the size of internal (plug gage) or external (ring gage) threads.

SUMMARY

- Machine dies are used for cutting external threads with the drill press.
- The solid type machine die permits adjusting to produce a thread of a specified depth.
- The spindle direction is reversed to remove the die when the releasing type die holder is used.
- Retractable dies (chasers) are mounted in either stationary or revolving types of holders. The chasers are held in a closed position for threading. They are opened after a thread is cut.
- A revolving die head rotates about the axis of a workpiece to cut an external thread. The head may be opened or closed manually or automatically.
- The reversing tap driver is used on a reversable spindle drill press for machine tapping.

- The nonreversing tap driver applies the required clockwise force on a tap for thread cutting. Raising the spindle lever causes a reverse clutch to engage. The tap then turns counterclockwise out of the threaded hole.
- A friction clutch permits adjusting to the torque required to turn a thread-cutting tool. The amount of torque is affected by the thread size (pitch and diameter), percent of full depth, kind of material, and cutting fluid.
- Heavy-duty tapping attachments contain speed-reducing units. These increase the capability (force) of a drilling machine to produce threads. In contrast, micro-tapping attachments are used for small size taps and fine threads.

UNIT 22 REVIEW AND SELF-TEST

1. Describe how a revolving die head works.
2. Compare the operation of *torque-driven* and *heavy-duty tapping attachments*.
3. a. List three factors the craftsperson must consider for machine tapping.
   b. List three conditions that govern the selection of cutting speeds for machine tapping.
4. State what effect the use of a correct cutting fluid has on machine threading.
5. Suggest three general safety precautions that must be followed to cut quality external threads and to prevent damage to a tap, die, or workpiece.
6. Tell what the *best metric equivalent* of an American Unified thread size means when it is specified on a drawing.
7. Indicate what *dual dimensioning* provides for a threaded part.

Unit 23
COUNTERSINKING, COUNTERBORING, SPOTFACING, AND BORING

The outer surface of a hole may be cut away at an angle, recessed with a shoulder, or machined with a flat circular area. The general machining processes that produce these shapes are called: *countersinking*, *counterboring*, and *spotfacing*. A hole may also be enlarged by *boring*. These processes and the cutting tools and accessories that are used in each instance are treated in this unit. The step-by-step procedures for actually performing each operation on a drilling machine are also covered.

COUNTERSINKING AND COUNTERSINKS

Countersinking refers to the machining of a cone-shaped opening or a recess in the outer surface of a hole. The angle of the *recessed indentation* corresponds with the angle of a screw, a rivet head, or other tapered object. These seat against the angular surface. A hole that is machined at an angle is called a *countersunk hole*. The tool that produces the shape is a *countersink*.

Screw holes are countersunk so that the angular head of the screw is level (*flush*) with the work surface. Fig. 23-1A shows the cone shape of the countersunk hole. In Fig. 23-1B the hole is countersunk to the same depth as a flat-head screw. Standard countersinks are manufactured with an included angle of either 60° or 82° (Fig. 23-2). Flat-head screws have an 82° angle.

Countersinks may be made of carbon tool steel, high-speed steel, or carbides. Holes may be countersunk by chucking and revolving the countersink. The depth of cut is controlled with a stop gage or by measuring across the outside (large) diameter. The mating part may also be tried in the countersunk hole. Countersinking should be performed at about one half the speed used in drilling the hole.

Chatter is produced when the speed is too great, the workpiece is insecurely held, or the countersunk is forced. A finer surface and smoother cutting action is produced when a cutting fluid is used. This is flowed or brushed on the countersink so that it reaches the cutting tool and work area.

Fig. 23-2. A standard countersink. (Courtesy of Do All Company)

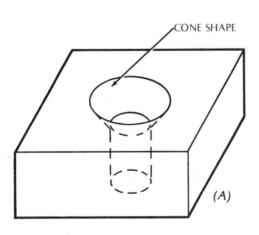

Fig. 23-1. The shape and depth of a countersunk hole.

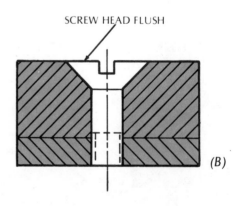

264

COMBINATION DRILL AND COUNTERSINK

One widely used shop practice for accurately guiding a larger drill is to drill a combination pilot and countersunk hole (Fig. 23-3). The pilot drilled and countersunk hole also provides a bearing surface for 60° angle machine centers. The countersunk hole keeps the workpiece centered and offsets the cutting force that is applied.

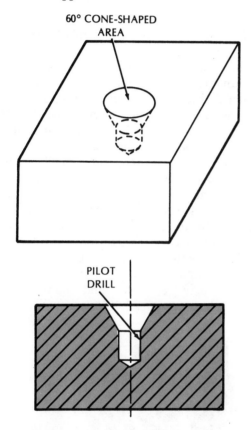

Fig. 23-3. A combination pilot and countersunk hole.

A *combination drill and countersink* is used for this purpose. The point of the cutting tool is positioned in a center-punched hole. The small-diameter point end produces a pilot hole. The angular surface of the countersink portion enlarges the hole. A 60° cone-shaped surface is formed.

The countersunk hole is drilled to the depth at which the outside diameter is smaller than the required finished hole size. Combination drill and countersinks are available either with a single cutting end or with double cutting ends.

Center drill tips are also available. These fit into adapters from which they may be easily removed and replaced. A single-point center drill tip and an adapter are illustrated in Fig. 23-4. Center drills are made of carbon tool steel, high-speed steel, or carbide. The small-diameter cones are fragile. Care must be taken not to exert excessive force on them during the drilling process. The feed per revolution is less on the countersinking portion than for normal drilling.

Fig. 23-4. A center drill tip and adapter. (Courtesy of Do All Company)

Combination drills and countersinks are produced with many sizes of points and outside diameters. A double-end solid-carbide combination drill and countersink is shown in Fig. 23-5. The cutting speeds for this tool fall within the carbide range.

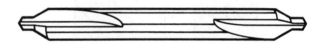

Fig. 23-5. A solid-carbide combination drill and countersink. (Courtesy of Morse Cutting Tools Division)

COUNTERBORES AND COUNTERBORING

Another shape of an enlarged and recessed hole has square shoulders. This is known as a *counterbored hole* (Fig. 23-6). Holes are counterbored to permit pins with square shoulders, screw heads, or other mating parts to fit *below* the surface of the workpiece. The cutting tool that is used for enlarging a hole is called a *counterbore*. A *counterbore* consists of a shank, body, and pilot.

The shank may be ground straight or to a standard Morse taper. An example of each type is illustrated in Fig. 23-7. The pilot is slightly smaller than the diameter of the hole to be counterbored. The pilot serves to center the counterbore. On some counterbores, like those in Fig. 23-7, the pilot is interchangeable. Different sizes of pilots may be used with one counterbore. Other counterbores that are used for only one size of screw head are made solid. Counterbores are usually three or four fluted. Counterbores are available in outside diameter sizes that vary, by increment of 1/16", from 1/4" to 2". Similar metric sizes range from 6 mm to 50 mm.

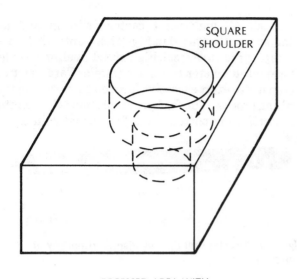

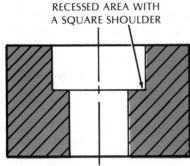

Fig. 23-6. A counterbored hole.

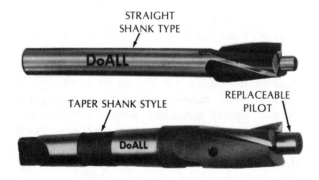

Fig. 23-7. A straight and a standard Morse taper counterbore. (Courtesy of Do All Company)

The cutting action of the counterbore is produced by the cutting lips. These are formed on the end (face) of each land. The cutting lips are radial. The lips are relieved to form the cutting angle. The flutes are spiral. These provide for the flow of chips out of the workpiece.

Counterbores made of high-speed steel are widely used. Carbide-tipped counterbores are practical where tough materials are to be counterbored. *Rough operations*, where the outer surface of the workpiece is uneven or hard (as in the case of scale on castings), require a carbide-tipped counterbore.

General counterboring operations are performed at a slower speed than that used for drilling. Cutting fluids must be used on materials that require a lubricant. The heat produced by the cutting action must be dissipated by the coolant properties of the cutting fluid.

DRILL/COUNTERBORE COMBINATION

Drilling and counterboring for socket head and other standard fasteners may be performed in one operation. A short-length drill/counterbore combination is designed for such operations. The drill/ counterbore may be a solid piece with a body of regular length or it may be a *tip*.

The tip type fits into an adapter body. The body has a slightly tapered hole that holds and secures the tip. The adapter has a hole drilled through the body. A *knockout rod* is inserted into the hole and tapped against the tip to remove it. A set of combination drill/ counterbore tips is illustrated in Fig. 23-8. This particular set accommodates socket head cap screw sizes #5, 6, 8, and 10, and 1/4″, 5/16″, 3/8″, 7/16″, and 1/2″. Holes may be drilled and counterbored to diameter tolerances of $+0.005″$.

Fig. 23-8. A set of drill counterbore tips in a holder. (Courtesy of Do All Company)

SPOTFACERS AND SPOTFACING

Spotfacing refers to the machining of a flat surface that is at right angles (90°) to a drilled hole.

Spotfacing produces an area against which a square-shouldered part may fit accurately (Fig. 23-9). The operation may be performed on a drilling machine. The end cutting tool is called a *spotfacer*. The surface is said to be *spotfaced*.

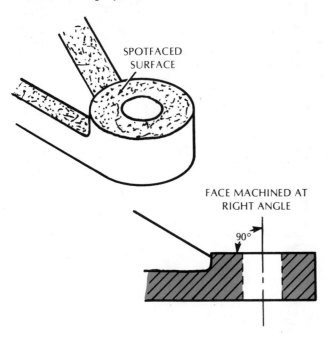

Fig. 23-9. A spotfaced surface permits correct seating of a mating part.

A counterbore may be used for spotfacing when the diameter is large enough to produce a spotfaced area of the required dimension. Another common type of spotfacer has an end-cutting tip, a shank, and a pilot end. The cutting end is removable. With this type of spotfacer it is possible to machine a large circular area. The cutting face may be sharpened easily. These features make this type of cutter inexpensive.

BORING AND BORING TOOLS

Boring describes the process of enlarging a hole by using a *single-point cutting tool* (Fig. 23-10). The cutting edge may be adjusted to accommodate different work sizes. Holes that have been formed by casting, punching, drilling, or any other process may be bored.

The cutting tool may be held in a *boring bar* or a *boring head*. The amount the tool is adjusted is indicated on a graduated scale. The boring process permits the taking of heavier finish cuts than is possible with reamers. The single-point cutting tool

is adjustable to bore any diameter within the range of the boring head.

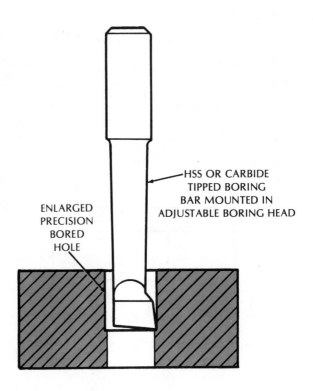

Fig. 23-10. Boring to enlarge a hole.

The taper shank of a boring head may be inserted directly in the taper bore of the drill press spindle. The cutter is secured in the offset section of the boring head. Small holes are usually bored on drilling machines, using the same tables of recommended cutting speeds that are used for similar boring operations on lathes and boring mills. The cutting speed (sfpm or m/min) is governed by:
- The kind of material to be cut
- The nature of the operation
- The type of cutting tool that is used
- The rigidity of the setup
- The design of the drilling machine.

REPRESENTATION AND DIMENSIONING OF BORED, RECESSED, AND SPOTFACED HOLES

There are drafting standards for representing and dimensioning countersunk, counterbored, spotfaced, and bored holes. The centerline is common for each operation. It designates the axis of the surfaces that are concentric.

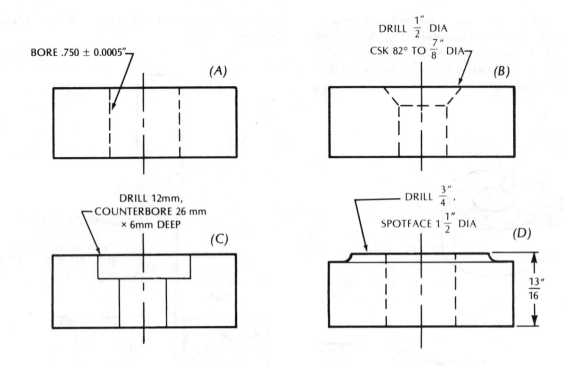

Fig. 23-11. Representation and dimensioning of different holes to be produced on a drilling machine.

Common drafting practices used for representing and dimensioning these forms of holes and surfaces in the inch-standard and metric systems are illustrated in Fig. 23-11. (A) shows the dimensioning for a bored hole. A drilled and countersunk hole is dimensioned in (B). (C) shows a drilled and counterbored hole. A drilled hole and spotfaced surface are specified in (D). The notations in each case guide the machine operator. The sizes, forms, and angles of the tools that are to be used, the required depth, and the degree of accuracy all are given.

HOW TO COUNTERSINK A HOLE

Step 1   Mount the workpiece in a holding device. Drill a hole of the required diameter.

Step 2   Select a countersink with the correct cutting angle and outside diameter.

Step 3   Determine the cutting speed and rpm. Set the spindle speed (rpm).
Note: The speed is largely governed by the maximum outside diameter of the countersunk hole.

Step 4   Replace the drill with the countersink. Bring the cutting edges carefully to the drilled hole. Note the reading on the graduated *quill* (spindle) of the machine. Set the depth stop.

Step 5   Start the machine. Brush or flow on the cutting fluid. Slowly cut to the required depth (Fig. 23-12).

Step 6   Stop the machine. Clean the workpiece. Check the depth.
Note: The mating part may be placed in the countersunk hole. The depth is checked by placing the flat edge of a steel rule across the mating part and workpiece.

Step 7   Remove any burred edge. Use a triangular scraper or a flat file.

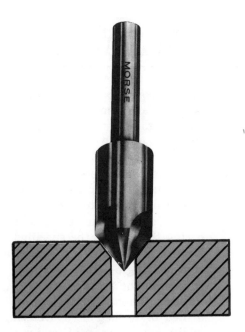

Fig. 23-12. Countersinking to a required depth. (Courtesy of Morse Cutting Tools Division)

HOW TO DRILL A CENTER HOLE

Step 1 Select the required size of combination drill and countersink. Chuck the cutting tool.

Step 2 Position the center-punched hole in line with the pilot drill point.

Step 3 Determine the cutting speed and rpm for the pilot drill. Set the spindle speed (rpm) to accommodate this diameter.
Note: This spindle speed may need to be reduced when the full depth of the point is reached and the angular (countersink) portion begins to cut. This is done to avoid chattering and to produce a finer surface finish.
Note: Brush on a small quantity of cutting fluid. This helps produce a better cutting action and a smooth-finished angular surface.

Step 4 Continue to drill and countersink to the required depth. Remove any burrs.

HOW TO COUNTERBORE A HOLE

Step 1 Select a counterbore with a pilot that is a few thousandths of an inch smaller than the drilled or reamed hole. The outside diameter of the counterbore must be the required size (Fig. 23-13).

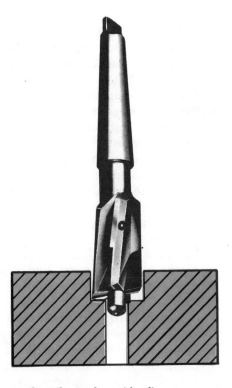

Fig. 23-13. The pilot and outside diameter meet required dimensions (carbide-tipped counterbore). (Courtesy of Morse Cutting Tools Division)

Step 2 Determine the cutting speed, spindle rpm, and feed. These depend on the outside diameter and the kind of material to be counterbored. Set the machine speed and feed.
Note: Reamer speeds for the same diameter may be used.

Step 3 Position the pilot in the hole. Bring the cutting edges carefully to the work surface. Set the spindle stop to the required depth.

Step 4 Apply cutting fluid to the cutting area. Start to counterbore. Continue until the hole is counterbored to the required depth.

Note: It is good practice during the process to periodically withdraw the counterbore and check the cutting action and quality of surface finish.

Step 5   Clean the workpiece. Burr the edge of the counterbored hole.

HOW TO SPOTFACE A HOLE

Step 1   Select a spotfacer with a pilot that is slightly smaller than the hole in the workpiece. The outside diameter of the cutter must meet the job requirements.

Step 2   Set the machine speed and feed. Since there is a large cutting area, the speed should be estimated by the largest outside diameter to be cut. The feed should be less than that for drilling a hole to this same diameter.

Step 3   Spotface to just *clean the surface* of the workpiece. Examine the quality of finish. Determine the depth dimension. Set the machine stop.

Step 4   Continue to spotface to the required depth. Apply a cutting fluid, either by nozzle or brush, where required.

CAUTION: If chatter marks appear, reduce the cutting speed and examine the angle of the cutting edges. These may be ground at too steep an angle.

Step 5   Clean the workpiece and burr it.

HOW TO BORE A HOLE USING A DRILL PRESS

Step 1   Select a boring bar and boring head. The boring bar diameter must be small enough to permit adjustments of the cutting tool.

Step 2   Examine the cutting edge of the boring tool. Sharpen if necessary.

Note: Check the diameter that the combination of cutting tool and boring head will bore. The diameter should permit boring the hole to the specified size.

Step 3   Determine the rpm for the recommended sfpm of a single-point cutting tool. Set the spindle at the required rpm.

Step 4   Position the cutting tool over the pre-drilled or formed hole. Determine the additional amount of material to be removed.

Step 5   Move the cutting tool to depth. Take a light cut for about 1/16". Stop the machine. Measure the diameter.

Step 6   Adjust the cutter if necessary. Take the cut. Bore through to the required depth.

Step 7   Continue to take successive cuts.

Note: If the hole is to be machined to close tolerances, more than one roughing cut may be required. The final finish cut is taken using a finer feed and a light cut of a few thousandths of an inch.

Step 8   Burr the bored hole.

SAFE PRACTICES IN COUNTERSINKING, COUNTERBORING, BORING, AND SPOTFACING

- Fit the spindle taper and cutting tool shank and tang properly. Securely seat them. Use a soft-faced hammer to carefully drive the tapers together.
- Rotate all right-hand cutting tools in a *clockwise* cutting direction.
- Examine the cutting edges of the cutting tool and stone any burrs.
- Stop the machine if there is chattering. Correct the causes of chatter.
- Use goggles. Cover the workpiece and tools if compressed air is used. Direct the air blast downward and away from other workers.

- Feed the small-diameter cutting point of a combination drill and countersink slowly and carefully into a workpiece. The point may be fractured if a heavy or jarring force is applied or if the speed is too slow.
- Use a brush to apply cutting fluid to the point where the cutting action is taking place.
- Remove wiping cloths from the work area whenever a cutting tool or spindle is to revolve.
- Secure loose clothing before operating any rotating machinery.

<table>
<tr><td colspan="2">TERMS USED IN COUNTERSINKING, COUNTERBORING, BORING, AND SPOTFACING WITH THE DRILL PRESS</td></tr>
<tr><td>Countersinking</td><td>The process of machining a conical surface at the end of a hole.</td></tr>
<tr><td>Countersink</td><td>A cutting tool with flutes and cutting edges ground at a specified included angle. An angular cutting tool used to produce a countersunk hole.</td></tr>
<tr><td>Counterboring</td><td>The process of enlarging the diameter of a hole. Recessing a hole with a shoulder to a required depth.</td></tr>
<tr><td>Counterbore</td><td>A multiple-fluted end-cutting tool. The outside diameter accommodates the diameter of a mating part that is to be inserted in a counterbored hole. A piloted end-cutting tool which produces a concentric, shouldered, recessed hole.</td></tr>
<tr><td>Spotfacing</td><td>The process of producing a surface of a given diameter that is square with the axis of a hole. A double-edge or multiple-fluted cutting tool of a required diameter. An end cutting tool that produces a flat, round surface. The surface is at a right angle to the axis of the hole.</td></tr>
<tr><td>Center drilling</td><td>The process of drilling a pilot hole and angular surface for positioning a drill or producing a center hole. A pilot and countersunk hole produced with a combination pilot and center drill.</td></tr>
<tr><td>Boring</td><td>The process of enlarging a hole with a single-point cutting tool. Machining a hole round. Accurately positioning the location of a hole.</td></tr>
<tr><td>Boring head</td><td>An accessory for holding and accurately positioning a cutting tool. Used in machining a hole to a close-tolerance dimension.</td></tr>
</table>

SUMMARY

- Shouldered pins, screw heads, and other parts often must be sunk into the workpiece. The top surface must often be flush.
- Recessed areas for angular and shouldered screws, bolts, and pins are produced by countersinking or counterboring.
- Spotfacing produces a flat, circular machined surface. The surface is perpendicular to the axis of a hole.
- Holes may be enlarged by boring. Boring tools may be adjusted to produce holes of various diameters. Bored holes may be machined to precise limits; for example, $\pm 0.0005''$ and ± 0.02 mm. Boring is a practical machining process and is adapted to drilling machines.
- A combination drill and countersink (center drill) produces a small pilot hole. This is followed by a cone-shaped section. The angular shape and pilot hole guide a drill in a fixed position or provide a bearing surface for a 60° center.
- The 60° and 82° included angle of the countersink cutting lips are common. Such countersinks accommodate screw heads, angular rivets, and parts that are manufactured to these standard angles.

- Counterbores are made of high speed steel for general machining. Tungsten carbide tips are used for heavy duty counterboring and on abrasive, hard, tough materials.
- Counterbores are designed solid or with a removable pilot. The pilot is slightly undersized to avoid scoring a previously drilled or reamed hole. The pilot guides the counterbore. A recessed hole is produced concentric with a reamed hole.
- The spotfacer is an end-cutting tool. A multiple-cutting lip counterbore of the required outside diameter may be used. Another common spotfacing tool has a two-lip cutter that is secured to the cutting bar.
- The cutting speed for spotfacing and counterboring depends on the outside diameter and the material in the workpiece.
- The cutting speed for countersinking and counterboring is slower than that for drilling.
- Cutting fluids are used in countersinking, counterboring, and spotfacing materials that require a lubricant and/or cutting compound.
- A countersunk hole is dimensioned to indicate the included angle and the outside diameter.

- The outside diameter and height of a spotfaced surface are indicated on a working drawing.
- The drill or reamer size and the outside diameter and depth are the dimensions normally provided for counterboring operations.
- General safety precautions relating to rotating spindles, cutting tools, burrs, and the use of eye protection devices must be observed.

UNIT 23 REVIEW AND SELF-TEST

1. State two main functions served in the machine shop by holes that are countersunk to either 60° or 82°.
2. Differentiate between a *combination drill and countersink* and a *center drill tip*.
3. Indicate two functions that are served by an interchangeable pilot on a counterbore.
4. Tell what the difference is between *spotfacing* and *boring* with respect to the cutting tools that are used for each process.
5. State the nature of the dimensioning information that is provided on drawings of holes that are to be countersunk, counterbored, spotfaced, or bored.
6. List the steps for boring a hole when using a drill press.
7. Give three safety precautions to observe when working with countersinks, counterbores, spotfacers, and boring tools.

PART FIVE

LATHE TECHNOLOGY AND PROCESSES

SECTION ONE

LATHE CHARACTERISTICS AND PROCESSES

The engine lathe is a precision machine tool. It is used for external and internal cylindrical machining processes. In most of these processes, material is removed from the internal or external surface of the workpiece by holding or moving the cutting edge of a formed cutting tool against the revolving workpiece. Following a brief look at the history of lathe development, this introductory section describes:

- Fundamental designs and functions of lathe components, systems, accessories, and attachments
- Machining processes commonly performed on lathes
- Categories and types of lathes and their typical applications
- General characteristics of and criteria for selecting the single-point cutting tools used in lathe processes
- General lathe operating and maintenance procedures and precautions.

Unit 24

LATHES: FUNCTIONS, TYPES, AND CONSTRUCTION FEATURES

The engine lathe is a basic machine tool. Machine processes that are fundamental to manufacturing and production are performed on the lathe. Lathe work skills, the principles of machining, and related technology may also be applied to other machine tools. Turret lathes, hand production and automatic screw machines, boring machines, and vertical turret lathes (some of which are numerically controlled) are adaptations of the lathe.

BASIC LATHE WORK PROCESSES

The lathe is used for *external* and *internal cylindrical machining processes*. The basic external operations include: *facing, straight turning, taper turning, shoulder turning, turning grooves, knurling, cutting off,* and *threading.* Basic internal operations relate to: *center drilling, drilling, boring; turning internal straight, taper, and undercut surfaces; countersinking, counterboring, reaming, tapping* and *threading.*

All of these internal and external operations are covered in detail in this part. Principles and applications of cutting tools, work-holding and driving devices, and machine setups are treated. Cutting speeds, feeds, cutting fluids and other machining practices also are included.

This unit serves as an introduction to the lathe as a machine tool. A few of the most significant historical developments are described first. The concepts of cylindrical machining that were incorporated in the early lathes led to the modern design of machine tools and production techniques. The functions of major lathe mechanisms, principal design features, and basic accessories are introduced. With this foundation, successive units in this part deal with lathe work practices and technology. These are applied to specific processes like straight turning, boring, threading, and others.

SIGNIFICANT HISTORICAL DEVELOPMENTS

The All-Iron Screw-Cutting Lathe

The term *lathe* was derived from one feature of an early *tree-turning device.* In this primitive machine, the workpiece was held between the centers of two tree trunks. The part was rotated by a rope mechanism that was foot powered. Later, the other end of the rope was attached to a flexible wood *lath* on the tree. The lath provided tension on the rope. This made it possible to produce a uniform cutting speed by foot power.

English instrument-maker Henry Maudslay built upon the experiences of earlier inventors and successively improved the lathe. In 1800 Maudslay succeeded in combining six important design features. These were incorporated in a lathe that was completely made of *iron.* The Maudslay lathe is illustrated in Fig. 24-1. This *all-iron screw-cutting lathe* combined a *spindle, slide rest, V-ways, lead screw, change gears,* and a *work support (tailstock),* and provided for a *power drive.* These same fundamental features are incorporated in the most sophisticated precision lathes of today.

Maudslay's screw-cutting metal lathe laid the groundwork for the design and manufacture of other machine tools and the mass production of interchangeable parts. The metal-cutting lathe met the growing need to industrialize. The capability of the lathe was increased. More complex parts were machined to ever higher dimensional tolerances and qualities of surface finishes.

Basic Features Of the Maudslay Lathe

The power-driven *spindle* rotated the work. The *slide rest* supported cutting tools so that they could be positioned and moved along the workpiece. The turning of the *lead screw* produced movement of the slide rest at a constant rate of feed. The *change gears* made it possible to vary the distance the slide rest moved in relation to the spindle rpm. This feature was also important in changing the nature of a *cut.* Various feeds were needed for rough and/or finish machine operations and to produce screw thread pitches of different sizes. The *tailstock* provided a *center* on which to support workpieces. One of the accessories on the Maudslay lathe was a *follower rest.* This attachment, which is not shown in Fig. 24-1, further supported long workpieces.

275

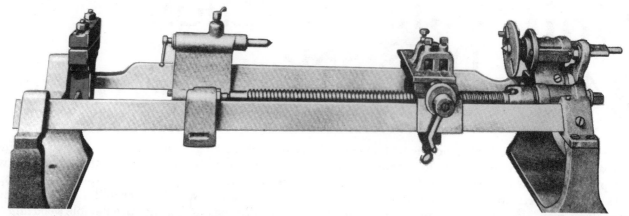

Fig. 24-1. The early Maudslay screw-cutting lathe.

Made all of iron and of heavier construction than previous lathes, the Maudslay lathe increased the speed and the accuracy with which parts could be machined, and the cutting capability to machine metal parts. The term *engine lathe* was derived from the fact that lathes at that time were power driven by steam engines.

The Hand-Fed "Eight-Station" Turret Lathe

From Maudslay's developments came the need for other machines. The milling machine, boring mill, and production screw-cutting machines are examples. The demands for metal fasteners like screws, bolts, and other threaded parts led to an adaptation of the early metal lathe.

In 1845 Stephen Fitch of Middlefield, Connecticut, designed and built an *eight-station turret lathe* (Fig. 24-2). Eight tools were mounted on and positioned by a *turret mechanism*.

Fig. 24-2. Fitch's eight-station turret lathe.

Each tool or positioning device was formed by a skilled craftsman. The tools and devices were held in the turret. Once the machine was *set up,* a lesser-skilled worker operated the turret lathe and mass-produced interchangeable parts.

During the same period, Gay and Silver of Chelmsford, Massachusetts, and James Hartness of Windsor, Vermont, also developed turret lathes. The Pratt and Whitney, and Brown and Sharpe machine tool companies were founded to manufacture small turret lathes during the Civil War. In the 1800s the northeast became the center of many significant machine tool developments.

The Automatic Turret Lathe (Screw Machine)

Still greater production demands for metal machine screws led to the invention of an *automatic turret lathe* in the late 1860s. This combined the features of Maudslay's screw-cutting lathe and Fitch's turret lathe. An added feature was called by its designer, Christopher Spencer, the *brain wheel.* This was a cam-activated device. The *brain wheel* (cam) positioned a cutting tool to perform a particular operation, like straight turning. At the end of the cut, a cam moved the cutting tool away from the workpiece. The die, mounted in a turret station, was brought to the workpiece. The die cut the threads in the turned workpiece. After withdrawing the die, other cams positioned cutting tools and stops in the different turret stations.

Since the *automatic turret lathe* was designed principally to completely machine screw threads, it became known as the *automatic screw machine.* Spencer later developed the first *multiple-spindle lathe.* Finally, he was able to mass-produce small screws by feeding coiled wire through a *three-spindle automatic screw machine.*

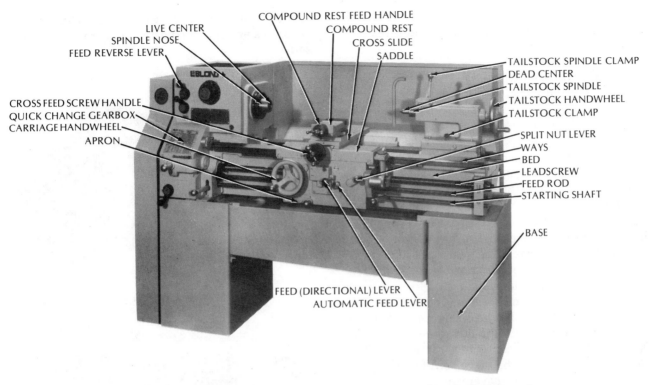

Fig. 24-3. The major features of a conventional lathe. (Courtesy of LeBlond Machine Tool Company)

Successive improvements and additions have been made to the work of Maudslay, Fitch, Spencer, and the early machine tool companies in the northeast and other parts of the United States, and throughout the industrialized nations. Each has contributed significantly to the present highly sophisticated mass-production capacity to make precision parts and mechanisms. Today, the *engine lathe* still remains a basic machine tool. A modern conventional lathe with major features identified is shown in Fig. 24-3.

Reduced to a simple explanation, the lathe is a mechanism for removing material. A formed cutting tool is moved against a revolving workpiece. The parts and mechanisms of a lathe serve one of the following three major functions, all of which are interrelated:

- To provide power to all moving parts
- To hold and rotate the workpiece
- To position, hold, and move the cutting tool.

THE LATHE BED

The body to which other components of a lathe are fitted is called the *lathe bed*. It consists of a ruggedly designed casting. *V-ways* and other *ways* are machined on the top surface of the bed. A headstock, gear box, lead screw and feed rod drive, and coolant system are usually attached to the lathe bed. The carriage and tailstock mechanisms are aligned on the ways of the bed and may be moved longitudinally.

COOLANT

The purposes for which a cutting fluid is used in lathe work are the same as those for other machining processes. The fluid serves:

- As a *coolant* to reduce the machining temperature of the cutting tool and workpiece
- As a *lubricant* to improve the cutting action and the quality of the surface finish
- To reduce the amount of *force* required during the cutting process.

Many lathes have a built-in coolant system. The system requires a *chip pan*, which is attached to the bed and base. Both the chips and the cutting fluid flow into the chip pan. The cutting fluid drains from the chip pan into a reservoir in the coolant system.

The coolant system consists of a circulating pump, a reservoir, piping, and a nozzle. The cutting fluid is pumped through a flexible hose to a nozzle. The nozzle is mounted on the carriage and is positioned to deliver the fluid where it is needed. The fluid follows the cutting action. As it returns to the reservoir, it flows through cleaning meshes. The fluid is recirculated *after* being cleaned.

POWER SOURCE AND MOTION – PRODUCING MECHANISMS

Motion for all moving parts of a lathe is transmitted from a power source, such as a *motor*. The motor is connected by belts or gears to a *headstock*, through which a spindle is driven. The speed and feed of each moving part of the lathe is directly related to the spindle movement.

End gears are used to establish precisely the rpm of the spindle in relation to the *gear train* in the *quick-change gear box*. The gear ratios, in turn, control the rpm of the *lead screw* and *feed screw* according to the spindle rpm.

The rotary motion and speed and feed of the lead screw are converted in the *apron* (Fig. 24-4). A linear movement is produced for the *carriage* or *cross* slide. On some lathes, the lead screw serves the dual function of controlling the movement for threading and feeding. These same movements of the carriage and cross slide also may be performed by hand feeding.

Fig. 24-4. The principal components of the apron and carriage. (Courtesy of LeBlond Machine Tool Company)

WORK-HOLDING AND WORK-ROTATING COMPONENTS AND ACCESSORIES

The *headstock* supports a hollow *spindle*. The *lathe spindle nose* has a ground internal taper and a special outside taper or threaded section. Details of a headstock and spindle construction are shown in Fig. 24-5.

Fig. 24-5. The construction of the headstock and the spindle. (Courtesy of South Bend Lathe, Inc.)

Work holders are mounted in or on the spindle nose. They are secured either by a threaded ring nut or by *cam-locking*. Collets and chucks are common types of work holders.

The back side of a lathe chuck has a special flange. This permits mounting the chuck directly on the lathe spindle and concentric with the spindle axis. The jaws on the front face of the chuck are adjustable. The jaws hold large-diameter workpieces, of regular or irregular shape, either *on center* or *off center*.

The *faceplate* is a work-holding and driving device. It is mounted on and secured to a spindle nose. Workpieces may be fastened directly to the faceplate with straps. Other parts may be positioned and held for machining operations by specially designed fixtures or setups.

The *driver plate* is still another work-rotating

GRADUATED
TAILSTOCK
SPINDLE

TANG SLOT FOR
EASY REMOVAL OF
TAPER SHANK TOOLS

SPINDLE
CLAMP

PRECISION
GRADUATED
COLLAR

BASE

CLAMP FOR
BASE AND MOVABLE
SECTION

MOVABLE UPPER
SECTION

Fig. 24-6. The features of a tailstock. (Courtesy of South Bend Lathe, Inc.)

device. This cylindrical plate has elongated slots. It is attached to the spindle nose. A centered workpiece is held between the headstock *live center* and the tailstock *dead center*. The tail of a lathe dog, which is clamped to the workpiece, fits one of the slots. Spindle motion is transmitted through the driver plate and the lathe dog to the workpiece.

The *tailstock* (Fig. 24-6) is a work-holding, tool-positioning, and feed mechanism. A tailstock spindle may be moved longitudinally to accommodate different lengths of work. Cutting tools mounted in the tailstock may also be fed into or out of a revolving workpiece.

The tailstock may be aligned for straight cylindrical turning. It may also be moved crosswise and *offset* for turning tapers. The whole mechanism may be slid along the ways to any position on the lathe bed. A tailstock handwheel moves a spindle in which a cutting tool is held. Many tailstock spindles are graduated to measure the depth of cut.

LATHE COMPONENTS FOR PRODUCING TOOL MOVEMENTS

The *end gears, quick-change gears,* and *carriage* mechanism are all related to tool movement functions. The *headstock and* the *spindle movement* serve the *work-rotating* function.

The spindle speed of small cone-pulley belt-driven lathes is controlled by the position of the belt on one of the step-cone pulleys. The drive may be direct. The range of speeds may be increased by the addition of different *gear trains.* Gear combinations (trains)

provide double, triple, quadruple, and other spindle speed ranges.

Heavier powered lathes have completely geared headstocks. The gears are engaged in a manner similar to shifting an automobile transmission. Newer machines are being designed for still greater increases in spindle speeds. Fluid and electromagnetic devices, variable-speed motors, and other mechanisms are being used to meet increased machine tool requirements.

The *headstock* is located at the *head,* or left-hand end, of the lathe bed. The headstock contains the spindle drive mechanism and the *lathe spindle.*

Power and motion from a motor is transmitted through the headstock drive mechanism to the lathe spindle. The spindle may be driven by multiple V-belts, a variable-speed drive, or by gear transmission. Fig. 24-7 is a phantom view of a gear drive headstock.

Speed changes on belt drives are made by shifting the belt to a different step-cone pulley combination. Additional speeds are obtained on small lathes through a two-step cone pulley drive between the motor and a countershaft. Other designs provide for a wider range of speeds by engaging a set of *back gears.* Still other variable-speed drives may be controlled electrically or mechanically while the lathe is in motion.

The gear ratio on a geared-head lathe is changed by speed-change levers. These engage different sets of gears to produce the required speed. Spindle speed changes must be made when the gears are *not* in motion, to prevent *stripping the gear teeth.*

On the end of the headstock there is an *end gear*

279

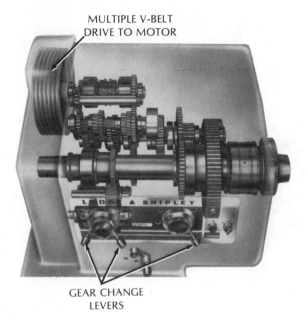

MULTIPLE V-BELT
DRIVE TO MOTOR

GEAR CHANGE
LEVERS

Fig. 24-7. A phantom view of a gear drive headstock. (Courtesy of The Lodge & Shipley Company)

train. These gears are enclosed by modern guards that are provided with a safety electrical disconnect switch. The switch shuts down the machine when the end guard is removed. The gears within the geared headstock are of special alloy steel. The gears are induction hardened and precision ground. All parts are lubricated with an oil bath system. The oil level is sighted on an oil level indicator.

The required spindle speed is obtained by setting the speed selector levers. The spindle is actuated by a *foward/stop/reverse* lever. This lever is located on the apron to provide *on-the-spot control* by the operator from any operating position.

The Headstock Spindle

The dominant feature of the lathe is the spindle. The condition, accuracy, and rigidity of the spindle affects the precision of most machining operations and the quality of surface finish. The headstock spindle is mounted in opposed, preloaded, precision anti-friction bearings. *Spindle runout (eccentricity)* is reduced to a minimum. The spindle is hollow to permit small parts, bar stock, and lathe attachments to pass through it. The face end of the spindle, called the *nose*, is accurately ground with an internal taper. Adapters are fitted to reduce the size of the spindle hole. This permits accessories, like a sleeve for the live center, to be inserted. Spindles are hardened and ground.

Spindle Nose Types

The spindle nose is accurately ground to receive nose mounting plates. These align chucks and face plates in relation to the spindle. The three basic designs for the outside of a spindle nose are shown in Fig. 24-8:
- The older threaded nose with a squared shoulder (A)
- The long taper and key-type spindle nose (B)
- The cam-lock type spindle nose (C).

*(A) THREADED
SPINDLE HOSE*

*(B) LONG TAPER
AMERICAN STANDARD*

(C) CAM-LOCK

Fig. 24-8. Three basic types of lathe spindle noses. (Courtesy of the Clausing Corp., and South Bend Lathe, Inc.)

280

The *threaded spindle nose* receives a threaded mounting. The shoulder helps align the mounted device. The threads hold and secure the device in place and transfer the spindle driving force.

The *long taper spindle nose* provides a tapered surface. This surface receives the corresponding taper of a spindle nose mounting plate. A key transfers the spindle driving force to the nose-mounting. The threaded ring on the spindle screws onto a matching thread. The threaded ring draws and securely holds the mounting on the taper. A spanner wrench is used to tighten or loosen the threaded ring nut.

The *cam-lock spindle nose* has a short taper. This accurately positions the spindle mounting. The spindle mounting has cam studs which fit into a ring of holes on the face. The spindle mount is held securely on the taper and against the face of the spindle lock nose by the cam studs. These studs are secured by turning a chuck key. The spindle mounting is driven by the cam studs.

End Gears And Quick-Change Gearbox

The gears on the *end* of the headstock between the spindle gear and the drive gear on the gearbox are called *end gears* (Fig. 24-9). End gears serve two functions:

Fig. 24-9. The end gear train. (Courtesy of Do All Company)

- To transmit motion from the spindle to the lead screw and feed screw through the gearbox
- To change the direction of the lead and feed screws.

The relationship between the spindle speed and the lead and feed screws is established through a series of gears. The gears in a *quick-change gearbox* (Fig. 24-10) are engaged by levers. One lever is positioned to permit the rapid selection of gears to produce a required *thread pitch*. The pitch may be in the Unified American/British system or SI Metrics. The direction of rotation of the lead screw and the feed screw may be changed from clockwise to counter-clockwise, to move the apron toward either the headstock or the tailstock.

Fig. 24-10. Change levers for spindle speeds and right- and left-hand feeds and thread pitches. (Courtesy of the Clausing Corp.)

Another gear-change mechanism for thread pitches (English and Metric) is pictured in Fig. 24-11. Gear shifting is done by positioning one or more of the three knobs *when the spindle is stopped*. The range of English threads that may be cut on this lathe is from 11 to 100 threads per inch. The Metric thread combinations may be set from .275 mm to 2.7 mm pitches.

Fig. 24-12 shows an index plate on a quick-change

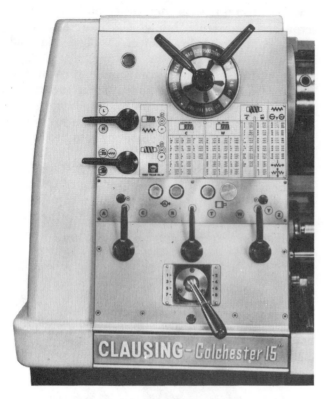

Fig. 24-11. Feed and thread pitch change knobs (English and metric systems). (Courtesy of Hardinge Brothers, Inc.)

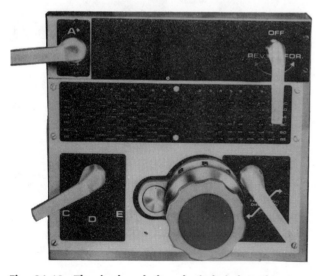

Fig. 24-12. The feed and thread pitch index plate on a quick-change gearbox. (Courtesy of the Clausing Corp.)

gear box. The index plate indicates the various lever positions to set the gearbox for either feed or thread cutting. Note that the pitches and feeds are given on the quick-change gearbox plate in terms of both inch and metric standard measurements.

The Carriage

The lathe cutting tool is secured, moved, and controlled by different attachments and mechanisms on the lathe *carriage* (Fig. 24-13). The *carriage* consists of a *saddle, cross slide,* and *apron.* With this combination, a cutting tool may be fed into or away from a workpiece. The rotary motion of the lead screw and the feed screw is converted to a horizontal motion. The movement may be across the length of a workpiece and/or at a right angle to it. A *compound rest* is attached to the cross slide to produce small tapers and other angle surfaces.

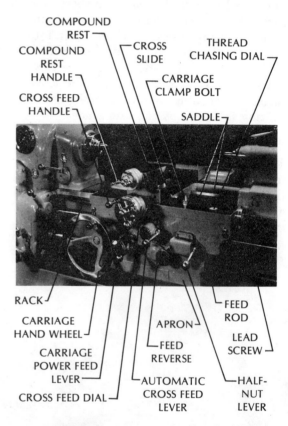

Fig. 24-13. The controls and major mechanisms of the carriage. (Courtesy of Do All Company)

The Apron—The controls and mechanisms for all movements of the carriage are housed in the apron. The principal parts of the apron are labelled in Fig. 24-13. A *selector lever* positions the proper gears for longitudinal and cross feeds. A *friction clutch* engages the feed. A *half nut lever* permits engaging the lead screw for thread cutting.

There are built-in safety features that prevent engagement of the half nuts and power feed at the same time. A large handwheel provides for rapid traverse of the carriage and for longitudinal feeding by hand. The crossfeed handle provides a complimentary motion that moves the cross slide at right angles. This combination permits tool placement and motion in two directions. The lead screw transmits motion through the half nuts for thread cutting. A *spline* is cut over the length of the lead screw. This design feature is used for thread cutting and feeding. A key rides in this groove. The motion of the lead screw is transmitted through the key to a *worm*. The worm slides over the lead screw as the carriage moves. The worm turns a *worm wheel* (gear).

When the feed clutch is engaged, the motion of the worm wheel is transmitted through a series of gears in the apron to a small *pinion gear*. The pinion gear teeth mesh with and revolve along a *rack*. The rack is mounted beneath the *ways* on the front side. This action produces the longitudinal movement of the carriage. Similarly, the power feed can be transmitted to the crossfeed screw.

In review, the threads on the feed screw are used for thread cutting. The split (half) nut engages or disengages the carriage when cutting threads. When the lead screw has a keyway cut along its length, it is designed for also transmitting power for feeding. A clutch engages or disengages the power feeds. Heavier lathes have one or more feed bars for transmitting power and a separate lead screw for thread cutting.

The Saddle—The saddle reaches across all of the *ways,* as shown in Fig. 24-14. This construction permits the saddle to hold the moving tool support

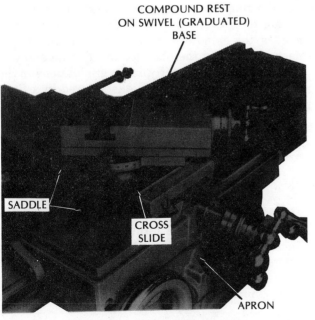

Fig. 24-14. The major design units of a lathe carriage. (Courtesy of LeBlond Machine Tool Company)

and tool feed mechanisms. The saddle also provides the necessary rigidity for all cutting actions. The saddle supports the apron and cross slide, on which a *compound rest* is mounted. These are identified in Fig. 24-14. The cross slide mechanism provides tool movement along the work axis or at right angles to it. This *crosswise motion* is produced whenever the cross slide feed screw is turned either by hand or power. *The Cross Slide And Compound Rest*—The compound rest is mounted on the cross slide. The compound rest may be swiveled though 360° to any angular position. The base is graduated in degrees to permit such adjustments. The compound rest may also be swung to conveniently position a cutting tool. The compound rest is fed by hand.

The basic functions of the compound rest include:
● Cutting angles and short tapers
● Moving the cutting tool to face to close tolerances (when set parallel to the ways of the lathe)
● Feeding the tool in thread cutting.

Cross Slide (Traverse) And Compound Feed Dials—The amount the cross slide is moved may be measured on a cross slide feed dial. Similar measurements for compound rest movements are read on the compound feed dial.

Each feed dial is graduated in relation to the distance the cross slide or compound rest (as the case may be) is advanced for one complete revolution. This distance depends on the pitch of the cross slide or the compound rest screw.

The feed dials are graduated according to the precision built into the lathe and the general job requirements for which it was designed. Some have readings in thousandths of an inch (0.001") or two-hundreths of a millimeter (0.02 mm). Higher precision lathes have dials (collars) that read to one ten-thousandth of an inch (0.0001") and one one-hundreth of a millimeter (0.01 mm).

The circumference of the feed dial collar is divided into an equal number of graduations. Different measurements are established from readings of the graduations on the dial. For example, if the cross slide feed screw has a pitch of five threads per inch, the cross slide moves 0.200" for each complete revolution. By dividing the circumference of the collar into two hundred equal parts, a movement from one graduation to the next represents 0.001".

The feed dials on lathes in the metric system are graduated in a similar manner. For instance, if the metric screw pitch is 5 mm (approximately 0.200") and the circumference of the collar is divided into two hundred equal parts, the movement between graduations is 0.025 mm. This is almost the equivalent of 0.001".

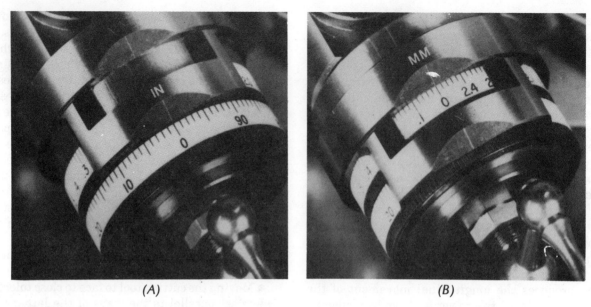

(A)   (B)

Fig. 24-15. Cross slide and compound rest feed dials (English and metric). Courtesy of Hardinge Brothers, Inc.)

A crossfeed graduated dial and collar is illustrated in Fig. 24-15. This design is used on a precision toolroom lathe. The metric graduations read in 0.1 mm on the setting collar (Fig. 24-15B). For ease in reading, every tenth graduation is numbered—1 (mm), 2, 3, etc. The dial that indicates the movement of the crossfeed (screw) is graduated in 0.01 mm. Every fifth graduation is marked—0.05 mm, 0.10 mm, etc.

On the opposite side, the English (inch) standard collar (Fig. 24-15A) is graduated in 0.001″. Each tenth graduation is marked—10 (thousandths of an inch), 20, etc.

Many of the newer lathes have dual inch standard and SI Metric standard dials. These are located on the cross slide and compound rest feed screws and on other lathe micrometer measuring devices. The dual dial is a convenient device for lathe measurements. It is possible with the dual dials to machine workpieces that must fit with other parts in either the inch standard or SI Metric system.

The Tailstock

There are two units to a tailstock: a *base* and a *head*. (Refer again to Fig. 24-6.) The base may be positioned quickly and secured at any longitudinal position along the bed of the lathe. The head may be moved *transversely* at right angles to the lathe central axis. This action positions the tailstock either *on center* (the headstock and tailstock axes aligned) or *off center*, for turning tapers. A *zero index line* and other graduations are machined on the base and head sections. These graduations are used to *center* or *offset the tailstock*. In either of these positions, the accuracy of the setting must be checked with more precise instruments.

The *head* includes the tailstock spindle. The spindle has a standard Morse internal taper. A *dead center* is secured in the tapered spindle. The center provides support for a centered workpiece and other cutting tools that are guided by the center. Movement of the spindle is controlled by a handwheel. Measurements of the distance the spindle and tool travel may be read directly on the graduations on the tailstock spindle. The spindle is secured in a fixed position by moving a *locking lever*.

WORK- AND CUTTING TOOL-HOLDING ACCESSORIES

Four different groups of work-holding accessories are common to general lathe practices. These are: *chucks, collets, faceplates and driver plates*, and *centers*.

Lathe Chucks

There are four common types of mechanical hand-adjusted chucks: *three-jaw universal, four-jaw*

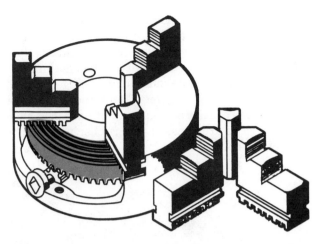

Fig. 24-16. A three-jaw universal chuck with a second set of jaws.

independent, combination, and *two-jaw chucks.*

The *three-jaw universal* chuck is adaptable to the *chucking* of *symmetrical workpieces.* These may be chucked on the inside or outside. The three jaws move together concentrically. One set of jaws is used for inside chucking; the other, for outside chucking. Fig. 24-16 shows some of the construction details and the second set of chuck jaws. In all chucking operations

the workpiece is held as close as possible to the chuck to avoid *overhanging.*

The *four-jaw independent chuck* has four jaws that move independently (Fig. 24-17). The jaws are reversable to permit inside and outside chucking. This type of chuck is practical for chucking nonsymmetrical workpieces, rough castings, and square, octagonal, and irregular shaped parts. Work may also be positioned off center. The construction details of a four-jaw chuck are illustrated in Fig. 24-17.

The *combination chuck* combines the distinguishing features of both the universal and the independent chuck. The general-purpose four-jaw combination chucks permits the jaws to be adjusted either universally or independently.

One of the less commonly used chucks is the *two-jaw chuck.* This type of chuck is especially adapted to hold irregular and nonsymmetrical workpieces. Many applications are made of the two-jaw chuck in turret lathe operations and in production work.

Collets

Collets are called by four different names: *spring collet, draw-in collet, collect chuck,* or just plain *collet.* Regardless of name, the experienced craftsperson knows each accessory and its function. A collet is used for holding small semifinished or finished parts so that additional operations may be performed.

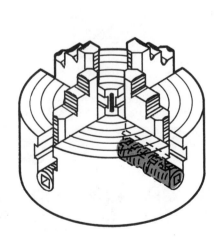

CONSTRUCTION
DETAILS

HEAVY
DUTY
TYPE

Fig. 24-17. A four-jaw independent chuck. (Courtesy of Cushman Industries, Inc.)

Collets are precisely machined. Their design features make collets a practical device for quickly and accurately chucking symmetrical workpieces.

Collets are available in several shapes, like round, square, and hexagonal (Fig. 24-18). The inside shape depends on the nature of the workpiece. The collets illustrated in Fig. 24-18 are of the *draw-in type*. One end is tapered on the outside and is slotted. The other end is threaded and has a keyway. This prevents the collet from turning in the collet sleeve.

A tapered collet sleeve is inserted into the spindle nose. The collet fits and is drawn into the sleeve by a hollow draw bar that fits through the spindle. Collet action in this setup is controlled by either a *lever (pull-type)* or a *handwheel (screw-type)*. These extend through the spindle from the back end. The outside taper of the hardened jaws (sections) of the collet fit accurately against the taper of the sleeve. The body is spring tempered. This permits the jaw sections to be drawn together against the work surface, to hold the workpieces securely and concentrically.

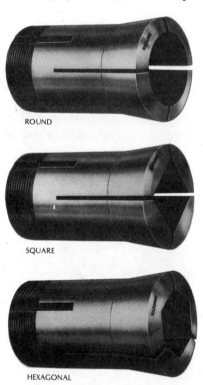

ROUND

SQUARE

HEXAGONAL

Fig. 24-18. Common shapes of spring type collets. (Courtesy of Hardinge Brothers, Inc.)

Collets are furnished in sets. Every steel spring collet varies in size in steps of 1/64″ for inch standard sizes and 0.5 mm for metric sizes.

Collets may also be operated from the spindle nose. The two handwheel types illustrated in Fig. 24-19 fit

directly on the spindle nose. The two collet designs, however, are different. The collets that are used on the Sjogren handwheel type collet chuck (Fig. 24-19A) are of the conventional all-steel design. The second type of spindle nose chuck (Fig. 24-19B) requires a *rubber flex collet*. The rubber flex collet holding areas consist of a number of hardened steel jaws. The jaws are encased in a rubber mount. The taper of the chuck ring is forced against a similarly

(A) SJOGREN COLLET CHUCK

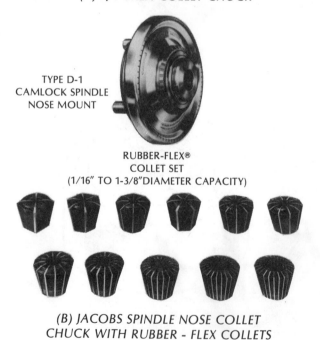

TYPE D-1
CAMLOCK SPINDLE
NOSE MOUNT

RUBBER-FLEX®
COLLET SET
(1/16″ TO 1-3/8″DIAMETER CAPACITY)

(B) JACOBS SPINDLE NOSE COLLET
CHUCK WITH RUBBER - FLEX COLLETS

Fig. 24-19. Two handwheel type collet chucks. (Courtesy of Hardinge Brothers, Inc., and The Jacobs Manufacturing Co.)

formed surface on the collet. At the same time, the back taper of the collet is drawn against the inner tapered surface of the collet chuck. This action causes the jaws to tighten against the work surface. Rubber flex collets have the advantage of being able to accommodate a wider range of work diameters than the all-steel type.

Tool Holders (Tool Posts)

Cutting tools are held and positioned in holders by three general types of cutting-tool attachments. These are called *tool posts*. Types 1, 2, and 3 are illustrated in Fig. 24-20.

● Type 1 is designed for a single operation at one setting. The cutting tool is held in a holder. The holder is positioned in a *tool block* (rectangular) or *tool post* (round) holding device.

● Type 2 provides for multiple operations. Each tool is adjusted in the holding device. Up to four successive operations are performed by a quick-release and positioning lever. This type of cutting-tool holder and positioning device is often called a *four-way turret*.

● Type 3 is a precision block and toolholder that permits rapid tool changes. The cutting tools are preset in each holder. The cutting tool and holder are accurately positioned on a locating pin and in an accurately slotted tool block.

The toolholders are furnished in sets. The sets provide for work positioning, turning, boring, cutting off, knurling with different knurling wheels, and other operations.

Each of the three basic types of toolholders has a T-shaped base. The base fits the T-slot of the cross slide. The base is tightened and held securely in the T-slot.

High-speed, removable and fused inserted carbide tips, and other hardened tool bits are widely used as cutting tools. These are firmly attached to a *shank* or held in a toolholder. The shanks or holders are then positioned and held in the tool post or tool block.

Lathe Centers

Lathe centers are available with tapered bodies. The headstock *(live)* and tailstock *(dead)* centers may be of the solid, hardened type. However, it is often desirable to use a *revolving dead center* like the one illustrated by the cutaway section in Fig. 24-21. The revolving dead center reduces friction. Otherwise, the friction makes it necessary to adjust the center to compensate for expansion of the workpiece. The *dead center* is mounted in the tailstock spindle. The

TYPE 1
HEAVY DUTY TOOL BLOCK
AND ROUND TOOL POST

TYPE 2
4—WAY STATION TURRET

TYPE 3
PRECISION BLOCK AND
PRESET TOOL HOLDERS

Fig. 24-20. Basic types of tool post holders. (Courtesy of Armstrong Bros. Tool Co., Do All Company, and South bend Lathe, Inc.)

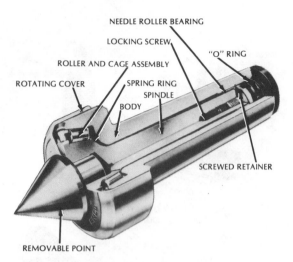

Fig. 24-21. A revolving dead center. (Courtesy of Enco Manufacturing Company)

workpiece revolves on the dead center. Carbide-tipped dead centers are available. These centers have the ability to withstand high temperatures. The *live center* has a tapered adapter or sleeve. The center and sleeve are held directly in the tailstock spindle.

The Faceplate

A *faceplate* is another common work-holding device. A faceplate (Fig. 24-22) is a round plate that fits on the spindle nose. The faceplate has a series of elongated *T-slots*. These slots are cut radially into the face. The T-slots accommodate *T-head bolt heads*.

Highly skilled machinists, toolmakers, and instrument makers must often perform lathe processes other than between centers. Sometimes, because the work part is irregular it must first be positioned on an angle plate. The angle plate is then strapped to the faceplate to permit machining the part at a particular axis. Because of their shape, other workpieces that require facing, drilling, boring, reaming, or threading may be screwed to the faceplate. The accurate positioning of different work centers is often done by using *toolmakers' buttons* or *special centering instruments*.

Care must be taken in faceplate work to use balancing weights. These offset any centrifugal forces that may be created by the unequal distribution of the mass. Any slight out-of-balance produces an unsafe condition. Unless the workpiece and work-holding devices are secured and balanced, the spindle speed may cause the parts to fly out and cause personal injury and damage to the machine.

The Driver Plate

A great deal of work is machined on the lathe *between centers*. The centered workpiece is driven by a *lathe dog* and *driver (dog) plate* (Fig. 24-23). The driver plate is a cylindrical plate that attaches to the spindle nose. The driver plate usually has two through, elongated radial slots. The *tail* of the lathe dog fits into a slot. The body of the lathe dog is

Fig. 24-22. A face plate. (Courtesy of Do All Company)

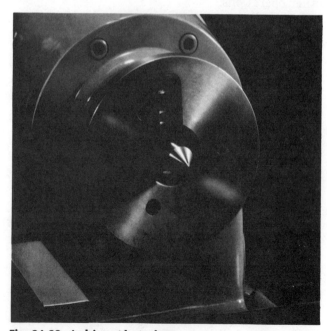

Fig. 24-23. A driver (dog) plate.

clamped over the workpiece. As the spindle and driver plate turn, the motion is transmitted through the lathe dog to the workpiece.

LATHE ATTACHMENTS

There are five general lathe attachments. Their names indicate their prime functions: *taper attachment, threading attachment, micrometer stop, center rest (steady rest)* and *follower rest,* and *rapid traverse mechanism.*

The Taper Attachment

The *taper attachment* (Fig. 24-24) is one of the most accurate, fastest, and practical methods of cutting a taper on the lathe. An adjustable angle attachment is clamped against one of the back ways. The mechanism consists of an adjustable block that clamps to the cross slide.

Fig. 24-24. A taper turning attachment. (Courtesy of Do All Company)

The block moves at a fixed rate according to the angular (tapered) setting of the attachment. This produces the same movement in the cross slide. As the longitudinal feed is engaged and the work turns, a taper is produced. There are two sets of graduations on the tapered attachment. These read in *degrees of taper* and *inches of taper per foot,* or the metric equivalent. On the model illustrated in Fig. 24-24, a graduated dial at the headstock end provides an accurate vernier setting.

Threading Attachments

The *thread chasing* dial is one type of threading attachment. It is fitted on the side of the apron. The chasing dial rotates freely when the split nut is not engaged for threading. The dial stops rotating when the split nut is engaged.

On some lathe models the dial is divided into eight parts, of which four are numbered and four are not. The operator uses these features to engage a thread-cutting tool so that it *picks up a thread* at any time during the thread-cutting process.

The split nut handle on the apron is engaged or disengaged as the cutting tool is withdrawn at any point along the thread. For even threads, the split nut handle is engaged when any line on the chasing dial is opposite the zero (0) index line. The handle is engaged for an odd number of threads when the index line is opposite any numbered line.

Another common device is the *high-speed threading attachment.* The attachment shown in Fig. 24-25 fits on the side of the apron. It contains its own half nut and a *tripping mechanism.* An adjustable stop is preset for the length of thread that is to be cut. The split nut automatically disengages each time that point is reached.

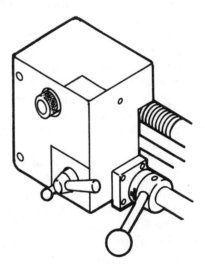

Fig. 24-25. A high-speed threading attachment. (Courtesy of Do All Company)

The Micrometer Stop

The *micrometer stop* is a convenient device consisting of a micrometer head and an attachment for securely clamping it on one of the front ways. The micrometer setting provides an accurate stop. The carriage is brought to the micrometer stop when duplicating multiple pieces or when finishing a part or surfaces to exact lengths. A modification of the micrometer stop is illustrated in Fig. 24-26. Four micrometer heads are attached to an indexing cylinder. Each head is set at a particular dimension. A *stop bar* is housed in the headstock. This provides a positive stop for each micrometer head setting.

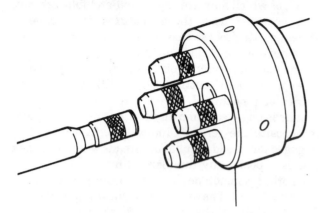

Fig. 24-26. Multiple, positive-length micrometer stops. (Courtesy of Do All Company)

Fig. 24-27. An application of a steady rest. (Courtesy of Do All Company)

Steady (Center) Rest And Follower Rest

Long workpieces must be supported at regular intervals and continuously while a cut is being taken. The *steady rest (center rest)* shown in Figure 24-27 is an accessory that performs this function. It helps prevent springing of the workpiece under the cutting action of a cutting tool. The steady rest is secured to the ways at a point along the workpiece that provides maximum support.

The *follower rest* performs a similar function except that it *follows* the cutting action. The follower rest (Fig. 24-28) is attached to the saddle and therefore moves along the work. The *work-supporting shoes* of the follower rest support the revolving workpiece at two places. The shoes prevent the forces of the cutting tool against the workpiece from springing it.

Rapid Power Traverse

Considerable time is saved, particularly on the heavier models of lathes, by the addition of a *rapid traverse* attachment (Fig. 24-29). The term *rapid traverse* relates to increased speed in making forward and reverse longitudinal movements of the carriage and the cross slide.

Rapid power traverse is controlled through longitudinal and crossfeed friction levers. A safety clutch is incorporated to automatically disengage one of the feeds, to prevent overloading.

FOLLOWER REST
ATTACHED TO
CROSS SLIDE

Fig. 24-28. A follower rest. (Courtesy of LeBlond Machine Tool Company)

Fig. 24-29. A rapid traverse attachment. (Courtesy of Do All Company)

COMMON LATHE MACHINING PROCESSES

Turning, facing, undercutting, internal recessing, cutting off, knurling, and *threading with a single-point cutting tool* are considered regular lathe work operations. Reaming, boring, spotfacing, counter-boring, and threading with taps and dies are internal operations common in lathe work. These operations were covered in earlier units on drilling machines. Reaming and threading are also bench work processes.

Turning

Turning relates to the removal of materials from the outer surface (diameter) of a workpiece. Turning is used to form one or more surfaces to specified dimensions. The machined surface may be *straight* (one continuous diameter), *tapered,* or *contoured* (as a concentric but irregularly-shaped surface).

Facing

Facing is the process of machining a flat surface *across* the face of a workpiece. The *faced surface* is at right angles to the lathe axis and the part itself.

Shoulder Cutting

The *face* which adjoins two surfaces of unequal diameter is known as a *shoulder*. Shoulders are turned on the lathe. Shoulders may be *square, filleted, angular, or a combination* of any of these forms. Shoulders may be formed by combining straight turning and facing, or by angle cutting, or by using a formed cutter.

Undercutting, or Internal Recessing

The process of cutting a concentric section into the inside diameter of a hole is called *undercutting,* or internal recessing. Such an operation is usually performed when a thread must *bottom squarely* or a mating piece must fit solidly against a machined surface.

Thread Cutting

Thread cutting may be done with formed cutters like taps and dies. When precisely machined threads are required, they are often *cut on a lathe* with a single-point cutting tool. The lathe is adapted to the cutting of inside or outside threads that are either straight or tapered.

Knurling

Knurling is the *displacing of metal* to form a particular pattern on the external surface of a part. Knurling provides a good gripping surface, enhances the appearance of a part, and increases the diameter. *Knurled surfaces* are produced by forcing hardened rollers of a particular pattern into a revolving workpiece.

Polishing

One of the final operations on a workpiece is that of *polishing*. Polishing produces a fine-grained and smooth surface finish and helps remove tool marks. A polished surface is produced by holding and moving an abrasive cloth across a work surface as it rotates at high speed. A very limited amount of material is removed in the process. Filing is also used to produce an evenly-grained but less highly polished surface.

These lathe work processes are combined with other common machining processes like drilling, boring, reaming, and threading. Lathes may also be used for such special purposes as simple grinding, milling, and spinning, and for roll flowing processes for producing contours on thin-gage materials.

LATHE SIZES AND SPECIFICATIONS

Lathes are usually designated by *model, swing over bed,* and the *center-to-center distance between the headstock and tailstock* (Fig. 24-30). For example, a 13″ (330 mm) by 25″ (635 mm) general-purpose lathe

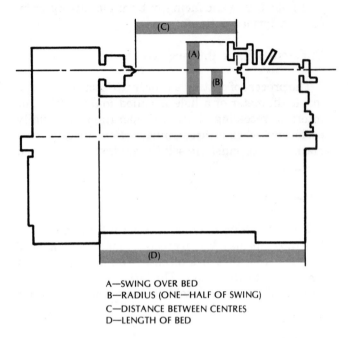

A—SWING OVER BED
B—RADIUS (ONE—HALF OF SWING)
C—DISTANCE BETWEEN CENTRES
D—LENGTH OF BED

Fig. 24-30. Designations of lathe size and swing.

will swing a 13″ (330 mm) diameter workpiece over the bed. However, only an 8¼″ diameter (210 mm) part will clear the cross slide. The 25″ (635 mm) center distance is the maximum length that can be accommodated between the centers of this lathe.

There is a tremendous range to other specifications such as:
- The spindle mounting; diameter, bore and nose shape; and headstock gear trains for spindle feeds
- The cross slide and compound rest construction details and carriage travel
- The gear combinations and number of thread and feed changes in the quick-change gearbox
- The series of inch standard and metric pitches of the lead screw
- The gearing and mechanisms for feeding and threading and for forward, neutral, and reversing directions.

Specifications of three basic models of lathes have been selected from different manufacturers of machine tools. Only some of the design features and specifications appear in Table 24-1. This purposely limited technical information does not indicate the available wide range of machine tool sizes, work capacities, and speeds, feeds, and threading possibilities for different work processes.

CATEGORIES OF LATHES

Lathe models may be grouped in three broad categories:
- All-purpose lathes
- Manufacturers' and production lathes
- Special production lathes.

All-Purpose Lathes

Small sizes of *all-purpose* lathes are mounted on benches. These are designated as *bench lathes.* Lathes of still smaller size that are used by instrument makers are called *instrument lathes.* Larger size (diameter and length) lathes like 10″ to 48″ (250 mm to 1200 mm) are known as *engine lathes.*

A high-precision model may be specified as a *toolroom* (or *toolmaker's*) *lathe.* The toolroom lathe has a wide range of speeds, especially the higher speeds, and is usually equipped with a complete set of accessories. The tolerances of all parts of the toolroom lathe are held to closer limits than those of a standard lathe. The toolroom lathe makes it possible to machine to the high degree of accuracy required in tool making, instrument making and other precision work.

A lathe that has a swing of 20″ or larger and is used for roughing *(hogging)* and finishing cuts is often referred to as a *heavy-duty lathe.* The *swing* (capacity) of a lathe may be increased. A section of the lathe bed adjacent to the headstock may be cut away to permit larger diameter workpieces to be turned. In other words, there is a *gap* in the bed next to the headstock. This model is referred to as a *gap lathe.*

Manufacturing Lathes

Any of the lathe models may be converted for multiple operations and limited production. A turret tool post may be mounted on the compound rest, and a turret head may be secured in the tailstock. One of the newer developments in lathe tooling is the eight-position tool-changing head. This head may be mounted on the cross slide.

All operations on an engine lathe may also be performed on a turret lathe. In addition to standard tooling, other tools are used with manufacturing lathes. These tools are broadly classified as bar tools, chucking tools, and cross-slide tools.

Table 24-1. SELECTED SPECIFICATIONS OF BASIC ENGINE LATHE MODELS

	Model	13" (330 mm) General-Purpose Lathe		(250 mm) 10" Precision Toolroom Lathe		48" (1220 mm) Heavy-Duty Turning Lathe	
	Design Features	Inch Standard	Metric Standard (millimeters)	Inch Standard	Metric Standard (millimeters)	Inch Standard	Metric Standard (millimeters)
Capacity (General)	Swing over bed	13"	330 mm	12½"	315 mm	48"	1220 mm
	Swing over cross slide	8¼"	210 mm	7¼"	180 mm	36"	915 mm
	Center distance	25"	635 mm	20"	510 mm	60"	1520 mm
Headstock	Hole through spindle (diameter)	1½"	38 mm	1 13/32"	35 mm	4 1/16"	103 mm
	Spindle range	40 to 2500 rpm		High 40 to 4000 rpm / Low 8 to 800 rpm		6 to 750 rpm	
	Number of spindle speeds (forward and reverse)	24		Infinite		36	
	Taper of spindle center	#3 Morse taper		#2 Morse taper		#7 Morse taper	
	Motor (two speed)	3 hp (2.25 kw)		5 hp (3.75 kw)		60 hp (45 kw)	
Cross-slide	Cross slide travel	7½"	190 mm				
	Compound rest travel	5½"	140 mm	2"	50 mm	7"	45 mm
Thread and Feed Ranges	Thread changes (UN)	35		60		48	
	Thread range (TPI)	2 to 56		3 to 184		½ to 28	
	Number of Metric pitches	39				48	
	Range of Metric pitches	0.2 to 14 mm					0.75 to 42.0 mm
	Range of feeds (longitudinal)	0.001" to 0.040"	0.03 to 1.0 mm	0.0005" to 0.016"	English Gear Box	0.0035" to 0.196"	0.09 to 5 mm
	Range of feeds (cross)	0.005" to 0.020"	0.013 to 0.50 mm				
	Leadscrew	1⅛" × 4 TPI	M 28 × 4 TPI	1" × 8TPI		2½" × 2 TPI	M63 × 12.7
Tail-stock	Spindle travel (traverse)	4⅜"	110 mm	3½"	85 mm	12"	300 mm
	Spindle taper		#3 MT	#2 MT			#7 Morse Taper
	Setover	±½"	±12 mm	±½"	±12 mm	±½"	±12 mm

Production Lathes

Large quantities of duplicated parts, which require that lathe operations be performed at high-production rates, are produced on other adaptations of the lathe. *Turret lathes, automatic lathes,* and *vertical turret lathes* are three common types of production lathes. The design of these machine tools permits multiple tool setups and simultaneous cutting with several tools. The cutting tools are preset. This eliminates individual tool setup each time. Turret lathes may be hand operated or automatic.

Ram And Saddle Turret Lathes—The two basic types of turret lathes are the *ram* and the *saddle*.

The ram type turret lathe includes all of the

features of the engine lathe. In addition, it has permanent accessories like the tailstock turret and a cross slide. A square turret tool post may be mounted as a front tool post. There is also a rear tool post for holding other cutting tools.

There are usually two aprons on the turret lathe. One apron controls the functions of a cutting tool on the saddle and cross slide. The other apron handles the requirements of the tools in the turret. The feed rates are controlled by selective gear trains in each operation.

The *saddle type* is larger and heavier than the ram type and is used for machining large parts. The turret is mounted on the saddle. The cutting tools may be automatically fed toward the headstock. The tools are rigidly held and supported by the saddle because of the weight and the heavy cuts that are taken. *Bar And Chucking Turret Lathes*—Bar and chucking turret lathes are two other common designs of turret lathes.

The *bar turret lathe* takes bar stock in a collet chuck mounted in the spindle nose. The bar is fed, by the bar feed, against a positive stop on one of the turret faces. The collet chuck is closed to hold the workpiece. The lathe operations then are performed on the part which extends from the chuck. The workpiece is cut off, and the cycle is repeated.

Chucking turret lathes require a chucking device for positioning and holding the workpiece. Three- or four-jaw chucks may be used. The part may also be mounted in a fixture on a faceplate. Usually, chuck work is performed on the saddle type turret lathe. *Automatic Bar And Chucking Turret Lathe*—The features of the bar and chucking turret lathes are combined in the *automatic bar and chucking turret lathe*. The customary practice in turret lathe work is to set up a *machining program*. The sequence and design of tools follows. Once the tools are produced, they are *set up* (positioned and secured). The machine set up permits the manufacture of required workpieces according to specifications.

The cutting tools on turret lathes may be positioned for *overhead turning* and *facing* as shown in Fig. 24-31A. Other simultaneous cutting operations may be carried on, like turning and drilling (Fig. 24-31B). The wide range of turret lathe operations has also been further extended by programming using numerical control and multiple tool heads.

Special Lathe Examples

Duplicating Lathes—The duplicating, or *contour, lathe* is used in machining irregular contours and blending sections. The cutting tool movement is guided by air, hydraulic, mechanical, or electrical devices. The shape that is produced is the result of a tracing device. This follows the contour of a *template*

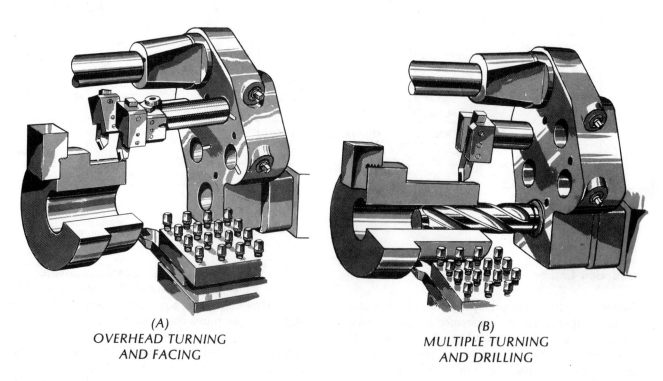

(A)
*OVERHEAD TURNING
AND FACING*

(B)
*MULTIPLE TURNING
AND DRILLING*

Fig. 24-31. Multiple processes on an automatic bar and chucking turret lathe. (Courtesy of The Warner & Swasey Company)

and controls the movement of the cross-slide tool.

All measurements are usually related to the first diameter that is machined, after that diameter is turned. Other contour lathes are guided by a tracer that scans the template. Any variation of air pressure between the tracer and template is amplified and converted to a hydraulic/relay valve motion. This movement controls the tool slide and moves it in the proper direction.

The Floturn Lathe—The floturn lathe is a departure from regular lathe work. No chip-producing tool is used and no chips are produced. This machine has the usual features of a lathe: bed, headstock, tailstock, saddle and cross slide, and work-holding devices.

The floturn lathe is used for *roll flowing* as a method of cold-forming metal. A heavy pressure is applied spirally with two hardened rollers against a metal block. The lathe and the floturn process may be used with flat plates, forgings, and castings, and for extruded and other shapes.

The metals may range in hardness from semihard steels to softer ferrous metals. In operation, the mandrel and workpiece turn. A roll carriage forces the metal ahead of it. The squeezing and stretching of the metal causes the workpiece to take the contour of the mandrel. The wall thicknesses can be held accurately to within ± 0.002″ (0.05 mm).

The two examples of a duplicating and a floturn lathe indicate the additional processes that may be performed by lathes. General features of the basic lathe are adapted to meet special requirements.

GENERAL SAFETY PRECAUTIONS FOR LATHE WORK

- Remove nicks and burrs from the bore and shank of each machine accessory and cutting tool.
- Check the overall diameter of work that is mounted in a chuck, on a faceplate, or other-work holding device. All objects must swing clear of the lathe bed and cross slide.
- Secure all machine guards in place before starting the lathe.
- Remove chips and foreign particles carefully from all lathe parts. Use a chip remover and brush. No chips should be pulled by hand. Chips are razor sharp.
- Wipe the ways and other machined surfaces clean after chip removal. Foreign matter can scrape and score a machined surface.
- Make sure the machine is stopped. Then, oil the lathe ways by hand before using.
- Check all tool and workpiece setups for rigidity. This helps ensure that no part will move due to cutting and/or centrifugal forces.

- Keep cutting and measuring tools and other setup tools off the machined surfaces of the lathe.
- Wipe accessories and all tools clean after use. Replace each item in a special tool rack or machine cabinet.
- Raise a heavy chuck or workpiece only with a cradle and portable hoist. Follow this practice to mount and remove a chuck from the spindle or for chucking heavy workpieces.
- Protect the lathe bed with a cradle block (wooden chuck support) whenever a chuck is being mounted or removed.
- Use cutting fluids where high-speed cutting is required and a great deal of heat is generated.
- Do not wear loose, hanging clothing near moving parts. Also, all wiping cloths must be removed from the machine.
- Check all machine lubricating levels *before* the lathe is started.
- Wear a face shield or eye protection device during all lathe operations.

COMMON LATHE TERMINOLOGY

Screw-cutting lathe   A machine tool with a precisely controlled rate of movement between a revolution of the spindle and the feed.

Basic lathe components   The grouping of units (components) of a lathe to perform a major function. The power source and motion producing mechanism for the workpiece; a work-holding and work-rotating device; and the motion producing mechanism for feeds, speeds, and threads.

Geared head lathe A headstock design that uses gear sets to produce a wide range of spindle speeds. Gear sets are engaged by sliding gears into position. One or more speed-change levers control the speed range.

Lathe carriage A mechanism for supporting, positioning, and securing cutting tools for longitudinal and transverse movements. A combination of apron, saddle, cross slide, and compound rest.

Threading attachment A device that permits the exact positioning and engagement of a thread-cutting tool to track in a previously cut thread form.

Taper attachment A control device that is set at a required taper angle and strapped to the lathe bed. A mechanism to transfer the angular movement of a slide block to the cross slide. Control of the cutting tool movement to produce a taper.

Basic lathe processes A series of fundamental external turning processes: facing, straight and taper turning, shoulder cutting, grooving, cutting off, knurling, and threading. In addition, a series of internal machining processes, including: center drilling, drilling, countersinking; boring, counterboring and spotfacing; reaming; recessing or undercutting, and tapping.

Engine lathe A general term for bench and floor models of lathes. A range of lathes from small bench type instrument lathes to larger precision toolroom lathes, to heavy-duty lathes.

Production lathe A machine tool on which all basic lathe processes may be performed. A lathe that uses a multiple-holder tool post, multiple back carriage-position tool holder, and a multiple-station tailstock turret head for production.

Long taper and cam-lock spindle nose Two standard types of spindle nose. A tapered spindle nose in which work-holding accessories and tools may be mounted, secured, and positioned concentrically with the spindle axis.

Longitudinal feed, or movement The movement (feed) of the carriage between the headstock and tailstock. A movement along the ways produced either by hand or by engaging the power feed.

Lead Screw A precisely machined screw thread running the length of the lathe bed. A screw which may be engaged by an apron mechanism. A screw that changes rotary motion to longitudinal motion to reproduce a thread of a required pitch.

Lathe chucks Work-holding devices for regular- or irregularly-shaped workpieces. Chucks permit mounting a workpiece on center or off center. A device that holds a workpiece securely. A term commonly applied to lathe work holding devices such as independent and universal jaw chucks, collets of various regular shapes, and fixtures that serve as chucks.

SUMMARY

- The Maudslay all-iron screw-cutting lathe refined and incorporated the spindle, slide rest, change gear, and tailstock features of earlier lathes. These were combined with a precision lead screw, which he produced.
- The addition of the multiple-station turret head led to higher levels of production. Individual tool setups were eliminated.

- The automatic screw machine combined design features of the engine lathe and turret lathe. The machining processes are cam-actuated.
- The three major functions of a lathe relate to:
  ○ Transmitting power to produce motion
  ○ Holding and rotating a workpiece
  ○ Positioning, holding, and moving a cutting tool at a fixed rate of feed or pitch.

- The headstock mechanism controls the work speed. The headstock spindle is used for holding and rotating the workpiece.
- The quick-change gearbox contains gear trains. Different gear ratios are combined to obtain different feeds and leads (pitch).
- The thread-chasing dial on the apron permits the engagement of the lead screw at a proper location. With it, successive thread cuts are positioned accurately in the previously cut thread groove.
- Straight, taper, and shoulder turning; facing; grooving; cutting-off; knurling; threading; and polishing all are common external lathe operations.
- Chuck work operations include facing, center drilling, drilling, straight and taper boring, reaming, countersinking and bevelling, counterboring, spotfacing, recessing, and threading.
- The steady rest is positioned at a fixed location to support a workpiece. A follower rest also supports the workpiece. It follows the cutting action to prevent springing of the workpiece.
- The apron of the carriage contains a half nut mechanism, gear sets, and control levers. The levers are used to engage the longitudinal feed of the carriage, the transverse feed of the cross slide, and the lead screw when thread cutting.
- The cross slide produces transverse motion (toward or away from the center). The compound rest on a cross slide makes it possible to take angular cuts and turn a short taper.

- The tailstock provides a work- or tool-supporting surface. Cutting tools may also be held in and moved by a tailstock spindle.
- Chucks may have jaws to hold work inside and/or outside. The jaws may be operated independently or universally. Collets may be opened or closed by lever, drawbar, or handwheel.
- The live center rotates with the work. The dead center may be of a solid type or have a 60° cone-shaped point which rotates on antifriction bearings.
- Inch standard and metric standard measurements may be read directly on graduated feed collars.
- Tapers may be cut by offsetting the tailstock. A taper attachment may also be used. The attachment is set at the required degree, taper per foot, or metric taper. Short tapers may also be machined by turning with the compound rest set at the required angle.
- Three simple groups of lathes are: the basic engine lathe, production turret lathes (ram, saddle, bar, and chucking machines), and special lathes for contour turning and flowing.
- Safety practices must be followed with respect to:
  - Lifting and positioning a heavy workpiece or chuck
  - Revolving machinery and the related need for gathering and tying of loose clothing
  - Placing all tools, parts, etc., on the work stand or tool cabinet,
  - Removing chips with a chip-removing rod (rake)
  - Cleaning the lathe only when the machine is turned *OFF*.

UNIT 24 REVIEW AND SELF-TEST

1. Indicate the fundamental differences between the construction of the spindle, change gears, and slide rest of Henry Maudslay's lathe of 1800 and those of a modern geared-head lathe.
2. Describe (a) the functions served by Christopher Spencer's *brain wheel* and (b) its important contribution to mass production.
3. Indicate the relationship of *end gears* to the spindle rpm of a lathe.
4. Identify the design feature of a tailstock that permits offsetting for taper turning.
5. Indicate a modern safety feature that is incorporated in the design of guards which cover *end gear trains*.
6. State the functions served by *external and internal tapers* on a lathe *spindle nose*.
7. Give two functions that are served by the *index plate* on a *quick-change gearbox*.
8. Describe the difference between graduations on *feed dials* (collars) in the inch standard and the metric standard systems.
9. Give two types of applications in which a *four-jaw independent chuck* is required instead of a *three-jaw universal chuck*.
10. List two advantages of the *four-way turret type 2* cutting toolholder and positioning device over a type 1 rectangular tool block for producing a number of identical parts.
11. Differentiate between the function of a *faceplate* and a *driver plate*.
12. Describe briefly how a *multiple-head micrometer stop* works.
13. Differentiate between a toolroom lathe and a heavy-duty lathe.
14. State when a *saddle type chucking turret lathe* is used instead of a *ram type bar chucking machine*.
15. Give two general safety precautions to take before operating a lathe.

Unit 25

SINGLE-POINT CUTTING TOOLS: TECHNOLOGY AND PROCESSES

The lathe operator has the responsibility of selecting the best available cutting tool, forming it to a required shape, and determining the correct cutting speed and feed. To make these judgments, the craftsperson must know:

- The materials of which the cutting tools are made
- The shapes of the cutting tools
- How chips are formed and controlled
- Causes of tool wear.

This information must be supported by knowledge of the factors which influence cutting speeds and feeds.

The single-point cutting tool is treated in this unit. This type is the most widely used for engine lathe processes. Later units deal with the actual work processes.

CUTTING TOOL MATERIALS

Cutting tools should possess three properties: *hardenability, abrasion resistance,* and *crater resistance. Hardenability* relates to the property of a steel to harden deeply. The cutting tool retains hardness at the high temperature developed at the cutting edge. *Crater resistance* is the ability to resist the forming of small indentations, or craters. These form on the side or top face of a cutting tool. A *crater* tends to undercut in back of the cutting edge. A crater weakens the cutting edge and, in some cases, causes it to break away under heavy cutting forces.

General cutting tools used for lathe work fall into the following six groups:

- carbon tool steel
- high-speed steel
- cast nonferrous alloy
- cemented carbide
- ceramic
- diamond.

Carbon Tool Steel

Tools that meet the increased requirements for higher speeds and tougher, harder materials have largely displaced carbon tool steel cutting tools. Vanadium and chromium are alloyed with carbon steels to increase their toughness and hardness depth. However, because of the concentration of heat at the cutting point and the lower temperatures at which the temper of a carbon tool steel cutter is drawn, such tool bits are seldom used for general lathe processes.

High-Speed Steel

High-speed steel is used in single-point cutting tools for the same reasons it is used in multiple-point cutting tools like machine drills, reamers, counterbores, and spotfacers.

High-speed steel is the most extensively used material for cutting tools. Its usage is based on factors such as tool and machining cost, tool wear, work finish, availability, and ease of grinding. High-speed steel single-point cutting tools are capable of withstanding machining shock and heavy cutting operations. A good cutting edge is maintained during red heat conditions.

Grades Of High-Speed Steel

There are many *grades of high-speed steel.* Each grade depends upon the chemical composition of the steel. The grades are identified according to the American Iron and Steel Institute *(AISI)* designation. Some of the commonly used "T" and "M" grades are indicated in Table 25-1. The addition of cobalt increases the ability of the cutting tool to retain its hardness during heavy roughing cuts and at high temperatures. Industrial designations (trade names) of these steels, their chemical compositions, and applications are given in Table 25-1.

Cast Alloys

Cast alloy tool bits are used for machining processes at extra high speeds. These speeds range from almost one quarter to two times more than those for high-speed steel. Such cutting tools have better wear resistant qualities than high-speed steel. Cast alloys, like *Stellite,* make it possible to efficiently cut hard-to-machine cast iron and steel. However, cast alloy tool bits are brittle. They are not as tough as high-speed steel.

Table 25-1. COMMON GRADES OF HIGH-SPEED STEEL FOR CUTTING TOOLS

Industrial Designation		Chemical Composition* (percent)						Application
AISI	Trade Name	C	W	Mo	Cr	V	Co	
T-1	High-Speed Steel	0.70	18.00	. . .	4.00	1.00	. . .	Single-point cutting tools; general purpose
T-4	Cobalt high-speed steel	0.75	18.00	. . .	4.00	1.00	5.00	Machining cast iron
T-6	Cobalt high-speed steel	0.80	20.00	. . .	4.50	1.50	12.00	Machining hard-scaled surfaces and abrasive materials
T-15**	Cobalt high-speed steel (Super HSS)	1.50	12.00	. . .	4.00	5.00	5.00	Machining difficult-to-machine materials, stainless steels, high-temperature alloys, and refractory materials
M-2	General high-speed steel	0.85	6.00	5.00	4.00	2.00	. . .	Single-point and multiple-point cutting tools: milling cutters, taps, drills, and counterbores

*Percent of chemical composition of: (C) carbon, (W) tungsten, (Mo) molybdenum, (Cr) chromium, (V) vanadium, (Co) cobalt.
**Extremely hard; requires slow grinding using a light grinding force

Cast alloys are manufactured as solid tool bits, as tips that may be brazed on tool shanks, and as cutting tool inserts. Cast alloy tool bits are popularly known by such trade names as: *Stellite, Rexalloy, Armaloy* and *Tantung*. These tool bits have the advantage of maintaining a good cutting edge at high cutting speeds

Each trade name tool bit relates to a specific composition of chromium, tungsten, carbon, and cobalt. These combinations produce tool bits with a high degree of hardness and high wear resistance. The tool bits also have the ability to maintain a cutting edge while operating at red heat cutting temperatures below 1400° F.

Cemented Carbides

Cemented carbides are powdered metals. They are compressed and *sintered* into a solid mass at temperatures around 2400° F. The composition and application of a few selected carbides appear in Table 25-2. Cemented carbides possess the following qualities:

- Excellent wear resistance
- Greater hardness than high-speed steels
- The ability to maintain a cutting edge under high temperature conditions
- Tips and inserts may be fused on steel shanks and cutter bodies, for economical manufacture of the cutting tool
- Cutting speeds may be increased three to four times those of high-speed steel cutters.

Table 25-2. APPLICATIONS OF SELECTED CEMENTED CARBIDE CUTTING TOOLS

Composition (Designation)	Application
Tungsten carbides	Gray cast iron, malleable iron, stainless steel, nonferrous metals, die-cast alloys, and plastics.
Tungsten carbide and cobalt (with titanium and tantalum carbide added)	Medium and light cuts on carbon and alloy steels, alloy cast iron, monel metal, and tool steels.
Titanium carbide and nickel or molybdenum	Light precision, and precision boring cuts requiring a high-quality surface finish.

Ceramic Cutting Tools

The sintering of fine grains of aluminum oxide produces a ceramic material that may be formed into *cutting tool inserts*. The inserts are particularly adapted for machining highly abrasive castings and hard steels. Small quantities of titanium, magnesium, or chromium oxide are added. The ceramic cutting tools are harder than the cemented carbides and may be used at extremely high cutting speeds on special high-speed lathes. Ceramic cutting tools may be used in turning plain carbon and alloy steels, malleable iron (pearlitic), gray cast iron, and stainless steel.

Industrial Diamonds

Industrial diamonds may be designed as single-point cutting tools. Diamonds are used on workpieces that must be machined to extremely close dimensional and surface finish tolerances Diamond-tip cutting tools are used for light finish-cut turning and boring operations. The depth of cut may range between 0.003″ to 0.030″ (0.08 mm to 0.8 mm) using a feed of from 0.0008″ to 0.005″ (0.02 mm to 0.1 mm) per revolution.

Industrial diamonds may be operated at extremely high speeds for machining nonferrous and nonmetallic materials. Single-point diamond cutting tools are especially adapted for finish machining hard-to-finish metals like soft aluminum. The factors of additional cost and brittleness of industrial diamonds must be considered against the advantages.

CUTTING SPEEDS FOR SINGLE-POINT TOOLS

Cutting speeds for metal-cutting operations are usually stated in manufacturers' tables, handbooks, and production plans. The cutting speeds are given in terms of *feet per minute (fpm)* or *meters per minute (m/min)*. In all cases, the craftsperson must judge whether there are conditions that require any modifications of the recommended speed. Associated with cutting speeds is concern for *tool life*. Tool life relates to the effective cutting time before a tool becomes dull and requires replacement.

The cutting speed is influenced by eight major factors:

- The material in the cutting tool
- The material in the workpiece (its heat treatment, hardness, abrasiveness, internal hard spots, and other surface conditions)
- The rake, the relief, and clearance angles and shape of the cutting areas

- The use of a chip breaker to form and to clear away chips
- The desired economical, efficient tool life
- The use of a cutting fluid where required
- The type of cut (roughing or finishing) and the depth
- The rigidity of the work setup, the capability and construction of the lathe, and the nature of the tool-holding device.

Tool Life

Since tool life is related directly to cutting speed, a reduction in speed always increases tool life. One objective in cutting is to select a cutting speed that provides both a fast, efficient cutting rate *and* maximum tool life. This fact means that there is a fair range of cutting speeds for most metals.

A greater amount of tool wear is permissible for roughing cuts where a precise surface finish is normally not required. Tool wear, however, is extremely important on finishing cuts. These are usually held to close dimensional and surface finish tolerances. Turning conditions, feed, and depth of cut also affect tool life.

Cutting Speed Tables And Calculations

With experience, a craftsperson always automatically considers all of the factors which influence cutting speed. From these, a judgment is made about the correct spindle speed. The speed-change levers or rpm controls on the headstock are then adjusted for the spindle speed.

In other instances, where different materials are to be cut or workpieces are to be produced in quantity, reference is made to a *Table of Recommended Cutting Speeds*. The cutting speed and spindle speed (rpm) may be calculated by formulas. The tables were developed after years of study. Consideration was given to different machining processes, speeds, feeds, depth of cuts, other factors affecting tool life, and the machined end product.

Cutting speed for lathe work, as in all other machining operations, refers to the rate at which a workpiece revolves past a fixed point of a cutting tool. In lathe work, the cutting point is at the turned diameter. The cutting speed is measured in surface feet per minute (fpm, or sfpm) or meters per minute (m/min). The cutting speed (CS) is equal to the product of the circumference of the workpiece at the bottom of the cut multiplied by the spindle speed in revolutions per minute (rpm). Thus,

$$CS \text{ (sfpm)} = \frac{\text{circumference} \times \text{rpm}}{12}$$

$$= \frac{(\text{diameter} \times \pi)^* \times \text{rpm}}{12}.$$

The metric equivalent of this formula is:

$$CS \text{ (m/min)} = \frac{(\text{circumference (mm)} \times \text{rpm})}{1000}$$

$$= \frac{(\text{diameter (mm)} \times \pi)^* \times \text{rpm}}{1000}.$$

*The value of π generally is rounded off to 3.14

In actual practice, the operator usually determines the rpm at which to set the spindle speed. The recommended cutting speed is found in a handbook table. The data provided is related to:

- The material to be cut
- The nature of the cutting tool
- The lathe process
- Whether roughing or finishing cuts are required
- Other specifications.

Part of such a table is reproduced in Table 25-3. A more complete table of cutting speeds, combined with cutting tool angles, is included in the Appendix. Two examples follow. These show how cutting speed tables and the cutting speed and rpm formula are used.

Example 1: Determine the cutting speed (sfpm) of a 2″ diameter brass workpiece that is being rough turned at 300 rpm.

$$CS \text{ (sfpm)} = \frac{(\text{diameter} \times \pi) \times \text{rpm}}{12}$$

$$= \frac{(2 \times 3.14) \times 300}{12}$$

$$= 157 \text{ feet per minute.}$$

The following simplified formula is commonly used in the shop: In this instance, the cutting speed would equal 150 feet per minute.

$$CS \text{ (sfpm)} = \frac{\text{diameter} \times \text{rpm}}{4}$$

Example 2: Refer to a table of cutting speeds (for example, Table 25-3). Calculate the lathe spindle

Table 25-3. CUTTING SPEED (sfpm AND m/min) USING HIGH-SPEED STEEL CUTTING TOOLS
(Partial Table)

Material to Be Machined		Cutting Speed (sfpm and m/min) for Lathe Operations*		
		Turning and Boring		Screw Thread Cutting
		Roughing	Finishing	
Low-carbon (C) steel 0.05 to 0.30%C	sfpm	90	100	35-40
	m/min	27	30	10-12
High-carbon (C) tool steel 0.6 to 1.7%C	sfpm	50	70	20-25
	m/min	15	21	6-8
Brass	sfpm	150	300	50-60
	m/min	45	90	15-18
Aluminum	sfpm	200	350	50-70
	m/min	60	105	15-21

*Without using a cutting fluid

speed required to rough turn a 60 mm diameter high carbon tool steel (0.60C) workpiece to a diameter of 50 mm. No cutting fluid is to be used.

In Table 25-3, the recommended cutting speed (CS) for rough turning high carbon tool steel without the use of a cutting fluid is 15 meters per minute (m/min). Substituting the diameter and sfpm values in the formula,

$$\text{rpm} = \frac{CS \text{ (m/min)} \times 1000}{\text{diameter (mm)} \times \pi}$$

$$= \frac{15 \times 1000}{50 \times 3.14} =$$

$$= 96 \text{ rpm (approximately)}.$$

The spindle speed is set to the lower rpm closest to the computed 96 rpm.

The simplified formula for rpm in English units of measure is equal to four times the surface feet per minute divided by the diameter, or:

$$\text{rpm} = \frac{4 \times \text{sfpm}}{\text{diameter}}.$$

CUTTING FEEDS AND DEPTH OF CUT

The *cutting feed* is the *distance* a cutting tool moves *across* a workpiece for each revolution. The cutting feed depends on whether a roughing or finishing cut is to be taken. Coarse feeds are used for rough machining; finer feeds are used for finish machining. The finer feeds produce a higher surface finish to a close dimensional tolerance. The feed is affected by the same eight factors that influence cutting speeds. In principle, a workpiece should be reduced to the required size and finish by taking the *least number of cuts*. Roughing cuts should be as deep as possible, using a coarse feed. Usually ten thousandths to fifteen thousandths of an inch (0.010" to 0.015" or 0.2 mm to 0.4 mm) is allowed for a finish cut. The workpiece is reduced in diameter by twice the depth of a cut. The simple line drawing (Fig. 25-1) shows a 3" diameter blank being reduced by 1/2". The cutting tool is set for a 1/4" roughing cut (depth of cut).

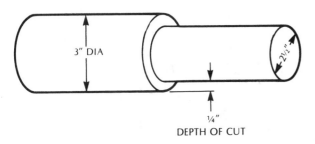

Fig. 25-1. Outside diameter reduced two times the depth of cut.

SINGLE-POINT CUTTING TOOL FEATURES AND TERMS

There are a number of common design features and terms for single-point cutting tools (Fig. 25-2). Each cutting tool has two cutting edges: a *side-cutting edge* and an *end-cutting edge*. The two cutting edges are connected by a *nose*. The nose is especially important because it produces the final finish turned surface.

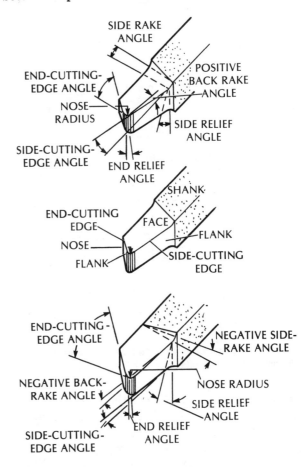

Fig. 25-2. Design features and terms for a single-point cutting tool.

The cutting edge is supported by the *flank*. The flank is *relieved at an angle*. The relief permits the cutting edge to have clearance for cutting action to take place. The cutting tool *face*, at the top of the tool, provides a surface for forming a chip. The chip then rides over this surface. A *shank* supports the cutting area. The shank also provides a body (mass) by which the tool may be positioned and held securely.

Cutting tools are classified as *right-or left-hand* and *side-cutting or end-cutting*. Typical side- and end-cutting operations are illustrated in Fig. 25-3. The cutting edge on a right-hand tool bit is on the left

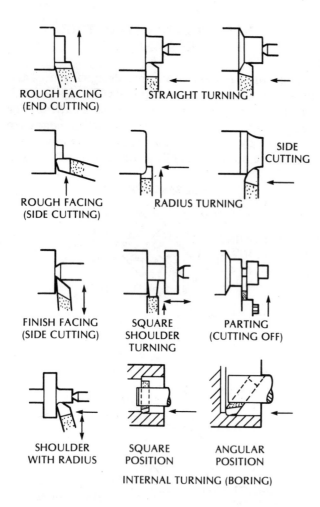

ROUGH FACING
(END CUTTING)

STRAIGHT TURNING

SIDE
CUTTING

ROUGH FACING
(SIDE CUTTING)

RADIUS TURNING

FINISH FACING
(SIDE CUTTING)

SQUARE
SHOULDER
TURNING

PARTING
(CUTTING OFF)

SHOULDER
WITH RADIUS

SQUARE
POSITION

ANGULAR
POSITION

INTERNAL TURNING (BORING)

Fig. 25-3. Typical side- and end-cutting operations with high-speed steel tool bits.

side. The *left-hand tool bit* has the side-cutting edge ground on the right side. The end-cutting edge tools cut with the end-cutting edges. This cutting tool is not provided with a side-cutting edge. Side-cutting tools cut as the carriage moves longitudinally. End-cutting tools cut by the cross (transverse) movement of the cross slide.

In addition to the nose radius, there are six principle tool angles:
- End- and side-relief angles
- Back- and side-rake angles
- End- and side-cutting edge angles.

Relief Angles

A *relief angle,* whether for the side-cutting or end-cutting edge or the nose, permits the cutting tool to penetrate into the workpiece. Tool life and tool performance depend on the relief angles. If the relief angle is too large, the cutting edge is weakened and may wear quickly, chip, or break away. Too small a relief angle may limit proper feeding. It may also cause rubbing and generate considerable heat, with limited cutting action.

High-speed steel cutting tools are ground with relief angles of 8° to 16°. The relief angle for general lathe processes is 10°. The smaller relief angles are used with harder materials. The harder and more brittle cemented carbide cutting tools require greater support under the cutting edge. The relief angle range for cemented carbides is from 5° to 12°. The relief angle of the nose radius is a combination and blending of the end- and side-relief angles.

Back-And Side-Rake Angles

The side-rake angle indicates the number of degrees the face slants from the side-cutting edge. The side-rake angle is shown in Fig. 25-4A. Similarly, the back-rake angle represents the number of degrees the face slants, as shown in Fig. 25-4B. The angle may slant downward (*positive rake*) or upward (*negative rake*) in moving along the cutting edges.

The side-rake may also be *positive* or *negative.* Negative side- and back-rake strengthen the cutting edge. This added strength is important when cutting hard and tough materials under severe machining

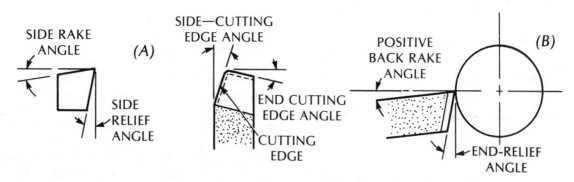

Fig. 25-4. Cutting edge, relief, and rake angles of a general-purpose tool bit.

Table 25-4. RECOMMENDED RAKE ANGLES FOR SELECTED MATERIALS (HIGH-SPEED STEEL AND CARBIDE CUTTING TOOLS)

Material	Hardness Range Bhn	HSS Rake Angle (°)		Carbide Rake Angle (°)	
		Back	Side	Back	Side
Carbon Steel (plain)	100 to 200	5 to 10	10 to 20	0 to 5	7 to 15
	400 to 500	−5 to 0	−5 to 0	−8 to 0	−6 to 0
Cast iron (gray)	160 to 200	5 to 10	10 to 15	0 to 5	6 to 15
Brass (free- (cutting)	. . . . . .	−5 to 5	0 to 10	. . .	. . .

conditions. Negative rake angles are used on throw-away carbide cutting tools. These angles provide an economical way of using the top and bottom faces of the insert for machining.

Manufacturer's recommendations and trade handbooks are available for determining correct rake angles. These are provided for high-speed steel and carbide cutters and inserts for use with different materials of varying degrees of hardness. The sample listings in Table 25-4 are from a more complete listing in Appendix Table A-24.

Side- And End- Cutting Edge Angles

The *side-cutting edge* is ground at an angle to the side of the tool bit or tool shank. The range is from 0° to 30°, depending on the material to be cut and the depth, feed, etc. A 20° angle is common for general lathe turning processes.

The *end-cutting angle* is formed by a perpendicular plane to the side of the tool bit and the end-cutting surface. This angle should be as small as possible and still permit the position of the side-cutting edge to be changed. Changes in the position of a cutting tool are common shop practice. These changes permit use of the cutting tool for shoulder turning and other operations close to a chuck, faceplate, other workholding device, or the workpiece itself.

In addition to clearing the workpiece, the end-cutting angle provides the backup area for the nose and adjoining end-cutting area of the cutting tool. A 15° end-cutting edge angle may be used for rough turning; an angle of 25° to 30° may be used for general turning processes.

Lead Angle, Chip Thickness And Cutting Force

The positioning of the side-cutting edge in relation to the workpiece influences the direction the chip flows, the thickness of the chip, and the required cutting force. The *lead angle* also affects these items. The lead angle is the angle the side-cutting edge forms with a plane that is perpendicular to the axis of the workpiece. Two different cutting tools are illustrated in Fig. 25-5. These are set to cut at the same depth. The *chip thickness* at (A) is equal to the feed. The chip thickness at (B) is less than the feed, but is longer. With the thinner chip, the cutting speed and/or the depth of cut and production rate may be increased.

A large lead angle has these disadvantages:
● A greater force is required to feed the cutting tool into a workpiece
● There is a tendency for the workpiece to spring away from the cutting tool and produce chatter
● The vibration of the workpiece may cause the cutting edge of the tool to break down.

The Nose Radius

Cutting tools are rounded at the nose to machine a smooth, continuous, straight surface. Otherwise, with a sharp point, a fine helical groove may be turned in the surface. The size of the radius depends on the rate of feed and the required quality of surface finish. There is a tendency to produce chatter when too large a radius or too fast a speed is used. A nose radius of from 1/64″ to 1/8″ is common.

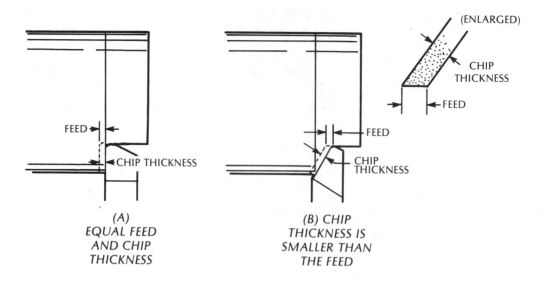

(A)
EQUAL FEED
AND CHIP
THICKNESS

(B) CHIP
THICKNESS IS
SMALLER THAN
THE FEED

(ENLARGED)

CHIP
THICKNESS

FEED

Fig. 25-5. Relationship of lead angle to chip thickness.

CHIP FORMING AND CONTROL

Chip Formation

The cutting action of a cutting tool produces three major types of metal chips:
- *continuous chip*
- *discontinuous chip*
- *segmental chip.*

The *continuous chip* represents the *flow* of a ductile metal (like steel) over the cutter face. The chip is in the form of a continuous spiral ribbon (Fig. 25-6A).

The *discontinuous chip* is produced when cutting brittle metals like cast iron. Since the metal does not flow, it fractures into *discontinuous particles* (Fig. 25-6B).

The *segmental chip* (Fig. 25-6C) has a file tooth appearance on the outer side. The inner side that flows over the cutting tool face has a smooth, burnished surface. The segmental chip is produced when taking a heavy feed on ductile materials. The internal metal fractures spread to the outer surface of the chip. The particles are welded together by the heat of the cutting force.

Chip Control

Chips that form a continuous or segmental chip at high speeds must be controlled. *Chip Control* means the disposing of chips in a safe manner. The accumulation of chips near the revolving workpiece, along the bed, carriage, and cross slide, or on cutting tools is dangerous. Chips may become wrapped

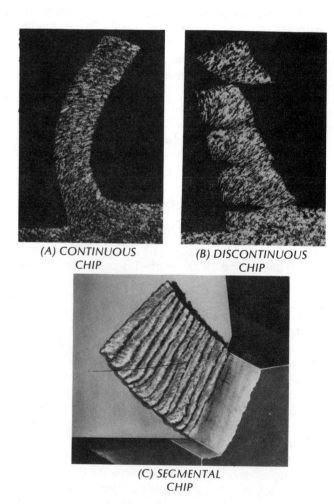

(A) CONTINUOUS
CHIP

(B) DISCONTINUOUS
CHIP

(C) SEGMENTAL
CHIP

Fig. 25-6. Basic types of metal chips. (Courtesy of Cincinnati Milacron)

around the work or chucking device. This often causes damage to the workpiece, cutting tool, or lathe. More important is the hazard of personal injury. Revolving chips or hot metal particles may fly in the work area. Operational dangers are increased with carbide cutting tools. A great quantity of chips forms and accumulates rapidly when heavy cuts are made at high speeds.

A *chip breaker* curls the chip. As the chip strikes against the toolholder, it breaks into short lengths. Chip breakers are of three basic types: parallel, angular and groove. Fig. 25-7 shows line drawings of the three types. Any one of these types may be ground into high-speed steel or formed in cemented carbide cutting tools.

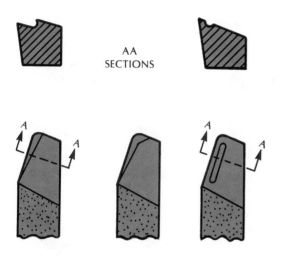

Fig. 25-7. Three basic types of chip breakers ground on high-speed steel tool bits.

CUTTING TOOL HOLDERS

A lathe cutting tool bit may be supported in a heat-treated toolholder. The cutting tool and holder are then secured in a tool post. Cemented carbide inserts are usually brazed on a steel shank. The shank is held in a tool post or tool rest. Disposable carbide inserts are held and secured in still another type of carbide toolholder. Holders for four of the eleven general styles of cemented carbide inserts are shown in Fig. 25-8.

The square, diamond-point, or parallelogram type of insert has a total of eight cutting edges. When the four top cutting edges become dull, the insert is reversed. Each of the four top cutting edges is then used. The insert is thrown away when the eight cutting edges are dull.

A toolholder for disposable carbide or ceramic inserts is designed as a *negative-* or *positive- rake* toolholder. The inserts for the positive-rake toolholder must be ground with a side relief angle. Thus, only the top face edges may be used. The general shapes of inserts are square, rectangular, triangular, round, and diamond. Fig. 25-9 illustrates various kinds and styles of ceramic and carbide inserts.

Different types and shapes of toolholders and cutting tools are treated in later units in relation to a particular work process.

DISPOSABLE CEMENTED CARBIDE INSERT APPLICATIONS

Manufacturers of disposable cemented carbide inserts use a series of letters to designate the *style* of

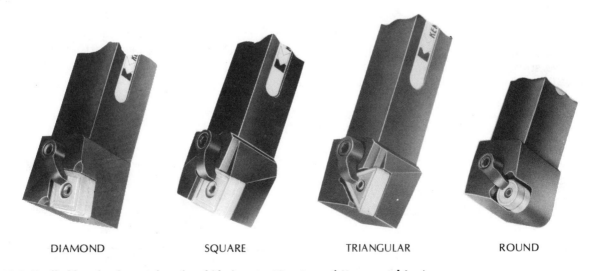

DIAMOND   SQUARE   TRIANGULAR   ROUND

Fig. 25-8. Toolholders for four styles of carbide inserts. (Courtesy of Kennametal Inc.)

307

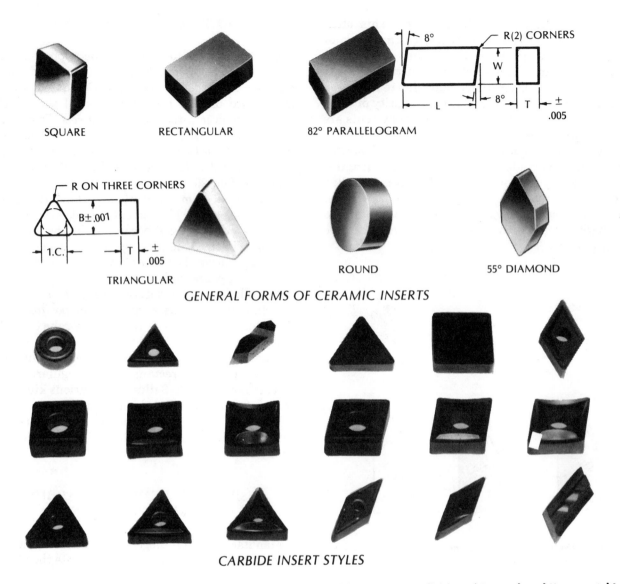

GENERAL FORMS OF CERAMIC INSERTS

CARBIDE INSERT STYLES

Fig. 25-9. Kinds and styles of ceramic and carbide inserts. (Courtesy of the VR/Wesson division of Fansteel, and Kennametal Inc.)

the insert. The style refers to the form and cutting design features. There are eleven general-purpose styles. These start with the letter *A*. Six different styles are illustrated in Fig. 25-10. Applications of each style are given. The smaller line drawings beside each identify the basic cutting edge and rake angles of the inserts.

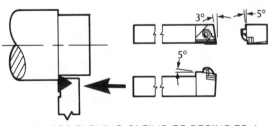

STYLE A FOR TURNING, FACING OR BORING TO A SQUARE SHOULDER.

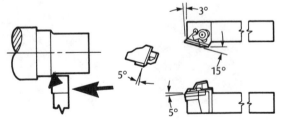

STYLE B FOR ROUGH TURNING, FACING OR BORING WHERE A SQUARE SHOULDER IS NOT REQUIRED.

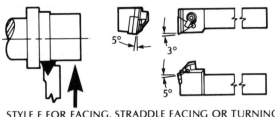

STYLE F FOR FACING, STRADDLE FACING OR TURNING WITH SHANK PARALLEL TO WORK AXIS.

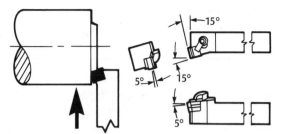

STYLE K FOR LEAD ANGLE FACING OR TURNING WITH SHANK PARALLEL TO WORK AXIS.

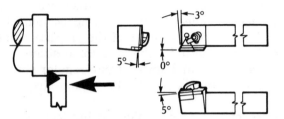

STYLE G FOR TURNING CLOSE TO CHUCK OR SHOULDER, OR FACING TO A CORNER.

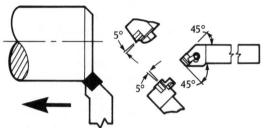

STYLE S FOR CHAMFERING AND FACING. 45° LEAD ANGLE.

Fig. 25-10. Six styles of disposable cemented carbide inserts and their general machining applications. (Courtesy of Kennametal Inc.)

SAFE PRACTICES IN THE USE OF SINGLE-POINT CUTTING TOOLS

- Restrict the use of carbon tool steel cutting tools to work processes requiring slow cutting speeds. The heat generated should not soften the hardened cutting edge.
- Select the grade of high-speed steel tool bit according to the conditions and properties of the workpiece, the nature of the operations, and the required cutting speeds and feeds.
- Use ceramic, cemented carbides, and industrial diamond cutting tools for work processes in which there is minimum shock. (Shock can fracture and destroy these cutting inserts.)
- Keep the spindle speed within the calculated range. The speed must be appropriate for the kind of material in the cutting tool and the characteristics of the workpiece.
- Use the smallest relief and rake angles possible. Too steep a relief angle or relief and rake angle weakens the cutting area. This can cause the cutting edge to break away. Too small a relief angle prevents the cutting edge from penetrating into the work surface. Too small a relief angle also reduces the cutting action, generates excessive heat, and may cause tearing instead of cutting.

- Grind the cutting edge with a nose radius. A nose radius prevents throwing up a fine-ridged surface. The nose radius helps produce a smooth, accurately finished surface.
- Check the lead angle, cutting edge, and speed. Too steep a lead angle, too long a cutting edge, and too high a speed can produce chatter.
- Use a chip breaker on continuous and segmental chips. The chips must be broken away and not allowed to accumulate or wrap around the workpiece.
- Stop the lathe to remove chips. Use a chip rake.
- Wipe cutting fluid spots as soon as practical. Cutting fluid dropped on the floor may create a hazardous slipping condition.
- Secure all loose clothing. Remove wiping cloths from the lathe *before* starting the machine.

SINGLE-POINT CUTTING TOOL TERMS

Single-point cutting tool	A cutting tool that is formed so that only one process or a single set of processes is performed with the one tool at each setting.
Hardenability	The capacity of a cutting tool to be hardened uniformly and deeply throughout its mass.
Crater resistance	The ability to resist wear caused by abrasion. Resistance to wear caused by heat generated at the cutting edge and the force of chips moving over the cutting tool.
AISI designation	A standardized system of a letter and numeral for designating a cutting tool material with a specific chemical composition. A trade identification of the properties and applications of a cutting tool material according to its chemical composition.
Ceramic inserts	Fine grains of aluminum oxide that are formed and sintered into regular shapes (inserts). Diamond, square, rectangular, triangular, and other round shapes of ceramic cutting tools.
Disposable insert styles	Manufacturers' designations, by letters, of cemented carbide cutting tool inserts. Throwaway inserts designated according to design features and the application of each style to specific cutting processes.
Tool life	The period over which a cutting tool maintains its ability to cut efficiently. The time span for producing dimensionally accurate surfaces.
Cutting speed	The rate at which a workpiece revolves past the cutting edge of a cutting tool. The speed of cutting materials, expressed in either surface feet (sfpm) or meters per minute (m/min).
Side-cutting and end-cutting edges	The edges of a single-point lathe cutting tool. Tool cutting edges that are ground to permit the tool to advance into a workpiece longitudinally (side-cutting) or crosswise (end-cutting).
Cutting tool angles (single-point, lathe)	A combination of three angles: relief angle, rake angle, and cutting edge angle.
Relief angle	An angle ground on the end and/or side. An angle which permits the cutting tool to advance into the moving workpiece.
Rake angle	An angle ground on the top face of a cutting tool. An angle that slopes away from the end- and/or the side-cutting edge.
Positive and negative rake angles	A rake angle that provides a smaller angle (positive rake) or a large angle (negative rake) at the cutting edge.
Cutting edge angles	The angles at which the side- and the end-cutting edges are ground.
Lead angle (cutting tool)	The angle formed by the side-cutting edge of a single-point cutting tool in relation to the axis of the workpiece.
Continuous or segmental chip	Chips that flow over the cutter as a continuous spiral or as a welded file-tooth-shaped ribbon (segmented chip).
Chip breaker	An external form secured with a cutting tool, or a groove ground near the cutting edge. A form to curl a chip so that it breaks continuously instead of flowing as a ribbon.

SUMMARY

- Three important properties of cutting tools are: hardenability, resistance to abrasive action, and resistance to the formation of a crater.
- Single-point cutting tools are generally made of high-speed steel, cast nonferrous alloys, cemented carbides, ceramics, and diamonds. These materials have largely replaced the old carbon tool steel cutter bit.

- High-speed steel cutting bits are designated by a trade name and AISI letter and number. Varying the amounts of carbon, tungsten, vanadium, chromium, molybdenum, and cobalt affects the cutting properties of high-speed steel.
- The original T-1 high-speed steel cutting tool is an excellent general-purpose, single-point cutting tool.
- M-2 high-speed steel provides the tool life qualities essential for single- and multiple-point cutting tools.
- Tungsten carbides and ceramic cutting tools are particularly adapted to the machining of hard and abrasive materials at exceptionally high cutting speeds.
- Industrial diamond cutting tools are especially suited for ferrous metals and nonferrous materials that are hard to finish to close dimensional accuracy. High-quality surface finish is possible.
- The lead angle of a cutting tool influences the chip thickness for a particular depth of cut.
- Chatter in lathe operations may be caused by excessive speed, too large a cutting area, vibration due to the springing of the workpiece, and improper rake, relief, or cutting angles.
- A surface finish depends on the size, shape, and the condition of the nose radius of the cutting tool.
- Lathe cutting tools produce three general types of cutting chips: continuous, discontinuous, and segmental.
- Chip breakers are employed for safety in machining. The chip breaker controls the shape, the size, and the flow of chips for rapid disposal.
- Disposable cemented carbide and ceramic inserts are available in round, diamond, square, rectangular, triangular, and other forms. Tool holders are designed to permit side or end mounting of the insert.

- Cutting speeds are influenced by the material of the cutting tool and workpiece; the cutting, clearance, and rakes angles of the tool bit; the depth of cut; the speed and other conditions of the lathe; and the required tool life.
- Cutting speeds in sfpm or m/min may be found in tables or computed by formula:

$$CS \text{ (sfpm)} = \frac{\text{Circumference (") } \times \text{ rpm}}{12}.$$

or,

$$CS \text{ (m/min)} = \frac{\text{Circumference (mm) } \times \text{ rpm}}{1000}.$$

- The lathe spindle speed (rpm) may be calculated in either metric or inch-standard values by the formulas:

$$rpm = \frac{CS \text{ (sfpm) } \times 12}{\text{Circumference (")}}$$

$$rpm = \frac{CS \text{ (m/min) } \times 1000}{\text{Circumference (mm)}}.$$

In general practice, the following simplified formula is used:

$$rpm = \frac{4 \times CS \text{ (sfpm)}}{\text{diameter}}.$$

- Cutting feeds are influenced by the same workpiece, cutting tool, and machine factors and requirements that apply to cutting speeds.
- The single-point cutting tool (right- or left-hand) is a side- or end-cutting tool having:
  ○ a positive or negative side or back rake
  ○ side, end, and nose relief
  ○ end-cutting and side-cutting edge angles.
- Cutting fluids must be correctly applied at the point where there is the greatest concentration of heat. This practice prolongs tool life and helps produce a dimensionally accurate workpiece and a higher quality surface finish.
- The increased speeds at which lathes are operated when using cemented carbides, ceramic inserts, and diamond cutters (as well as newly developed high-speed steel and nonferrous alloy cutting bits) requires the worker to wear eye protection devices and/or a face shield.

UNIT 25 REVIEW AND SELF-TEST

1. State what effect *crater resistance* of a lathe cutting tool has on its cutting ability.
2. a. Give an application of a cast alloy tool bit.
   b. Tell why it is preferred in this case to a standard high-speed steel tool bit.
3. Distinguish between the manufacture of cemented carbide and ceramic inserts.
4. Cite a practical application of single-point diamond cutting tools in lathe operations.
5. State why a greater amount of tool wear is permissible in rough machining than for finishing cuts.
6. Indicate what relationship exists between the rpm of a lathe spindle and the cutting speed (sfpm).
7. List the nature of the technical information generally found in handbook tables of cutting speeds for lathe work.
8. Identify (a) the two cutting edges of a single-point cutting tool and (b) the purpose of relieving the flank.
9. Tell what effect too small a *relief angle* on the side-cutting edge has on the cutting action.
10. Describe the effect of *negative side-* and *back-rake angles* on the cutting edge of a single-point cutting tool.
11. a. Define *lead angle* of the cutting tool in relation to the workpiece.
    b. State what influence the lead angle has on a cut.
12. Indicate what safety precaution must be taken to control the accumulation of continuous or segmental chips at high speeds.
13. a. List five common shapes of cemented carbide and ceramic inserts.
    b. Give an advantage of disposable cemented carbide or ceramic inserts.
14. Cite two cutter precautions to take when selecting diamond, ceramic, or cemented carbide cutting tools.

Unit 26

LATHE CONTROLS AND MAINTENANCE

The lathe is a precision machine tool. Its accuracy and life span depend upon the worker. The lathe must be operated following safe practices and procedures. These must be within the limitations of the machine tool, the work-holding accessories, cutting tools, and the workpiece. All movements, controls, and adjustments must be maintained carefully. The lathe must be level to ensure continued accuracy.

Equally important is the care that must be taken to reduce and compensate for wear. All mating surfaces must be properly and regularly lubricated. The cutting fluid system must be kept clean. This permits a continuous flow of coolant under hygienic and safe conditions.

The operating controls, alignment, care, and maintenance of an engine lathe are covered in this unit.

LATHE ALIGNMENT AND ADJUSTMENTS

One of the requirements in installing any precision equipment is *leveling*. Leveling is the process of positioning a machine surface so it falls in a true horizontal (0°) plane or a right-angle (90°) vertical plane. When a machine is designed and constructed, the mating and moving surfaces are machined in relation to these planes. When installed, leveling affects all of the machine parts, mechanisms, and the actual accuracy in machining a workpiece. The leveling of a lathe is shown in Fig. 26-1.

A *precision level* is used to level the lathe bed. The level is placed lengthwise. The adjusting screws in the base or leg are turned until the *bubble* of the level is centered in the level sight.

The level is turned 90° and placed across the ways.

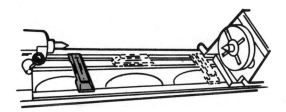

Fig. 26-1. Leveling a lathe with a precision level.

Sometimes, parallels are used between the flat ways and the level. The leveling screws are again adjusted. The lengthwise leveling process is repeated to ensure that the longitudinal leveling has not been changed. Once leveled, shims are placed under the legs and the machine is bolted or nested in position. The levelness is again checked. Further adjustments are made as necessary.

HEADSTOCK CONTROLS FOR SPINDLESPEEDS

Spindle speeds are adjusted on the headstock. To review, there are four principle types of speed-change mechanisms:
- The step-cone pulley drive
- The back-gear drive
- The variable-speed drive
- The geared-head drive.

Slow speed ranges are produced by incorporating *back gears* in the design of each type.

Step-Cone Pulley Drives

Spindle speed (rpm) changes on lathes with step-cone pulley drives are made by stopping the lathe. The belt tension lever is then moved to loosen the belt. A higher speed is produced by changing the belt to a larger diameter on the driver pulley. The belt is moved also to the corresponding smaller step pulley on the spindle. The reverse process is followed for reducing the spindle speed.

Further changes may be made if the motor is two speed or if there is a two-step drive between the motor and the drive shaft. Before starting the lathe, the belt is rotated by hand. This test is to make sure the belt tracks in the right pulley combination. The manufacturer's spindle speed table is usually mounted on the headstock cover. The table (index plate) gives the full range of spindle speeds.

Back-Gear Drives

On some designs of lathes there is a *back-gear* drive for slower speeds. The back gears increase the force the spindle may exert for heavier cuts. The speed is reduced to provide a slower range of rpm. Back gears should be changed only when the power is *OFF*.

A *pin* engages and disengages the back gears for direct drive. Back-gear reduction is produced by a smaller gear on the driving unit turning a larger gear. In turn, a smaller gear on the same solid shaft turns a larger bull gear on the spindle. After the bull gear is engaged, it is checked for proper meshing of the gears. The lathe spindle is rotated slightly by hand. When the drive pin is *IN*, the spindle is driven directly. The back gears are engaged when the pin is in the *OUT* position.

Another model of a back-gear drive is illustrated in Fig. 26-2. The back-gear mechanism is engaged or disengaged by moving the lever as indicated. The right position produces a direct drive. The left position engages the back gear.

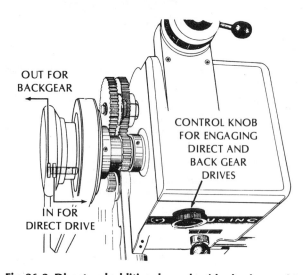

Fig. 26-2. Direct and additional speeds with a back-gear drive.

Variable-Speed Drive Controls

Spindle speed changes are made on variable-speed drives while the motor is running. The speeds are changed hydraulically by positioning the speed-control dial at a desired speed. The spindle rpm is read directly from the dial at the *index line*. Fig. 26-3 shows a variable-speed control dial. The speed range for this back-gear drive is from 52 to 280 rpm. The direct-drive range (open belt) is from 360 to 2000 rpm.

Back-gear (low spindle) speed combinations on variable-speed drives are engaged by a second lever on the headstock. The lathe spindle is stopped. The back-gear lever is positioned for direct drive. When the lever is moved in an opposite position, the back gears are engaged.

Fig. 26-3. A variable-speed control dial for high and low spindle speed ranges. (Courtesy of the Clausing Corp.)

Geared-Head Controls

The *geared-head* lathe incorporates all the features of direct spindle drive and back-gear reduction. Spindle rpm changes are made when the spindle is stopped and different sets of gears are engaged. The gears are moved into position by adjusting one or more gear-change levers on the headstock.

The spindle speed changes appear on an index plate that is attached to the headstock. The positions of the gear-change levers are indicated for each speed.

The interior view of an 18-spindle-speed geared headstock is shown in Fig. 26-4. On this model, spindle speeds from 35 to 1500 rpm are obtained by using the two speed-change levers.

MACHINE CONTROLS FOR FEEDS AND THREADS

Quick-Change Gearbox Controls

The *quick-change gearbox* serves the two-fold

Fig. 26-4. Interior view of an 18-speed geared-head engine lathe. (Courtesy of South Bend Lathe, Inc.)

function of turning the feed screw and the lead screw. An index plate on the gearbox indicates the range of feeds per revolution and the pitch of inch standard or metric threads.

The positions and combinations of the quick-change gear levers for a particular feed or screw thread are given on the index plate. Another lever is known as the *feed-reverse lever*. This lever changes the direction of rotation of the lead and feed screws. Fig. 26-5 shows the quick-change gearbox control levers. Two levers control the feed and thread pitch ranges. A third lever is identified as the feed-reverse lever.

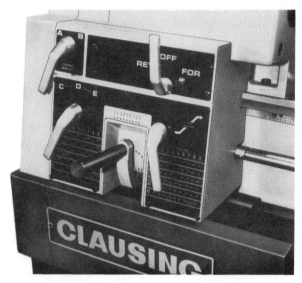

Fig. 26-5. Quick-change gearbox control levers. (Courtesy of South Bend Lathe, Inc.)

Apron Controls

The actual automatic feeding of the cutting tool on the carriage is controlled by a feed lever or automatic feed knob. Usually, when the lever is in the *up position*, the cutting action is toward the headstock. The lever is moved to a neutral position to stop the feed. The *downward position* of the feed lever reverses the direction of feed. Similarly, a second feed lever or knob engages the automatic crossfeed. The direction of the crossfeed may be changed by moving the *feed-reverse lever.*

Thread-Cutting Controls

A *split nut lever* is used for thread cutting. Before this lever is engaged, the feed-change lever must be moved to a neutral position. Again, the *feed-reverse lever* is used to change the direction of the lead screw to right- or left-hand threads.

A high-speed threading attachment is illustrated in Fig. 26-6. The attachment fits in the right side of the apron. This model is designed with an adjustable stop. This automatically disengages the half nut at the same location at the end of each cut.

Fig. 26-6. Automatic adjustable stop high-speed threading attachment. (Courtesy of Do All Company)

COMPENSATING FOR WEAR

The ways on the lathe bed are designed to provide for the *continuous seating* of the carriage. This seating compensates for wear. The cross slide and compound rest are each provided with an *adjustable gib*. This permits adjustment for wear on the mating surfaces.

In principle, a gib is adjusted by loosening a jam nut or screw. A tapered gib is drawn in. When the slide may be moved without *play*, the gib is locked in position with the jam nut.

A flat gib is adjusted by loosening the jam nuts. Adjusting screws are turned until the gib has *taken up the wear*. The jam nuts are then tightened, and the adjusted sliding surfaces are checked. Before any adjustments are undertaken, the bearing surfaces must be free of foreign particles and thoroughly cleaned. Fig. 26-7 shows how a gib adjusting screw is turned on one make of lathe.

Fig. 26-7. Gib adjustment of the cross slide. (Courtesy of the Clausing Corp.)

LOST MOTION

All machine tools and instruments have *lost motion*. Mating parts, like screw threads and gears, require a small difference in size in order to move. This minute variation increases with wear. The craftsperson recognizes that this condition produces lost motion. Unless lost motion is compensated for, a workpiece will be machined to a wrong dimension.

Lost motion is the distance a male or female screw thread or a gear tooth turns before the mating slide, gear, or other mechanism moves. Lost motion is also known as *backlash, slack, play, or end play*. The lost motion on mating screw threads is diagrammed in Fig. 26-8.

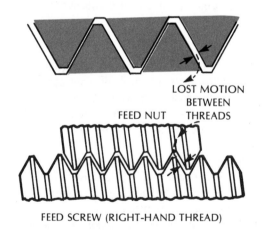

Fig. 26-8. Lost motion between two threads.

Lost motion may be corrected by withdrawing or turning *(backing off)* the screw thread or gear away from a required point. The direction is then reversed. The cutting tool is advanced by feeding it in one direction only. For example, a cutting tool that is

mounted in a tool post on a compound rest is advanced by turning the crossfeed handwheel clockwise. If the cutting tool is moved too far, the correct diameter cannot be machined by simply reversing the direction of motion and bringing the cutting tool out of the workpiece a given distance. Instead, the tool must be backed away from the workpiece. The required distance is at least a part of a turn beyond the graduated measurement to which the tool is to be set. The crossfeed screw direction is then changed (reversed). The cutting tool is advanced to the required graduated reading on the handwheel dial. A trial cut is taken for a short distance. The part is measured. All subsequent cuts may be accurately taken by continuously advancing the cutting tool in the one direction. Checks are made with a micrometer or other measuring instrument.

OVERLOAD SAFETY DEVICES

There is always the possibility of overloading gear trains, lead and feed screws, and other machine parts. Manufacturers use *shear pins* and *slip clutches* to prevent damage.

A *shear pin* is designed to be of a particular diameter and shape. The shear pin withstands a specific force. The pin shears and the machine motion stops when the design force is exceeded. The shear pin in the end gear train (Fig. 26-9) is used to prevent overload and the stripping of gear teeth. A shear pin makes it almost impossible to apply an excessive force on a shaft, rod, or lever.

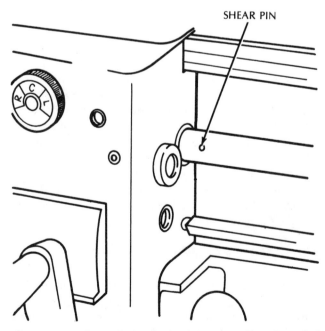

Fig. 26-9. A shear pin in the lead screw and gearbox shaft protects them from overload. (Courtesy of the clausing Corp.)

A *slip clutch* protects the feed rod and connecting mechanisms. The slip clutch is designed to release the feed rod when a specific force is exceeded. The slip clutch mechanism has an added feature that permits the feed rod to be automatically reengaged when the force is reduced. Fig. 26-10 provides an interior view of a spring-ball slip clutch applied to a feed rod.

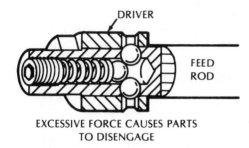

Fig. 26-10. Application of a spring-ball clutch to the feed rod.

TAILSTOCK ALIGNMENT

The tailstock should be tested for correct alignment. This is necessary to ensure accuracy in turning between centers or for positioning, holding, or other machining operations.

If great precision is not required, the alignment of the headstock and tailstock spindles may be checked by sight (Fig. 26-11). The live and dead centers are aligned by adjusting the tailstock head until the dead center axis coincides with the live center axis.

Greater precision may be obtained by using a centered *test bar* or by turning a centered workpiece. A dial indicator, mounted in the tool post, is used with the test bar. Precision alignment by the test bar and dial indicator method is pictured in Fig. 26-12. The tailstock is adjusted until the dial readings are equal at both ends of the test bar.

The third method of testing is to take a light cut to the same depth at both ends of a centered piece of stock (Fig. 26-13). Each diameter is measured by

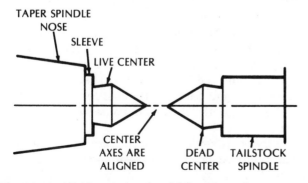

Fig. 26-11. Aligning centers by sighting along the axes.

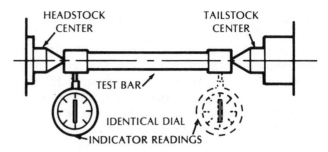

Fig. 26-12. Precision alignment of the centers with a test bar and dial indicator.

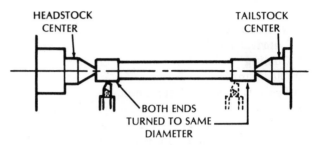

Fig. 26-13. Aligning the tailstock and spindle axes.

micrometer. The tailstock is adjusted until the turned diameters are precisely the same dimension.

LATHE MAINTENANCE

Cleaning The Lathe

Lathe cleaning means the safe removal of cutting chips and foreign particles from the machine, accessories, cutting tools, and workpiece. It also involves cleaning the cutting fluid reservoir and system. The purpose of cleaning, as explained before, is to protect the worker, the machine, and the workpiece. Chips should be removed when the machine is stopped. A chip rake should be used when continuous chips are formed into a large mass.

When chip breakers are applied, the broken chip fragments may be brushed away from the machine areas. Comparatively large quantities of chips are shoveled out of the chip tray. A chip disposal container should be readily available so that none of the cutting fluid is spilled on the floor.

After the chips are removed, the headstock, tailstock, and carriage are then brushed. This is followed by wiping with a clean cloth. The painted surfaces are wiped first; the machine surfaces, last. All accessories and cutting tools must be brushed and wiped clean.

The lead screw must also be cleaned. In this instance the machine may be started with all feeds disengaged. A hard, short-fiber brush may be moved in the threads along the entire length of the lead screw as it turns slowly.

With continuous use, gum and oil stains form on the exposed machined surfaces of the lathe. After these surfaces are wiped with a clean cloth, they should be wiped again with a cloth saturated with a small amount of kerosene or other gum or nontoxic stain-cutting solvent. Both the painted and finished surfaces are wiped clean in this manner. Solvents are available in spray containers for easy application.

Lubricants And Lubrication

Lubricants serve four main functions:
- Lubricating
- Cooling
- Cleansing
- Protecting against corrosion.

Lubricating Function—A lubricant provides a *hydrostatic fluid shim* between all mating parts when the machine is in motion. In other words, the lubricant fills all clearance spaces. The metal surfaces are cushioned and slide with the least amount of friction. Less force is required to move the mating parts, and friction-generated heat is decreased. The finer surface movement conditions produce smoother operation. There is less power consumption, and wear is reduced. The fluid shim also serves the important function of helping to maintain the precision of the machine tool.

Special *way lubricants* are available for reciprocating mating surfaces. These lubricants have a high surface tension. The way lubricants may be used alone for heavy-duty operations where there are severe abrasive conditions. The way lubricants may also be added to other lubricating oils. The way lubricant additive increases the load-carrying capabilities and lubricating properties of the other oils.

Cooling Function—The heat that normally is generated in the machine tool itself is carried away by the flow of the coolant. Another cooling effect results from the reduction of friction. Cooling also helps prevent seizing due to expansion.

Cleansing Function—The lubricant washes away minute foreign particles. These are produced by the abrasion and wearing of surfaces that move against each other. Cleansing reduces wear and helps maintain the precision of the machine tool.

Protection Against Corrosion—Exposure of unprotected ferrous metals to air, use of certain water-soluble cutting fluids, and turning of corroded surfaces may produce rust. Rust on a machined part results in a pitted and rough surface that has abrasive properties. This condition increases wear, affects machining accuracy, and decreases machine tool life.

Corrective measures require the covering of exposed machine surfaces wherever possible. Rusted chips should be removed quickly. The coolant system requires more frequent system checks and replacement of the cutting fluid. The machined surfaces must be wiped clean and protected with a rust-inhibiting solution.

Rust preventive mists are available to protect unfinished parts or machine surfaces. The rust preventive is sprayed as a mist. Workpieces and machine parts are thus protected during idle periods or when operating without a cutting fluid. The rust preventive mist has a water-displacing property. It is effective on wet or dry surfaces.

MAINTAINING THE LUBRICATION AND COOLANT SYSTEMS

Care Of The Lubricant System

Most lathes are furnished with pressure lubricating systems. Lubricants are forced between all gears, the spindle and bearings, and other moving parts. Where necessary, sight level indicators show the amount of lubricant in an oil reservoir. These must be checked before the lathe is operated. The indicator level must be maintained continuously. Each manufacturer provides information about the type and grade of lubricant to use. This is important. An improper grade may cause overheating, impair the machining accuracy, and cause damage to various parts of the lathe.

In addition to machine lubrication from a pressurized automatic system, the ways and other oil holes may need lubrication with a hand oiler. These parts require daily cleaning and lubricating.

Care Of The Cutting Fluid System

Recommendations for completely draining the cutting fluid reservoir and system must also be followed carefully. Sludge sediment and foreign particles must be removed at regular intervals.

The cutting fluid is pumped out of the reservoir into a disposable container. The sediment-collecting trays are removed and cleaned. The collection of chips and foreign particles that settle in the bottom are shoveled out. The inside container is washed clean.

The trays are replaced and the reservoir is refilled to the required level with fresh cutting fluid. The cutting fluid must be tested and appropriate for the material to be turned and the work processes.

Visual, on-the-spot tests of fluid concentration may be made with an *industrial fluids tester*. Such a tester is illustrated in Fig. 26-14. This hand-held instrument measures the refractive index (concentrations in the mixture) of the cutting fluid.

The instrument compensates for temperature variations in the cutting fluid. A few drops of the fluid

Fig. 26-14. An industrial fluid tester. (Courtesy of Do All Company)

provide the test sample. The cutting fluid concentration is quickly established by a direct reading on the fluid tester.

FACTORS AFFECTING MACHINING CONDITIONS

There are a number of general problems the lathe operator faces. These affect the accuracy and quality of the machined part and the machining efficiency. Being able to recognize and correct each problem is essential to maintaining a lathe in an excellent and safe working condition.

A number of common lathe problems are listed in Table 26-1. The most probable causes of each problem and the corrective actions the worker may take are also given.

Table 26-1. FACTORS AFFECTING MACHINING CONDITIONS AND CORRECTIVE ACTION

Problem	Probable Cause	Corrective Action
Machine vibration	● Work-holding device or setup is out of balance	● Counterbalance the unequal force ● Reduce the spindle speed
	● High cutting speed	● Reduce the spindle speed
	● Floor vibration	● Secure the base firmly to a solid footing
	● Overhang of cutting tool or compound rest	● Shorten the length the tool holder extends beyond the tool post ● Position the tool holder so that the compound rest has maximum support from the bottom slide surfaces
Vibration of the cross slide and compound rest	● Compound rest slide extends too far from the bottom slide support surfaces	● Reset the cutting tool so that there is greater area in contact between the top and bottom slide surfaces
	● Loose gib	● Adjust the gib to compensate for wear, and lock it in the adjusted position
	● Overhang of cutting tool or compound rest	● Shorten the length of the cutting tool in the holder and the tool holder in the tool post
Play in feed screw	● Wear on the feed screw and nut	● Back the cutting tool away from the work. Take measurements by maintaining continuous contact, turning in one direction only
Chatter	● High rate of cutting speed or feed	● Reduce the spindle speed and rate of feed
	● Incorrect rake or relief angles, or form or cutting angles	● Regrind the cutting bit to the recommended rake, relief, and cutting angles and form
	● Overhang of cutting tool or tool holder	● Shorten the length the cutting tool extends from the holder and the tool post
Misalignment of headstock and tailstock axes	● Tailstock offset	● Move the tailstock head in its base until the spindle axis aligns with the headstock spindle axis

HOW TO ALIGN THE TAILSTOCK

Aligning By Sight

Method One:

Step 1   Locate the index lines on the base and head of the tailstock.

Step 2   Align the index lines (witness marks).
Note: Unloosen the tightening bolt. Move the head on the base until the two index lines coincide. At that point, the centers are aligned.

Step 3   Lock the tailstock head and base in this position.

Method Two:

Step 1   Clean the headstock and tailstock center spindles. Insert the live and dead centers.

Step 2   Move the tailstock toward the headstock until the centers almost touch.

Step 3   Look down on the centers. Sight across the points.

Step 4   Move the head section of the tailstock so that the live and dead centers are aligned.
Note: Loosen the nut that binds the head and base sections of the tailstock. Then move the head section until the dead center is aligned with the live center.

Step 5   Tighten the binding nut to hold the tailstock head and base sections securely.

Checking Alignment With A Test Bar

Step 1   Mount the live and dead centers.

Step 2   Place and hold a centered cylindrical parallel test bar between centers.
CAUTION: Clean the center holes of the test bar before mounting between centers.

Step 3   Mount a dial indicator in the tool post. Position the dial indicator so that the spindle is in a horizontal position.

Step 4   Bring the point end into contact with the test bar. Note the reading.

Step 5   Move the dial indicator toward the second end (near the headstock).
CAUTION: It is good practice to move the indicator point away from the test bar, using the index finger. The point is then carefully released at the second position.

Step 6   Note any variation in reading. If the reading is greater at the headstock end, it means the tailstock must be moved toward the operator. The distance to move the tailstock is equal (approximately) to the difference in the two dial readings.

Step 7   Return the dial indicator to the tailstock portion of the test bar. Loosen the binding nut and adjust the tailstock until the dial reading is the same as the reading at the headstock end.

Step 8   Secure the head and base sections of the tailstock. Recheck the alignment by taking a dial reading at both ends of the test bar. Make whatever further adjustments may be required.
Note: The centers are aligned when the dial readings on both machined ends of the test bar are the same.

Checking Alignment By Machining

Step 1   Mount the workpiece between centers.

Step 2   Position the cutting tool. Take a light cut for a short distance. Note the reading on the graduated dial of the cross feed.

Step 3   Back the cutting tool away from the workpiece.

Step 4   Reposition the cutting tool near the headstock end of the workpiece. Move the cutting tool to the same depth, according to the graduated dial reading. Take a cut for a short distance.

Step 5   Measure both turned diameters.
Note: If the tailstock diameter is larger, it means the tailstock must be moved forward toward the operator.

Step 6   Position a dial indicator so the point makes contact with the tailstock spindle.

Step 7   Loosen the binding nut. Move the tailstock head the required distance toward the operator if the turned portion at the tailstock end is larger in diameter than the headstock end. Move the tailstock in the opposite direction if the reverse condition exists.

Step 8   Secure the head and base sections of the tailstock. Take a second cut on the centered workpiece. Recheck the turned diameters. Readjust the tailstock (if necessary) until the same diameter is produced on both ends of the workpiece.

HOW TO CLEAN A LATHE

Step 1   Cut *OFF* all power. Shut down the lathe.

Step 2   Rake or brush away all chips from the machine, the accessories, and the cutting tool.

Step 3   Wipe each accessory and cutting tool.

Step 4   Moisten a clean wiping cloth with a thin film of oil or rust inhibitor. Wipe over all machined surfaces. '

Step 5   Replace each accessory and cutting tool in its appropriate place in the machine stand or tool room.

Step 6   Clean all painted and machined surfaces of the lathe with a wiping cloth.
Note: Thick gums and grease may be removed with kerosene or a manufactured solvent. These cut the grease without damaging the adjacent surfaces.

HOW TO LUBRICATE A LATHE

Step 1   Check the level of the lubricant in the geared headstock. The level must be at or above the index line on the sight level.

Step 2   Add lubricant by removing the filler cap. Bring the level to the index line.
CAUTION: The lathe manufacturer's technical manual must be consulted for the correct lubricant to use.

Step 3   Repeat the steps with the quick-change gear box and any other gear train mechanism.

Step 4   Fill any other oil holes with a hand oiler. Use the kind of lubricant recommended by the lathe maker.
Note: The skilled worker checks the technical manual to determine the number and types of lubrication systems. *Non-forced systems* are lubricated before operating the lathe.

HOW TO CLEAN THE COOLANT SYSTEM

Step 1   Remove the chip and sediment trays from the coolant reservoir.

Step 2   Clean the sludge that has accumulated. Use a trowel or metal scoop.
Note: The trays are washed in a cleaning and degreasing tank when one is available in the shop.

Step 3   Siphon off the cutting fluid from the reservoir into a disposal or separating container.

Step 4   Remove the fine particles and gum that have been deposited in the base. These may be shoveled out. Flush the reservoir.

Step 5   Remove the coolant nozzle. Clean and reinstall it.

Step 6   Remove the coolant pump if it is portable. Flush the impellor blades or immerse them in a cleansing bath.

Step 7   Determine the cutting fluid that is to be used. Fill the reservoir to the indicated height.

Step 8   Check the system. Turn on the coolant pump. Open the nozzle valve and check to see that the cutting fluid is circulating properly.

SAFE PRACTICES IN THE CARE AND MAINTENANCE OF THE LATHE

- Protect the lathe bed and mechanisms against warpage and distortion. This requires proper positioning and leveling.
- Make sure the lathe spindle is stopped when making speed changes on step-cone pulley, geared-head, and back-gear drives.
- Change the spindle speeds of variable-speed motor drives when the spindle is in motion.
- Replace overload shear pins with shear pins of the same material.
- Check for excessive force if there is continuous slipping of a slip clutch on a feed rod. Damage may result if the condition is not corrected.
- Provide compensation for lost motion when setting a cutting tool. Otherwise, an incorrect dimension may result. Excessive lost motion requires adjustment of a gib or replacement of a screw thread or nut.

- Position the axes of the tailstock and headstock spindles so that they coincide. This alignment is necessary to accurately machine a straight surface.
- Remove chip fragments only when the lathe is shut down. Use a brush or a chip rake.
- Place chips in a chip container. Avoid splashing cutting fluid on the shop floor.
- Clean lead screw threads carefully. A hard, short-fiber brush may be held against a slowly turning screw.
- Check the manufacturer's recommendations for lubricating the lathe parts.
- Use machine lubricants in forced-feed systems that meet the lathe manufacturer's specifications. Lubricants must be maintained at the level indicated on the sight gage.
- Learn the location of hand oil cups. Oil the lathe before operating it.
- Coat exposed machine surfaces with a rust preventive film to protect against possible corrosion, particularly during a machine shutdown period.
- Drain and clean the cutting fluid system regularly.
- Check the composition of the cutting tool to be sure it maintains the same properties when a coolant and lubricant are used.
- Check to see that the feed rod and lead screw may not be engaged simultaneously. This condition can cause damage to both mechanisms.
- Analyze any chatter. Chatter is a sign that the workpiece, cutting tools, or setup are incorrect, or the speed or feed are excessive. These unsafe conditions can cause injury to the operator or damage to the lathe or workpiece.
- Wear safety goggles when cleaning the lathe and when performing lathe operations.

LATHE CONTROLS AND MAINTENANCE TERMS

Leveling (lathe)	A process of positioning a machined surface of the lathe so it always lies in the same horizontal or vertical plane. A process designed to overcome lathe distortion and prevent machine inaccuracies.
Spindle speed drives	Combinations of pulley and gear drives which provide for the change of spindle speeds. Mechanisms which permit speed changes to accommodate different machining processes and materials.
Back gears	A simple gear train that, when engaged, provides a slow-speed spindle range.
Speed- or feed-change levers	A series of levers mounted on the headstock or quick-change gearbox. Positioning internal gear combinations to produce a specified spindle speed (rpm), feed, or correct pitch.
Starting, stopping, and reversing levers	Levers mounted on the apron to control the longitudinal and crossfeed direction and movement.
Slip clutch	An overload prevention safety device. A spring-activated safety device that disengages the feed rod whenever an excessive force is applied.
Lost motion, backlash, end play	Excessive clearance between two movable mating parts. A condition in which movement of one member of a mechanism fails to simultaneously move the other member, because of excessive clearance between them.
Lubricating function	The filling of a void area (clearance space) between two mating parts with a hydrostatic fluid film. Reduces friction between mating moving parts.

SUMMARY

- A precision level is used to position the lathe bed in a horizontal plane. Leveling prevents machine distortion and permits accurate machining.
- The four basic types of speed drive mechanisms are:
  - Step-cone pulley
  - Back gear
  - Variable speed
  - Geared head.
- The use of back gears extends the normal speed range to obtain slower speeds and increased torque.
- Pulley- or gear-driven speed and feed mechanisms should be stopped before changes are made in the spindle rpm.
- The range of spindle speeds and the related gear lever combinations are given on an index plate. This is attached to the headstock.
- Feed and pitch changes are made through the quick-change gearbox.
- Overload safety devices include shear pins and slip clutches. These are designed to stop the movement of mating parts whenever excessive forces are applied.
- The tailstock may be aligned by sighting, with a test bar, or by machining a workpiece.
- Machined surfaces must be cleaned of gummy substances and protected against rust and corrosion.
- Vibration of the cross slide may be caused by loose fitting or overhang of the compound rest.

- Machine vibration may be caused by excessive speed, poor support by the work-holding device, or unequal turning forces.
- Lost motion may be compensated for by backing a tool away from a workpiece. The direction is then reversed and the tool is fed to the required depth.
- Chatter results when there is excessive speed, tool overhang, or the work is not held rigidly. Incorrect rake, relief, cutting angles, or nose radius may also produce chatter.
- The lathe manufacturer's specifications must be followed. The proper lubricant must be used for all movable mating parts, such as gear trains and feed and speed mechanisms.
- Machine lubricants must be checked daily. These must be held at or above the index line on the sight gage.
- The lathe must be cleaned by removing all chips, wiping all machine surfaces, and giving them a protective coating.
- Hygienic conditions must be maintained in coolant reservoirs and systems. The concentration and condition of the cutting fluid must be checked regularly and the sediment removed.
- Operator safety requirements include the use of a safety shield or goggles. The lathe must be shut down during cleaning. All guards must be replaced *before* the machine is turned on. The operator should secure all loose clothing before operating any machine.

UNIT 26 REVIEW AND SELF-TEST

1. State the prime function that is served by leveling.
2. Describe briefly the operation of (a) *variable-speed drive controls* and (b) *geared-head controls* on an engine lathe.
3. Explain how *lost motion* in mating threads or gears on a lathe is (a) produced and (b) corrected.
4. Name two general *protective overload devices* on a lathe.
5. a. Explain the effect of corrosion on machined surfaces that slide on each other.
   b. Give a preventive step that may be taken to prevent corrosion during shut-down periods.
6. State two functions that are served by an *industrial fluid tester*.
7. List the steps for checking the alignment of the headstock and tailstock spindle axes by machining an already centered workpiece.
8. Name five terms that are associated with the cleaning of the coolant system on a lathe.
9. State three safety precautions to observe in the care and maintenance of a lathe.

SECTION TWO

LATHE CUTTING TOOLS AND TOOLHOLDERS

Lathe work processes begin with the selection of the most suitable cutting tool and toolholder. This section presents:
- The design characteristics and specifications of basic lathe cutting tool bits and the coding systems used to identify them
- The design characteristics and general applications of lathe toolholders
- The procedures for grinding high-speed steel tool bits.

Unit 27

BASIC LATHE CUTTING TOOLS AND HOLDERS

The basic cutting tool for lathe work is called a *tool bit* or an *insert*. Both forms are held securely by a toolholder during machining processes. Other cutting tools are brazed onto a hardened steel shank, as in the case of cemented carbides. The toolholder or solid shank is then positioned in relation to the work and process in a tool post, turret head, or other holding device. The basic cutting tools and toolholders for lathe processes are covered in this unit.

CUTTING TOOL BITS

The efficiency of the cutting tool depends on the following factors:
- Forming the correct relief, clearance, rake, and cutting edge angles. (These depend on the cutting

requirements and necessary dimensional accuracy.)
- Positioning the cutting edges properly with respect to the workpiece and processes
- The rigidity of the setup used to support the cutting action
- The properties of the cutting tool to maintain effective tool life
- Speeds and feeds which permit the processes to be performed economically and safely.

A single-point, high-speed steel tool bit that has a square cross section is commonly used for lathe operations. The tool bits are usually ground on a bench or pedestal grinder. An aluminum oxide abrasive grinding wheel is used for grinding. Diamond-impregnated grinding wheels are used for cemented carbides. The high-speed steel tool bit is usually ground by hand. The tool is held and guided

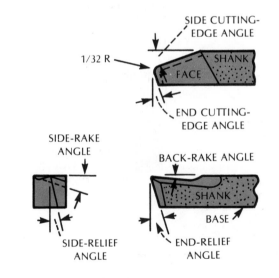

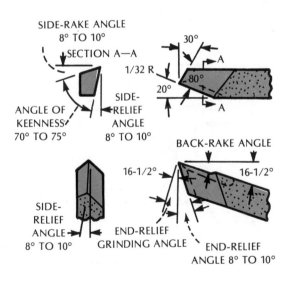

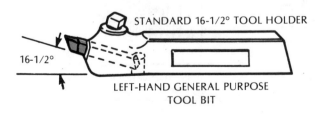

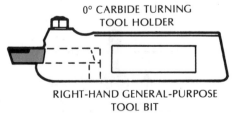

Fig. 27-1. General-purpose single-point turning tools and toolholders and their angles and related terms.

325

by hand to form cutting edges that have a particular shape and appropriate clearance (relief) and rake angles.

SINGLE-POINT CUTTING TOOL TERMS

The particular shape and the angles to which a tool bit are ground are determined by the process to be performed and the material to be cut. Other conditions that relate to speed, feed, and finish also are considered. General terms used in the shop in relation to the tool bit are illustrated in Fig. 27-1. The various angles and their designations that were treated in the preceding unit are reviewed for further development.

● The *cutting edge* designates the portion of the tool bit that does the cutting.

● Chips flow over the top surface of the tool bit. This surface is called the *face*.

● The surface which is adjacent to and supports the cutting edge is the *flank*.

● The end- and side-cutting edges are joined at the *nose*.

● The *point* refers to all the surfaces on the face

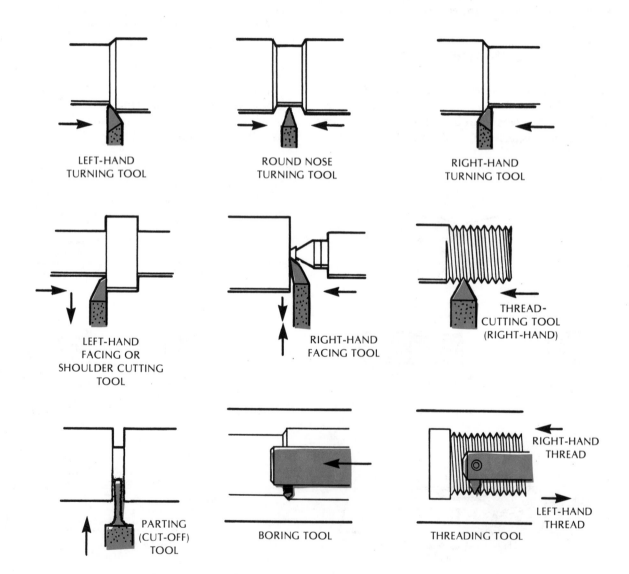

Fig. 27-2. Common shapes of tool bits and their general-purpose applications.

that produce the cutting edges.

● The term *right-hand* or *left-hand,* when used, precedes the name of the cutting tool or operation. For example, a right-hand rough cutting (roughing) tool. The cutting action of this single-point tool bit is from right to left. The movement is toward the headstock or from the center of the workpiece outward. A left-hand cutting tool cuts from left to right.

Some common shapes of tool bits and their applications are illustrated in Fig. 27-2. In external turning processes, rough and finish cuts are taken with either *right- or left-hand turning tools.*

Shoulders are turned to a particular radius with a modified form of right- or left-hand turning tool. *Right-hand* and *left-hand facing tools* are adapted to facing processes where the work is held between centers. The *end-cutting tool bit* is used for cutting-off operations. Another form of end-cutting tool produces a high-quality surface finish.

SYSTEM FOR DESIGNATING BRAZED-TIP CARBIDE TOOL BITS

Carbide brazed-tip single-point tool bits (Fig. 27-3)

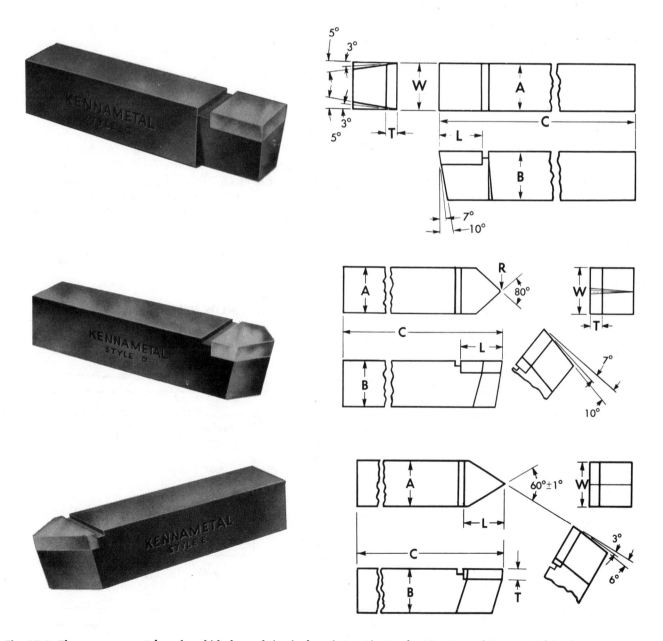

Fig. 27-3. Three common styles of carbide brazed-tip single-point cutting tools. (Courtesy of Kennametal Inc.)

Table 27-1. SYSTEM FOR DESIGNATING BRAZED-TIP CARBIDE TOOL BITS

SAMPLE DESIGNATION:

| B | R | — | 6 |

TOOL CHARACTERISTICS

Shape	Applications
(A) 0° lead	Shouldering, turning
(B) 15° lead	Forming to lead angle
(C) Square nose	Grooving and facing
(D) 30° lead	Undercutting and chamfering
(E) 45° lead	Threading

HAND

R—right hand

L—left hand

SHANK SIZE

(Expressed in $\frac{1}{16}$″)

are designated by a system of two letters and a numeral. As shown in Table 27-1, the first letter indicates the tool shape and the general process(es) for which it may be used. The second letter identifies the side-cutting edge and the direction of cut. The numeral gives the size of the square shank in sixteenths of an inch.

The designation of the carbide tool bit used in the example (BR-6) in Table 27-1 indicates that it has a lead angle of 15° (B). The brazed tip is ground as a right-hand cutting tool (R). The tool bit may be used for any general purpose. The (6) shows that the shank is 6/16″, or 3/8″, square.

Table 27-2. STANDARD INDUSTRIAL DESIGNATIONS FOR THROWAWAY CARBIDE INSERTS

SAMPLE DESIGNATION:

| S | N | E | F | — |
| 1 | 2 | 3 | 4 | |

SHAPE	CLEARANCE	CLASS		TYPE
R Round	N 0°		Cutting Point	Thickness	A With hole
S Square	A 3°	A	.0002″	.001″	B With hole and one countersink
T Triangle	B 5°	B	.0002″	.005″	
P Pentagon	C 7°	C	.0005″	.001″	C With hole and two countersinks
D Diamond 55°	P 10°	D	.0005″	.005″	
C Diamond 80°	D 15°	E	.001″	.001″	F Clamp-on type with chipbreaker
O Octagon	E 20°	G	.001″	.005″	
H Hexagon	F 25°		.002″		G With hole and chipbreaker
L Rectangle	G 30°	M**	to	.005″	
M Diamond 86°			.005″		H With hole, one countersink and chipbreaker
A Parallelogram 85°			.005″		
B Parallelogram 82°		U**	to	.005″	
E Parallelogram 55°			.012″		J With hole, two countersinks and chipbreaker
F Parallelogram 70°					
V Diamond 35°					

*Used only when required
**Exact tolerance is determined by size of insert

SPECIFICATIONS OF THROWAWAY CARBIDE INSERTS

Carbide cutting tool inserts (Fig. 27-4) are pre-formed. A system for identifying eight design features of throwaway inserts provides complete information. The system consists of letters and numerals. For example, a throwaway carbide insert may be specified as SNEF-334E. Table 27-2 gives the standard industrial designations for each symbol. The SNEF-334E sample designation indicates that the insert:

1) is square (S)
2) has a 0° clearance angle (N)
3) has a cutting point and thickness class of 0.001″ (E)
4) is a clamp-on type with a chip breaker (F)

The second group of symbols shows that the insert:

5) is 3/8″ square (3)
6) is 3/16″ thick (3)
7) has a cutting point of 1/16″ radius (4)
8) is unground and honed (E).

Fig. 27-4. Four basic forms of throwaway carbide inserts.

CHIP BREAKERS AND CUTTING ACTION

There are three prime forces that act during the cutting process: *longitudinal, radial* and *tangential*. As the cutting tool feeds into the workpiece, a *longitudinal force* is produced. This is shown in Fig. 27-5. The force of the workpiece against the front face of the cutting tool is a *radial force*. A *tangential force* results from the downward action of the workpiece against the top of the tool.

Materials are removed during machining by these three forces. The forces *wedge* or *shear away* the material.

On soft materials the shearing action produces a continuous ribbon or chip. The shape of the top face helps form a spiral chip.

SIZE	THICKNESS	CUTTING POINT Radius of Flats	OTHER CONDITIONS*
For size ¼″ and over, use number of ⅛ths in size For Rectangular and Parallelogram, use two digits: number of ⅛ths in width and ¼ths in length	For size ¼″ and over, use number of ¹⁄₁₆ths in thickness	0 Sharp corner 1 ¹⁄₆₄ Radius 2 ¹⁄₃₂ Radius 3 ³⁄₆₄ Radius 4 ¹⁄₁₆ Radius 6 ³⁄₃₂ Radius 8 ⅛ Radius A Square insert with 45° chamfer	A Ground all over —light honed B Ground all over —heavy honed C Ground top and bottom only— light honed D Ground top and bottom only— heavy honed E Unground insert —honed F Unground insert —not honed T Chamfer-cutting edge

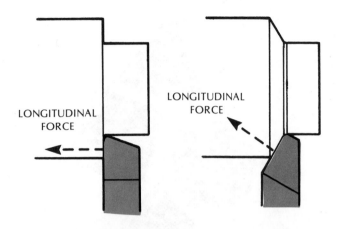

Fig. 27-5. Longitudinal cutting force perpendicular to the side-cutting edge.

On harder materials the wedging forces cause the material to compress until the shearing point is reached. The material is then separated at the cutting edge, forming a chip. As stated earlier, continuous spiral chips create a hazard for the operator. They may also cause damage to the workpiece or machine.

Continuous chips are broken into small, safe sizes with a *chip breaker*. This feature may be ground as an area on the top of the tool bit or preformed on an insert, or added as a chip groove. Examples of parallel, angular, and groove type chip breakers for brazed carbide cutting tools are shown in Fig. 27-6. The preformed chip grooves for carbide inserts are also illustrated.

Provision is made to accommodate chip breakers

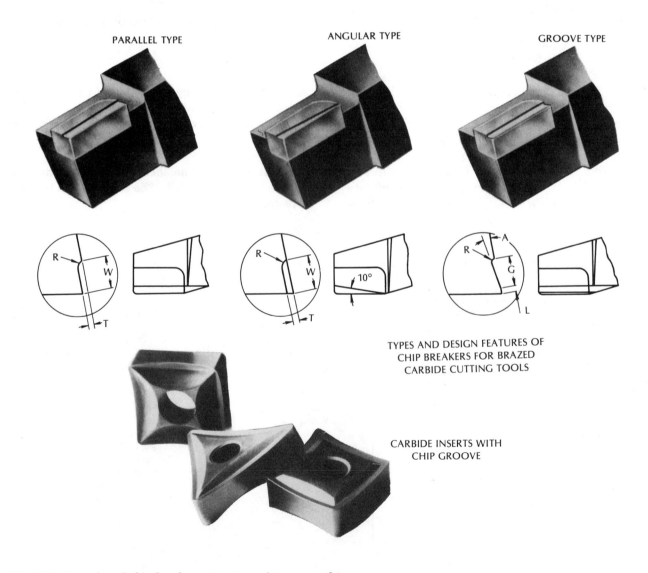

PARALLEL TYPE

ANGULAR TYPE

GROOVE TYPE

TYPES AND DESIGN FEATURES OF
CHIP BREAKERS FOR BRAZED
CARBIDE CUTTING TOOLS

CARBIDE INSERTS WITH
CHIP GROOVE

Fig. 27-6. Examples of chip breakers. (Courtesy of Kennametal Inc.)

on toolholders. Chip breakers for throwaway insert toolholders are commercially available. Chip breakers are designed for positive-, neutral-, and negative-rake toolholders and for general, light, medium, and heavy feeds.

ANGLE GRINDING GAGES FOR SINGLE-POINT TOOL BITS

Tables of recommended relief and rake angles for single-point cutting tools provide guidelines for the craftsperson. (See Appendix Tables A-24 and A-25.) The angles may be checked for accuracy by using a flat metal gage. Two side edges of the gage are ground to conform to specified end- and side-relief angles. Two V-grooves, one in the top and one in the bottom of the gage, are used to measure the cutting tool angles. Fig. 27-7 shows how a tool bit held in a toolholder is measured for side relief, end relief, and the tool angle.

TOOLHOLDERS

There are four basic types of toolholders that are widely used for holding high-speed tool bits (Fig. 27-8):

- *Standard* toolholder
- *Cutting off* and side toolholder
- Spring type *side-cutting* toolholder
- *Threading* toolholder.

Each of these toolholders is designed as a *right-hand* or *left-hand (offset)* or *straight type*. The type that is used is determined by the nature of the operation. The shape of the toolholder must permit the cutting tool and all parts of the machine setup to *clear without hitting*.

A separate form for the threading toolholder permits the use of a circular tool blade that is especially ground for thread cutting.

Standard Tool Bit Holders

A standard toolholder is designed to hold the tool bit at a fixed angle of either 0° or 16½°. Fig. 27-9 shows the true back rake angle. This includes the rake angle to which the tool bit is ground plus the 16½° angle of the toolholder. The end relief angle is also increased to offset for the 16½°.

A standard *zero degree angle* toolholder is shown in Fig. 27-10. This angle represents the angle formed by the bottom of the slot in the toolholder and its base. The zero degree toolholder is used principally with brazed carbide-tipped tool bits. Carbide cutting tools require little or no back rake.

The 0° and 16½° toolholders are furnished in five sizes to accommodate tool bits that range from 1/4" to 5/8" square.

Toolholders For Carbide And Ceramic Inserts

The *throwaway (disposable) carbide or ceramic insert* is held in a solid shank by any one of many holding devices. The insert may be *locked* securely by a cam action. The insert may also be strapped in position with a clamp. The chip breakers that are widely used with carbide and ceramic inserts are locked in position over the insert (Fig. 27-11).

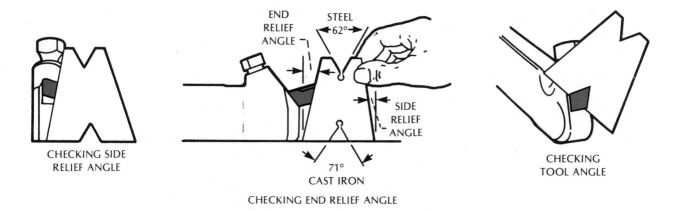

Fig. 27-7. Applications of a relief-and tool-angle gage.

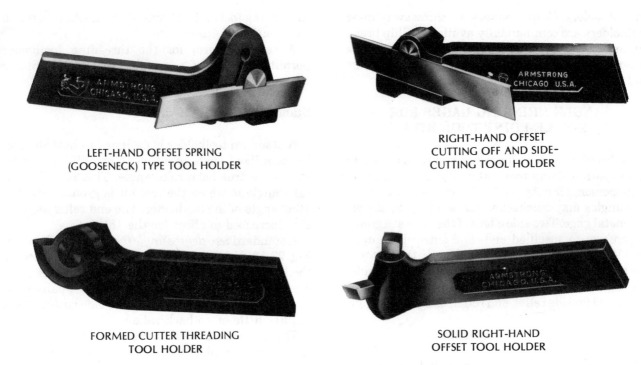

LEFT-HAND OFFSET SPRING
(GOOSENECK) TYPE TOOL HOLDER

RIGHT-HAND OFFSET
CUTTING OFF AND SIDE-
CUTTING TOOL HOLDER

FORMED CUTTER THREADING
TOOL HOLDER

SOLID RIGHT-HAND
OFFSET TOOL HOLDER

Fig. 27-8. Basic types of holders for HSS cutting tools. (Courtesy of Armstrong Bros. Tool Co.)

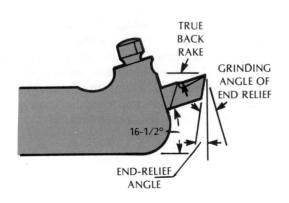

Fig. 27-9. True back-rake angle (16½°-angle toolholder).

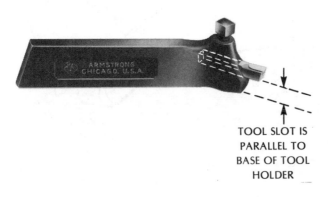

TOOL SLOT IS
PARALLEL TO
BASE OF TOOL
HOLDER

Fig. 27-10. Left-hand offset 0°-angle solid toolholder. (Courtesy of Armstrong Bros. Tool Co.)

Fig. 27-11. Chip breaker secured over carbide insert in a solid right-hand offset toolholder. (Courtesy of Do All Company)

Regular square tool bits and insert toolholders are available for use with conventional tool posts, turret type, and heavy-duty tool posts.

Quick-Change Tool System

Quick-Change Toolholder—The quick-change toolholder is designed with a dovetail slide. This type and the features of a facing and turning toolholder are shown in Fig. 27-12. This permits the holder to fit and be held in a main dovetailed section of a tool post. The holder provides for cutting tool adjustments. The unit may, thus, be *preset*. Vertical adjustments are also possible.

Fig. 27-12. Quick-change toolholder for turning and facing tools. (Courtesy of Armstrong Bros Tool Co.)

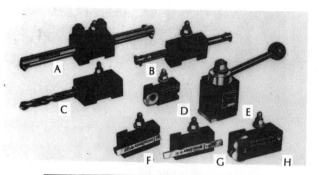

NATURE OF TOOL HOLDER	
A	HEAVY-DUTY BORING
B	LIGHT-DUTY BORING
C	#3 MORSE TAPER SHANK DRILL
D	THREADING
E	QUICK-CHANGE TOOL POST
F	TURNING
G	PARTING (CUT OFF)
H	KNURLING

Fig. 27-13. A quick-change tool system.

The advantage of this setup is that, in quantity production, several cutting tools may be ground and preset. A preset unit may be positioned quickly to replace a worn cutting tool. The unit is held securely by a clamp.

Quick-Change Tool Post—A quick-change tool post combines the features of the quick-change toolholder and the multiple-station turret. An individual toolholder is secured to one side of the quick-change tool post. When the tool setup is completed, each tool is positioned for a particular operation. Quick-change tool posts are held by simply turning to the next station and tightening the clamping lever.

Fig. 27-13 illustrates a *quick-change tool system.* The *system* consists of the quick-change tool post and a number of toolholders for different cutting tools. Each cutting tool is mounted quickly, aligned, and secured in position. The combination illustrated is used for performing general external and internal lathe operations.

Heavy-Duty, Open-Side Tool Block (Holder)

The heavy-duty, open-side tool block, or toolholder, is used to rigidly hold the cutting tool. The holder supports the cutting tool against the tremendous forces that are exerted when heavy cuts are taken. One tool is held at one time. The C-shaped block (Fig. 27-14) fits in the compound rest slide. It is secured by a T-slot clamp. This type of cutting tool holder is also adapted for operations requiring a rear cross slide for a cutting tool. Fig. 27-14 shows an application of the heavy-duty tool block mounted on the compound rest.

Fig. 27-14. A heavy-duty C-shaped block secured on a compound rest.

HOW TO GRIND HIGH-SPEED STEEL TOOL BITS

Example: Right-Hand Turning Tool (Roughing Cut)

Step 1   Determine the grade of high-speed steel that is recommended for the machining processes to be performed.

Step 2   Consider the shape and angles of the side-, end-, and nose-cutting edges.

Step 3   Select the appropriate toolholder.

Note: The 16½° standard toolholder is used for general lathe processes.

Step 4   Determine the correct end- and side-relief angles and side- and back-rake angles.

Note: These are based on the nature of the finish required, the composition of the workpiece, the nature of the process, the speed, and the feed.

Step 5   True and dress both the coarse-grit (roughing) and fine-grit (finishing) aluminum oxide grinding wheels on the bench or floor (pedestal) grinder.

Grinding Side-Cutting Edge And Relief Angles

Step 1   Position the tool rest. Hold, position, and guide the cutting tool with the fingers of both hands (Fig. 27-15). The bottom (heel) is tilted toward the grinding wheel.

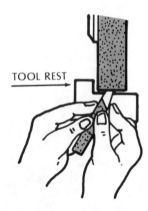

Fig. 27-15. Holding and guiding a tool bit while grinding.

Step 2   Rough grind the side-cutting edge and relief angles (Fig. 27-16).

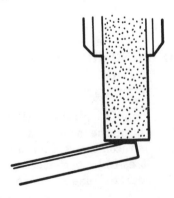

Fig. 27-16. Grinding the side-cutting edge and relief angle.

Step 3   Check the side-relief angle. Use a bevel protractor or a side-relief angle gage.

Note: Rough grind to within approximately 1/16″ from the top of the tool bit.

CAUTION: Move the tool bit across the *entire* face of the grinding wheel. This is to avoid uneven wearing of the wheel face. Quench the tool bit regularly. Quenching permits safe handling and protects the properties of the cutting tool.

Grinding The End-Cutting Edge And Relief Angles

Step 1   Raise the front end of the tool bit to increase the angle of contact with the face of the grinding wheel (Fig. 27-17).

Fig. 27-17. Grinding an end-cutting edge and relief angle.

Step 2   Rough grind the end-cutting edge and end-relief angles to within 1/16″ of the tool bit point.

Note: The formed surface must blend the end-cutting and side-cutting angles, the flanks, and the nose radius (Fig. 27-18).

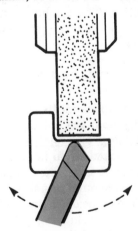

Fig. 27-18. Grinding a nose radius.

Grinding Side And Back Rake

Step 1   Hold the tool bit against the face of the grinding wheel so that the side-cutting edge is horizontal.

Step 2   Tilt the tool bit surface adjacent to the cutting edge inward toward the grinding wheel to grind the back-rake angle.

Step 3   Rough grind the face (Fig. 27-19) to within 1/16″ of a sharp edge.

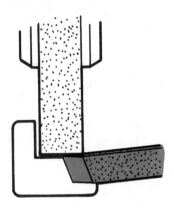

Fig. 27-19. Grinding the side- and back-rake angles

Finish Grinding

Step 1   Repeat all of the previous grinding steps. Use the fine-grained wheel. Take light finishing cuts.

Step 2   Grind just below the roughly ground surfaces. The finish grinding produces a fine-grained edge.

Honing (Stoning) The Cutting Edges

Step 1   Select a medium/fine-grain abrasive stone *(oil stone)*. Apply a small amount of kerosene or cutting fluid to the face of the stone.

Step 2   Hold the medium-grain side of the stone against the flank of the tool bit.

Step 3   Move the stone parallel across the side (face) of the cutting tool (Fig. 27-20).

Step 4   Repeat the honing process. Use the fine-grained side of the stone.

Note: Some workers perfer to move the tool bit when honing the cutting edges. In this instance, place the cutting tool flank on the stone. Rub the tool bit over the stone lengthwise until a fine cutting edge is produced.

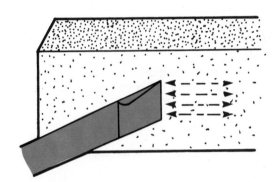

Fig. 27-20. Honing the cutting edge.

Step 5   Hone the end-cutting edge in the same manner.

Step 6   Repeat the honing steps for the end-relief and top-rake angles.

Step 7   Lay the round nose on the sharpening stone. Swing the cutting tool in an arc while moving it along the stone.

Step 8   Wipe the face of the sharpening stone to clean it and to remove the fine metal particles.

SAFE PRACTICES IN USING CUTTING TOOLS AND TOOLHOLDERS

- Grind high-speed tool bits and cemented carbide and ceramic inserts to the specified clearance, rake, and cutting-edge angles. These affect the cutting efficiency, tool life, quality of surface finish, and dimensional accuracy.
- Grind a chip breaker into the cutting tool bit or place a separate chip breaker adjacent to the cutting edges. The continuous breaking away of small chips provides safe working conditions and prevents damage to the lathe and workpiece.
- Select lathe speeds and feeds which permit turning processes to be performed economically and safely.
- Pay attention during rough turning (taking hogging cuts) to the effect of the longitudinal, radial, and tangential cutting forces and the heat generated on the workpiece and cutting tool.
- Select a straight-, right-, or left-hand type toolholder that permits the cutting tool to cut to the required dimension without interference.
- Consider the base angle of the toolholder when grinding a relief or rake angle.
- Secure the grinder machine guards and position the tool rest close to the grinding wheels before starting the machine.
- Position the machine safety shields and use safety goggles during all grinding operations.
- True and dress the faces of the rough- and finish-cut grinding wheel before grinding.
- Quench the high-speed tool bit regularly when dry grinding.

LATHE CUTTING TOOL TERMS

Cutting tool efficiency	The ability of a cutting tool to machine to required specifications and still maintain its cutting effectiveness.
Tool bit	A common term used in the machine shop to designate a solid, single-point cutting tool. A general cutting tool of square or rectangular cross-sectional area.
Cutting forces (lathe-work)	Longitudinal, radial, and tangential forces. Forces produced by the cutting tool as it feeds into a revolving workpiece.
Shearing point	The place where the material that is compressed above the cutting edges of the tool is caused to be separated (sheared).
Throwaway insert	Commercially available carbide or ceramic inserts. Inserts that may be positioned and held securely in a tool holder. Inserts that are economical to replace when worn.
Right- and left-hand toolholders	A cutting tool holder that is offset at the tool-holding end. A toolholder that extends the range of the cutting action beyond that which normally may be performed with a straight toolholder.
Right- and left-hand turning tools	Lathe tool bits that are ground to cut from right to left (right-hand) or left to right (left-hand).
Carbide insert designation system	An eight-position series of letters and numerical symbols. A system that provides full specifications for carbide inserts.
Quick-change tool holder	A toolholder designed to fit a dovetailed tool post. The cutting tool is preset, and the whole unit may be replaced quickly.
Quick-change tool system	A combination of multiple-station, quick-change cutting tool holders.
Heavy-duty toolholder	An open-side toolholder with a T-base that may be secured to the T-slot of the compound rest. A toolholder designed to withstand the forces produced by heavy roughing cuts (speeds and feeds).
Stoning or honing a tool bit	A hand process of moving each cutting edge against the face of an abrasive stick (stone). The producing of finely-finished cutting edges on a tool bit.

SUMMARY

- High-speed steel tool bits and carbide and ceramic inserts must be ground to specified rake, clearance, and cutting-edge angles. The tool bits must be positioned accurately and held securely.
- *Cutting edge, face, flank, nose point,* and *right-* and *left-hand* are terms used daily in connection with tool bits.
- The four basic tool bit shapes deal with turning, facing, cutting off, and threading. Tools bits are ground with the side-cutting edge on the left side for regular turning processes. The feed is from right to left. Left-hand cutting tools are fed from left to right.
- Square-cross-sectional-area high-speed steel and carbide-tipped tool mounts may be held in conventional square-opening toolholders.
- Large square and rectangular tool bits, tool shanks, and toolholders are held in a single- or multiple-type toolholder.
- Carbide and ceramic inserts are mounted and held to a solid shank by either a pin-block, cam-action, or clamp-type tool post.
- Carbide inserts are specified according to a combination of letters and numerals. The eight symbols designate shape, clearance angle, class of cutting point and clearance, type, size, thickness, cutting point, and other conditions.
- The action of a cutting tool as it is fed into a revolving workpiece produces longitudinal, radial, and tangential forces.
- Shearing results from compressing the material over the point of the cutting tool. The chip flows in a continuous ribbon.
- The quick-change tool post provides a series of accurately dovetailed slides. Several quick-change toolholders may be accurately positioned and held in these slides. The multiple head permits several cutting tools to be set up and moved into position.
- The open-side, heavy-duty toolholder is used for roughing cuts that produce excessive forces.
- Flat tool bit gages provide an accurate measurement of the relief, rake and other cutting angles.
- High-speed steel tool bits are rough ground, finish ground, and then hand stoned. This produces a smooth, clean cutting edge.
- Provision should be made for chip breakage by either grinding or adding a chip-breaker insert.
- Grinding wheel guards and machine safety shields should be in position. A coolant for the high-speed steel tool bit should be available for quenching during the grinding process.

UNIT 27 REVIEW AND SELF-TEST

1. State three factors that affect the efficiency of lathe cutting tools.
2. Distinguish between the following terms as they apply to single-point cutting tools: *cutting edge, face,* and *point.*
3. List four of the eight design features that provide specifications for throwaway carbide inserts.
4. Explain the meaning of *tangential force* during a cutting process on a lathe.

5. a. Describe a *quick-change tool system.*
   b. List the advantages of such a system over a *multiple-station turret.*
6. Tell how to grind the side- and end-cutting edges and the relief angles and the face of a single-point cutting tool.
7. State two safe practices to observe in selecting, grinding, or cutting with a single-point cutting tool for lathe work.

SECTION THREE

LATHE WORK BETWEEN CENTERS

Many lathe work processes are performed with the workpiece mounted between the lathe centers. This section describes the principles of and procedures for performing the following basic between-centers processes:

- Facing and straight turning
- Shoulder turning, chamfering, and rounding ends
- Grooving, form turning, and cutting off
- Filing, polishing, and knurling
- Taper and angle turning.

Unit 28

CENTER WORK AND LATHE SETUPS

All basic lathe turning processes require a workholding device and a method of turning the workpiece. Common lathe practices 'may be grouped as *chuck work, faceplate work,* and *fixture work.* Other work parts may be *turned between centers.*

The lathe must also be set up in terms of mounting an appropriate chuck, faceplate, or centers. The cutting tools must be selected, checked for cutting qualities, positioned, and secured. Speeds, feeds, and the appropriate cutting fluid must be determined. These factors influence the desired degree of dimensional accuracy and surface finish. The worker must also make other decisions based on economical tool life and cutting efficiency. Five basic turning processes, using carbide inserts, are illustrated in Fig. 28-1.

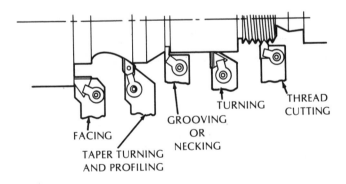

Fig. 28-1. Basic turning operations using carbide inserts and toolholders.

A turned part usually is machined from a solid bar, forging, or casting. Before any turning process may be performed, the part must be *prepared.* The end may need to be *faced.* This provides a flat plane reference surface. A cone-shaped (center) hole may be required. The center hole permits the workpiece to be *mounted between the lathe centers* or *supported on one end.*

This unit deals with center drilling and the preparation of the lathe for external turning processes.

CENTER WORK AND CENTER DRILLING

Parts that are to be machined between centers require center holes. These permit the workpiece to be machined, removed, and then accurately reset for subsequent operations. The center holes provide an accurate bearing surface. The outer cylindrical surface remains concentric with any previously turned surface.

Functions Of Center Holes

A *center hole* is a *cone-shaped* or *bell-mouthed* bearing surface for the live and/or dead centers. Each center hole must be of a particular size and depth. These depend on the size of the part to be machined and the nature of the operation (Fig. 28-2A). An improperly drilled center hole limits the bearing surface, wears away rapidly, and may result in inaccurate turning (Fig. 28-2B). The center drilled hole should have a fine surface finish to reduce friction on the cone- or bell-shaped bearing area.

Types And Sizes Of Center Drills

Table 28-1 gives recommended sizes of plain center drills for different diameters of workpieces. Note in each case that the diameter to which the hole is countersunk is always less than the body diameter of the combined drill and countersink (center drill).

A radial type of center drill has an 82° angle above the 60° bearing band. This type of center drill automatically produces a *safety center.* The radial center hole is lubricated easily. The radial type is available in sizes ranging from #00 (0.025" drill and 5/64" body diameters) to #18 (1/4" drill and 3/4" body diameters). Similar metric sizes and styles of center drills are available. The sizes ranging from #00 to #8 are adapted for precision tool, gage, and instrument work. The heavy-duty type ranges from #11 to #18. These sizes are used for heavy production parts, especially forgings and castings. The features of a plain- and a radial-type center drill are shown in Fig. 28-3A and B.

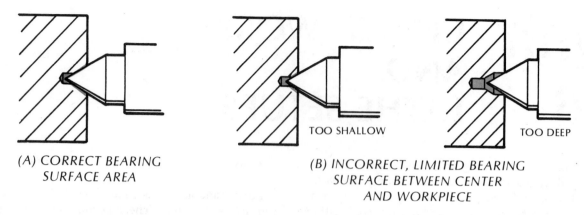

(A) CORRECT BEARING SURFACE AREA

(B) INCORRECT, LIMITED BEARING SURFACE BETWEEN CENTER AND WORKPIECE

TOO SHALLOW

TOO DEEP

Fig. 28-2. A correctly prepared center hole and two common center hole problems.

Table 28-1. RECOMMENDED CENTER DRILL SIZES (PLAIN TYPE) FOR DIFFERENT DIAMETER WORKPIECES

Size No.	Diameter (inches)			
	Workpiece	Countersink	Drill Point	Body
00	³/₃₂ to ⁷/₆₄	³/₆₄	0.025	⁵/₆₄
0	⅛ to ¹¹/₆₄	⁵/₆₄	¹/₃₂	⅛
1	³/₁₆ to ⁵/₁₆	³/₃₂	³/₆₄	⅛
2	⅜ to ½	⁹/₆₄	⁵/₆₄	³/₁₆
3	⅝ to ¾	³/₁₆	⁷/₆₄	¼
4	1 to ½	¹⁵/₆₄	⅛	⁵/₁₆
5	2 to 3	²¹/₆₄	³/₁₆	⁷/₁₆
6	3 to 4	⅜	⁷/₃₂	½
7	4 to 5	¹⁵/₃₂	¼	⅝
8	6 and over	⁹/₁₆	⁵/₁₆	¾

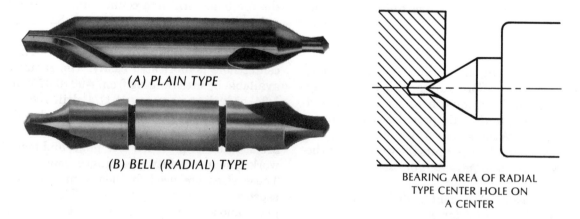

(A) PLAIN TYPE

(B) BELL (RADIAL) TYPE

BEARING AREA OF RADIAL TYPE CENTER HOLE ON A CENTER

Fig. 28-3. Two types of center drills and a radial center hole. (Photos courtesy of Do All Company)

Common Center Layout Techniques

The workpiece is generally gripped in a lathe chuck and faced. The center drill is held in a drill chuck in the tailstock. The center drill is fed carefully into the revolving workpiece. Sometimes, the process is reversed. The workpiece is fed into a revolving center drill that is mounted in the lathe spindle.

Standard diameter, square, and hexagonal workpieces may also be held in collets or fixtures. The end is faced. Then, the center hole is drilled using the center drill, drill chuck, and tailstock method.

PREPARING THE LATHE FOR TURNING BETWEEN CENTERS

The Live Center

The accuracy to which a part may be machined on centers depends on the *trueness* of the headstock (live) center and the alignment of the headstock and tailstock centers. The headstock spindle taper should be clean and free of burrs. Similarly, the tapered shank of the live center and the tapered surfaces of the center sleeve should be burr-free and wiped clean. The live center (and sleeve, if required) should be inserted in the spindle. The live center is *driven home* by sharply and rapidly pushing the center into the spindle. The center, sleeve, and spindle often have *matching lines* for accurately replacing the center in the same position.

The trueness of the live center must then be checked. A dial indicator may be used. The indicator point is brought into contact with the surface of the live center. The spindle is revolved by hand. If the center *runs out*, it either may be ground in place or another live center must be inserted and tested.

Turning A Center

When a chuck is already mounted, it is common practice to chuck a piece of steel and turn a center. The compound rest is swung to a 30° angle with the spindle axis. A center is *roughed out* and then *finish turned*. The side of the chuck jaw then serves to drive the lathe dog and workpiece.

Aligning The Dead Center

The three basic methods of checking the alignment of centers were described earlier. The centers may be aligned visually if the workpiece does not require a high degree of accuracy. A more accurate alignment is produced with a test bar and dial indicator or by

turning two similar diameters. Another fast, accurate method of aligning centers is to use a *micro-set adjustable center* in the tailstock. One model of this center is illustrated in Fig. 28-4. Since this type of center revolves with the workpiece, it is identified as a live center. The micro-set adjustable center is used on *light* work.

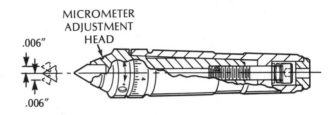

Fig. 28-4. A micro-set adjustable center. (Courtesy of Enco Manufacturing Company)

The amount the tailstock is out of alignment is determined by the test bar and dial indicator method or by the trial cut (two turned diameters) method. The micro-center is adjusted. The center is moved by turning the graduated dial the required amount.

Workpieces that are to be faced accurately from the center hole to the outside diameter often require the use of a tailstock center that is *relieved*. This permits taking a cut across the entire face. The cut is started slightly inside the center drilled hole.

Preparation Of The Workpiece

A lathe dog is customarily used to drive a round, square, hexagonal, or other odd-shaped workpiece. Some types of dogs have a movable V- or parallel jaw. On other dogs, a setscrew forces the workpiece against the V-shape. Machined surfaces require that a soft metal band be placed around the workpiece. The band prevents scoring the finished surface.

Before placing the workpiece between centers, the center holes are cleaned. They are rechecked to ensure that there are no chips in the drilled areas. A drop of lubricating oil is placed in the dead center drilled hole.

General Practices In Setting Up The Machine And Cutting Tool

● Position the cutting tool so that there is the least amount of overhang between the workpiece, tool bit, toolholder, and tool post.

● Set up the tool post as close to the center of the compound rest as possible. Avoid excessive overhang.

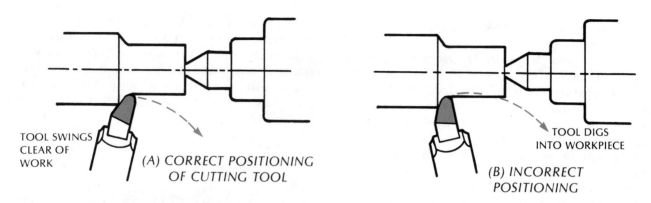

TOOL SWINGS CLEAR OF WORK

(A) CORRECT POSITIONING OF CUTTING TOOL

TOOL DIGS INTO WORKPIECE

(B) INCORRECT POSITIONING

Fig. 28-5. Effects of excessive cutting forces on the position of the cutting tool.

• Position the toolholder as illustrated in Fig. 28-5A. If the cutting forces cause the cutting tool to move, it will not *dig in*. This condition is shown in Fig. 28-5B.

• Set the cutting tool point on the centerline for general machining.

• The position of the lathe dog must be checked to see that the tail does not bind in the driver plate. The centered workpiece must seat properly on the live center.

• Feed the dead center slowly into the workpiece. Adjust it until there is a *snug fit*. Lock the tailstock spindle at that point. (*Snug fit* means that the workpiece is free to turn without any end play.)

• Check the cutting tool and machine setup. The carriage and cross slide are moved to the furthest cutting position, with the cutting tool close to the workpiece. The spindle is then revolved by hand to see that the workpiece, cutting tool, carriage, and machine accessories all clear each other.

• Set the lathe spindle speed at the recommended rpm and feed for the material to be turned, the type of cutting tool, the nature of the operation, and the condition of the machine.

• Set the lathe spindle speed at the lower speed when the calculated rpm falls between two spindle speeds on the lathe.

• Take the deepest cut using the coarsest feed possible. Allow about 1/32″ for a finish cut.

HOW TO LAY OUT A CENTER HOLE

Laying Out A Center With A Center Square

Step 1   Cut off the workpiece. Allow sufficient

stock to face the ends. File a slight chamfered edge to break the outside edge. Coat the end with layout dye.

Step 2   Bring the V-section of the center head against the outer surface on one end of the workpiece. Scribe a line close to the blade. Fig. 28-6 shows this step.

Step 3   Turn the center square head about a third of a revolution. Scribe a second line.

Step 4   Turn the center head another one third and scribe a third line.

Step 5   Place the sharp point of a center punch at the intersection of the scribed lines.

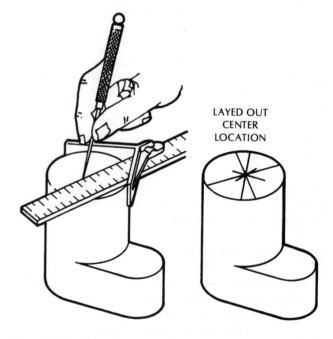

LAYED OUT CENTER LOCATION

Fig. 28-6. Laying out a center location with a center head.

Step 6   Strike the center punch squarely to produce a center-punched indentation.

Step 6   Repeat steps 1 through 5 on the opposite end of the workpiece.

Note: The workpiece may also be layed out by using dividers. A height gage and V-blocks (Fig. 28-7) are sometimes used. The center may also be located by simply positioning and striking the punch of a bell-shaped cup. The cup centers over the end of the workpiece.

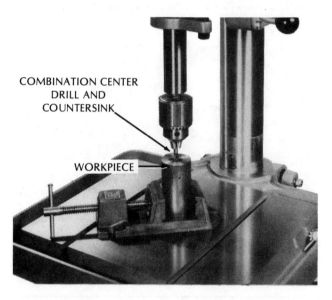

COMBINATION CENTER DRILL AND COUNTERSINK

WORKPIECE

Fig. 28-7. Drilling a center hole on a drill press. (Courtesy of the Clausing Corp.)

Checking The Layout Of A Center Hole

Step 1   Set the legs of a divider to the radius of the workpiece.

Step 2   Insert one leg in the center punch mark. Swing the other leg around the circumference.

Note: If the center mark is not centered, the divider leg will not swing concentrically in relation to the circumference.

Correcting An Off-Centered Punch Mark

Step 1   Place the center punch in the center mark. Tilt the punch at an angle.

Step 2   Tap the center punch so that the blow drives the center mark toward the accurate center.

Note: The workpiece may also be checked for accuracy of center hole layout by

placing it between centers. As the workpiece is rotated by hand past a fixed point, the amount of variation (off-center) is noted. The workpiece is then removed. The center hole is repunched to move it to an exact location.

HOW TO DRILL CENTER HOLES

Drilling On A Drill Press

Step 1   Position the workpiece using a V-block or the V-groove of a drill press vise.

Step 2   Select the number of center drill that will produce a large enough center hole for the job requirements.

Step 3   Chuck the center drill. Set the spindle speed. Align the drill point with the center punch mark.

Step 4   Brush on a thin film of cutting fluid. Bring the drill carefully into contact with the workpiece (Fig. 28-7). Drill to depth.

Note: Use a fine feed and a cutting fluid to finish the center drilling. A highly finished surface is required to produce an excellent bearing surface.

Drilling A Center Hole On A Lathe

Method One: Center Drilling From A Lathe Spindle

Step 1   Mount the drill chuck with a center drill in the headstock spindle.

Step 2   Position and hold the center-punched hole on one end of the workpiece against the dead center.

Step 3   Start the lathe. Guide the center drill in the second center-punched hole on the opposite end. Feed the workpiece toward the center drill by turning the tailstock handwheel.

Step 4   Feed to the required depth. Use a fine feed and a cutting fluid for the final cutting action.

Method Two: Center Drilling Work
Mounted In A Chuck

Step 1   Chuck the workpiece so that the outside surface runs concentric. For accurate centering, use a dial test indicator.

Step 2   Mount the drill chuck and center drill in the tailstock spindle.

Step 3   Bring the tailstock up toward the work-piece and tighten it. Put a light pressure on the tailstock spindle by partially clamping it. Feed the center drill to start the process (Fig. 28-8).
Note: In chuck work, the workpiece is usually faced before center drilling.

Step 4   Continue to feed carefully to the required depth. Use a cutting fluid.

Step 5   Feed the center drill point carefully. Continue to feed until the required outside diameter of the countersunk portion is reached. Fig. 28-9 shows the steady rest setup for center drilling long workpieces. Use a cutting fluid and a fine finish feed.

Step 6   Repeat Steps 1-5 to center drill the other end.

Fig. 28-9. Using a steady rest to center drill a long workpiece. (Courtesy of LeBland Machine Tool Company)

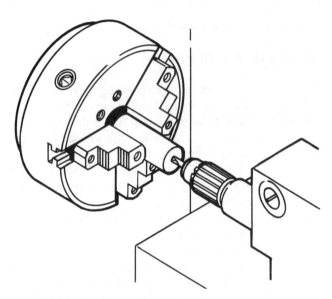

Fig. 28-8. Center drilling work held in a chuck.

Center Drilling Long Workpieces

Step 1   Chuck one end of the long workpiece so that it runs true.

Step 2   Support the other end with a steady rest. Open the steady rest jaws and adjust each one as required. The scribed center on the end of the workpiece must align with the axis of the tailstock center.
Note: The workpiece may also be checked for alignment by moving a dial test indicator along the top surface. The steady rest jaws are moved independently until there is a zero reading on the dial test indicator.
CAUTION: A surface gage and scriber point should be used when the end of the workpiece has a rough scale surface.

Step 3   Face the end. Insert a center drill and chuck in the tailstock spindle.

Step 4   Set the spindle speed. Apply a lubricant between the steady rest jaws and the workpiece. Start the lathe.

HOW TO MOUNT WORK BETWEEN CENTERS

Step 1   Secure a driver plate on the spindle nose. Align the headstock and tailstock centers. Aligning may be done visually or by one of the more precise methods using a dial indicator.

Step 2   Check the trueness of the live center.

Step 3   Select a lathe dog that will accommodate the size and shape of the workpiece. Slide it over the workpiece to the headstock end.

Step 4   Clean the center holes and the center. Apply a center lubricant on the tailstock center if a solid dead center is used.

Step 5   Adjust the center so that the workpiece turns freely without end play.

Step 6   Insert the lathe dog tail in a driver plate slot. Check to see that it moves freely and does not bind in the slot when the safety set screw is tightened.

CAUTION: The center hole must be checked for adequate lubrication during each successive cutting step. The heat that is generated may cause the workpiece to expand and tighten against the centers.

Step 7   Check the center height of the cutting tool and the position of the holder and compound rest. Be sure all parts are clear. Then, proceed with the machining process.

HOW TO TRUE A LIVE CENTER

Truing A Soft-Steel Live (Headstock) Center

Step 1   Mount the soft-steel live center and sleeve in the headstock spindle.
Note: Align the witness marks on the center, sleeve, and spindle.

Step 2   Set the compound rest at a 30° angle to the axis of the lathe spindle. Position a turning tool at the center height.

Step 3   Start the lathe. Take shallow cuts along the angular face until the 60° point runs true. The final cut should be a finish cut.

Truing A Hardened-Steel Live (Headstock) Center

Step 1   Mount the hardened-steel live center and sleeve in the headstock spindle.

Step 2   Mount a portable tool post grinding attachment in the tool post on the compound rest. Set the axis of the grinding wheel at a 30° angle to the lathe spindle axis.

Step 3   True and dress the grinding wheel face.

Step 4   Start the lathe spindle. Feed the grinding wheel into the center using the crossfeed screw. Feed the grinding wheel along the center by turning the compound rest handwheel.

CAUTION: Use protective cloths or pads to prevent the abrasive particles from reaching any machined surface.

CAUTION: The guard on the tool post grinding attachment must be in place, and safety goggles must be worn.

Step 5   Continue to take successive grinding cuts until the center runs true.
Note: The hardened tailstock center must be inserted in the headstock spindle and ground in the same manner.

SAFE PRACTICES IN CENTER WORK AND LATHE SETUPS

● Check the depth of a center drilled hole. It must provide an angular bearing surface adequate to support the workpiece and withstand the cutting action.

● Use a rotating dead center when turning workpieces that revolve at high speeds. Otherwise, considerable friction may be produced between the center drilled surface and the solid dead center.

● Check the trueness of the live center and the alignment of both centers. The accuracy of a turned workpiece depends on the trueness and alignment of the centers.

● Reduce the overhang of the cutting tool and the toolholder to a minimum. Support the tool post rigidly and as close as possible to the center of the compound rest T-slot.

● Position the cutting toolholder so that if there is

any movement the cutting point swings away from the workpiece.

● Use lathe dogs equipped with safety screws or safe locking devices.

● Check the bent tail of the lathe dog to see that it clears the bottom or sides of the driver plate slot. The dog position must permit the workpiece to ride on the center.

● Support long centered workpieces with a steady rest.

● Lubricate the angular or radial bearing surface of a workpiece that revolves on a solid dead center.

● Rotate the workpiece and tool setup by hand to ensure that all parts clear without touching.

● Cover the lathe bed and other machined surfaces whenever a tool post grinding attachment is used.

- Adjust the centers between successive cuts when the amount of heat generated is sufficient to produce *binding*.

- Place a protective soft material around a machined surface to protect it from being scored by the lathe dog screw or parallel strap.

TERMS USED IN CENTER WORK AND LATHE SETUPS

Centering	The process of drilling a small pilot hole and countersinking an angular surface in the end of a workpiece. Drilling a pilot hole and countersinking are usually done in one operation.
Center hole	A drilled hole with a cone- or radial-shaped outer end. A cone- or radial-shaped recess that may be used to center a drill or to provide an angular bearing surface for a lathe center.
Radial center drill	A center drill that has the added feature of an 82° angle above the 60° bearing band. A center drill that produces a safety center.
Overhang	The portion of a cutting tool, toolholder, or tool post that extends beyond a supporting surface.
Checking the layout	A general precautionary step to ensure that a workpiece has been laid out accurately. Steps taken to correct any layout inaccuracy.
Mounting work (lathe)	The process of positioning a workpiece so that it may be machined. Securing a workpiece in a chuck or between centers.
Witness marks	A light prick punch mark or scored line on a center, sleeve, or spindle. Position marks on mating parts. Marks made for the purpose of assembling parts in the same relative position each time.
Portable tool post grinding attachment	A complete grinding unit with a shank attachment that fits into a standard tool post.

SUMMARY

- A faced end provides a flat reference surface. Measurements may be made accurately from a faced end.
- Center holes provide a 60° angle bearing surface for turning work between centers. The cylindrical surfaces that are produced are concentric with the axis of the lathe centers.
- Center drills of the plain type have the countersunk portion ground to an angle of 60°. The radial type of center drill has an added countersunk angle of 82°.
- Center drills are ordered by number sizes. These sizes range from #00 (0.025″ drill and 5/64″ body diameters) to the heavy-duty #18 size.
- Center holes are generally drilled with center drills on a drill press or lathe.
- The drill point and outside diameter of the countersunk area of a center drilled hole are determined by (a) the material and size of the workpiece and (b) the nature of the required turning operations.

- The live center must be checked for trueness. Any inaccuracy must be corrected. The center is turned if it is soft, and ground if it is hardened. The included angle of a live and dead center is 60°.
- Misalignment of the centers produces inaccuracies in a turned part.
- Workpieces that are to be turned between centers are usually driven by a lathe dog. When secured to the workpiece, the dog must permit the centers to be positioned to provide a correct bearing surface.
- The setting up, positioning, and securing of a tool bit, toolholder, and tool post must provide maximum support with the least possible overhang.
- The machine and tools must be set up so that all movable parts rotate freely and clear each other at the farthest cutting position.
- Workpieces that extend a distance from a chuck must be supported at the tailstock end with a steady rest. The steady rest jaws must be positioned to align the axes of the workpiece and tailstock spindle.

● *Before* the lathe is started, all loose clothing must be secured, wiping cloths must be removed, and all parts must be checked to see that they are free to rotate.

UNIT 28 REVIEW AND SELF-TEST

1. Identify (a) the functions and (b) the characteristics of a precision radial center drilled hole.
2. State three practices the skilled mechanic follows when using a live center.
3. Cite two advantages of the *micro-set adjusting center* for light turning processes as compared to using a solid dead center.

4. List five general practices for setting up a lathe and taking basic cuts.
5. State the method by which long workpieces are supported for center drilling.
6. Tell why a center is often turned from a mild-steel chucked piece rather than by using a standard, hardened live center.
7. Give four precautionary steps to take before starting to turn a workpiece between centers.

Unit 29

FACING AND STRAIGHT TURNING

Precision machining processes require an allowance of excess material on the *rough stock*. The rough size of the workpiece should be as close as possible to the finished size and shape. This practice conserves material, reduces machining time, and cuts costs.

Two of the most common lathe processes deal with linear (length) and concentric (diameter) measurements. The process of machining a flat plane surface on the end of a workpiece on a lathe is referred to as *facing*. The turning of one or more diameters is known by such terms as *straight turning, parallel turning, cylindrical turning*, or *taking a straight cut*.

Facing and turning processes require differently formed cutting tools and different work-holding devices and tool setups. Some of the common shapes of turning tools are illustrated in Fig. 29-1. The machined surfaces produced are represented on drawings. These give the dimensions and show the features so that each may be uniformly and accurately interpreted.

This unit deals with the theory of and step-by-step practices for facing and straight turning, and the representation and dimensioning of faced and turned surfaces on drawings.

SETTING UP FOR FACING OR STRAIGHT TURNING

Cutting Speed And Feed

The engine lathe must be set up for efficient machining. The cutting speeds and the required spindle speed may need to be calculated. These may also be established from handbook tables, as explained in an earlier unit. The lathe spindle speed is set as close as possible to the calculated rpm.

The feed for general lathe work processes usually must be determined by the operator. To repeat, consideration is given to such factors as the kind of material, the processes and cutting tools, the rigidity

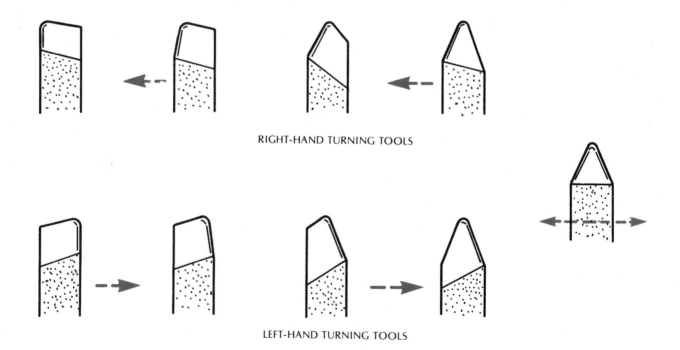

RIGHT-HAND TURNING TOOLS

LEFT-HAND TURNING TOOLS

Fig. 29-1. Common shapes of turning tools.

of the workpiece, the power (capacity) of the lathe, and the use of cutting lubricants. The selected feed is then obtained by positioning the feed-change levers in the quick-change gearbox.

Positioning The Compound Rest

Compound Rest Set At 30°—The compound rest is graduated on the base and may be rotated through 360°. Angular settings such as 30°, 45° and 60° are common. The 30° angle provides a convenient method of producing a fixed cutting-tool movement. For instance, the amount of side (longitudinal) movement of a cutting tool is equal to one half the distance the compound rest is moved. This movement is shown in Fig. 29-2.

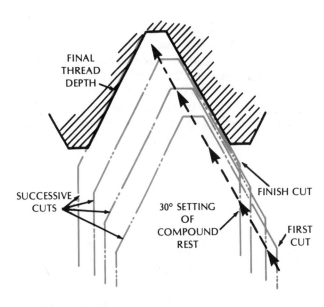

Fig. 29-3. Feeding a thread-cutting tool to thread depth by angular setting of the compound rest.

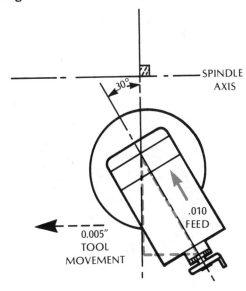

Fig. 29-2. At a 30° setting, longitudinal movement equals one half the feed of the compound rest.

When the compound rest is set at 30° and the handwheel is turned 0.010" (or 0.2 mm on a metric-graduated dial), it produces a side motion of 0.005" (0.1 mm). The graduated collar on the compound rest feed screw is often used to set the tool depth. The 30° angular setting is applied to straight and to 60° included angle turning processes. This setting is especially important when cutting standard 60° thread forms. The feed per cut and the final thread depth are calculated. The cutting tool is fed to depth as shown in Fig. 29-3. The graduated collar on the compound rest is used to feed the thread-cutting tool to depth.

The compound rest may be set at any angle to the axis of the cross slide. The three positions (*quadrants*) in which an angle may be set are indicated by quadrants I, II, and III in Fig. 29-4.

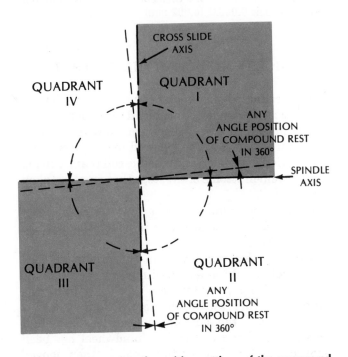

Fig. 29-4. Range of angle position settings of the compound rest.

Compound Rest Set At 90°—A 90° angular setting (parallel to the work axis) of the compound rest is practical for facing and shoulder turning. The amount the cutting tool is fed is read directly from the compound rest graduated collar setting. If the compound rest is moved 0.5 mm (0.020"), the cutting tool advances the same distance (0.5 mm or 0.020"). *Compound Rest Set At 84° 16'*—Although not a common practice, precise tool settings to approximately 0.0001″ (0.002 mm) may be made with the compound rest set at an angle of 84° 16′. Fig. 29-5 shows this movement. The cutting tool feeds into the workpiece 0.0001″ (0.002 mm) for each 0.001″ (0.02 mm) movement of the compound rest feed screw. Because the angle graduations on compound rests are not graduated in minutes, the 16′ must be estimated.

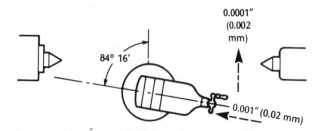

Fig. 29-5. Compound rest movement of 0.001″ (0.02 mm) at 84° 16′ equals 0.0001″ (0.002 mm) transverse movement.

PRACTICES FOR SETTING THE DEPTH OF CUT

The graduated micrometer collar on feed screws is a practical device for economically producing accurately machined parts. There are, however, certain procedures that must be followed to compensate for lost motion (backlash) and possible collar and cutting tool movement. Some of the steps that must be taken are:

- Turn the compound rest feed screw at least one-half turn in the direction in which the depth of cut is to be set
- Set the locking device on the graduated feed collar before taking any readings
- Feed the cutting tool into (not away from) the workpiece when setting the depth of cut
- Back out the cutting tool at least one-half revolution whenever the feed handwheel has been moved too far. The cutting tool is then moved clockwise. This removes the backlash and permits feeding the cutting tool to the correct setting
- Avoid handling a friction-held graduated collar when setting the cutting tool.

HOW TO SET A CUTTING TOOL FOR DEPTH OF CUT

Step 1   Select an appropriate toolholder. It must

accommodate the workpiece and the nature of the cutting operation.

Step 2   Tighten the tool bit in the toolholder. Position the tool bit for the machining operation. The cutting edge is set at center height. Solidly secure the setup to the compound rest.

Step 3   Determine the correct spindle rpm and feed. Check the lubricating systems. Start the lathe.

Step 4   Turn the crossfeed handle until a light cut is taken and forms at least half a circle around the workpiece.

Step 5   Move the carriage handwheel so that the tool clears the workpiece. Stop the lathe without disturbing the handwheel setting.

Step 6   Measure the turned diameter section. Determine how much stock is still to be removed with one or more roughing cuts (Fig. 29-6).

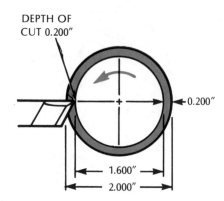

Fig. 29-6. An outside diameter is reduced by twice the depth of cut.

CAUTION: The amount the workpiece diameter is reduced is *double* the depth of cut.

Step 7   Continue to move the cutting tool to the depth of cut. The graduated collar provides an accurate measurement.

Step 8   Start the lathe again. Take a trial cut for a short distance (usually 1/8″). Stop the lathe. Measure the diameter.

Step 9   Take the roughing cut across the workpiece. Note the reading on the graduated collar.

Step 10   Continue to take successive roughing cuts. Each cut is set for depth by feeding the crossfeed screw in the same direction. The distance is measured by the reading on the graduated collar.

CAUTION: If the motion of the feed screw is reversed, compensation must be made for lost motion.

Step 11   Increase the speed and decrease the feed for the finish cut.
Note: Precision turning may require replacing the roughing tool bit with a finishing one.
Note: The same procedure is followed in taking a trial final cut for a short distance. Check the diameter for accuracy and surface finish at each setting before the entire surface is turned.

FACING TOOLS: SETUP AND PROCESSES

A workpiece is faced for three main reasons:

● To produce a flat plane. This provides a reference surface for other measurements and operations

● To machine a flat-finished area that is at a right-angle to the axis of the workpiece

● To turn a surface to a required linear dimension.

The Cutting Tool

Facing may be done with a side-cutting or end-cutting tool. The cutting point has a radius. The cut may be started on the outside diameter. The cut may be fed either toward or away from the center. The cut may be a roughing or a finishing one. The workpiece may be held in a chuck, between centers, on a faceplate, or in a holding fixture.

Although the high-speed steel tool bit is commonly used, all factors and conditions must be considered. The tool bit that is selected must be the most productive for the job. Cemented-tip carbides with solid shank and carbide and ceramic inserts cut more efficiently at high speeds and on harder and tougher materials than do high-speed steel tool bits.

Two shapes of cutting tools are generally used for facing. The side-cutting tool with a 55° cutting edge angle permits the facing of work between centers. The end-cutting shape (Fig. 29-7) is adapted to hogging cuts. A considerable amount of material may be removed. A hogging cut should be taken below the scale of a casting or the outer hard or rough surface of a workpiece.

Another common rough-facing process requires the use of a regular turning tool. Rough cuts are taken lengthwise to successively reduce the outside diameter of the workpiece. These cuts are brought almost to the layout line marking the linear length. The same cutting tool can then be used to *square up the end* by feeding it to or away from the center (transversely).

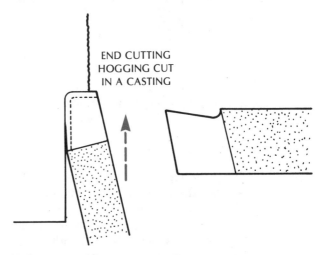

END CUTTING HOGGING CUT IN A CASTING

Fig. 29-7. End-cutting tool bit used for a rough facing operation.

When an end-cutting tool with a slightly rounded nose is used, hogging cuts are taken first. These cuts start at the outside diameter. A side-cutting tool is then set on center. A light finishing cut is taken with the side-cutting tool, starting at the center.

The facing tool bit with a 55° cutting angle and a slight radius nose is used when facing the end of a workpiece that is mounted between centers (Fig. 29-8). The final finish-facing cut is started at the center. If a dead center that is not cut away is used, it is carefully backed off from 1/32" to 1/16" with the tailstock handwheel (Fig. 29-8A). At the same time, the facing tool is fed by hand toward the center. This turning action cuts away the burr that is formed around the center. A final finish cut is started at the center (Fig. 29-8B). The dead center is moved back in the center hole. A continuous cut is then taken across the face.

HOW TO FACE A WORKPIECE

Chuck And Face Plate Work

Rough Facing
Step 1   Layout the required length.

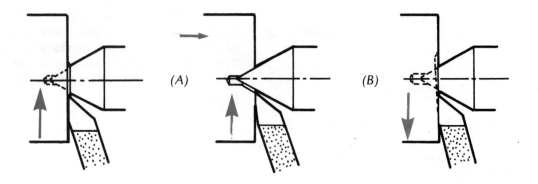

Fig. 29-8. Removing a center section burr and finish facing an end.

Step 2 Chuck or mount the workpiece. Select and set the spindle rpm. This speed is based on the largest diameter to be faced.

Step 3 Use an end-cutting or plain side-cutting rough-turning tool. Set the depth of cut. Use a coarse feed. Feed the cutting tool across the face almost to the layout line. *Note:* The end-cutting tool is fed directly into the workpiece.
Note: When a considerable amount of material is to be faced off of a workpiece, a series of longitudinal cuts are first taken. The diameter is reduced for each successive cut. The cutting tool is fed to almost the required length.

Step 4 Feed the cutting tool across the face. This evens the steps that are formed on the end of the workpiece.

Finish Facing

Step 1 Replace the end- or side-cutting tool with a facing tool. Position the cutting point on the center line.

Step 2 Increase the speed. Decrease the feed. Take a light finishing cut. Start at the center and feed to the outside diameter.

Facing Work Between Centers

Rough Facing

Step 1 Check the depth of the center drilled hole.
CAUTION: The center hole must be deep enough to permit the dead center to still seat properly when the workpiece is faced to the required length.

Step 2 Select a facing tool with a 55° cutting angle. Mount the tool bit so that the point is at center height.

Step 3 Advance the cutting tool into the workpiece to the required depth.

Step 4 Feed the tool from the center outward across the face of the work.

Removing The Center-Section Burr,
And Finish Facing

Step 1 Unclamp the tailstock spindle. Move the spindle out from the workpiece from 1/32" to 1/16".
CAUTION: The unclamping and slight moving of the tailstock spindle must be done carefully.

Step 2 Feed the end-cutting tool toward the center. This action causes the burred center edge to be faced off.
Note: Reset the center in the center hole as soon as the burred edge has been removed and the facing cut has been started.

Step 3 Start a final finish cut at the center.

Step 4 Feed the facing tool across the face to the outside diameter.
Note: Precise linear measurements may be taken with a depth micrometer. Additional facing cuts may be set accurately by using a micrometer stop. The movement may also be measured with the graduated collar of the compound rest when it is set parallel to the axis of the lathe.

GENERAL RULES FOR STRAIGHT TURNING

Straight, or *parallel,* turning means that a workpiece is machined to produce a perfectly round cylinder of a required size. The general rules for all lathe processes apply.

● The cutting tool should be fed toward the headstock wherever possible. This is especially important on work between centers. The cutting forces should be directed *toward* the live center.

● The number of cuts should be as few as possible. Sufficient stock should be left for a final finish cut if a fine surface finish is required.

● The size and condition of the lathe, the shape and material of the workpiece and cutting tool, and the type of cut all affect the depth of cut.

● The roughing cut must be deep enough on castings, forgings, and other materials to cut below the hard outer surface or scale.

● Deep roughing cuts should be taken, using the coarsest possible feed.

● The cutting tool and holder must be set in relation to the workpiece. Any movement of the holder produced by the cutting forces should cause the cutting tool to swing *away* from the work.

● Trial cuts are taken for a short distance. Each trial cut permits the workpiece to be measured before a final cut is taken. Any undersized positioning of the cutting tool may be corrected before the part is machined.

TRIAL CUT METHOD OF STRAIGHT TURNING

The most widely used method of straight, or cylindrical, turning involves the taking of a *trial cut*. The tool is positioned with the crossfeed handwheel to take a shallow cut that still leaves the diameter oversize. The automatic longitudinal power feed is engaged. A cut is taken for a distance of 1/8″ to 1/4″. The carriage and tool are moved longitudinally away from the workpiece. The lathe spindle is stopped and the work is measured.

The cutting tool may be adjusted for depth using the precision graduated feed collar. The carriage and cutting tool are again brought into position to take the cut. The longitudinal power feed is engaged when the cutting tool is set for depth. The cut is then taken to the required length.

The amount of stock to be left for finish turning of the diameter and the shoulder length must be determined in advance. In general practice, the diameter allowance for turning varies from 1/64″ (0.4 mm) on small diameters to 1/16″ (1.6 mm) on large diameters. The allowance for finish turning of shoulders is about half this amount. Shoulder turning allowances range from 0.010″ to 0.030″ (0.3 mm to 0.8 mm).

The same trial cut procedure may be followed for the final finish-turning processes. The cutting tool may be changed from a general-purpose roughing tool to one ground for the finish-turning process. Again, the cutting speed should be increased and the power feed rate decreased for finish turning. This is particularly important with cemented carbide cutting tools. These tools must be operated above a *critical cutting range,* generally between 150 to 200 sfpm.

HOW TO DO STRAIGHT (PARALLEL) TURNING

Taking Roughing Cuts

Step 1 Select the kind and shape of tool bit that is appropriate for rough turning the particular part.

Step 2 Use a straight or left-hand toolholder that will clear the workpiece and the machine setup.

Step 3 Position and solidly secure the tool bit and toolholder to the compound rest.

Step 4 Determine and set the spindle speed at the required rpm.

Step 5 Set the feed lever for a roughing cut.

Step 6 Move the cutting tool to the end of the workpiece. Adjust the depth of cut until at least half of a true diameter is turned.

Step 7 Take the trial cut for a distance of from 1/8″ to 1/4″. Leave the tool set. Return it to the starting end. Stop the lathe.

Step 8 Measure the turned-diameter portion only.

Step 9 Note the reading on the graduated collar. This is the reference point. Subtract the measured diameter from the required diameter. An oversize allowance is made for the finish cut.

Step 10 Move the cutting tool to depth. Start the lathe. Engage the power feed. Take the roughing cut (Fig. 29-9).

Note: When more than one roughing cut is required, each successive cut should be as deep as conditions permit. The cutting tool must be set to cut through any scale.

CAUTION: Close attention must be paid to the cutting tool and cutting action. If the cutting tool tends to dig in or move, immediately withdraw it. Recheck the cutting edge and the tool height and position. Then, securely retighten the tool.

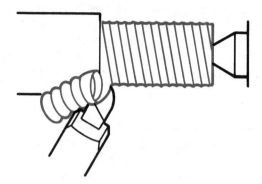

Fig. 29-9. Taking a roughing coarse feed) cut.

Rough Turning An Entire Length Between Centers

Step 1   Rough turn the workpiece to about half the required length. Turn the workpiece end-for-end.

Step 2   Take successive roughing cuts along the remaining portion of the workpiece.

Step 3   Continue to the same depth of reading on the feed collar as for the first portion.

Taking Finishing Cuts

Step 1   Recheck the condition of the center holes. Lubricate the dead center hole. Remount the workpiece.

Step 2   Replace the roughing tool with a finish turning tool. Mount and position the tool bit at center height and to form the shoulder.

Step 3   Increase the spindle speed and decrease the cutting feed for finish turning.

Step 4   Start the lathe. Take a light trial cut. Move the finishing tool clear of the workpiece.

Step 5   Stop the lathe. Measure the turned diameter. Subtract this diameter from the required diameter.

Step 6   Note the graduated collar reading. Move the cutting tool in for a depth of cut that equals one half of the distance between the turned and the required diameter.

Step 7   Retake the trial cut. Stop the lathe. Check the accuracy of the turned diameter. Adjust further for depth. Engage the power feed. Turn the required diameter

to the specified length (Fig. 29-10). *Note:* If the diameter is smaller than required, the cutting tool must be backed out (away) from the workpiece. Reposition the depth. Attention must be paid to overcome the lost motion of the crossfeed screw.

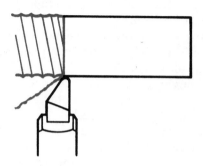

Fig. 29-10. Taking a finishing cut.

REPRESENTATION AND DIMENSIONING OF CYLINDRICAL PARTS

Most cylindrical machined parts have diameters that are symmetrical. These features are often represented in one-view drawings. A centerline denotes that the diameters are symmetrical about the axis that the centerline represents. The diameter may be abbreviated as DIA, as shown in Fig. 29-11.

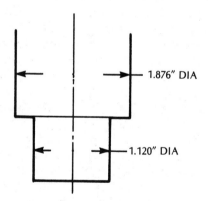

1.876" DIA

1.120" DIA

Fig. 29-11. Designation of a diameter on a drawing.

Decimal Tolerances

When machining interchangeable parts, the worker must know the dimensional accuracy to which each feature must be machined. The finished features must be held to certain *tolerances*. These are usually specified on drawings under *Notes*. Such a note appears in Fig. 29-12. The tolerances may be given as decimal values in either the inch- or metric-standard systems, or in both, depending on the system(s) in which the dimensions are given.

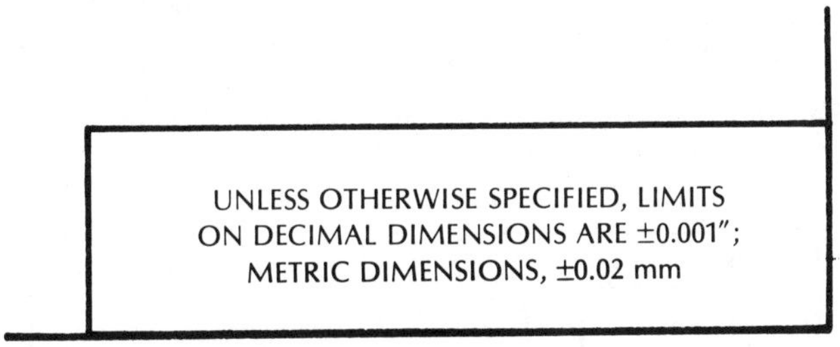

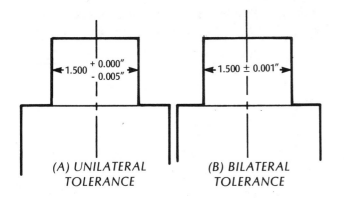

UNLESS OTHERWISE SPECIFIED, LIMITS ON DECIMAL DIMENSIONS ARE ±0.001"; METRIC DIMENSIONS, ±0.02 mm

Fig. 29-12. Drawing note which specifies dimensional accuracy.

The precision to which a part must be machined to function correctly in relation to a mating part is known as a *decimal tolerance*. The decimal tolerance gives the *upper* and *lower* (the largest and smallest) *diameter* to which the workpiece may be machined and still be accepted.

Unilateral And Bilateral Tolerances

Tolerances may be *unilateral* or *bilateral*. A *unilateral tolerance* is in *one* direction only. It may be *either* (+) or (−). The tolerance illustrated in Fig. 29-13A is unilateral. The basic diameter (1.500″) must be held in one direction (−) only. The part may be turned between 1.500″ and 1.495″.

The dimension in Fig. 29-13B is *bilateral*. It means the dimensional accuracy may vary above (+) or below (−) the basic dimension of 1.500″. It may range from 1.501″ to 1.499″ diameter. A part that is machined in this range meets the specified tolerances.

Upper And Lower Limits Of Dimensions

A note is widely used on drawings to indicate that uniform tolerances are applied to all or most dimensions. Since the three diameters that are dimensioned in Fig. 29-14 are all in millimeters, the mm may be omitted. It is obvious that this is the unit of measurement. The note indicates for the drawing that the (+) and (−) tolerances are different. Each dimensional measurement (25.4 mm, 33.4 mm and 38.8 mm) must be kept within the +0.01 mm and −0.02 mm ranges to be accepted.

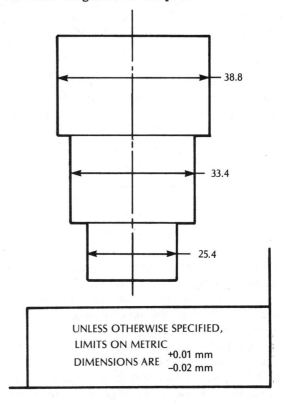

Fig. 29-14. Upper and lower limits specified in a note on a drawing.

The *larger* diameter is referred to as the *upper limit* of a dimension. The *smaller* diameter is the *lower limit*. These limits may also be given on a drawing by placing the *upper* limit *above* a line and the *lower* limit *below* the line. Fig. 29-15 shows the diameter of a workpiece with an upper limit of 25.41 mm and a lower limit of 25.38 mm.

Diametral Dimensions (Equal And Unequal Tolerances)

Drawings are further simplified by the use of

Fig. 29-13. Dimensioning with unilateral and bilateral tolerances.

symbols. The symbol ℺ following a dimension indicates that a particular feature represents a diameter. The one-view drawing of the cylindrical blank in Fig. 29-16 gives the *diametral dimensions* of the two turned sections. The dimensions are stated in both the metric- and inch-standard systems.

The *NOTE* in the lower right corner of Fig. 29-16 gives the permissible range of tolerances: ± 0.05 mm for the metric dimensions; ± 0.002″ for the inch-standard sizes. The two outside diameters may be turned from 50.85 mm to 50.75 mm and from 76.25 mm to 76.15 mm. The equivalent inch-standard dimensions range from 2.002″ to 1.998″ and 3.002″ to 2.998″.

Since both the (+) and (−) tolerances are equal, they are referred to as *equal bilateral tolerances*. When the upper (+) and lower (−) limits vary, they are identified as *unequal bilateral tolerances*.

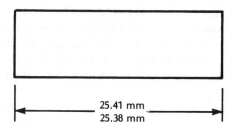

Fig. 29-15. Upper and lower limits specified directly on the drawing.

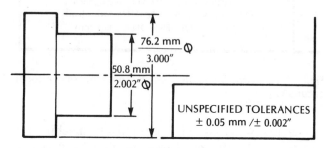

Fig. 29-16. A one-view drawing with tolerances given in a note.

SAFE PRACTICES IN FACING AND STRAIGHT TURNING

- Check the edges of the compound rest for clearance between it and the revolving workpiece or work-holding device. Special care must be taken when the compound rest is positioned at an angle to the spindle axis.
- The remaining amount of material (depth of cut) to be removed from a workpiece is equal to one half the difference between the machined (reference surface) outside diameter and a required diameter.
- Take up all lost motion in the crossfeed screw. Then, lock the graduated collar at the starting point of a measurement.
- Continue to feed the cutting tool in the same direction once the lost motion has been taken up.
- Turn the crossfeed handwheel *clockwise* to apply a force to set the depth of cut. The inward force causes the tool to remain at the depth of cut even when cutting forces are present.
- Position the cutting tool and holder so that any movement during the cutting process moves the cutting edge *away* from the workpiece.

- Increase the sfpm beyond the critical point range to produce a high-quality surface finish when using a keen, sharp-edged carbide cutting tool.
- Take a trial cut for a short distance and measure the diameter *before* turning the required length.
- Allow sufficient stock for removal during the finish cut. The workpiece must *clean up* to the required degree of dimensional accuracy and quality of surface finish.
- Move the tailstock center out carefully so that a facing tool may remove the burred ring around the center hole.
- Return the center back to full bearing position while feeding the facing tool outward.

	COMMON TERMS USED IN FACING AND STRAIGHT TURNING

Turning (straight, parallel, cylindrical)	Common shop terminology used to describe the machining of one or more external diameters. Advancing a cutting tool longitudinally. Taking a cut across the outside diameter at a desired depth while the workpiece is revolving.
Equivalent feed of compound rest	The distance a cutting tool advances longitudinally or transversely when fed by the handwheel of a compound rest that is set at an angle.
30° compound rest setting	The position of the compound rest at 30° from either the center or cross slide axis. An angular setting that produces a longitudinal or transverse movement equal to one half the movement of the compound rest.
Locking the graduated collar	A precautionary step of securing the cross slide or compound rest graduated collar at a desired graduation. Securing a graduated collar so that its position is not changed when other adjustments of the cutting tools or work are made.
Depth of cut (lathe turning)	A setting for a lathe cutting tool. Reducing the outside diameter by twice the actual depth of cut.
Trial cut	A common shop practice of turning a concentric area for a length of about 1/8″ to 1/4″. Establishing a reference surface (a first cut) from which other cuts may be planned in reducing a workpiece to a required size.
Reference layout points	Identifying lines or marks to show a linear dimensional limit to which a surface is to be machined.
Tolerance	A dimensional limit that guides a craftsperson in making or fitting a part in relation to a mating surface or part. The maximum variation permitted in the finished size of a machined part.
Bilateral tolerance	A two-direction tolerance. A tolerance above (+) and below (−) a basic dimension.
Unilateral tolerance	A single-direction tolerance. A tolerance above (+) *or* below (−) a basic dimension, as specified on a drawing.
Upper and/or lower dimensional limit	An ultimate dimension for a feature of a part. The largest (upper) and smallest (lower limit) dimension to which a part may be machined to keep within a required degree of accuracy.
Unequal tolerances	A difference in the (+) and (−) tolerance values.
Allowance	An intentional difference allowed between mating parts. A positive or negative difference. Tolerance and allowance may not be used interchangeably.

	SUMMARY

- Compound rest settings of 30° and 60° are widely used in turning and facing. The compound rest feed movement may be converted into either a transverse or longitudinal movement.

- Movements of 0.0001″ may be made by setting the compound rest at an angle of 84° 16′ to the cross slide axis. As the compound rest is advanced 0.001″, the tool is fed 0.0001″.

- A trial cut permits measuring a workpiece. If the cut produces an incorrect diameter, it may be corrected. A trial cut on a tough abrasive casting or forging must be taken below the hard outer scale or surface.

- The amount a workpiece is reduced in diameter is equal to *twice* the actual depth of cut.

- A workpiece that is to be reduced in length by facing for a considerable distance should be rough turned longitudinally. Each cut may be stepped off. Finally, the cutting angle is changed and the uneven steps are faced off to produce a square, flat end.

- The dial-graduated micrometer collar is used on modern lathes for direct measurement of the cross slide and the compound rest movements. Graduations on one band are in decimal parts of an inch. On the opposite band the graduations are in decimal values of millimeters.

- Once lost motion is compensated for, the graduated collar is locked to the feed screw. The graduation line becomes the reference point. New measurements are related to the reference point.

- The burred ring at the center hole is removed by carefully moving the dead center away from the workpiece for about 1/32″. The facing tool then is fed against the end of the workpiece. As the cutting tool is fed outwardly, the center is again brought into the full bearing position.
- The length of a round workpiece is sometimes scribed on a color dyed surface, for visibility during machining.
- Workpieces should be rough turned leaving 1/64″ of stock for finish turning on small diameters and 1/16″ on large diameters. The spindle speed is increased and the feed is decreased for finish turning.
- A faced surface provides a flat reference plane at a required distance from another surface.
- Chuck and faceplate work may be faced with a regular side- or end-cutting tool. Centered work is usually faced with a 55° cutting point in a rightside facing tool.
- Round workpieces are generally represented by a single-view drawing or sketch. The tolerances specify the acceptable dimensional range within which a part is to be machined.
- Tolerances may be unilateral (one direction, either + or −) or bilateral (both + and − values).
- The largest diameter to which a feature may be machined is its upper limit; the smallest diameter is the lower limit.
- Bilateral tolerances may be equal (± 0.001″) or unequal (+ 0.02 mm/− 0.01 mm).
- Take a first cut through the scale of a casting or irregular, rough surface of a forging. The final roughing cut must permit the workpiece to be finish turned to the required degree of accuracy.
- Turn the workpiece end-for-end when the entire length is to be turned between centers. When a quantity of parts are to be machined, the roughing operations are completed first on all pieces. This is followed with the final finish cut.
- All lubricating systems and moving parts of the lathe must be checked. Cloths and waste must be removed. Loose clothing is to be secured. General safety precautions for machine operation must be followed.

UNIT 29 REVIEW AND SELF-TEST

1. State why the compound rest is generally set at a 30° angle for straight turning processes.
2. List the steps to take to turn a steel part from an outside diameter of 100 mm down to 80 mm.
3. Identify two general forms of facing tools.
4. State four practices the lathe operator follows that apply to all straight turning processes.
5. Tell what a dimension with an *unequal bilateral tolerance* (like 3.500″ $^{+0.0005}_{-0.0010}$) means to the lathe operator who is machining the part.
6. Give the (a) *upper* and (b) *lower limits* and (c) the tolerance of a part that is dimensioned 50.80 mm ±0.005 mm.
7. Indicate three safe practices to follow in straight turning.

Unit 30

SHOULDER TURNING, CHAMFERING, AND ROUNDING ENDS

A *shoulder* is the area that connects two diameters. *Shoulder turning* is the process of machining this connecting area to conform to a particular size and shape. There are three basic shoulder shapes (Fig. 30-1):

- The square shoulder
- The filleted (round) shoulder
- The beveled (angular) shoulder.

A *chamfer* is an outside edge or end that is beveled or cut away at an angle. A round end or corner differs from the chamfer in that the shape is a radius.

This unit covers the laying out and turning of shoulders and rounded or beveled ends, their representation on drawings, and the steps for producing each shape.

PURPOSES OF SHOULDERS, CHAMFERS, AND ROUNDED ENDS

Chamfered And Rounded Ends

Chamfered and rounded ends are turned for appearance. They also provide clearance so that the face of a mating part may seat properly against an internal shoulder (Fig. 30-2). Chamfered and rounded ends are also machined for safety and comfort in handling.

A chamfered end is especially cut on workpieces that are to be threaded. The chamfer angle provides a good face against which a single-point threading tool or die head may be started. As stated before, the chamfer reduces burrs that form easily on a sharp edge. A 30° chamfered end permits a 60° formed thread to be engaged easily.

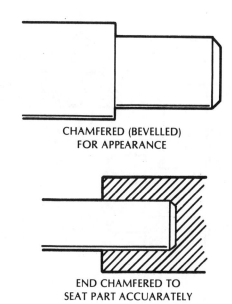

CHAMFERED (BEVELLED)
FOR APPEARANCE

END CHAMFERED TO
SEAT PART ACCUARATELY

Fig. 30-2. Functions served by chamfered ends.

The size of a chamfered or rounded end is usually given on a drawing. However, the craftsperson often has to judge the angle or size of radius. Many times there is a note on the drawing to BREAK ALL SHARP EDGES. In such cases, just a light cut may be taken with a mill file against a revolving workpiece. The end may be filed to a radius or angle.

Shoulders

The particular shape of a shoulder depends on its

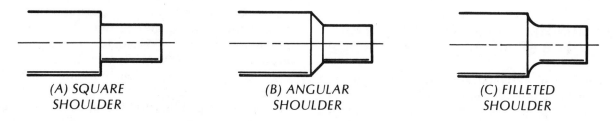

(A) SQUARE SHOULDER   (B) ANGULAR SHOULDER   (C) FILLETED SHOULDER

Fig. 30-1. Three basic shoulder shapes.

use. For example, the underside of the head of a bolt and the body form a square shoulder. This shape permits the head to seat flat so that there is maximum contact of the surface areas. A square shouldered part may serve also as an end bearing or stop, as shown in Fig. 30-3.

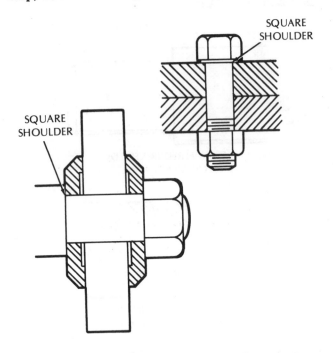

Fig. 30-3. Use of a square shoulder as an end bearing surface.

The shoulder on a forming punch may be round or beveled to form a part with a corresponding radius (fillet) or angle. When a sharp corner is to be avoided, the filleted shoulder strengthens the part without increasing its size. Beveled shoulders eliminate sharp corners, add strength, and improve the appearance of the part. Beveled shoulders are cut mostly at 30°, 45°, and 60° angles.

Regardless of the design, the production of each shoulder involves three basic processes:

- Laying out the location of the shoulder
- Machining the two diameters
- Forming the shoulder to the required shape and length.

COMMON MEASUREMENT PROCEDURES FOR SHOULDER TURNING

The length of a shoulder is usually marked by cutting a light groove around the workpiece. The cutting tool is positioned by measuring the required shoulder length with a steel rule.

Shoulders that are to be machined to a precision

linear dimension may be measured with a gage or depth micrometer. In such instances, the final cut is positioned using a micrometer stop for the carriage. The compound rest feed screw may also be moved the required distance. The depth is read on the graduated collar. The diameter and shoulder length are generally turned to within 1/64″ of the finished size. This leaves sufficient material at the shoulder to permit turning to the required shape and dimension.

Sometimes, the length of a square shoulder is located by cutting a groove in the workpiece with a necking tool. After grooving, the body diameter is turned to size. One or more cuts is taken until the required dimension is reached.

Another shoulder-turning technique is to *block out the shoulder,* leaving enough material at the shoulder to finish turn it to the required form. The adjacent cylindrical surface is then turned to size.

APPLICATION OF THE RADIUS GAGE

The *radius gage* provides a visual check on the radius of the cutting tool or the filleted shoulder. The accuracy of the cutting tool is checked by placing the correct radius portion of the gage over the ground radius. A radius gage and five general applications of it are pictured in Fig. 30-4.

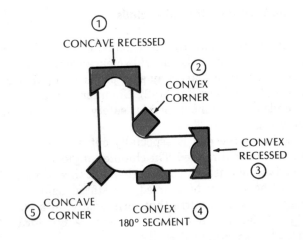

Fig. 30-4. Applications of the five gaging edges of a radius gage.

Low spots are sighted by any background light that shines between the gage and the cutting tool. The dark, high spot areas are ground or honed if there are only a few thousandths to be removed. When no light is seen along the radius, the cutting point is formed to the correct shape and depth.

One common practice that is repeated in turning a shoulder to a large radius is to *step off* the shoulder. Successive facing or straight turning cuts are taken.

These end in a series of steps. The steps, or excess corner material, are rough turned. A stepped-off section is shown in Fig. 30-5.

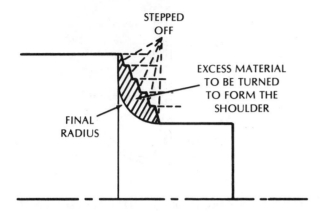

Fig. 30-5. Stepping off a large radius.

The lathe operator often *contour forms* the radius. This involves moving the longitudinal and crossfeed screws simultaneously by hand. This action *sweeps* the cutting tool through the desired radius.

The shape and size of the radius are checked with a radius gage. A great deal of machining skill is required to blend all of the cuts made with a round-nose cutting tool until the desired radius is formed. A tool ground to the correct radius is often used.

Commercial flat, metal radius gages (as previously illustrated) are available. The general sizes range from 1/32″ R to 1/2″ R. There are usually five different gaging areas on each blade. These permit gaging the radius of a corner, a shoulder, or a recess in a circular workpiece.

REPRESENTING AND DIMENSIONING SHOULDERS, CHAMFERS, AND ROUNDED ENDS

The techniques of representing and dimensioning the three general types of shoulders and the rounded and beveled ends are illustrated in Fig. 30-6. The square shoulder (1) is turned to an outside body diameter of 1.5000″ for the 1 1/2″—12UNF thread. The length or distance to the square shoulder is 1 1/4″.

The radius of the filleted shoulder (2) is indicated as ¼ R. The beveled shoulder (3) joins the 1.250″ + 0.000″/− 0.002″ and 1.500″ ± 0.001″ diameters at a 45° angle. The distance to the shoulder is 2″.

The beveled end (5) angle corresponds to one half of the thread angle and is 30°. While no depth is given, there is a note on the drawing: CHAMFER TO THREAD DEPTH.

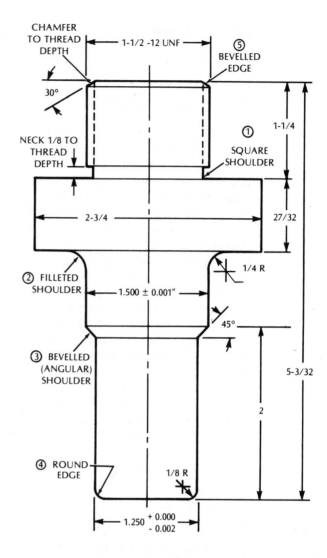

Fig. 30-6. Representation and dimensioning of general types of shoulders and edges.

This indicates the chamfer to be cut at least to the thread depth. The Unified Fine Thread (1 1/2″— 12UNF) base diameter is 1.400″.

The chamfered depth is one half the difference between the outside and base diameters of the thread, or 0.050″. In practice, the lathe operator cuts slightly beyond the thread depth. The drawing shows the rounded edge (4) is dimensioned 1/8 R. The end is turned to a 1/8″ radius.

HOW TO TURN SHOULDERS

Note: The procedure for turning each type of shoulder starts with a workpiece that has been laid out and rough turned to diameter, length, and shape.

Turning The Square Shoulder

Method 1—Feeding Outward

Step 1 — Reduce the excess material at the shoulder. This permits taking a finishing cut.

Step 2 — Position a right-side-cutting facing tool. Set the crossfeed graduated collar at "0".

Step 3 — Finish turn the outside diameter. Take the finish cut with the automatic power feed. Turn almost to the shoulder. Continue to feed the facing tool into the shoulder by hand. Then, take a facing cut outward.

Step 4 — Stop the lathe. Measure the length to see whether the cut *splits the layout line*.

Step 5 — Remove any additional material on the shoulder face. Bring the facing tool back to the "0" setting each time. The tool then is again fed toward the outside diameter. *Note:* The cutting tool may also be moved to the required length either against a micrometer stop or with the compound rest. For a less precise measurement, the operator often estimates, from experience, how much more the facing tool should be moved into a workpiece.

Method 2—Feeding Inward

Step 1 — Move the end-cutting facing tool so that the cutting edge just touches the turned diameter of the body portion of the shoulder. Set the cross slide graduated collar at "0".

Step 2 — Back the cutting tool away from the workpiece. Start a trial cut for length on the outside diameter. Check the length with a steel rule or by micrometer measurement, depending on the accuracy needed. Lock the carriage for facing the shoulder portion.

Step 3 — Feed inward to the "0" setting on the graduated collar.

Step 4 — Feed the cutting tool in this position to the right to remove any excess material (Fig. 30-7). Blend the diameter at the shoulder with the body. *Note:* Some operators move the cutting point slightly beyond the "0" setting for the final shoulder finishing operation. Just the corner of the shoulder is turned at that setting.

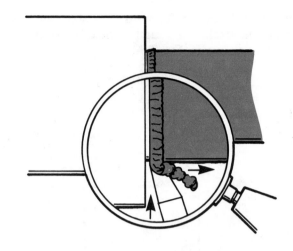

Fig. 30-7. Feeding an end-cutting tool to cut a square shoulder.

Step 5 — Repeat the last two steps, if required, to machine the shoulder length to size. *Note:* End-feeding is a quicker way of establishing the length and then facing. However, a smoother finished surface is produced by feeding the cutting tool outward on the shoulder face.

Step 6 — Remove the tool marks. Speed up the spindle rpm to about twice the speed used for turning. Polish with an abrasive cloth.

Turning A Filleted (Round) Shoulder

Step 1 — Position a right-side round-nose turning tool.

Step 2 — Rough face the shoulder. Allow sufficient material to finish the body diameter, shoulder radius, and shoulder face (Fig. 30-8).

Step 3 — Rough turn any excess material left on the rounded portion.

Step 4 — Set the cutting tool at "0" on the crossfeed graduated collar when the final diameter is reached.

Step 5 — Finish turn the diameter. Disengage the longitudinal automatic power feed before the shoulder is reached.

Step 6 — Feed the carriage slowly by hand. Move the tool a slight distance into the shoulder face.

Step 7 — Lock the carriage. Take a facing cut outward.

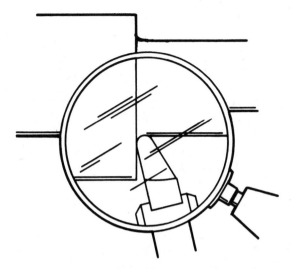

Fig. 30-8. Blending (turning) the body diameter, shoulder radius, and face.

Note: Decrease the spindle speed when cutting a radius. This is to prevent chatter and digging in of the tool at the shoulder.

Note: A cutting fluid is used to produce a fine surface finish.

Step 8 Measure the shoulder length. Take any required additional cuts by bringing the cutting tool point to the "0" setting each time. Move the tool by hand into the shoulder face to the required depth. Lock the carriage, and reface.

Turning A Beveled Or Angular Shoulder

Method 1—Setting A Side-Cutting Tool

Step 1 Position the side-cutting edge at the required angle. The turning of an angular shoulder is shown in Fig. 30-9.

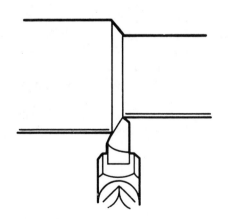

Fig. 30-9. Turning an angular shoulder.

Step 2 Remove excess stock from the angular shoulder.

Step 3 Take a trial cut on the body diameter. Measure the diameter, then move the cutting tool to the required depth. Set the crossfeed graduated collar at "0".

Step 4 Use a power feed and take the cut almost to the shoulder. Disengage the power feed.

Step 5 Decrease the spindle speed for a large deep-cut shoulder. This tends to reduce chatter. Continue to slowly feed the cutting tool by hand into the angular shoulder face.

Step 6 Feed to the required length. Use a cutting fluid.

Method 2—Setting The Compound Rest

Step 1 Position the compound rest at the required angle.

Step 2 Set up a side-cutting tool.

Step 3 Take cuts from the center outward. Feed the cutting tool by hand by turning the compound rest feed screw.

Step 4 Continue to machine the bevel. The final beveled surface starts at the required length of the small diameter. It continues at the required angle to the larger diameter.

HOW TO TURN A CHAMFER (BEVEL)

Step 1 Set a right-side or end-cutting tool at the required angle. Three setups are shown in Fig. 30-10.

Step 2 Lay out the width of the chamfer on the end of the workpiece.

Step 3 Move the carriage by hand so that the side- or end-cutting tool cuts into the revolving workpiece.

Note: Reduce the spindle speed on wide chamfers, to prevent chatter.

Step 4 Continue to slowly feed the cutting tool until the chamfer (bevel) is cut to the required width.

Note: A cutting compound should be used. The tool may also be fed by locking the carriage when the cut is almost to final depth and then feeding to depth with the cross slide handwheel.

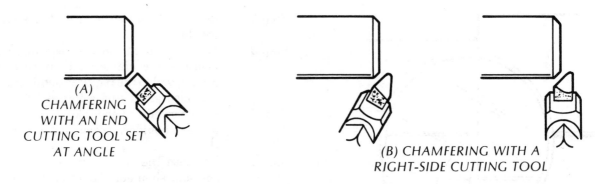

(A)
CHAMFERING
WITH AN END
CUTTING TOOL SET
AT ANGLE

(B) CHAMFERING WITH A
RIGHT-SIDE CUTTING TOOL

Fig. 30-10. Setups for turning a chamfered (beveled) edge.

HOW TO ROUND A CORNER

Using A Radius-Cutting Tool

Step 1   Grind an end-cutting tool to the specified inside radius. Check the accuracy with a radius or other gage.

Step 2   Set the cutting tool point on center. Position the tool so that the radius will blend into the face. This setup is shown in Fig. 30-11.

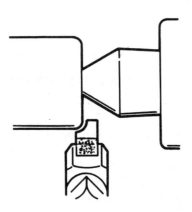

Fig. 30-11. Turning a round corner with a radius tool.

Step 3   Reduce the spindle speed. The larger the radius, the slower the spindle speed.

Step 4   Bring the inside edge of the cutting point close the face of the workpiece. Lock the carriage. Flow on a cutting fluid. Feed the radius tool with either the cross slide or compound rest feed screw. Continue until the required radius is formed.

Step 5   Stop the lathe. Remove all chips, and wipe the workpiece.

Step 6   Increase the spindle speed. Start the lathe. Remove the tool marks by finish filing. Polish with an abrasive cloth.

Filing A Radius

Step 1   Select a clean mill-cut file.

CAUTION: Check to see that the handle securely fits the tang.

Step 2   Set the lathe at the high spindle speed. Hold the file as for filing a flat surface.

CAUTION: Left-hand filing is recommended. There is less chance of hitting any rotating part of the lathe or workpiece.

Step 3   Take long, regular file strokes. Turn the file during each forward stroke to form the radius.

Note: Use a file card and brush to remove chips and to keep the file clean.

Step 4   Check the radius with a template or radius gage.

Step 5   Remove the file marks by polishing with an abrasive cloth. The abrasive cloth should be held by hand and pressed firmly against the workpiece.

Note: An 80-100 grain size is used for general finish. A finer finish is obtained by polishing with a finer grain size.

SAFE PRACTICES IN TURNING SHOULDERS, CHAMFERS, AND ROUNDED ENDS

- Blend the stepped sections of a filleted or square shoulder to the desired form before a final round-nose or blunt angular-nose facing tool is used for the finishing cut.
- Lock the graduated collar on the crossfeed or compound rest feed screw. This prevents any movement that might produce an incorrect reading.
- Align the cutting point of a facing tool on center.
- Reduce the spindle speed and use an appropriate cutting fluid whenever a large radius, angular shoulder, or a rounded or beveled end are to be machined. The speed reduction helps prevent chatter.
- Check the accuracy of a turned concave or convex radius with a radius gage.
- Use the left-hand method of filing a workpiece on the lathe. This is a safety precaution to prevent hitting the revolving workpiece, lathe jaw, or chuck.

- See that the file handle fits solidly on the tang.
- Check the condition of the center hole on work that has been turned between centers. If necessary, take a *cleaning up cut* with a center drill to smooth the angular bearing surface.
- Add a lubricant periodically to the dead center. This is usually done between finish turning, filing, or polishing operations.
- Check all lubricant levels. Oil the bearing surfaces of the lathe by hand *before* beginning any process.
- Apply a cutting fluid where one may be used, particularly on rough turning. This will help control the temperature of the workpiece and improve the cutting efficiency.

TERMS USED IN TURNING SHOULDERS, CHAMFERS, AND ROUNDED ENDS

Filleted shoulder (round)	The joining of two different diameters with a rounded end section. A shoulder formed by a radius. A radius turned to an inside corner.
Beveled shoulder (angular)	The joining of two different diameters with an angular surface. A shoulder formed at an angle.
Square shoulder (turned)	A plane radial face. A face that is at right angles to the axis of a cylindrically turned workpiece.
Rounded end (turned)	A round curved surface that connects a face with an adjoining cylindrically turned portion of a workpiece. A radius turned to an outside corner.
Rounding a corner (turning)	The process of forming a radius on the end of a workpiece. Feeding a cutting tool with a round cutting face of a specified radius into a revolving workpiece.
Radius gage	A flat, L-shaped metal plate with circular internal and external shapes of a specified radius. A flat plate applied in checking the form and dimensional accuracy of inside or outside radii.
Stepping off	The process of taking a series of successive cuts. Excess material is left at the end of each cut. A shoulder is then formed to a particular shape and size.
Reducing the stepped off shoulder	Removing the excess material left after rough forming a shoulder. Finish turning a shoulder to shape and size.
End-feeding	Hand or power feeding a cutting tool from an outside diameter inward toward the center.
Chamfering (turned workpiece)	Forming an angular (beveled) end. An angular surface adjoining two adjacent flat surfaces. Positioning the cutting tool at a required angle and feeding the cutting edge into the revolving workpiece.

Chamfer to thread depth A drawing note that means: (a) the chamfer angle is the same as the thread angle, and (b) the minimum depth of the chamfer is equal to the depth of the thread.

Representing shoulders and end shapes Graphic techniques for communicating information about the shape and dimensions of different kinds of shoulders and turned ends.

SUMMARY

- Angular and filleted shoulders are turned to strengthen the area adjoining two diameters of a workpiece or to improve the general appearance.
- A square shoulder permits the seating of two mating faces. The square shoulder also serves as a bearing or positioning surface.
- Shoulder turning requires three basic processes: (1) layout (marking the length), (2) turning both diameters, and (3) machining the connecting area to a particular shape and size.
- Shoulder lengths may be precision measured with a depth micrometer.
- A 1/64″ material allowance is usually left for finish turning both diameters and the shoulder.
- A radius gage is used to measure inside and outside turned forms. These forms are circular and of a specified radius.
- Angular shoulders are usually cut by positioning the face of the cutting tool at the required angle. The tool is fed lengthwise into the shoulder.
- Turned shoulders and ends are dimensioned in terms of length, diameter, and the shape and size of the round or beveled area.
- A chamfered thread end helps prevent the forming of a burr. The chamfer also permits a thread form and the mating thread to be started accurately.
- Turned shoulders are sometimes finished by hand filing. Their diameters and shoulder portion may be polished by increasing the spindle speed and using an abrasive cloth.

- Deep filleted (round) shoulders are rough turned by stepping off each successive cut. The excess material is then rough turned to the required radius.
- Filleted shoulders may be rough and/or finish turned with a formed-radius cutting tool.
- A large radius is often turned by hand feeding. The crossfeed and longitudinal feed handwheels are moved at the same time.
- The final cut in shoulder turning is usually begun with the graduated collar set at "0". Subsequent shoulder forming or facing cuts are brought to or started at the "0" setting.
- The spindle speed is reduced when using a radius-forming shoulder-cutting tool for the finish facing cut. A cutting lubricant helps produce a finer cutting action and surface finish.
- The sharp edge of a round workpiece is often broken by taking a light angular or rounded file cut.
- A handle must be properly fitted to the tang of a file, for safe use. The handle permits convenient manipulation of the file for the correct cutting action.
- The file teeth must be kept clean with a file card and brush.
- Left-hand filing on the lathe is recommended.
- The work center must be clean, checked for quality of bearing surface, and lubricated.
- Machine parts should be protected against filing and abrasive particles.

UNIT 30 REVIEW AND SELF-TEST

1. Indicate what the worker is expected to do when a drawing of a part has the note:

   BREAK ALL SHARP EDGES.

2. Explain the meaning in shoulder turning of the term: *Block out the shoulder*.

3. List three lathe work applications of a *radius gage*.

4. Describe the meaning of a note on a shop sketch that reads:

   CHAMFER TO THREAD DEPTH.

5. State two common methods of (a) turning a square shoulder, (b) turning a beveled or angular shoulder, and (c) rounding a corner.

6. Tell what safety precautions to take in turning a large radius or an angular (beveled) shoulder.

Unit 31

GROOVING, FORM TURNING, AND CUTTING OFF

Turning processes that produce a specially shaped cutaway section around a workpiece are referred to as *grooving* (Fig. 31-1). A groove that is cut into an outside diameter is called a *neck*. A groove that is formed inside a hole is known as a *recess* or an *undercut*. The process is called *grooving, necking, recessing,* or *undercutting*.

Form turning is used to produce a regular or irregular shape around a cylindrical surface. *Cutting off (parting)* is the separating of material by cutting through a revolving workpiece. The purposes, general cutting tools, and setups for grooving, form turning, and cutting off, together with the representation and dimensioning of these processes, are treated in this unit.

FUNCTIONS SERVED BY GROOVES

A part is usually grooved for one of the following six reasons:
- To permit a cutting tool to end a cut in the workpiece. The tool may be withdrawn from this position without damage to the cutting point. One application of necking is in threading to a shoulder, as shown in Fig. 31-1A. The thread-cutting tool emerges from the last thread into the groove. It may then be safely withdrawn and positioned for subsequent cuts
- To provide a groove in which a round, spring-metal slit washer may be inserted and held. This washer design is used in assembling and holding parts in a fixed location on a shaft
- To permit a square-shouldered mating part to seat squarely on the shoulder face
- To simplify other work processes, such as turning a square shoulder. A square groove is often cut to depth and to the required length. A regular side-cutting tool may then be used to turn the diameter to size
- To form a V-shaped groove in which a belt may track and exert a driving force (Fig. 31-1B)
- To improve the appearance, as in the case of a knurled section
- To provide a round recess between turned sections of a workpiece (Fig. 31-C).

COMMON FORMS OF GROOVES

There are three basic forms of grooves:
- Square
- Round
- Angular (V-shaped).

The cutting tools that produce these shapes are named: square-, round-, and V-nosed grooving tools.

Square Groove

The *square groove* (Fig. 31-2) may be cut at a shoulder or at any point along the workpiece. As stated earlier, the square groove provides a channel. This permits a threading tool to ride into the groove

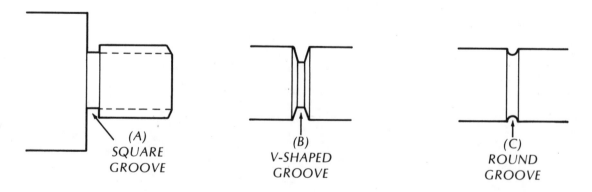

(A) SQUARE GROOVE

(B) V-SHAPED GROOVE

(C) ROUND GROOVE

Fig. 31-1. Three common forms of grooves.

or a mating part to shoulder squarely. Parts that are to be cylindrically ground are often undercut at a shoulder. The undercut makes it possible for the face of the grinding wheel to grind the diameter to size along the entire length.

The square groove may be turned with a cutoff (parting) tool. A standard high-speed steel tool bit (Fig. 31-2) or a carbide-tipped bit ground to the required width are commonly used.

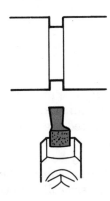

Fig. 31-2. A regular tool bit ground to cut a square groove.

Round Groove

The *round groove* (Fig. 31-3) eliminates the sharp corners of the square groove. This feature is particularly important on hardened parts or on workpieces that are subjected to severe stresses. Under these conditions, a sharp corner may crack and fracture easily.

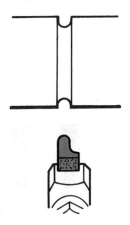

Fig. 31-3. A form-ground tool bit for turning a round groove.

The round groove also provides a pleasing appearance for ending a work process, like a knurled area.

Round grooves may be produced by regular cutting tools that are formed to the required shape and size. This setup is pictured in Fig. 31-3.

V-Groove

The *V-groove* (Fig. 31-4) has two major applications: (1) to form the sides against which a V-belt may ride, and (2) as an angular channel into which a thread cutting tool may end. The angle provides a better surface for finishing the last portion of a thread. Deep V-grooves are usually turned by offsetting the compound rest at the required angle. One angular side of a V-groove is cut at a time. Shallow V-grooves may be cut by using a formed V-cutting tool like that in Fig. 31-4.

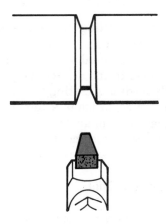

Fig. 31-4. A form-ground tool bit for turning a V-groove.

Special care must be taken when cutting a deep groove in a long small-diameter workpiece or in work that has been turned between centers. A follower rest should be positioned to support the workpiece behind the cutting edge. This setup overcomes the tendency of the cutting tool to dig in and the workpiece to *ride* on the tool face. Either condition causes the work to spring out of shape.

Grooving tools are end-cutting tools. They are ground with relief for the tool end and sides. Cutting speeds are decreased in grooving because of the greater cutting area. Cutting fluids and fine feeds are required to produce a fine surface finish.

FORM TURNING

Methods Of Turning Forms

Convex, concave, and other irregular forms are

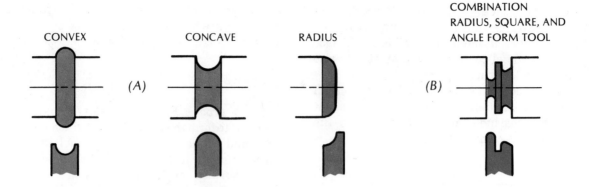

Fig. 31-5. Examples of form tools for producing regular shapes.

generally turned by one of the following four methods:

- Manually controlling the movements of the carriage and cross feed
- Turning to a required depth with a cutting tool that is preformed to the desired shape
- Manually feeding a cutting tool until the shape that is produced conforms to a template
- Producing the form automatically using a template and a tracer attachment

The first two methods are covered in this unit.

Preformed Cutting Tools

Form turning is a practical way of producing regular shapes as well as intricate contours. Form tools that are precision ground to a required shape and size are used. Such tools are sharpened on the top cutting face to maintain the accuracy of the form. One advantage of using an accurately ground form tool is that multiple pieces may be reproduced to the same shape and size. Gaging and measuring each part thus may be eliminated.

The contours produced are usually a combination of a round, angular, or square groove, or a raised section. The radius may be concave or convex (Fig. 31-5). The convex radius requires the workpiece to be turned first with a raised collar section. The collar is then reduced to shape and size with a convex radius form tool. Fig. 31-5A provides two examples of forms that are turned with concave and convex radii. Fig. 31-5B shows a contour consisting of concave radii, square, and angular sections.

The cutting action with formed turning tools takes place over a large surface area. Accordingly, the spindle speed must be reduced to about one half the rpm for regular turning. On larger and deep contours, a *gooseneck* tool holder is used. This holder helps eliminate chatter and produces a fine surface finish. The same safe practices must be observed in form turning as are observed for grooving and cutoff operations.

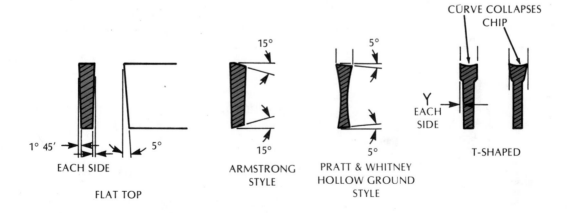

Fig. 31-6. Types of HSS, carbide and alloy cutoff blades.

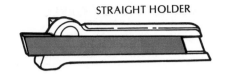

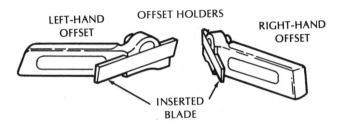

Fig. 31-7. Three standard styles of cutting-off and side-cutting toolholders with inserted blades.

Producing a Grooved Form Freehand

Another common practice in form turning is to form the groove *freehand*. This technique is used when one or just a few parts are to be machined and no template is available. The form is produced by the operator. A partially formed cutting tool is fed into the revolving workpiece. The longitudinal and transverse feeds are applied manually and simultaneously. The amount the carriage moves in relation to the movement of the cross slide helps to govern the shape that is turned.

CUTTING-OFF (PARTING) PROCESSES

Cutting-off processes are usually performed with a specially formed thin blade that is rectangular in shape. The sides of the cutting tool are shaped so there is a relief angle. The blade is held solidly against a corresponding angular surface of a straight or offset tool holder. Cutoff blades are made of high-speed steel, carbides, and cast alloy. Four common shapes are shown in Fig. 31-6.

The blade holders are commercially available in straight and right- and left-hand offset styles (Fig. 31-7).

Small workpieces are often cut off with a standard tool bit that is ground as an end-cutting parting tool. The depth to which the narrow blade or point enters the workpiece, the extent the blade must extend, and the fragile nature of this type of cutting-off tool requires that special precautions be taken.

PROBLEMS ENCOUNTERED IN CUTTING OFF STOCK

The lathe operator must recognize four major problem areas in cutting off (parting) stock:
- Chatter
- Rapid dulling of the cutting tool
- Cutting tool digging into the work and the *work climbing on the tool*
- Cutter riding on the work.

A simplified statement of each problem, the probable cause, and the recommended correction of each condition is given in Table 31-1.

REPRESENTING AND DIMENSIONING GROOVES AND TURNED FORMS

The three common grooves are generally represented on drawings as shown in Fig. 31-8. The

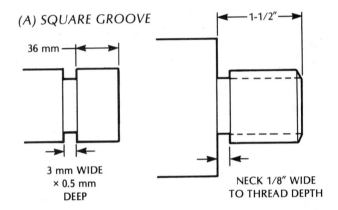

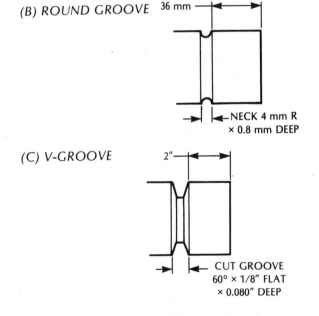

Fig. 31-8. Representation and dimensioning of common grooves.

Table 31-1. CUTTING OFF PROBLEMS: CAUSES AND RECOMMENDED CORRECTIONS

Cutting Off Problems	Probable Causes	Recommended Correction
Chatter	Speed too great	Decrease the cutting speed
	Tool extended too far	Clamp the cutting tool *short*
	Play in cross slide and compound rest	Adjust the gibs on cross slide and compound rest
	Backlash in compound rest feed screw	Back the compound rest away from the work. Turn the feed screw to the left
	Play in lathe spindle	Adjust the end play in the spindle bearings
Rapid dulling of cutting tool	Cutting speed too high	Decrease the cutting speed
	Too much front clearance	Decrease the front clearance angle
	Lack of cutting lubricant	Apply lard oil or required cutting compound for part being turned
	Overheating and drawing the tool temper due to insufficient clearance	Grind front, side, or back clearance
	Overheating due to excessive cutting speed	Decrease spindle speed
Cutting tool digging into work and climbing on cutting tool	Tool set too low	Set tool at height of lathe center
	Play in compound and cross slide	Adjust gibs on cross slide and compound rest
	Backlash in compound rest feed screw	Back the compound rest away from the work. Turn the feed screw to the left
	Too much front clearance	Decrease the front clearance angle
	Too much top rake	Decrease the top rake
Cutting tool riding on work	Tool too high	Set tool at the height of the lathe center
	Insufficient front clearance	Increase the amount of the front clearance angle

dimensions give the width and depth and provide other information about the form.

The square groove (Fig. 31-8A) is dimensioned for width and depth. A notation is used when this groove is cut at a shoulder or to provide a channel at the end of a thread. The notation gives the width of the groove and directs that the groove be cut "TO THREAD DEPTH." In some instances, the width and depth may be given as "⅛" WIDE × ⅛" DEEP".

In Fig. 31-8B, the round groove (neck) is 36 mm from the end of the workpiece. The 4mm radius is cut

to a depth of 0.8 mm. This same groove may be dimensioned by the note: UNDERCUT 4 mm R × 0.8 mm DEEP.

The V-groove (Fig. 31-8C) is formed with a 60° included angle form cutter. The width of the flat end-cutting edge is ground ⅛". The groove is cut 2" in from the end and to a depth of 0.080".

Sometimes, a part drawing, like that in Fig. 31-9, shows the width (6.4 mm) of the groove in relation to a reference plane (20.0 mm). The root diameter to which the groove is turned is indicated by the symbol Ø (31.8 mm Ø).

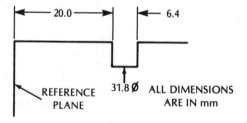

Fig. 31-9. A part drawing that illustrates the dimensioning of a square groove.

HOW TO TURN GROOVES

Step 1   Grind the tool bit to the desired shape and size. Hone the point end. Test a round-nose tool with a radius gage; test a

V-shaped groove with a gage or bevel protractor.

Note: The grooving processes are similar even though the shape of the cutting points differ.

Step 2   Mount the workpiece in a chuck or between centers. Secure the tool bit in a tool holder. Position the cutting point at center height and at a right angle to the axis of the workpiece.

Step 3   Lay out the location of the groove.

Note: The position of a square groove may be measured by placing the steel rule against the side of the square-nose tool. The length is read directly.

Step 4   Set the spindle speed at approximately one half the rpm used for turning.

Step 5   Start the lathe spindle. Carefully bring the cutting point to the workpiece. Keep the point within the boundaries of the groove to be cut.

Step 6   Set the cross-feed screw graduated collar at zero (0).

Step 7   Feed the square, round, or V-shaped formed grooving cutter to the required depth. The depth is read on the graduated collar. Use a fine feed and a cutting fluid.

Note: Many deep grooves are cut or widened by feeding the formed tool against the sides of the groove. This relieves the force produced if the whole point is cutting, and prevents chattering.

Note: The skilled craftsperson usually checks the root diameter of the groove with a caliper, knife-edge vernier, or other precision measuring tool. Fig. 31-10 shows the root diameter being checked with an outside caliper. The workpiece must be stopped during this check.

CAUTION: Use a follower rest to support the workpiece when a deep groove is to be cut or the part is long and slender.

Step 8   Widen a square groove by bringing the cutting point clear of the workpiece. Move the carriage to the right or left within the scribed lines. Again, feed to the required depth.

Note: The graduated collar may be used for repositioning the cutting tool for width when the compound rest is set parallel with the center.

Step 9   Break the sharp corners of the groove to remove any burrs. Use a smooth cut mill file.

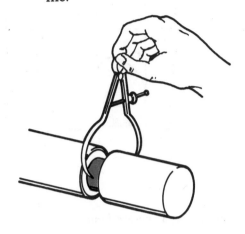

Fig. 31-10. Checking the diameter of a square groove.

HOW TO TURN A FORM

Machining A Concave Form By Hand Feeding

Step 1   Mount the workpiece. Set the spindle speed at about three fourths the speed for turning. Mark the location and the centerline of the groove.

Note: This speed is selected because only part of the cutter face will be cutting at one time.

Step 2   Position and secure a round-nose cutting tool bit.

Note: The tool bit should be ground to the largest possible radius.

Step 3   Grasp the carriage handwheel with one hand and the cross-feed handwheel with the other.

Step 4   Starting at one end of the groove, carefully feed the tool bit to cut into the workpiece. Feed to the opposite end of the groove.

Step 5   Continue to feed in the tool. Take successive cuts from each side to step off the concave form (Fig. 31-11).

CAUTION: Start the roughing cuts from each side and work toward the center.

CAUTION: Take *light* roughing cuts to get the correct coordination between the carriage feed and cross feed in relation to the concave radius to be formed.

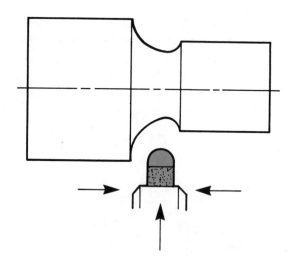

Fig. 31-11. Machining a concave form.

Step 6   Check the form with a radius gage. Measure the root diameter of the groove with a caliper.
Note: A round piece of stock of the required diameter is often used as a radius gage when such a gage is not available.

Step 7   Continue to *sweep* the radius tool into the concave surface. Take *light* finishing cuts for the final cutting of the radius.

Step 8   Use a smooth-cut round or half-round mill file. Finish form the radius, removing the cutting tool marks.

Step 9   Step up the spindle rpm. Polish with an abrasive cloth.

Turning A Raised Convex Section By Hand Feeding

Step 1   Lay out the width of the section.

Step 2   Turn both sides of the raised radius section to the required diameter.

Step 3   Use a round-nosed tool and rough turn the radius by stepping-off.
Note: The roughing cuts are usually started at the center of the convex section and fed to the outside face. One side is roughed out, then the other.

Step 4   Form the roughed-out steps into a continuous curve. Start the finish cut either at the width of the groove or at the center. Then, sweep the form by simultaneously feeding the tool while moving the carriage and the cross slide.

Step 5   Stop the workpiece periodically. Check the convex section for size and shape. Use a radius gage or template.

Step 6   Take a fine finish cut to smooth out the surface of the radius.

Step 7   Finish file the convex section to remove the tool marks. The radius may be polished with an abrasive cloth.

Form Turning With A Formed Cutting Tool

Step 1   Check the form cutting tool for shape and size.

Step 2   Reduce the spindle speed to one half the rpm usually used for turning.

Step 3   Secure and position the form cutting tool on center. Mark off the length of the turned section.

Step 4   Start the cut at the required length. Feed the lead end of the form cutter until it just touches the revolving workpiece.

Step 5   Set the cross feed graduated collar at zero (0). Start the flow of cutting fluid.

Step 6   Feed in to the required depth.
Note: The worker often carefully moves the carriage a slight amount. This action breaks the chips and gives the form tool a little more freedom.

Step 7   Hand file any sharp edges that result from turning.

HOW TO CUT OFF (PART) STOCK

Step 1   Select the cutoff tool blade that is recommended for the workpiece material.

Step 2   Place and secure the blade in either a straight or right-hand offset tool holder. Locate the tool holder so that the cutting point is on center and the blade is at 90° to the work axis. These settings are shown in Fig. 31-12.
Note: The blade should extend beyond the holder about ⅛" more than the radius of the part to be cut.

Step 3   Locate the cutting-off tool at the required workpiece length (Fig. 31-13).

Step 4   Check the tool setting to be sure the

workpiece can be parted without interference.

Step 5 Set the spindle speed at one half the rpm used for straight turning.

Step 6 Feed the cutoff tool continuously (slowly but constantly) into the workpiece (Fig. 31-14). Use a cutting fluid for materials that require it.

Note: Move the parting tool to the right and left a few thousandths with the carriage handwheel. This action produces a slightly wider groove. This keeps the blade from binding, particularly when cutting off to depths of ¼″ and more.

Step 7 Continue to feed until the workpiece is almost cut off.

Note: Before the work is parted, remove the burrs from both sides by filing.

CAUTION: Use a *light* feed as the cutoff point approaches. The cutoff part should not be touched; it may be hot.

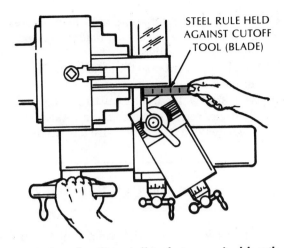

Fig. 31-13. Locating the cutoff tool at a required length.

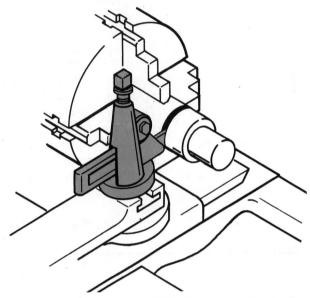

Fig. 31-14. Feeding a parting tool inward to cut off a workpiece.

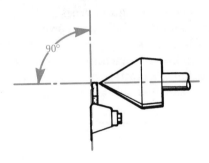

Fig. 31-12. Setting a cutting-off blade on center and square.

SAFE PRACTICES IN GROOVING, FORM TURNING, AND CUTTING OFF

- Set the spindle speed at about one half the rpm that is used for straight turning with high-speed steel tool bits or cutoff blades. Apply this general rule to grooving, form turning with a preformed cutting tool, and cutting off.
- Back up the workpiece with a follower rest when turning deep grooves that may weaken the workpiece.
- Stop the lathe each time the groove is to be calipered.
- Move the carriage *slightly* to the right and left as the form cutting tool is fed into the workpiece

during the roughing cuts. This helps eliminate chatter and binding.
- Grip the workpiece securely in a collet, chuck, or other holding fixture.
- Use a straight or right-hand offset tool holder. Hold the stock so that the least amount extends beyond the holding device. Cut off as close to the holding device and spindle nose as possible.
- Chuck the largest diameter of the workpiece. Cut off the smaller diameter. Otherwise, the workpiece may bend.
- Extend the cutoff blade beyond the holder about

⅛" more than the radius (one half the diameter) of the workpiece. Grip the tool holder as close to the tool end as possible.

- Set the blade or center so that the top is flat and has no back rake.
- Apply a steady feed. The cutting edge dulls rapidly under too light a feed or an intermittent feed. Two heavy a feed may cause tool breakage.

This is caused by digging in or jamming.
- Select the cutting fluid that is most appropriate for the material and process and direct it, so that it reaches the cutting point.
- Use a protective shield or goggles. The hot chips tend to *spatter*. Avoid handling hot pieces, particularly at the final stage of parting.

GROOVING, FORM TURNING, AND CUTTING OFF TERMS

Grooving, (necking-undercutting)	The process of turning a cutaway surface of a particular shape and size.
Recessing	The process of boring an internal, circular, grooved surface to a specific size and shape.
Square, round, and angular grooves	Three basic shapes for external grooves. The cross-sectional shape of a circular indentation on the outside of a workpiece.
Form turning	The turning of an external or internal contour of a specified shape and size.
Form turning by hand	A turning process in which a formed cutting tool and manually-controlled movements of the carriage and cross slide are used to produce a desired form.
Cutting off (Parting, turning)	The severing of a section by cutting a groove through a revolving workpiece. The process of using a mounted parting tool to cut through a workpiece.
Cutoff (parting) blade	A comparatively thin cutting tool, commonly of rectangular cross section. The side faces taper (or are hollow ground) or are formed into a T-shape to provide clearance.
Representing and dimensioning of grooves	Graphically providing a full description of the shape, size, and other features of a groove. A line drawing with full specifications for machining a groove. Sample notations that appear on drawings. Information to a craftsperson to turn a groove of a particular shape to at least the depth of a specified thread.

SUMMARY

- Grooving, forming, and cutting-off processes all involve turning a circular indentation or an external collar to a desired shape and dimension. A specifically formed cutting tool is used in each process.
- Grooving, forming, and cutting-off tools are made of various grades of high-speed steel, alloys, and carbides. The nature and size of the workpiece and the production requirements determine the cutting tool material.
- High-speed steel tool bits and cutoff (parting) blades are used for general purposes.
- Cobalt and vanadium high-speed steels, alloys, and carbide blades are practical when (a) a great amount of material must be removed as rapidly as possible, (b) the material is tough to machine, or (c) there is considerable abrasive action.
- A groove:
  - provides a recess for ending a process

  - relieves the material around a shoulder so that a mating part will seat properly
  - improves the appearance of a workpiece
  - provides an angular surface on which a drive belt may ride.
- Grooves are usually square, circular, angular, or a combination of these shapes.
- Grooves are represented and dimensioned on drawings so that the width, depth, location, and shape are specified.
- Grooves may be cut with standard tool bits. These are especially ground to a required square, round, or angular shape, or a combination of these shapes.
- The cutting speed for grooving with formed cutters and for cutting-off operations is usually one half the speed used for regular turning.
- Deep angular grooves are usually turned by feeding the cutting tool with the compound rest set at the required angle.

- A cylindrical form may be turned by:
  (a) using a preformed cutting tool
  (b) automatically feeding a cutting tool by the tracer-template method
  (c) manually feeding a cutting tool. The turned form is checked with a template.
- A thin rectangular-section blade with relieved sides is used for cutting off stock. The blade is set at 90° to the lathe axis, on center, and with no back rake.
- On deep cuts, the cutoff blade is moved by hand a few thousandths to the right and left to prevent binding.
- Cutting fluids are used for grooving, form turning, and cutting-off processes on all materials that should be machined wet.
- Concave and convex radii may be measured for shape and size with a radius gage. Like the square

and V-shoulder, the root diameter may be measured with an outside caliper or a vernier. Combinations of square, angular, and round shapes require checking against a template.
- The length that a form cutting or cutoff tool extends from the holder should not exceed ⅛″ more than the depth of cut.
- Form turning and cutting-off operations should be performed as close as possible to the chucking device and spindle.
- Special precautions must be taken to see that chips do not accumulate in the groove. An accumulation of chips during a deep cutoff operation may cause the blade to bind.
- Grooving, form turning, and cutting-off operations require lower spindle speeds and a finer feed than those used for straight turning.

UNIT 31 REVIEW AND SELF-TEST

1. List four reasons for grooving.
2. Identify two methods of form turning on a lathe.
3. Explain why a *gooseneck toolholder* is used for form turning large, deep contours.
4. State four common cutting-off problems in lathe work.
5. Explain what the drawing notation *CUT TO THREAD DEPTH.* means in relation to a groove.
6. Indicate the main work processes to turn a groove.
7. List the steps to follow in cutting off a workpiece to a specified length.
8. List four safe practices that must be observed in grooving, form turning, and cutting off.

Unit 32 LATHE FILING, POLISHING, AND KNURLING

Filing and polishing are two lathe processes that produce a smooth, polished surface finish. A minimum amount of 0.002″ to 0.003″ is left on the workpiece outside diameter for filing and polishing.

FILING TECHNOLOGY AND PROCESSES

Types Of Files

The single-cut bastard mill file is a general-purpose file for lathe work. A 10″ to 12″ long-angle lathe file is also commonly used. As stated in an earlier unit, this file has sides (width and thickness) that are parallel. It has two uncut (safe) edges which permit finish filing to a shoulder. The long-angle lathe file has features that partially eliminate chatter, provide for rapid cleaning of chips, and reduce the probability of scoring the workpiece. The bastard, second-cut, and smooth-cut mill files produce a fine-grained surface. The two common types of lathe files used on ferrous metals are shown in Fig. 32-1.

Fig. 32-1. Two common types of lathe files. (Courtesy of the Cooper Group)

Three other types of files are used on nonferrous metals. These are the *aluminium file*, the *brass file*, and the *super-shear file*. The first two files are shown in Fig. 32-2.

The teeth of the *aluminium file* are designed to quickly clear chips and to eliminate clogging. The combination of *deep undercut* and *fine overcut* features produces a tool form that tends to break up the file particles. The chips are cleared during the filing process. The tooth form also helps overcome chatter.

Brass, bronze, and other tough, ductile metals are finish filed with a *brass file*. The teeth on this type of file have a *deep, short upcut angle* and a *fine, long-angle overcut*. This design breaks up the chips so that the file clears itself.

Aluminium, copper and most other nonferrous metals may be filed on the lathe with a *super-shear*

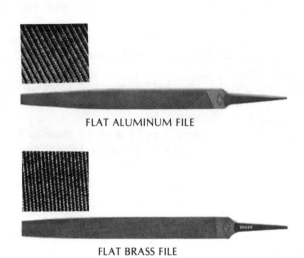

FLAT ALUMINUM FILE

FLAT BRASS FILE

Fig. 32-2. Files for filing aluminum and brass on a lathe.

file. The teeth are cut in an arc (Fig. 32-3). The arc is off-center in relation to the file axis. The milled double-purpose tooth permits free cutting, easy removal of chips from the teeth, and produces a smooth surface finish.

Fig. 32-3. The teeth of a super-sheer file are cut in an arc.

The Filing Process

The cutting is always done on the forward stroke when filing by hand. The strokes are overlapped to about one half the width of the file, the same as for bench filing. The spindle speed used for filing should be about twice as fast as that used for turning. A revolving dead center is preferred for filing and polishing work that is held between centers. A speed of 30 to 40 strokes per minute provides good control of the file in removing the limited amount of material that is left for filing. Filing is also intended to remove light tool marks.

Only a *slight* force should be applied to the file. Excessive force produces an out-of-round surface and tends to clog the teeth and score the workpiece. The file must be cleaned frequently with a file card and brush. The teeth are sometimes *chalked* to prevent clogging and to simplify cleaning. The extreme sharpness of the fine cutting edges of a new file are removed by filing a flat surface of a cast iron block.

POLISHING ON THE LATHE

After completion of the filing process, a smoother polished surface is obtained by using an abrasive cloth. Polishing is done by pressing and moving the abrasive cloth across the revolving workpiece. Aluminum oxide abrasive cloths are used for polishing steels and most ferrous metals. Silicon carbide is widely used for nonferrous metals. Grit sizes of 80 to 100 are generally used. Finer grained and more highly polished surfaces are produced with grit sizes of 200 or finer.

High spindle speeds are required for polishing on the lathe. A still higher gloss finish is usually produced by using a worn abrasive cloth and applying a few drops of oil. During the initial polishing, greater force may be applied by holding the abrasive strip against a file. Again, long, regular strokes are used against the revolving workpiece.

Polishing straps are common when polishing large diameters. The abrasive cloth is held by two half sections of a form *(polishing strap)*. Force is applied on the polishing strap to press it against the surface that is turning.

KNURLS AND KNURLING

A *knurl* is raised area that is *impressed* upon the surface of a cylindrical part. *Knurling* is the process of forcing a pair of hardened rolls to form either a diamond- or straight-line shaped pattern. The knurls are pressed into a slowly revolving workpiece, causing the material to flow (material displacement). The surface patterns produced may be *diamond shaped* or *straight-line*.

There are three general purposes for knurling:

● To provide a positive gripping surface on tools, instruments, or work parts. This permits ease of handling and more precise adjustment

● To raise the surface and increase the diameter to provide a press fit or an irregular surface. The knurl produces a gripping surface. This prevents two mating parts from turning, as in the case of a plastic handle on the shank of a screwdriver

● To add to the appearance of the work.

The *diamond pattern* is formed by overlapping the shape produced by two hardened rolls. Each roll has teeth or ridges. One set of teeth has a right-hand helix (lead). The teeth on the other roll are cut with a left-hand helix. The general patterns and pitches of knurls are shown in Fig. 32-4. One roll forms a series of right-hand ridges; the other forms left-hand ridges. These two series of ridges cross to form the diamond-shape pattern.

The *straight line pattern* is formed by using two hardened steel rolls that have grooves cut parallel to the axis of each roll.

Knurling is a *displacement process*. It requires great force for the knurls to penetrate the workpiece surface and impress the pattern.

Knurl Sizes And Types Of Holders

Knurls are available in three basic sizes, which are identified by their *pitch*. The three sizes and their pitches are: *coarse* (14-pitch), *medium* (21-pitch), and *fine* (33-pitch). The pitch refers to the number of teeth

HELICAL RIDGE KNURL
FORMS DIAMOND PATTERN
(THREE PITCH SIZES)

THREE PITCHES OF STRAIGHT
LINE PATTERN KNURLS

Fig. 32-4. Knurl rolls for diamond and straight-line pattern knurls. (Courtesy of the J.H. Williams Division of TRW Inc.)

per linear inch. The knurl disc is heat treated and rides on a hardened steel pin.

A single set of knurls is usually held on a self-centering-head toolholder (Fig. 32-5A). Some holders are made for a single set and pitch. A multiple-head holder contains three sets of coarse, medium, and fine knurls, as shown in Fig. 32-5B. The knurls are mounted in a revolving self-centering head that pivots on a hardened pin. The multiple heads are also designed to fit a dovetailed quick-change tool post (Fig. 32-5C).

Forming A Knurled Pattern

Knurling is performed by securing the toolholder so that the faces of the knurls are parallel to the work surface. A low spindle speed and a steady flow of cutting fluid are required. At least one third to one half the width of the knurl should be in contact with the workpiece surface and fed inward. The depth of feed depends on the pitch of the knurl and the type of material to be knurled. Fine-pitch knurls may be fed to full depth. A coarse pitch knurl and the knurling of hard materials requires a series of *cuts*. Generally, the knurl is fed to a depth of 0.025″ (0.6 mm) on the first pass.

In cases where the knurled surface is narrower than the width of the knurl, the knurls are fed in slowly to the total depth. Surfaces that are wider than the width of the knurls require the power longitudinal feed to be engaged immediately. The knurls are moved across the required length of the workpiece. The knurls are fed inward *before* the direction of feed is reversed. The process is continued until a reasonably sharp diamond-point or a straight-line knurl pattern is produced.

The following practices must be used to avoid a few problems that are common in knurling:

● The knurls must be kept in continuous contact with the workpiece until the process is completed.

● The automatic feed must be engaged continuously to traverse the full length of the knurled surface. Otherwise, a *ring* is formed, which produces a knurled surface with a varying pattern. This condition is impractical to correct.

● A *double impression* knurled pattern is produced when uneven pressure is applied on both rolls (Fig. 32-6). This condition may be corrected by:

   ○ Positioning the knurl rolls on center

   ○ Tracking the teeth carefully

   ○ Infeeding the knurl rolls again so that the teeth penetrate evenly.

● The knurl teeth must be cleaned *before* the operation and *while the machine is stopped*.

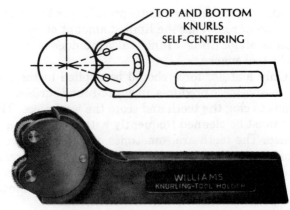

(A) KNURLING TOOL HOLDER FOR SINGLE KNURL SET AND PITCH

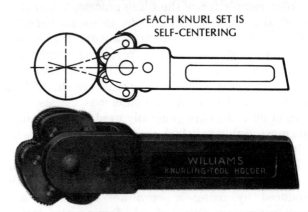

(B) COMBINATION MULTIPLE (REVOLVING) HEAD KNURLING TOOL HOLDER

(C) REVOLVING HEAD WITH SIX KNURLS MOUNTED IN DOVETAIL SLIDE FOR QUICK-CHANGE TOOL POST

Fig. 32-5. Knurl sets and knurl holders for general and quick-change tool post mounting. (Courtesy of the J.H. Williams Division of TRW Inc., and Do All Company)

● The flow of cutting fluid must be adequate to wash any particles away from the work surface and the rolls.

● A follower rest and/or a steady rest may be required. The location of the knurl and the size and shape of the workpiece determine when to use these accessories. The follower rest jaws offset the force applied during the forming of the knurl. The steady rest provides added support of the workpiece along its length.

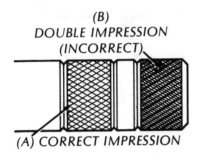

Fig. 32-6. Starting the knurl.

Calculating The Turned Diameter For A Knurled Section (Medium Pitch)

The raising of the crests of the knurled surface increases the outside diameter of the part. Consequently, when the finished outside diameter of the knurled surface must be held to a specified size, the workpiece must be turned to a smaller diameter before knurling. The displacement of metal during knurling then forms the surface to the required outside diameter. The size is calculated for medium-pitch knurls only by using the following two simple formulas. These involve the knurl pitch and the number of knurled *serrations* that are formed around the workpiece.

● Number of Serrations = (medium pitch × π) × (required finished diameter of workpiece minus 0.017″)

● Outside Diameter = (number of serrations) × (0.015″)

Example: Calculate the outside diameter to which should be turned a workpiece that is to be knurled with a medium (21-pitch) diamond knurl. The finished diameter is to be 2″.

Step 1 Number of Serrations = (21) × (3.1416) × (2″ − 0.017″)

= 130.62 (round off to 131)

Step 2 Outside Diameter = (131) × (0.015″)

= 1.965″.

If the workpiece first is turned to 1.965″, the raised outside diameter produced by a 21-pitch knurl will be the required 2″.

REPRESENTATION AND DIMENSIONING OF A KNURLED SURFACE

The width, outside diameter, and location of a knurled surface appear on a drawing or print as dimensions. The pattern (diamond or straight), pitch, the shape of the knurled ends, and the finished outside diameter (when it is to be held to a close tolerance) appear as notes.

Two common examples are provided in Fig. 32-7. The straight knurl in (A) is formed to depth on a 38-mm diameter. The knurled surface is narrower (6 mm) than the width of the knurl. Thus, the knurls are fed directly to depth without moving them longitudinally across the workpiece.

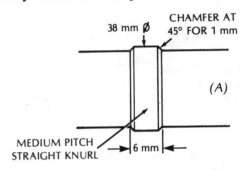

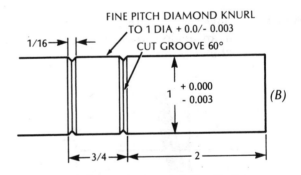

Fig. 32-7. Representing and dimensioning a knurled surface.

The fine pitch diamond knurl in (B) is 3/4″ wide and begins 2″ from one end. The knurl ends in two 60° V-grooves that are 1/16″ wide. The outside diameter must be held to 1″ + 0.000″/− 0.003″. This means the diameter of the section to be knurled first must be turned undersize. The raised knurled surface then increases the diameter to within the required tolerance range.

HOW TO FILE A TURNED SURFACE

Step 1   Select a mill or lathe file of the size and cut suited for the job. Clean the file teeth. Chalk the file face. Check the handle.

Step 2   Set the spindle speed at twice the rpm used for turning.
CAUTION: Move the carriage out of the way. Disengage the lead screw and feed rod *before* starting the lathe.

Step 3   Grasp the file handle in the left hand, as shown in Fig. 32-8. Hold the point of the file with the right hand (to file *left-hand*).

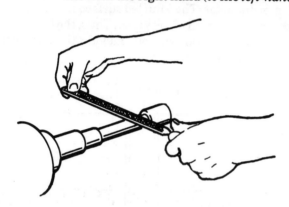

Fig. 32-8. Correct positions of the hands and file for filing left-handed on a lathe.

Step 4   Move the file with a slight pressure and at an angle across the workpiece (Fig. 32-9).

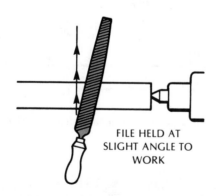

FILE HELD AT
SLIGHT ANGLE TO
WORK

Fig. 32-9. The correct position of the file for filing on a lathe.

Note: Take full strokes. Maintain a file rate of 30 to 40 strokes per minute.

Step 5   Release the pressure on the return stroke and raise the file slightly off the work.

Step 6   Move the file laterally about half of its width so that each stroke overlaps. This action is illustrated in Fig. 32-10. Continue across the entire face in this manner until the surface is filed.

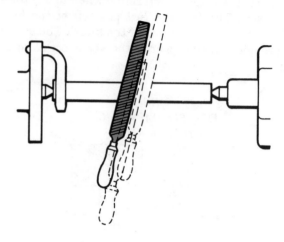

Fig. 32-10. Overlapping file strokes.

CAUTION: Clean the file teeth frequently to remove chips and to prevent scoring.

Step 7   Stop the lathe at intervals and check the diameter of the work.

CAUTION: Check the center to see that it is properly lubricated and that the center hole is not scored.

Step 8   Continue to file. Allow from 0.0005″ to 0.001″ for polishing.
Note: Avoid touching the outside diameter by hand. Otherwise, a *slick surface* will be produced.

HOW TO POLISH ON THE LATHE

Step 1   Set the spindle speed at three to four times the rpm used for turning.
Note: It is assumed that the workpiece has just been filed.

Step 2   Recheck the tailstock center and center hole, and lubricate them.
Note: If possible, use a revolving tailstock center.

Step 3   Select the appropriate abrasive grain and grit size. Tear a strip slightly wider than the width of the file and from 10" to 12" long. Place one end of the abrasive strip around the file point end. The abrasive cutting face must be *toward* the workpiece. Hold the abrasive at the point end with the right hand. Pull the abrasive cloth along the file. Hold it to the file with the left hand (Fig. 32-11).

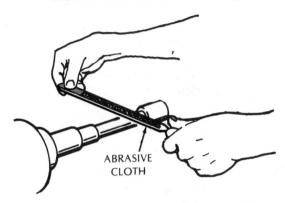

ABRASIVE CLOTH

Fig. 32-11. The position of an abrasive cloth for polishing on the lathe.

Step 4   Start the lathe. Take long, overlapping strokes. Check the quality of surface finish and measure the diameter.
Note: Use successively finer grit sizes to obtain a finer surface finish.

HOW TO KNURL

Step 1   Lay out the boundaries or the position of the knurled section. Check the turned diameter on any knurled surface that must be held to a required finished dimension. Cut grooves or bevel or round the ends as required.

Step 2   Check the condition of the knurls. Clean the teeth. Mount the knurls on center with the faces parallel to the workpiece (Fig. 32-12).

Step 3   Set the spindle speed at about one half the rpm used for turning. Select a carriage feed of from 0.020" to 0.030" (0.5 to 0.8 mm). The amount of feed depends on the pitch of the knurl. Start the lathe.

Step 4   Move the knurling tool until about one half of the face bears on the workpiece. Feed in the tool to a depth of 0.020" to 0.025" (0.5 mm to 0.6 mm).

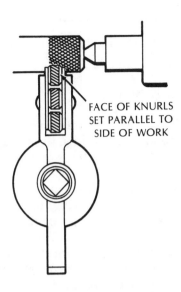

FACE OF KNURLS SET PARALLEL TO SIDE OF WORK

Fig. 32-12. Positioning the knurls and holder.

Step 5   Stop the lathe. Check the correctness of the pattern.
Note: A double impression indicates that the rolls are not centered. One roll is cutting deeper than another. It may be necessary to move the knurls to a new location to correct a double impression.

Step 6   Position the cutting fluid nozzle to flow a small quantity of cutting fluid on the knurls.

CAUTION: Use a follower rest or steady rest when support is required. This prevents springing and damage to the workpiece.

Step 7   Engage the automatic feed. Start the lathe and feed for the full length of the knurled section.

CAUTION: Stop the lathe. Use a stiff brush to remove particles of the workpiece material from the knurls.

Step 8   Check the knurled pattern. Continue to feed the rolls in to depth. Reverse the automatic feed. Take successive passes across the workpiece until a clean, smooth crest pattern is produced.

CAUTION: A damaged knurl pattern results when the feed is stopped and the knurls keep impressing while the workpiece revolves.

HOW TO SUPPORT A WORKPIECE

Using A Steady Rest

Step 1   Clean the lathe ways and the base of the steady rest. Position the steady rest. The location along the workpiece must provide maximum support and permit the greatest possible movement of the carriage and the cutting tools.

Step 2   Tighten the clamping bolt on the steady rest. Adjust one jaw. Bring it lightly into contact with the workpiece.
Note: A true surface or spot may need to be turned in advance on the workpiece. This provides a bearing surface for the jaws of the steady rest.

Step 3   Use a dial test indicator. Position it 180° from one of the lower jaws.

Step 4   Adjust the steady rest lower jaw until the pointer moves a few thousandths from the "0" setting. Then, *back away the jaw* until the pointer retuns to the original "0" setting.

Step 5   Follow the same procedure with the other two steady rest jaws. Position the indicator in relation to the jaw to be set. Secure each jaw.

Step 6   Check to see that the workpiece revolves freely within the three-jaw setting.
Note: A centered workpiece that extends a considerable distance from a chuck or holding device is first positioned with the dead center. The steady rest is then located on the turned section and the jaws are adjusted.

Step 7   Move the carriage and cutting tool to make certain that the position of the steady rest provides clearance for the cutting action.

Step 8   Perform the required operations. When they are completed, open the upper section of the steady rest frame. Swing it away from the workpiece. Remove the workpiece.

Step 9   Clean the lathe and steady rest. Return each accessory to its proper storage section.

Using A Follower Rest

Step 1   Attach the follower rest to the saddle.

Step 2   Turn a small area of the workpiece.

Step 3   Adjust the rear and top jaws of the follower rest so they just contact the workpiece. Apply a lubricant between the jaws and the workpiece.

Step 4   Proceed with the required operations.
CAUTION: The follower rest jaws may need to be adjusted and lubricated for each successive cut.

Step 5   Clean the lathe and follower rest. Replace all equipment and cutting tools.
Note: Long or thin-sectioned workpieces are often supported with a steady rest. A follower rest is used sometimes in combination with a steady rest. This provides additional support against the forces exerted by a cutting or forming tool.

SAFE PRACTICES IN FILING, POLISHING, AND KNURLING

- Be sure the file handle fits correctly *before* beginning the lathe filing. A revolving workpiece tends to force the file *toward* the operator.
- File left-handed by holding the file handle in the left hand. This method provides maximum safety in clearing all revolving parts.
- Chalk the file teeth of a new file to remove excessive sharpness from the tooth edges. Chalking and using a light force prevents scoring and produces a fine, smooth finish.
- Apply force only on the forward stroke. Take long, slow file strokes. Out-of-roundness is produced by taking fast, short strokes.
- File dry. Avoid rubbing your hand over the filed surface. Body moisture and oil produce a difficult surface to file and may cause scratches.
- Chalk the file teeth to easily clean them and to prevent clogging.
- Check the workpiece to see that it turns freely on centers. Periodic lubrication and adjustment may be needed to compensate for expansion. This is caused by the friction-generated heat produced by revolving the workpiece at a high speed.
- Set up and support the workpiece to withstand the forces necessary to impress a knurled

pattern.

- Knurl rolls must be cleaned and set on center before starting the process. The amount of cutting fluid must be adequate to wash away particles and to lubricate the rolls. Lubrication permits the material to flow easier during the knurling process.
- Keep the fingers, wiping cloth, or brush away from the revolving workpiece and the knurl rolls.

- Check the support of the workpiece. A follower and/or steady rest may be needed to prevent bending or distortion of the workpiece during knurling. Considerable force is required to flow the material into a knurled pattern.
- Disengage the lead screw and feed rod when filing and polishing. This precaution prevents the lead screw or feeds from being accidentally engaged at a high spindle speed.

FILING, POLISHING, AND KNURLING TERMS

Filing (lathe)	A finishing process for removing fine tool marks and reducing a diameter a few thousandths of an inch. Reducing a round part to a required size within a specified tolerance.
Polishing (lathe)	Producing a smoothly finished and polished surface. The cutting (abrading) action of abrasive cloth against a revolving workpiece. A final finishing process.
Knurling	The raising of a series of diamond- or straight-line patterned surfaces. Causing a material to flow by exerting force on a revolving workpiece with a set of specially formed rolls.
Diamond knurl pattern	A series of raised areas with a square pyramid shape. Uniformly formed areas around the surface of a workpiece. A shape produced by impressing overlapped right- and left-hand helixes into a revolving cylindrical surface.
Straight-line knurl pattern	A uniform series of grooves of triangular cross section with a slightly flat crest. A series of uniformly shaped surfaces that are impressed parallel to the work axis of a cylindrical workpiece.
Representing (dimensioning) a knurled surface	A graphic description such as a sketch, drawing, or print that provides the craftsperson the accurate information needed to accurately produce a knurl. Describing the features of a knurled surface. Pitch, location, size dimensions, and other information required for grooving or forming the ends of a knurled section.
Displacement of metal (knurling)	The flowing of metal. Impressing a set of formed knurls into a revolving workpiece. Forcing metal to flow and form uniform series of diamond (triangular-shaped) or straight-line raised areas.
Serrations (knurling)	Markings or formed grooves. A uniform shape (pattern) impressed into the outside diameter of a cylindrical workpiece.
Overlapping file cuts	A series of successive file cuts. Moving the file about one half its width over a new area during each stroke. The blending of a newly-filed diameter with the preceding file cut.

SUMMARY

- A turned workpiece is often filed to produce a finely-finished surface without tool marks. From 0.002″ to 0.003″ (0.05 mm to 0.08 mm) is left for both finish filing and polishing.
- Bastard and second- or smooth-cut mill files, long-angled lathe files, or other special metal-cutting files are commonly used for lathe filing of ferrous metals. Aluminum, brass, and super-shear files are used for filing nonferrous metals and other materials. File lengths of 10″ to 12″ (or 250 mm to 300 mm) are common.
- The spindle speeds that should be used for filing average two times those used for turning. A speed of 30 to 40 full-length strokes per minute provides good control of the file.
- A finer surface finish may be produced by using an

abrasive cloth of 80 to 120 grit. Still finer grain surfaces may be produced with 200 and finer grit sizes.

- Extremely high speeds are used for finish polishing. A high-gloss surface is produced by applying a small amount of oil and using a slightly worn abrasive cloth.
- Knurling causes a flow of material on the outside diameter of a revolving workpiece. Diamond and straight-line knurl patterns are general. The common pitches are coarse (14), medium (21), and fine (33).
- Spindle speeds for knurling are about one half to two thirds those required for straight turning.
- Knurling: (a) provides an easier-to-grip surface, for more positive handling and more precise adjustments; (b) adds to the appearance of a part; and (c) enlarges a diameter for a press fit.
- Diamond and straight-line patterns are produced by positioning the axis of the knurl rolls parallel to the work surface. The knurl rolls are impressed into the workpiece.
- The flutes of the diamond or straight-line knurls must be cleaned before and after each successive pass, while the machine is stopped.
- A cutting fluid should be flowed or brushed on the knurls. The fluid flows away chips or other foreign particles.

- A double-impression knurl may be corrected by centering the work roll to cut uniformly or by starting in a new location.
- The knurls must be moved across the work face without stopping. Otherwise, a varying knurl pattern is produced.
- Knurled surfaces are represented on drawings by a combination of dimensions and notes. These specify the pitch and pattern, the length, width, location, and finished diameter, and the nature of the grooves or rounded edges at the ends of the knurled section.
- The outside diameter to which a medium-pitch knurled area should be turned is calculated by first determining the number of serrations that will be formed around the finished diameter and then multiplying this number of serrations by 0.015".
- A follower rest may be used to provide a bearing surface. This withstands the forces produced by knurling and cutting processes.
- Personal and machine safety precautions must be observed. Slower speeds are required for knurling. High speeds are used in filing and polishing. These require constant checking and lubrication of the dead center.
- Particles may be removed by flowing a cutting fluid or by brushing with a stiff-bristle brush.

UNIT 32 REVIEW AND SELF TEST

1. State the functions served by the *deep undercut* and the *fine overcut* features of an *aluminum file.*
2. Indicate the characteristics of a *brass file.*
3. Describe the shape and functions served by *super-shear file teeth.*
4. State what purpose is served by filing a workpiece on a lathe.
5. Indicate the kind of abrasive cloth to use (a) to polish steels and other ferrous metals and (b) to polish nonferrous metals.
6. State three guidelines to follow in polishing a workpiece on a lathe.
7. Give two reasons for knurling.
8. Specify the basic *knurl sizes and patterns.*
9. Identify three common knurling problems and their causes.
10. Tell what dimensions appear on a drawing to provide the specifications of a knurled area.
11. Differentiate between the use of a *steady rest* and a *follower rest* to support a workpiece.
12. List three safety precautions to observe when filing on a lathe.
13. Explain why the *lead screw* and *feed rod* on a lathe must be disengaged when filing or polishing.

Unit 33 TAPER AND ANGLE TURNING: TECHNOLOGY AND PROCESSES

A *taper* is a cone-shaped form. A *taper* increases or decreases uniformly in diameter along the length of a workpiece. *Taper turning* is the process of machining this uniform change in diameter. A taper may be internal or external.

A steep taper is often referred to in machine shops as either an *outside* or *inside angle*. Internal angles are *bored*. External angles are *turned*. A steep taper angle may be set and the cutting tool may be fed with the compound rest.

This unit deals with different types of external tapers and angles, how tapers are calculated and represented on drawings, taper- and angle-turning processes, and methods of checking these for dimensional accuracy. Machine tool and personal safety precautions are also covered.

PURPOSES AND TYPES OF TAPERS AND TAPER SYSTEMS

The taper is a commonly used design feature in machine and metal products industries. Cutting tools like twist drills, machine reamers, and end mills have taper shanks. The spindles of lathes, milling machines, boring mills, and other machine tools, have internal and external tapers. Many mating parts, such as sleeves, sockets, centers, and spindles, have tapers that correspond with the tapers of the parts into which they fit.

Tapers serve four main purposes:

● A taper provides a method of accurately locating two mating parts concentrically about a common axis.

● *Self-holding* tapers are designed to securely hold together two mating parts.

● *Self-releasing* tapers accurately align parts in relation to an axis. However, they require some other device to securely hold them together.

● A taper provides for accurate positioning and holding of a cutting tool, holder, spindle, or other part. These revolve as a single unit around a particular axis.

Tapers permit interchangeability, accuracy in centering, and an efficient way of quickly changing a tool or machine setup. The specifications of standar-

dized tapers are given in trade and engineering handbooks and in manufacturers' technical manuals. There are many series of standard tapers. These meet the extensive design needs within and among different industries such as machine, metal manufacturing, aerospace, maritime, and a host of others.

The six standard taper systems that are widely used in the machine and metal industries are:

● The American Standard (Morse) Taper
● The Brown and Sharpe Taper
● The Jarno Taper
● The American Standard Self-Releasing Steep Taper
● The American Standard Taper Pin
● The Jacobs Taper.

American Standard Taper Pin Series

American Standard Taper Pins are widely used when two mating parts must be accurately positioned and held in a fixed position. The taper pin permits easy assembling and disassembling. Taper pins are furnished in standard lengths in relation to the taper size. The taper pins have a standard *taper per foot* of either 0.250″ or 0.0208″ per inch.

The general sizes are numbered from 7/0 to 2/0 and 0 to 11. Taper pin holes are usually reamed. The reamer sizes are designed so that one size overlaps the next succeeding one. The diameters at the small end of the taper range from approximately 1/16″ (for the 7/0 size) to 41/64″ (for the #10 size). American Standard Taper Pin tables give the large and small diameter, length of each pin, and the drill size. Table 33-1 shows the manner in which this information is presented. The complete table is included in Appendix Table A-35.

The Jacobs Taper

The *Jacobs Taper* is often used with a standard Morse or Brown and Sharpe taper which fits standard machine spindles. The Jacobs is a short, self-holding taper. This corresponds with a short taper bore. These bored holes are used on drill chucks or external tapers of a shaft, shank, or spindle. The Jacobs taper is widely applied on portable power-driven tools.

Table 33-1. SELECTED STANDARD TAPER PINS
(Diameters and Drill Sizes)

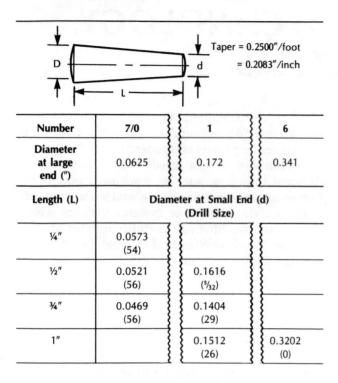

Taper = 0.2500"/foot
= 0.2083"/inch

Number	7/0	1	6
Diameter at large end (")	0.0625	0.172	0.341
Length (L)	Diameter at Small End (d) (Drill Size)		
¼"	0.0573 (54)		
½"	0.0521 (56)	0.1616 (⁵/₃₂)	
¾"	0.0469 (56)	0.1404 (29)	
1"		0.1512 (26)	0.3202 (0)

American Standard (Morse) Taper

The *American Standard (Morse) Taper* has a small taper angle. This produces a wedging action and makes the Morse a *self-holding* taper. As stated before, the Morse is the most extensively used taper in the machine industry. It is used on common cutting tools features such as the shanks of twist drills, reamers, counterbores, and countersinks, and on machine tool features such as drill press spindles and lathe spindles.

The Morse taper series is numbered from #0 to #7. The small-end diameters of this series range from 0.252" to 2.750". The lengths vary from 2" to 10". Unlike the other taper series, the amount of taper per foot varies in the Morse series with each number except #0, #4½, and #7. The approximate taper per foot is 5/8".

Table 33-2 lists the kind of information that is required for designing and machining an American Standard, or Morse, taper.

Brown And Sharpe Taper

The *Brown and Sharpe (B & S) Taper* is another self-holding taper. The B & S taper series is numbered from #1 to #18. Within this series the diameters at the small end of the taper range from 0.200" to 3.000". The B & S taper is 0.502" per foot (0.0418" per inch) for #1, #2, #3 and #13. The taper

Table 33-2. SAMPLE DIMENSIONS: AMERICAN STANDARD (MORSE) TAPER

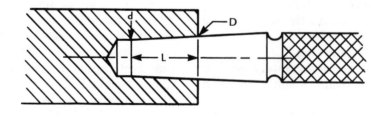

Number of Taper	Diameter at Small End (d)	Standard Plug Length (L)	Diameter at Gage Point (D)	Tpf	Tpi
#1	0.369"	2"	0.475"	0.5986"	0.0499"
#3	0.778	3³/₁₆	0.938	0.6024	0.0502
#7	2.750	10	3.270	0.6240	0.0520

varies from 0.5024″ to 0.4997″ for numbered tapers #4 through #14. The taper for numbers 14 through 18 is 0.5000″.

Jarno Taper

The *Jarno Taper* series (#1 to #20) uses a uniform taper per foot of 0.600″. The number of a Jarno taper is directly related to the small diameter, the large diameter, and the length. The small-end diameters of the series range from 0.100″ to 2.000″. The following simple formulas may be used to calculate the sizes:

$$\text{Diameter at Small End} = \frac{\text{\# of Jarno Taper}}{10}$$

$$\text{Diameter at Large End} = \frac{\text{\# of Jarno Taper}}{8}$$

$$\text{Length of Jarno Taper} = \frac{\text{\# of Jarno Taper}}{2}$$

For example, a #6 Jarno taper is 6/10″ (0.600″) at the small end of the taper, 6/8″ (0.750″) at the large end, and is 6/2″ (or 3″) long.

American Standard Self-Releasing (Steep) Taper

The *American Standard Self-Releasing Steep Taper* series is similar to the earlier *Milling Machine Taper* series. The taper of 3½″ per foot ensures easier release of arbors, adapters and similar accessories from the spindles of milling and other machines. Because the steep tapers are not self-holding, they require slots or keys to drive an accessory (Fig. 33-1A). The mating parts are usually drawn and held together with a draw bolt, cam-locking device, or nut. The lathe type-L spindle nose (Fig. 33-1 B) has a steep taper, a key drive, and a threaded ring nut. These features accurately position, hold, and drive a faceplate, driver plate, or chuck that is mounted on the spindle nose.

The short steep taper (3″ per foot) of the type D-1 lathe spindle nose (Fig. 33-1C) centers the arbor or accessory. The cam-lock device holds the arbor or accessory accurately and firmly on the spindle nose.

TAPER DEFINITIONS AND CALCULATIONS

The setting up of a lathe to cut a taper requires that the taper per foot or taper per inch be known. The taper may then be turned either by setting the taper attachment, by offsetting the tailstock, or by determining the angle and then positioning the compound rest. The short taper or angle may then be turned by feeding with the compound rest handwheel.

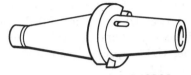

(A) MILLING MACHINE ARBOR

(B) LONG TAPER, TYPE L   *(C) STEEP SHORT TAPER, TYPE D-1*
(TAPER 3½″ PER FOOT)   *(TAPER 3″ PER FOOT)*

Fig. 33-1. Examples of steep, self-releasing tapers.

The general terms and simple formulas for calculating the various features of a taper are presented here.

Taper Formula Terms

- T designates *taper*. This is the difference in size between the large diameter and the small diameter.
- T_{pi} is the *taper per inch*. This is the amount the workpiece diameter changes over a 1″ length.
- T_{pf} is the *taper per foot*. It indicates the change in diameter in a one-foot length.
- D refers to the diameter at the *large* end.
- d denotes the diameter of the taper at the small end.
- L_t is the *length* of the taper in inches.
- L_o is the *overall length* of the workpiece.

Calculating Taper Values

The taper (T) equals the difference in the large (D) and small (d) diameters:
$$T = D - d.$$
If a tapered part has a large diameter of 1.500″ and a small diameter of 1.250″, the taper is 0.250″:

$$T = D - d$$
$$T = 1.500 - 1.250''$$
$$= 0.250''.$$

The taper per foot (T_{pf}) is equal to the difference between the large (D) and small (d) diameters in inches multiplied by 12 and divided by the required length of taper (L_t). Expressed as a formula:

$$T_{pf} = \frac{(D - d) \times 12}{L_t}$$

Also,

$$T_{pf} = T_{pi} \times 12.$$

Example: Calculate the T_{pf} from the dimensions given in Fig. 33-2.

$$T_{pf} = \frac{(D - d) \times 12}{L_t}$$
$$= \frac{(1.500 - 1.250) \times 12}{6}$$
$$= \frac{0.250 \times 12}{6}$$
$$= 0.500''.$$

The taper per inch (T_{pi}) equals the difference between the large diameter (D) and small diameter (d) divided by the length of the taper in inches (L_t).

$$T_{pi} = \frac{(D - d)}{L_t}$$

or,

$$T_{pi} = \frac{T_{pf}}{12}.$$

In the example,

$$T_{pi} = \frac{(D - d)}{L_t}$$
$$= \frac{(1.500'' - 1.250'')}{6''}$$
$$= 0.0417''.$$

When the taper per inch (T_{pi}) is known, the taper per foot (T_{pf}) is found by simply multiplying by 12:

$$T_{pf} = T_{pi} \times 12.$$

In the example in Fig. 33-2,

$$T_{pf} = T_{pi} \times 12.$$
$$= 0.0417'' \times 12$$
$$= 0.500''.$$

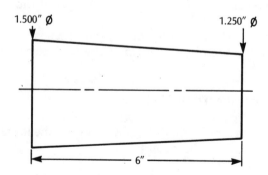

Fig. 33-2. A tapered workpiece.

Calculating The Diameters

The large diameter (D) may be calculated by multiplying the taper per inch (T_{pi}) by the length of the taper in inches (L_t) and adding the small diameter (d):

$$D = (T_{pi} \times L_t) + d,$$

or if the taper per foot (T_{pf}) is given:

$$D = \frac{(T_{pf} \times L_t)}{12} + d.$$

Example: Calculate the large diameter (D) of the tapered workpiece using the dimensions given in Fig. 33-3.

$$D = \frac{(T_{pf} \times L_t)}{12} + d$$
$$= \frac{(0.600 \times 4.000'')}{12} + 1.250''$$
$$= 1.450''.$$

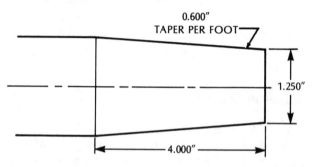

Fig. 33-3. A taper problem.

The small diameter (d) may be calculated by using the formula:

$$d = D - (L_t \times T_{pi}),$$

or if the taper per foot (T_{pf}) is given:

$$d = D - \frac{(L_t \times T_{pf})}{12}.$$

Example: Given the large diameter (D) of 1.450'' in the previous example instead of the small diameter (d),

$$d = D - \frac{(L_t \times T_{pf})}{12}$$
$$= 1.450'' - \frac{(4.000'' \times 0.600)}{12}$$
$$= 1.250''.$$

Calculating The Tailstock Offset

The offset tailstock method is used to cut external shallow tapers. The amount to offset a tailstock depends on the overall length of the workpiece (L_o) and the amount of taper (T_{pi} or T_{pf}). Simple formulas are used to compute the required tailstock offset (T_o).

● When the taper is given as Taper Per Inch (T_{pi}),

$$(T_o) = \frac{(T_{pi}) \times (L_o)}{2}$$

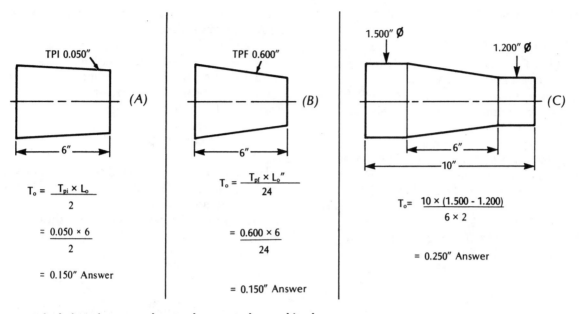

$$T_o = \frac{T_{pi} \times L_o}{2}$$

$$= \frac{0.050 \times 6}{2}$$

$$= 0.150'' \text{ Answer}$$

$$T_o = \frac{T_{pf} \times L_o''}{24}$$

$$= \frac{0.600 \times 6}{24}$$

$$= 0.150'' \text{ Answer}$$

$$T_o = \frac{10 \times (1.500 - 1.200)}{6 \times 2}$$

$$= 0.250'' \text{ Answer}$$

Fig. 33-4. Calculations for tapered parts that are to be machined.

• When the taper is expressed as Taper Per Foot (T_{pf}),

$$(T_o) = \frac{(T_{pf}) \times (L_o)}{24}$$

• When a taper is dimensioned with a small diameter (d), large diameter (D), length of taper (L_t), and the overall length (L_o''),

$$(T_o) = \frac{(L_o \times (D - d)}{(L_t) \times 2}.$$

Examples: See Figs. 33-4 A, B, and C for examples of how to use the three formulas to calculate the tailstock offset required to turn three differently dimensioned tapered parts.

REPRESENTING AND DIMENSIONING TAPERS AND ANGLES ON CYLINDRICAL PARTS

Tapers On Cylindrical Parts

The taper on a round workpiece is represented by two symmetrical tapering lines. The amount of taper is usually specified in terms of TAPER PER FOOT or TAPER PER INCH. For example, a standard taper pin that has a taper of one-quarter inch per foot is indicated by the note: 0.250″ (or 1/4″) TAPER PER FOOT.

A drawing or sketch may include the diameter at either the small or large end of the taper, or at both. The length of the taper may be given or computed.

Figs. 33-5A and B show how tapered parts are represented and dimensioned on drawings. The diameters and length are given as dimensions. The taper per foot or taper per inch appears as a note.

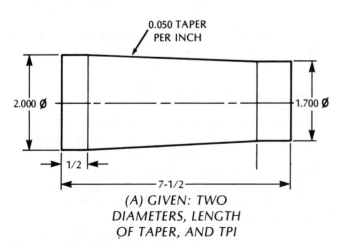

(A) GIVEN: TWO DIAMETERS, LENGTH OF TAPER, AND TPI

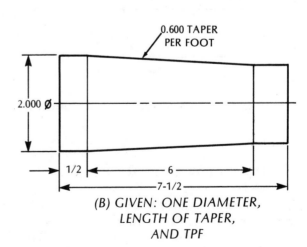

(B) GIVEN: ONE DIAMETER, LENGTH OF TAPER, AND TPF

Fig. 33-5. The dimensioning of tapers.

Angles On Cylindrical Parts

An angular surface of a turned part is represented by the angle of the adjacent sides. The angle is generally expressed in degrees and minutes. The dimension may show the total (included) angle between the two angular surfaces in relation to a center (reference) line.

The two drawings in Fig. 33-6 show how an angle may be represented and dimensioned. The centerline shows that the angle is symmetrical. The 60° ± 15′ in Fig. 33-6A indicates that the machined angle of 60° must be held to within ± 15′. The limits, therefore, are from 59° 45′ to 60° 15′.

The dimension in Fig. 33-6B gives the number of degrees of the angular surface in relation to the centerline (30°) and the required tolerance (± 5′).

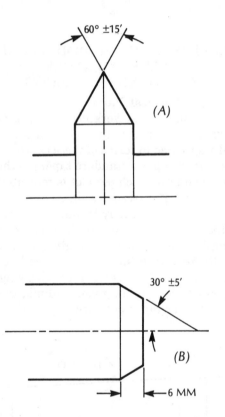

Fig. 33-6. Representing and dimensioning a turned angle.

MEASURING AND GAGING TAPERS

Tapered surfaces must be precision machined for two mating parts to fit and align accurately and to provide maximum holding ability for small-angle tapers. There are four common methods of measuring the accuracy of the taper angle and/or the size of the large and small diameters.

- Gaging with a taper plug or ring gage
- Measuring with a standard micrometer
- Measuring with a taper micrometer
- Testing with a sine bar setup.

Each method is described briefly.

Taper Plug Gage Method

The taper plug gage (Fig. 33-7A) is hardened and precision ground. The face and a ground area are identified as steps (A) and (B). When the small end of a correctly turned taper falls between these steps, the part is turned to size within the allowable taper limits.

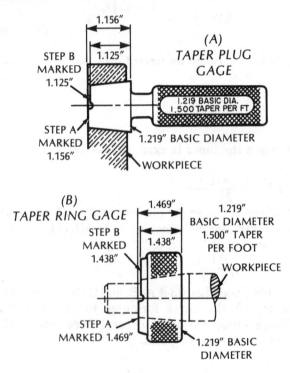

Fig. 33-7. Features of taper plug and ring types of go-not-go gages (step style).

Taper Ring Gage Method

A hardened *taper ring gage* (Fig. 33-7B) has a precision-ground tapered hole. Part of the outside of the body is cut away. One or more index lines are marked on the flat section. When the taper is correct and the end of the tapered part cuts an index line, it indicates that the outside diameters are accurately machined to size.

The taper must first be checked for accuracy. Usually, a fine coating of Prussian blue is applied to the machined part. The gage is then placed carefully on the taper. If the machined part is turned slightly and then removed, the high areas on the machined tapered surface are identified easily. Once the taper fits accurately, the diameters may then be checked by gage or micrometer measurement.

A few simple precautions must be taken. The mating part must be clean and free of burrs and nicks. Since a small-angle taper is self-locking when a small force is applied, the workpiece should be turned *clockwise* carefully. A counterclockwise withdrawing motion makes it easy to turn and remove the tapered plug or ring gage. Excessive turning causes undue wear on the gage

Standard Micrometer Method

When a taper ring gage is not available, the standard outside micrometer is often used to measure the accuracy of the taper and the diameter. This method is shown in Fig. 33-8.

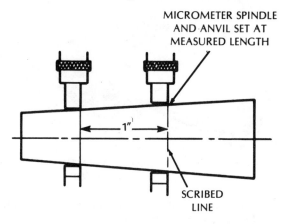

MICROMETER SPINDLE AND ANVIL SET AT MEASURED LENGTH

1″

SCRIBED LINE

Fig. 33-8. Measuring the diameter of a taper with a standard micrometer.

The turned taper is given a coating of a colored dye. Lines are scribed 1″ apart. The diameters at these reference points are measured with a micrometer. The difference between the two is then compared with the required taper per inch. It is often necessary to convert taper per foot to taper per inch and to multiply this value by the number of inches between the two scribed lines.

Any oversize variation may be corrected by adjusting the taper attachment or tailstock offset. Another light cut is then taken to depth. If the variation is around. 0.002″, the tapered surface may be finish filed. Considerable skill is required to machine and file to precision measurements by this method.

Taper Micrometer

The *taper micrometer* provides a more reliable and accurate measuring instrument and method than does the standard micrometer. The taper micrometer includes an adjustable anvil and a 1″ sine bar that is attached to the frame. The sine bar is adjusted by the movement of the spindle.

The accuracy of the taper is measured by placing the taper micrometer over the workpiece. The thimble is adjusted until the jaws just touch the tapered surface. The reading on the spindle indicates the taper per inch. This reading may need to be converted to taper per foot or angle of taper. Tables or simple formulas may be used to obtain these values.

Sine Bar Method

The *sine bar* (Fig. 33-9) is a precision tool. It is used to check and measure angles to a high degree of accuracy. A sine bar consists of a steel bar to which two *plugs* of the same diameter may be mounted. The bar and plugs are hardened and machined to close tolerances. The center distance between the plugs is held accurately. Center distances of 5″, 10″, or 20″ are used for ease in making calculations. The edge or edges (depending on the design) of the sine bar are parallel with the reference axis of both plugs.

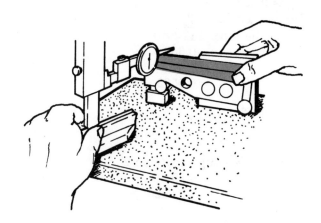

Fig. 33-9. Using the sine bar method of accurately measuring a taper.

The sine bar requires the use of handbook tables and trigonometry. The *sine value* must be calculated for the angle of the tapered surface. The sine bar is used with gage blocks, a dial indicator, and a flat reference surface like a surface plate.

If the included angle of the taper is not given, it must be calculated by one of two formulas. When the taper per foot is specified, the tangent of the included taper angle is equal to the taper per foot divided by 12.

When the end diameters and length of taper are given, the tangent of the included angle of the taper is equal to the large diameter minus the small diameter, divided by the length of taper.

The gage blocks that are required to position the sine bar so that any variation in taper may be read on a dial indicator may be calculated. This setup is shown in Fig. 33-9. The gage block height is equal to the sine of the angle multiplied by the center distance of the sine bar plugs (i.e., 5", 10", 20", etc.).

The taper angle is measured by placing the workpiece axis parallel to the axis of the preset sine bar. The taper angle is correct when there is no movement of the dial indicator point across the length of the workpiece.

The sine bar method is not as widely used as other simpler methods. These require less skill to set up and to measure the taper angle. Angle gage blocks are simple to use. They are more accurate than the sine bar method. With a set of sixteen angle gage blocks, over 350 thousand angles in increments of one second may be set to an accuracy within millionths of a second.

Example: Determine the gage block build up required to measure the accuracy of a No. 10 American Standard Taper Plug. The taper per foot of the No. 4 taper is 0.602".

$$\text{Tangent angle of the taper} = \frac{T_{pf}}{12}$$
$$= \frac{0.602}{12}$$
$$= 0.0502$$

$$\text{Tangent of the angle} = 0.0502$$
$$= 2° 52'$$

$$\text{Gage Block setup} = 10 \times \text{sine } 2° 52'$$
$$= 10 \times 0.0502$$
$$= 0.5020".$$

TAPER TURNING PROCESSES

Taper Attachment

The taper attachment provides a quick, economical, practical way of accurately turning internal and external tapers. In principle, the taper attachment guides the position of the cutting tool in an angular relation to the lathe center axis. A sliding block moves along the guide bar of the taper attachment. The angular setting of the guide bar is transmitted by the sliding block to produce a similar movement of the cutting tool. There are two basic types of taper attachments:

- Plain taper attachment
- Telescopic taper attachment.

Plain Taper Attachment Features—The plain taper attachment requires the removal of the binding screw that connects the crossfeed screw to the cross slide. The cross slide is then moved by securing the sliding block to the extension arm on the cross slide. The compound rest feed handwheel is turned to set the depth of cut with the plain taper attachment.

Telescopic Taper Attachment—The cross slide and crossfeed screw and nut are not disengaged with the *telescopic taper attachment* (Fig. 33-10). The depth of cut may be set directly with the crossfeed handwheel.

When a tapered workpiece must be duplicated, the plain and telescopic taper attachment may be set accurately by using a dial indicator. The guide bar is adjusted until there is no dial movement when the indicator point is moved along the workpiece.

MAXIMUM TAPER
4"/FT. (100 mm/300 mm)

Fig. 33-10. A permanently-mounted telescopic taper attachment. (Courtesy of South Bend Lathe, Inc.)

Advantages Of Turning A Taper With A Taper Attachment

Tapers may be turned accurately, economically, and with less danger of work (center) spoilage by using a taper attachment. There are also many other advantages of turning a taper by this method.

- Alignment of the live and dead centers, required by other methods, is eliminated.
- The workpiece is supported by the full angular bearing surface at the center holes.
- Tapers may be cut on work held in a chuck, between centers, or on a faceplate, fixture, or other work-holding setup.
- A greater range of inch- and metric-standard taper sizes may be turned.

• Internal and external tapers may be cut with the same setting of the taper attachment. This one setup permits machining the matching parts accurately.

• The taper attachment may be positioned directly at the required taper or in degrees of taper. The base of the taper attachment has a scale graduated in taper per foot, millimeters of taper, and degrees of taper. Calculation of these dimensions is thus eliminated.

• The angle of taper is not affected by any variation in the length of the workpiece. The one taper setting may be used to produce multiple parts with the same taper, regardless of any variation in the length of the workpiece.

ANGLE TURNING WITH THE COMPOUND REST

Description Of The Compound Rest

The carriage of the engine lathe has a secondary slide mounted on the cross slide. The secondary slide is called a *compound rest*. The compound rest may be swiveled on its base. This permits a cutting tool to be set and fed at a required angle.

The compound rest, as generally used in the machine shop, includes three major parts. These are shown in Fig. 33-11.

• The compound rest *top slide* contains a T-slot for holding a tool post or multiple-head toolholder. Movement of the cutting tool is controlled by a handwheel on the slide.

• The compound rest *swivel slide* is graduated in degrees. This makes it possible to set the cutting tool at a required angle. The swivel slide is clamped to the cross slide.

• The *lower, or cross, slide* provides for transverse movement of the compound rest. This means that the cross slide moves at right angles to the ways of the lathe.

The feed screw of the top slide has a graduated collar. This permits micrometer adjustment for either inch-standard or metric dimensions. The compound rest is hand-fed. The feed is independent from the power crossfeed.

Graduations On The Compound Rest Swivel Slide

While all compound rest swivel slides are graduated in degrees, the arrangement is not standardized. Fig. 33-12 shows a graduated swivel slide marked from 90° to 0° to 90°. Other lathe manufacturers graduate from 0° to 90° to 0°. Two zero (0) index lines on the cross slide provide reference points.

Angles are usually stated as an *included angle* or as an *angle with the centerline*. This last designation is one half the amount of the included angle.

Compound Rest Setup For Angle Turning

The angle at which the compound rest is set usually corresponds to the angle with a centerline. Thus, the angle produced is the included angle. The degree reading on the swivel slide depends on the axis from which the compound rest is set and the arrangements of the numbered graduations.

Short, steep angles are usually turned by swiveling the compound rest. Three common setups are illustrated in Fig. 33-12. If an included angle is shown on a drawing, the compound rest is positioned at one half of this required angle. To turn the 60° center shown in Fig. 33-12A, the compound is swiveled to 30°. The 45° angle in Fig. 33-12B may be produced by swiveling the compound rest 45° to the transverse axis. The two sides of the 70° angle in Fig. 33-12C are turned by swiveling the compound rest to

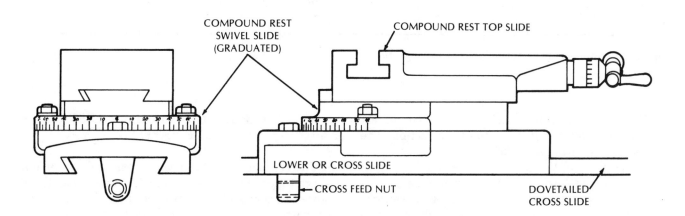

Fig. 33-11. Three major parts of a lathe compound rest.

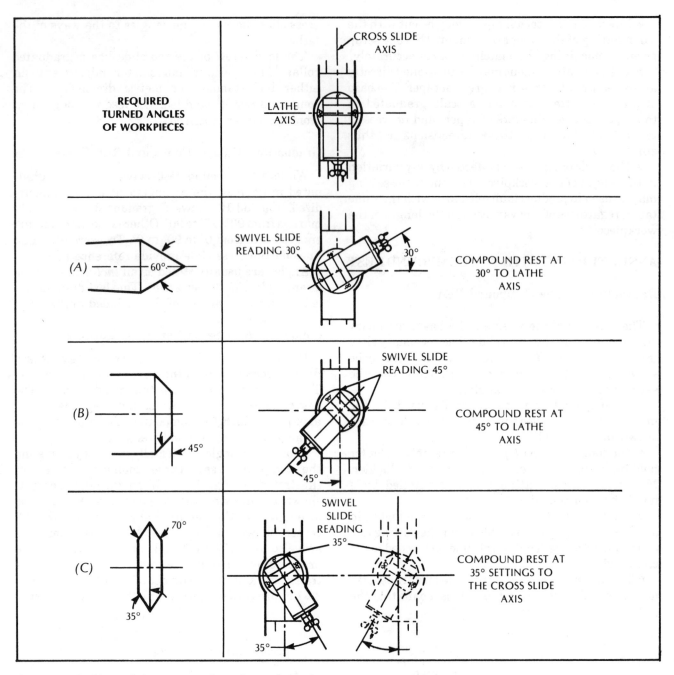

Fig. 33-12. Positions of the compound rest for angle turning.

a 35° angle and turning one side. The compound rest is then swiveled 35° in the opposite direction. The turned included angle is 70°.

The position of the compound rest for feeding in relation to the lathe center or cross slide axis is influenced by safety and convenience considerations. Feeding with the compound rest handwheel from a position in back of the workpiece is not as safe or as convenient as working in front of the workpiece.

COMPUTING COMPOUND REST ANGLE SETTING

The compound rest is used occasionally to turn a short taper. The dimension of the taper usually is stated in terms of taper per foot (T_{pf}). This must be converted to degrees and minutes, to correspond with the graduations on the swivel slide. The angle may be computed by one of two methods:

Converting T_{pf} To Angle With Centerline Setting

Calculate the angle to which the compound rest must be set in relation to the centerline by using the following formula:

Compound Rest Angle $= T_{pf} \times 2.383$.

Example: Calculate the compound rest angle in relation to the centerline for turning a short taper of 1″ T_{pf}.

$$\begin{aligned} \text{Compound Rest Angle} &= T_{pf} \times 2.383 \\ &= 1'' \times 2.383 \\ &= 2.383°. \end{aligned}$$

Note: The graduations on the compound rest are in divisions of one degree. In this case, the worker must estimate 3/8 of one degree (0.383°). After the compound rest is set, a trial cut is taken. The T_{pf} is measured. Further adjustment of the compound rest then is made if needed.

Tangent Of The Angle With The Centerline

Use the following formula to calculate the tangent (tan) of the angle to which the compound rest must be set in relation to the centerline:

$$\text{Tan} = \frac{T_{pf}}{24}$$

Example: Using a T_{pf} of 1″, the tangent is equal to 1/24 or 0.04167. A *table of natural tangent values* shows that 0.04167 represents an angle of 2° 23′.

Note: Since the swivel slide graduations are in degrees, the 23′ must be estimated. Consequently, after a trial cut the taper must be checked to determine if the compound rest setting requires further adjustment.

HOW TO TURN AND MEASURE A TAPER

Turning With A Taper Attachment

Setting The Telescopic Taper Attachment

Step 1 Mount the workpiece. Set the cutting tool on center.

Step 2 Check the guide bar and sliding block on the taper attachment for *free play*. Adjust the gibs as needed.

Step 3 Clean and lubricate all sliding and mating surfaces.

Step 4 Adjust the guide bar. Read the taper per foot (or taper angle in degrees) on the graduated base plate.

Step 5 Tighten the lock screws. These hold the guide bar to the base plate.

Step 6 Center the base plate with the cross slide. Position the taper attachment.
Note: The cutting tool must be able to traverse the workpiece the length of taper that is to be turned.

Step 7 Secure the taper attachment in position. Tighten the clamping bracket to the lathe bed.
Note: Steps 1 through 7 are used for setting either the telescopic or the plain taper attachment.

Setting The Plain Taper Attachment

Step 1 Set the compound rest at zero. This aligns it with the cross slide axis.

Step 2 Remove the binding screw. This disconnects the cross slide and the crossfeed screw nut.

Step 3 Tighten the binding screw. This secures the slide block and cross slide extension arm.

CAUTION: The binding screw hole on the cross slide must be covered. This prevents chips or dirt from reaching the crossfeed screw.

Step 4 Take out the end play in the taper attachment setup. Move the carriage clear of the workpiece after each cut. Then, move it back to feed longitudinally into the work at the beginning of each new cut.

Step 5 Set the cutting tool for a light trial cut for about 1/8″.

Step 6 Measure the small diameter. Set the depth for a roughing cut. Use the same cutting speed and feed as for regular turning. Take a tapered cut for approximately 1¼″.

Step 7 Check the accuracy of the taper and the diameter. The taper usually is measured with a standard micrometer.

CAUTION: The workpiece must be clean and burr free. This is a necessary step before gaging or measuring the tapered surface.

Note: Readjust the taper attachment if necessary. Take a very light cut and recheck the taper.

Step 8 Finish turn the taper to size. The usual 0.002″ to 0.003″ should be left for fine finish filing and polishing.

Taper Turning By The Offset Tailstock Method

Step 1   Determine the required tailstock offset. Check the centers for alignment.

Step 2   Position a dial indicator pointer so that it touches the side of the tailstock spindle. Set the indicator at zero.

Step 3   Loosen the tailstock binding nut. Turn the screw to move the head on the base. *Note:* The tailstock setting is sometimes positioned by the direct reading. The amount of offset is determined from the graduations on the base and the index line on the head of the tailstock.

Step 4   Continue to move the tailstock head to the required offset. The amount is represented by the number recorded on the dial indicator.
     CAUTION: The range of movement of the dial indicator pointer must be within the limits of the instrument.

Step 5   Tighten the binding nut. Position and secure the cutting tool on center.

Step 6   Check the centers on the workpiece. Lubricate the dead center hole. Test the setup to see that the workpiece rotates freely.

Step 7   Take a trial cut. Check the large and small diameters against the required taper.
     Note: This step is a particularly important check because the length of the workpiece affects the taper. Adjust the tailstock with a dial indicator if further adjustment is required.

Step 8   Take a roughing cut. Test the taper for accuracy. Use a gage, micrometer, or other method for checking the taper and the accuracies of the small and large diameters.

Step 9   Take a finish cut. Allow 0.002″ to 0.003″ on a workpiece that ranges up to 1¼″ diameter. The part may then be brought to size by finish filing and polishing.

HOW TO TURN AN ANGLE FOR A SHORT TAPER USING THE COMPOUND REST

Step 1   Determine the required angle for setting the compound rest.

Step 2   Loosen the lock nuts on the compound rest.

Step 3   Swivel the compound rest to the required angle. Lock the compound rest in this position.
     Note: The compound rest is set at one half of the included angle.

     CAUTION: Position the compound rest so that the cutting tool may be fed toward the headstock.

Step 4   Set the cutting tool on center. Start the lathe. Move the carriage longitudinally so that the cutting tool is positioned to take a trial cut.

Step 5   Hand-feed the tool bit by turning the compound rest feed screw.

Step 6   Check the turned angle with a plug or angle gage, a protractor, or a template.

Step 7   Take a roughing cut. Check the angle for dimensional size and accuracy.

Step 8   Follow with other roughing cuts as required and a final finish cut.
     Note: The angular surface may be filed and polished to meet the required surface finish.

When two angles are to be cut, as for a bevel gear:

Step 9   Reset the compound rest to turn the complimentary angle. Repeat the steps for rough and finished turning and angle checking (Steps 1-8).

A short taper may be turned with a compound rest. The compound rest is set at the computed angle to the centerline. The angle corresponds to the taper per foot or metric equivalent. Steps 2 through 8 are then followed.

SAFE PRACTICES IN TURNING TAPERS AND ANGLES

- Check the size of the bearing surface of the center holes. They must be large enough to permit the workpiece to be moved out of alignment when using the offset tailstock method.
- Lubricate the dead center hole. Carefully adjust the workpiece so that it may turn freely.
- Test the lathe dog to see that the tail is free to ride in the driver plate slot.
- Move the cutting tool away from the workpiece and then position it for the next cut. This movement takes up the lost motion of the taper attachment and setup.

- Turn the tapered workpiece clockwise carefully when testing for accuracy. To release the tapered surfaces, reverse the turning direction (counterclockwise) and gently pull outward. This action helps release the tapered surfaces without damaging them.
- Use a plastic or soft metal bar to gently tap and dislodge a self-holding taper.
- Remove any burrs and clean the work surface of a taper *before* testing.
- Use a soft metal collar around a finished surface when a part requires additional machining. This prevents scoring the workpiece.

TERMS USED IN TAPER AND ANGLE TURNING

Taper	The uniform increase or decrease of a diameter.
Outside or inside angle	A steep turned taper along an outside or inside angle.
Self-holding taper	A slightly tapered part. Two mating tapered parts that are held securely together by the tapered surfaces.
Self-releasing taper	A steeply tapered part that requires a clamping device to hold it securely on a mating taper.
Standard taper systems	Different series of tapers that are nationally or internationally accepted as a specific standard. The tapers within a series extend over a wide range of diameters.
Taper per foot (T_{pf})	The amount a large diameter varies from a small diameter for each 12″ length.
Large diameter (D)	The diameter at the large end of a taper.
Tailstock offset (T_o)	The amount a tailstock is moved from a central alignment (zero index) position.
Taper ring gage	A gage for measuring the accuracy of an external taper. A tapered ring that is precision ground to a close tolerance.
Taper micrometer	A reliable, accurate, precision measuring instrument. A sine bar design feature brought into a micrometer. Micrometer spindle movement that provides for minutely measuring a taper.
Plain taper attachment	An accessory fastened to a lathe bed. An attachment that produces an angular movement of the cross slide.
Telescopic taper attachment	A mechanism that permits the transfer of motion. Transferring the angle setting of the taper attachment directly through the cross slide.
Steep taper (angle) turning	The process of setting the compound rest at one half of the included angle. Feeding the cutting tool at a preset angle by turning the compound rest handwheel.
Compound rest	A combination of a cross slide, an angle-graduated swivel slide, and a top slide. A mechanism on which one or more tools are positioned and fed.

SUMMARY

- Tapers provide for:
  - The accurate alignment of mating parts
  - Holding mating taper surfaces securely or for easy release
  - Positioning and revolving of cutting tools and other machine units.
- Self-holding short tapers are used to align and hold mating tapers rigidly.
- Self-releasing steep tapers are designed for accurate alignment of mating parts. Some form of locking device is required. These tapers are freed easily.
- The American Standard Taper Pin series ranges from #7/0 to #11. The standard taper per foot is 0.2500″ or 0.0208″ per inch. The pins come in a variety of standard lengths and are commercially produced.
- The American Standard (Morse) Taper falls in the small-taper-angle, self-holding taper group. The amount of taper per foot varies within the series from almost 0.600″ to 0.625″.
- The Brown and Sharpe (B & S) Taper series ranges from #1 to #18, with small-end diameters from 0.200″ to 3.000″. The taper per foot varies among the different numbers. The T_{pf} is approximately 0.5000″.
- The Jacobs Taper is usually a short, self-holding taper. A principle application is the centering and holding of a drill chuck.
- The Jarno Taper is a 0.6000″-per-foot, self-holding taper. The number of the Jarno taper, divided by 10, 8, or 2, determines the small and large diameters and the length, respectively.

- The American Standard Self-Releasing Steep Taper combines features of quick, precision alignment and easy release. The long steep taper, like the type-L spindle nose, has a 3.5000″ taper per foot. The steep taper of the type D-1 spindle nose is 3.000″ taper per foot.
- The taper per foot is equal to the difference between the large and small diameters multiplied by 12 and the resultant product divided by the required taper length.

$$T_{pf} = \frac{(D-d) \times 12}{L_t}.$$

- The large diameter of a taper is equal to the taper per inch multiplied by the length of the part in inches, plus the small diameter.

$$D = (T_{pi} \times L_t) + d.$$

- Tapers may be turned by offsetting the tailstock or by using a plain or telescopic taper attachment, or with the compound rest.
- The amount to offset a tailstock is equal to the taper per foot multiplied by the overall length of the workpiece in inches and the resultant product divided by 24.

$$T_o = \frac{T_{pf} \times L_o}{24}$$

- Tapers and steep angles for cylindrical parts are specified on shop drawings by dimensions, notations, or both.
- Tapers may be gaged with a taper ring gage. They may also be measured with a standard or taper micrometer. The accuracy may also be checked precisely by using a sine bar and dial micrometer setup.
- The compound rest set at the required angle provides a practical method of turning steep angles.

UNIT 33 REVIEW AND SELF-TEST

1. Identify four main purposes that are served by tapers.
2. a. Give the general range of numbered sizes of *taper pins*.
   b. State two major applications of American Standard Taper Pins.
3. Distinguish between a *Jacobs taper* and an *American Standard* (Morse) *Taper*.
4. State the *taper per inch* for (a) Brown and Sharpe tapers and (b) Jarno tapers.
5. Differentiate between the holding devices for an American Standard *self-releasing steep taper* and a *Type D-1 short steep taper* (3.000″ per foot).
6. Define each of the following taper terms or give the formula for calculating each value:
   a. taper
   b. taper per foot
   c. large diameter
   d. small diameter.
7. a. Identify two common methods of turning shallow tapers.
   b. Tell how a taper is represented and dimensioned on a drawing.
8. List three general methods of measuring the accuracy of the taper angle and/or the large and small diameters.
9. State four advantages of turning a taper with a *taper attachment* compared to the *offset tailstock method*.
10. List the steps required to turn an included angle of 60° on a workpiece between centers, using a compound rest.
11. Tell how to turn two tapered mating parts when testing the accuracy of the tapers.
12. State two safe practices the lathe worker must observe when turning tapers.

SECTION FOUR

LATHE WORK HELD IN A CHUCK

Boring, reaming, drilling, and internal thread cutting processes on the lathe generally require that the workpiece be mounted in a chuck on the spindle. This section opens with descriptions of the types of chucks typically used for these processes and the methods for truing and centering the work. Subsequent units describe the principles of and procedures for performing the following lathe processes:

- Drilling, countersinking, and reaming
- Straight hole boring, counterboring, recessing (undercutting), and taper boring
- External surface finishing with the work mounted on a mandrel
- Cutting internal and external threads.

Unit 34

CHUCKS AND CHUCKING: TECHNOLOGY AND PROCESSES

This unit covers the mounting and removing of standard chucks and collets from the lathe spindle nose. Step-by-step processes are included for basic chuck setups. Some are for machining concentric surfaces. Other setups deal with off-center (eccentric) surfaces. Safety practices are stressed to prevent personal injury or damage to the machine, accessories, or tools.

THREADED AND STEEP TAPER ANGLE SPINDLE NOSES AND CHUCK ADAPTER PLATES

Type-L Spindle Nose

As noted in Unit 24, there are three main designs of spindle noses. Of these, the threaded spindle nose (Fig. 34-1) is the oldest type. The chuck flange and the chuck itself are centered by the accuracy of the threads, a straight turned section, and the spindle nose shoulder.

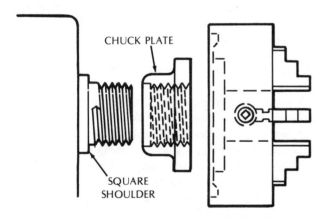

Fig. 34-1. An older style of threaded spindle nose.

The American Standard Type-L Spindle Nose (Fig. 34-2) has a taper of 3.500" per foot. The taper bore of the chuck flange centers on the corresponding taper of the spindle nose. Note, also, that the chuck flange has a keyway. This fits the key in the tapered nose of the spindle. The key and keyway prevent the chuck from turning on the tapered surface of the spindle nose.

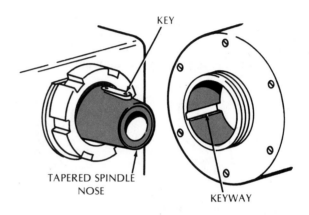

Fig. 34-2. The American Standard Type-L spindle nose with a key.

The chuck is held securely on the taper and concentric with the spindle by a *lock ring*. The lock ring fits the threaded flange on the chuck. When the lock ring is tightened, the chuck is drawn tightly onto the tapered spindle. A spanner wrench is used for both tightening and loosening the lock ring.

There are a number of advantages of the type-L and D-1 steep taper spindle noses over the threaded type.

● A chuck or accessory with the same taper may be aligned to run true to the spindle.

● Driver plates, faceplates, and accessories from other machine tools may be interchanged and run true.

● The type-L and D-1 spindle noses (and corresponding tapers in the accessories) may be cleaned readily. Accessories can be mounted or removed easily.

Type D-1 Cam-Lock Spindle Nose

There are five main design features of the D-1 cam-lock spindle nose. These are identified in Fig. 34-3. The short steep taper of the D-1 nose is 3.000" per foot. There are from three to six *cam-lock studs*, depending on the chuck size. These extend from the chuck or other accessory. The studs fit into corresponding holes in the face of the spindle nose. There are, also, eccentric *cam-locks* that match. As the eccentric cam-lock turns against the stud, the accessory is drawn firmly onto the taper and against the *face* of the *spindle flange.*

There are *register lines* for each of the cam-locks. The *index line* of each cam-lock must match with the *register position* to mount or remove the accessory. The cam-locks are given a partial turn *clockwise* to mount a chuck; *counterclockwise,* to remove it.

A *chuck cradle* of hard wood (Fig. 34-4) is usually used to install or to remove a heavy chuck. The cradle is grooved to fit and slide on the ways and under the chuck. The cradle protects the operator from injury and prevents damage to the lathe ways.

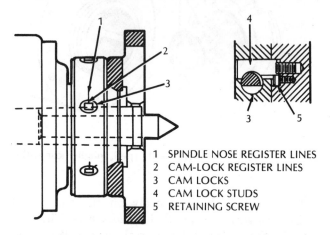

1 SPINDLE NOSE REGISTER LINES
2 CAM-LOCK REGISTER LINES
3 CAM LOCKS
4 CAM LOCK STUDS
5 RETAINING SCREW

Fig. 34-3. The features of a cam-lock spindle nose.

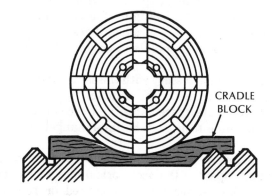

Fig. 34-4. Supporting a heavy chuck with a cradle block.

BASIC CHUCK WORK SETUPS

The Universal Three-Jaw Chuck

The name of the *universal three-jaw chuck* indicates that there are three accurate, self-centering jaws. These are controlled by a bevel gear-driven scroll, as shown in the cutaway section in Fig. 34-5.

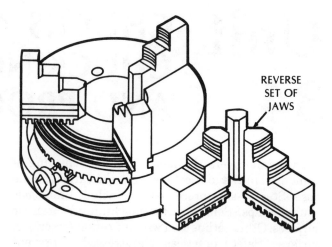

REVERSE SET OF JAWS

Fig. 34-5. A cutaway section of a three-jaw universal chuck.

All three jaws may be *actuated* (moved) by turning any one of the adjusting sockets.

Due to the shape of the screw thread on their back sides, the jaws are not reversible. Other jaw sets are available to accommodate large-diameter work and work that requires inside chucking.

With heavy use and wear, the jaws may spring. This condition affects the accuracy with which a workpiece is held concentric. Under these conditions, the jaws are checked for accuracy, and reground if needed. Some chucks are designed so that the jaws may be adjusted to compensate for any wear or out-of-true condition.

To restate, the universal chuck is self-centering. The jaws do not require individual setting, as do independent-jaw chucks.

Changing Universal Three-Jaw Chuck Jaws

A number is stamped on the face of the chuck at each slotted section. Each jaw has a particular number. Each jaw must be replaced in a slot having the same number. The jaws are started on the *scroll,* one at a time. The jaws must be assembled in the chuck body in correct sequence to be concentric. Jaw #1 is started first. The scroll is turned slowly until jaw #2 is engaged. This is followed by continuing to turn the scroll until jaw #3 is engaged.

Four-Jaw Independent Chuck

Each jaw of the four-jaw independent chuck (Fig. 34-6) is adjusted independently. The jaw is moved by the action of the *chuck screw thread.* This thread meshes with a corresponding partial thread on the underside of each jaw. The phantom section in Fig. 34-6 shows one of the *chuck screws* engaged with the

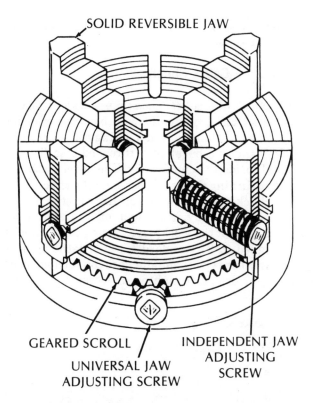

SOLID REVERSIBLE JAW

GEARED SCROLL

UNIVERSAL JAW
ADJUSTING SCREW

INDEPENDENT JAW
ADJUSTING
SCREW

Fig. 34-6. A phantom section of a four-jaw combination chuck.

threaded jaw. There are four chuck screws, one for each chuck jaw.

The chuck jaws are *stepped* to take work of small and large diameters. They may be used to hold workpieces on either an inside or an outside diameter. The jaws are also reversible. The four-jaw independent chuck can accommodate a wider range of shapes and sizes than the universal types.

The four-jaw chuck may be used to grip square, round, or irregular-shaped pieces. The jaws may be positioned concentric or off-center, depending on the job requirements. The work surfaces may be finished or rough, as are the outer surface of castings, forgings, or surfaces produced by flame cutting or other processes.

The face of the chuck has a series of concentric rings cut into it at regular intervals. These rings, or grooves, aid the operator in chucking a workpiece. The chuck jaws are opened in relation to the concentric rings. These are a guide that the chuck will take a particular size workpiece. The jaws are then *set* (tightened) against the workpiece. The position of each jaw is observed in relation to the concentric grooves. Adjustments are made by loosening one jaw and tightening the opposite jaw. This *rough setting* is then checked by other approximate or precise methods of truing a workpiece.

The three general positions for chucking work are shown in Fig. 34-7. In (A) the jaws are in a normal position. To accommodate a large diameter, the chuck jaws are reversed. The piece then is chucked at one of the jaw steps as shown in (B). Parts that require facing, other internal operations, or turning across the outside diameter are *chucked on the inside* as shown in (C). When the jaws are positioned as in (C), an outward force is exerted on the workpiece.

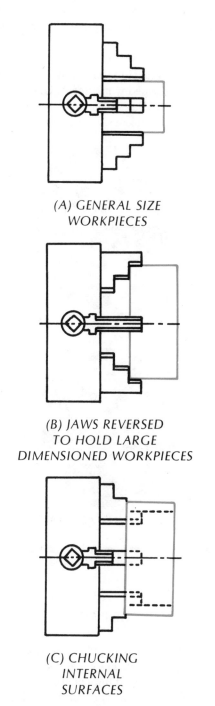

*(A) GENERAL SIZE
WORKPIECES*

*(B) JAWS REVERSED
TO HOLD LARGE
DIMENSIONED WORKPIECES*

*(C) CHUCKING
INTERNAL
SURFACES*

Fig. 34-7. Chucking workpieces in a three-jaw chuck.

Power-Operated Self-Centering Chucks

Power-operated two-and three-jaw chucks with adjustable and nonadjustable jaws are widely used in production. Both types may be *self-centering power chucks*.

The nonadjustable-jaw power chuck (Fig. 34-8A) is recommended for general manufacturing service. It is particularly suited for repetitive operations, especially on heavy workpieces. This chuck is balanced to eliminate chatter and vibration at high spindle speeds (rpm).

The adjustable-jaw type (Fig. 34-8B) provides independent jaw action. The jaws are designed to hold irregular-shaped workpieces. Once the jaws are adjusted, the initial accuracy of positioning is maintained with successive workpieces.

(A) NONADJUSTABLE JAWS

(B) ADJUSTABLE JAWS

Fig. 34-8. Two power-operated three-jaw self-centering chucks. (Courtesy of Cushman Industries, Inc.)

COLLET CHUCKS

Collet chucks are simple, accurate, practical work-holding devices. They are used principally for holding regular-shaped bars of stock and workpieces that are finished on the outside. The work is generally of round, square, or hexagonal shape. Two common types of collets are the *spring* and the *rubber-flex collet*.

Spindle Nose Collet Chuck

The *spindle nose collet chuck* takes a wider range of work sizes than does the draw-in bar type of collet chuck. The phantom view of a spring collet chuck in Fig. 34-9 shows a bevel gear threaded disc. This is mounted inside the chuck body. The bevel teeth are moved by turning the bevel socket on the chuck body with a chuck wrench. As the spring collet is drawn into the chuck body, the taper on the split jaws tightens against the workpiece. Spring collet chucks are designed to fit standard types of lathe spindle noses. Another type of spindle nose collet chuck uses a spur gear design to tighten or loosen the collet.

Fig. 34-9. A spring collet chuck and a standard collet. (Courtesy of Cushman Industries Inc.)

Jacobs Spindle Nose Collet Chuck

The *Jacobs spindle nose type chuck* takes rubber-flex collets. This chuck is also made to fit different types of spindle noses. The usual range of each collet is 1/8″ (1/16″ over and under a nominal size). A single rubber-flex collet and Jacobs type of spindle nose chuck serves a wide range of sizes.

The work is first inserted in the collet. The chuck handwheel is turned *clockwise* to tighten the collet jaws against the workpiece. If the workpiece is not long enough to extend into the rubber-flex collet for at least 3/4″, then a plug of the same diameter is needed. This plug is placed in the back of the collet. The plug helps ensure that the collet grips the work securely. It prevents the work from springing away if a heavy force is applied during the machine operation.

The Drill Chuck

The standard *drill press chuck* (Fig. 34-10) is

commonly used for drilling, reaming, tapping, and other operations on the lathe. This *chuck* is fitted with a taper shank. The shank fits the spindle bore of the tailstock. Another sleeve or taper socket may be used to accommodate the chuck in the spindle headstock. As for all mating parts, the taper fitting sections should be clean and free of burrs.

Fig. 34-10. A standard drill chuck fitted with a taper shank. (Courtesy of The Jacobs Manufacturing Co.)

The drill chuck is also designed with a hollow core. This permits the holding of long bars of small cross section that extend through the spindle. The drill chuck in Fig. 34-11 is designed to fit a threaded spindle nose.

Fig. 34-11. A drill chuck designed to fit a threaded spindle nose. (Courtesy of The Jacobs Manufacturing Co.)

Supporting Workpieces

Many times a long slender workpiece, such as the one in Fig. 34-12, is held in a collet chuck. The other end must be supported in a special tailstock chuck or with a steady rest. In addition, it is impossible to machine accurately without further supporting the

workpiece. A follower rest is used near the cutting area. This helps prevent springing of the workpiece by the cutting force.

FOLLOWER REST PROTECTS WORKPIECE AGAINST CUTTING FORCES

STEADY REST SUPPORTS THE WORKPIECE AND SERVES AS A CENTER

Fig. 34-12. Use of a steady rest and follower rest to support a workpiece. (Courtesy of the Clausing Corp.)

WORK TRUING METHODS

Approximate Methods

There are a number of approximate and precise methods of truing chuck work. The concentric rings on the chuck face provide the first *approximate method*.

Once chucked, the work may be further tested by revolving the spindle slowly. A piece a chalk is held in one hand. The hand is steadied and the chalk is brought to the workpiece so that any *high spot* produces a chalk mark (Fig. 34-13A). The jaw (or jaws) opposite the high spot is moved out. The opposing jaw (or jaws) is moved in. The process is repeated until the work is concentric (no high spots).

The trueness may also be checked by using a tool holder (Fig. 34-13B). Reduction of the light (space) showing between the back end of the tool holder and the slowly turning workpiece indicates the high spot. A piece of white paper may be placed on the cross slide under the tool holder. This helps show the light (space) and identify which jaws must be adjusted.

Precision Truing By The Dial Indicator Method

Out-of-trueness may be corrected by a more accurate method. A dial indicator and holder are positioned and held in the tool post (Fig. 34-14). The indicator point is carefully brought into contact with the workpiece. The work is revolved slowly by hand. The amount that the work runs out-of-true is indicated by movement of the dial indicator pointer. The difference between the high and low spots is noted. The jaw opposite the high spot is moved *out* one half the distance. The opposing jaw is moved *in* the same distance. The workpiece is again revolved and further adjustments are made. Each chuck jaw is then checked for tightness. The work is concentric when there is no movement of the dial indicator pointer.

USE OF CONCENTRIC
RINGS ON CHUCK FACE

(B) CHECKING WITH
THE BACK END
OF THE TOOL
HOLDER

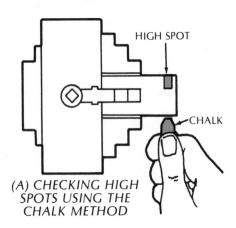

HIGH SPOT

CHALK

(A) CHECKING HIGH
SPOTS USING THE
CHALK METHOD

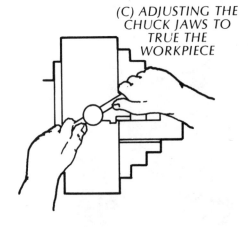

(C) ADJUSTING THE
CHUCK JAWS TO
TRUE THE
WORKPIECE

Fig. 34-13. Approximate work truing methods.

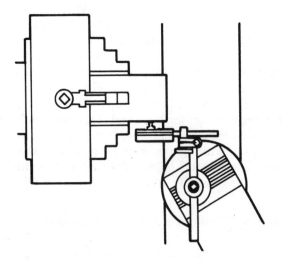

Fig. 34-14. Dial indicator method of precision truing.

HOW TO REMOVE LATHE SPINDLE ACCESSORIES

Removing A Live Center

Step 1   Insert a solid-metal *knockout bar* in the spindle bore (Fig. 34-15).

Step 2   Hold the end of the live center with the right hand.

Step 3   Move the bar with a quick stroke to firmly tap the center. This frees the tapered surface.

Note: The live center is sometimes left in place if it does not interfere with any of the chuck work processes.

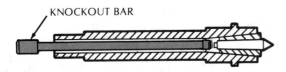

Fig. 34-15. Driving a live center with a knockout bar.

Removing A Faceplate, Driver Plate, Or Chuck

Note: The following two steps apply to the mounting and removal of faceplates, driver plates, and chucks that are large and heavy:
● Engage the back gears
● Slide a wooden cradle of a size and form suitable for the particular accessory to be mounted or removed.

Threaded Spindle Nose Faceplate or Driver Plate

Step 1   Select a flat or square metal bar that fits the slots of the faceplate or driver plate. Remove burrs. Then, center the bar in one of the plate slots.

Step 2   Tap the bar firmly in a *counterclockwise* direction. Retap if the mating shoulders of the accessory and the spindle nose are not freed.

Step 3   Continue to slowly turn the faceplate or driver plate by hand in a counterclockwise direction.

    CAUTION: Extra care must be taken as the last threads are reached. This is to avoid having the driver plate or faceplate move off the spindle with the full weight riding on just one or two threads.

Threaded Spindle Nose Universal or Independent-Jaw Chuck

Step 1   Extend a chuck jaw nearly to the outside diameter of the chuck.

Step 2   Place a hardwood block over the back ways as shown in Fig. 34-16.

Step 3   Reverse the direction of the spindle to revolve *counterclockwise*. Turn the spindle so that the extended jaw strikes the hardwood block sharply.

    CAUTION: The hardwood, plastic, or soft metal block must be strong enough to withstand the force of the jaw striking it.

Step 4   Move the cradle under the chuck if it is large and heavy. Continue to turn the spindle counterclockwise. Exercise additional caution as the last few threads are reached.

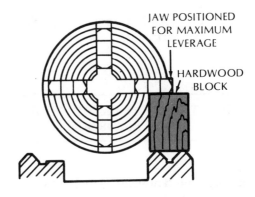

Fig. 34-16. Removing a chuck from a threaded spindle nose.

Steep Taper Type-L Spindle Nose Plates And Chucks

Step 1   Select the correct size of C-spanner wrench. Place it around the lock ring. Rotate the lathe so that the lock ring slot permits the spanner wrench to be held in an upright position.

Step 2   Engage the back gears. Place a wooden cradle under the accessory if it is heavy.

Step 3   Place the spanner wrench around the lock ring. Hold it in place with one hand.

Step 4   Tap the wrench handle sharply in a *counterclockwise* direction to *free up* the threads. This unseats the tapered surfaces. Continue to turn the lock ring counterclockwise until it disengages.

Step 5   Use the cradle on a heavy chuck. Carefully slide the accessory away from the spindle nose.

Cam-Lock Spindle Nose Plates And Chucks

Step 1   Slide the wooden cradle under the accessory.

Step 2   Turn each cam-lock in a *clockwise* direction. Stop when the index line and the register marks coincide.

Step 3   Tap the chuck sharply but lightly to break the taper contact, if necessary.

Step 4   Slide the accessory on the cradle until it clears the lathe spindle nose.

Step 5   Store the accessory in its proper compartment.

Removing A Draw-In Collet Attachment

Step 1   Remove the spring collet and the drawbar.

Step 2   Turn the spindle nose cap *counterclockwise*. This frees the taper sleeve from the spindle bore.

Step 3   Place each part in its proper storage compartment.

HOW TO MOUNT LATHE SPINDLE ACCESSORIES

Assembling The Threaded Spindle Nose Draw-In Type Collet Attachment

Step 1   Remove any nicks or burrs from the spindle bore or the taper sleeve and other parts. Wipe all surfaces to remove any foreign particles.

Step 2   Place the spindle nose cap in position.

Step 3   Insert the taper sleeve. Move the sleeve in with a quick motion. This sets the mating tapered surfaces. The parts are aligned and the sleeve is also held securely in the spindle bore.

Step 4   Lock the lathe spindle by engaging the back gears. On small bench lathes, engage the spindle stop pin.

Step 5   Select a collet of the shape and size of the workpiece.
Note: Accuracy and efficiency may be impaired if the collet segments are not able to bear evenly along the length of the workpiece.

Step 6   Insert the collet in the taper sleeve. Turn the collet until the key in the taper sleeve engages the keyway of the collet.

Step 7   Extend the drawbar through the lathe spindle.

Step 8   Press against the face of the collet with the palm of the right hand. This prevents the collet from being pushed out of the sleeve.

Step 9   Turn the handwheel of the drawbar slowly *clockwise*. This engages the threaded end of the collet. Continue to turn the handwheel *clockwise* until most of the end play is taken up.

Step 10   Insert the burr-free workpiece. Extend it the distance necessary to perform the lathe operations.

Step 11   Continue to turn the drawbar *clockwise* by hand. The force produced by the handwheel is adequate for the collet to hold the workpiece securely.

Assembling A Steep Taper Spindle Nose Collet Chuck

Step 1   Select a spring collet chuck with an adapter plate. These must fit either the type-L or D-1 spindle nose on the lathe.

Step 2   Remove burrs and nicks. Wipe all surfaces clean.

Step 3   Proceed to assemble the collet chuck on the tapered spindle nose. Secure the collet chuck to the spindle nose. Tighten either the lock ring or the cam-locks, as the case may be.

Step 4   Insert the collet, aligning the keyway and key. Engage a few threads.

Step 5   Place the burr-free workpiece in the collet. It should project only far enough to permit performance of the required operations.

Step 6   Turn the chuck handwheel. Draw the collet against the taper portion until the work is held securely.

Mounting Universal And Independent-Jaw Chucks

Threaded Spindle Nose

Step 1   Clean the threads on the spindle and chuck. Check to see that there are no burrs. Wipe all surfaces to remove foreign particles. Apply a few drops of oil on the threads.
Note: The threads on the chuck plate are usually cleaned with a spring thread cleaner.

Step 2   Place the chuck on a cradle if it is large and heavy.

Step 3   Slide the straight-turned portion of the chuck plate over the threads. At the same time, slowly turn the spindle by hand to engage the first thread.

CAUTION: If the chuck binds, stop turning the spindle. Reverse the spindle direction and disengage the thread. Check to see that the thread has not been damaged. Then, start again to properly engage the first thread.

Step 4   Screw the chuck on the spindle nose until it shoulders.

CAUTION: The chuck must not be jammed or forced too tightly against the shoulder. When this happens, it is difficult to remove the chuck.

American Standard Taper Spindle Nose (Type L)

Step 1   Turn the lathe spindle by *hand* until the key on the nose and the keyway in the chuck adapter plate are aligned.

Step 2   Slide the cradle so that the tapered bore fits onto the spindle nose.

Step 3   Turn the lock ring. This engages the threads on the chuck adapter plate.
Note: If the threads are not engaged easily, turn the lock ring in the reverse direction and remove the chuck. Remove any dirt or chips that were not cleaned out the first time.

Step 4   Remount the chuck. Tighten the lock ring by hand.

Step 5   Place a spanner wrench around the lock ring. *Set* the tapered and shoulder surfaces by striking the spanner wrench sharply with the palm of one hand.

Cam-Lock Spindle Nose

Step 1   Align the index line on each cam-lock stud with the register line on the spindle.

Step 2   Slide the cleaned and burr-free chuck on a cradle. Position it on the nose spindle.

Step 3   Turn the lathe spindle slowly by hand. Align the cam-lock studs with the clearance holes in the spindle nose.

Step 4   Slide the chuck onto the tapered portion of the spindle and up to the shoulder.

Step 5   Turn each cam-lock in a *counterclockwise* direction. Apply equal force to each cam-lock. The chuck should be drawn tightly against the spindle shoulder. The short tapered surfaces ensure that the chuck is correctly aligned and runs concentrically.

HOW TO CHANGE THE JAWS ON A UNIVERSAL CHUCK

Step 1   Remove the installed set of jaws. Clean and store them in an appropriate container.

Step 2   Select a matched set of jaws that will accommodate the work size and the nature of the operations.

Step 3   Start assembling each jaw. Begin with jaw #1. Turn the scroll with the chuck key until the jaw is engaged.

Step 4   Insert jaw #2 while continuing to turn the scroll. As soon as this jaw is engaged, proceed with jaw #3 and the remaining jaws in the set.
Note: The jaws must be inserted and engaged in numerical sequence. Otherwise, they will not be concentric.

Step 5   Check the concentricity of the jaws. Chuck a piece of finished stock. Test its trueness with a dial indicator.

Step 6   Correct any inaccuracy either by grinding the jaws or by adjusting them.
Note. An adjusting screw is provided on each jaw of some universal chucks. The inaccuracy is corrected by turning the adjusting screw. The jaws may then be moved in or out as needed.

HOW TO MOUNT AND TRUE WORK IN A CHUCK

Approximate Methods

Setting By The Concentric Chuck Face Grooves

Step 1   Determine which set of concentric grooves will accommodate the workpiece size.

Step 2   Locate the jaws at the same distance from the circular grooves.

Step 3   Insert the workpiece. Tighten each chuck jaw. Recheck to see that each jaw is still the same distance from the groove.

Truing with Chalk

Step 1   Mount the workpiece as concentrically as possible.

Step 2   Set the spindle to turn at a slow speed. Hold a piece of chalk in one hand. Steady the hand against the back of a toolpost or on the compound rest.

Step 3   Bring the chalk to the revolving work. A chalk mark will appear at the *high spot*.

Step 4   Release (open) the jaw that is opposite the high spot.

Step 5   Tighten the opposite jaw.
Note: If the high spot is between two jaws, the two jaws opposite the mark are opened. The jaws opposite these then are tightened. This moves the workpiece in toward center.

CAUTION: One set of jaws should be adjusted at one time. This practice prevents the part from falling out of the chuck.

Step 6   Continue to test and adjust the jaws until the workpiece runs true. Then, apply equal force to tighten each jaw.

Truing The Face Of A Workpiece

Step 1   Mount and true the outside surface using the preceding steps.

Step 2   Measure the distance the work projects from the chuck face. Tap the workpiece in at any point where it projects more than it should.

Step 3   Start the workpiece revolving slowly. Hold a piece of chalk so that it may be steadied as it is brought to the work face. The chalk spots indicate the high spots.

Step 4   Use a plastic or soft-faced hammer. Firmly but carefully tap the workpiece to move the high spot in toward the chuck face.

Step 5   Continue to locate the high spots. Tap and adjust the workpiece until the face is even.

Note: A square or flat homemade block is sometimes used to chuck a flat workpiece. The block is placed between the work and the chuck. The part is tapped until the back face rests against the block. The block is then removed and the process is repeated at each jaw.

CAUTION: The parallel or block between the work and chuck must be removed before starting the lathe.

Testing For Trueness By The Toolholder Method

Step 1   Mount and center the work in the chuck. Use the concentric grooves on the chuckface as a guide.

Step 2   Turn the toolholder backward in the tool post. Place a sheet of white paper on the cross slide.

Step 3   Revolve the workpiece slowly by hand. Bring the tool holder toward the work.

Step 4   Note whether the amount of light (space) showing is even. The darker areas (reduced spacing) are high spots.

Step 5   Adjust the chuck jaws until the workpiece runs concentric.

Note: Square- or irregular-shaped workpieces may be checked for trueness by taking a light trial cut. The jaws are then adjusted to move the part to the correct position.

Precision Testing For Trueness With A Dial Indicator

Step 1   Chuck the workpiece. Check it for trueness by one of the approximate methods.

Step 2   Clamp the holder of a dial indicator in the tool post. Position the movable plunger of the indicator so that the point may continuously touch the workpiece.

Step 3   Turn the spindle slowly by hand. The dial indicator should clear the workpiece.

Step 4   Move the indicator in until the contact point touches the part as it is turned by hand. Take the reading on the dial.

Step 5   Note any out-of-trueness, as indicated by movement of the dial pointer. Adjust only two opposite jaws at one time.

CAUTION: The dial indicator is a relatively fragile and expensive instrument. It can be damaged easily. The internal mechanism should be moved slowly and gently in relation to a workpiece.

Step 6   Adjust the other two jaws. Continue until there is no dial movement over the entire workpiece.

SAFE PRACTICES WITH CHUCKS AND CHUCK WORK

- Shut off the power to the lathe when mounting or removing a chuck.
- Obtain assistance when handling a heavy chuck or other machine accessory.
- Use a wooden cradle to slide a heavy chuck onto or off a spindle nose.
- Keep the fingers out from under the chuck and the cradle.

- Stone away any burrs on the spindle nose, chuck adapter plate, or other machined part.

- Wipe all mating surfaces with a clean wiping cloth. Carefully move the palm and fingers over the parts to feel if there are any foreign particles left. Apply a drop of oil on the taper or thread of the spindle nose and the shoulder.

- Clamp irregular or rough surface parts in an independent jaw chuck. Universal chucks become inaccurate when the jaws are tightened against out-of-round workpieces.
- Engage the threaded chuck plate by turning the lathe spindle *counterclockwise* by hand. The threads are disengaged (unscrewed) by reversing the direction.
- Use the correct size of chuck wrench. The force applied by hand should be adequate for the jaws to hold a workpiece securely.
- Avoid excessive force on thin-sectioned areas of a workpiece. The additional force may cause the part to spring out of shape.

- Revolve the spindle by hand before starting any operation. Check to see that the chuck jaws and the workpiece clear the carriage. Also, check the tool setup.
- Tighten a collet only when there is a workpiece between the jaws. Otherwise, the jaws may become sprung.
- Use protecting soft shims between a finely machined surface and the chuck jaws.
- Protect the spindle bore against chips and dirt. When the lathe is stopped, a piece of waste cloth may be packed in to prevent chips from entering.

TERMS USED WITH CHUCKS AND CHUCK WORK

Chucks	A work-holding device for mounting and holding a workpiece in a particular position. A device that transfers motion and force from a lathe spindle through the workpiece.
Chuck adapter plate	A circular flanged plate designed to fit a particular type and size of spindle nose. A plate fastened to the back side of a chuck. A plate that accurately positions and holds a chuck on a lathe spindle.
Type-L spindle nose	A steep-angle taper lathe spindle nose. A spindle nose with an American Standard taper of $3.500''$ T_{pf}. A tapered lathe spindle nose having a key and a lock ring. The lock ring draws and holds a chuck securely on the taper and against the shoulder of the spindle.
Type D-1 cam-lock spindle chuck adapter	A short steep-angle taper-bored adapter plate. A chuck plate bored to a $3.000''$ T_{pf}. An adapter plate designed to secure a chuck on a cam-lock spindle nose.
Universal chuck (lathe)	A work-holding device with jaws that move radially and together. A device to position and hold a workpiece concentric.
Independent jaw chuck	A work-holding device for positioning a workpiece centrally or off-center. A chuck with jaws that are moved independently.
Inside chucking	The process of positioning a chuck to hold a workpiece from an inside diameter or relieved area. Exerting an outward force against a workpiece.
Spindle nose collet chuck	A work-holding device mounted on a lathe spindle. Usually, a hand-operated chucking mechanism that moves a metal spring collet or rubber-flex collet. A device for adjusting a collet to hold a workpiece so that it may run concentric.
Approximate methods of work truing	Positioning a workpiece by sighting the high, or out-of-true, spots. Adjusting chuck jaws to move a part so it runs concentric.
Precision testing for trueness	Truing a finished machined surface to run concentric within $\pm 0.0005''$ (0.01 mm) or to a finer tolerance. Using a dial indicator to true a workpiece until no movement (out-of-trueness) is recorded on the instrument.
Spindle nose accessory	A chuck, driver plate, faceplate, live center, or other device that is secured to the spindle nose. A device which positions and holds a workpiece or a cutting tool in a spindle nose. The accessory serves to transmit the turning force of the lathe spindle to the area of cutting action.
Cradle block	A hardwood block that fits across the ways. The center area is curved to slide (cradle) under a chuck. A moveable safety device on which a chuck is rested for mounting or removal.

SUMMARY

- Chuck adapter plates are designed for mounting universal, independent-jaw, and combination chucks.
- Spindle nose accessories are accurately positioned on a type-L American Standard steep taper (3.500″ T_{pf}) by a lock ring. The lock ring draws and holds together the mating taper and shoulder surfaces.
- The universal three-jaw chuck is a common shop chuck. It is recommended for bar stock and other workpieces that are produced with a fairly accurate round or hexagonal finished surface.
- The independent jaw chuck requires more setup time than a universal chuck. However, it is used for both regular finished work or for rough-surfaced and irregular-shaped parts. The independent jaw chuck is practical for holding workpieces that require concentric or off-center machining.
- The combination chuck combines the features of the independent-jaw and the universal chuck. A workpiece may be set in a desired location using the independent jaws. Duplicate parts may then be chucked by turning the sockets of the universal jaws.
- Cam-lock studs fit corresponding holes in the D-1 short steep-taper (3.000″ T_{pf}) spindle nose. The taper and shoulder surfaces are drawn and held securely by turning each cam-lock.

- Taper and shoulder contact may be broken by tapping the chuck firmly but lightly. A soft-faced hammer is used.
- A set of universal jaws must be inserted and engaged in numerical sequence on the scroll. Start with the #1 jaw.
- Collets provide a quick, accurate, dependable method for holding machined bars and other finished-surface products.
- The spring collet may be drawn in with a draw bar attachment. A spindle nose collet chuck may also be used with the spring collet.
- Rubber-flex collets are designed for a spindle nose chuck. The tapered face of the chuck cap forces the metal blades (inserts) against the inside tapered walls of the chuck. This movement causes the blades to close against the workpiece.
- Standard or cored drill press chucks are used for chucking and driving small-diameter cutting tools and workpieces.
- Work may be roughly trued by noting the high spots. Truing to precise limits is usually done with a dial indicator.
- Heavy-work accessories require extra care. Machine or tool damage and personal strain or other injury may be avoided by safe handling.
- General precautions about machine care and safe practices must be followed.

UNIT 34 REVIEW AND SELF TEST

1. Explain how a *D-1 cam-lock spindle nose* operates.
2. List the basic differences in the operation of a *universal three-jaw chuck* and that of an *independent four-jaw chuck*.
3. Tell why *rubber-flex collets* mounted in a Jacobs or Sjogren spindle-nose collet chuck are more flexible than *spring collet chucks*.
4. Identify (a) two rough and (b) one precision method of truing work held in a chuck.
5. List the steps in removing a chuck or plate from a *steep taper (Type -L) spindle nose*.
6. State two precautions to take when changing the jaws on a universal chuck.
7. List four safe practices to observe when handling chucks and doing chuck work.

Unit 35 CENTERING, DRILLING, COUNTERSINKING, AND REAMING ON THE LATHE

Multiple-point cutting tools like the center drill, countersink, twist drill, and machine reamer are commonly used for producing holes. The tool design features, terms, applications, cutting speeds and feeds, and materials of which the tools are manufactured were treated in earlier units. The cutting tools and processes were related to bench work and drill press work.

This unit deals with the same cutting tools, but as applied to lathe work. One main difference is that in lathe work the cutting may be done by holding the cutting tool stationary (nonrevolving). It is then fed into a revolving workpiece. The same process may also be carried on by feeding a stationary workpiece into a revolving cutting tool. These lathe hole-producing methods, additional toolholders, and machine setups are covered in this unit.

CENTERING WORK HELD IN A CHUCK

Center holes provide a bearing surface for lathe centers. A center hole may also be used as a pilot hole for subsequent operations. A common and accurate method of drilling a center hole is to use a combination center drill (drill and countersink). The small pilot drill point is especially important for work that is to be mounted between centers.

Sometimes just a central starting hole needs to be *spotted* in the end of a workpiece. In such a case, either a specially ground *tool bit* or a *flat drill* may be used.

Work to be centered is usually held in a universal chuck or collet chuck. Irregular and rough surfaces are centered using an independent jaw chuck. The center drill is held in a standard drill chuck. This chuck has a tapered shank to fit the tailstock spindle. The chuck, center drill, and workpiece setup for drilling a center hole are shown in Fig. 35-1.

After the workpiece end has been faced, the center drill is fed into the revolving workpiece. Although the cutting speed remains the same, the diameter of the angular body increases. Care must be taken in feeding the center drill. This is particularly important when feeding the small-diameter pilot drill portion at the start.

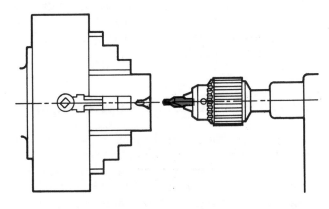

Fig. 35-1. Center drilling a workpiece held in a chuck.

A center may also be *spotted* by grinding a tool bit to a steep-angle point (Fig. 35-2). The area in back of the cutting face is ground away sharply. This permits the cutting edge to cut without interference from rubbing against the angle hole.

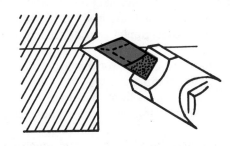

Fig. 35-2. "Spotting", truing, or centering with a specially ground tool bit.

A special 60° countersink may also be used when a center hole needs to be trued quickly and fairly accurately. Only one lip of the countersink is ground to do the cutting. The countersink is held in a drill chuck and is mounted in the tailstock. A high spindle speed is used. The tailstock spindle is brought back as far as possible. This cuts down on the *overhang*. The spindle clamp screw is tightened lightly. The countersink is then fed slowly into the revolving workpiece. These steps help prevent *play* in the tailstock spindle. Play may be caused by the interrupted, uneven cut that exists until a centered hole is produced.

DRILLING PRACTICES ON LATHE WORK

A drill is held on a lathe in a number of different holding devices. Straight-shank drills may be held in a drill chuck. This is usually the case if they are smaller than one-half inch in diameter. The drill chuck with taper arbor may fit directly into the matching tapered tailstock spindle. With an adapter, the chuck may also fit the spindle nose. Large-diameter, straight-shank drills are usually held from turning by a *drill holder*. At other times, a drill is gripped in a regular lathe chuck.

REAMING PRACTICES ON LATHE WORK

The jobber's, shell, fluted-chuck, expansion, and adjustable reamer are all used in lathe work. Two or more step reamers and taper reamers are also used. All of these types were described in Unit 21, which covered machine reamers and reaming processes.

The cutting angles and characteristics of reamers, the cutting speeds and feeds, and cutting fluids are similar to those used for drilling, boring, and other machine tool processes. Tables of recommended operating conditions for reaming different materials are included in the appendix. Holes to be reamed on the lathe are uniformly represented and demonstrated on drawings and sketches.

The selection of a high-speed drill, cobalt high-speed steel, carbide-tipped, or other machine reamer depends on:
- The material of the workpiece
- Production requirements
- Whether the hole is interrupted
- Other factors similar to those that affect hand reamers.

The workpiece requirements and machine conditions determine whether a straight-flute or right-hand or left-hand spiral-flute reamer is the most practical. The reamer may be of solid, expansion, or adjustable design. The machine reamer may have a straight shank that permits direct chucking. Or, it may have a taper shank, which is adaptable for holding in a taper socket or sleeve or directly in a tailstock or headstock spindle. The same machine and tool safety precautions described in Unit 21 must be followed to prevent breakage and excessive reamer wear.

Extremely concentric holes require that the hole first be bored, leaving a minimum amount for reaming. Precise dimensional accuracy also may require both boring and *hand* reaming to meet the specified tolerances.

Another type of carbide-tipped reamer is illustrated in Fig. 35-3. The flutes are carbide tipped for the full length. These provide a good bearing surface in the reamed hole. An inserted plug may be driven into the body. This permits fine adjustments of the reamer to within 0.0001″. The reamer may also be reground to size numerous times.

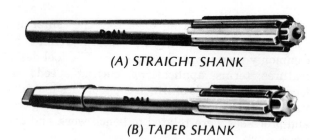

(A) STRAIGHT SHANK

(B) TAPER SHANK

Fig. 35-3. Full-length carbide flutes on a plug-type expansion reamer. (Courtesy of Do All Company)

REPRESENTING AND DIMENSIONING COUNTERSUNK HOLES

A hole that is recessed with a cone-shaped section is a *countersunk hole* (Fig. 35-4). Such a hole may receive a cone-shaped flat-head bolt, screw, rivet, or other part with a corresponding included angle.

Fig. 35-4A shows how countersunk holes are represented in a cross-section view. The conventional drawing in Fig. 35-4B indicates how the countersunk hole is represented on a flat plate. Fig. 35-4C illustrates a countersunk hole in a round workpiece.

As shown in Fig. 35-4B, countersunk holes are dimensioned by: (1) the diameter of the drilled or reamed hole, (2) the angle at which the hole is countersunk, and (3) the outside diameter at the large end of the countersunk hole. Multiple countersunk holes are dimensioned by indicating the number of holes to be countersunk (4 in Fig. 35-4B).

> ### HOW TO CENTER DRILL, COUNTERSINK, AND DRILL HOLES ON THE LATHE

Cutting Tools Mounted In A Tailstock Spindle

Center Drill And Countersink Held In A Drill Chuck

Step 1   Select the center drill or countersink that meets the job specifications. Secure a drill chuck with a taper shank that fits

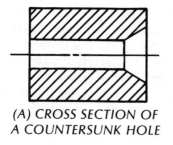

(A) CROSS SECTION OF
A COUNTERSUNK HOLE

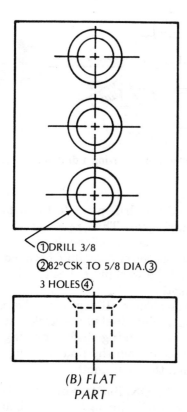

① DRILL 3/8
② 82°CSK TO 5/8 DIA. ③
3 HOLES ④

(B) FLAT
PART

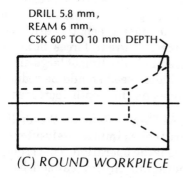

DRILL 5.8 mm,
REAM 6 mm,
CSK 60° TO 10 mm DEPTH

(C) ROUND WORKPIECE

Fig. 35-4. Representation and dimensioning of countersunk holes.

the tailstock spindle. Check the shank for burrs. Remove burrs and wipe the taper surface clean.

Step 2   Mount the chuck in the tailstock spindle.

Sharply snap the chuck into position to *set the tapers.*

Step 3   Insert the center drill or countersink in the drill chuck and tighten it. Determine the required spindle rpm. Set this speed.

Step 4   Move the tailstock spindle back as far as possible. Slide the tailstock up toward the workpiece until the point of the center drill or countersink nearly touches the workpiece end (Fig. 35-5). Clamp the tailstock.

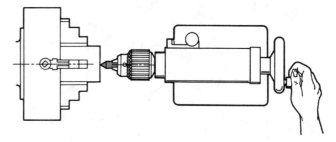

Fig. 35-5. The setup for starting a center drill.

Step 5   Start the lathe. Turn the tailstock hand-wheel and feed slowly. Check the center drill point or the countersink to see that the cutting tool is centered.
Note: Since the workpiece was faced prior to center drilling, the facing tool may not have cut completely through the center. The facing tool may need to be brought back to the center to make a small concentric indentation for center drilling. The same step may be taken to ensure that a countersink is correctly centered.

CAUTION: The point of the center drill may fracture if it is fed into the workpiece off-center.

Step 6   Apply a few drops of cutting fluid. Continue to feed slowly to the required depth.
Note: Many long parts must be drilled or reamed on the lathe. In such instances, the end of the workpiece is supported by a steady rest. The drill is usually held in the tailstock spindle. The setup of the workpiece and drill and the drilling operation are shown in Fig. 35-6.

Drill Held In A Drill Chuck
Step 1   Examine the shank of the twist drill.

417

Stone any burrs. Wipe it clean. Insert and tighten it in the drill chuck.

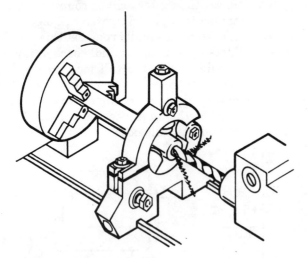

Fig. 35-6. Drilling a long workpiece supported by a steady rest, with the drill mounted directly in the tailstock spindle.

Step 2   Determine the correct cutting speed (rpm) for the material to be drilled, and the cutting conditions (wet or dry) of the drill. Set the lathe spindle speed to the correct rpm.

Step 3   Check the position of the tailstock spindle. It should be back far enough to permit drilling the hole to depth. Bring the tailstock toward the workpiece. Clamp it in position.

Step 4   Feed the drill into the workpiece until the point enters to the full diameter. Note the reading on the graduated tailstock spindle (Fig. 35-7).
Note: The drill point must be started carefully. It must cut concentrically with the center hole or center spot.

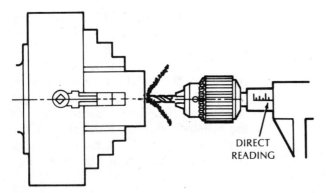

Fig. 35-7. Measuring the drill depth on the graduated tailstock spindle.

CAUTION: An out-of-true center spot must be corrected before proceeding further with the drilling operation.
Note. To start the drill true, turn the back end of the tool holder toward the drill. Move the tool holder in until it just touches the drill body (Fig. 35-8). Slowly

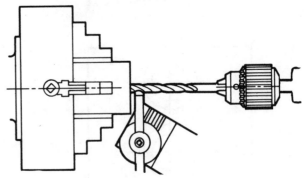

Fig. 35-8. A method of steadying a drill point to start a true hole.

feed the drill point to the depth of the cutting lips. Then, remove the tool holder from the side of the drill.

Step 5   Direct the cutting fluid between the drill and the workpiece. Feed to the required depth.
Note: The drill depth may be measured on the graduated tailstock spindle. The initial reading (before drilling) is subtracted from the final reading to establish the drill depth.

CAUTION: Back the drill out of the work repeatedly during deep-hole drilling. The flutes may need to be brushed to remove the chips. A continuous flow of cutting fluid must reach the drill point at all times. If great forces build up during deep-hole drilling of large-size holes, the spindle speed should be reduced several times as the depth increases.

CAUTION: Remove the drill from the drill chuck immediately after completing the lathe operation. This is done to avoid personal injury caused by brushing against the drill point.

Cutting Tools Held In A Quick-Change Toolholder

Step 1   Select the correct type and size of twist drill for the hole to be drilled. Stone any burrs from the shank.

ADJUSTABLE HOLDER
SETUP PERMITS
USE OF LATHE
CARRIAGE FEED
FOR DRILLING

ADJUSTABLE MORSE TAPER
DRILL HOLDER

Fig. 35-9. A holder mounted on a quick-change toolholder slide, for power feed drilling. (Courtesy of Do All Company)

Step 2   Select a quick-change drill holder that will accommodate the drill taper shank (Fig. 35-9). Insert and tighten the drill in the holder.

Step 3   Slide the drill holder on the quick-change tool post. Lock it in position with the drill point at center height.

Step 4   Determine the required spindle speed (rpm). Set the spindle speed. Start the lathe.

Step 5   Set the feed rod at the recommended cutting feed. Start the flow of cutting fluid.

Step 6   Move the drill into the center spot or center drilled hole. Feed the drill by hand until its outside diameter contacts the workpiece.

Step 7   Stop the lathe if a blind hole is to be drilled. Locate and adjust a carriage stop at the required depth.

Step 8   Start the lathe. Engage the power feed. Feed a through hole automatically. Discontinue the feed on a blind hole when the drill has almost reached full depth.

Step 9   Feed to final depth by hand on a blind hole. Clear the chips. Test for depth with a depth gage.

Note: Deep hole drilling requires the same precautions for chip removal as in drilling by other methods.

Step 10   Remove burrs caused by the drilling operation.

HOW TO REAM HOLES ON A LATHE

Note: Setups for machine reaming are similar to drilling and countersinking. The machine reamer may be held in the headstock spindle and turned. Or, the work may rotate and the reamer may be held stationary in the tailstock spindle or in a quick-change tool post. Holes that are to be reamed concentric, parallel to a fixed axis, and to within precise tolerances should first be bored.

Step 1   Select a reamer that will produce a hole within the specified tolerance.

Step 2   Chuck the reamer if it is a straight-shank type. Mount a taper-shank reamer in the headstock or tailstock spindle. Fig. 35-10 shows a typical setup for machine reaming on a lathe.

Step 3   Determine the cutting speed and cutting feed. These depend on the material in the workpiece, the amount of material to be removed, the reamer type, and the use of a cutting fluid.

Step 4   Set the spindle speed (rpm). Set the feed if the setup permits machine feeding. Direct the flow of cutting fluid. Start the lathe.

Step 5   Bring the reamer to the workpiece so that it starts to cut. Take a cut for about 1/8″. Withdraw the reamer. Stop the machine. Clean the workpiece. Measure the hole size with a gage or micrometer.

Note: If an adjustable machine reamer is used, make a (+) or (−) adjustment as needed. If a solid reamer is being used

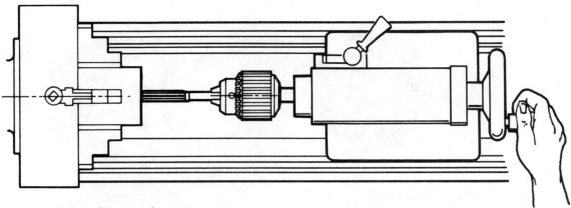

Fig. 35-10. A typical machine reaming setup.

and it is cutting oversize, it should be replaced.

Step 6   Continue to ream to the required depth.

Step 7   Withdraw the reamer. Stop the lathe. Remove the fine sharp burr. Clean the workpiece. Check for final size. Reream if necessary.

CAUTION: Cut off the power when cleaning a reamed hole. The outer edge of a reamed hole always has a fine sharp burr. Remove all burrs. Wrap the wiping cloth around a rod to clean a deep hole.

HOW TO SPOT AND TRUE A CENTER HOLE

Truing With A Centering Tool Bit

Step 1   Grind a 60° centering tool. The top face should be flat.

Step 2   Set the tool bit on center. Align the cutting side at 30° to the axis of the lathe.

Step 3   True the workpiece in a four-jaw chuck. Use a dial indicator to test for trueness.
Note: Use a steady rest to support the end of a workpiece that projects a considerable length from the chuck.

Step 4   Set the lathe for a slow spindle speed (around 200 rpm).

Step 5   Position the centering tool in the center hole by hand. Take light cuts with the full face of the centering tool. Continue until the hole is concentric.
Note: Once centered, the center hole is usually finished with a 60° countersink.

Truing With A Special Countersink

Step 1   Mount and true the workpiece. Use a steady rest to support a long workpiece.

Step 2   Mount the special 60° countersink in a drill chuck. Secure the drill chuck and countersink in the tailstock spindle.

Step 3   Position the tailstock spindle so that it projects as little as possible. Bring the tailstock close to the workpiece.

Step 4   Set the spindle speed according to the size of countersink and length of cutting face.

Step 5   Turn the tailstock spindle clamp until it applies a slight force on the spindle.

Step 6   Apply cutting fluid. Bring the special countersink into the center hole so that it cuts on the high side, or eccentric surface.

Step 7   Feed slowly. Continue to feed only as far as is needed to make the center hole run true.

SAFE PRACTICES IN CENTER DRILLING, DRILLING, COUNTERSINKING, AND REAMING ON THE LATHE

- Guide the cutting tool to cut a concentric hole by center drilling or spotting a center.
- Position the center hole of a cutting-tool holder against the dead center. Maintain a continuous force against the dead center with the back end of a regular tool holder. This action controls the feed of the cutting tool. It also prevents the cutting tool from *breaking through*.
- Clamp a workpiece that is to be drilled or reamed. This prevents it from turning. Workpieces should *not* be held by hand.
- Remove any raised point that may be left from facing at the center, *before* attempting to center drill.
- Touch the back end of the tool holder against the body of a drill. This helps position the drill so that it cuts concentrically.
- Continue to use force to hold a drill or reamer against the dead center when withdrawing the centering tool.

- Reduce the speed and feed when drilling large-diameter holes to a depth of more than twice the drill size.
- Remove a tool bit or other cutting tool when it is not in use. This prevents injury caused by brushing against the cutting point and edges.
- Grind the area steeply beyond the cutting point of a center-hole-spotting tool bit. The angular cutting edge must be able to cut to center hole depth without rubbing.
- Stop the lathe to clean out a drilled or reamed hole. Use a wiping cloth around a rod if the hole is deep.
- Burr the edges of each drilled or reamed hole if the job permits.

TERMS FOR CENTERING, DRILLING, COUNTERSINKING, AND REAMING ON THE LATHE

Spotting a center	Cutting an indentation at an angle in the center of a workpiece. Machining a center concentric with the axis of a revolving workpiece.
Special centering countersink	A 60°- included-angle countersink with only one lip ground to cut a center hole.
Drill holder	A device for securely holding a drill.
Straight-shank holder	A holding device that clamps on the straight shank of a drill or reamer. The holder permits the drill to be centered, fed into, and removed from the workpiece.
Threading into the work (breaking through)	The action of a cutting tool as it cuts through a workpiece. A drill or other cutting tool that feeds into a workpiece at the rate of the helix angle. The rapid movement through a workpiece when a drill or reamer cuts faster than the cutting feed.
Truing a center	Recutting a damaged center hole so that it is concentric with the axis of a workpiece. The process of truing or centering a hole with a center tool bit or specially ground countersink.

SUMMARY

- Center drills, drills, countersinks, and machine reamers for lathe work are the same as those used on drilling and other machine tools. The size, classification system, design features, and materials of construction of hole-producing cutting tools are similar regardless of the machine tool.
- Center drills and smaller sizes of straight-shank drills, countersinks, and reamers may be held in a drill chuck. A taper arbor permits mounting in a headstock, sleeve, socket, or in the tailstock spindle.
- Center drilling, drilling, countersinking, and reaming may be performed by rotating either the cutting tool or the workpiece.
- Holes may be produced on the lathe by feeding either the workpiece or the cutting tool to the required depth.
- Cutting tools used for producing holes on the lathe may be mounted and driven from the headstock spindle.
- Cutting tools also may be positioned, held in, and fed with the tailstock spindle.

- Taper-shank center tools may be held in taper sleeves, sockets, the taper spindle nose, or a tailstock. A taper-shank cutting-tool holder is used to hold and position the cutting tool.
- The cutting speeds and feeds for center drilling, drilling, countersinking, and reaming are similar to those used with drilling machines.
- An out-of-round or damaged center hole may be trued with a specially ground centering tool bit. A countersink ground so that one lip cuts the angular section is another center-hole-cutting tool.
- The cutting action is easier, friction and heat are reduced, and a finer surface finish is produced when the correct cutting fluid is used.
- Workpieces that extend a distance from the spindle nose or chuck should be supported with a steady rest. The steady rest should be located near the right-hand end of the workpiece for facing, centering, drilling, and reaming operations.
- Safe operating procedures must be observed for holding the workpiece, setting up the cutting tools, and performing the machining operations.

UNIT 35 REVIEW AND SELF-TEST

1. Tell what functions are served by (a) the *pilot drill* and (b) the *angle cutting faces* of a center drill.
2. State two differences between drilling and reaming on a lathe.
3. Identify three different (a) reamer types, (b) materials of which reamers are made, and (c) factors influencing the selection of the reamer.
4. State (a) how a 60° countersunk hole is represented on a drawing and (b) what dimensioning information is to be found in a note.
5. Tell how an out-of-true center spot may be corrected before a hole is drilled.
6. Indicate the advantages of a *quick-change toolholder* for holding cutting tools.
7. List two cautions to observe with reamed holes.
8. List three functions that are served by a cutting fluid for deep-hole drilling.
9. State two safety precautions to follow in *through drilling* or *reaming*.

Unit 36

BORING PROCESSES AND MANDREL WORK

Boring is the enlarging of a hole by internal turning processes. In general practice, boring is done on the lathe with a single-point cutting tool. A boring tool is mounted in a holder. This is secured in the tool post. As the cutting tool is fed into a revolving workpiece, a hole or internal formed surface is cut to a required size. Boring is a comparatively expensive and time-consuming process. Boring is used primarily when the concentricity of a hole is important.

This unit deals with common types and forms of boring tools, boring bars, and holders. These are especially designed for lathe work. A few other adaptations of boring processes, like counterboring, recessing, and taper boring, also are described. Another part of the unit concentrates on applications of bored holes to other work processes. The arbor press and mandrels are covered in relation to force fits and mandrel work.

Sometimes the shape and size makes it impractical to turn the workpiece. While almost an obsolete practice, a work-holding fixture is used in the carriage. This replaces the compound slide and rest. The boring tool is mounted on an adjustable boring head in the spindle. The workpiece is secured in the fixture. The workpiece is fed longitudinally into the cutting tool by carriage feed. Longitudinal and vertical boring mills are specially designed for such operations. Holes are usually bored in irregular-shaped workpieces using a vertical milling machine. Such holes may be bored easier, faster, and more accurately.

PURPOSES OF BORING PROCESSES

Boring serves four main functions:
- To enlarge the diameter of a hole, particularly very large holes
- To true up a hole. (The surface is bored concentric and straight in relation to the axis of the workpiece.)
- To produce an accurate, high-quality surface finish in an odd-size hole
- To machine a true hole as a pilot for subsequent cutting tools.

Holes that are to be reamed concentric and to size with a hand or machine reamer should be bored whenever possible. A hole is usually bored to within 0.005″ (0.1 mm) to 0.007″ (0.2 mm) for machine reaming. This tolerance may be increased to 0.010″ (0.3 mm) for ½″ (12.5 mm) – diameter reamed holes; 0.016″ (0.4 mm) for 1″ (25 mm) diameter; 0.030 (0.8 mm) for 2″ (50 mm) diameter; and up to 0.045 (1.1 mm) for a 3″ (76 mm) diameter hole. Tolerances for hand reaming on diameters up to 1″ (following a boring operation) range from 0.002″ (0.5 mm) to 0.005″ (0.1 mm).

While called by different terms, there are a number of other boring operations. Counterboring of large diameters is usually done with a single-point cutting tool. The cutting of a groove or undercut is another internal boring process. Threads and tapers are cut internally. However, in general practice, the term *boring* is applied to the enlarging of a diameter.

BORING TOOLS, BARS, AND HOLDERS

Four groups of cutting tools are in common use for boring processes. Forged boring bits, standard-size square high-speed steel cutting tool bits, carbide-tipped bits, and ceramic inserts are widely used. Each may be ground or formed for internal boring, grooving, threading, or form turning.

Forged Boring Tool And Holder

The forged single-point boring tool (Fig. 36-1) has an offset end. It is generally used for light boring operations. The cutting tool may be made of high-speed steel, or the cutting end may be tipped with a carbide insert. The forged type is secured in an offset V-grooved toolholder. The yoke on the toolholder type illustrated in Fig. 36-1 is reversible. This permits the holder to be used as both a right- and a left-hand toolholder. A single toolholder can accommodate a number of different cutting tool diameters. The forged cutting tool is used in small-diameter holes, particularly for sizes from ⅛″ (3 mm) to ½″ (12 mm).

The cutting end of a forged boring bit is ground similar to that of a left-hand turning tool. The cutting end must have front clearance, side clearance, and

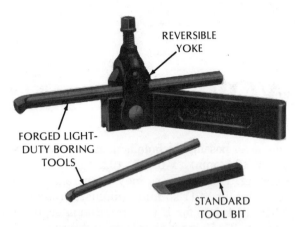

Fig. 36-1. A reversible-yoke holder and boring tools. (Courtesy of Armstrong Bros. Tool Company)

side rake like a turning tool. Fig. 36-2 illustrates these characteristics. The amount of front clearance increases sharply for smaller diameters. The front clearance angle must be adequate to permit the cutting edges to cut freely without rubbing at the heel. However, the amount of front clearance should not be excessive. Otherwise, the cutting edge will not be properly supported, and it will break away.

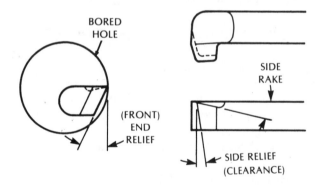

Fig. 36-2. Rake and relief angles of a solid, forged boring tool.

The shape of the cutting edges depends on the operations to be performed. Fig. 36-3A shows a forged boring bar ground for shoulder turning. The shape is similar to that of a left-hand facing tool. The cutting edge of a cutting tool for boring a straight hole is shown in Fig. 36-3B. It is shaped like that of a regular left-hand turning tool.

Boring Bars And Holders

A *boring bar* is a round steel bar. The bar positions and holds a tool bit. Boring bars may be positioned for maximum rigidity. One type has a broached square hole in one end. This accommodates a regular-size square high-speed steel or carbide-tipped tool bit.

Boring bits may be held in a boring bar at 90°, 45°, and 30° angles.

Fig. 36-4A is a *plain boring bar*. The tool bit may be secured in either end by means of a set screw.

Another general-purpose boring bar is shown in Fig. 36-4B. One end is slotted to position and hold the cutting tool at 90°, 45° or 30°.

End-Cap Boring Bar

A third type is called an *end-cap* boring bar (Fig. 36-4C). The cutting tool is held in position by the wedging action of a hardened plug. Three interchangeable ends are shown in Fig. 36-4C.

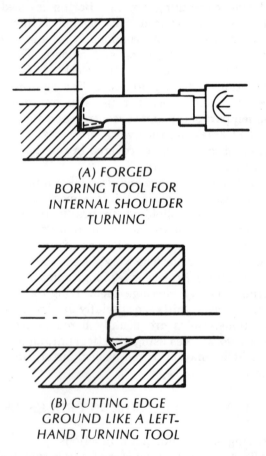

(A) FORGED BORING TOOL FOR INTERNAL SHOULDER TURNING

(B) CUTTING EDGE GROUND LIKE A LEFT-HAND TURNING TOOL

Fig. 36-3. The correct cutting edge shape of a boring tool is determined by the boring process.

Web-Bar Boring Toolholder

The *web-bar* boring toolholder is slotted on one end. The cutting tool is slipped into this slot. The nut on the opposite end is tightened. This draws a tapered draw bar that transmits a tremendous clamping force on the boring bar slots. These grip the cutting tool so securely that the cutting tool will break before it moves in the bar.

Fig 36-5 shows a heat-treated web-bar boring bar. The line drawings illustrate a few common shapes of cutting tools.

Correct adjustment of a boring tool is particularly important. The boring bar and/or cutting tool must be clamped as short as possible. This minimizes the overhang. Greater tool rigidity is thus provided for the cutting process.

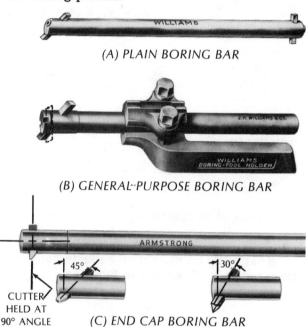

(A) PLAIN BORING BAR

(B) GENERAL-PURPOSE BORING BAR

45°   30°

CUTTER HELD AT 90° ANGLE   (C) END CAP BORING BAR

Fig. 36-4. Common types of boring bars for straight and angle positioning of a tool bit. (Courtesy of the J.H. Williams Division of TRW Inc., and Armstrong Bros. Tool Co.)

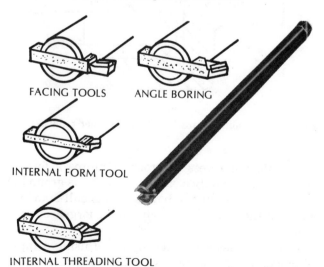

FACING TOOLS   ANGLE BORING

INTERNAL FORM TOOL

INTERNAL THREADING TOOL

Fig. 36-5. A web-bar boring toolholder and common shapes of cutting tools. (Photo courtesy of Do All Company)

Heavy-Duty Boring Bar Set

This set consists of a combination toolholder/tool post and three sizes of boring bars. These permit the operator to use the largest size possible for a particular job. With greater strength in the boring bar, it is possible to take longer cuts at increased speeds.

Fig. 36-6 shows a heavy-duty boring bar set. Note that the cutter may be positioned at 90°, 45°, and 30° in any one of the three bars. The boring bar and cutting tool are positioned and secured by turning the nut. This clamps the boring bar in the holder body. The body is secured at the same time in the compound rest T-slot.

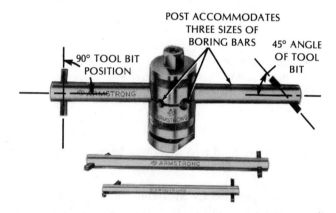

POST ACCOMMODATES THREE SIZES OF BORING BARS

90° TOOL BIT POSITION

45° ANGLE OF TOOL BIT

Fig. 36-6. A three-bar, heavy-duty boring toolholder and boring bars. (Courtesy of Armstrong Bros. Tool Co.)

Boring Bar And Quick-Change Toolholder

A boring bar may easily be set in a quick-change toolholder. One design of bar has three angular ends. These permit boring with the tool bit held at an angle or at 90° to the work axis. The quick-change toolholder may be slid quickly in the dovetailed slots. The cutting tool may be adjusted vertically and locked in this position.

Boring Bars For Production Work

Different types of boring bars are used with turret lathes and other semi-automatic turning machines. Such boring bars require throw-away carbide or ceramic inserts. Two fixed-head types are illustrated in Fig. 36-7. The type on the left uses square inserts for through boring. The boring bar on the right may be used for both threading and grooving by just changing the precision insert. It is shown here with a grooving insert installed.

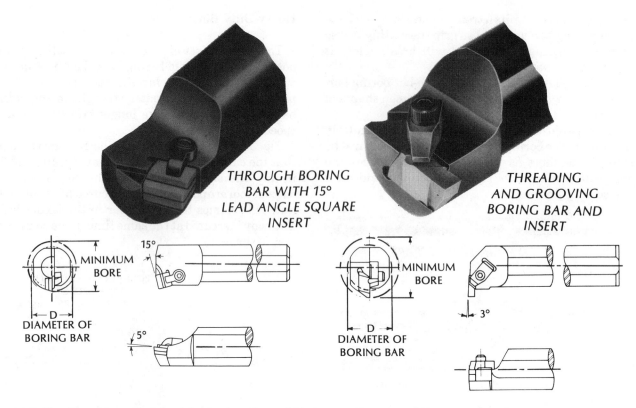

Fig. 36-7. Fixed-head type steel-shank boring bars for carbide inserts. (Courtesy of Kennametal Inc.)

Other types also are available. Holes may be bored to a square shoulder by using boring bars with triangular inserts. The shanks of these boring bars are made of steel or tungsten carbide. The latter provides greater rigidity than the standard steel boring bars.

CUTTING TOOL SETUPS FOR BORING

The 90° slot position for the cutting tool is used for straight through boring. The cutting tool is set on center. Feeding is (right-hand) toward the spindle. A side rake of 5° to 7° provides a good cutting-edge angle. The amount of end relief depends on the inside diameter. The side-relief angle should be between 12° and 15°.

Counterboring, Recessing, And Boring Tapers

Three other common boring operations, in addition to internal threading, are counterboring, recessing (undercutting), and taper boring.

For most workpieces, counterbored holes are formed with a counterbore. On larger diameters, the counterbored hole is formed by boring. The shoulder of a counterbored hole is machined either square or with a small corner radius. The cutting tool and

boring bar setup for machining a counterbored hole with a radius is illustrated in Fig. 36-8.

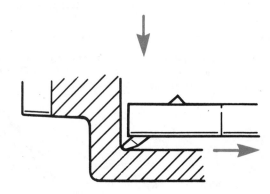

Fig. 36-8. A counterboard hole with a corner radius.

An internal recess or groove requires a formed cutting tool. Usually a standard tool bit is ground to shape and used with a boring bar. Small-diameter recesses may be formed with solid forged cutting tools.

Tapered holes are generally bored using a taper attachment. The process is similar to regular taper turning, but with two major exceptions: (1) Although the tool is fed toward the spindle, it is ground for turning left-hand. (2) The end-relief angle on smaller diameters must be increased to prevent the end of the cutting tool from rubbing.

The boring of short steep tapers combines angular turning with boring. The compound rest is positioned at the taper angle (Fig. 36-9A). The carriage is set in position. Angular feeding is done with the compound rest handwheel. The length of taper that can be bored is limited to the movement of the compound rest slide. Large steep tapers are often cut using a regular toolholder. Fig. 36-9B shows the setup with a right-hand offset toolholder for boring a large-diameter angle surface.

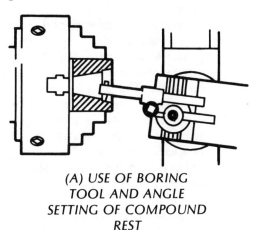

(A) USE OF BORING TOOL AND ANGLE SETTING OF COMPOUND REST

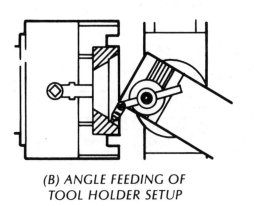

(B) ANGLE FEEDING OF TOOL HOLDER SETUP

Fig. 36-9. Boring short steep tapers.

A standard toolholder provides greater rigidity. It is preferred wherever the work size and operation permit boring in this manner. Other short tapers are bored using a boring bar or a solid, forged tool bit. The tapers are usually checked for machining accuracy of the taper angle and size with either a taper gage or a mating angle part.

HOW TO BORE A HOLE

Setting Up The Lathe And Workpiece

Step 1   Mount the workpiece.
Note: Position the workpiece out from the chuck or faceplate to bore a through hole.

Step 2   Face the end. Center drill to within 1/32″ (0.8 mm) of the finished diameter.
Note: Larger diameter holes are often cast or roughly cut out with a torch. Such surfaces are often machined directly by boring. In the case of cast iron, the first roughing cut must be below the scale.

Step 3   Set the spindle speed. The same rpm should be used as for outside turning the same diameter.

Step 4   Set the feed. For rough boring, the feed is decreased because of the nature of the boring operation and the tendency of the boring tool to spring under heavy cuts.

Straight Hole Boring

Step 1   Select the largest size boring bar that can be accommodated in the hole. Grind the cutting tool according to the nature, size, and shape of the boring operation.
Note: The angle settings of the boring bar are used for undercutting and shoulder boring. The 90° setting is used for straight boring.

Step 2   Mount the boring bar in the toolholder, with the cutting tool as close to the toolholder as possible. The boring bar length should permit clearing the depth of the hole to be bored.

Step 3   Set the cutting edge at center height (Fig. 36-10). Check to see that the heel of the cutting tool clears the diameter.

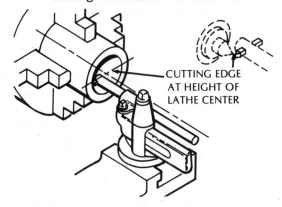

CUTTING EDGE AT HEIGHT OF LATHE CENTER

Fig. 36-10. Tool and holder setup for straight boring.

Step 4   Start the lathe. Move the cutting tool to take a trial cut for a short distance.
Note: Cutting must take place around at least half the circumference of a hole that is not concentric. This is necessary for taking a first measurement.

Step 5   Stop the lathe. Measure the bored diameter. Measurements may be taken with a steel rule, inside caliper, telescopic gage, inside micrometer, or a vernier caliper. Fig. 36-11 shows the correct handling of a vernier caliper to measure an inside diameter.

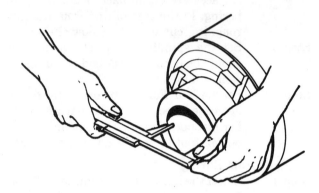

Fig. 36-11. Measuring the diameter of a bored hole with a vernier caliper.

Step 6   Position the boring tool to take the deepest possible cut with the coarsest feed.
Note: If chatter marks are produced, check the correctness of the cutting and relief angles of the boring tool. Reduce the speed, feed, and depth of cut, if necessary.

Step 8   Use a cutting tool that has been honed for a fine finish cut.

Step 9   Take a trial cut with a fine feed for about 1/8". Measure the diameter. Adjust the depth of cut if required.

Step 10   Apply cutting fluid and take the finish cut.
Note: The diameter of the bored hole should be checked at several places along its length. This is done to be sure the hole is not *bell-mouthed*. If this happens, it indicates the tool is springing away from the work. Take an additional cut or two at the same setting. This should produce a parallel bored hole.

Step 11   Break the sharp edge of the bored hole with a hand scraper.

Boring A Counterbored Hole

Step 1   Use the largest practical diameter of boring bar. Grip the boring bar short.

Step 2   Insert a left-hand boring tool. Rough bore the counterbored area.
Note: The cutting tool is usually held in the 30° or 45° position. This permits the cutting tool to cut close to the shoulder.

Step 3   Step off the area adjacent to the shoulder with a smaller radius tool point.
Note: Stepping off reduces the amount left for final machining of the shoulder area.

Step 4   Measure the diameter and depth of the counterbored hole.
Note: A steel rule, depth gage, depth micrometer or a vernier depth gage may be used. The measuring tool to use depends on the required tolerance.
Note: The depth is sometimes reached using a carriage stop. Also, the compound rest axis may be set parallel to the lathe axis. The boring bar is then fed by the compound rest handwheel to the specified depth.

Step 5   Stone the nose and cutting point for the final cut. Increase the speed; decrease the feed. Take a light finish cut along the inside diameter (Fig. 36-12). Disengage the power feed before the shoulder is reached.

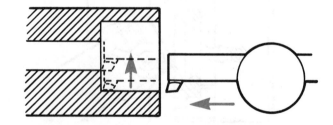

Fig. 36-12. Finish boring the diameter and face.

Step 6   Feed the tool to the shoulder by hand. Then, feed inwardly to take a light finish cut across the face of the shoulder.

Step 7   Remove the burr on the outside edge. Recheck the diameter and the depth.

Undercutting (Internal Recessing)

Step 1   Grind a boring tool to the shape of the required groove or recess.

Step 2   Swivel the compound rest to 0°. Check the setting. The compound rest should be parallel to the lathe axis.

Step 3   Secure the tool bit in the boring bar. The boring bar should project far enough from the holder to clear the workpiece.

Step 4   Set the spindle speed for the diameter of the workpiece and material to be bored. Use the same cutting speed as for a cutting-off process.
Note: The speed may need to be decreased further if the boring bar or solid forged tool bit overhangs the workpiece for any distance. There is a tendency under these conditions to produce chatter.

Step 5   Position the left side of the cutting tool so that is almost grazes the flat face of the workpiece.

Step 6   Set the graduated collar on the compound rest at zero. Lock the carriage at this position.

Step 7   Start the lathe. Move in the compound rest the required distance.

Step 8   Apply a cutting fluid. Feed the cutting tool until it touches the inside diameter. Set the cross slide micrometer dial to 0°. Move the cutting tool in to depth.
Note: The cutting tool is moved sideways 0.002″—0.003″ in the groove. This *frees up* the sides the same as in cutting an external groove.

Step 9   Use a triangular hand scraper. Remove any burrs formed on the edges of the undercut section and the outside diameter.

MANDRELS AND MANDREL WORK

Many workpieces that are bored or reamed require further machining on the outside surfaces. Such parts may be positioned accurately in relation to the work axis by using a *mandrel*. A mandrel is a hardened cylindrical steel bar that is pressed into the finished hole of a workpiece.

Two common types of mandrels are the *solid* and the *expansion mandrel*. Other types include *gang*, *threaded*, and *taper shank mandrels*. Each of these general types is discussed.

The Solid Mandrel

A mandrel is a hardened cylindrical steel bar. The *solid mandrel* (Fig. 36-13) has a recessed center hole in each end. The ends are also turned smaller than

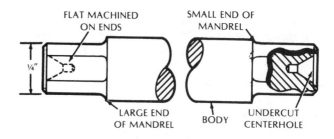

Fig. 36-13. The features of a solid mandrel.

the body size. A flat is machined to provide a positive clamping surface. The set screw of a lathe dog may be tightened against this surface. The large end has the mandrel size stamped on it. This marking also indicates the end on which the lathe dog should be clamped.

The body is ground with a slight taper of 0.0005″ per inch of length. The small end of mandrels under 1/2″ diameter (12 mm) is usually a half thousandth under the standard diameter. On large-size mandrels the small end is ground up to 0.001″ undersize. Due to the taper, the large end is a few thousandths larger than the normal diameter. The accuracy of a mandrel depends on the condition and accuracy of the center holes.

The Expansion Mandrel

The *expansion mandrel* (Fig. 36-14) accommodates a wider variation in hole sizes than the solid mandrel. The expansion mandrel consists of a taper mandrel and a slotted sleeve. The taper bore of the sleeve corresponds to the taper of the mandrel. The expansion mandrel may be expanded from 0.005″ to 0.008″ (0.1 mm to 0.2 mm) over the nominal size for

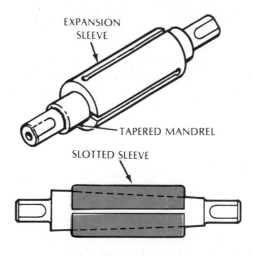

Fig. 36-14. Two types of expansion mandrels.

diameters up to 1″ (25 mm). The slotted sleeves come in different diameters. The same taper mandrel may be used with more than one diameter of sleeve.

The Gang Mandrel

The *gang mandrel* (Fig. 36-15) provides for the machining of multiple pieces. This type differs from the solid and expansion mandrels. The mandrel has a flanged, parallel-ground body, and a threaded end. The workpieces are placed side by side. A collar and nut are tightened to hold the parts in place for machining.

The gang mandrel illustrated in Fig. 36-15 is mounted between centers. A lathe dog, tightened against the flat surface on one end, provides the drive force.

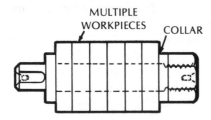

Fig. 36-15. A gang mandrel.

Threaded Mandrel

The *threaded mandrel* is used for mounting threaded parts that are to be turned. The workpiece is screwed onto the threaded end of the mandrel. The recessed thread and square flange permit the parts to be fed squarely.

The threaded mandrel may be designed for machining work between centers. This style is shown in Fig. 36-16A. The taper shank mandrel in Fig. 36-16B is another style. This mandrel may be fitted to the headstock spindle by using an adapter. An adaptation of the taper-shank mandrel is illustrated in Fig. 36-16C. One end is turned straight and is slotted. A special flat-head screw applies a force against the slotted segments. These, in turn, hold the workpiece.

THE ARBOR PRESS

The mandrel is pressed into the finished hole of a workpiece. The force is sufficient to permit the machining of the workpiece outside surfaces. Sometimes, the arbor is driven into position by a soft-face hammer. In such cases, the workpiece is placed on a

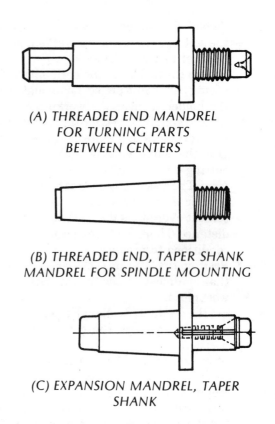

(A) THREADED END MANDREL FOR TURNING PARTS BETWEEN CENTERS

(B) THREADED END, TAPER SHANK MANDREL FOR SPINDLE MOUNTING

(C) EXPANSION MANDREL, TAPER SHANK

Fig. 36-16. Threaded and taper-shank mandrels.

mandrel block (Fig. 36-17) and the mandrel is driven into it.

When turning work on a mandrel, the cutting tool is set so that the cutting force is directed toward the large end of the mandrel. Light cuts are taken on large-diameter work. This is to prevent the work from turning on the mandrel.

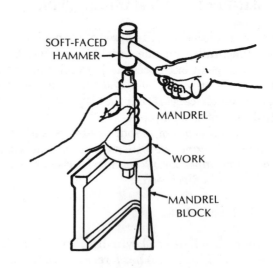

Fig. 36-17. Driving a mandrel into a workpiece.

An arbor press is designed for mounting work on arbors. Fig. 36-18 shows one common type. The workpiece is mounted on the table plate. Force is applied to the lever. This force is multiplied through a pinion gear to the rack teeth on the ram. The force is applied by the ram to move the mandrel into the bored or reamed hole.

The arbor press is also used widely in the assembling of shafts, pins, bushings, and other parts that require a force fit.

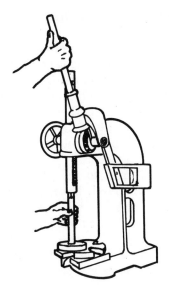

Fig. 36-18. A bench type of hand arbor press.

HOW TO MOUNT AND REMOVE WORK ON A MANDREL

Pressing A Mandrel And A Workpiece

Step 1   Select the type and size of mandrel most suited for the job and the size of the bored hole.
Note: The solid mandrel is preferred for standard hole sizes where a single part is to be machined. The expanded-bushing type is adapted for nonstandard hole sizes.

Step 2   Wipe the workpiece hole and the mandrel. Examine the surfaces. Remove any burrs or score marks.
Note: The hardened surface of the mandrel requires the use of an oilstone on any burrs.

Step 3   Apply a thin film of lubricant on the mandrel body and the bored hole.

CAUTION: The lubricant prevents seizing and scoring when pressing the mandrel into or out of the workpiece.

Step 4   Insert the small end of the mandrel body into the workpiece. As the mandrel is pressed in, the force of the slight taper against the bored hole securely holds the two parts.

Step 5   Place the mandrel and the workpiece upright between the ram and table plate of the arbor press.
Note: If a mandrel block and soft-face hammer are used, the workpiece is placed on the flat surface of the mandrel block. The mandrel is then driven by the soft-face hammer.
Note: Parallel bars or a disk are generally placed between the work face and the table plate. This setup provides better support and prevents scoring of a finished surface.

Step 6   Hold the mandrel with the left hand. Lower the ram by pulling down on the lever with the right hand. Check to see that the workpiece rests securely on the table. The mandrel must be square with and centered under the ram.

Step 7   Adjust the lever arm for maximum leverage.

Step 8   Apply force to the lever. Press the arbor into the workpiece until the parts feel secured.

CAUTION: The amount of force required depends on the material in the workpiece, its size, and the nature of the cuts that are to be taken. A part may be distorted or a casting may crack if too much force is exerted.

Step 9   Reexamine and clear the center holes of the mandrel before mounting between centers.

Step 10   Fasten the lathe dog on the large (stamped) end of the mandrel.

Removing A Mandrel

Step 1   Place the workpiece on the table plate. Use parallels or a disk between the work face and table plate if added support is needed.

CAUTION: The large diameter (stamped size end) of the mandrel must face *downward.*

431

Step 2   Center the small end of the mandrel under the ram. Check to be sure that the mandrel and workpiece are positioned squarely.

Step 3   Hold the mandrel with the left hand.

Apply force on the lever with the right hand.

Note: A drop or two of lubricating oil should be applied where the mandrel and hole fit together.

SAFE PRACTICES IN INTERNAL BORING

- Select the largest tool bit-and-holder combination that meets the job requirements. This will provide maximum rigidity with adequate clearance.
- Clamp the boring bar so that the cutting tool is as close to the holder as possible. Check to see that the head end clears without rubbing into the chuck or the end of the workpiece.
- Grind the cutting point (front) relief angle steep enough so that the heel will not rub on the workpiece.
- Position the cutting tool at an angle when counterboring and facing against a shoulder.
- Move the workpiece out from the face of a chuck or faceplate for through-hole boring.
- Take trial cuts. Measure the diameter (and depth if necessary) before setting for the final cut.
- Stop the lathe before taking a measurement.
- Use cutting fluids when boring, to improve the quality of the surface finish.

- Position the nozzle so that the lubricant flows *inside* the bored hole.
- Check the condition of the bored hole. This is particularly important in a long bored hole or if a light forged boring tool is used.
- Shorten the holder length if chatter is produced. Also, decrease any or all of the following: speed, feed, and depth of cut.
- Burr the edge of the bored hole or recessed area with a triangular scraper.
- Apply sufficient force to hold the workpiece securely on the mandrel. Excessive force may expand the bore or crack a casting.
- Stop the lathe to remove chips and clean a bored hole.
- Use a protective shield or safety goggles, and observe all machine safety practices.

TERMS APPLIED IN BORING PROCESSES AND MANDREL WORK

Boring (lathe work)	A process of enlarging a hole by advancing a single-point cutting tool into a revolving workpiece.
Boring tool	A standard tool bit ground for an internal turning process. A solid shank forged cutting tool that is offset and ground to form a cutting tool.
Boring bar	A round or square-shaped steel bar. A bar for positioning and holding high-speed steel, carbide-tipped, or ceramic inserts. A steel bar that is secured in a tool post or other holder. A steel bar that is slotted or has a square-shaped hole to receive a cutting tool.
Bored tapered hole	A hole that regularly increases in diameter a given amount over a specified distance. A steep angle or a small taper angle produced by turning the inside of a hole.
Counterboring (turning)	Producing an enlarged bored hole. A bored hole that is separated by a shoulder from a smaller but concentric hole.
Undercutting	Boring a groove or recess internally on a workpiece.
Mandrel	A work-holding device on which a bored part may be mounted and secured for subsequent turning operations. Generally, a hardened, ground, and centered tool having a body that tapers 0.0005″ per inch.

Gang threaded mandrel   A straight, ground, cylindrical device. A mandrel that has a shoulder end and a threaded end. A device for holding a number of bored workpieces for subsequent machining operations.

Taper shank threaded mandrel   A mandrel that is threaded at one end to receive a threaded part. The other end is tapered to fit a standard spindle nose adapter.

Arbor press   A device consisting of a ribbed frame, table, table plate, and movable ram. A device for applying force to assemble or disassemble mating parts.

Pressing a mandrel   The process of applying force on a mandrel to hold or to remove a bored or reamed workpiece.

SUMMARY

- Boring is the process of enlarging and producing a concentric and straight hole.
- Boring produces a quality finished internal surface. This may be machined accurately to any standard or odd-size diameter.
- Forged and regular square-shaped high-speed and carbide-tipped tool bits are widely used for general-purpose boring processes. The forged type solid cutting point and shank are largely used for boring small diameters.
- Most tool bits are ground for left-hand turning. However, the boring tool is usually moved horizontally toward the spindle.
- The end of the standard boring bar holder is broached to receive a tool bit. This may be positioned at a 30°, 45°, or 90° angle to the boring bar.
- Boring bars and holders are fitted with interchangeable ends. The bars are of different diameters.
- Boring bars may also be positioned for rapid tool changes. The cutter and bar may be held in a quick-change tool holder.
- Heavy-duty boring bars have a clamping device on what is called a fixed head. Square, triangular, and rectangular inserts may be inserted and clamped in position.
- Solid tungsten carbide boring tools provide maximum rigidity.
- Counterboring, recessing, internal taper turning, and internal threading are adaptations of the boring process.
- Cutting speeds and feeds normally used for external turning processes are reduced for boring, except for light cuts.

- Long tapered holes are bored with a taper attachment. Short and steep tapers are bored by swinging the compound rest to the required angle. The cutting tool is fed by the compound rest handwheel.
- The largest practical size of boring bar is used. The overhang from the tool post is reduced to a minimum. Workpieces should be rough bored using the deepest cut, coarsest feed, and fastest speed possible. These are governed by the material, size, strength, and shape of the workpiece, and the type of operation and cutting tool setup.
- Chatter may be eliminated by reducing the speed, feed, and overhang, and by establishing the correct relief and rake angles, and accurately positioning the cutting tool on center.
- The stepped-off sections of a counterbored hole are roughed out. The surface between the turned diameter and face is then formed. The boring tool is fed longitudinally in to depth. The tool is then fed toward the center to face the end.
- Internal recessing and undercutting require cutting tools with shapes similar to those used for external processes.
- The external surfaces of a bored part may be machined by first mounting the part on a solid or expansion mandrel. Multiple pieces may be machined on a gang mandrel. Threaded mandrels are used for mounting finish threaded parts.
- The arbor press is used to force a mandrel into the bored hole of a part. The force fit must be sufficiently tight to prevent the part from turning on the mandrel. Care must be taken, particularly with castings, to not create excessive holding force.
- Goggles or other protective devices should be worn in the machine area. All safety precautions must be observed.

UNIT 36 REVIEW AND SELF-TEST

1. Determine and state the amount of material to be allowed for machine reaming the following bored holes: (a) 25 mm diameter, (b) 2″, and (c) 90 mm.
2. Name four groups of cutting tools that are used for boring processes.
3. Give two advantages of a heavy-duty boring bar set compared to a plain boring bar.
4. a. Identify three different boring operations performed with *throwaway carbide* or *ceramic inserts*.
   b. Name the shape of the insert that is used for each application.
5. State two differences in grinding a cutting tool for taper boring instead of for regular turning.
6. List the steps for boring a counterbored hole.
7. Describe *undercutting (internal recessing)*.
8. a. List five common types of lathe mandrels.
   b. Cite the advantage of a *multiple-piece workholding mandrel* compared to one used for an individual part.
9. Indicate two different applications of arbor presses.
10. State two safe practices that must be followed to prevent damage to work that is mounted on a mandrel.
11. Indicate corrective steps to take if chatter is produced during boring.

Unit 37

CUTTING AND MEASURING EXTERNAL AND INTERNAL SCREW THREADS

Thread cutting was related in earlier units to the hand tapping of internal threads and the cutting of external threads with a die. Many other threads are machined on a lathe using a single-point thread-cutting tool. This unit covers the gages needed to measure the different cutting tool angles or the thread pitch. The lathe setups and tools, the cutting of single-pitch right- and left-hand (internal and external) threads, and new terminology are also included. Thread measurements are applied. Formulas are used to compute required thread dimensions.

GAGES, CUTTING TOOLS, AND HOLDERS

Center Gage

A small, flat gage is used to check the accuracy of the sides of a thread-cutting tool (Fig. 37-1). Called a *center gage*, is has a series of 60° angles. The parallel edges have a number of fractional graduations. The graduations are used to measure thread pitches. The center gage is also used to position the thread-cutting tool in relation to the axis of the workpiece.

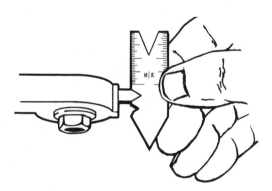

Fig. 37-1. A center gage being used to check a 60° thread-cutting tool.

Screw Thread Gages

One type of *screw thread gage* is a circular disk with a series of V-shaped (thread form) openings around the circumference. The thread sizes on this type (Fig. 37-2A) conform to the standards for American National or Unified Threads. Another type is a flat plate with a 29° included angle (Fig. 37-2B). It is used for checking Acme threading tools. An Acme screw thread gage is used for measuring the accuracy of the thread-cutting tool, including the flat point. A similar flat plate gage is available for checking SI Metric threads.

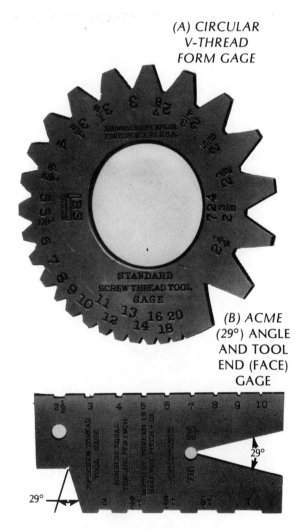

(A) CIRCULAR V-THREAD FORM GAGE

(B) ACME (29°) ANGLE AND TOOL END (FACE) GAGE

Fig. 37-2. Examples of 60° and 29° thread form gages. (Courtesy of Brown & Sharpe Manufacturing Co.)

Threading (Thread-Cutting) Tool

A *threading tool* is a single-point cutting tool that has an included angle and a point shape that meet a specific thread form standard. The thread-cutting tool requires that the flank of the side cutting edge be ground with additional clearance. The cutting edges are ground at an angle equal to the thread helix angle *plus* the regular relief angle. This clearance angle prevents the cutting tool from rubbing against the threads as the tool advances along the workpiece (Fig. 37-3). The cutting tool on the left in Fig. 37-3 has sufficient side clearance. The tool on the right has inadequate clearance. The thread cannot be cut accurately. The flank rubs against the side of the thread, preventing the tool from cutting.

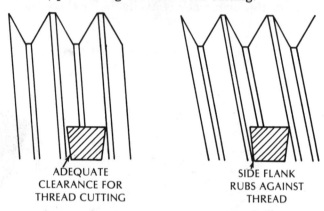

ADEQUATE CLEARANCE FOR THREAD CUTTING

SIDE FLANK RUBS AGAINST THREAD

Fig. 37-3. Providing side clearance to permit thread cutting.

Fig. 37-4A shows a correctly ground, standard high-speed steel thread-cutting tool. This is for a 60° right-hand thread. The point is rounded slightly for a Unified thread. The front clearance angle permits the cutting tool to clear the diameter of the revolving workpiece. The top face is ground at the angle of the cutting tool holder. When secured in the tool holder at center height for threading, the top face is horizontal. The clearance angle for cutting a left-hand thread is formed on a left (side) cutting flank (Fig. 37-4B).

Formed Threading Tool

A circular blade ground to a particular thread form is called a *formed threading tool*. The blade has appropriate relief and clearance angles. Fig. 37-5 shows both a formed cutter and a holder. The adjustment of the cutting edge at center height is illustrated in Fig. 37-6. The circular cutter is held securely against the side face of a toolholder. A hardened stop screw is used to adjust the cutter at center height. This screw also prevents the blade from being moved by the cutting force.

Spring Head Thread-Cutting Toolholder

The *Spring head* thread toolholder is generally used with a standard high-speed steel square tool bit. The spring head may be tightened with a locking nut. This permits the taking of heavy and roughing cuts. When loosened, the holder has a spring feature. This is especially desirable for finish threads.

Fig. 37-7 shows an offset spring head thread toolholder. (A) shows that the square hole in the holder is broached at a 30° angle. Just one side of a high-speed tool bit requires grinding to cut coarse or fine 60° threads. The toolholder in (B) shows such a tool bit. Another design feature is the 4° (from horizontal) angle of the square slot. This is illustrat-

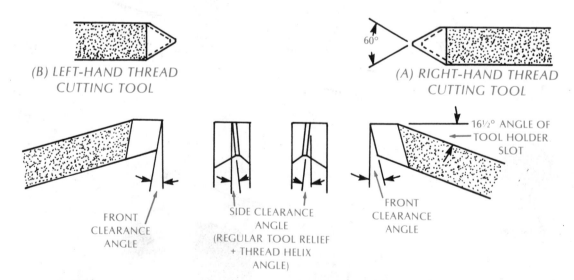

(B) LEFT-HAND THREAD CUTTING TOOL

60°

(A) RIGHT-HAND THREAD CUTTING TOOL

16½° ANGLE OF TOOL HOLDER SLOT

FRONT CLEARANCE ANGLE

SIDE CLEARANCE ANGLE (REGULAR TOOL RELIEF + THREAD HELIX ANGLE)

FRONT CLEARANCE ANGLE

Fig. 37-4. Additional clearance needed for thread cutting.

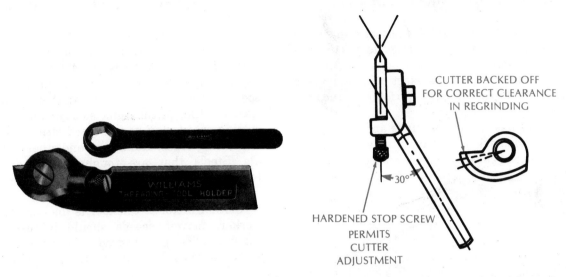

Fig. 37-5. A common circular-form threading cutter and holder. (Courtesy of the J.H. Williams Division of TRW Inc.)

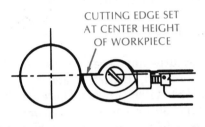

Fig. 37-6. The position of a formed tool for thread cutting.

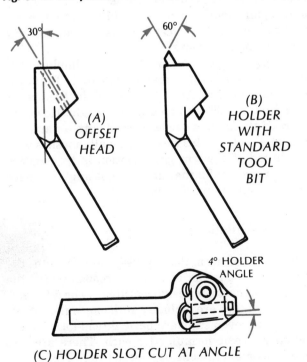

(C) HOLDER SLOT CUT AT ANGLE

Fig. 37-7. Design features of an offset spring head holder for a thread-cutting tool.

ed in (C). The tool bits are usually ground with a 15° clearance angle and a slight top rake.

MULTIPLE THREADS

Up to this point only regular *single-lead* threads have been considered. As a threaded part is turned one revolution, the mating part advances a distance equal to the thread pitch. However, there are times when a thread must be advanced quickly. *Multiple threads* are designed to increase the *lead*. The *lead* is the distance a threaded part moves longitudinally for one revolution.

Lead and *pitch* are the same on a *single-lead thread*. On a *double-lead thread*, the lead is *twice* the pitch. A *triple-lead thread* advances *three* times the distance of a single-lead thread per revolution. A *quadruple-lead thread* moves *four* times the distance of a single-lead thread. The double-, triple-, and quadruple-lead threads are known as *multiple-lead* threads, or *multiple threads*.

The cutting of multiple threads requires a different set of calculations for depth, pitch, and other dimensions. The lathe setup, the use of the thread-cutting dials, and the thread-cutting process for multiple threads differ from those used for cutting single-lead threads.

SCREW PITCH GAGE

The *screw pitch gage* is a gaging tool with a series of flat blades. One edge of each blade has a threaded form of a specific pitch. The blades are used to determine or to check the pitch of a thread. Gages are available with thread forms and sizes for American National, Unified, and SI Metric threads.

THREAD-CHASING ATTACHMENT

A *thread-chasing attachment* is a threading device that is attached to the lathe carriage. Its function is to locate a position at which a lead screw may be engaged or disengaged. The exact point of engagement permits the threading tool to follow in the helix of the previously cut groove.

The device has a dial with lines and numbers. A simple thread chasing attachment is shown in Fig. 37-8. The thread-chasing dial is connected to a worm

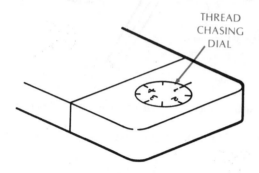

THREAD CHASING DIAL

Fig. 37-8. A thread-chasing attachment mounted on the carriage.

wheel. This is turned by the lead screw. The split-nut lever in the carriage is engaged when the chasing dial reaches a particular line. This stops the rotation of the dial. Threads in the American National and Unified thread series have an even or odd number of *half threads* per inch. Fig. 37-9 gives the even and odd lines on the chasing dial. The correct engagement of the split nut on the lead screw is indicated for even, odd, and half threads.

SETTING THE LEAD SCREW

The amount a lead screw moves in relation to each revolution of the lathe spindle is controlled by gear combinations in the quick-change gearbox. (The design features of gear boxes were described in an earlier unit.) The correct lead screw setting may be made by positioning the gears as indicated on the index plate. There are usually two levers to be positioned. Another lever controls the direction the lead screw turns. There is one position for cutting right-hand threads. A second position disengages the lead screw. The third position reverses the direction of rotation, to cut left-hand threads.

CAUTION: The feed rod must be disengaged when cutting threads. Otherwise, the feed may be accidentally engaged. This produces excessive forces on the feed screw. Damage may result.

SPINDLE SPEEDS FOR THREAD CUTTING

The spindle speeds required for threading are slower than those used for turning. The spindle speed for cutting coarse threads on 3/4″ (18 mm) and larger diameters is one fourth that for turning. A faster speed, about one third to one half that for turning, is used for fine pitches and smaller diameters. The cutting speed may be increased still further when machining brass, aluminum, and other soft materials. Steels that are tougher and harder than low-carbon (soft-cutting) steels require slower speeds.

Another factor to consider is the expertise of the worker. Slower speeds should be used by the beginner. The speed is gradually increased within a speed range as skill and coordination are developed.

DESIGN FEATURES AND THREAD FORM CALCULATIONS

Additional Thread Forms

Design features have been covered for American National Standard, British Standard Whitworth, Unified, and SI Metric threads. There are three other common forms of threads: *square, Acme,* and *Brown and Sharpe* (B&S) *worm threads.* Each of these threads has a coarser pitch than the four previously covered types. Such threads are designed principally to transmit motion and to apply a great force. Applications include feed screws, vises, steering mechanisms, and jacks.

The square thread, as the name implies, has square, parallel sides and a flat root. Both the American National Acme thread and the B&S worm thread have a 29° included thread angle. However, the B&S thread is a deeper thread with a narrower crest and root. A B&S thread normally meshes with the thread of a worm gear. Motion is transmitted between two shafts that are at right angles to each other and in different planes.

Thread Calculations

A series of formulas is used to compute the different dimensions of screw threads. While the letters designating a design feature may be different, the same values are included in the formulas. Table 37-1 shows seven standard basic thread forms and a few of the basic formulas for each. There are five other important dimensions not shown in Table 37-1: major, minor, and pitch diameters, and normal and actual size.

NATURE OF THREADS PER INCH TO BE CUT	POINT TO ENGAGE SPLIT NUT		POSITION OF THREAD CHASING DIAL
EVEN NUMBER OF THREADS	ENGAGE AT ANY GRADUATION ON THE DIAL	½ 1 1½ 2 2½ 3 3½ 4	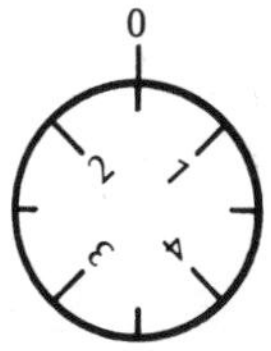
ODD NUMBER OF THREADS	ENGAGE AT ANY NUMBERED GRADUATION	1 2 3 4	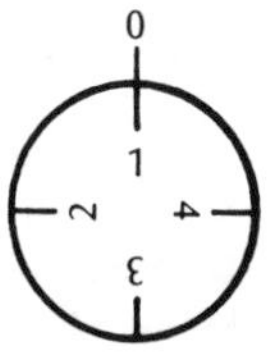
FRACTIONAL NUMBER OF THREADS	ENGAGE AT EVERY OTHER ODD OR EVEN GRADUATION	1 OR 2 3 4	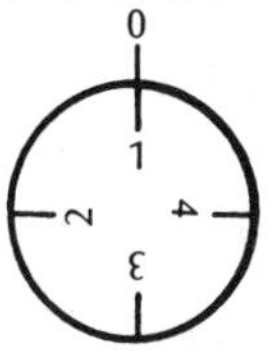
THREADS WHICH ARE A MULTIPLE OF THE LEAD SCREW PITCH	ENGAGE AT ANY POINT WHERE THE SPLIT NUT MESHES		

Fig. 37-9. Positions of the thread-chasing dial to engage the splitnut for cutting threads.

The *pitch diameter* is important in design and measurement. It represents the diameter at that point of the thread where the groove and the thread widths are equal. The pitch diameter is equal to the major diameter minus a single thread depth. Thread tolerances and allowances are given at the pitch diameter.

The *minor diameter* (formerly called *root diameter)* is the smallest thread diameter. Minor diameter applies to both external and internal threads.

ALLOWANCE, TOLERANCE, LIMITS, AND SIZE

Screw threads are machined to various specifications. These depend on the application, the material from which the part is made, the method of generating the thread form, and other factors. The size and fit depend on *allowance, tolerance,* and *limits.*

Allowance refers to the difference allowed between the largest external thread and the smallest internal thread. Allowance is an intentional difference between mating parts. The allowance may be positive *(clearance)* so that parts will fit freely. A negative allowance is specified for parts that must be assembled with force *(force fit).* In other words, a given allowance produces the tightest acceptable fit. Reference tables are used to establish allowances. Maximum and minimum pitch diameters are given for any classification of fit.

Example: The allowance for a 1″—8UNC class 2A (outside) and 2B (inside) fit is the difference between the minimum pitch diameter of the inside thread and the maximum pitch diameter of the outside thread:

Minimum pitch diameter
of 2B, (the inside thread) = 0.9188″
Maximum pitch diameter
of 2A, (the outside thread) = 0.9168

Allowance = $\overline{0.0020″}$

Tolerance, as stated earlier, is the acceptable amount a dimension may vary in one direction or in both. For example, if a workpiece is to be turned to a 25.4 mm ± 0.02 mm, the total tolerance is 0.04 mm. A part turned to 1.000″ ± 0.001″ has a tolerance of 0.002″. The tolerance for threads in the American

Table 37-1. PARTIAL SET OF FORMULAS FOR COMMON THREAD FORMS

Symbols

P = Screw Thread Pitch
N = Number of Threads Per Inch
D = Single Depth of Thread
W = Width of Groove and Ridge

C = Width of Flat at Crest
R = Width of Flat at Root
r = Radius at Crest or Root

$D = 0.6495 \times P$ or $\dfrac{0.6495}{N}$

$F = 0.125 \times P$ or $\dfrac{0.125}{N}$

American National

$D = 0.7035\ P$ (maximum)
$\quad 0.6855\ P$ (minimum)
$F = 0.125\ P$
$R = 0.0633\ P$ (maximum)
$\quad 0.054\ P$ (minimum)

SI Metrics

D for (external thread) = $0.6134 \times P$ or $\dfrac{0.6134}{N}$

D for (internal thread) = $0.5413 \times P$ or $\dfrac{0.5413}{N}$

F for (external thread) = $0.125 \times P$ or $\dfrac{0.125}{N}$

F for (internal thread) = $0.250 \times P$ or $\dfrac{0.250}{N}$

$D = 0.500\ P$
$F = 0.500\ P$
$C = 0.500\ P + 0.002$

Square

$D =$ (minimum) $0.500\ P$
$\quad =$ (maximum) $0.500\ P + 0.010$
$F = 0.3707\ P$
$C = 0.3707\ P - 0.0052$
$\quad$ (for maximum depth)

Acme

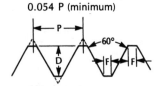

Unified

$D = 0.6866\ P$
$F = 0.335\ P$
$C = 0.310\ P$

Brown and Sharpe Worm

$D = 0.6403 \times P$ or $\dfrac{0.6403}{N}$

$R = 0.1373 \times P$ or $\dfrac{0.1373}{N}$

Whitworth

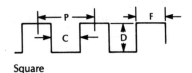

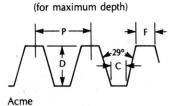

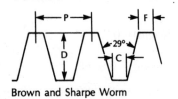

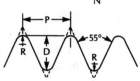

National and the Unified thread systems are plus (+) on outside threads and minus (−) on inside threads. With this rule, when a thread varies from the nominal (basic) size, it will have a freer fit.

Example: The tolerance for the 1″-8 UNC class 2A (external) thread used in the previous example is the difference between the maximum and minimum pitch diameters. These pitch diameters may be obtained from reference tables.

Maximum pitch diameter = 0.9168″
Minimum pitch diameter = 0.9100″
Tolerance (or
  acceptable variation) = 0.0068″

Limits represent maximum and minimum dimensions. In the previous example, the *upper limit* of the pitch diameter for a class 2 fit is 0.9168″. The *lower limit* is 0.9100″.

The *basic size* is the theoretical exact size of the designated thread. It is from the basic size that size limitations are made. The basic size of a 1 ⅜″—6NC thread is 1.375″. The thread notation on a part drawing indicates the 1 ⅜″ size.

The *actual size* is the measured size. This may differ from the basic size, depending on the accuracy of machining.

THREAD MEASUREMENT AND INSPECTION

Measurement standards must be followed to ensure that each threaded part functions as designed. This means that the surface finish, thread profile, and dimensions must be checked. On *rough threads* with a class 1A or 1B fit, the thread may be checked for size with a corresponding size of bolt, nut, or other mating part.

Where a more accurate fit is required, like a class 3A or 3B fit on a precision part, any of the following inspection and measurement techniques may be used.

- Thread ring and/or thread plug gage
- Thread snap gage
- Screw thread micrometer
- Thread comparator micrometer
- Optical comparator
- Toolmaker's microscope
- One-wire (Acme) or three-wire (60° thread) method (form angle).

Screw Thread Micrometer

The screw thread micrometer (Fig. 37-10A) measures the pitch diameter. These micrometers are designed to measure 60° angle threads in the inch-standard or SI Metric standard systems. Screw

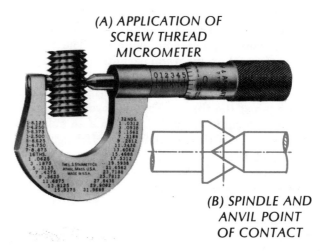

(A) APPLICATION OF SCREW THREAD MICROMETER

(B) SPINDLE AND ANVIL POINT OF CONTACT

Fig. 37-10. Using a screw thread micrometer to measure the pitch diameter of a screw thread.

thread micrometers have a 1″ (25 mm) range. Different micrometers are required in measuring American National and Unified threads. For instance, four micrometers cover the range from 8 to 40 threads per inch (T_{pi}):

- 8 to 13 T_{pi}
- 14 to 20 T_{pi}
- 22 to 30 T_{pi}
- 32 to 40 T_{pi}

The screw thread micrometer has a 60° point spindle. The swivel anvil has a corresponding cone shape. The line drawing in Fig. 37-10B shows these shapes and the point of contact. The micrometer measurement is taken at the pitch diameter of the thread, along the helix plane. This produces a slightly inaccurate measurement. For extremely precise threads, the helix angle must be considered. In these cases, the micrometer is usually set with a master thread plug gage. The threaded part may then be measured precisely.

Thread Comparator Micrometer

The *thread comparator micrometer* compares a thread measurement against a thread standard. The micrometer has a 60° cone-shaped spindle and anvil. The micrometer reading is established by first gaging a standard thread plug gage. The advantages of the thread comparator micrometer are the speed, ease, and accuracy with which a thread may be measured.

Optical Comparator

The common optical comparator magnifies a part from 5 to 250 times and projects it upon a screen.

Design features of a thread are measured in one step. The measurements include form, major, minor, and pitch diameters, pitch, and lead error. The surface illuminator of the comparator permits examining the surface finish of the thread.

The optical comparator casts a high-intensity light beam against the thread. The shadow produced by the thread form is magnified. It forms an enlarged silhouette of the thread when projected against a ground-glass screen.

The thread to be measured is positioned between centers or on a V-block on the table. The table may be moved in three directions: vertically, parallel to the lens, and perpendicular to it. Measurements are established by using a micrometer or end measuring rods on the comparator.

Angles are measured by rotating the hairlines on the viewing screen. A vernier scale permits the making of angular measurements to an accuracy of one minute of one degree. Sometimes a chart is substituted for the viewing screen.

Toolmaker's Microscope

The toolmaker's microscope makes it possible to measure the same thread features as can be measured with the optical comparator. In this case, the table is moved by precision lead screws. A micrometer thimble is used to turn each lead screw. Measurements, read directly from the graduations on the thimble, are within 0.0001″ (0.002 mm). The eyepieces of the microscope have hairlines in the form of the thread.

Three-Wire Method

The *three-wire* method (Fig. 37-11) is used to check the pitch diameter of 60° angle screw threads in the American National, Unified, or SI Metric systems. The standard micrometer may be used to measure threads that require an accuracy of 0.001″ (0.002 mm). Threads may be measured to greater precision with an electronic comparator. The optical comparator is used for checking the thread form and size against an accurate thread template.

The three-wire method is recommended by the National Bureau of Standards and the National Screw Thread Commission. The method provides an excellent way of checking the pitch diameter. Any error in the included thread angle has a very limited effect on the pitch diameter.

Three wires of the same diameter are required. Two of these are placed in thread grooves on one side. The third is placed in a thread groove on the opposite

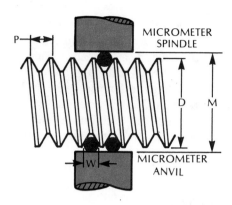

Fig. 37-11. Symbols, wire placement, and position of the micrometer spindle for three-wire measurement.

side. The wires and threads are positioned between the anvil and spindle of a standard micrometer. Fig. 37-11 shows the related dimension symbols, the placement of the three wires, and the position of the micrometer anvil and spindle.

The Best Wire Size—The size wire to use depends on the thread pitch. There are three possible wire sizes that may be used: *largest, smallest,* and *best.* The *best wire size* is recommended. With the best wire size, the wires come in contact with the thread angle at the pitch diameter. This condition exists when a thread, machined at the correct angle, is cut to thread depth.

Calculating Wire Size And Thread Measurement—Four formulas for measuring American National and Unified threads are given in Table 37-2. The first relates to the measurement over the wires (M). The other three formulas may be used to calculate the wire size.

Example: Using the formulas in Table 37-2 calculate (a) the best wire size and (b) the micrometer measurement over the wires for a 1″—8UNC threaded part.

(a) Best wire size:
$$W_b = \frac{0.57735}{N}$$
$$= \frac{0.57735}{8}$$
$$= 0.0722″ \text{ diameter (Answer)}$$

(b) Measurement over the wires:
$$M = D + 3W_b - \left(\frac{1.5155}{N}\right)$$
$$= 1.000 + 3(0.0722) - \left(\frac{1.5115}{8}\right)$$
$$= 1.0272″ \text{ (Answer).}$$

One-Wire Method (Acme Thread)

The one-wire method of measuring an Acme thread requires one wire or pin of a particular size (Fig.

Table 37-2. FORMULAS FOR WIRE SIZE AND THREE-WIRE MEASUREMENT

Dimension	Formula	Symbols
Measurement over wires (M)	$M = D + 3W_b - \dfrac{1.5155}{N}$	D = Major thread diameter
Largest wire size (W₁)	$W_1 = \dfrac{1.010}{N}$ or $1.010P$	W_1 = Largest diameter of wire
Best wire size (Wb)	$W_b = \dfrac{0.57735}{N}$ or $0.57735P$	W_b = Best wire diameter
Smallest wire size (Ws)	$W_s = \dfrac{0.505}{N}$ or $0.505P$	W_s = Smallest wire diameter N = Number of threads per inch M = Three wire measurement

37-12). The wire or pin is placed in the thread groove. The outside (major) diameter is then measured with a micrometer. The thread is cut to the correct depth when the micrometer reading over the wire is the same as the specified major diameter. The reading should be taken with the wire tight in the thread. An accurate reading is not possible unless burrs are first removed from the outside diameter.

The correct wire size may be determined by a single formula:

Wire size = 0.4872 × pitch.

The pitch equals one (1) divided by the number of threads per inch (N).

Example: Calculate the wire size for an Acme thread having two threads per inch.

Wire Size = 0.4872 × pitch
= 0.4872 × ½
= 0.2436″ (Answer).

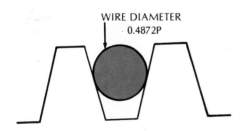

WIRE DIAMETER
0.4872P

Fig. 37-12. One-wire method of measuring the depth of an acme thread.

CALCULATING THE LEADING AND FOLLOWING SIDE ANGLES FOR CUTTING TOOLS

Tables are usually used to establish the clearance for the leading and following side cutting edges for all forms of threads. There are times, however, when these must be calculated. The helix angle of the thread may be found by considering a right triangle (Fig. 37-13).

The helix angle of the leading side is represented as the tangent of the lead divided by the circumference of the minor diameter (Figure 37-13A).

$$\text{Tan of leading side angle} = \frac{\text{lead of thread}}{\text{circumference of minor diameter}}$$

The tangent of the following side (Fig. 37-13B) is equal to the lead divided by the circumference of the major diameter.

$$\text{Tan of leading side angle} = \frac{\text{lead of thread}}{\text{circumference of major diameter}}$$

The helix angle is found by using a table of natural trigonometric functions to convert the numerical value of the tangent to the angle equivalent. This is usually given in degrees (°) and minutes (′).

Usually, a relief angle of 1° is *added to* the helix angle for the *leading* side. One degree is *subtracted* from the helix angle for the *following* side.

Example: A 1¼″ – 4 square thread with the following specifications is required:

Lead = 0.250″
Single depth = 0.125″
Double depth = 0.250″
Major diameter = 1.250″.
To calculate the minor diameter:

443

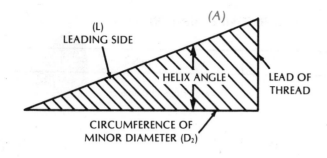

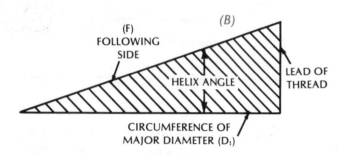

Fig. 37-13. Helix angles of the leading and following sides.

Minor diameter = major diameter − double depth
= 1.250″ − 0.250″
= 1.000.

To calculate the leading side angle of the tool bit:

$$\text{Tan of leading side} = \frac{\text{lead of thread}}{\text{circumference of minor diameter}}$$

$$= \frac{0.250}{(\pi \times 1.000)}$$

$$= 0.0796 \text{ Tan.}$$

The tangent value of 0.0796, established from a table of natural trigonometric functions, equals 4° 33′. The leading side angle of the tool bit is 4° 33′ + 1° relief angle, or 5° 33′ (Answer).

The following side angle of the tool bit is found in a similar way:

$$\text{Tan following side angle} = \frac{\text{lead}}{\text{circumference of major diameter}}$$

$$= \frac{0.250″}{\pi \times (1.250″)}$$

$$= 0.0637 \text{ Tan}$$

$$0.0637 \text{ Tan} = 3° 38′.$$

The following side angle of the tool bit is (3° 38′) − the 1° relief angle, or 2° 38′ (Answer).

DEPTH SETTINGS FOR AMERICAN NATIONAL FORM (60°) THREADS

The feeding of a thread-cutting tool is usually done by turning the compound rest handwheel. Some threads are cut by directly infeeding the cutting tool with the compound rest set at 0″. In other cases where a precision form thread is to be produced, the thread-cutting tool may be fed with the compound rest set at 30°. The tool is fed at the 30° angle to thread depth. However, common shop practice is to set the compound rest at 29°. This provides a slight angular clearance. The right side of the cutting tool just *shaves the thread* to produce a fine finish.

Table 37-3 gives the depth-of-feed settings for a thread range of 4 to 64 threads per inch. The depth (in thousandths of an inch) for each thread is shown for compound rest settings of 0°, 29°, and 30°.

CAUTION: The thread depths are based on the correct width (0.125 P) of the flat cutting-tool point. Unless this width is accurate, the depth readings will not be correct.

HOW TO SET UP A LATHE FOR THREAD CUTTING

Preparing The Workpiece

Step 1   Machine the workpiece to the major diameter of the thread.
Step 2   Chamfer the end to a 30° or 45° angle.
Step 3   Measure and mark the length of the threaded portion.
Step 4   Undercut the length (if possible). Cut the groove to thread depth (minor diameter). *Note:* Sometimes the groove is cut a few thousandths deeper. The groove simplifies the threading process. Ending the thread in a groove ensures that the mating part fits over the required length or shoulders properly.

Setting Up The Lathe

American National And Unified (60° Angle) Thread Forms
Step 1   Set the compound rest at 0°, 30°, or 29°. The angle depends on the selected method of cutting. *Note:* Set the compound rest *counter-clockwise* to cut a right-hand thread.
Step 2   Determine the cutting speed and spindle rpm. Set the spindle at this speed.
Step 3   Engage the lead screw gears. Position the quick-change gear box levers. This is

Table 37-3. DEPTH SETTINGS FOR AMERICAN NATIONAL FORM THREADS

Threads Per Inch (tpi)	Compound Rest Angle Setting and Depth of Feed* (in thousandths of an inch)		
	0°	30°	29°
4	0.1625	0.1876	0.1858
6	0.108	0.1247	0.1235
7	0.093	0.1074	0.106
8	0.081	0.0935	0.092
9	0.072	0.083	0.082
10	0.065	0.075	0.074
11	0.059	0.068	0.0674
13	0.050	0.0577	0.057
14	0.0465	0.0537	0.0525
16	0.0405	0.0468	0.046
18	0.036	0.0417	0.041
20	0.0325	0.0375	0.037
24	0.027	0.031	0.0308
28	0.0232	0.0270	0.0265
32	0.0203	0.0234	0.0232
36	0.0180	0.0208	0.0206
40	0.0162	0.0187	0.0185
44	0.0148	0.0170	0.0169
48	0.0135	0.0156	0.0154
56	0.0116	0.0134	0.0133
64	0.0101	0.0117	0.0115

*Based on: D = 0.64952P, and width of flat = 0.125P

indicated on the thread cutting section of the index plate.

Step 4 Position the lead screw lever to turn in the correct direction for a right- or left-hand thread.

Step 5 Disengage the feed rod by moving the feed-change lever to the *neutral* position.

Acme And Brown & Sharpe 29° Thread Angle

Step 1 Set the compound rest (14½°) at one half the included thread angle.

Step 2 Check and set the cutting speed, thread pitch, and thread direction.

Step 3 Disengage the feed rod by moving the lever to the *neutral* position.

Step 4 Check the angle and size of the flat point on the cutting tool. Use an Acme thread gage.

Step 5 Select an appropriate type of toolholder. Position the tool bit as close to the holder as possible. Set the tool bit at center height. Apply a *slight* turning force on the tool post screw to secure the tool bit.

Step 6 Place one edge of the Acme thread tool gage against the workpiece. Carefully feed the point of the cutting tool into the appropriate size of thread form on the gage.

Step 7 Tap the toolholder until no spacing (light) is visible between the cutting tool and the thread gage. Securely tighten the tool post screw. Recheck the setting.

Square Threads

Step 1   Set the compound rest axis parallel to the lathe axis. This permits readjusting the tool bit to *pick up the thread,* if necessary.

Step 2   Grind the tool bit to one half the thread pitch.

Note: Coarse-pitch square threads require a roughing tool and a finishing threading tool. Grind the roughing tool at least 0.015″ (0.4 mm) smaller than the required thread groove (P/2).

Note: The variation in width depends on the coarseness of the thread and the required class of fit.

Step 3   Mount the cutting tool in a toolholder. Position the cutting point on center and the edges at right angles to the work. Secure this setup.

Step 4   Set the speed levers at about one fourth the rpm used for turning. Position the change gears to produce the specified thread pitch. Disengage the feed rod. Engage the lead screw to turn in the right direction.

Positioning The Thread-Cutting Tool

Step 1   Check the angle of the thread-cutting tool with a center gage (60°), an Acme thread gage (29°), or a square thread template (90°).
Note: The size of the flat cutting point is checked against the thread form on a screw thread gage.

Step 2   Check the clearance angles with a gage or protractor.
Note: The leading and following sides must be ground to compensate for the helix angles.

Step 3   Select the appropriate type of thread-cutting tool and holder.
Note: The single-point square tool bit and circular formed cutter, and the solid and spring head threading tool holders each have advantages.

Step 4   Secure the cutting tool in a holder. Clamp the tool bit as close to the holder as possible.

Step 5   Set the cutting edge at center height.

Step 6   Place the center gage against the workpiece.

Step 7   Position the thread-cutting tool in the center gage (Fig. 37-14). This sets the point at right angles to the workpiece.
Note: A light paper placed on the cross slide makes it easier to see any spacing (light) between the cutting edges and the center gage. If necessary, gently tap the tool to center it.

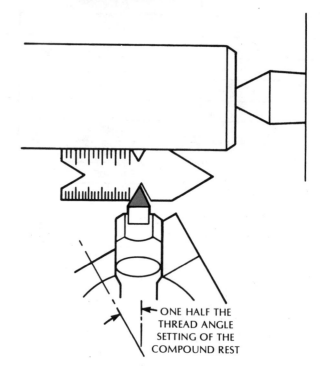

ONE HALF THE THREAD ANGLE SETTING OF THE COMPOUND REST

Fig. 37-14. Positioning a 60° angle thread-cutting tool using a center gage.

Step 8   Tighten the toolholder in the tool post. Recheck the accuracy of the tool setting. Then, move the setup to see that all moving parts clear the cross slide and the tooling setup.

HOW TO CUT THREADS

Checking In Preparation For Thread Cutting

Each of the following major steps require checking before a thread is cut:
- The correct positioning of the cutting tool
- The clearance of all moving parts, which must be adequate to prevent interference at both the start and the finish of each cut

- The positions of the speed, feed, and screw levers, for correct engagement or disengagement
- The mounting of the workpiece, to see that it is secure

  Note: Long slender workpieces must be supported with a follower rest. Otherwise, the work will spring away from the threading tool. This produces an inaccurately formed thread.
- The maximum diameter, chamfered end, and undercut for thread length and depth.

Cutting An Outside Right-Hand Thread

Step 1   Back the thread-cutting tool away from the workpiece.

Step 2   Start the lathe. Bring the cutting tool in until it just touches the workpiece.

       CAUTION: A beginner should use a slower spindle speed until all steps are coordinated.

Step 3   Set the micrometer collars on the crossfeed and compound rest feed screws at "0". Lock the collars at this setting.

Step 4   Move the carriage until the cutting tool clears the workpiece.

Step 5   Turn the compound rest handwheel to feed the tool to a depth of 0.002″ to 0.003″.

Step 6   Place the right hand on the split-nut lever. The left hand is placed on the crossfeed handwheel.

Step 7   Determine which lines on the chasing dial are to be used. Engage the half-nut when the correct line on the chasing dial reaches the index line.

Step 8   Turn the crossfeed handwheel *counterclockwise* as quickly as possible at the end of the cut. This clears the tool of the thread groove.

Step 9   Disengage the split-nut as the crossfeed handwheel is turned counterclockwise.

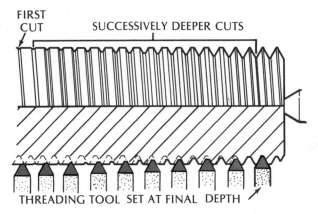

Fig. 37-16. The sequence of cuts to machine a thread.

       Note: Steps 8 and 9 may be reversed when the thread ends in an undercut (grooved) section. The split-nut is disengaged at the undercut with the right hand. The tool is brought out away from the workpiece by the left hand.

       Note: When these steps are coordinated, the spindle speed may be increased within the recommended range.

Step 10   Stop the lathe. Return the carriage to the starting position. Check the pitch.

       Note: A thread pitch gage, rule, or center gage may be used. These applications are shown in Fig. 37-15A, B, and C.

Step 11   Start the lathe. Feed the threading tool for a roughing cut. Set the depth of cut with the compound rest.

       Note: Apply a cutting fluid over the work surface with a brush.

Step 12   Continue to take roughing cuts. The depth of each successive cut should be decreased. Fig. 37-16 provides an example of decreasing an initial feed of 0.020″ (0.5 mm) to 0.015″ (0.4 mm), 0.010″ (0.3 mm), and 0.005″ (0.1 mm).

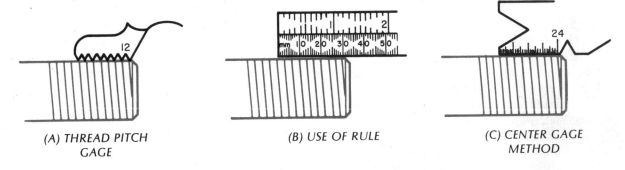

(A) THREAD PITCH GAGE

(B) USE OF RULE

(C) CENTER GAGE METHOD

Fig. 37-15. Common methods of checking thread pitch.

Note: Determine in advance the final depth reading on the compound rest micrometer collar.

Step 13   Take the last two cuts as finish cuts. The depth may be set for 0.003″ (0.08 mm), then 0.002″ (0.05 mm).

Note: Both rough and finish thread-cutting tools are sometimes used. These speed up the roughing out and finish turning processes. In some instances, the finish thread-cutting tool is fed in the last 0.001″ (0.025 mm) by infeeding at 30° or 90° (Fig. 37-17).

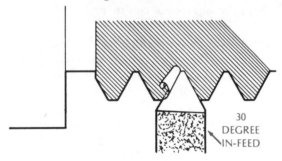

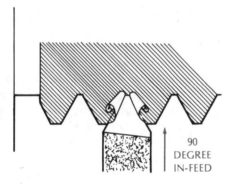

Fig. 37-17. Infeeding at 30° and 90° on a final cut.

Step 14   Check the thread size. A nut or other mating part may be used for checking most threaded workpieces. Final precision parts require measurement and checking with a thread micrometer, comparator, or by the wire method.

CAUTION: Mark the slot on the drive plate before removing a workpiece that is mounted on centers. This ensures that if the workpiece is removed for checking the thread size, it may be returned to its original position (slot). In no event should the dog be removed from the workpiece. Otherwise, *crossed threads* may be produced if additional finish cuts are required.

Cutting Square Threads

Step 1   Proceed to cut a groove at the end of the thread.

Step 2   Rough and finish turn the square thread. Follow the same steps used for the 60° form threads. The exception is that the cutting tool is fed inward at right angles to the workpiece. The *crossfeed hand-wheel* is used to feed the cutting tool.

Note: Brush on a continuous supply of cutting fluid where applicable.

Step 3   Rechase the thread with the finishing tool.

Note: The worker often *touches up* the edges of the square threads. This removes the sharp corners and burrs. A single-cut smooth file produces a fine finished edge.

Cleaning Up The Back Side Of The Thread

As stated earlier, the cutting tooth is fed to depth for each successive cut by turning the compound rest handwheel. Since the compound rest is set at a 29° or 30° angle, one side cutting edge and the nose of the tool do the cutting. One chip is formed and flows freely over the top of the tool. If both side cutting edges and the point of a cross slide infeed were used, the cutting tool would tend to tear the thread and produce a rough surface finish.

Often, the side opposite the thread flank that is being cut (the back side) is rough. The second side cutting edge of the threading tool is then used to *clean up the back side.* The process of cleaning up the back side of a right-hand thread is shown in Fig. 37-18.

Step 1   Feed the tool at the 29° or 30° angle almost to thread depth (Fig. 37-18). The larger the diameter and the coarser the pitch, the greater the amount left for finish cutting the thread to depth.

Step 2   Back out the cutting tool from 0.010″ to 0.015″ (0.2 to 0.4 mm).

This is shown at Position B in Fig. 37-18. *Note:* The tool is backed out and *fed back in to this depth.* This is done to reverse the lost motion of the compound rest feed screw.

Step 3   Infeed the cutting tool for the next several cuts. This infeeding of the cross slide screw is represented in Fig. 37-18 by positions C and D.

Note: In position E the tool nose has been fed to within a few thousandths (0.04 to

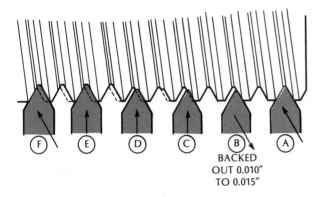

Fig. 37-18. Successive infeeding cuts taken to clean up the back side of a thread.

0.06 mm) of the required depth.

Step 4   Brush or flow on an adequate amount of cutting fluid where required. This is essential to produce a good surface finish.

Step 5   Infeed the cutting tool with the compound rest until the thread is brought to size. Position F in Fig. 37-18 shows the final infeeding.

Resetting The Thread-Cutting Tool

Whenever the workpiece is removed from a chuck, holding fixture, or lathe dog, or the cutting tool is changed, the thread-cutting tool must be reset.

Step 1   Repeat the steps for setting up the workpiece and tool in preparation for cutting the thread.

Step 2   Close the split nut at the correct chasing dial position. Feed the carriage a short distance along the workpiece.

Step 3   Stop the lathe. Move the crossfeed and compound rest handwheels. Keep adjusting the cutting tool position until the sides and point are congruent with (match) the previously cut thread.

CAUTION: Do not let the cutting edges touch the thread surfaces. If the tool is brought onto the surfaces, the new cutting edges will be damaged.

Step 4   Reset the compound rest and crossfeed micrometer collars at zero. Back the cutting tool out of the thread groove. Disengage the split-nut. Return the cutting tool to the beginning of the thread.

Step 5   Continue to take roughing and/or finishing cuts.

Note: The skilled craftsperson often determines the depth of cut by revolving the workpiece and turning the compound rest handwheel until the point of the cutting tool just grazes the major diameter. The micrometer collar reading is then subtracted from the "0" to which the tool was set for depth. The resultant micrometer collar reading represents the depth within an accuracy of approximately 0.002″ to 0.003″ (0.05 to 0.08 mm).

Cutting An Outside (External) Left-Hand Thread

Step 1   Prepare the workpiece and set the cutting tool as for cutting a right-hand thread.

Step 2   Set the compound rest at an angle of 30° (or 29°) for a 60° angle thread. Use a 14½° angle for an Acme thread. Set the compound rest axis at 0° (parallel to the lathe axis) for a Square thread. The compound rest is set to the left of the cross slide.

Step 3   Position the lead screw lever for *reverse*. As the split nut is engaged, the cutting tool feeds from left to right.

Step 4   Position the cutting tool to take a light trial cut. Set the graduated collars at "0". Move the cutting tool to the left end of the threaded section.

Step 5   Engage the half-nut at the appropriate mark on the chasing dial. Take the trial cut.

Step 6   Proceed to rough out and finish cut the thread. Check it with the same procedures used for checking a right-hand thread.

LATHE SETUP FOR CUTTING SI METRIC THREADS

The quick-change gear boxes of most modern lathes are designed for cutting threads in both the American National Unified system and the SI Metric system.

Changing Quick-Change Gearboxes To Cut SI Metric Threads

Lathes that are geared according to the inch-standard system of measurement may also be used to cut SI Metric threads. Two change gears having 50 and

127 teeth are needed. This ratio represents the relationships between the inch-standard and Metric systems of measurement. The change gear ratio of 50/127 represents a ratio of 1 inch to 2.54 centimeters.

$$\frac{1}{2.54} \times \frac{50 \text{ (teeth)}}{50 \text{ (teeth)}} = \frac{50}{127}.$$

These two gears are placed on the end gear train. The 50-tooth gear is mounted on the spindle. The 127-tooth gear is mounted on the lead screw. The lathe is now geared to cut threads per centimeter.

To use the quick-change gearbox combinations, the millimeter pitches in the SI Metric system must be converted to the *number of threads per centimeter*. For example, a screw thread having a 5-mm pitch (almost 0.200″) is equal to 2 threads per centimeter. The quick-change gearbox combination to use is 2 threads per inch (tpi). With the 50-to-127 ratio, the lathe is set to cut 2 threads per centimeter.

The major difference between cutting Metric and inch-standard threads is that for Metric threads the split-nut is engaged continuously. The Metric threads bear no relation to the lines on the thread-chasing dial. Thus, to track each successive cut, the threading tool is backed out at the end of each cut. The lathe is reversed, and the cutting tool is returned to the starting position. The tool is fed to the next depth. The cut is again started, with the spindle turning in the original direction. This procedure continues until the thread is completed.

Modern Inch-Standard And SI Metric Lathes

The ratios in the gear trains of late-model lathes permit thread cutting for the inch-standard and SI Metric systems. The spindle and lead screw gears in the end gear train do not need to be replaced to establish the 50-to-127 ratio.

The levers used to obtain different spindle speeds, feeds, and lead screw movements, and to reverse the direction, are mounted on the head stock and quick-change gear box. The lever for starting, stopping, and reversing the lathe is located on the apron.

**HOW TO CUT METRIC THREADS
RIGHT- OR LEFT-HAND ON THE MODERN
INCH/METRIC STANDARD LATHE**

Step 1   Position the quick-change gear box levers to produce the required SI Metric pitch.

Step 2   Prepare the workpiece and set up the cutting tool. Follow the same procedures used for cutting right- and left-hand American and Unified National Threads.

Step 3   Take the usual light trial cut. At the end of the cut, back the cutting tool out of the workpiece.

           CAUTION: Keep the half-nut engaged throughout the thread-cutting process.

Step 4   Reverse the direction of the spindle to return the cutting tool to the starting position. Stop the lathe.

Step 5   Feed the cutting tool to take successive roughing and finishing cuts. Follow the same steps used for cutting and measuring nonmetric threads.

INTERNAL THREADING

Two common methods of cutting internal threads include tapping and machine threading. The larger and coarser thread sizes are sometimes roughed out with a single-point thread-cutting tool. A tap is then used to cut the thread to size.

When a hole is to be tapped on the lathe, it is first drilled or bored to the required tap drill size. This must take into account the *percent of full thread depth* required. In no instance may the hole size be smaller than the minor diameter of the thread. The larger size hole provides thread clearance. The 75 percent depth of thread is widely used, as explained in hand tapping (Unit 13).

Threading With A Single-Point Cutting Tool

A center gage or Acme thread gage, boring bar or other form of toolholder, and a forged or standard cutting tool are needed for internal threading. A bored hole is preferred because it is concentric. A drilled hole might vary a few thousandths of an inch or a fractional part of a millimeter (0.08 to 0.1 mm) from being perfectly round, straight, and concentric.

A groove (recess) should be cut a few thousandths of an inch below the maximum thread diameter for blind holes. The groove width should be about 1½ times the thread pitch. If the part permits, a shallow recess may be cut to the same depth on the end of the workpiece. The outside of the bored or recessed hole should be chamfered 14½° to accommodate the Acme thread form and 30° for 60° angle thread forms. The chamfer reduces burrs. Also, the mating threads fit easier. The inside and end recesses and the chamfered end of a bored workpiece are illustrated in Fig. 37-19.

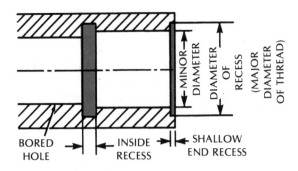

Fig. 37-19. Recessing for internal threading.

The Internal Thread-Cutting Tool Bit

The cutting tool is ground to the required included angle and width. Both the leading and following cutting edges must have regular relief plus clearance, to compensate for the thread helix angle. The required clearances vary with the major and minor diameters of the threads. The top face is flat, without back rake.

The largest size boring bar possible is used. This provides maximum rigidity, particularly when cutting long threads. However, the boring bar diameter must be small enough to permit backing the tool out of the recess and returning it to the starting position.

HOW TO CUT AN INTERNAL THREAD

Tapping An Internal Thread

Step 1 — Drill the hole to the required diameter. *Note:* If thread concentricity is required, the thread should be chased, not tapped.

Step 2 — Mount a square (or angular) cutting tool in the boring bar. Position the tool at thread depth. Start the lathe. Cut an inside recess to about 0.003″ larger than the major diameter.

Step 3 — Turn a 30″ chamfer on the end of the workpiece. (Turn a 14½° chamfer for Acme threads.)

Step 4 — Position the taper tap, guiding it against the tailstock center.

Step 5 — Start the taper tap by turning the tap holder and feeding the dead center.

Step 6 — Turn the chuck slowly by hand for a partial turn. Continue to thread, and turn. Use a cutting lubricant throughout the threading process where possible.

Note: Whenever possible, the tap should be turned with a tap wrench. A chuck does not provide the *feel* of the tapping process that a tap wrench does.

Step 7 — Repeat these steps until the threads are cut to full depth.

Note: On through holes, the spindle may be run at a very slow speed. The dead center is kept continuously in the center hole of the tape.

CAUTION: The tailstock handwheel should be moved with the right hand. At the same time, the left hand should be kept on the control lever. The lever then can be moved quickly to instantly stop all motion in an emergency.

Step 8 — Use a plug tap as a second tap for through holes. Use a bottoming tap to cut threads that *shoulder*.

CAUTION: Power should *not* be used when tapping a blind hole.

Step 9 — Turn the tap in the reverse direction, to remove it. At the same time, move the dead center at the same rate. The dead center should continue to serve as a guide.

Step 10 — Turn off any burrs from the ends. Clean the workpiece. Test for fit with the mating part or a gage.

Cutting An Internal Thread With A Single-Point Threading Tool

Step 1 — Check the diameter of the bored hole for size.

Step 2 — Undercut a recess (equal to the major diameter) at thread length. Cut a chamfer on the end of the workpiece. If possible, bore to the major diameter for about 1/16″.

Step 3 — Set the lathe at the correct speed for threading.

Step 4 — Position the lead screw levers to produce the specified thread pitch.

CAUTION: Be sure the feed rod is *disengaged*.

Step 5 — Select an appropriate size of boring bar. Grind the tool bit to accommodate the thread size. Secure the tool bit in the boring bar.

Note: Be sure the tool bit extends a short distance beyond the full depth of thread.

CAUTION: The length of the tool bit must be short enough to clear the inside bore when the bit is moved out of the thread groove.

Step 6 Set the compound rest at the correct number of degrees (one half the included angle).

Note: The compound rest is positioned to the left for right-hand threads and to the right for left-hand threads.

Step 7 Secure the boring bar in a holder. Avoid excessive overhang. Position the cutting edges at center height.

Step 8 Select a thread gage appropriate for the thread form. Place one side of the gage against the workpiece or a parallel straight edge.

Step 9 Bring the tool cutting edges into the gage. Sight between the mating edges of the gage and the tool bit. Gently tap the tool holder until no spacing (light) is seen between the mating surfaces. Secure the setting.

Note: Recheck the center height and the squareness of the threading tool.

Step 10 Place a mark on the boring bar at the length of the threaded portion.

Note: This is important. The boring bar mark indicates when to disengage the split nut.

Step 11 Turn the compound rest handwheel to take out all play. Set the micrometer collar at "0".

Step 12 Start the lathe. Turn the crossfeed handle to remove end play. Reset the micrometer collar at "0" when the cutting tool just grazes the bored hole.

Step 13 Move the compound rest to take a roughing cut.

Step 14 Hold the split nut lever with the right hand and the crossfeed handwheel with the left.

Step 15 Engage the split nut at the correct index line on the chasing dial.

Step 16 Cut the thread. Watch the boring bar. Disengage the split nut and move the boring bar away from the internal thread when the specified thread length is reached.

Step 17 Move the boring bar out of the workpiece. Continue to chase the thread at successive depths until the required depth is reached.

MACHINE SETUP FOR TAPER THREAD CUTTING

The recommended method of cutting an accurate taper thread is to use a taper attachment. Taper threads that are cut by offsetting the tailstock center are not as accurate as those cut by the taper attachment method. The angularity between the lathe dog and driver plate during each revolution causes variation in the speed of the workpiece. Other inaccuracies are produced by the wear on the work centers caused by cutting with the centers offset. The amount of error depends on the distance the centers are offset.

The thread-cutting tool is set perpendicular to the axis of the workpiece (Fig. 37-20). If there is no

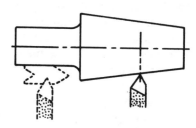

Fig. 37-20. Correct positioning of thread-cutting tool.

cylindrical solid surface on the workpiece from which the cutting tool can be set squarely, it may be necessary to use a parallel cylinder test bar. Once the thread-cutting tool is positioned and secured, the test bar is replaced with the workpiece. The tailstock spindle also can be used for squarely setting the cutting tool.

The same steps are followed in taper threading as for cutting regular straight threads.

Step 1 Take a roughing cut. Return the cutting tool to the starting position.

Step 2 Check the quality of the cut and the pitch.

Note: Brush on a cutting fluid if it is required.

Step 3 Continue to take roughing and one or two finishing cuts. Proceed to reach the thread depth in the same manner as for cutting an external thread.

Note: If the front end is turned (recessed) to the major diameter, this may be used as a guide for thread depth.

Step 4 Stop the lathe. Clean the workpiece. Remove burrs. Check with a mating part or gage, depending on the required class of fit.

SAFE PRACTICES IN CUTTING INTERNAL AND EXTERNAL THREADS

- Disengage the feed rod *before* engaging the feed screw. Otherwise, the feed and lead forces opposing each other may damage apron parts and the accuracy of the lead screw.
- Use slower speeds for thread cutting than used for other turning operations.
- Avoid excessive overhang of the cutting tool or boring bar.
- Grind the thread-cutting tool with clearance adequate to compensate for the thread helix angles.
- Use one side cutting edge and the point if possible. *Clean the threads* by infeeding with 0.002″ to 0.003″ (0.04 mm to 0.06 mm) finish cuts.

- Set the thread-cutting tool on center.
- Cut an internal recess at the thread length, to thread depth, and about 1½ times as wide as the thread.
- Keep fingers and wiping cloths away from the revolving workpiece and cutting tool.
- Reposition a changed cutting tool in the thread groove by moving both the compound rest (at the angle setting) and the cross slide.
- Grind the flat or round cutting point accurately. Thread depth measurements depend on the width of the point.

TERMS USED IN EXTERNAL AND INTERNAL THREADING

Lead	The axial movement (distance) of one mating threaded part from another for one revolution. Lead and pitch are equal for a standard single-lead screw thread.
Multiple lead	The number of times a male and female thread may be engaged around the circumference. Common leads are double (2P), triple (3P), and quadruple (4P). The depth of multiple-lead threads is only a fraction of a single-lead thread.
Thread chasing	The process of engaging the split nut at a particular position. Precisely locating a threading tool for successive cuts to follow in the same thread groove.
Chasing dial	A rotation dial on the threading attachment. The dial is graduated to indicate the position at which a half nut may be engaged to chase a thread.
Acme thread	A 29°-included-angle thread form with a flat crest and root. A coarse angular-sided thread that is used where great forces are to be exerted.
Brown & Sharpe (B&S) worm thread	A thread form similar to the 29° Acme. The thread depth is greater. The crest and roots are smaller. A thread form adaptable for transmitting motion between two shafts at right angles and in different planes.
Pitch diameter	An imaginary diameter midway between the major and minor diameters of a thread. The diameter where the thread and thread groove dimension is the same. An important thread diameter from which tolerances are measured to provide a specified fit.
Nominal size	A close approximate size of a thread. For example, the nominal size of a 1″—8NC thread is 1″. (By contrast, the *basic size* is 1.000—8NC.)
Limit	The upper and/or lower acceptable dimension.
Optical comparator	A device by which each design feature of a form may be compared against a master and with which measurements may be taken.
Three-wire method of measuring threads	The use of three precision-ground wires of a special size to measure the pitch diameter of a screw thread.
Best-wire size	The diameter of three wires that contact the thread sides (angle) at the pitch diameter.
Tangent of thread helix angle	The trigonometric value of the thread lead divided by the circumference of the thread diameter.

| *Side (edge) cutting angle of a threading tool* | An angle equal to the regular clearance angle plus the additional helix angle of the thread. |
| *End gears* | The train of gears between the spindle and the lead screw. Older model lathes require gear changes to accommodate inch- and metric-standard thread pitches. |

SUMMARY

- Many outside and inside threads are cut on a lathe with a single-point cutting tool.
- The common thread forms include those with 60° and 29° included angles, and square threads. These may be in the American National, Unified, and SI Metric systems.
- The regular clearance angle for each cutting edge of a thread-cutting tool must be increased. The additional clearance must compensate for the thread helix angle.
- The spring head toolholder may be adjusted for roughing and finishing cuts. The spring feature is important for taking finish cuts to produce a high-quality surface finish.
- The distance (lead) a double thread travels is 2P. Triple threads have a lead of 3P; quadruple threads, 4P.
- Threads are cut on the lathe using the thread-chasing dial and threading attachment. This device permits engaging the split nut. The thread-cutting tool is tracked in the same groove for successive cuts.
- The spindle speeds for thread cutting are influenced by the skill of the operator, the specifications of the material, and the conditions under which the threads are machined. Speeds of one quarter to one half those used for turning are applied to thread cutting.
- Successive roughing cuts are taken as deeply as possible. Allow about 0.006″ (0.01 mm) for finish cuts on diameters up to 1″.
- Thread measurements are taken at the pitch diameter. Internal threads are cut to the specified major diameter; outside threads are cut to the minor diameter.
- Allowance, according to class of fit, is the specified (intentional) difference between mating parts or surfaces.
- The tolerance given on a drawing establishes the acceptable amount a dimension may vary from the basic size.
- A thread may be checked for fit with a bolt, nut, or other mating part. Ring and plug thread gages may also be used. Other precise measurements may be

- taken with thread micrometers and optical comparators.
- Precision wires are used for V-thread (60° angle) measurements. The best-wire size in the three-wire method permits the mesurement to be taken at the pitch diameter.
- Mathematical formulas are used for calculating wire sizes and the measurement over wires.
- A single wire is used for measuring thread depth for Acme and Brown & Sharpe worm threads.

- The helix angle of threads is found in handbooks or it may be calculated. A cutting tool must include additional clearance for the cutting angle.
- The compound rest is set at 30° to the *right* for American National, Unified, and SI Metric *outside right-hand* threads.
- The angle setting is to the left of the perpendicular axis of the compound rest for cutting internal threads.
- Where possible, the threading tool is fed to depth with the compound rest handwheel. Thread depths may be read on the handwheel micrometer collar.
- The roughing tool for a square thread is ground undersize. The width of the finishing tool is ground a few thousandths oversize.
- The cutting speed is reduced for cutting coarser threads to greater depths. The depth for each successive roughing cut may be reduced.
- The cutting tools for standard threads are generally positioned with a thread gage. These tools are set on center.
- The crossfeed and compound rest micrometer collars are set at "0" when the cutting edge is just set to cut.
- A thread-cutting tool is reset by turning the handwheels on the cross slide and the compound rest. These movements permit changing the depth and position.
- The directions of the lead screw and the threading tool are reversed for cutting left-hand threads. End play must be taken out to permit feeding inwardly.

- Metric threads may be cut on an inch-standard system lathe. A ratio of 50:127 must be established between the spindle and lead screw.
- Internal threads may be tapped or machine cut on the lathe. A single-point cutting tool ground to the required thread form is used for machine cutting.
- The largest size boring bar that the diameter of the hole will accommodate is preferred.

UNIT 37 REVIEW AND SELF-TEST

1. Cite the different functions served by a *center gage* and a *circular screw pitch gage*.
2. Tell how the *cutting edge clearance angles* of a thread cutting tool are established.
3. Indicate the relationship of *lead to pitch* (to a single thread) for each of the following multiple threads:
   a. double-lead
   b. quadruple-lead
   c. triple-lead.
4. Explain the function of a *thread-cutting attachment*.
5. Indicate conditions where square, Acme, and Brown and Sharpe threads are more functional than American National Unified and SI Metric 60° form threads.
6. State why the *pitch diameter* is important in thread calculations and measurements.
7. Tell what the differences are between the terms *allowance, tolerance,* and *limits* as applied to screw threads.
8. Identify two thread (a) measuring tools and (b) inspection instruments.
9. State how the *one-wire method of measuring an Acme thread* differs from the use of three wires for measuring 60° form threads.
10. Explain why a *relief angle* of 1° is *added to the helix angle* for the *leading-side* and is *subtracted from the helix angle* of the *following-side* in thread cutting.
11. Indicate the setting of the *compound rest* for cutting the following threads:
    a. 60° SI Metric
    b. Brown and Sharpe
    c. square.
12. List four settings or dimensions that the lathe operator must check before cutting a thread.
13. Explain the meaning of *cleaning up the back side of a thread* (that is rough).
14. Explain how the cross feed and compound rest handwheels are used to reset a thread-cutting tool.
15. Explain the relationship between the *50-tooth spindle gear* and the *127-tooth gear on the lead screw*.
16. State two characteristics of internal threads cut with a single-point threading tool that make such threads superior to die-cut threads.
17. Indicate why a taper thread may be cut more accurately by using a taper attachment than by offsetting the tailstock.
18. List three safety precautions to observe when thread cutting on a lathe.

PART SIX

MILLING MACHINES

SECTION ONE

MACHINES, ACCESSORIES, CUTTING TOOLS, AND MAINTENANCE

A milling machine shapes and sizes a part by feeding the workpiece past a rotating multiple-tooth cutter that removes the required amount of material. After a brief look at the history of milling machine development, this section presents the following introductory information about milling machines and milling processes:

- The designs and functions of principal features and accessories
- Maintenance procedures
- Devices for holding milling cutters and workpieces
- The design features and applications of standard milling cutters
- Speeds, feeds, and cutting fluids used for milling processes.

Unit 38

MILLING MACHINES: FUNCTIONS AND MAINTENANCE

Milling is the process of cutting away material by feeding the workpiece past a rotating multiple-tooth cutter. The cutting action of the many teeth around the *milling cutter* provides a fast method of machining. The machined surface may be flat, angular or curved. The surface may also be milled to any combination of shapes. The machine for holding the workpiece, rotating the cutter, and feeding it is known as a *milling machine*.

GENERAL APPLICATIONS OF MILLING AND MILLING MACHINES

One or several milling cutters may be mounted on an arbor at the same time. Multiple cutters increase the rate of cutting. Several surfaces may also be machined simultaneously. Two or more work-positioning and work-clamping fixtures may be used. These permit one part to be loaded while another is being machined. The milling machine is practical for machining a single part or for mass producing a number of interchangeable parts.

Milling machines are also designed with quick-change gear boxes or variable-speed drives. These permit a wide range of cutting speeds and feeds. Rapid traverse feeds are available to reduce setup time in bringing the work to the cutter.

Workpieces may be held in a regular or rotary vise, strapped directly to the table, nested and secured in a fixture, or positioned between centers.

Standard milling cutters are available with various diameters, widths, and shapes. These vary from regular straight-tooth plain milling cutters to keyseat cutters, form cutters, and end mills. These and other cutters and forms are described and applied in later units.

The cutter names normally indicate the kind of shape they produce. Flat surfaces, slots, keyways, irregular profiles, and parts like spur, spiral, and bevel gears and cams that require equal divisions may be milled to a high degree of accuracy.

SKILLS AND KNOWLEDGE REQUIRED

Throughout the units that follow, basic processes are performed on standard milling machines. To perform these basic processes, a competent worker must have the following expertise:

● A working knowledge of the design features of the machines, cutters, and accessories
● The ability to make cutter and work setups
● Good judgement related to proper speeds, feeds, and cutting action
● An understanding of the corrective steps to take to adjust the machine and cutter. These actions are based on preliminary measurements or settings that are made.

This background of knowledge and machine skill may then be applied to the operation of any size and type of milling machine.

BRIEF HISTORY OF SIGNIFICANT MILLING MACHINE DEVELOPMENTS

The rifle manufacturing industry of the early 1800's is responsible for many developments and improvements in the early designs of milling machines. In 1818, Eli Whitney promoted the manufacture of interchangeable rifle parts. Whether Whitney built the milling machine in America at that time is not known.

Whitney's contributions were followed in the years prior to 1840 by those of Robbins and Lawrence of Windsor, Vermont. Their miller used a rack-and-pinion feed mechanism. This design was later improved by George Lincoln of Hartford, Connecticut. Lincoln's worm and worm wheel were incorporated in the feed mechanism. This design feature reduced chatter. It also added significantly to the accuracy and quality of surface finishes and made possible further applications of the milling machine. The Lincoln machine was produced in quantity.

During the civil war, F. A. Pratt and Amos Whitney formed the "P&W" Machine Tool Company. The Pratt and Whitney Company still bears their names. The main products at the time were small turret lathes and milling machines.

Concurrently, Brown and Sharpe also produced these two basic machine tools. Their added contribution was the design and construction of a *universal milling machine*.

459

The Brown and Sharpe and Pratt and Whitney Companies have contributed significantly to the design and development of machine tools and instruments and other precision tools, and manufacturing and production technology.

Essentially, the milling machines of today incorporate features that date back to those of Robbins and Lawrence. Another major improvement was the contribution of the *omniversal miller* by the Brown and Sharpe company. The knee of this machine may be rotated on an axis perpendicular to or around (parallel with) the column face. This movement makes it possible to machine tapered spiral grooves, bevel gear teeth, etc. This machine was manufactured for years and was widely used in tool rooms for complex milling machine processes.

More recent milling machine developments include:

- Preloaded antifriction spindle bearings, which maintain greater dimensional accuracy of the spindle
- Functional coolant systems with more efficient coolants
- Force-feed lubrication systems
- Greater accuracy of all moving and mating parts
- Wider ranges of speeds and feeds, with a greater degree of control
- Applications of carbides and ceramic cutting tools
- Single micrometer collars graduated for direct reading in either the inch-standard or the Metric measurement system.

INDUSTRIAL TYPES OF MILLING MACHINES

The milling machine is used widely in the machining of a single part as well as for mass production. The machine designs range from those of simple hand-operated millers to those of automated milling machines for multiple· processes that are numerically controlled.

The major types of general-purpose and common manufacturing machines are:

- Column-and-knee type
- Fixed bed
- Rotary table
- Planetary
- Tracer controlled.

These types and the design features of numerically-controlled machines are described briefly.

There are other types of milling machines *(millers)* for particular machining processes. For example, the *thread miller* is used for milling lead screws. The *cam miller* is designed for machining disk cams. The *skin*

miller is adapted to special milling processes in the aircraft industry. The outer covering (skin) of wings is tapered to accommodate the stresses within each section. The taper of one inch or more along the skin may be milled on this machine.

The range of milling machines sizes is also extensive. The table lengths and movements extend from those of the bench miller to bed lengths of 100 to 150 feet. Some spindles are positioned stationary. On other millers, the spindle carriers may be swiveled about a vertical or horizontal plane. Together with the cross feed (tranverse), longitudinal, and vertical movements, this type of miller has five *axes of control*. Many milling machines are *programmed* and are numerically controlled.

Knee-And-Column Types Of Milling Machines

There are five basic types of *knee-and-column* milling machines.

- Hand
- Plain
- Universal
- Omniversal
- Vertical spindle.

Hand Milling Machine—As the name implies, the hand milling machine is small. It is entirely hand operated. The hand miller pictured in Fig. 38-1 may be mounted on a bench or on a floor base. This machine is particularly useful in small milling machine manufacturing operations. Slotting, single- and multiple-cutter milling, direct indexing and other simple milling operations may be performed economically. The hand feed permits a rapid feed approach and withdrawal after a cut. The hand miller is less expensive and easier to operate than a heavier plain or universal miller.

Plain Milling Machine—The plain miller has three principal straight-line movements. The table moves in a longitudinal path. The movement is parallel to the face of the machine column. The table may be moved (cross fed) in or out and/or up and down (vertically). The spindle is mounted in a horizontal position. The cutter is rotated by the spindle movement. The feed is in a straight line. The plain miller is a general-purpose, standard machine tool. The features and operation of the plain milling machine are illustrated and described in detail later in this unit.

Universal Milling Machine—The universal milling machine has the added movement of a swivel table. A standard universal miller is shown in Fig. 38-2. The table is mounted on a saddle. The base of the table is graduated and designed for setting it either *straight* (0°) or at an angle. The angle setting permits the table to travel at an angle to the column face. The universal miller is especially adapted to produce straight and spiral cuts. Common examples are the spiral flutes on reamers, drills, and other cutters, and the spiral teeth on helical form gears.

Omniversal Milling Machine—The features of this machine were described earlier. It is primarily a precision tool room and laboratory miller. The added rotation of the knee about the column face is in a perpendicular axis to the knee. This makes it possible to machine *tapered spirals* and *tapered holes*.

Vertical Spindle Milling Machine—The spindle on this machine is mounted vertically. The two basic types of vertical spindle millers are classified as: (1) fixed bed and (2) knee-and-column. A knee-and-column type of universal toolroom model vertical spindle machine is shown in Fig. 38-3. The machine

Fig. 38-1. A hand milling machine. (Courtesy of the W.H. Nichols Company)

Fig. 38-3. A vertical milling machine. (Courtesy of Bridgeport Machines Division of Textron Inc.)

Fig. 38-2. A universal milling machine. (Courtesy of Cincinnati Milacron)

has an adjustable overarm that is attached to the column. The base of the overarm is graduated to permit angular adjustment. A *head* is mounted on the overarm. This head may be swiveled in an arc parallel and/or transversely to the table. The three

461

angular adjustments make it possible to machine at a compound angle.

The head also includes the spindle drive mechanism and controls for speeds, feeds, and positioning of the cutting tool. On the fixed-bed type the workpiece is positioned and moved longitudinally or is crossfed by the table movement. The spindle is set at a fixed distance from the column. The spindle is adjusted to depth.

On the knee-and-column type the depth of cut is set by moving the knee vertically. The vertical spindle milling machine is versatile. Positioning the workpiece and cutter, observing the cutting action, and inspection are more easily done than on a horizontal milling machine. The vertical milling machine is also adapted to precision hole-machining operations. Accurate center distances are easily attained. The workpiece may be positioned accurately by direct measurement with the micrometer collars on the table, crossfeed, and knee-elevating screws.

Fixed-Bed Milling Machine

The fixed-bed miller is usually a heavier, more rigidly constructed machine tool than the knee-and-column type. A fixed-bed milling machine has an adjustable spiral head. This permits adjustment of the cutter in relation to the fixed position of the workpiece and table. Only the table travels longitudinally along the fixed ways of the bed.

The fixed-bed milling machine is primarily a manufacturing type of machine. Milling cutters such as face mills, shell end mills, and other cutters mounted on arbors are used. Face milling, slotting, and straddle milling are common operations performed on this machine.

An automatically cycled fixed-bed milling machine is illustrated in Fig. 38-4. The workpiece is positioned and secured in a fixture or other holding device on the table. The automatic cycle includes the following functions:

- Starting
- Rapid traverse feeding of the work to the cutter
- Machining at the required rate of feed
- Reversing.

Other fixed-bed machines are equipped for *automatic rise and fall* of the spindle head. This permits the milling of surfaces at different levels. These machines are also called *rise and fall millers*. The spindle head mechanism is synchronized with the automatic traverse of the table. The cutter may thus be raised over any projection in the workpiece or work-holding device. Once over the obstacle, the spindle head is again positioned for depth in the same plane or a different plane.

Bed-type milling machines may be equipped with multiple spindle heads. The heads may be adjusted vertically. Some are designed for horizontal adjustments.

There are many adaptations of the principles of fixed-bed milling machines to other classes of machine tools. The horizontal boring mill and the vertical boring mill are examples.

Rotary-Table Milling Machine

The design features of the *rotary table milling machine* differ considerably from those of the flat-bed, plain, and universal millers. The table revolves. The worker loads and unloads workpieces in fixtures attached to the table. The table rotates under a face-type cutter. Usually, there are two spindle heads. A roughing cutter is mounted in one spindle. The finishing cutter is driven by the second spindle. This machine is built ruggedly to face large surfaces in a production setup. For instance, the rotary table machine is particularly adapted to facing operations on marine and automotive engine cylinder heads.

Planetary Milling Machine

The workpiece is held stationary on the *planetary milling machine*. One or more cutters may be held in vertical and/or horizontal spindles. Production units are also designed with magazine loading. This machine performs many operations that normally are done on the lathe. The planetary milling machine is used for workpieces that are heavy, difficult to machine, or are so unbalanced or delicate that they cannot be rotated.

The versatility of the machine is shown by naming a few typical operations.
- Facing
- Internal and external threading of different pitches simultaneously
- Milling radial crankcase bores in aircraft engines
- Milling elongated slots.

The operations indicate that plain and form cutters and thread cutters are used. Operations are performed on the inside or outside of a workpiece, or on both surfaces at the same time.

Tracer-Controlled Milling Machine (Duplicator Die Sinker)

There are two basic types of tracer-controlled

Fig. 38-4. A fixed-bed, manufacturing type milling machine with a variable cycle controller. (Courtesy of Cincinnati Milacron)

milling machines. One is called a *profile miller (profiler)*. The second type has a number of names, including *duplicator, die sinker,* and *Keller machine.*

Profile Miller—This miller is similar to the vertical spindle milling machine. Machining is usually done by small-diameter end mills. The cutting path is controlled by a *tracer (stylus)* with the same diameter and form as the end mill. The tracer follows a template of the required size and shape. The movement of the cutter is controlled by hand or automatic feed.

The combination of movements in two dimensions produces the required form in *one plane.* On the automatic feed type, the tracer mechanism disengages whenever contact is broken with the surface of the template. When a profile miller has an optical scanner, line drawings may be used in place of the template when a high degree of accuracy is not needed.

Die Sinking Machines (Profiler, Automatic Tracer-Controlled Miller)—Many form dies and formed parts require three-dimensional machining. *Die sinkers* are designed for operation manually or in combination with electronic and hydraulic controls. A template is used. The template is a replica of the work to be produced. Fig. 38-5 shows a setup for duplicating a convex form. The tracer provides three-dimensional information as it is moved over the entire surface. This is duplicated by the movement of the corresponding profiler head and form cutter.

There are three major components to an automatic profiler: *control (command), processing,* and *drive.* The command component includes a tracer head, a stylus, and a control panel. As the stylus moves over the template the movements are converted to electrical signals. These movement signals are transmitted to the *processing component.*

The movement signals are processed and used to develop electrical *demand signals,* which are sent to the *actuating mechanisms.* Very simply, these are the motors, hydraulic cylinders, and control valves.

463

These mechanisms establish the relative position of the cutting tool at a point along any one or a combination of the three axes.

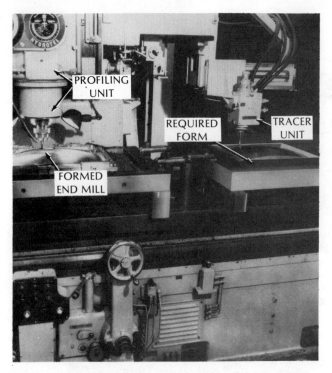

Fig. 38-5. Three-dimension profiling. (Courtesy of Cincinnati Milacron)

FUNCTIONS AND CONSTRUCTION OF MAJOR UNITS

The knee-and-column milling machine (Fig. 38-6) is a widely used, general-purpose type. The name is derived from two basic units: the *knee* and the *column*. The terms commonly used to describe the major units of all milling machines are given in Fig. 38-6. The longitudinal, crosswise (or transverse), and vertical movements are shown in Fig. 38-7. The major units of a horizontal milling machine include the *knee, column, saddle, table, overarm, table controls, spindle, arbor support, overarm braces, drives, speed* and *feed controls*, and the *coolant system.*
A description of each of these units follows.

The Column

The *column* (Fig. 38-8) supports the knee, saddle, table, and coolant system. The column is a hollow casting frame. Space is provided internally to house the driving motor and the gear trains or variable drives. These provide for speed and feed changes. The dovetailed ways at the top support the overarm.

Some models also house the oil reservoir and the pumping system. These supply oil to the spindle, gears, and other parts through a pressure controlled

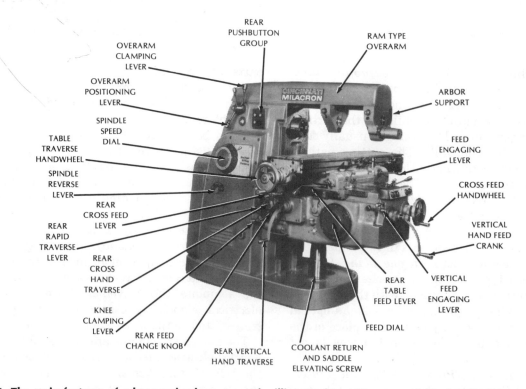

Fig. 38-6. The main features of a knee-and-column type of milling machine. (Courtesy of Cincinnati Milacron)

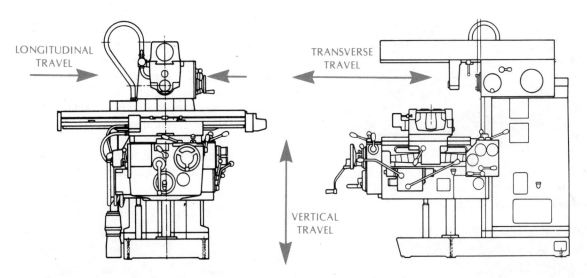

LONGITUDINAL TRAVEL

TRANSVERSE TRAVEL

VERTICAL TRAVEL

Fig. 38-7. Three basic movements on a horizontal milling machine.

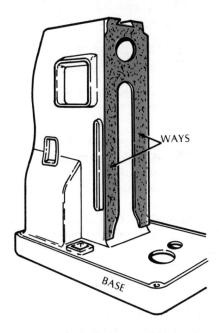

WAYS

BASE

38-8. The column of a horizontal milling machine.

38-9. The knee of a universal horizontal milling machine. (Courtesy of Cincinnati Milacron)

system. A second reservoir supplies the cutting fluid that flows over the cutter and work.

The front face of the column is flat. The face sides are dovetailed and are called *ways*. The face and ways provide the bearing surfaces that keep the knee aligned and permit accurate height adjustments. The bearing surfaces are precision finished and are sometimes scraped.

The Knee

The *knee* (Fig. 38-9) has two sliding surfaces. These are accurately machined at right angles to each

other. The vertical face on the back fits the ways of the column. The knee thus may be moved vertically. The horizontal face on top of the knee provides the bearing surface and horizontal plane for the *saddle*. Adjustable *gibs* (metal strips) are included for the dovetailed areas of the column, knee, and table. These are adjusted to compensate for wear of the bearing surfaces and to maintain the original sliding fit.

The Saddle

Universal Milling Machine—The *saddle* of the universal milling machine (Fig. 38-10) is made in two parts. The bottom half slides over the dovetailed ways on the top of the knee. The top half of the saddle is also dovetailed, to accommodate the table. The mating surfaces of the top and bottom sections of the saddle are machined with circular ways. This

465

Fig. 38-10. The swivel housing and saddle of a universal horizontal milling machine. (Courtesy of Cincinnati Milacron)

provides a swivel joint that permits the table to be set parallel or swung to a desired angle. The two halves are clamped by bolts that extend from circular T-slots in the base portion.

Plain Horizontal Milling Machine—The *saddle* of this milling machine is also attached to the knee. Its function is to provide movement in two directions in relation to the saddle: crosswise (transverse) and longitudinal. The saddle dovetails slide along the angular ways on the top of the knee. This provides the bearing surfaces on which the saddle may be moved crosswise.

The dovetailed slide on top of the saddle is machined at right angles to that on the bottom. The top dovetailed slide accommodates the ways of the table. The table may thus be moved lengthwise (longitudinal movement).

The Table

Milling machine accessories, work-holding devices, and workpieces are all strapped to a *table*. The underside of the table is dovetailed to fit and slide longitudinally in the ways on the top section of the

saddle (Fig. 38-11). The top face of the table is ground in a perfect horizontal plane. This provides a smooth, accurate, flat plane. The sides are accurately machined at right angles to the face. Precision-machined T-slots are accurately located along the length of the table. They are parallel to the sides of the table. On some models, a T-slot is machined in the table side that faces the column. This design feature permits fixtures, attachments, and other parts that are fitted with *keys* to be aligned quickly and accurately.

Fig. 38-11. A table designed to fit the saddle of a milling machine. (Courtesy of Cincinnati Milacron)

Other grooves run parallel with the table slots. These carry the coolant to a deeper trough at one end of the table. The coolant flows through this trough as it returns to the coolant reservoir, from which it is recirculated.

These combined features of the table, saddle, and knee of the universal milling machine are shown in Fig. 38-12.

Table And Knee Controls And Graduated Dials

Table and knee movements are made with feed screws and specially designed nuts. These are incorporated in the knee, saddle, and table design.

The knee is moved by an *elevating screw*. This extends from the inside of the knee to the base. The motion of the screw is controlled by the *vertical hand feed crank*.

A second feed screw is provided to move the saddle crosswise, toward or away from the column. The *crossfeed handwheel* is used to turn this screw.

A third screw, known as the *table lead screw*, runs the length of the table. It is turned by a handwheel located at the end of the table.

There are micrometer collars (dials) on each of the three screw-movement controls. The collars are graduated to read in thousandths of an inch (0.001″) or in two-hundredths of a millimeter (0.02 mm). Most

Fig. 38-12. The major units of a universal milling machine. (Courtesy of Cincinnati Milacron)

modern machine tool collars are designed to permit readings in both English and Metric units of measure. A precision graduated inch/metric micrometer collar for such readings is shown in Fig. 38-13.

Since the lead and feed screws and mating parts on milling machines are precision machined, the screw movements indicated on the micrometer collars are accurate to within 0.001″ (0.02 mm).

Overarm, Arbor Support, And Braces

The Overarm—The dovetailed ways on the top of the milling machine column accommodate the *overarm*. This member may be moved in a horizontal plane. The overarm axis is parallel to the spindle axis. The location of the overarm and an assembly of an overarm and outer arbor support are shown in **Fig.**

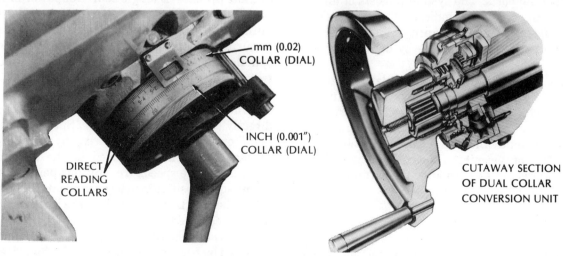

Fig. 38-13. Precision graduated inch/metric micrometer collars for table, knee, and saddle movements. (Courtesy of Kearney & Trecker Corp.)

38-14. The overarm slides toward (into) or away from the column. On heavy-duty millers, a screw-and-cam arrangement permits moving and securely locking the overarm in position.

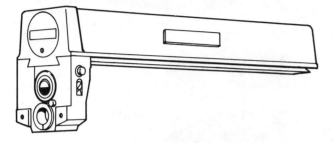

Fig. 38-14. Overarm and outer arbor support.

Arbor Supports—*Arbor supports* are designed to slide on the overarm. Two common designs of arbor supports are available.

The *outer arbor support* (Fig. 38-14) extends down below the milling machine arbor. The lower section of the support contains a bearing into which the pilot end of the arbor fits. The arbor support and overarm are used together to keep the arbor aligned and to prevent it from springing. Otherwise, chatter marks, poor surface finish, and an inaccurate workpiece may be produced.

The *intermediate support* has a large center hole. A bearing sleeve rides in this center hole. The arbor fits through and is supported by the bearing sleeve. The intermediate support and bearing sleeve are used when roughing cuts and multiple cuts are to be taken.

A self-contained oiling system is incorporated in many arbor supports. The system consists of an oil reservoir, a sight gage that indicates the supply of oil, and a plunger. A small quantity of oil is forced directly into the bearing sleeve by depressing the plunger.

Overarm Braces—On the heavier models, the overarm is rigidly supported for heavy cuts by using *overarm braces*. Although the designs vary, all overarm braces are attached to the knee, have slotted arms, and extend and are secured to the arbor support. The slots permit the knee to be adjusted. The overarm braces are secured after the cutter and the work are positioned.

The Spindle

The *spindle* is housed in the upper section of the column. Four main functions are served by the spindle:

- Aligning and holding cutting tools and arbors
- Transmitting motion and force from the power source for cutting
- Ensuring that the machining processes are dimensionally accurate
- Driving a slotting or vertical spindle attachment.

The spindle itself is hollow bored. The front end has a tapered hole. Standard arbors, adapters, and cutting tool shanks fit the tapered hole. A draw-in bar is usually used to secure these tools against the taper. The front end of the spindle is fitted with two set-in *lugs* that extend beyond the face. Arbors, adapters, and cutting tools have identical slots. These match the positions of the lugs. The lugs fit into the slots and provide a positive drive.

Four other holes are threaded in the spindle face. These are used for mounting face mills.

Around the shell of the spindle are end support bearings. These are housed in corresponding bearing supports in the column assembly. The bearing supports are precision machined and are adjusted to produce a minimum of friction with maximum sturdiness and accuracy. The spindle assembly includes gear combinations or other variable speed mechanism. Each provides a wide range of spindle speeds.

The direction of the spindle rotation is reversed by one of two common methods. Some models have two motor switches, for clockwise rotation and a second for counterclockwise rotation. Other machines are equipped with a *spindle direction change lever*.

Milling Machine Drives

Spindle Speeds—A single source of power is used on the smaller, all-purpose horizontal and vertical milling machines. Larger, heavy-duty, and manufacturing models use individual motors to power the spindle, knee, and table for regular feeding and rapid traverse.

Some spindles are driven by a belt or a silent chain. The motor is usually housed in the base. Others are directly connected to the motor. Speed changes are made on a constant-speed drive by a sliding gear transmission. This is housed in the column. The various spindle speeds (revolutions per minute) are obtained by positioning the speed control levers. The index chart on the column gives the lever settings for a required speed.

On variable-speed motor drives, speed changes are obtained by turning the speed control dial. Step-cone pully drives require changing the belt position.

Feeds And Feed Levers—Feed is the rate at which a

workpiece is fed to a cutter. The rate is specified as so many thousandths of an inch or millimeters per revolution of the spindle. The feed rate may also be given in inches or millimeters per minute when the milling machine is equipped with a constant-speed drive.

A feed index chart on the knee or column gives the range of feeds. The position of the feed change levers and a dial design are illustrated in Fig. 38-15. The lever or dial settings regulate the lengthwise movement of the table, the crossfeed movement of the saddle, and the vertical travel of the knee when the automatic feed is applied.

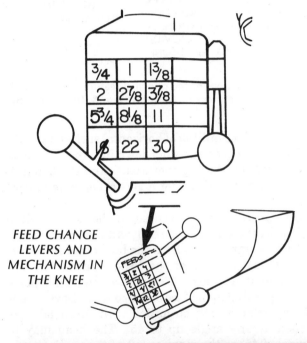

FEED CHANGE LEVERS AND MECHANISM IN THE KNEE

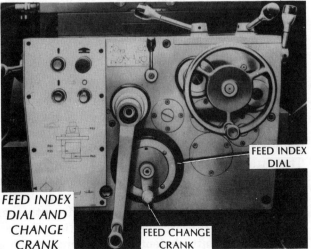

FEED INDEX DIAL AND CHANGE CRANK

FEED INDEX DIAL

FEED CHANGE CRANK

Fig. 38-15. Types of feed controls. (Courtesy of Cincinnati Milacron)

Other levers are used to manually engage or disengage the automatic feed mechanism. These levers are located on the knee and saddle. Some machines are also designed for rapid traverse. Once the starting point is reached, the feed lever is moved to the normal feed position. The work is then fed at the required feed past the cutter.

Trip Dogs—Trip dogs are simple adjustable guides. They control the length of movement. The trip dog moves against a pin, causing it to move and disengage the feed. Trip dogs may be placed on the front of the table, under the saddle, and behind the knee. The location determines the position at which the movement of a particular unit is disengaged. Trip dogs are also a safety device. They prevent a cutting tool from moving so far that it causes damage to the setup or workpiece.

PRINCIPAL MILLING MACHINE ACCESSORIES

Two principal milling machine accessories are found in most training centers, machine shops, and tool rooms. These are the *dividing head* and the *rotary (circular) table*. While described briefly at this point, the dividing head and work indexing processes are dealt with in considerable detail in a unit by itself.

The Dividing Head

The *dividing head* (Fig. 38-16) is also called an *index head*. Its function is to turn the workpiece through a precise series of arcs. Thus, the workpiece may be positioned for any required number of cuts or measurements to be taken around the periphery. The indexing may be for regular, *even spacing*. Design features may also be irregularly spaced. The angle between each feature may be different.

The dividing head may be *plain, standard,* or *universal*. The principal parts of the universal dividing head are pointed out in Fig. 38-16. Note that this is a more complex dividing head.

Once the relative position of the workpiece is indexed, the dividing head spindle may be turned continuously. This movement is necessary for such processes as machining splines, flutes, and gear teeth that are spiral. The ratios of the gears between the lead screws of the table and dividing head are determined according to the required helical angle. These gears are inserted in the gear train between the table and the dividing head.

The plain *index center* has plates with equally spaced holes. After each cut, the part is moved directly to another hole, depending on the required

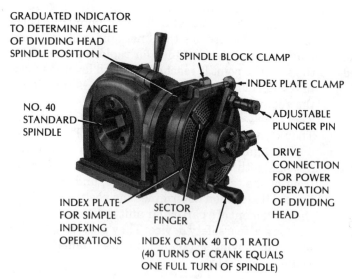

GRADUATED INDICATOR TO DETERMINE ANGLE OF DIVIDING HEAD SPINDLE POSITION

SPINDLE BLOCK CLAMP

INDEX PLATE CLAMP

NO. 40 STANDARD SPINDLE

ADJUSTABLE PLUNGER PIN

DRIVE CONNECTION FOR POWER OPERATION OF DIVIDING HEAD

INDEX PLATE FOR SIMPLE INDEXING OPERATIONS

SECTOR FINGER

INDEX CRANK 40 TO 1 RATIO (40 TURNS OF CRANK EQUALS ONE FULL TURN OF SPINDLE)

Fig. 38-16. The main features of a universal dividing head. (Courtesy of Kearney & Trecker Corp.)

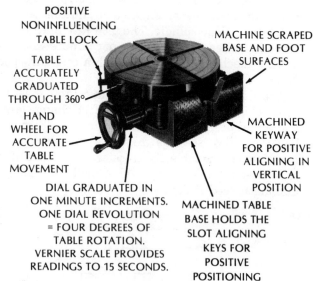

POSITIVE NONINFLUENCING TABLE LOCK

TABLE ACCURATELY GRADUATED THROUGH 360°

HAND WHEEL FOR ACCURATE TABLE MOVEMENT

MACHINE SCRAPED BASE AND FOOT SURFACES

MACHINED KEYWAY FOR POSITIVE ALIGNING IN VERTICAL POSITION

DIAL GRADUATED IN ONE MINUTE INCREMENTS. ONE DIAL REVOLUTION = FOUR DEGREES OF TABLE ROTATION. VERNIER SCALE PROVIDES READINGS TO 15 SECONDS.

MACHINED TABLE BASE HOLDS THE SLOT ALIGNING KEYS FOR POSITIVE POSITIONING

Fig. 38-17. Functions of the principal features of a precision rotary table. (Courtesy of Universal Vise & Tool Company)

number of spaces. The plain index center plates are divided around the circumference with 28 and 48 or 30 and 36 holes or notches. These combinations are divisible by a great many common numbers from 1 through 48.

The *standard index head* is used for hand indexing a wider range of divisions. This dividing head and general dividing head work are covered later. A *footstock* or *tailstock* cutter is provided to support work between centers.

Rotary Table

The *rotary table* permits the workpiece to be moved in a circular path. It consists of a base and a rotary table. The base of a comparatively simple direct-setting rotary table, like that shown in Fig. 38-17, is graduated in 1-minute increments through 360°. The rotary table is provided with a T-slotted face plate and a chuck. Irregular-shaped workpieces may be strapped to the face plate or held in a fixture.

A precision type of rotary table is geared and moved by a handwheel. The micrometer collar on the lead screw mechanism reads in degrees, minutes, and seconds. Some rotary tables are power operated where continuous motion is needed. Milling a circular groove is an example.

VERTICAL MILLING HEAD AND SLOTTER HEAD ATTACHMENTS

Vertical Milling Head Attachment

There are some instances when vertical milling is

preferred. Face milling of horizontal surfaces and the cutting of grooves and slots down into the workpiece are typical machining operations. An attachment is available to substitute for a vertical spindle milling machine.

The attachment is secured to the column. It is driven by the spindle and really is a vertical gear box. Motion is transmitted from the horizontal spindle to the vertical spindle of the head attachment. The attachment may be positioned at an angle in a plane that is parallel to the face of the column. Beveled and other surfaces may be milled with the head positioned at any angle up to 45°. The head may be swiveled 45° to the right or left of the vertical axis.

Slotter Head Attachment

A less commonly used attachment is the *slotter head*. This unit is mounted on the column face. The slotter head converts the rotating motion of the milling machine spindle to a reciprocating motion. The attachment is adapted to machine keyways, slots, and grooves, and square and other irregular-shaped holes. These forms all require a reciprocating motion. The cutting tools are shaped differently than those that require a circular cutting motion. The *ram* of the slotting attachment has a normal range of strokes up to four inches. The attachment may also be positioned at an angle. This provides extra capability to machine regular and irregular angular surfaces.

MAINTENANCE OF THE LUBRICATING SYSTEMS

Preserving Accuracy And Smoothness Of Operation

Almost all of the parts on the milling machine are made of metal. The mating moving parts are machine finished and precision ground. Some flat bearing surfaces are scraped. The scraped surfaces fit precisely and provide smooth, efficient operation of the mating parts.

The bearing surfaces on the saddle of larger millers carry the weight of the table and workpiece. The column and knee support even heavier weights. The minutely intermittent cutting action of straight-tooth milling and other cutters produces additional forces. The turning or sliding of any two metallic surfaces against each other produces wear. As the speed increases, there is increased wear. An extremely thin film of lubricant between the surfaces reduces friction and maintains the dimensional accuracy. The film also reduces heat caused by friction. Heat produces expansion that, in turn, further increases friction. If not prevented by adequate lubrication, this cycle of friction-heat-expansion-friction may require added force to move the parts.

Oiling systems are incorporated into the design of plain, universal, and other types of milling machines. The systems maintain a continuous film of oil between the mating surfaces of moving parts. This provides good, regular lubrication and at the same time maintains the precision of the mating parts.

Pressure, gravity, and splash systems are widely used. The oil is forced through tubing to the surfaces, threads, and bearings that require lubrication. Glass-faced (sight) gages, like the one shown on the saddle in Fig. 38-18, indicate whether the oil is circulating. Other sight gages show the amount of oil in the reservoir and indicate the level at which it should be maintained.

Fig. 38-18. Glass-faced sight gage on a machine unit.

Force Feeding The Spindle, Knee, Table, And Column Systems

Oil is distributed to the spindle, spindle keyways, and gear train by pumping from the reservoir in the column.

The column, the knee, and the saddle are three units that are usually oiled by pressure and splash systems.

The gears and screws housed in the knee and the bearing surfaces in the column are usually force fed oil from the knee reservoir. A mechanically operated pump delivers the oil under pressure.

The saddle parts and the table ways are lubricated by a *single-shot oiling* device. This is located at the front of the saddle. The hand-operated plunger force feeds the oil from the reservoir to the bearing surface.

As indicated before, the pilot and bearing sleeve on the arbor are oiled directly from the arbor support reservoir. *Oil cups* that require filling by hand are used for other surfaces that require lubrication. Some of the smaller machines are lubricated almost entirely by hand. The type and the size of the *oiler* are indicative of the quantity of lubricant required. For instance, the cross feed screw requires a few drops of oil. Larger bearing surfaces and the gear train require a continuous flow over the gears and sliding parts.

It is important on all machine tools that the operator follow the recommended oiling procedures and routine. In practice, after the machine is set up, an inspection is made of all sight gages. There must be an adequate supply of oil in each of the oil wells. The machine is started only after all lubricating systems have been checked.

Regular Procedures For Machine Lubrication

Technical manuals identify the lubricating systems and points and specify the procedures to follow. The machine tool builder recommendations center around four basic considerations that the machine operator should follow. These include:
- Lubricant specifications
- Required quantity of each lubricant
- Frequency
- Routine (Fig. 38-19).

Frequency—This refers to the work time intervals at which the various systems are checked and maintained. Plant and machine tool maintenance require occasional *shut down* of a machine tool while the reservoirs are drained, cleaned, and refilled. Fine metal particles and other sediment collect in the bottom of the reservoir. If the fine metal particles are

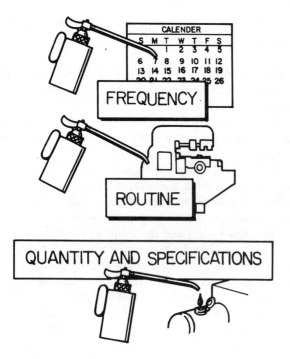

Fig. 38-19. Basic considerations in regulating machine lubrication.

left to flow within the lubricant supplied to revolving and sliding parts, they produce additional wear. *Routine*—All systems and hand oiling stations must be lubricated. To avoid missing a reservoir or lubrication point, a specific sequence of checks must be followed. It is important to first know the number and location of each lubrication point. Then start at one lubrication point and proceed to every other point in the established order.

Lubricant Specifications And Quantity

The specifications (type) and quantity of lubricant depends on the application. Geared spindles are in constant rotation and produce the forces necessary for cutting. The spindle lubricant must withstand variations in temperature and must have the required viscosity to flow, cool, and lubricate under varying machining conditions. The lubricant on the column ways requires different properties than those of the oil used to meet the heat-removing needs of the spindle.

The quantity of lubricant to use is established by the manufacturer according to the conditions of the system. The indicated range between *full* and *add oil* on a sight indicator, for example, provides the operator with the designer recommendations. The tendency in hand oiling is to use too much lubricant. Any excess may cause a safety hazard if it flows onto

the floor. Particles of materials also accumulate around the overflow and produce a dirty work station. Excess oil should be wiped off immediately.

HOW TO LUBRICATE AND MAINTAIN THE MILLING MACHINE

Systems Requiring Daily Lubricant Checks

Step 1   Read the lubricant chart supplied with the machine. Note all of the lubricating systems and the types of oiling devices. Locate each lubrication point and reservoir.

Step 2   Check the lubricant level. Use the sight gages on the column, saddle, and knee reservoirs (Fig. 38-20A).

Step 3   Locate the reservoir filler cap if the lubricant level requires that oil be added.

CAUTION: Remove all dirt particles and wipe the area around the cap until it is clean (Fig. 38-20B). It is important to prevent dirt, chips, and grit from accumulating and entering an oil reservoir.

Step 4   Check the recommended viscosity and other specifications if additional lubricant is needed. Add the lubricant with a spout oiler until the correct level indicated on the sight gage is reached (Fig. 38-20C).
Note: Pour in a small quantity at a time to allow the oil level to rise to give a true indication.

CAUTION: Wipe up any lubricant that may have spilled on the floor or onto the machine.

Step 5   Replace the reservoir cap. Again, wipe up any lubricant on the cap and surrounding area.

Step 6   Start the milling machine. Check the flow gage to determine that the oil pumps are operating correctly.
Note: A fine stream of lubricant appears in the flow gage when the oil pump is functioning. Periodically during the machine operation, recheck to see that the pump is circulating the lubricant.

Step 7   Hand oil the oil cups as specified by the machine manufacturer.
Note: Determine the number and loca-

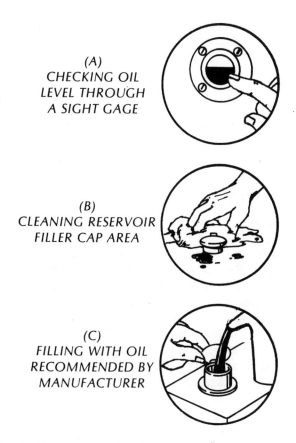

(A)
CHECKING OIL LEVEL THROUGH A SIGHT GAGE

(B)
CLEANING RESERVOIR FILLER CAP AREA

(C)
FILLING WITH OIL RECOMMENDED BY MANUFACTURER

Fig. 38-20. Basic procedures for machine lubrication.

tion. Then oil all stations in an established sequence.

Step 8   Brush all chips from the table and T-slots. Remove chips from between the table and column. Wipe the machined surfaces until they are clean and free of chips and cutting compound. These include the overarm and arbor supports.

Step 9   Lubricate all machined sliding surfaces that are not automatically oiled. These include the knee, column, and table ways and the overarm.
Note: Some workers apply a small quantity of oil on the machined surfaces. This is then spread to produce a thin film of lubricant over the whole surface area. A clean wiping cloth is preferred to spreading by hand.
Note: Use a spray cleaner on surfaces where the lubricant has congealed.
CAUTION: During a shut-down period of a few days or longer, the machined surfaces should be sprayed or coated with a rust inhibitor.

Step 10   Clean and lubricate the cross slide, table, and knee elevating screws. These are not included in an automatic lubricating system on some models.

Systems Requiring Periodic Lubrication Checks

The condition of oil filters, driving chain (if one is included in the machine design), motors, and reservoirs should be checked periodically.

Step 1   Check the type and condition of the oil filters. Use a cleaning bath if the filter pad or screen requires cleaning.

    CAUTION: Replace a *loaded* filter pad. Otherwise, the filter will not remove metal and other foreign particles.

Step 2   Test the condition of the lubricant in each reservoir.
Note: A viscosity tester may be used to determine whether the lubricant has *thinned out.*

    CAUTION: Discoloration of the lubricant is a sign that the lubricant is dirty. It has picked up and is circulating dirt and other fine foreign particles. Under conditions of reduced viscosity and discoloration, the lubricant is not able to maintain a correct, clean film between the moving and sliding parts.

    CAUTION: The power supply must be cut off before opening the motor compartment. On many models this safety feature is an integral part of the machine construction.

Step 3   Check the lubricating level of the motor bearings. If lubricant is to be added, check the specifications. Then, add the right lubricant.

Step 4   Examine the driving chain if there is one. Use the driving chain lubricant specified by the manufacturer.
Note: If the driving chain is lubricated by a *splash system,* check the oil supply level. Add lubricant as required.

Step 5   Examine the condition of each reservoir. If the lubricant requires changing, remove the drain plug. Drain the lubricant. Cleanse the reservoir to remove the *sludge* (thickened, gummy lubricant and foreign particles).
Note: A solvent is used to dissolve the sludge for easy removal.

Step 6   Follow the manufacturer's recommendations and refill the reservoir with clean lubricant. Bring the level to the line indicated on the sight gage.

SAFE PRACTICES IN LUBRICATING AND MAINTAINING MILLING MACHINES

- Shut down the milling machine when examining, cleaning, and lubricating moving parts.
- Avoid leaning or rubbing against the machine. Any slip against a lever or a handwheel may cause the machine to start or a cut to be taken accidentally.
- Treat the table top and top flat surface of the knee as a precision surface plate. Layout tools, wrenches, hammers, excess straps, etc., should be placed on an adjacent stand.
- Remove any nicks or burrs from the table face and the sides of the longitudinal slots.
- Check the bases and keys of vises, dividing heads, and other attachments. These also must be free of burrs. The keys and slots must be fitted accurately.
- Use a T-slot metal cleaner to remove the extremely sharp chips from the T-slots. A stiff brush and small chip pan should be used to clear the chips from the table, saddle, knee, and base.
- Keep the coolant strainer on the end of the table clear. This permits recirculating the cutting fluid.

- Check all sight gages. Maintain the proper level of lubricant in each reservoir.
- Follow a daily routine for checking and maintaining all systems and moving surfaces.
- Examine the face, keys, and base of the spindle. Carefully remove any burrs or nicks.
- Check the pilot end of the arbor and an intermediate bearing (if required) for burrs and correct fit. Scrape or stone any internal burrs in the arbor support bearing.
- Clean the area around an oiler to remove foreign matter before lubricating. Wipe up any excess oil after lubricating.
- Wipe up any lubricant that has been spilled or has overflown onto any machine surface or the floor.
- Place oily and dirty wiping rags or cloths in a *metal* container.
- Observe the standard personal safety precautions against loose clothing and the use of wiping cloths around moving parts.

TERMS APPLIED TO MILLING MACHINE DESIGN FEATURES AND MAINTENANCE

Milling	The process of removing material with a revolving multi-tooth circular cutter. Machining flat, angular, circular, and irregular straight, tapered, and spiral surfaces with a milling cutter.
Knee-and-column type hand, plain, and universal horizontal milling machines	Milling machines designed around three major components: a column, a knee, and a table. A classification of milling machines. Usually, a single horizontal spindle machine on which a workpiece on the table is fed past a revolving cutter.
Fixed-bed milling machine	A milling machine adapted for manufacturing. The table moves longitudinally parallel to the face of the column. The dovetailed ways of the table slide along fixed bed ways.
Planetary milling machine	Performs typical lathe and milling machine operations. Especially adapted to heavy, intricately designed, or hard-to-handle workpieces that cannot be turned or machined readily from a vertical position.
Tracer-controlled milling machine	Two-and three-dimensional duplicators and profilers. Styli and duplication heads synchronized to duplicate either two- or three-dimensional irregular forms.
Automatic profiler	A complementary system of control, processing, and drive mechanisms for pneumatically, electronically, and mechanically actuated units. A profiler that

	automatically reproduces a desired form.
Saddle (universal)	A two-section mechanism that moves on dovetailed slides of the knee.The top slide nests the table. The two sections permit setting the table at an angle.
Overarm, arbor support, and braces	Collectively the machine features that rigidly support a spindle, arbor, and cutters.
Feed change levers (milling machine)	Levers that are positioned in a fixed relationship to geared, V-belt, or variable-motor drives.
Trip dogs	Movable metal guides that are part of a tripping mechanism. A method of controlling distance and stopping a cut.
Dividing head	A mechanism designed to position and form a workpiece. A mechanism used to divide the circumference of a workpiece into a given number of spaces.
Rotary table	A round milling machine table on which a workpiece is secured and moved in a circular path. A table for rotating workpieces at a desired angle in relation to a milling cutter.
Force feeding (lubrication)	A system of forcing a lubricant from a reservoir to rotating or sliding members in the machine.
Lubrication system	A composite of a reservoir for the lubricant, a gravity or force-feed pump, piping to areas that are to be lubricated, and a sight gage.
Sight gage	A glass insert in a lubricating system. A visible device that indicates the amount and condition (color) of the lubricant in the system. A device that shows the required level of a lubricant.

SUMMARY

- The first milling machines were built in Europe in the mid to late 1760s. Developments were accelerated by Eli Whitney's concept of interchangeable parts and other needs of the rifle industry.
- The replacement of the rack-and-pinion feed mechanism with the worm and worm wheel provided better feed controls and a higher degree of machining accuracy.
- The angle adjustment feature of the universal mill extended the range of milling machine operations.
- Some of the newer design features include: preloaded bearings, force feed, lubrication systems, hardened and precision ground moving parts, greater speed and feed ranges, direct micrometer measurements, and readings in both inch and metric standards.
- There are six broad categories of milling machines: knee-and-column, fixed-bed, rotary table, planetary, tracer-controlled, and numerically controlled. Milling machines are also designed for thread, cam, skin and other specific machining processes.
- The general, all-purpose knee-and-column type millers include hand, plain, universal, and vertical spindle milling machines.
- Fixed-bed millers are a production machine. One or more spindles are adjusted in relation to the table. Table movement is longitudinal in a fixed position on the bed. Standard milling operations are performed. All movements may be automatically cycled for production.
- Rotary table milling machines are adaptable to large workpieces that are mounted on a revolving table. These are fed past one or more rotating spindles with cutters.
- The planetary milling machine performs a combination of lathe and milling machine operations. It is used for machining many parts that are difficult to hold or need to rotate vertically. The operations include the simultaneous threading of internal and external thread pitches.
- The profile miller is a vertical spindle, two-dimensional milling machine. A tracer follows the contour. A similarly shaped end mill cuts the part to size and shape.
- The die sinking (profiler) machine forms to three dimensions. Vertical, longitudinal, and transverse movements of a tracer over the desired form are duplicated by a cutter head. The cutting tool movement may be manually operated. Some profilers are designed for electronic-hydraulic automatic tracer controls.
- The major structural members of general-purpose milling machines are the column, the knee, the standard or universal saddle, and the table.

- Table and knee movements are made by lead screws. Graduated micrometer collars provide direct measurements.
- The arbor and cutting tools are supported by the overarm, arbor supports, and overarm braces.
- The spindle bearings are preloaded. The spindle is designed to align, hold, and drive cutting tools and attachments. Gears and sliding parts mesh with mating members in the gear box.
- Spindle speeds and feeds are controlled by positioning gear change levers and power levers. Trip dogs control the table movement.
- The dividing head with footstock center, the vertical milling head, and the slotter head are three general attachments.
- Milling machines are designed for pressure, gravity, and splash lubrication systems.
- Sight gages provide a visual check of lubrication. Lubrication systems are incorporated for force feeding the knee, column, and saddle.
- Hand lubrication is required on the flat or angular sliding surfaces of some machines.
- Regular lubricating procedures involve frequency (daily and longer periods), a routine, and knowledge of the specifications and quantity of lubrication needed.
- Reservoirs also require checks on the viscosity of the lubricant and the amount of sediment.
- The milling machine must be shut down when power drives are examined.
- General machine tool and personal safety precautions must be followed.

UNIT 38 REVIEW AND SELF-TEST

1. Name (a) two early (1800-1900) developers of milling machines, and (b) state the contribution made in each case.
2. List five developments since 1900 that have added to the versatility and efficiency of milling machines.
3. Distinguish between the principal straight-line table movements of *plain* and *universal horizontal millers* and *plain vertical milling machines*.
4. Classify three nonstandard milling machine types.
5. Identify the major construction parts and mechanisms of a milling machine.
6. Describe the functions of (a) a *dividing head* and (b) a *rotary table*.
7. Explain the functions of the *automatic lubrication systems* of modern milling machines.
8. Indicate four precautionary practices to follow when lubricating and maintaining a milling machine.

Unit 39
MILLING CUTTER- AND WORK-HOLDING DEVICES

Precision machining depends on securely holding the cutting tool and the workpiece in a fixed position. The number of shapes, sizes, and functions of cutting tools requires a variety of holders to accommodate different operations. Similarly, workpieces may be held in vises, between centers, or in chucks or fixtures. They may also be strapped to the table.

This unit concentrates on general-purpose devices for holding milling cutters and workpieces. Step-by-step procedures for mounting and removing these devices are presented.

CUTTER-HOLDING DEVICES

The spindle of the milling machine provides a bearing surface. This aligns the arbor, adapter, or tool shank. The power to drive the cutter is transmitted from the spindle to the cutter.

Modern milling machine spindle noses are ground to one of the National Milling Machine Taper sizes. This standard has a steep taper of 3½" per foot, or an included angle of 16° 36'. The angle of taper makes it *self-releasing*. A locking device is needed to secure an arbor or adapter to the spindle taper. A *quick-releasing collar* or *draw-in bar* is used for this purpose.

Milling machine cutters are generally held by four methods:

- Mounting the cutter on the nose of the spindle
- Inserting a shank directly in the spindle bore
- Using adapters and collets
- Mounting the cutter on an arbor.

Spindle Nose

The four common steep taper sizes of spindle noses are #30, #40, #50, and #60. A steep taper spindle nose is shown in Fig. 39-1. The range of diameter of the common taper sizes are given in Table 39-1.

Older milling machine spindles were designed with a threaded nose. This was used to hold a milling machine chuck and other cutting tools. The inside bore was machined to a standard self-locking Brown and Sharpe taper. The arbors and cutting tools were held securely by the taper and a draw-in bar. Fig. 39-2 shows a threaded spindle nose.

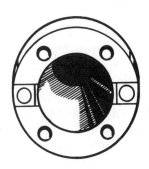

Fig. 39-1. A steep taper spindle nose.

Table 39-1. COMMON MILLING MACHINE SPINDLE TAPER SIZES

National Milling Machine Tapers	
Taper #	Large Diameter
30	1¼"
40	1¾"
50	2¾"
60	4¼"

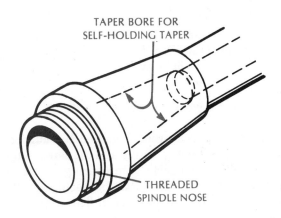

TAPER BORE FOR SELF-HOLDING TAPER

THREADED SPINDLE NOSE

Fig. 39-2. An older style threaded spindle nose.

Another old type had a steep outside taper and a Brown and Sharpe internal taper. The end had two recessed slots. The lugs on the arbor or driver plate transmitted power to the cutter.

Face milling cutters were mounted directly on the spindle nose. The cutter was drawn against the steep taper by the *driver plate* and the draw-in bar.

Arbor Styles

There are three *styles of arbors: A, B* and *C.*
Style A—The *style A arbor* (Fig. 39-3A) has a pilot. The pilot fits into a corresponding bearing in the arbor support. Style A permits the workpiece to be brought up close to the arbor. The advantage is that smaller diameter cutters may be used with style A arbors than can be used with style B. Where needed, a bearing sleeve and an intermediate arbor support are added. These provide additional bearing support and prevent a long arbor from vibrating.

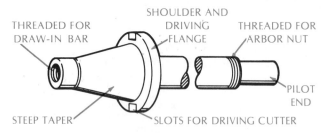

*(A) STYLE A ARBOR WITH PILOT
FOR END SUPPORT*

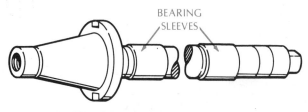

*(B) STYLE B ARBOR SUPPORTED BY
INTERMEDIATE AND OTHER
ARBOR SUPPORTS*

Fig. 39-3. Style A and B milling machine arbors.

Style B—The style B arbor (Fig. 39-3B) is supported by one or more *bearing sleeves* and the same number of intermediate supports. The bearing sleeves are *keyed* to the arbor. The arbor is machined with a *keyway* along most of its length. The keyway is a square slot.

The large diameters of the bearing sleeve and arbor support bearing permit the arbor to be supported anywhere along the length of the arbor.

Multiple milling operations often require the use of two arbor supports for heavy-duty milling. This setup provides maximum rigidity, eliminating chatter. The bushing and bearing sleeves have a freerunning fit.

Style B arbors are used where a minimum of clearance is needed between the cutters, workpiece, and holding device.

Style C—The style C arbor (Fig. 39-4) is a short arbor. It is sometimes called a *shell end mill arbor.* The cutter is mounted on a shouldered end. It is held against the face of the arbor by a screw. A special wrench is used to reach into the counterbored face of the shell end mill. The cutter is driven by two lugs on the arbor. These mate with corresponding slots in a shell end mill.

Keyway and Key

A metal *key* with a rectangular cross section is used between the arbor keyway, spacing collar, cutter, and bearing sleeve. The key and keyway are shown in Fig. 39-5. The keyway and key prevent the cutter, collar, and sleeve from rotating on the arbor. The key also provides the contact surface for driving the cutter.

Spacing Collars, Bearing Sleeve, And Arbor Nut

Spacing collars (Fig. 39-6A) are used for spacing or locating purposes. They also hold one or more cutters on an arbor. Spacing collars have parallel faces ground to accuracies as close as $\pm$ 0.0005" ($\pm$ 0.01 mm).

The *bearing sleeve* (Fig. 37-6B) also fits on the arbor and is keyed. Usually, the ends of the sleeve are not ground to the same limits of accuracy as are those of spacing collars. However, when the bearing sleeve is used as a spacer between cutters, it must be machined to the same precision as that of a spacing collar. The outside diameter of the bearing sleeve must be ground accurately to fit the arbor support bushing.

The *arbor nut* is a round nut with two flat surfaces. These are machined along part of the length. The milling machine arbor wrench fits this nut. The arbor nut is used to tighten the collars, cutter, and arbor support bushing. The threads direction is opposite that of the milling operation. For instance, if an arbor normally turns clockwise, the nut has a right-hand thread. The nut therefore tends to be tightened by any vibration in cutting. The right-hand nut is the most common.

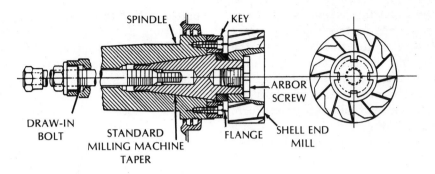

Fig. 39-4. A style C milling machine arbor.

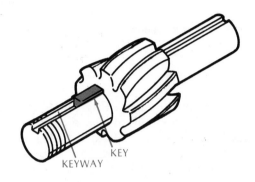

Fig. 39-5. The key and keyway transmit cutting force to the cutting tool.

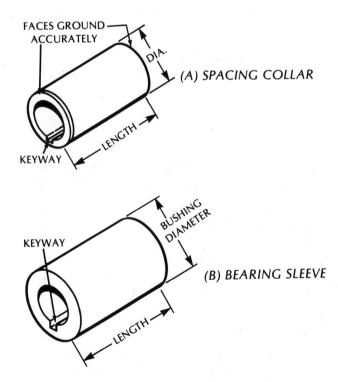

Fig. 39-6. Features of spacing collars and bearing sleeves.

The Draw-In Bar

As stated earlier, the steep taper is self-releasing. The mating surfaces must be drawn together for proper alignment and to provide direct contact between the spindle and the cutter-holding device. The driving lugs on the spindle fit in the slots of the work-holding device and power the cutting tool.

A *draw-in bar* has two threaded ends, a knob, and a threaded collar (Fig. 39-7). The draw-in bar extends almost the length of the spindle. One threaded end screws into the tapered end of the arbor or adapter. The draw nut is adjusted against the left (outside) face of the spindle. The threaded collar is tightened to draw the tapered arbor and spindle surfaces together.

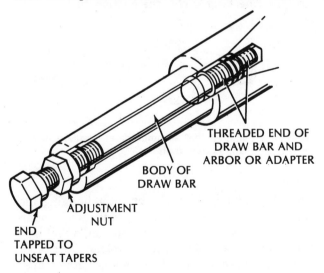

Fig. 39-7. Assembly and functions of a draw-in bar.

The process is reversed to release the surfaces and remove the cutter-holding device. The end of the draw-in bar is tapped gently but firmly with a soft-faced hammer to release the tapers.

Some milling machines are still equipped with the self-holding Brown and Sharpe taper. The draw-in bolt is also designed to mount and hold arbors and adapters in Brown and Sharpe taper spindles. Care must be taken to tighten the nut only until the surfaces are mated and secured. The self-holding taper requires less force to secure than the self-releasing steep taper. The arbors and adapters are released by turning the thread collar about one turn. If the draw-in bolt is not designed to push against a retaining collar, the end should be tapped to release the taper surfaces.

Adapters

An *adapter* is a holding device. As the term indicates, it adapts the design features of different sizes and types of cutters to the spindle nose. An adapter permits use of a large variety of tools and accessories that otherwise would not fit a milling machine spindle.

Adapters are still available to accommodate older types of arbors and cutting tools that have self-holding tapers. Fig. 39-8 shows an adapter for holding a Brown and Sharpe or a Morse taper shank cutting tool. The adapter is secured in the steep taper nose by a draw-in bar. The lugs on the face of the spindle provide the driving force. The taper and tang of the cutting tool fit the inside bore of the adapter. The self-holding taper aligns and secures the tapered surfaces. The tang drives the cutting tool.

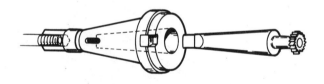

Fig. 39-8. An adapter that accommodates standard B&S or Morse taper-shank cutting tools.

Another adapter is made for interchanging face milling cutters from one type of spindle to a standard spindle. With this adapter, shell end mills that have an internal taper designed for the older short-taper-nose spindle can be mated to standard steep-taper-nose spindles. A special long draw-in bar is used to pull the tapered surfaces together and to secure the face milling cutter.

The *cam-lock adapter* (Fig. 39-9) is designed to hold cutting tools that have steep taper shanks. As the cam-lock is turned against a corresponding surface on the tool shank, there is a positive lock. This prevents the cutter or holder from turning. There is a positive drive and grip so that the cutter will not be

pulled out of the adapter by the cutting action or vibration. End mills and other cutters and holders may be released quickly. The cutter or holder is secured and released by a partial turn of the cam-lock with a wrench.

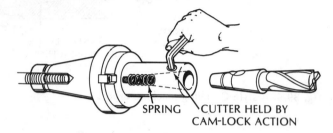

Fig. 39-9. A cam-lock adapter.

Sleeves And Collets

The milling machine *sleeve* (Fig. 39-10) serves the same function as a standard sleeve used on a drill press, lathe, or other machine tool. A sleeve has an internal and external taper. The outside taper surface fits into an adapter. The taper on the cutting tool corresponds with the inside taper of the sleeve. The sleeve makes it possible to use smaller-taper-diameter cutters in a large spindle bore.

The assembly of a sleeve, cutter, and adapter are shown in Fig. 39-10. This particular style has a tang end. This fits onto a corresponding tang slot in the adapter. The end of the adapter in Fig. 39-10 is threaded. The adapter is secured by a draw-in bar. The cutter is released from the sleeve in the same manner as a drill and drill sleeve. A drift is tapped a striking blow to release the tapers.

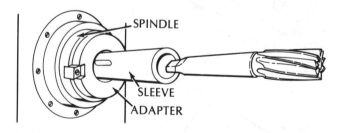

Fig. 39-10. An adapter, sleeve, and cutter assembly.

Collets

A *collet* serves a function similar to that of a sleeve. A collet is a precision-ground cylindrical sleeve (F and H in Fig. 39-11). A series of slots are cut partly along the length. This splits the cylinder to form a collet. As the collet is held in a tool holder and

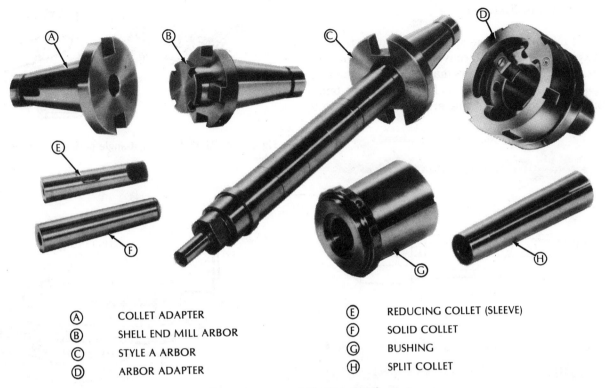

Ⓐ	COLLET ADAPTER	Ⓔ	REDUCING COLLET (SLEEVE)
Ⓑ	SHELL END MILL ARBOR	Ⓕ	SOLID COLLET
Ⓒ	STYLE A ARBOR	Ⓖ	BUSHING
Ⓓ	ARBOR ADAPTER	Ⓗ	SPLIT COLLET

Fig. 39-11. Standard arbors, adapters, and collets. (Courtesy of Cincinnati Milacron)

tightened, force is applied to the cutting tool. Collets are usually used with small-diameter, straight-shank end mills. Collets provide a comparatively inexpensive method of accommodating a wide range of cutting tool shanks.

A combination of arbors, adapters, and collets are shown in Fig. 39-11. These are some of the everyday types used for general-purpose milling operations.

Cutting Tool Holders

Straight-shank end mills and other cutting tools may be held directly in a cutting tool holder. These holders are available to accommodate different shank sizes. Fig. 39-12 shows three general tool holder designs. The cutting tool is secured in any one of the three holders by a set screw. The shanks of holders (A) and (B) are ground to the Brown and Sharpe self-holding taper. However, the ends are machined to different forms. The end of holder (A) is threaded to receive a draw-in bar. The end of holder (B) has a tang that fits the tang slot of an adapter.

The steep-angle taper of holder (C) fits directly into the standard steep-taper-angle bore of the spindle.

WORK-HOLDING DEVICES

Plain Milling Machine Vise

One of the most widely used and practical holding devices is the *milling machine vise*. Usually, these are heavier than the all-purpose vises used with drill presses. The milling machine vise base is slotted (grooved) in two directions. Keys fit the vise grooves and the table slot. These align the vise jaws either parallel or at right angles to the column face. The parts of a plain milling machine vise are labeled in Fig. 39-13. The right-angle slots and the strapping of a plain vise in the parallel and right-angle positions are shown in Fig. 39-14.

Swivel Vise

The swivel vise is another common milling machine vise. It has the added feature of a movable, graduated base (Fig. 39-15). The base is strapped to the table. The vise may be turned to position the workpiece. The workpiece may be machined parallel to the column face or at a required angle. Clamping bolts on each side secure the vise and base sections.

A third type, which is especially adapted to machining multiple pieces, is the cam-action vise.

Vise Jaws And Fixtures

Most milling machine operations require that the workpieces be held with conventional (straight) vise jaws. However, special jaws may be added to the vise. Two examples are shown in Fig. 39-16A. In one a workpiece is held at an angle. In the other a round

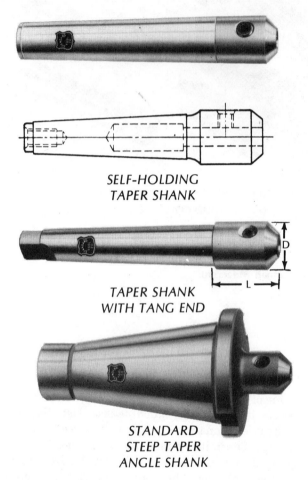

SELF-HOLDING
TAPER SHANK

TAPER SHANK
WITH TANG END

STANDARD
STEEP TAPER
ANGLE SHANK

Fig. 39-12. General designs of toolholders for straight-shank end mills and other cutting tools. (Courtesy of Standard Tool Div./Lear Siegler, Inc.)

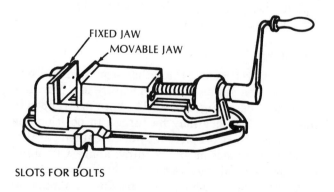

FIXED JAW
MOVABLE JAW

SLOTS FOR BOLTS

Fig. 39-13. A plain milling machine vise.

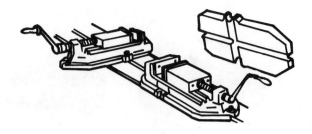

Fig. 39-14. Parallel and right-angle positions of a vise on the table.

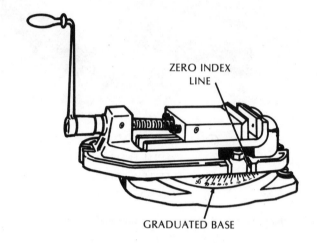

ZERO INDEX
LINE

GRADUATED BASE

Fig. 39-15. A graduated-base swivel vise.

workpiece is positioned and secured in a V-groove.

Fig. 39-16B shows a set of jaws that serve as a fixture. Irregularly shaped parts may be positioned, secured, and machined accurately with this simple setup. For production of a large quantity of irregularly shaped parts that require precise locating, a fixture is usually used. After the fixture is set up and the cutting tools are positioned, multiple pieces may readily be produced. The fixture cuts down the setup time and reduces machining costs.

Other Work-Holding Devices

A universal vise is used when workpieces must be held at a compound angle. This vise was described in an earlier unit.

Accessories and work-holding devices are strapped to the table with standard straps and clamps. A number of shapes and sizes are available to accommodate different workpieces and milling processes. Flat, finger, adjustable, and the V-strap clamp are examples.

Standard milling machine bolts, like those used on drilling machines, are of three common types. The T-heads that fit the T-slot of the table may be

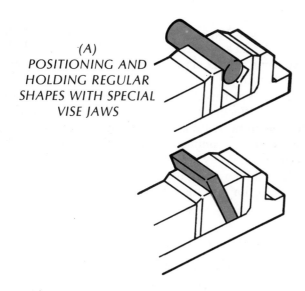

(A)
POSITIONING AND
HOLDING REGULAR
SHAPES WITH SPECIAL
VISE JAWS

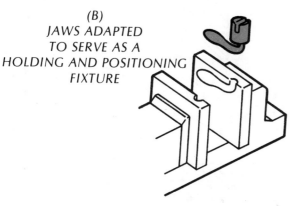

(B)
JAWS ADAPTED
TO SERVE AS A
HOLDING AND POSITIONING
FIXTURE

Fig. 39-16. Uses of special jaws and inserts to hold regular and irregular forms.

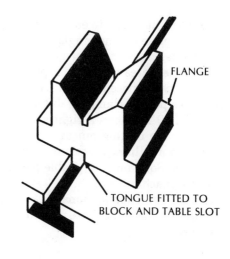

FLANGE

TONGUE FITTED TO
BLOCK AND TABLE SLOT

Fig. 39-17. A milling machine V-block.

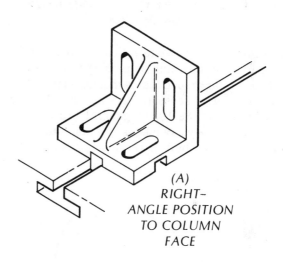

(A)
RIGHT–
ANGLE POSITION
TO COLUMN
FACE

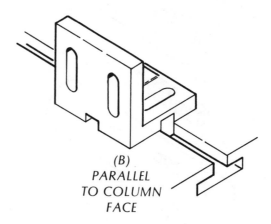

(B)
PARALLEL
TO COLUMN
FACE

Fig. 39-18. Parallel and 90° positions of an angle plate on the milling machine table.

designed with a square-head, cutaway T-head, or a tongue block with a stud.

Flat and step *blocks* support the ends of clamps. Adjustable jacks are used to accommodate different work heights. Wedges and shims are used to compensate for unevenness. Rough castings require shims and wedges for leveling.

Round work is generally held in *V-blocks* (Fig. 39-17). These differ from the V-blocks described earlier. The V-opening at the top is machined at 90°. The base is flanged for convenience in strapping the V-block to the table. The underside of the base is grooved. When fitted with a tongue and mounted over a table slot, the V-block faces are parallel to the table. The V-groove, tongues, and keyed slot are shown in Fig. 39-17. V-blocks are made and used in pairs.

Milling machine *angle plates* (Fig. 39-18) provide a surface that is at 90° to the horizontal plane of the table. The base of the angle plate has two grooves.

These are cut at right angles to each other. A tongue is fitted to the groove. The angle plate may be positioned parallel with or at a right angle to the column face. Angle plates are ribbed to provide better support for the two adjacent sides. The elongated slots in the sides permit clamping different sizes of workpieces. Fig. 39-18 shows a right-angle plate in two positions on the milling machine table.

Parallels serve auxiliary functions in layout and work positioning. Solid and adjustable parallels are used in milling machine work.

The types, sizes, and applications of C-clamps and parallel clamps are similar to those used in bench and drill press work.

REPRESENTING KEYWAYS AND KEYS ON DRAWINGS

The term *keyway* is used to indicate a groove that is cut lengthwise into a hub or mating part. *Keyseat* refers to a similar groove that is machined in a shaft. The features of a keyseat, keyway, and key are illustrated in Fig. 39-19. In general practice, the phrase *cutting a keyway* means the machining of a groove in either a shaft or the mating part.

A *key* is a metal part that fits in and between the keyway and keyseat. As stated earlier, the key locks two parts in a fixed position. The key also provides a positive drive.

There are three common shapes of keys: square, flat, and Woodruff (Fig. 39-20). The *square key* has a square cross section. The *flat key* is rectangular. A groove that accommodates one half the height of the square or flat key is machined in the shaft. The other half of the key height is accommodated by a groove cut into the mating part.

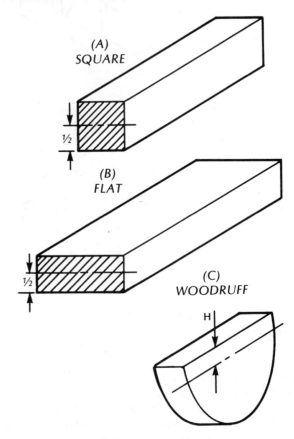

Fig. 39-20. Three common shapes of keys.

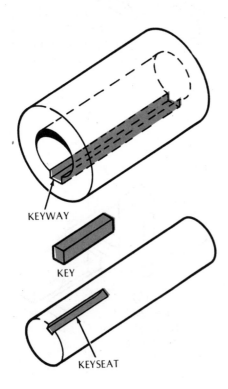

Fig. 39-19. Features of a key, keyway, and keyseat.

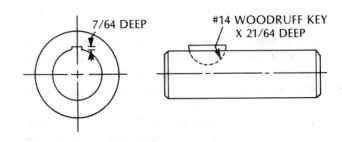

Fig. 39-21. A shop note for a Woodruff keyway and key.

The *Woodruff key* has a semicircular shape. This key is identified by a numbering system. The number also relates to the circular milling cutter that is used to mill the keyseat. The circular depth of the keyseat and the height of the key in the mating part are usually given in shop notes on a drawing.

Fig. 39-21 shows a #14 Woodruff key as an example. The circular groove is cut $\frac{21}{64}''$ deep. The slot in the mating part is cut $\frac{7}{64}''$. This fits the Woodruff key area that extends beyond the shaft diameter.

Square and flat keyways are represented on a drawing in the manner shown in Fig. 39-22. The dimensions appear as shop notes. The width of the keyway is given first, followed by the depth and the length of the key. The length of the keyway is usually indicated in a second view.

HOW TO MOUNT AND REMOVE A MILLING MACHINE ARBOR

Preparing The Machine Arbor

Step 1   Remove chips from the table top and T-slots.
CAUTION: Safety goggles or a protective shield must be worn for all machining processes.
CAUTION: The machine must be *off*. Use a T-handle cleaner, stiff brush, and pan to remove the chips. Wipe the table with a clean cloth.

Step 2   Clean the nose end of the spindle with a wiping cloth.

CAUTION: If it is necessary to use an air gun, place a cloth over the back end of the spindle.

Step 3   Feel and visually examine the taper bore and spindle nose to be sure the surfaces are clean. They must also be free of burrs and nicks.
Note: Burrs and nicks may be removed by carefully using a round abrasive stick.

Step 4   Clean the draw-in bar, particularly the threaded front end and the collar.

Step 5   Insert the draw-in bar part way into the spindle.

Step 6   Lock the spindle to prevent it from turning.

Step 7   Clean the arbor support. Check the bearing for nicks and burrs. Remove the arbor support if the arbor cannot be installed with it in place.

Step 8   Select the required style and size of arbor.

Step 9   Recheck the arbor. Be sure the arbor, spring collars, bushing, and nut are clean. All parts, especially the tapered end and bushing, must be free of burrs and nicks.
Note: For inspection and cleaning, the arbor should be placed on a clean cloth on a portable machine stand.

Mounting A Milling Machine Arbor

Step 1   Hold the tapered-shank end of the arbor in one hand and the shaft portion in the other (Fig. 39-23A).

Step 2   Insert and guide the tapered shank carefully into the spindle nose. Align the

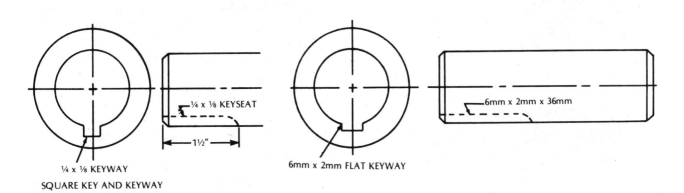

¼ x ⅛ KEYSEAT

1½"

¼ x ⅛ KEYWAY

SQUARE KEY AND KEYWAY

6mm x 2mm FLAT KEYWAY

6mm x 2mm x 36mm

Fig. 39-22. Dimensioning of square and flat keys and keyways.

slots (Fig. 39-23B). Move the arbor in as far as possible.

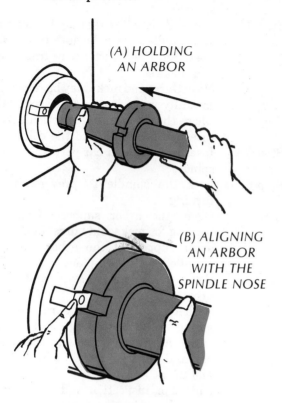

(A) HOLDING AN ARBOR

(B) ALIGNING AN ARBOR WITH THE SPINDLE NOSE

Fig. 39-23. Mounting a milling machine arbor.

Step 3   Move in the draw-in bar and turn it to engage the threads of the arbor.

Step 4   Turn the draw-in bar nut by hand as far as possible (Fig. 39-24A).

Step 5   Use an open-end wrench to turn the draw-in bar nut. Tighten the nut until the arbor is held securely in the taper and against the spindle face (Fig. 39-24B).

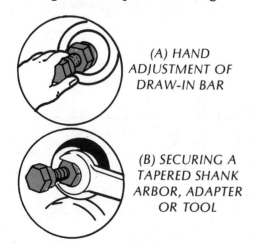

(A) HAND ADJUSTMENT OF DRAW-IN BAR

(B) SECURING A TAPERED SHANK ARBOR, ADAPTER OR TOOL

Fig. 39-24. Tightening the draw-in bar.

Step 6   Unlock the spindle. Start the machine. Check the end of the arbor to see that it runs perfectly true.
Note: If the arbor does not run true, two conditions may exist. First, the arbor may not be seated correctly and is improperly mounted. Second, the arbor may be *sprung out of shape.* The arbor should be removed and reexamined. If it is not sprung, it can be remounted. If it is sprung, it must be replaced.

Removing A Milling Machine Arbor (Styles A, B, And C)

Step 1   Stop the machine and clean it.

Step 2   Remove the cutter, workpiece, and arbor support. Then lock the spindle.

Step 3   Fit an open-end wrench to the nut on the draw-in bar.

Step 4   Apply a force on the wrench in a *counterclockwise* direction (Fig. 39-25A). Turn the nut almost one turn.

Step 5   Tap the end of the draw-in bar with a soft-faced hammer. Deliver a sharp blow to release the steep-taper arbor from the spindle (Fig. 39-25B).

(A) LOOSENING THE DRAW-IN BAR

(B) UNSEATING THE SPINDLE NOSE AND ARBOR, ADAPTER OR CUTTING TOOL TAPERS

Fig. 39-25. Removing a milling machine arbor.

Note: It is often more difficult to release an arbor from a self-locking taper spindle. In such a case, remove the draw-in bar. Insert in its place a bar that is just slightly smaller in diameter than the spindle hole. Gently but firmly tap the end of this bar.

CAUTION: Hold the arbor with one hand during this freeing process. This prevents the arbor from jumping out of the spindle and falling when the tapers are freed.

Step 6   Unscrew the draw-in bar with the left hand while supporting the arbor with the right hand.

Step 7   Use *both* hands to remove the arbor when the draw-in bar is free.

Note: The smaller style C arbors can usually be handled with one hand.

CAUTION: Handle the arbor carefully. Do not allow it to hit any part of the machine.

Step 8   Return the arbor (with the collars, bearing sleeve, and nut in place) to the storage rack or cabinet (Fig. 39-26).

Step 9   Remove the draw-in bar and store it.

Step 10   Unlock the spindle.

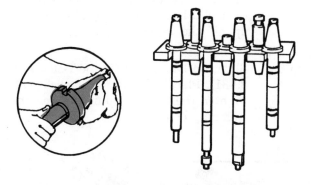

Fig. 39-26. Cleaning and storing milling machine arbors.

Mounting A Cam-Lock Adapter

Step 1   Select the adapter that fits the spindle nose taper and the shank of the cam-lock toolholder (or cutting tool).

Step 2   Lock the spindle. Be sure all surfaces are clean and free of burrs.

Step 3   Mount the adapter (Fig. 39-27). Follow the same steps as used for mounting an arbor.

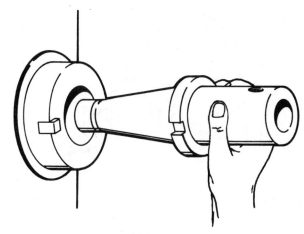

Fig. 39-27. Mounting an adapter.

Step 4   Recheck the taper hole of the adapter and the taper shank of the arbor. Each must be clean and free of burrs.

Step 5   Insert the cam-lock arbor in the adapter (Fig. 39-28). Match the cam lock and slot.

Step 6   Turn the setscrew clockwise with a setscrew wrench.

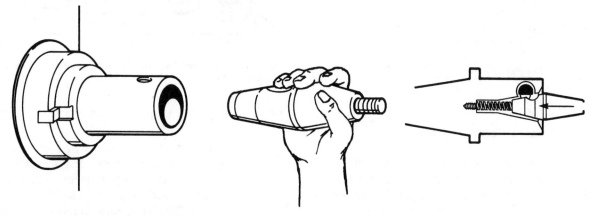

Fig. 39-28. Positioning a cam-lock arbor.

Note: Make sure that as the cam is turned it seats itself in the arbor shank groove.

Step 7   Continue to turn the setscrew until the arbor is securely seated and the tapers are locked together (Fig. 39-29).

Step 8   Unlock the spindle. Start the machine and test the arbor for trueness.

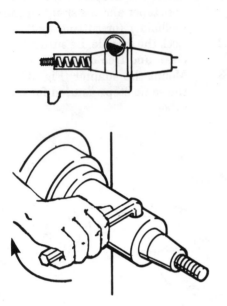

Fig. 39-29. Tightening the cam-lock arbor in the adapter.

Removing A Cam-Lock Adapter

Step 1   Stop the machine and clean it.

Step 2   Remove the cutter, workpiece, and arbor support. Lock the spindle.

Step 3   Turn the setscrew counterclockwise. This releases the cam lock (Fig. 39-30). **CAUTION:** Hold the arbor in one hand when releasing it. This is done to prevent the arbor from falling on any part of the machine.

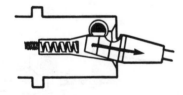

Fig. 39-30. Releasing a cam-lock arbor

Note: If the arbor does not run true, stop the machine. Remove the arbor and adapter as described in *Removing A Cam-Lock Adapter.* Check for burrs and dirty surfaces.

HOW TO MOUNT AND DISMOUNT A PLAIN VISE

Mounting The Vise

Step 1   Clean out the table slots with a slot cleaner. Wipe with a clean cloth.

Step 2   Examine the slots and remove any burrs or nicks.

Step 3   Place the vise on a bench or tool stand. Wipe it clean. Remove any burrs from the base.

Step 4   Insert the tongues in the appropriate set of slots. These depend on whether the vise is to be set with the jaws parallel or at 90° to the arbor.

Step 5   Place the vise carefully on the table. The tongues should fit snugly into the table slots without being forced.
Note: Position the vise as near to the middle of the table as possible.

CAUTION: Be sure no chips fall between the vise base and table when the vise is turned. It is good practice to lift the vise up off the table a short distance to examine it.

Step 6   Slide two T-head bolts in the T-slot (Fig. 39-31). Place a washer over each bolt. Tighten the two nuts uniformly on the clamping bolts.

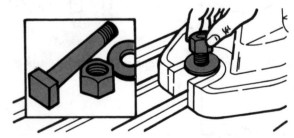

Fig. 39-31. Securing a work-holding device with a T-bolt, washer and nut.

Dismounting The Vise

Step 1   Take the same steps and observe the same safety precautions for removing chips and wiping all parts clean as listed under *Mounting The Vise.*
CAUTION: Avoid brushing over or touching the fine, razor-sharp chips. The

cutter and arbor should be removed first.

Step 2   Loosen the nuts on the clamping bolts. Move the clamping bolts clear of the vise slots.

Step 3   Lift the vise clear of the table. Place it on a tool stand or in its storage area.
Note: The vise is usually placed on a piece of clean wood to protect the base from damage.

Step 4   Remove the T-bolts. Clean and return them to the proper rack or drawer.

HOW TO POSITION SWIVEL VISE JAWS

Using Graduations On The Vise

Step 1   Loosen the clamping bolts that hold the vise to the base.

Step 2   Swing the vise so that the zero index line is aligned with the required angle graduation on the base.

Step 3   Tighten the clamping nuts by hand until a slight force is applied.

Step 4   Recheck the angle setting.
Note: A magnifying glass is often used to see that the index line *splits the graduation*. If necessary, tap the vise gently with a soft-face hammer until the required setting is reached. This procedure is for approximate alignment.

Step 5   Tighten the clamping nuts uniformly with a wrench. Recheck the angle setting.

Using A Steel Square (90° To The Column Face)

Step 1   Swivel the vise so that the jaws are 90° to the column face. This setting is read on the graduated base.

Step 2   Tighten the clamping nuts by hand to apply a slight force.

Step 3   Move the vise crosswise to bring it as close as possible to the column.

Step 4   Clean the beam and blade of the steel square. Hold the beam firmly and carefully against the machined face of the column.

Step 5   Slide the blade slowly up to the fixed jaw of the vise.

Step 6   Sight down between the edge of the blade and the fixed jaw.
Note: If light can be seen, the vise is out of alignment.

Step 7   Move the blade away from the vise jaw. Tap the vise lightly with a soft-face hammer.

Step 8   Retest. This time place two narrow pieces of tissue paper between the blade of the square and the fixed vise jaw.
Note: When the same force is required to remove each piece of tissue paper, the vise jaw is set 90° to the column face.

Step 9   Tighten the clamping nuts securely. Recheck the vise setting.

Using A Dial Indicator

A dial indicator is used to accurately align vise jaws either parallel or at right angles to the table movement. The accuracy of the setting is limited by the dimensional accuracy of the dial indicator. The fixed jaw can thus be set to within 0.0005″ (0.01 mm) or less, depending on the indicator.

Parallel With The Column Face

Step 1   Set the vise so that the index line is aligned with the 90° graduation. The vise jaws should be parallel to the column face. Tighten the clamping nuts by hand.

Step 2   Assemble the indicator, with an adjustable universal joint, a clamp, and an extension rod. These are illustrated in Fig. 39-32.

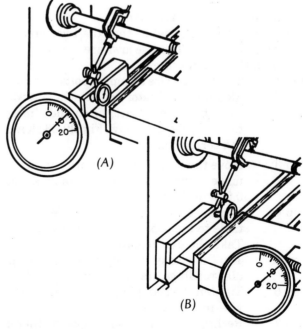

Fig. 39-32. Checking the accuracy of the alignment of a solid vise jaw (parallel to column).

Step 3   Position the indicator assembly so that

the contact point almost touches the fixed jaw of the vise.

Step 4   Clamp the indicator assembly on the arbor in this position.

Step 5   Move the table out slowly until contact is made and the indicator dial begins to move. Continue to move the table about ten one-thousandths of an inch (0.010", or 0.2 mm). This should provide a good contact (Fig. 39-32A).

Note: Some craftspersons turn the dial face so that the dial registers *zero* (0).

Step 6   Move the table carefully toward the opposite end of the vise. Note if there is any difference in the two readings (Fig. 39-32B).

Step 7   Tap the vise gently (if the jaws are not parallel) in a direction away from the higher indicated reading. The movement should be half the difference between the two indicator readings.

Step 8   Move the dial indicator to each end of the jaw. Adjust further until there is no variation in reading.

Step 9   Tighten the clamping nuts securely.

Step 10   Recheck the dial indicator readings at both ends of the vise jaw.

Step 11   Remove the indicator and attachments. Store them in the indicator case.

Parallel With The Spindle Axis

In this case the vise is set so that the two zero (0) graduations coincide. The vise jaws are then perpendicular to the column face.

Step 1   Tighten the vise clamping bolts by hand.

Step 2   Mount the indicator and attachments on the arbor. The contact end should almost touch the fixed vise jaw. Fig. 39-33 shows the correct setup.

Step 3   Move the table carefully until contact is made. Continue to move the table about 0.010" (0.2 mm).

Step 4   Use the cross feed handwheel. Move the table to the opposite end of the vise jaw. Take the dial reading.

Step 5   Tap the vise gently, if needed, until the dial readings are the same.

Step 6   Tighten the clamping nuts to secure the vise in this position.

Step 7   Recheck for *parallelism.* Adjust further if necessary.

Step 8   Disassemble the dial indicator attachments and store them in the indicator case.

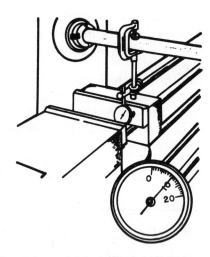

Fig. 39-33. Checking vise jaw alignment parallel to the spindle axis.

HOW TO USE WORK-HOLDING ACCESSORIES

Using Stops, Strap Clamps, And Bolts

Step 1   Clean the table slots, table, and workpiece. Wipe with a clean cloth. Check and remove any burrs or nicks. Place the workpiece on the table.

Step 2   Select two *stops.* These metal pieces are the width of the table slots. Insert each stop into the table slot. Position each one near the end of the workpiece.

Step 3   Move the workpiece against the two stops.

Step 4   Slide the T-bolts in the table slots as close as possible to the work.

Note: The cutaway T-head bolt may be dropped into position and turned.

Step 5   Use a clamp that accommodates the size of the workpiece and the nature of the milling machine operation.

Step 6   Place a block under the end of the clamp. This location provides maximum clamping leverage on the workpiece.

Note: The block size should be as near as possible to the thickness of the clamping surface.

Step 7   Tighten the nut securely.

Note: If the workpiece requires that clamps be placed on both ends or in different sides, tighten the nuts uniformly.

Step 8   Check to see that the correct force is being exerted. The workpiece must be strapped *flat* on the table.

Using V-Blocks And Step Blocks

Step 1   Select the size of V-block required to adequately nest the workpiece.
Step 2   Clean the table, workpiece, and V-blocks. Remove any burrs or nicks.
Step 3   Position the V-blocks to provide maximum support for the workpiece. The blocks and clamps must also clear the milling operation.
          Note: A general practice is to position the workpiece as close as practical to the center of the table and column.
Step 4   Select a flat or other appropriate type of clamp and step block.
Step 5   Insert the T-bolts in the table. Slide each one to the middle of the V-block.
Step 6   Protect the surface finish if necessary. Place a metal shim between the clamp and the workpiece.
Step 7   Support one end of the strap clamp on the step block. The height should be the same as the workpiece setup.
Step 8   Tighten each nut securely.

Using An Angle Plate, Parallel And C-Clamp, And Jack

Step 1   Select an angle plate that accommodates the size of the workpiece. It also must permit securely clamping the part.
Step 2   Clean all machined surfaces of the table and angle plate. Remove any burrs or nicks.
Step 3   Position the angle plate on the table. Locate it as near to the center as possible. The angle plate and workpiece should be as close as practical to the column. Secure the angle plate in this position.

Step 4   Select either a C-clamp or a parallel clamp.
          Note: C-clamps are generally used to support larger workpieces where heavier cuts are taken.
Step 5   Adjust the C-clamps or the jaws of the parallel clamps. Place the workpiece against the vertical face of the angle plate. Tighten each clamp. Use only enough force to correctly position the workpiece.
          CAUTION: A protecting shim of metal, plastic, or cardboard is placed between any rough surface and the machined face of the angle plate or under a strap clamp.
Step 6   Select one or more support jacks. Place them under any overhanging surface that is to be milled (Fig. 39-34).
          Note: Sometimes the overhanging portion is also clamped. In such cases, the jack should be located under the workpiece, with one end of the clamp directly above it.

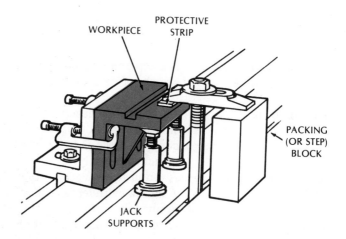

Fig. 39-34. The clamping setup for machining a workpiece mounted on an angle plate.

SAFETY PRECAUTIONS WITH MILLING CUTTER-AND WORK-HOLDING DEVICES

• Use the shortest style arbor possible. This permits machining operations to be performed closer to the column.
• Remove burrs or nicks from the spindle face and nose and the arbor adapter taper, shaft, bearing, and spacing collars.

• Tighten the draw-in bar with enough force to pull together the tapered and face surfaces of the spindle and arbor.
• Use a key in the keyway of the arbor, milling cutter, and support bearing.

- Locate milling cutters on an arbor as close to the column as possible. This provides maximum support for the cutting tool during machining.
- Disengage steep tapers by striking the draw-in bar a sharp blow. The bar must not be loosened more than one thread, to prevent damage to the thread.
- Withdraw the draw-in bar of a tightly secured self-holding taper. Use the largest rod the spindle hole will accommodate to apply the force needed to free the tapered surfaces.
- Clean and remove burrs from the base and working surfaces of a vise, angle plate, packing blocks, and other mating finished parts. Check the tongue size to see that it accurately fits a table slot.
- Strap work-holding devices and position the workpiece and cutter for maximum rigidity and strength.

- Use protecting shims between the finished surface of a workpiece and a clamp.
- Mount and remove a style A arbor by holding it with two hands.
- Store arbors vertically in a rack to prevent them from hitting together or being bent.
- Move the table slowly to position a dial indicator pointer against a vise jaw or workpiece. Normally, a 0.010″ (0.2 mm) movement is adequate.
- Use the table of the moveable cabinet to clean and check heavy work-holding devices or arbors.
- Remove chips from table slots with a T-shaped metal cleaner, stiff brush, and tray.
- Wipe up any coolant that has been spilled on the floor.
- Place a cloth covering over the air gun to prevent chips from flying up toward the operator.
- Use a properly fitted protective shield or safety goggles during all milling operations.

TERMS USED WITH MILLING CUTTER-AND WORK-HOLDING DEVICES

American Standard Milling Machine Tapers	A steep-angle taper system of 16° 36′, or 3½″ taper per foot. A self-releasing taper. A system of designating milling machine spindle nose tapers (#30, #40, #50, etc).
Style A, B, and C Arbors	Three general designs of milling machine arbors. *Style A* is a long arbor supported on the end by a pilot. *Style B* is an arbor that is steadied by a bearing sleeve and an arbor support. *Style C* is an arbor to which shell end mills and other cutting tools are attached directly.
Spacing collars	A round sleeve with parallel faces. These are accurately ground to specified lengths.
Bearing sleeve	A hollow cylinder with a keyway cut lengthwise. The outer bearing surface is ground accurately to fit the arbor support bushing.
Draw-in bar	A round bar threaded at one end to screw into an arbor or adapter. The screw end permits a shouldered nut to draw in and secure tapered surfaces.
Adapter	A cutter-holding device. A device that adapts the various sizes of cutters and shanks to the milling machine spindle size.
Sleeve	A cylindrical toolholder. The outside surface fits into an adapter. The inside bore accommodates the shank of a cutting tool.
Collet	An adjustable, slotted sleeve. The sections move to permit holding a limited range of tool shank diameters.
Cam-lock adapter	A device for quickly and securely locking and unlocking a cutting tool in an adapter. An eccentric stud and corresponding slotted cutter. A device for applying the force needed to seat and hold a cutter in an adapter.
Tongue	A rectangular-cross-section metal piece. A part fitted into a machined groove in the base of a work-holding device. A rectangular block that aligns a work-holding device parallel or at right angles to the column face.
Woodruff keyway and key	A semicircular groove cut into a shaft to accommodate a circular key. A number system in which a milling cutter produces a special size of semicircular groove.

SUMMARY

- National Milling Machine Tapers are self-releasing. They are machined to a taper of 3½″ per foot, or at a 16° 36′ included angle. The tapers are designated by a two-digit numbering system. Common sizes are #30, #40, #50, and #60.
- Self-releasing tapers require a set of lugs. These transmit the spindle force to the cutter.
- Self-releasing tapered arbors are drawn against the mating spindle surface with a draw-in bolt.
- Style A arbors permit the use of smaller diameter cutters. The cutting action takes place closer to the milling machine arbor.
- Style B arbors require a bearing sleeve for support. This style of arbor is adaptable to heavy-duty milling operations.
- Style C arbors permit quick release of cutting tools so that they may be changed conveniently. This arbor is widely used for shell end mills. Sleeves, collets, and other toolholding accessories may be securely held in different designs of style C arbors.
- Style A and B arbors and bearing collars are machined with a keyway. A key is inserted between the arbor keyway and that of the cutter bearing sleeve. Cutting force is applied to the cutter by the forces exerted against the faces and by the key.
- Keyways and keys are usually represented and dimensioned on drawings with shop notes. Square and flat keys are machined to half depth in both the shaft and the mating part.
- Adapters are work-holding devices. Adapters make it possible to fit a great range of cutter designs and sizes on the standard spindle.
- Sleeves permit the use of regular self-holding-taper-shank tools. Morse and Brown and Sharpe tapers may be held in a steep-taper spindle adapter.
- Collets are slotted sleeves of various designs. The sections may be brought together or opened to accommodate a limited range of tool shank diameters.
- Collets are used for holding and changing from one size and type of round-shank cutting tool to another. The collet and cutter are held in an adapter.
- Plain and swivel vises, angle plates, and V-blocks are aligned by tongues. These fit snugly into the table slots. The graduations on the base of a swivel vise permit setting the vise jaws to a required angle.
- A dial indicator is used to set a fixed vise jaw or the vertical surface of an angle plate perpendicular to within an accuracy of ± 0.0005″ (0.01 mm). The surfaces may be checked for alignment parallel or at a right angle to the column face.
- Workpieces are commonly clamped to the table in V-blocks or on an angle plate. Flat, curved, and tongue strap clamps and C-or parallel clamps are used.
- Jacks support overhanging surfaces of workpieces that are held in a vise or on an angle plate.
- Standard personal and machine safety precautions and procedures are followed for cleaning the T-slotted table, machine, and all tools and instruments.

UNIT 39 REVIEW AND SELF-TEST

1. Describe *Style A, B,* and *C* milling machine arbors.
2. Explain briefly the function served by the *draw-in bar.*
3. List five work-holding devices for general milling processes.
4. Explain the meaning of the following dimension found on a drawing of a machine part with a flat keyway:

    $0.500″ \times 0.187″ \times 2″.$

5. Give the steps to follow in removing a *Style A, B,* or *C* milling machine arbor.
6. Describe three practices for positioning vise jaws on a milling machine.
7. Explain how V-blocks and other work-holding accessories are used to nest a round workpiece and hold it securely on a milling machine table.
8. State three machine or tool safety precautions the milling machine operator must observe.

Unit 40

STANDARD MILLING CUTTERS: TECHNOLOGY AND APPLICATIONS

A *milling cutter* is a cutting tool that is used on the milling machine. These cutters are available in many standard and special types, forms, diameters, and widths. The teeth may be straight (parallel to the axis of rotation) or at a helix angle. The cutter may be right-hand (to turn clockwise) or left-hand (to turn counterclockwise). Standards for milling cutters have been approved by the American National Standards Institute (ANSI).

This unit covers general features of milling cutters, forms of cutter teeth, materials used in milling cutters, and standard milling cutter types. Applications are included for each type. The actual milling processes are covered in later units.

GENERAL FEATURES OF MILLING CUTTERS

Milling cutters may be broadly grouped in two categories: standard and special. The *standard* types are manufactured according to the dimensional standards of the ANSI.

Some of the terms used to identify the major features of milling cutters appear in Fig. 40-1. The drawings illustrate three different forms of milling cutters. A staggered-tooth side mill is illustrated in (A). The features and terms associated with a form-relieved cutter are given in (B). A few added features of a helical plain milling cutter are shown in (C).

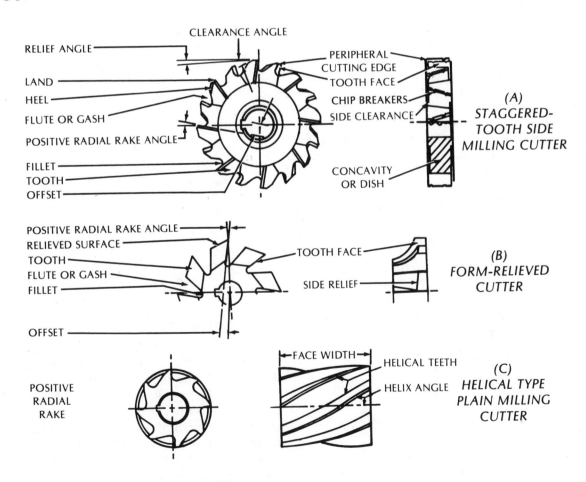

Fig. 40-1. General features of three forms of milling cutters.

There are a number of other terms that are used to describe a milling cutter or to order one from a toolroom.

• *Diameter*—The diameter refers to the outside diameter of the cutter (Fig. 40-2). The diameter of a new cutter will be the size stamped on it by the manufacturer. However, this diameter is changed each time the teeth are reground.

Fig. 40-2. Measuring the cutter diameter.

• *Cutter Width*—The cutter width is the measurement across the face of the cutter. This dimension is also stamped on the cutter by the manufacturer. The width may be measured with a steel rule or a micrometer.

• *Hole Size*—The hole size is represented by the diameter of the bore (Fig. 40-3). The bore and axis of the cutter are ground concentric. The hole size is usually held to within + 0.001″ of the nominal arbor size. The hole size is designed in proportion to the size and nature of the cutting operation.

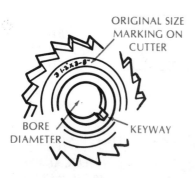

Fig. 40-3. The cutter bore diameter, keyway, and original size marking.

• *Keyway*—The keyway in the cutter (Fig. 40-3) matches the keyway in the arbor. A key is inserted between the arbor and the cutter. The key prevents the cutter from being turned on the arbor by the pressure of the cut. Keyway and key sizes are standardized.

• *Shank*—Shanks (Fig. 40-4) are an integral part of milling cutters that are not mounted on an arbor. The shank serves the same function as a drill or reamer shank. The shank provides a gripping surface for holding and accurately centering an end milling cutter. The shank also transmits force to the cutter. The shank may be straight or tapered. Three different styles are illustrated in Fig. 40-4. Straight shanks are ground to standard sizes. Tapered-shank milling cutters usually have either a Brown and Sharpe or a Morse self-holding taper. The milling cutter may be secured on the taper and driven by a tang end. Other taper shanks and milling cutters are drawn up tightly by the draw-in bar. The threaded end of the draw-in bar screws into the tapped end of the taper shank. Both the tang and threaded end designs are illustrated in Fig. 40-4.

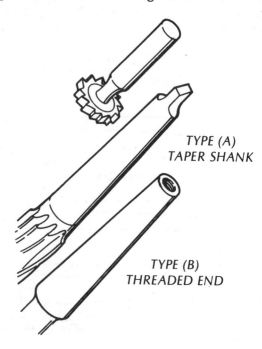

Fig. 40-4. Types of milling cutter shanks.

• *Cutter Teeth*—The cutter teeth provide the cutting edges. The shape and function of the cutter determine the design of the teeth. For instance, the teeth on a plain milling cutter cut on the end. The teeth are machined around the periphery of such a cutter. A side milling cutter has peripheral teeth and others on its sides. The teeth may be straight. That means the teeth are parallel to the axis of the cutter. Other teeth are helical. The cutting edges are at an angle (helix) to the cutter axis.

BASIC FORMS OF CUTTER TEETH

Circular milling cutters are designed with three basic forms of teeth.

- The saw tooth
- The formed tooth
- The inserted tooth.

The Saw Tooth Form

Some of the common design features of the *saw tooth* are illustrated in Fig. 40-5. The teeth are said to be *radial*. The tooth face lies along a line that cuts through the center of the cutter. Teeth are cut with a *radial tooth face* or at an *angle* to the radial line.

As illustrated in Fig. 40-5, the tooth face is machined at an angle. This produces the same cutting effect as grinding a rake angle on a single-point cutting tool. In this case, the cutter has a *positive radial rake*. The *rake angle* is the number of degrees formed by the tooth face and a radial line.

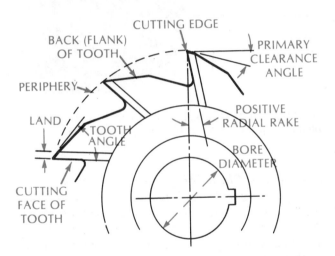

Fig. 40-5. The design features of and terms used with a standard milling cutter.

A cutting tooth has *zero rake* when the plane of the tooth face cuts through the cutter axis. A *positive rake* permits the metal to flow along the face of a high-speed steel cutter. Positive rake for these cutters improves tool life by lowering the temperature of the tooth. Less power is required and a higher quality of surface finish is produced.

The teeth may also have a *negative rake*. The tool life and performance of carbide-tipped teeth, which operate at exceedingly high speeds and coarse feeds, are improved when the teeth have negative rake.

The purpose of negative rake is to protect the cutting edges of carbide tools. Their edges are brittle

in comparison to those of high-speed steel tools. With a negative rake, the cutting forces fall within the cutter body. This is the reason why the cutting edges of carbide-typed cutters have a zero or negative rake. Surprisingly, carbide cutters are capable of producing a high-quality surface finish at high machining speeds.

There may be two or three clearances on the saw tooth form. The teeth are ground at a slight *clearance angle (primary clearance)* around the periphery. This clearance extends from the tooth face and produces a cutting edge. The narrow, flat surface directly behind the cutting edge is called the *land*. The angle formed by the cutting face and the land is the *tooth angle*. The body of the tooth is cut at a secondary angle and at a third angle. This design provides maximum support for the cutting edge. It also provides a *chip space* for the fast removal of chips.

The teeth on the saw tooth form are sharpened by grinding the land.

The Formed-Tooth Milling Cutter

As the name suggests, a *formed-tooth cutter* has a contour, or tooth outline, of a particular shape. A concave milling cutter with a specified diameter produces a round shape on a workpiece. A gear cutter machines a gear tooth that conforms to specific design requirements. The right- or left-hand radius cutter mills a round corner. The flute cutter cuts flutes on drills, reamers, taps, and other cutting tools. A single formed cutter is used for some applications. In applications that involve the milling of a wide contour, the form cutters may be set up in combination.

The outline of the teeth on a formed cutter is reproduced starting at the tooth face. The clearance, or *eccentric relief,* follows the same contour as the cutting edge. Fig. 40-6 shows a semicircular cutting face on a radial-tooth formed cutter. This cutter may be resharpened by grinding the tooth face. Grinding is *radial*. Although the teeth get thinner each time the cutter is ground, the cutting edge retains the original shape.

STANDARD TYPES OF MILLING CUTTERS

Plain Milling Cutters

The teeth of a plain milling cutter are formed around the periphery only. This type of cutter is used to produce either a narrow or a wide flat surface (Fig. 40-7). The surface is parallel to the axis of the cutter.

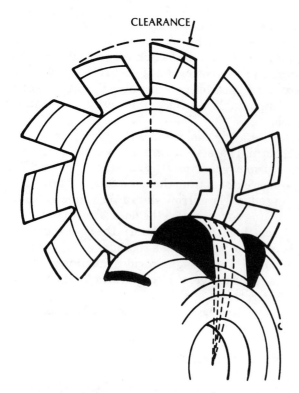

CLEARANCE

Fig. 40-6. The shape of a formed cutter face is retained by radial grinding.

Two examples of plain milling cutters are shown in Fig. 40-7. There is a wide variety of plain milling cutters. The diameter and the width of the cutter depend on whether a part is to be *slab milled* (milling a wide, flat surface) or requires a narrow-width slot. There are three broad classes of plain milling cutters: light-duty, heavy-duty, and helical.

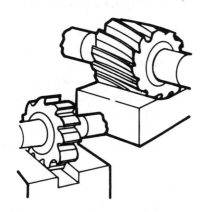

Fig. 40-7. Two examples of plain milling.

Light-Duty Plain Milling Cutters—Light-duty plain milling cutters up to 3/4″ (18 mm) wide generally have *straight teeth*. These are parallel to the cutter axis. High-speed cutters have from four to five teeth for each inch of diameter.

Cutters over 3/4″ (18 mm) wide usually have helical teeth. The helix angle ranges from 18° to 25°. These tooth angles produce a shearing, cutting action. Less force is required, vibration and chatter are reduced, and a good quality of surface finish is produced. These are the advantages of a helical-tooth over a straight-tooth cutter. The teeth of light-duty plain milling cutters (Fig. 40-8) are smaller and have a finer pitch than those of heavy-duty cutters. The light-duty plain cutter is designed for light cuts and fine feeds. Plain milling cutters up to 3/4″ wide with straight teeth are used for milling grooves and keyways that must have accurate widths.

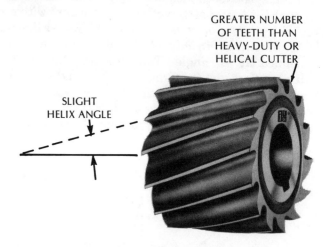

GREATER NUMBER OF TEETH THAN HEAVY-DUTY OR HELICAL CUTTER

SLIGHT HELIX ANGLE

Fig. 40-8. A light-duty plain milling cutter. (Courtesy of the Illinois/Eclipse Division of Illinois Tool Works Inc.)

Heavy-Duty Plain Milling Cutters—Heavy-duty plain milling cutters (Fig. 40-9) are also called *coarse-tooth cutters*. These cutters average two to three teeth for each inch of diameter. As an example, a 3″ heavy-duty plain milling cutter usually has 8 teeth. The helix angle of the teeth is also steeper than that of light-duty cutters. Different cutter manufacturers vary the helix angle of the teeth from 25° to 45°.

This class of cutter is designed for plain, heavy cuts. The cutting edges and wide flutes combine strength with sufficient space for easy flow of chips. However, the long, needle-like chips are difficult to remove from the workpiece and machine setup. Extra safety steps must be taken. A metal cleaner and stiff brush should be used to clear away the chips.

Wide, flat surfaces are usually slab milled with a particular type of heavy-duty cutter. This cutter is called a *slab mill*, *slabbing cutter*, or *roughing cutter*. This type of cutter has interrupted teeth. Part of each tooth is relieved at a different place along the length. This design breaks up the chip. A slabbing cutter is able to produce a good surface finish under heavy cutting conditions.

Fig. 40-9. A heavy-duty plain milling cutter. (Courtesy of the Illinois/Eclipse Division of Illinois Tool Works Inc.)

Helical Plain Milling Cutters—Helical plain milling cutter teeth are formed at a steep helix angle (Fig. 40-10). This ranges from 45° to 60°, or steeper. The teeth are coarser than those of the heavy-duty type. The teeth are designed so that they engage the work at a steep angle. The cutting force is absorbed in end thrust. The teeth are continuously engaged in comparison with the intermittent cutting action of a straight-tooth cutter. The helical plain milling cutter therefore eliminates chatter.

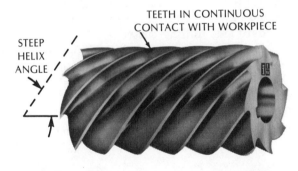

Fig. 40-10. A helical type plain milling cutter. (Courtesy of Illinois/Eclipse Division of Illinois Tool Works Inc.)

The direction of the helix (spiral) establishes the direction of the force. The angle of the helix determines the end forces that are generated. The helix may be right-or left-hand. A right-hand spiral on a cutter turning clockwise produces a force in the direction of the spindle (Fig. 40-11). The end bearings are designed to absorb such thrusts. This is preferred to the direction of force produced by a left-hand helix. This helix tends to pull the arbor from the spindle.

Wide surfaces are often milled by interlocking a right- and a left-hand helical cutter. The forces exerted by each cutter (Fig. 40-12) are thereby offset (canceled).

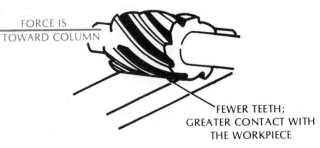

Fig. 40-11. The cutting force exerted by a helical cutter is toward the column.

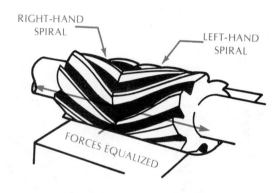

Fig. 40-12. Right- and left-hand spiral cutters equalize the cutting forces.

Helical plain milling cutters are efficient for wide, shallow cuts. They are not as practical as the heavy-duty cutters for deep cuts, slab milling, or coarse feeds.

Plain milling cutters are sharpened on a tool and cutter grinder. The land on each tooth is ground concentric with the cutter axis.

Side Milling Cutters

Side milling cutters are similar to plain milling cutters. However, in addition to teeth around the periphery, others are formed on one or both sides. Most of the cutting is done by the teeth around the periphery. The side cutting teeth cut the side of the workpiece. These cutters are mounted on and are keyed to the arbor.

The cutter may be used alone or in pairs, as for *straddle milling.* In such applications, spacing collars of a desired width are placed on the arbor between the two cutters. The cutters mill two parallel sides at the same time. Good machine finishes and dimensional accuracy are possible.

Side mills are not recommended for milling slots. There is a tendency for the side cutting teeth to mill wider than the specified cutter width.

Four types of side milling cutters are in general use. These are:

- Plain side milling cutters
- Half side milling cutters
- Staggered-tooth side milling cutters
- Interlocking side milling cutters.

Plain Side Milling Cutter—This cutter has teeth on the periphery and on both sides. The teeth on the sides taper slightly toward the center of the cutter. This *concavity* provides clearance, or side relief. The plain side milling cutter is adaptable for general-purpose side milling, slotting, and straddle milling.

The four milling cutters illustrated in Fig. 40-13 have different design features. The most commonly used cutter, (A), is made of high-speed steel. The three remaining plain side milling cutters (B, C, and D) have heat-treated alloy steel bodies. These are *stress relieved*. The teeth consist of carbide inserts. These are diamond tapered. The extremely smooth surface finish prevents abrasive chips from adhering to the cutting edges.

The style in (B) is used for nonferrous and nonmetallic materials. The four teeth on this cutter have a *positive radial rake*.

The (C) cutter has a *negative radial rake*. This cutter is designed for milling steels.

The (D) cutter has a *neutral radial rake* and a *neutral axial rake*. The cutter sides are not formed with clearance. This plain side milling cutter is used for milling cast iron and malleable iron parts.

The carbide-tipped cutters in Fig. 40-13 machine at considerably higher cutting speeds and feeds than do those made of high-speed steels.

Half Side Milling Cutters—One side of the half side milling cutter has teeth (Fig. 40-14). The other side is flat and resembles a plain milling cutter. The teeth around the circumference are helical. These produce a shearing, cutting action. This minimizes chatter. The top and side teeth are deep cut. This serves two purposes: (1) The wide spacing permits the free flow of chips. (2) Tool life is extended. The teeth may be reground a greater number of times than can those of a standard side milling cutter.

A picture of a right-hand half side cutter is presented in Fig. 40-14. Fig. 40-14 also shows a right-hand and a left-hand cutter positioned for straddle milling.

Half side milling cutters are designed for heavy-duty side milling. These cutters may also be used in

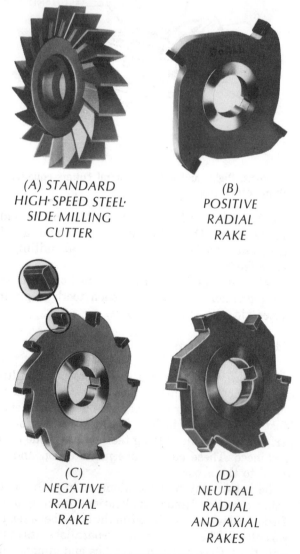

(A) STANDARD HIGH-SPEED STEEL SIDE MILLING CUTTER

(B) POSITIVE RADIAL RAKE

(C) NEGATIVE RADIAL RAKE

(D) NEUTRAL RADIAL AND AXIAL RAKES

CARBIDE TIPPED PLAIN SIDE MILLING CUTTERS

Fig. 40-13. High-speed steel and carbide-tipped plain side milling cutters. (Courtesy of Do All Company)

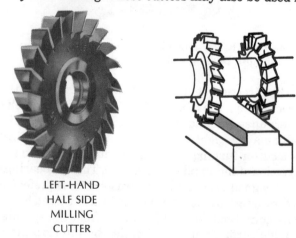

LEFT-HAND HALF SIDE MILLING CUTTER

Fig. 40-14. Straddle milling with right- and left-hand side milling cutters. (Photo courtesy of Morse Cutting Tools Division)

pairs for slotting to a required dimension. Spacing collars are used to separate the cutters.

Staggered-Tooth Side Milling Cutters—This cutter has cutting teeth around the circumference (Fig. 40-15). The teeth are cut with alternate right- and left-hand helixes. One side of each tooth forms a side tooth. The white section of each alternate tooth in Fig. 40-15 indicates a *land*. These lands are ground from 1/64" (0.4 mm) to 1/32" (0.8 mm) wide. No clearance is provided, and the side teeth are not designed to cut. There is a minimum back clearance of 0.001" per inch. This prevents the cutter from binding in the cut.

Fig. 40-16. Interlocking side milling cutters.

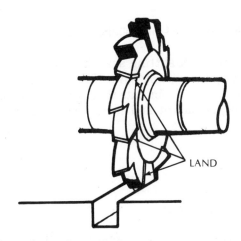

Fig. 40-15. A staggered-tooth side milling cutter.

The overall width of each tooth is narrower than the width of the cutter. The chip produced is thus prevented from wedging in the slot. The teeth are sharpened around the circumference. Occasionally, grinding is required on the side teeth.

The staggered-tooth side milling cutter is a heavy-duty cutter. It is designed to remove large amounts of metal, to mill with a minimum of vibration and chatter, and to produce a high-quality surface finish in deep cuts.

Interlocking Side Milling Cutters—This cutter is made in two halves. These are placed side by side and *interlock* (Fig. 40-16). The teeth around the circumference of the *interlocking cutters* are alternately long and short. The alternate teeth interlock. The cutters may be separated by using spacing washers. The amount of separation is controlled by the overlap range of the interlocking teeth.

Interlocking cutters are used for milling plain faces and other parallel surfaces to close tolerances. If the cutters are to be used as adjustable side milling cutters, it is practical to regrind the face and side teeth. The width is adjusted by inserting spacing washers on the arbor between the right- and left-hand cutters. The cutters may be reground so long as the teeth interlock.

Slitting Saws

Slitting saws are another type of plain or side milling cutter. They are designed for cutoff work on the milling machine and for cutting narrow slots. Three general-purpose types are:

- Plain metal-slitting saws
- Metal-slitting saws with side teeth
- Staggered-tooth metal-slitting saws.

The *screw-slotting cutter* is a modification of a slitting saw.

Plain Metal-Slitting Saw—This narrow plain milling cutter has teeth around the circumference. The number of teeth per inch of diameter is larger than on plain milling cutters. The rate of feed is a fraction of that used in plain milling (usually no more than one fourth).

The sides of the plain slitting saw are ground with a slight taper. This extends from the circumference to the hub. This hollow grinding provides clearance, particularly when very deep, narrow slots are being cut. The clearance prevents binding and the generation of excess friction. The saw is thus able to cut freely and accurately.

A slitting saw with side teeth and an application are illustrated in Fig. 40-17. Metal-slitting saws may be mounted for gang milling (slitting). Spacing collars of the required width are placed between the hubs of each saw.

Like other sawing processes, it is important to select a fine-pitch saw that permits two or more teeth to be engaged at the same time. This practice helps overcome the tendency of the cutter to dig into thin materials and also helps eliminate vibration.

Plain slitting saws are available to fit standard arbor diameters. The general widths range from

1/32″ (0.8 mm) to 3/16″ (5 mm). The outside diameters vary from 2½″ (64 mm) in the narrower widths to 8″ (200 mm) for the wider cutters.

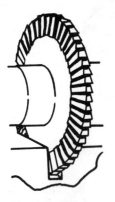 

Fig. 40-17. A slitting saw with side teeth. (Photo courtesy of Standard Tool Div./Lear Siegler, Inc.)

Screw-Slotting Cutters—This cutter is designed for cutting shallow slots and screw heads. Screw-slotting cutters look like plain metal-slitting saws. However, the nature of the operations they perform require different design features. The general cuts taken are short, the depth is shallow, and the material being cut is usually an easily machined steel. These cutters are also used on soft nonferrous metals and thin-sectioned parts.

Screw-slotting cutters have a fine pitch. For example, the 2¾″ diameter cutter has 72 teeth. A fine feed is required. There are three general-purpose (arbor-mounted) cutter diameters: 1-1/34″, 2-1/4″ and 2-3/4″. The sides are ground parallel so that there is no side relief. The thicknesses vary from 0.006″ (0.15 mm) to 0.182″ (4.6 mm).

Slitting Saws With Side Teeth—This design of slitting saw is recessed on both sides. The recess extends from the hub to the bottom of the teeth. Side teeth are cut from the circumference to the recessed portion. The side teeth are ground to prevent rubbing against the side and to provide chip clearance. All cutting is done by the cutting edges around the periphery of the cutter. The side teeth help remove the chips in deep slotting operations. This type of cutter is available in the same range of widths and diameters as the plain slitting saw.

Staggered-Tooth Slitting Saws—The principle of the staggered tooth is applied to slitting saws. Teeth are alternatively staggered as shown in Fig. 40-18. The teeth start at the circumference and extend to the recessed area of the cutter. Cutting is done by the teeth at the periphery of the cutter. Production of a chip that is slightly narrower than the width of the slot and the freedom to clear the cutter provide

maximum chip clearance. The wedging of chips, particularly in extra-deep cuts, is reduced.

Fig. 40-18. A staggered-tooth slitting saw.

Staggered-tooth slitting saws have coarser teeth than do regular slitting saws. The staggered-tooth type is best adapted for cuts that are 3/16″ (4.5 mm) and wider. Standard and heavy feeds are recommended.

Angle Milling Cutters

There are two basic types of angle milling cutters: *single-angle* and *double-angle*. Some single-angle cutters are of the hole type shown in Fig. 40-19A. These are mounted on a regular arbor. Others are threaded for mounting on a threaded adapter (Fig. 40-19B). Angle milling cutters are used for processes such as milling angular grooves, dovetails, serrations, and cutter teeth.

Single-Angle Milling Cutters—These cutters have teeth on the angular face and on the adjacent side (Fig. 40-19). They are used for milling a 45° angle or a 60° included angle.

Double-Angle Milling Cutters—The teeth of these cutter are V-shaped (Fig. 40-20). They are cut in the two angular faces. Double-angle milling cutters are available with standard included angles of 45°, 60°, and 90°. The angles on each side of a perpendicular (to the axis) centerline are equal. The angles are 22-½°, 30°, and 45°, respectively. Fig. 40-20 shows a double-angle milling cutter with a 90° included angle. The two angles with the centerline are equal (45°).

Double-angle milling cutters with unequal angles are also available. For instance, the angle of one side to the perpendicular centerline may be 12°. The adjacent angle may be 48°.

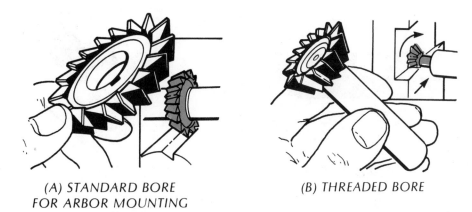

(A) STANDARD BORE
FOR ARBOR MOUNTING

(B) THREADED BORE

Fig. 40-19. Standard and threaded types of single-angle cutters.

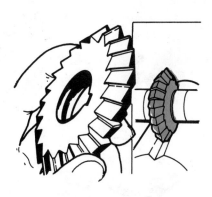

Fig. 40-20. A 90° double (equal-angle) milling cutter.

T-Slot And Woodruff Keyseat Cutters

T–Slot Cutters—This is a single-process cutter (Fig. 40-21). It is used to mill a T-slot after a groove has been milled with a side or end milling cutter. This is an unusual cutter. It mills five sides of a T-slot at the same time: the bottom, the two interrupted faces, and the right and left sides.

The T-slot cutter is made in one piece (Fig. 40-21). It consists of a shank which fits into an adapter. The other end has teeth. These are usually of staggered-tooth design, cut at a helix angle. The area between the teeth and shank is machined to a smaller diameter than the narrow width of the T-slot.

T-slot cutters are available to mill slots in tables, columns, and plates. These take standard T-bolts with diameters ranging from 1/4" (6 mm) to 1½" (38 mm). Because of the construction and simultaneous cutting on five surfaces, light feeds are required. High-speed steel and carbide-tipped T-slot cutters are available.

Woodruff Keyseat Cutters—As described earlier, *Woodruff keys* are widely used between two mating parts that are to be keyed together. The cutting tool for milling the semicircular keyseat is called a *Woodruff keyseat cutter*. The milling process and a standard Woodruff keyseat cutter are shown in Fig. 40-22.

There are two general types of Woodruff keyseat cutters. The smaller sizes, up to 1½" (38 mm) in diameter, have a shank. This fits into an adapter. These cutters have teeth on the circumference only. Side clearance is provided by hollow grinding toward

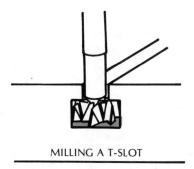

MILLING A T-SLOT

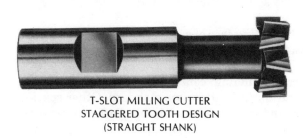

T-SLOT MILLING CUTTER
STAGGERED TOOTH DESIGN
(STRAIGHT SHANK)

Fig. 40-21. A T-slot cutter and its milling process. (Photo courtesy of Morse Cutting Tools Division)

Fig. 40-22. Woodruff (shank type) keyseat cutter and the keyseating process. (Photo courtesy of Morse Cutting Tools Division)

the center. The width (thickness) of the Woodruff keyseat cutter is held to close tolerances. On the smaller diameter cutters, the neck is reduced in diameter. This permits clearing the workpiece when the cutter reaches the full depth to which the keyseat is to be milled.

The cutter size is stamped on the shank of some cutters. This may be given in the form of a numerical code like 306, or as the actual dimensions, such as 3/32″ × 3/4″. The last two digits of the numerical code (06 in this case) indicates the diameter in eighths of an inch. The (3) digit preceding the last two in the example gives the width in 32nds of an inch. The 306 cutter is 3/32″ wide × 06/8″ in diameter.

The shank type of cutter is available in standard sizes. These range from 1/8″ to 3/8″ in width and 1/2″ to 1½″ in diameter. These are furnished in high-speed steel or as a carbide-tipped cutter. The model shown in Fig. 40-22 has a centered end that may be supported when milling at high speeds with a comparatively rough cut.

The arbor (hole) type of Woodruff keyseat cutter is more practical for larger diameter, deeper keyseats. The hole type is used for keyseats that are larger than two inches in diameter. These cutters are manufactured with staggered-tooth construction.

Formed-Tooth Milling Cutters

To review, there are three distinguishing design features of a formed-tooth milling cutter:
- The formed teeth are relieved behind the cutting face. The same profile is ground on the face of each tooth.
- The cutting face of each tooth is ground radially.
- Subsequent grinding to sharpen the cutter will not alter the tooth profile.

Corner rounding, concave and convex, gear milling, and fluting cutters are standard. Other special shapes and sizes are also available. Form cutters may be used alone or in pairs as indicated in Fig. 40-23. They may also be combined with other regular shapes of milling cutters.

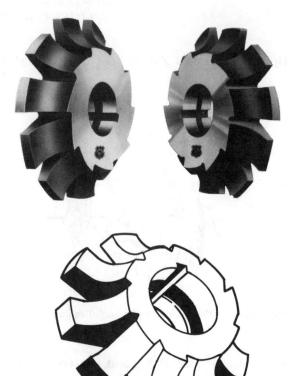

Fig. 40-23. Corner-rounding form milling cutters. (Photos courtesy of Standard Tool Div./Lear Siegler, Inc.)

The radial tooth faces may be ground with positive, zero, or negative rake. The original rake must be maintained to reproduce the exact profile. The cutting teeth may be straight or helical. The helical tooth produces a shearing action with reduced chatter.

End Milling Cutters

End milling is the process of machining horizontal, vertical, angular, and other irregular-shaped surfaces. The cutting tool is called an *end mill*. End mills are used to mill grooves, slots, keyways, and for other larger surface milling. They are also widely used for profile milling in die making. End mills are coarse-tooth cutters. The teeth cut on the periphery as well as on the face.

Standard End Mills (Solid Type)—Smaller size end mills are of a *solid type*. The teeth and shank are one piece. The larger diameter end mills are of the *shell type*. The cutter body and the shank are made in two pieces. A number of different designs of the solid type of end mill are identified in Fig. 40-24.

Solid end mills may have a *single* or a *double cutting end*. These may be two, three, four, or more flutes. The end may be straight or shaped to cut a round groove or radius. The teeth may be parallel to the axis or helical. Both right- and left-hand cutters

LEFT-HAND HELIX,
FOUR-FLUTE, SINGLE
END MILL

RIGHT-HAND HELIX,
FOUR-FLUTE, TAPER
SHANK END MILL

RIGHT-HAND HELIX,
FOUR-FLUTE, HEAVY-
DUTY, BALL END

FOUR-FLUTE, DOUBLE
END, END MILL

Fig. 40-24. Examples of single- and double-end and taper-shank multiple-flute end mills. (Courtesy of Morse Cutting Tools Division)

are available. End mills are called *spiral end mills* when the flutes are cut at an angle. The whole cutter may be made of high-speed steel, the teeth may be cemented carbide-tipped or solid cemented carbide.

End mills are made with either a straight or tapered shank. Some straight-shank end mills have a groove for locking in an adapter. They may also be chucked directly or held in a collet.

Taper-shank end mills have one of three possible tapers. The self-holding tapers include either a Brown and Sharpe or a Morse taper. These taper shanks are usually designed with a tang. The self-releasing taper shanks have the standard 3½" taper per foot. The cam-lock adapter described in an earlier unit is used with the self-releasing taper shank type.

The end teeth radiate from the center to the circumference. Two-fluted end mills have flutes cut across the bottom (end). These form two cutting lips. Each extends beyond the center axis. These cutting lips permit this type of end mill to cut into solid material without first drilling a hole.

The end mill may be right- or left-hand (Fig. 40-25). The end mill in Fig. 40-25A has a right-hand helix. The cutter rotates counterclockwise (right-hand, in the same direction as a drill). A right-hand end mill with a right-hand helix is standard. The end mill in Fig. 40-25B is a left-hand end mill. It has a left-hand helix. It must rotate clockwise to cut. It is also possible to order a right-hand end mill with a left-hand helix. The left-hand helix has the advantage of directing the thrust to the spindle. This feature is especially important when milling a *pocket*. The end mill does not feed itself into the pocket.

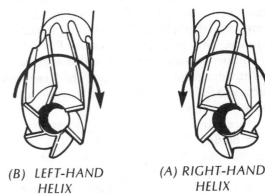

(B) LEFT-HAND HELIX　　　*(A) RIGHT-HAND HELIX*

Fig. 40-25. Right- and left-hand spiral end mills.

Shell End Mills—The shell end mill is a circular *shell* (Fig. 40-26). The shell has a bored hole that centers on and fits an adapter on the spindle. Two slots are milled on the back face. These fit the driving lugs on the adapter face. The center of the shell is recessed for a screw head or nut. This secures the cutter to a stub arbor.

HIGH SPEED STEEL TYPE

CARBIDE-TIPPED SHELL END MILL

Fig. 40-26. Shell end mills. (Courtesy of Morse Cutting Tools Division)

The teeth are designed around the circumference and front end of the shell. This permits face milling in a plane parallel to the column face, surface milling, and step or shoulder milling. Shell end mills are used for machining larger surfaces that require heavier cuts and coarser feeds than are possible with the smaller solid end mills.

Face Milling Cutter

This cutter (Fig. 40-27) is essentially a special form of end mill. The size is usually 6″ or larger in diameter. The teeth are beveled or rounded at the periphery of the cutter. Thus, only a small portion of the tooth is in contact with the workpiece. The remaining part (front face) of the tooth is ground with a clearance. This reduces the width of the blade, which eliminates scoring of the milled surface and reduces any drag against the cutter. The face of each tooth acts as a finish cutting tool and removes a small amount of material.

Some face milling cutters have both roughing and finishing blades. The finishing blades are set to a smaller diameter. These blades extend out a little more from the face of the body. The finishing blades also have a slightly wider surface on the cutting face. This produces a fine surface finish.

The body of a face milling cutter, like the one shown in Fig. 40-27, is usually made of an alloy steel. The blades (inserted teeth) may be of high-speed steel, cast alloy, carbide, or carbide-tipped. The blades are adjustable. They may be either reground and resharpened or replaced.

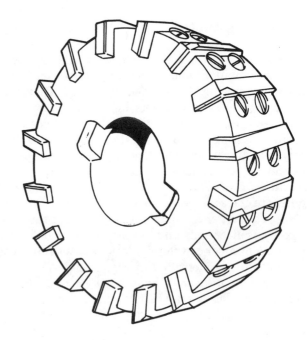

Fig. 40-27. Face milling cutter. (Courtesy of Cincinnati Milacron)

Face milling cutters are heavy-duty cutters. Heavy cuts, coarse feeds, and high cutting speeds are essential. Hogging cuts also require cutting below the scale on castings, rough edges on parts that are cut out by burning processes, and forgings.

The blades must extend far enough to permit free flow of the large quantity of chips that are produced. At the same time, the blades must be rigidly supported. The blades must cut without chatter and must be protected against breakage. Sometimes, grooves (*gullets*) are cut into the body. These are machined ahead of the blades. The gullets provide additional space for the chips.

The blades are serrated for easier adjustment and more secure locking in position. The body may be machined so that the blades are held straight or at an angle to the radial and axial planes. These angles give the face mill the same shearing action as the helical teeth on a plain milling cutter.

The blades are secured to the body by screws and other wedging devices. The body on some models is accurately positioned by the internally ground taper bore. The body is held on a stub arbor or other adapter. It may be held securely on the taper by a clamping strap and the draw-in bar. Face milling cutters may be either right- or left-hand.

Fly Cutters

A *fly cutter* consists of an arbor or other holding

device, one or more single-point cutters, and set-screws or other fasteners. The cutting tool is usually a tool bit ground to a desired shape. A fly cutter may be used for internal boring or for external operations. Plain, angular, and form milling are a few examples.

The fly cutter is practical for experimental work where an inexpensive spiral cutting tool is required. It is also used on materials like aluminium and magnesium. The trend is toward fewer teeth for the efficient milling of these and other metals. Fly cutters are made of high-speed steels, cast alloys, and cemented carbides.

Rotary Files And Burs

Although they are circular in shape like a milling machine cutter, rotary files and burs are not classified as such. The cutting edges consist of closely spaced, shallow grooves. These are cut around the circumference of the cutter.

Rotary files and burs are adapted to metal-removing applications using a flexible or other portable hand unit. Trimming a weld, preparing parts to be welded, and removing small amounts of metal in form dies are examples of the type of work performed with these cutting tools. Rotary files and burs are made of high-speed steel or cemented carbide.

SAFE PRACTICES IN THE CARE AND USE OF MILLING CUTTERS

- Stop the milling machine *before* setting up or removing a workpiece, cutter, or other accessory.
- Loosen the arbor support arm screw slightly. The arm should move but should still provide support while the arbor nut is loosened.
- Make sure that all chips are removed from the cutter and workpiece and that the cutter is not hot. A wiping cloth should be placed on the cutter to protect the hands. Use two hands when sliding the cutter off the arbor.
- Examine the cutting teeth. If they are dull, the milling cutter requires regrinding on a tool and cutter grinder.
- Remove any burrs on the milling cutter. Use an abrasive stone for this purpose.
- Store hole type milling cutters on a rack. Make sure the teeth do not hit other cutters or metal parts. *Note:* Solid-shank mills are usually stored in separate slots or compartments.

- Check the bore size of the milling cutter, the arbor diameter, and the key. The parts must fit with a sliding fit.
- Determine the *hand* of the cutter, the required direction of spindle rotation, and whether the arbor nut is right- or left-hand. Rotating the cutter in the wrong direction will ruin the teeth. Any movement of the arbor nut during cutting must be in the direction that tightens it.
- Select the cutting speed, feed, and cutting fluid (if required) that are most appropriate for the material, cutter, and milling process. Keep the workpiece and cutter as cool as possible.
- Protect the table surface with a wiping cloth or protective tray if setup and measuring tools are to be placed on it.

COMMON TERMS USED WITH MILLING CUTTERS

Diameter, width, hole and keyway sizes, shank type, cutter teeth form, rake, tooth pitch	A combination of features by which a craftsperson specifies a milling cutter.
Radial rake	The angle formed by the face of a cutter tooth and a radial plane. The plane extends from the outside tooth edge through the centerline.

Negative, zero, and positive rake	Radial tooth (rake) angles. These provide maximum tool life and cutting efficiency. A required rake angle for a specific workpiece, machining process, and cutter application.
Axial clearance	Clearance ground on the side of a cutter. Hollow grinding side teeth to prevent binding and to reduce friction on the sides of a cutter.
Plain milling cutter (light- and heavy-duty and helical)	A class of cutters designed for light, medium and slab (coarse, heavy) milling. Straight (parallel) teeth or teeth cut at an angle.
Side milling cutter	A combination of teeth machined on the face and side of a milling cutter. A cutter with teeth designed for side and/or face milling operations.
Staggered tooth	Teeth on a circular cutter that alternate with a right- and left-hand helix. Additional chip clearance and tooth relief is provided on the noncutting side.
Interlocking cutters	Milling cutters with sides recessed and alternate teeth ground back on the inside faces. Two cutters that are meshed. The teeth overlap and permit adjustment to a required width.
Slitting (slotting)	The process of milling a very narrow groove for purposes of cutting apart.
Single- and double-angle cutter	The included angle and angle form of a milling cutter. The number of angle faces on a cutter: one (single-angle); two (double-angle).
Woodruff keyseat cutter	A milling cutter designed to mill a circular groove (keyseat) to a specified diameter and width.
Formed tooth cutter	A tooth form that produces a desired profile. A cutter with teeth relieved to the same shape and size behind the cutting face. A cutter that retains its profile when reground parallel to the original plane of the cutting face.
End milling	Milling processes using the end and/or face of a cutter. Cutting with a solid-shank or adapter-held shell end mill.
Face milling	Usually, the process of machining the face of a workpiece. Using a milling cutter that cuts on the beveled or rounded cutting edges on the periphery of the cutter.
Fly cutting	Milling a face or boring with one or more single-point cutting tools (tool bits). These are mounted in an arbor or adapter.
Rotary filing	Removing minimum amounts of material with a portable (flexible) driving shaft and rotary file.

SUMMARY

- Common ANSI terms are used to describe the features of standard milling cutters.
- Milling cutters range from single-point fly cutters to multiple-tooth hole and shank types.
- Milling cutter forms are available to mill flat, angular, shoulder, round, and other contours and for slitting and sawing.
- A cutting tool face may be designed with a positive, zero, or negative radial rake. The positive and zero rakes are generally applied to high-speed steel cutters.
- Zero and negative rakes are used with carbide inserts and carbide-tipped cutters. The negative rake gives added strength to the cutting edges. A fine surface finish is also produced.
- Plain milling cutter teeth are ground with a primary clearance angle on the land.

- Formed-tooth milling cutters are relieved to the same tooth profile as the cutting face.
- Standard plain milling cutters include types for light- and heavy-duty flat (horizontal) milling. Helical cutters have a steep (45° to 60°) angle. The number of teeth per inch of diameter is also less.
- Side milling cutters are used primarily for milling steps (shoulders) and sides and for straddle milling. The teeth are machined around the circumference and on one (plain side) or two (half side) sides.
- Staggered teeth provide increased space for chip clearance. Each tooth in a set cuts to a narrower width. The teeth usually alternate in a right and left helix angle.
- Interlocking cutters are two cutters with teeth relieved on one face. The overall width of the two cutters may be adjusted with spacing washers.

Alternate teeth interlock and overlap.
- Slitting and slotting cutters have a finer pitch than other milling cutters. A minimum of two teeth should be engaged at all times. Slitting saws may have side teeth. These are hollow ground in the sides.
- Angles may be milled with single-angle or double-angle cutters. The teeth are ground to cut on the side and angular face. Double-angle cutters may have equal or unequal angles.
- T-slot milling involves first the machining of a slot, followed by forming of the remaining five surfaces of the T-slot.
- Woodruff keyseat cutters produce standard semi-circular slots (keyseats). The dimensions correspond with those of a fitted Woodruff key.
- Some of the common shapes of milling cutters include concave, convex, radius or corner, gear, and fluting forms. Formed tool cutters may be combined with other standard flat, angle, and face milling cutters.
- End mills of the solid type and smaller sizes are widely used on vertical spindle milling machines. The larger sizes are better suited for horizontal milling machine processes.

- Solid, straight-shank end mills are secured with a screw, a cam-lock fastener, or a collet. The hole type end mill is located and held securely in an adapter. Taper shank end mills are fitted with a self-holding taper.
- Face milling is a practical method of taking hogging and finishing cuts on work faces that are 6″ wide and larger.
- High-speed steel, cast alloy, and carbide-tipped blades are inserted and held in a rigidly constructed body. Usually, only the angular or round cutting tooth faces on the edge of the cutter do the cutting.
- Fly cutters are best adapted for milling operations using single-point cutting tools. Regular or special forms may be quickly and inexpensively produced.
- Safe practices must be followed to protect the hands when mounting or removing a milling cutter.
- Tool safety in relation to fits, burrs, and correctness of speeds and feeds, direction of cut, and cutting angles must be observed.

UNIT 40 REVIEW AND SELF-TEST

1. State the general specifications that are used to order milling cutters for Style A and B arbors.
2. Define (a) *rake angle* and the effect on cutting using milling cutters with (b) *positive rake* or (c) *negative rake*.
3. Explain the function that is served by providing *eccentric relief* on formed cutters.
4. Describe how milling cutter teeth that are formed at a steep helix angle affect: (a) the *direction of force* and (b) the *cutting forces*.
5. Differentiate between a *half side milling cutter* and an *interlocking side milling cutter*.
6. a. Name three different types of *slitting saws*.
   b. Give an application of each slitting saw type.
7. Give the dimensions of a *# 608 Woodruff keyseat cutter*.
8. Indicate the differences between a *solid end mill* and a *shell end mill*.
9. Describe (a) the function of a *face milling cutter* and (b) its major design features.
10. List three safe practices to follow in the care and use of milling cutters.

Unit 41

SPEEDS, FEEDS, AND CUTTING FLUIDS FOR MILLING

Speeds, feeds, and cutting fluids are widely used terms in milling machine work. These determine the rate at which parts are machined, the quality of surface finish, the way in which a workpiece is to be secured for machining, and the type of cutter to use. Speeds and feeds are further influenced by the size and condition of the milling machine.

This unit deals in depth with factors that influence cutting speeds and feeds and cutting fluids for basic milling machine processes. Simple mathematical formulas for determining spindle (rpm) and cutting speeds and feeds are applied. The remainder of the unit covers speed and feed controls and cutting fluids and systems.

CUTTING SPEEDS IN MILLING MACHINE WORK

Cutting Speed Defined

Cutting speed (for a milling machine) refers to the distance the circumference of a milling cutter revolves in a fixed period of time (Fig. 41-1). The *rate* is given in surface feet per minute (sfpm) or meters per minute (m/min). Too slow a cutting speed reduces cutting efficiency. Too high a speed may cause

damage to the cutter or workpiece. Operator judgment is necessary. The cutting speed and feed must be suited to the material and the work processes. The rate of speed expresses the distance a point on the circumference of a milling cutter travels in one minute.

Feed is the distance the workpiece moves in relation to the milling cutter. Feed may be expressed in terms of inches or millimeters per minute or per revolution of the cutter, or the rate of movement (cut) for each tooth on the cutter.

Determining Cutting Speeds

Parts materials vary according to hardness, structure, and machinability. The materials of which cutters are made also differ. High-speed milling cutters are practical for general-purpose operations. Cast alloy and cemented carbides are adapted to the machining of tough, abrasive materials. Cutting with carbide is done at speeds that are faster than those used in the high-speed steel cutter range.

Data has been gathered over many years on cutting speeds and feeds. Tables cover the ranges of cutting speeds and feeds for different milling cutters and workpiece materials. The tables represent *average* cutting speeds and/or feeds. The cutting speeds may be increased by one third to two thirds for finishing cuts. The cutting speed tables give the surface feet or meters per minute at which each material may be machined efficiently. Table 41-1 lists the cutting speeds for five different types of cutter materials. Soft cast iron is used as an example of the workpiece material. A more complete table of cutting speeds appears in Appendix A-26.

The operator determines the cutting speed from the work order or a table. This cutting speed must then be related to the diameter of the cutter. The cutter or spindle rpm is then computed.

Computing Spindle Speeds (RPM) And Cutting Speeds

The rate (rpm) at which a spindle should rotate may be calculated by one of the following simple

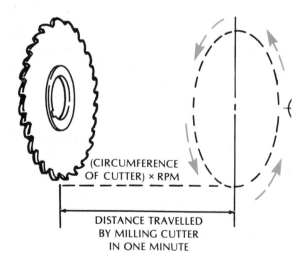

(CIRCUMFERENCE OF CUTTER) × RPM

DISTANCE TRAVELLED BY MILLING CUTTER IN ONE MINUTE

Fig. 41-1. The meaning of cutting speed.

Table 41-1 EXAMPLE OF CUTTING SPEED (sfpm)

Material to be Milled		High-Speed Steel	Super High-Speed Steel	Stellite	Tantalum Carbide	Cemented Carbides
				CUTTER MATERIAL		
Cast Iron	Soft	50 to 80	60 to 115	90 to 130		250 to 325

formulas:

Inch Standard:

$$rpm = \frac{(\text{Cutting Speed in sfpm}) \times (12)}{\text{Cutter Circumference (")}}$$

Metric Standard:

$$rpm = \frac{(\text{Cutting Speed in m/min})}{\text{Cutter Circumference (m)}}$$

The same basic formula is used whether the cutter diameter is expressed in inch standard or metric standard dimensions. The cutting speed in surface feet per minute is multiplied by 12. The product is divided by the circumference of the cutter in inches. If the cutting speed is given in meters per minute and the cutter diameter is measured in millimeters, the cutter circumference in terms of its meter value is used.

Example: Table 41-1 indicates that a high speed steel milling machine cutter may be operated within a range of 50 to 80 sfpm. Assume that the highest cutting rate (80 sfpm) is to be used on a solid, soft-cutting cast iron part. The part is to be face milled. A 3.500″ diameter, plain, high speed steel helical milling cutter is selected. Determine the spindle (and cutter) rpm. Use $\pi = 3$.

Step 1.　Cutter circumference $= \pi \times d$
$$= 3 \times 3.5$$
$$= 10.5'' \text{ (Answer)}.$$

Step 2.　Cutting speed of HSS cutter $= 80$ sfpm.

Step 3.　$rpm = \dfrac{(CS) \times 12}{(C'')}$

$$= \frac{80 \times 12}{10.5}$$

$$= 91.42 \text{ (Answer)}.$$

Step 4.　Select the closest (next lower) spindle speed to 91.42 rpm; for instance, 90 rpm.

Sometimes it is necessary to compute the cutting speed (sfpm or sfm) when the spindle speed and cutter diameter (d) are known.

Step 1　$CS \text{ (Sfm)} = \dfrac{(\pi \times \text{cutter diameter}) \times (rpm)}{12}$

Step 2　Using rpm $= 90$, the cutter diameter of 3.500″, and π rounded off to 3:

$$CS = \frac{(3 \times 3.5) \times 90}{12}$$

$$= 78.75 \text{ sfpm (Answer)}.$$

The variation between the initial cutter speed of 80 sfpm and the computed 78.75 sfpm is due to using 90 rpm as the closest spindle speed. In general shop practice, the following simplified formula is used in the inch standard system:

$$rpm = \frac{sfm \times 4}{dia}$$

or,

$$sfm = \frac{rpm \times dia.}{4}$$

FACTORS INFLUENCING CUTTING SPEEDS

Tables show, in general, that the harder or more abrasive the material, the slower the cutting speed. In addition to the materials in the workpiece and cutter, a number of the following factors also affect cutting speeds.

● Depth, width, and thickness of the cut
● Design and sharpness of the cutter teeth
● Method of holding and supporting the cutter and workpiece
● Nature and flow of cutting fluid
● Required degree of surface finish and dimensional accuracy
● Type and condition of the milling machine.

Depth And Width Of Cut And Chip Thickness

The cutting speed is generally reduced as the depth and width of the cut and the chip thickness are increased. Light, finish cuts may be taken at the higher speed within the specified range of cutting speeds. The cutting speed is also reduced for deep, narrow cuts. During such operations, it is difficult to flow a cutting fluid to the point of the cutting action.

Design Of The Cutter

The shape and design of the cutter and teeth affect the cutting speed. The cutting speed may be increased when less heat is generated. For example, the steep helix angle of the teeth of a coarse-tooth cutter permits chips to be washed out quickly. Therefore, a minimum of heat is produced by the cutting action and friction of the chips.

Cutters and circular saws with side teeth have a limited area of contact with the workpiece. These cutters may be run at higher surface feeds than may cutters that have a large contact area. The cutting speed must be reduced if there is excessive wear on the lands. Operations with formed cutters, like thread milling, require increased cutting speed.

Cutter And Work Support

The cutter should be supported close to the place where the greatest force is applied. This means machining as close as possible to the column and/or arbor support arm. The cutting speed is usually increased on light workpieces when a fine feed is used or when there is a nonrigid work-holding setup.

Nature And Flow Of Cutting Fluid

Restated, the cutting fluid reduces friction, removes heat, increases tool life, and helps produce a high-quality surface finish. The specifications of the fluid are determined by the material in the cutter and workpiece and the nature of the machining process. The more efficiently the cutting fluid serves these four basic functions, the higher the cutting speed.

Surface Finish And Degree Of Accuracy

Cutting speeds may be increased on soft materials and when light cuts are taken. Fine, accurate surface finishes are milled by increasing the cutting speed within the recommended range. The same quality of surface may be produced by decreasing the cutting feed. However, increasing the speed, where practical, is preferred because it saves time.

FACTORS THAT AFFECT CUTTING FEEDS

Most milling machines have a *constant-rate feed drive*. The feed is independent of the cutting speed. Once the required feed is set, it is independent of the spindle rpm.

The basis for calculating feed relates to the amount of material (thickness of the chips) removed by each succeeding milling cutter tooth. Feed is controlled by the same set of factors as those that apply to cutting speed.

Increasing And Decreasing The Rate Of Feed

Feeds per tooth for general milling processes are suggested in Table 41-2. The range extends from feeds for rough milling operations on hard, tough ferrous metals to light cuts on nonferrous metals. Feeds are given in 0.000″ and 0.00 mm. An average starting feed for each process, in both fractional inch and millimeter measurements, appears under the "Start" column.

Table 41-2. SUGGESTED FEEDS PER TOOTH FOR GENERAL MILLING PROCESSES

	Feeds			
	Start		Range	
Process	0.000″	0.00 mm	0.000″	0.00 mm
Face milling	0.008	0.20	0.005 − 0.030	0.10 − 0.80
Straddle milling	0.008	0.20	0.005 − 0.030	0.10 − 0.80
Channeling or slotting	0.008	0.20	0.005 − 0.020	0.10 − 0.50
Slab milling	0.007	0.20	0.005 − 0.020	0.10 − 0.50
End milling or profiling	0.004	0.10	0.002 − 0.010	0.05 − 0.25
Sawing	0.003	0.10	0.002 − 0.10	0.05 − 0.25
Thread milling	0.002	0.05	0.001 − 0.005	0.02 − 0.10
Boring	0.007	0.20	0.005 − 0.020	0.10 − 0.50

Table 41-3. RECOMMENDED FEED PER TOOTH (0.000″) FOR CEMENTED CARBIDE MILLING CUTTERS

Material to be Milled	Milling Machine Cutters and Processes					
	Face Mills	Spiral Mills	Side and Slotting Mills	End Mills	Form Relieved Cutters	Circular Saws
Malleable and Cast Iron (Medium hard)	0.016	0.013	0.010	0.008	0.005	0.004

Feeds may be increased for certain operations like slab milling; heavy roughing cuts; abrasive, scaled surface conditions; and easily machinable materials. On light cuts it is possible to increase *both* the feed and the speed.

Feeds should be decreased if the cutter chips at the cutting edges, for deep slotting cuts, for parts that cannot be clamped securely, or where a thin or fine-tooth cutter is used. Feeds are reduced when chips are not broken and continuously washed out of the workpiece.

Production depends on the *rate of feed*. The coarsest possible feed, consistent with the job requirements and machine setup, should be used. Cutters should be machine sharpened accurately. Good cutting qualities require less power and produce a higher quality of surface finish, with economy. The recommended practice is to begin with a finer feed per tooth. This feed is then increased depending on actual machining conditions.

Computing Cutting Feeds

Feed is a combination of distance and time. The time interval is expressed as *per minute*. The distance represents how far a workpiece moves during a milling process. A feed may be given in terms of *inches per minute* (ipm) or *millimeters per minute* (mm/min). The feed may also be stated as the thickness of the chip that is cut away by each tooth.

Some feed tables list recommended feeds (0.000″ or 0.00 mm) per tooth for selected materials and milling processes. Table 41-3 shows one line from the feed table in Appendix A-27. Six milling machine processes are recorded. The recommended feed per tooth is given for using cemented carbides to mill malleable and medium-hard cast iron parts. These feeds are reduced from one half to one third for finer surface finishes.

Example: A six-tooth spiral mill is used on a malleable iron casting. Each tooth cuts a 0.013″ thick chip. The cutter rotates at 200 rpm. The automatic feed (inches per minute) for roughing out is calculated in the following manner:

Step 1   The feed per revolution of the spiral mill is equal to the feed per tooth (0.013″) multiplied by the number of teeth, or 0.078″.

Step 2   The feed (inches per minute) equals the feed per revolution (0.078″) multiplied by the number of revolutions of the cutter per minute (200), or 15.6 inches per minute.

FEED AND SPEED FORMULAS (MILLING PROCESSES)

Table 41-4 gives the formulas for computing different speed and feed values. As mentioned earlier, cutting speeds (sfpm) and feed (per tooth) are specified in tables of technical information. Speeds and feeds are related to different cutting tools, materials in workpieces, and milling processes. Cutting speeds and feeds are influenced by many of the same conditions found in lathe, drill press, and many other machining processes.

Tables that show the maximum rate of cutting in *cubic inches per minute* (i³ pm) are often used in production milling. The feed in inches per minute is equal to the volume of material to be removed per minute divided by the depth of cut times the width of cut.

SPINDLE SPEED DRIVE AND CONTROL MECHANISMS

Spindle speed changes are usually made by arranging gears in specific combinations. Spindle speeds are indicated on the side of the column either

Table 41-4. FORMULAS FOR COMPUTING FEEDS AND SPEEDS FOR MILLING PROCESSES

Required Value	Formula	Symbols
Feed per tooth (F_t)	$F_t = \dfrac{F''}{(n) \times (rpm)}$	F'' = Feed in inches per minute
Feed per Revolution of cutter (F_r)	$F_r = (F_t) \times (n)$ $F_r = \dfrac{F''}{rpm}$	F_t = Feed per tooth
Feed in Inches per minute (F'')	$F'' = (F) \times (n) \times (rpm)$ $F'' = \dfrac{i^3 pm}{(D) \times (W)}$	n = Number of teeth in cutter
Revolutions of cutter per minute (rpm)	$rpm = \dfrac{(sfpm) \times (12)}{C \text{ (in inches)}}$ Simplified Formula: $rpm = \dfrac{(sfpm) \times (4)}{(d)}$	C = Circumference of cutter $.d$ = diameter of cutter in inches
Cutting speed (sfpm)	$sfpm \ (CS) = \dfrac{(C) \times (rpm)}{12}$ Simplified Formula: $sfpm = \dfrac{(d) \times (rpm)}{4}$	rpm = Revolutions of cutter per minute $sfpm$ = Surface feed per minute CS = Cutting speed i^3pm = Cubic inches of material removed per minute D = Depth of cut W = Width of cut

on an index plate or a dial. Gear combinations are controlled by levers. The levers are a mechanical means of sliding (shifting) gears in the transmission to produce a desired speed. Speeds in the geared head are usually arranged in two ranges: *low* and *high*. The low range uses a back gear reduction. The back gears are not engaged in the high range.

Fig. 41-2 shows how power and speed are transmitted from the motor to the spindle. This particular 7½ horsepower model has sixteen spindle speeds. The range is ample for general-purpose milling of ferrous and nonferrous metals. Either high-speed steel or cemented carbide milling cutters may be used. The back gears in the low range provide eight spindle speeds from 30 to 167 rpm. The change gears in the high range also produce eight spindle speeds. These range from 214 to 1200 rpm.

Lighter models of milling machines have eighteen spindle speeds. The low range on some is between 60 and 530 rpm. The higher spindle speeds range from 480 to 4150 rpm.

The spindle speed on some small milling machines is controlled by shifting belts. Other designs include a variable-speed drive. The speeds are changed

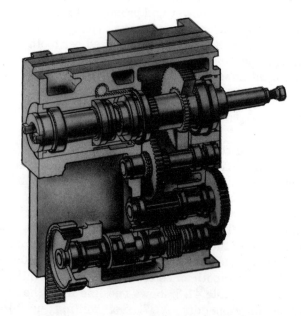

Fig. 41-2. Cutaway section of a 16-spindle-speed geared-head drive. (Courtesy of Cincinnati Milacron)

hydraulically. The larger and heavier machines require a gear box. The gear combinations are housed

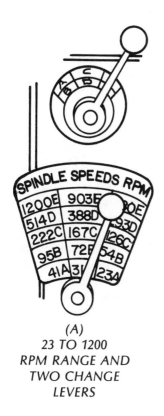

(A)
*23 TO 1200
RPM RANGE AND
TWO CHANGE
LEVERS*

*(B) 12 TO 1500 RPM RANGE SPEED
DIAL SELECTOR*

Fig. 41-3. A speed index plate and dial for setting spindle RPM. (Courtesy of Cincinnati Milacron)

inside the upper section of the column. The gear combinations are controlled on some models by a single *speed selector lever* or *crank*.

Another gear change setup for controlling spindle speeds is illustrated in Fig. 41-3A. This machine has two spindle speed selector levers. The lower lever may be moved in an arc to any one of three positions. The top lever may be set to one of five positions: A, B, C, D, or E. With the top lever in position A, the bottom lever may be set at 23, 31, or 41 rpm. The two levers provide fifteen spindle speeds, ranging from 23 to 1200 rpm on this model. By contrast, the single dial selector in Fig. 41-3B provides a series of spindle speed changes from 25 to 1500 rpm.

Spindle speed changes on geared-head milling machines are made when the spindle is stopped. *Forward* (clockwise) and *reverse* (counterclockwise) *spindle rotation* is controlled by a spindle starting and direction lever. A push button combination is used on machines equipped with a reversing motor.

HOW TO SET THE SPINDLE SPEED

Step 1 Determine the appropriate spindle speed

for the job at hand. It may be necessary to compute the rpm. The work order, blueprint, cutter specifications, and cutting speed tables provide the necessary information.

Step 2 Position the spindle rotation lever.

CAUTION: The spindle must rotate clockwise for a right-hand cutter; counterclockwise for a left-hand cutter.

Variable-Speed Hydraulic Drive

Step 1 Move the speed selector lever to the required rpm on the dial.
Note: Speed changes may be made on hydraulic drive machines while the spindle is rotating.

Step 2 Start the machine and the milling operation.
Note: The skilled worker often starts a cut and then determines whether the spindle speed should be increased, decreased, or left unchanged.

Geared-Head Spindle Drive

Step 1   Select the appropriate spindle speed (rpm) on the index plate. Determine the required position of the speed selector lever(s).

    CAUTION: The spindle must be stopped *before* any gear positions are changed.

Step 2   Proceed with the milling operation.

Step 3   Change the spindle speed if it is necessary for more efficient milling.

FEED MECHANISMS AND CONTROLS

The designs of feed mechanisms, like those of speed devices, vary. Some provide fixed feeds. Others provide variable feeds. These give an infinite number of feeds in ipm or mm/min. Some of the heavy milling machines have variable feed rates from 1″ to 150″ (25 mm to 3800 mm) per minute. This range permits the skilled operator to select the desired thickness of chip per tooth.

Rapid traverse speeds promote a fast operating cycle. Rapid traverse feeds up to 300 rpm (7600 mm/min) are used in some semiproduction machines.

Feed selection is made during the cutting cycle. The cutter is fed slowly into the workpiece. The feed is then increased or decreased. The feed is determined by the amount the cut is varied for depth and width.

Feed controls are usually independent of the spindle speed. Changes in spindle speed do not affect the rate of feed. The range of feeds may be indicated on a dial or index plate like those shown in Fig. 41-4. A single feed selector lever is usually used on the dial type (Fig. 41-4B). In other applications, two levers are used (Fig. 41-4A). Lever A is moved vertically to one of four positions. Lever B is positioned horizontally to one of three positions. These two levers control twelve gear combinations and therefore provide 12 different feeds. These range from 3/4″ (19 mm) per minute to 30″ (760 mm) per minute.

Other models have 16 change feeds. The horizontal feeds range, in *geometric progression,* from 1/2″ to 60″. The vertical (knee) feeds are from 1/4″ to 30″. Fig. 41-4B shows a single-crank feed control and dial.

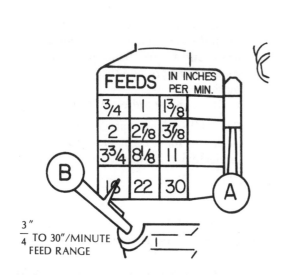

(A) OLDER MODEL FEED CHANGE INDEX PLATE AND LEVERS

$\frac{1}{2}$″ TO 60″/MIN. HORIZONTAL AND $\frac{1}{4}$″ TO 30″/MIN. VERTICAL FEED RANGES

(B) SINGLE CRANK-TYPE CONTROL

Fig. 41-4. Feed index plate and dial with feed controls. (Courtesy of Cincinnati Milacron)

Milling machines are provided with regular and rapid power feeds. The knee may be raised or lowered. The regular feed is engaged by a directional lever (up-neutral-down).

The saddle may be moved toward the column or away from it. The directional power feed (in machines that are so equipped) provides for cross feeding.

The table may be moved longitudinally by power. The feed is engaged by the directional feed control lever. When the lever is positioned to the *right,* the table *feeds to the right.* In the lever *neutral* (center) *position,* the table remains stationary. In the lever *left* position, the table *feeds to the left.*

The feed selector dial and lever and feed control levers are pointed out in Fig. 41-5. A single power (rapid traverse) lever permits rapid positioning of the workpiece. This feature provides for fast travel in any direction in which a feed is engaged. The workpiece may also be returned quickly to a starting position.

Milling machines are also equipped with *clamp levers.* The knee, saddle, and table may be clamped securely. The clamps may also be opened (relieved) so that the units may slide easily. In the locked position

of a clamp lever the units are held solidly together. A unit is secured when its movement is not required during a milling operation.

HOW TO SET THE FEED (TABLE, SADDLE, AND KNEE)

Step 1 Determine the correct feed.
Note: It may be necessary to compute the feed. One of the cutting speed and/or feed formulas may be used. All factors that affect cutting speed must be considered.

Step 2 Move the power feed control lever(s) to the required feed position(s). The lever position(s) is usually shown on the feed index plate or dial.

Step 3 Move either the longitudinal (table), transverse (saddle), or vertical (knee)

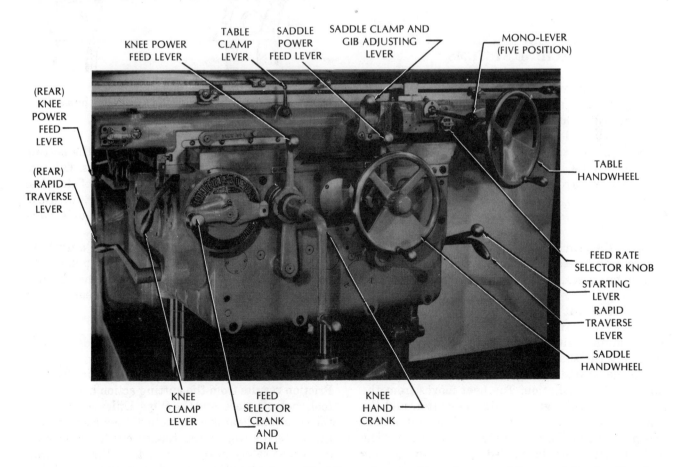

Fig. 41-5. The feed and traverse controls of a heavy-duty plain milling machine. (Courtesy of Kearney & Trecker Corp.)

feed control lever. The lever to use depends on whether the table, saddle, or knee is to be power fed. The direction in which the lever is moved depends on the required direction of power feed.

Step 4　Turn the appropriate clamp lever to permit the table, knee, or saddle to slide. The other clamp levers are tightened to securely hold the corresponding milling machine units.

CAUTION: Set the trip dogs if necessary. The dogs will trip (stop) the power feed at set positions. This prevents damage to the workpiece, setup, and machine.

Step 5　Use the rapid traverse lever to quickly position the workpiece and cutter.

Step 6　Disengage the power feed, and engage the handwheel for hand feeding.
Note: Hand feeding is used to start a cut. When the operator is assured that the milling process can be carried on safely, power feed is then engaged.

CUTTING FLUIDS AND MILLING MACHINE SYSTEMS

Cutting fluids and the principles of application to a milling machine are similar to those used on other machine tools. Cutting fluids may be liquid or air. These are applied to increase the quantity of work, to improve the cutting action and surface finish, to increase tool life, and to protect the finished surfaces from rust or corrosion.

Cooling Properties Of Cutting Fluids

Two of the prime functions of cooling and lubricating were discussed in an earlier unit. As applied to milling processes, the cooling properties of a cutting fluid help reduce the cutting temperature of the milling cutter. It is equally important to keep the workpiece cool. At cooler temperatures the part may be machined without changing its hardness or distorting the shape.

Heat generated during a milling process may be carried off by radiation. The heat moves from the cutting point through the chip, the cutter, and the workpiece (Fig. 41-6). However, the amount of heat dissipated by radiation is usually insufficient. The major portion must be carried away by the cutting fluid.

Fig. 41-6. Dissipation of heat by radiation.

The cutting fluid must have at least the following two important properties:

● A high specific heat factor. This relates to the ability to absorb heat. The greater the capacity to absorb heat, the faster the cooling action.

● Sufficiently low viscosity, which regulates the property of the cutting fluid to cling and to flow (Fig. 41-7).

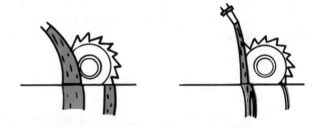

Fig. 41-7. The effect of viscosity on the flowing properties of cutting fluid.

The lower the viscosity, the quicker the cutting fluid reaches the cutter, work, and chips. Oil-and-water (emulsion) mixtures, the addition of chemical additives to make the cutting fluid *wetter,* the use of low-viscosity oils, and other cutting fluids were explained in an earlier unit.

In summary, the cutting fluid selected must have maximum fluidity so that it penetrates rapidly to the point of cutting. The cutting fluid must then be able to quickly absorb and dissipate the heat that is generated.

Lubricating Properties Of Cutting Fluids

The milling cutter is lubricated to reduce friction. Friction results from the cutting action between the tool, the workpiece, and the chips. Unless friction and the resultant heat are controlled, they may damage the cutter or workpiece, reduce cutting efficiency, and cause safety hazards. Fig. 41-8 illustrates that considerable force is required to shear material. The

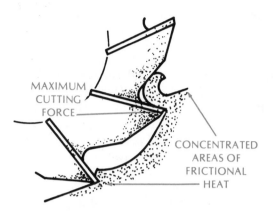

Fig. 41-8. The shearing forces of milling cutter teeth generate heat.

friction of the cutting action generates heat. A good cutting fluid reduces friction in the following three ways:

• By providing a film of lubricant between the contact surfaces and work

• By lubricating the chip, the contact surfaces of the workpiece, and the milling cutter. Lubrication reduces friction so that the chips and cutter can slide freely. A chip develops heat within itself. The chip is forced into the space between the cutter and the workpiece. As the chips curl or break up into smaller pieces, considerable heat is generated as the particles slide over one another. A good cutting fluid lubricates the chips and all contact surfaces.

• By maintaining a film of lubricant between the cutting edges and faces of the cutter and the chip. The frictional heat generated by chips sliding over a cutting face may become so intense that the chips fuse on the cutter teeth. This condition is shown in Fig. 41-9. The excessive heat and force may also dull the cutting edge and cause it to become ragged (Fig. 41-10).

Fig. 41-9. Fusion of chips to the face of a cutting tooth.

Fig. 41-10. The breakdown of sharp cutting edges.

Not all metals require a lubricant. Cast iron should be cut without a lubricant because it contains *free graphite*. The free graphite lubricates the cutting tool. Any additional oily lubricant produces a gummy substance. This retards chip movement. Under this condition, the chips tend to clog the chip spaces, and the cast iron becomes glazed.

Other Characteristics Of Cutting Fluids

There are two additional considerations in selecting a cutting fluid. The cutting fluid must be able to:

• Resist destructive cutting forces

• Resist deterioration resulting from excessive heat.

Tremendous forces are developed in shearing a chip from the workpiece with a milling cutter. The lubricant must be able to withstand both the cutting force and the intense heat generated at the cutting edge. Machining occurs by prying (shearing) a chip from the work slightly ahead of the cutting edge. The cutting edge is protected by fine particles of the material that fill the small space between the chip and the cutting edge. So long as these fine particles flow off with the chip, they do not become imbedded (welded) to the cutting edge. The cutting lubricant must be able to penetrate (flow) to the cutting edge, maintain lubricating properties under great force, and prevent chip particles from becoming welded to the cutting edge. In other words, the cutting fluid must have *antiweld* characteristics.

Chip Elimination

To produce ideal cutting conditions, a large enough quantity of cutting fluid must flow to do two things:

• It must move the chips off of the cutting edges and out of the cutter flutes (Fig. 41-11).

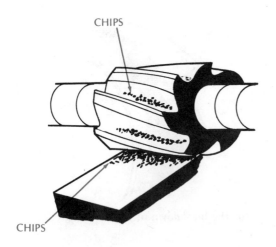

CHIPS

CHIPS

Fig. 41-11. Cutting fluid is needed to remove chips from the cutting tool edges and flutes and from the workpiece.

• It must flow chips away from the cutting area. If the chips are drawn back between the revolving cutter and the workpiece, additional cutting is required.

There are only a few teeth in continuous contact during a milling process. Therefore, the rate of flow and the selection of an appropriate cutting fluid should permit the chips to be removed during the noncutting time.

Increasing Tool Life

A good cutting lubricant prevents the development of frictional heat and resists the adhesion of chip particles to the cutting edges and faces. The cutting edge is also protected from wearing away. The use of an appropriate cutting fluid makes it possible to machine a greater number of pieces before the cutter teeth need to be reground. Thus, *cutter life* is extended because of production efficiency.

A great deal of milling with carbide cutters is done *dry*. However, when a cutting fluid is used, there must be a large, constant flow. Interrupted flow of the cutting fluid on a carbide tool causes *thermal cracking*. This, in turn, produces chipping of the tool.

Improving And Preserving Surface Finishes

A good surface finish may be produced by dry machining certain metals like cast iron and brass. The chips *flake* off and do not adhere to the cutting tool. Other materials produce chips that tend to cling to the tool face. When these pass between the work and the cutter, the chips tend to gouge or tear the work surface. The surface finish of these materials

can be improved by using a cutting fluid to remove the chips.

Preserving the surface finish means selecting a cutting fluid that will not discolor or rust a finished surface. Discoloration and corrosion of nonferrous metals are produced by additives in the cutting fluid. Rusting of ferrous parts may be avoided by rapidly evaporating the water in a cutting fluid solution. The remaining oil coats and protects the surface. Manufacturers' tables usually specify the proportions of water, other lubricants, and additives needed to meet specific requirements for milling ferrous and nonferrous metals and other nonmetallic parts.

KINDS OF CUTTING LUBRICANTS AND COMPOUNDS

Water Solutions

The use of a plain stream of water on a cutting tool makes it possible to increase the cutting speed. Water has excellent properties to rapidly absorb and carry off heat. Unfortunately, water has two drawbacks. First, rust forms on the workpiece and machine surfaces of the miller. Second, water thins the lubricating oils needed between mating surfaces.

Soluble oils are mixed with water to form an *emulsion*. The emulsion is milky in appearance. The soluble oils are usually a mineral oil, a vegetable oil, or an animal oil. An *emulsifying agent,* such as soap, is added. The consistency of the emulsion depends on the ratio of water to soluble oil. The emulsion consists of fine droplets of oil which are suspended in the water, soap, and oil mixture.

Other soluble oils are available in a clear, transparent mixture. This type is used when the operator must be able to see a layout or when the machining operation must be visible at all times. Soluble oils are applied when cooling is the prime consideration.

Cutting Oils

Generally, mineral, animal, and vegetable cutting oils are *compounded* (mixed). This blending is done for economical reasons as well as to improve certain cutting qualities. The straight cutting oils include lard oil and untreated mineral oils. Lard oil is mixed with mineral oils to prevent it from becoming rancid and clogging feed lines. Lard oil is still an excellent tool lubricant under severe cutting conditions.

When mixed with the cheaper mineral oils, the percentage of lard oil is determined by the nature of the cutting action and the hardness of the material to be cut. Mineral oils generally have better lubricating qualities than soluble oils. Lard oil is added when the cutting forces are too severe for straight mineral oils.

Sulphurized Oils

The addition of sulphur to a cutting fluid permits the cutting speed to be increased significantly. The cutting fluid is also able to withstand the greater forces that accompany heavy cuts on tough materials. Sulphurized oils are dark in color. They continue to darken as more sulphur is added, until all transparency is lost.

A pale yellow, transparent cutting fluid is produced by mixing mineral oil with a *base oil* (such as lard oil) to which sulphur is added. The amount of sulphur depends on the nature of the operation.

Kerosene

Kerosene is used primarily on nonferrous metals like aluminum, brasses, bronzes, magnesium, and zinc. The machinability ratings on these metals are above 100 percent.

Air As A Coolant

Although air is not a cutting fluid, it performs two similar functions. First, a flow of air under pressure helps remove chips from the cutter and workpiece. Second, the air cools the cutting tool, the chips, and the part. Air is used on cast iron and other metals and materials where a cutting fluid cannot be applied. The air stream may be produced by suction or as a blast.

APPLICATION OF CUTTING FLUIDS

The method of applying a cutting fluid is determined by the following conditions:
● The nature of the operation
● The width of the cutter and the depth of cut
● The design of the cutter.

Simple Hand Feeding

The same volume of cutting fluid is not adequate for all operations. In some cases, the fluid may be dripped on the cutter by a simple gravity feed. Where a small quantity is required, the cutting fluid may be brushed across a slowly revolving cutter. **Extreme caution must be exercised during this procedure.** A stiff brush must be used, and only as the teeth emerge from the cut.

Hand feeding is used when:
● It is the most practical method of applying a cutting fluid
● An operation requires the application of a limited amount of cutting fluid
● A light cut is taken and the surface finish is to be improved.

Circulating Pump System

A circulating pump system provides a continuous volume of cutting fluid. The hollow base of the milling machine usually contains the cutting fluid reservoir. A pump drains the cutting fluid from the reservoir and forces it through the feed lines. A flexible tube at the cutter end permits the nozzle (distributor) to be positioned to reach any point on any style of arbor. The cutting fluid flows through the nozzle and over the cutter, workpiece, and chips. The amount of flow is controlled by a shut-off valve near the nozzle. The design and shape of the nozzle shapes the stream. The stream may be confined to a single narrow cutter (Fig. 41-12C), or separate streams for multiple cutters (Fig. 41-12B), or a wide stream for applications such as slab milling (Fig. 41-12A).

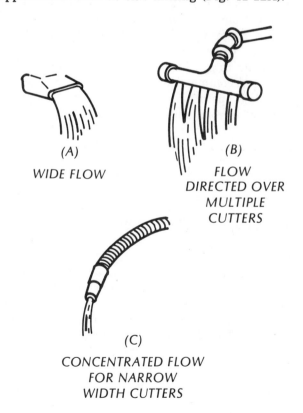

(A)
WIDE FLOW

(B)
FLOW DIRECTED OVER MULTIPLE CUTTERS

(C)
CONCENTRATED FLOW FOR NARROW WIDTH CUTTERS

Fig. 41-12. Three common types of cutting fluid distributors.

The cutting fluid and chips drain through the T-slots and channels in the table. The fluid drains through strainers into troughs at the end of the table. The fluid then passes through tubing back to the reservoir, to be recirculated. The rimmed base also has a screened opening. This permits any cutting fluid that reaches the base to be screened and returned to the reservoir.

Circulating pumps that operate at a constant flow rate are designed with an automatic relief valve. This operates when the volume of cutting fluid is increased. The excess fluid coming from the pump is diverted. The relief valve on the discharge side of the pump opens. The excess cutting fluid is returned to the reservoir instead of being sent into the feed line.

The distributor, flexible tube, and valve are generally attached to the overarm by a clamp. The clamp permits adjustment. Once it is set, the unit is secured in position.

Chips and other particles are first strained out by metal screens (sieves). Some are placed in the ends of the table or in the hose connection. Some machine designs include additional screening. The cutting fluid returns to the fluid tank, where it flows through a box-shaped strainer into the reservoir.

The whole circulating system and reservoir require regular cleaning. Sediment that accumulates in the bottom must be removed. The cutting fluid requires constant checking for composition and other properties. Once checked, the cutting fluid must be brought up to specifications. Appendix Table A-28 lists the recommended cutting fluids for various machining operations and various machinability ratings of ferrous and nonferrous metals.

HOW TO SUPPLY CUTTING FLUIDS

Using A Constant-Flow Circulating System

Step 1 Check the composition of the cutting fluid.
Note: The comparator shown in an earlier unit may be used. A fluid sample is drawn from the reservoir and tested. The sample should be representative of the solution in the reservoir.

Step 2 Select the distributor. The type should provide the kind and quantity of flow needed for the particular job.
Note: One of the three common types illustrated in Fig. 41-12 may be selected. The elongated type (A) provides a uniform flow across a wide cutter. Type (B) has a number of openings that direct the fluid over multiple cutters. Type (C) directs a limited flow over a narrow-width cutter.

Step 3 Attach the distributor to the flexible tube or pipe.

Step 4 Position the distributor bracket. Move it in the overarm until the distributor is centered with the cutter.

Step 5 Adjust the distributor in relation to the face of the cutter. Move it to within one inch of the cutter.

Step 6 Place splash guards to confine the fluid to the table area if a large volume of cutting fluid is to be used.

Step 7 Start the circulating system pump.

Step 8 Move the control valve slowly until the desired flow is delivered.

Step 9 Start the milling operation.
Note: Chips may need to be removed from the table during heavy cutting operations. The operator needs to check continuously to see that there is a free circulation of the cutting fluid.
CAUTION: Any cutting fluid that drops on the floor must be wiped up immediately. Oily cloths and rags should be placed in a metal container.

Step 10 Check during the cutting operation to see that there is a continuous flow of cutting fluid. Adjust the valve to control the flow.

Step 11 Turn off the control valve and the machine at the end of the operation or to set up a new workpiece.
Note: Standard machine cleanup procedures are followed.

- Use a cutting speed within the limits of the cutter and the job. Too high a speed may cause damage to the cutter and workpiece.
- Reduce the cutting speed for thin-sectioned parts or when the depth and width of cut are increased.
- Increase the cutting speed with coarse-tooth steep-angle milling cutters when the chips flow away easily.
- Machine as close as possible to the column or arbor support.
- Reduce the spindle rpm if there is excessive wear on the lands.
- Shift the speed selector lever to change the spindle speed in a geared head when the gears are stopped.
- Check the direction of spindle rotation. The spindle must rotate according to the hand of the cutter.
- Start a cut by hand feeding. Once it is established that the work and cutter are correctly positioned and there is good cutting action, either the speed or feed, or both, may be increased.
- Set the trip dogs before engaging a power feed.
- Check clearances between the cutter, arbor support, work-holding device, and workpiece before engaging the rapid traverse or regular power feed.
- Loosen the unit clamping lever before turning a handwheel or applying a power feed.
- Sample the cutting fluid by testing for composition, viscosity, and sediment. Bring the cutting fluid back to strength before using it on a cut.
- Select an appropriately designed distributor. The cutting fluid must flow over the entire cutter and workpiece. The volume must be adequate to wash out chips.
- Spray a rust inhibitor on the workpiece if the cutting fluid tends to produce corrosion or rust.
- Refer to a manufacturer's table on cutting fluids to determine the most efficient one to use on a particular job.
- Remove chips to permit recirculation of the cutting fluid. Use splash guards where practical.
- Wipe up any cutting fluid that spills on the floor. Dispose of oily waste and wiping cloths in a metal container.
- Avoid skin contact with cutting fluids. Direct contact with cutting fluids may cause skin infections.
- Wear a protective shield or safety goggles. Secure all loose clothing when working with or around moving machinery.

TERMS APPLIED TO MILLING MACHINE SPEEDS, FEEDS, AND CUTTING FLUIDS

Spindle speed (rpm)	The rpm a spindle should rotate for a milling cutter to cut at a required cutting speed.
Cutting speed (sfpm or m/min)	The distance the circumference of a revolving cutter travels in one minute. The surface feet or meters through which a milling cutter moves in a one-minute interval (sfpm or m/min). A specific rate established for a milling operation using a particular cutter on a specified material.
Constant-rate feed drive	Table, saddle, and knee feeds that are independent of the spindle speed (rpm).
Feed (per tooth)	The thickness of a chip removed by one tooth of a milling cutter. Usually the chip thickness in thousandths or metric equivalents.
Rate of feed (ipm or mm/min)	The distance a workpiece feeds into a milling cutter in one minute.
Rate of feed (i³pm)	The volume (mass) of material removed in one minute. The number of cubic inches of material machined in one minute. A production milling term.
Feed control levers	Separate levers that control the direction and movements of the knee, saddle, and table. A three-position lever for forward, neutral, and reverse longitudinal movements, and transverse and vertical movements.

Rapid Traverse	A mechanism for increasng feeds by moving the rapid traverse positioning lever. A rapid feed for speeding up the process of bringing the cutter and workpiece together.
Speed dial or index plate (geared head)	A circular cone-shaped or rectangular index plate on which speeds are indicated. A dial with speeds marked for different gear combinations.
Spindle direction lever	A lever mounted on the milling machine column. A lever for controlling the direction of rotation of the spindle.
Feed selector dial or index plate	A circular cone-shaped or rectangular index plate on which the ranges of feeds in rpm or mm/min are given. The dial or plate also shows the position(s) of the feed change lever(s) to obtain a particular feed.
Clamp levers	Levers that are locked to secure the knee, column, and/or saddle so that they are rigidly held during machining. The corresponding clamp lever is unlocked to permit moving one of the major milling machine units.
Heat removed by radiation	Heat that moves through a body and is dissipated in the surrounding area. The natural flow of heat through a milling cutter, chips, and workpiece.
Pitting of cutting tooth face	The welding of chip particles on the tool face near the cutting edge. Metal fusion caused by the intense heat generated by the cutting action.
Chip flow	Flowing a large volume of cutting fluid to wash away the chips from the cutter and workpiece.
Circulating pump system	A mechanism for flowing the cutting fluid on the cutter and workpiece at a constant rate. A reservoir, circulating pump, tubing, flexible nozzle, valves, and applicator. A system for circulating and recirculating a cutting fluid.

SUMMARY

- The cutting speed of a milling cutter is expressed in surface feet per minute (sfpm) or meters per minute (m/min). This represents the distance the circumference of a cutter travels in one minute.
- The cutting speed is converted to spindle speed; the rpm that a given diameter cutter must rotate.
- Cutting speed tables represent averages established by cutting tool and production specialists after extensive testing.
- Because of the many variable factors, the milling machine operator must apply personal judgement about speed and feed to the job at hand.
- Formulas are used to determine the feed per tooth, feed per revolution of a cutter, rpm of the spindle and cutter, and cutting speeds.
- Feeds are given for general milling processes in terms of ipm or mm/min. In production milling, feeds are stated as cubic inches of material removed in one minute (i³pm).
- Cutting speeds and feeds are influenced by the following factors:
  - Size of cut (depth and width)
  - Style and size of cutter
  - Sharpness of cutter teeth
  - Cutter and work supports
  - Type and quantity of cutting fluid
  - Surface finish and dimensional accuracy
  - Conditions of the milling machine.
- The higher cutting speeds in the recommended range may be used when:
  - Heat is dissipated quickly
  - Coarse tooth cutters with free-cutting helix angles provide fast removal of chips
  - Side teeth reduce the friction
  - Cutter support is at a maximum
  - Efficient cutting fluids reduce friction and lower the cutting temperature
  - Soft materials are to be cut or light cuts are to be taken.
- Milling cutter feeds are reduced for: deep slotting cuts, sawing with a fine-tooth narrow cutter, parts that cannot be clamped with great force, and operations where the chips are not broken and washed out readily.
- The range of spindle speeds is indicated on a dial or index plate on the column. Large machines have a high and a low range of speeds. These are controlled by a lever which sets the gear combinations on a geared-head machine. Some small machines are equipped with an infinite variable-speed drive.

- A lever usually controls the forward, neutral, and reverse direction of the spindle.
- Feeds are indicated on a dial or feed index plate. Usually, the longitudinal and transverse feeds are the same. Feed selector levers control the direction of the longitudinal, transverse, and vertical movements and feeds.
- When the saddle, table, and/or knee are to be held stationary during a milling operation, the respective unit is clamped securely.
- A trial cut is started by hand feeding. The feed is increased as soon as the operator determines that the operation is being safely performed. The power feed is then engaged.
- Cutting fluids are needed to remove heat that is not carried off by radiation.
- Cutting fluids reduce friction and high temperatures on the cutting edges. Excessive temperatures and forces tend to weld chip particles on the cutting face and rapidly wear away the cutting edges.
- Cutting fluids maintain a lubricating film between the cutter, chip, and workpiece.
- Cutting fluids are recommended by manufacturers according to the machinability and design of the material, the kind of milling operation, and the type and material in the milling cutter.
- The colors of cutting fluids vary from a clear transparent mixture, to milky and darker colors.
- Cutting fluids may be applied by hand on simple operations. A continuous flow is produced by using a circulating pump system. This is selected and positioned to flow the cutting fluid over the workpiece and cutter(s).
- Safety precautions must be considered in relation to the direction of cut, engagement of gears, cutting action, and removal of chips.
- Cutting fluids require regular testing. Hygienic conditions must be observed when using cutting fluids.
- The use of protective goggles and other standard personal and machine tool safety practices must be followed.

UNIT 41 REVIEW AND SELF-TEST

1. Express how cutting speed and feed rates are specified for milling machine work in inch and metric standard systems.
2. Calculate the cutting speed of a 4″ diameter cutter traveling at a spindle speed of 100 rpm.
3. Explain how each of the following conditions affects *cutting speeds* for general milling machine processes:
   a. Increasing the width and depth of cut and the chip thickness
   b. Changing from a standard tooth to a steep helix angle where a minimum amount of heat is generated
   c. Using fine feeds on thin-sectioned workpieces
   d. Improving the efficiency of the cutting fluid.
4. Indicate how each one of the following conditions affects *cutting feeds* for general milling processes:
   a. Cutting through scaled surfaces
   b. Roughing cuts
   c. Deep grooving cuts
   d. Chipping of the cutter teeth at the cutting edges.
5. Explain the meaning of *low-range* and *high-range spindle speeds*.
6. Cite the advantage of using *rapid traverse speeds* compared to using *regular traverse speeds*.
7. State three ways in which a quality cutting fluid (where applicable) reduces friction between a cutting tool, the workpiece, and the chips.
8. Give reasons why the use of an ideal cutting fluid increases productivity.
9. Provide a list of steps to follow to supply a cutting fluid with a *constant-flow circulating system*.
10. List three machine safety practices to follow in setting speed, feed, and coolant system controls.

SECTION TWO

TYPICAL MILLING PROCESSES AND PRACTICES

This section describes the principles of and procedures for performing three of the most common milling processes performed on the horizontal milling machine:

- Plain milling, in which the plane of the milled surface is parallel to the axis of the cutter
- Face milling, in which the plane of the milled surface is at a right angle to the face mill cutter
- Side milling, by which workpiece features such as ends, parallel steps, and grooves are produced

Unit 42 PLAIN MILLING ON THE HORIZONTAL MILLING MACHINE

Plain milling relates to the machining of a flat surface. The plane of the surface is parallel to the axis of the cutter. The surface is usually produced by a plain milling cutter mounted on an arbor. The terms *peripheral* and *slab milling* are also used. *Slab milling* refers to the machining of wide, flat surfaces.

When the plane of the milled surface is at a right angle to a face mill cutter axis, the process is referred to as *face milling*. This is covered in Unit 43.

This unit deals with plain milling. The competent operator must be able to apply the technology and procedures that relate to each of the following processes:

● Preparing and laying out the workpiece
● Determining the safest and most practical method of holding the workpiece
● Setting up the workpiece accurately for each successive operation
● Selecting and mounting a milling cutter appropriate to the job requirements
● Determining the correct spindle speed, cutting feed, and cutting fluid
● Setting up the machine
● Taking the necessary milling cuts and judging the cutting action of the cutter
● Measuring the workpiece for dimensional accuracy
● Cleaning and maintenance of the machine and accessories.

Applications are then made to plain milling, face milling, grooving, slotting, sawing, and form milling.

CONVENTIONAL AND CLIMB MILLING

The two common methods of removing metal are called *conventional* and *climb milling*. Since the milling machine was invented, operations have been performed by rotating the cutter in a direction opposite to that of the table feed. This direction of movement and table feed is necessary. Otherwise, the limited rigidity of some machines and the backlash of the table tend to draw the workpiece into the cutter. This causes damage to the part, the cutter, or the work-holding device.

Conventional (Up) Milling

Conventional milling is also referred to as *up milling*. Fig. 42-1 shows the direction the cutter turns in relation to the table feed. The cutter rotates clockwise. The workpiece is fed to the right. If the cutter rotates counterclockwise, the table feed is to the left (into the cutter).

The cutting action takes place from the bottom of the cut to the face of the workpiece. The part is fed into the revolving cutter. The chip starts as a very fine cut. It gradually increases in thickness, as shown by the black area in Fig. 42-1. The maximum thickness of the chip is at the point where each cutting tooth cuts through the top surface of the workpiece.

Theoretically, the cut starts at the perpendicular centerline of the cutter and work. In practice, the chip is started just ahead of the center, as indicated by the X in Fig. 42-1. This delayed action is caused by the teeth sliding over *(burnishing)* the material. This occurs momentarily. When a sufficient force is produced, the cutter teeth *bite into the surface* to start forming the chip. There is also a minute time interval between the cutting action of two successive teeth.

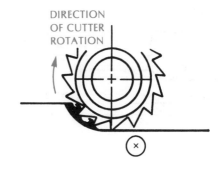

DIRECTION OF CUTTER ROTATION

Fig. 42-1. The directions of the cutter and work for conventional milling.

The combined cutting action produces *tooth marks* in a milled surface. The use of a fine feed and a sharp milling cutter traveling at high speed (within the specified range) produces a machined plane surface

527

that meets practical dimensional accuracy requirements.

The variations in the cross section of a chip and the tooth marks on a milled surface are exaggerated in Fig. 42-2. The burnishing rather than cutting action tends to dull the cutting edges of the milling cutter. The major forces in conventional milling are upward. These forces tend to lift the workpiece.

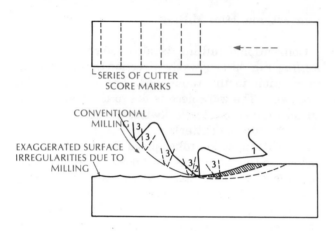

Fig. 42-2. The surface finish generally produced by the cutting action of a milling cutter.

In all milling it is important to use a *stop*. This prevents the part and/or work-holding device from moving on the table. Fig. 42-3 shows the use of a stop. It is bolted to the table immediately ahead of the workpiece. The workpiece in this example is secured directly to the table.

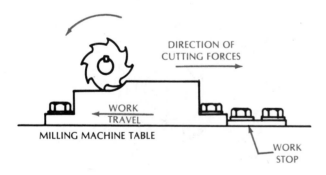

Fig. 42-3. A work stop prevents the cutting forces from moving the workpiece.

Climb (Down) Milling

In *climb milling,* the work is fed in the same direction as the rotation of the cutter teeth. In Fig.

42-4, the cutter rotates in a clockwise direction. The workpiece is fed into the cutter from right to left. If the cutter rotates in a counterclockwise direction, the cutting feed is from left to right.

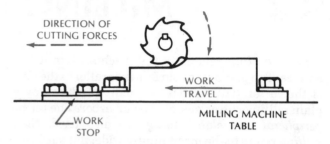

Fig. 42-4. Forces that must be overcome by a work stop in climb, or down, milling.

The cutting forces in climb milling are mostly downward. A stop is usually placed ahead of the workpiece to prevent any movement during a heavy milling operation. There is also a tendency to pull the work into the revolving cutter. For this reason, climb milling is avoided unless the milling machine is rigidly constructed and/or equipped with an automatic *backlash* eliminator. Once this device is engaged, it is automatically activated while climb milling.

The chip formation in climb milling is opposite that in conventional milling. Fig. 42-5 shows that the cutter tooth is almost parallel with the top surface of the workpiece. The cutter tooth begins to mill the full chip thickness. The chip thickness gradually diminishes. Finally, there is a disengagement between the teeth and work. During this cutting cycle the force is reduced to almost zero. This helps eliminate the revolution or feed marks that are normally produced by conventional milling.

Advantages Of Climb Milling:
● Climb milling tends to eliminate surface burrs that

Fig. 42-5. A full chip is milled from the top of the work surface in climb milling.

normally are produced during conventional (up) milling. The cutting force in up milling increases as the teeth penetrate deeper into the material. When the surface of the workpiece is no longer strong enough to resist the force of the cut, the material is torn away and turned to form a burr. As the cutter dulls, the burrs get heavier. In down milling, as just explained, the cutting force diminishes.

- Because revolution and feed marks are minimized, climb milling produces a smoother cut.
- Climb milling is more practical than conventional milling for machining deep, narrow slots. Narrow cutters and saws have a tendency to flex and crowd sideways under the force of a cut. The cutting action in climb milling permits these cutters to cut without springing under a heavy force.
- Climb milling forces the work against the table fixture or surface to which it is clamped. This feature makes climb milling desirable for machining thin or hard-to-hold workpieces and for cutting off stock.
- Laboratory tests indicate that less power is required for climb milling than conventional milling.
- Consistently parallel surfaces and dimensional accuracy may be maintained on thin-sectioned parts.
- Cutting efficiency is increased because more efficient cutter rake angles may be used.

Disadvantages Of Climb Milling:
- **Climb milling is dangerous.** The milling machine should be equipped with a backlash eliminator. All play must be removed between the lead screw and nut. The ways and sliding surfaces must also be free of lost motion. Any of these conditions causes the cutter to pull the work into the teeth.
- Climb milling is not recommended for castings, forgings, hot-rolled steels, or other materials that have an abrasive outer scale or surface. The continuous contact of the cutting teeth on a rough, hard surface causes the teeth to dull rapidly. Conventional milling is desirable in these cases. Once the cut is started, the teeth no longer come in contact with the outer scale (Fig. 42-6). As the chip and force become greater toward the outer surface, the scale breaks with a slight force.

MILLING FLAT SURFACES

Setting Up The Machine, Work-Holding Device, And Workpiece

Setting up the milling machine refers to the following functions:

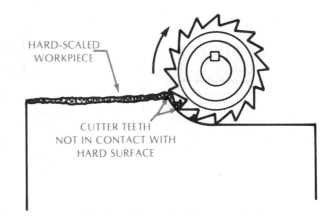

Fig. 42-6. Conventional milling is recommended for hard-scaled casting, forged, and rough rolled surfaces.

- Clamping of the work
- Mounting the cutter
- Setting the spindle speed
- Adjusting the feed mechanism
- Positioning the cutting fluid nozzle.

Small, regularly-shaped workpieces are usually mounted in a vise. Larger pieces may be fastened directly to the table. Irregularly-shaped parts are nested and secured in fixtures.

The depth of cut depends on the kind and the amount of material to be removed, the type of plain milling cutter, and the required degree of accuracy. The depth of the first cut on castings, forgings, and other rough-surface parts must be below the scale or hard outer surface.

Most jobs require a roughing cut and a finish cut. The roughing cut should be as deep as possible. The depth is determined by the rigidity of the setup, the capacity of the machine, and the surface condition. Usually 1/64″ (0.015″ to 0.020″, or 0.4 mm to 0.5 mm) is left for a finish cut. A finish cut is necessary for dimensional accuracy and to produce a high-quality surface finish.

Most workpieces may be conveniently held in a milling machine vise. The vise jaws are positioned accurately. The work is generally set on parallels. The height of the jaw must permit the workpiece to be held securely. Fig. 42-7 shows three setups. The workpiece in (A) is too low; in (B) it is too high. The setup in (C) provides maximum seating and holding power. It is good practice to seat the workpiece on two narrow parallels. These permit the operator to determine when the part is properly seated.

A protecting strip is placed between any rough surface, the ground jaws of the vise, and the ground faces of the parallels. Often when *seating the work,*

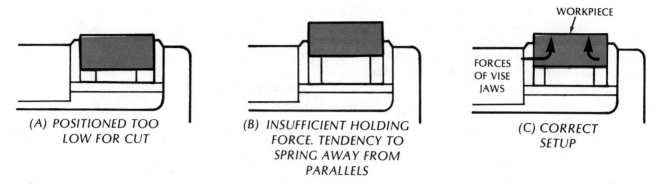

(A) POSITIONED TOO LOW FOR CUT

(B) INSUFFICIENT HOLDING FORCE. TENDENCY TO SPRING AWAY FROM PARALLELS

(C) CORRECT SETUP

FORCES OF VISE JAWS

WORKPIECE

Fig. 42-7. Incorrect and correct setups of the workpiece on parallels in a vise.

the movable jaw tends to lift the part slightly off the parallels. This condition is overcome by applying a slight force with the movable jaw. The workpiece is then gently but firmly tapped. (Too hard a blow tends to make the work rebound.) The vise is tightened further. The work is rechecked, and reseated if necessary.

The Milling Cutter And Setup

A light-duty, heavy-duty, or helical plain milling cutter should be selected. The correct type depends on the material, the nature of the operation, and the time within which the part is to be milled.

The cutter or combination of cutters should be wide enough to mill across the width of the workpiece. If two cutters are required, it is good practice to use interlocking cutters. These permit the cutting forces to be equalized.

Usually, the smallest diameter cutters possible are used. The larger the diameter, the longer the approach time required to travel across the same length of workpiece. Fig. 42-8 illustrates the added distance required when using a larger diameter cutter.

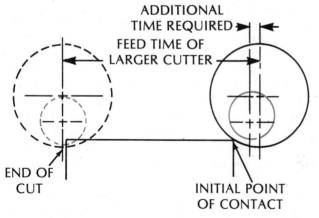

ADDITIONAL TIME REQUIRED

FEED TIME OF LARGER CUTTER

END OF CUT

INITIAL POINT OF CONTACT

Fig. 42-8. An illustration of the shorter feed time needed for a smaller diameter cutter.

When heavy slab cuts are to be taken, consideration must be given to selecting and mounting a cutter so that the cutting forces are directed *toward* the column. The arbor support must also be positioned to give the greatest rigidity possible.

Setup For Milling Surfaces Parallel And/Or At Right Angles

The surface that is milled first becomes a reference plane. If the opposite surface is to be milled parallel, the first surface is seated on parallels and secured. The usual procedure for removing burrs is followed prior to positioning. A trial cut is usually taken across the second face. The ends are measured for parallelism and thickness.

When all four sides are to be milled square, the first milled side is held against the solid vise jaw. The adjacent side, if it is straight, may be placed on parallels. Otherwise, the face to be milled may need to be set level by using a surface gage or other indicator. The third side is milled by seating the second side on parallels. A trial cut is taken for about 1/4″. The part is measured with a steel rule, caliper, or micrometer. The knee is adjusted to produce the required dimension.

If the fourth side is rough, a round bar may be placed above center between the movable jaw and the workpiece. When the vise is tightened, the force is toward the solid jaw and parallels. These forces tend to seat the workpiece. The workpiece is measured for size (Fig. 42-9). The knee is adjusted if required. The part is then milled to size.

A number of personal and machine safety precautions must be observed continuously. Once a setup is made, the knee and saddle must be locked. The cutter must rotate in the right direction. The spindle must be stopped before any measurements are taken or before the workpiece is moved. The part must be

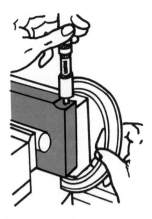

Fig. 42-9. Checking the workpiece size after a trial cut.

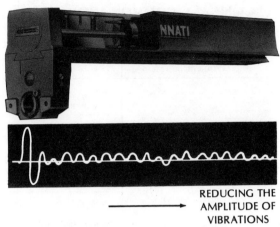

REDUCING THE
AMPLITUDE OF
VIBRATIONS

Fig. 42-10. Vibration damping unit inside the overarm.
(Courtesy of Cincinnati Milacron)

checked to see that it has not moved. There must be an adequate flow of cutting fluid.

Stopping the cutter part way through a cut causes the milling of a slight depression in the surface. Other unwanted indentations are produced if a revolving cutter is moved across a finished face.

COMMON MILLING PROBLEMS: PROBABLE CAUSES AND CORRECTIVE STEPS

Milling problems are similar to those that occur on other machines on which single- and multiple-point cutting tools are used. Chatter ranks as the number one problem. Other conditions make it difficult to mill parts to a required size or quality of surface finish. For example, some cutters dull prematurely, the teeth break or burn, or the cutter digs into the workpiece.

Vibration of the machine, cutter, or work (or of all three) causes unsafe machining conditions. Vibration also often results in damaged work. Some models have a *vibration damping unit* located inside the overarm. This feature is shown in the cutaway section of the overarm in Fig. 42-10. Its function is illustrated by the *vibration amplitude* line drawing. This vibration (chatter) damping capability permits greater depths of cuts and higher feeds and speeds. Cutting is more efficient and smoother. The vibration damping overarm may be used with both climb and conventional milling.

These problems and others are listed in the first column of Table 42-1. This table is a simple guide for analyzing common milling problems. The most probable causes of each problem are given in the second column. Corrective steps are suggested in the third column.

HOW TO MILL FLAT SURFACES (PLAIN MILLING)

Milling Parallel And Right-Angle Surfaces On Square And Rectangular Workpieces

Step 1 — Secure a workpiece, casting, or forging that has enough stock to permit milling to size. Lay out the part.

Step 2 — Locate the workpiece centrally in a vise, or on an angle plate or other holding device. Position it on parallels. Shim and support any uneven surface.

Step 3 — Select a sharp plain milling cutter appropriate for the job. Mount and key the cutter. Support the cutter as rigidly as possible.

Step 4 — Determine the cutting speed. Calculate and set the spindle to the required rpm and direction of rotation.

Step 5 — Determine the rate of feed for a roughing cut. Set the feed dial accordingly.

Step 6 — Position the workpiece so that it is centered under the cutter.
Note: The workpiece and cutter are brought to position by using rapid traverse feed. If the distance is short, the table, knee, and saddle handwheels are used.

Step 7 — Lock the knee and saddle. Set the trip dogs to the required limits of table travel.

Step 8 — Position the distributor nozzle to provide the necessary flow of required cutting fluid.

Step 9 — Start a trial cut. Turn the table hand-

Table 42-1. COMMON MILLING PROBLEMS: PROBABLE CAUSES AND CORRECTIVE STEPS

Milling Problem	Probable Cause	Corrective Steps
Chatter	• Lack of rigidity in the machine, work holding device, arbor or workpiece.	• Increase rigidity; machine support and secure workpiece, cutter, and arbor more effectively
	• Excessive cutting load	• Use cutter with smaller number of teeth
	• Dull cutter	• Resharpen
	• Poor lubrication or wrong lubricant	• Improve lubrication and check lubricant specifications
	• Straight tooth cutter	• Use helical tooth cutter
	• Peripheral relief angle too great	• Decrease relief angle
Inability to hold size of workpiece	• Excessive cutting load causes deflection	• Decrease the number of teeth in contact with the workpiece
	• Chip packing	• Increase fluid system pressure, redirect flow of cutting fluid to wash chips out of teeth
	• Chips causing misalignment of work	• Remove all chips and clean workpiece and holding device before mounting new workpiece
Poor quality of surface finish	• Feed too high	• Decrease feed or increase speed
	• Dull cutter	• Resharpen accurately
	• Cutting speed too low	• Increase sfpm (spindle speed) or decrease feed
	• Cutter has insufficient number of teeth	• Use finer tooth cutter
Burnishing the workpiece	• Cut is too light	• Increase depth of cut
	• Insufficient peripheral relief	• Increase peripheral relief angle
	• Land too wide	• Decrease width of land
	• Cutter is dull	• Regrind cutter teeth accurately

wheel by hand. Observe the cutting action and the layout line. Enough stock is to be left for other roughing cuts or a finish cut.

Step 10 Stop all spindle motion. Measure the thickness of the part.

Step 11 Unlock the knee clamp if a vertical adjustment in the depth of cut is required. Make the adjustment, and then relock the knee clamp.

Step 12 Restart the machine. Start the cut by hand. If the cutting action is satisfactory, engage the power feed.

Step 13 Disengage the power feed at the end of the cut. Stop the spindle. Return the table to the starting position.

Step 14 Decrease the feed. Increase the speed. Take a finish cut of at least 0.015″ (0.4 mm). This should produce a milled surface to the required degree of finish.

Step 15 Stop the spindle. Clean the workpiece, holding device, and parallels. Remove the burrs from the edges of the workpiece.

CAUTION: For personal safety, use a stiff brush and metal chip cleaner.

Step 16 Replace the part in the work-holding device. Machined side #1 is held against the solid vise jaw or the machined face of an angle plate.

Step 17 Use a surface gage or measure the layout line to be sure that both ends of the workpiece are the same height from the table or vise jaws.
Note: If necessary, shim or jack the under surface so that the cutting forces will not move the workpiece.

Step 18 Secure the workpiece. Position the cutter

(Table 42-1 continued)

Milling Problem	Probable Cause	Corrective Steps
Cutter dulls prematurely	● Cutting load too great	● Decrease number of teeth in contact with the workpiece or increase the spindle rpm
	Insufficient or improper cutting fluid	● Increase the flow of cutting fluid
		● Check cutting fluid and specifications; change cutting fluid if necessary
Burning of cutter	● Insufficient cutting fluid	● Increase flow of cutting fluid
	● Speed too fast	● Decrease speed
	● Improper cutting fluid	● Test composition and viscosity and bring up to the required specifications
Breaking of cutting edges on teeth	● Feed too high	● Decrease feed per tooth or increase the number of teeth
	● Cutter rotating in wrong direction	● Correct direction of spindle
Cutter digs (hogs) in	● Peripheral relief angle too great	● Decrease relief angle; use recommended angles
	● Rake angle too large	● Decrease rake angle
	● Improper speed	● Check recommended speed; adjust accordingly
	● Failure to tighten saddle or knee clamping levers	● Tighten saddle and knee clamping levers
Vibration	● Cutter rubs; insufficient clearance	● Use staggered tooth cutter or cutter with side relief teeth
	● Arbor size and support	● Use larger arbor and adjust arbor support arm

to depth for a trial cut. Start the flow of cutting fluid.

Step 19   Feed by hand a short distance. Then stop the spindle and check for depth against the layout line. Adjust the depth if necessary. Then, restart the machine, engage the power feed, and take the rough cut across side #2.

Note: If a finish cut is required, perform Step 14.

Step 20   Machine sides #3 and #4. Steps 9 through 15 are followed after the workpiece is seated on parallels (on side #1 and #2, respectively).

Note: Micrometer measurements are taken near each corner to ensure that the opposite faces are parallel and that the part is machined to the correct dimension. Since the solid jaw faces are accurately machined at 90° to the vise base, the milled faces of the workpiece should be at 90° to each other. The skilled operator always checks the accuracy of the faces (90°) with a solid steel square.

HOW TO MILL ENDS SQUARE

There are four common methods for milling the ends of a workpiece *square with the sides.*

Method 1.   The part is held in a vise that is parallel to the arbor axis. The ends are squared with a side milling cutter.

Method 2.   A part with a long rectangular area may be held in a vise that is parallel to the column face.

Method 3.   A part with a relatively small cross-sectional area may be positioned and held in vise jaws that are set parallel to the arbor axis.

Method 4.   A part with a small cross-sectional area may be strapped to an angle plate.

The first method is described in Unit 44 under *How To Mill Ends Square.* The other three methods are described here.

Vise And Workpiece Positioned Parallel To The Column Face

Step 1   Position the vise parallel to the column face.

Step 2   Place the workpiece on a parallel that rests on the base of the vise. Tighten the vise jaws so that the part may be tapped into position.
Note: Use a protective metal strip to prevent the hardened vise jaws from scoring any finished surface.

Step 3   Place the steel square on the base of the vise if possible. Otherwise, use a parallel. Slide the blade up to the workpiece. Sight along the blade to see if the workpiece side is positioned squarely.
Note: If any light shines through, move the square away. Then, tap the workpiece to square it. When it is squared, securely tighten the workpiece. Recheck it with the square.
Note: A surface gage may also be used if there is a layout line on the part.

Step 4   Take a roughing cut. Close observation of the milling operation is necessary.

CAUTION: If the cutting force causes the workpiece to shift, immediately move the cutter away and stop the machine. The workpiece then needs to be reset and held more securely.

Step 5   Continue with the roughing and finishing cuts. Follow the same steps as for plain milling flat surfaces.

Step 6   Setup and mill the opposite end in the same manner.

Vise Positioned Parallel With The Arbor Axis

Short workpieces that have a small cross-sectional area may be quickly and accurately machined while held in a vise. This is positioned parallel with the arbor axis.

Step 1   Square the side of the workpiece relative to the base of the vise. The steel square method may be used.
Note: A layout line may be checked using the surface gage method. A more accurate method is to use a dial indicator. The arm is secured to the arbor. The dial indicator point is brought into contact with the side of the workpiece. The knee is raised and lowered as required. The workpiece is tapped gently until the dial

indicates that the side is vertical (square with the vise base).

Step 2   Take a roughing and a finishing cut. Follow the same steps as for plain milling other flat surfaces.
Note: At the end of the roughing cut, the spindle is stopped. The workpiece is burred. The squareness of the end is checked with the steel square.

CAUTION: The cutting force must be directed *toward* the solid vise jaw. If necessary, the vise may need to be swung through 180° to cut in this manner. Also, the part should extend only a short distance above the vise jaws. The ends of long workpieces should be milled by side milling. Medium lengths are often held on an angle plate.

Step 3   Reposition and secure the workpiece to mill the opposite end. Proceed in the same manner as for milling and measuring two parallel surfaces.

Using An Angle Plate To Mill Ends Square

Step 1   Select an angle plate that will accommodate the length of the workpiece. Secure the angle plate so that the cutting force is toward the face and knee.

Step 2   Center the workpiece on the angle plate. Clamp the workpiece so that it may be squared.
Note: Limit the amount the workpiece extends beyond the top of the angle plate. If they are needed for support, use a packing block and shims between the bottom of the workpiece and the table.

Step 3   Square up the workpiece. The dial indicator, solid steel square, or surface gage method may be used. The selection depends on the accuracy required. Tighten the strap clamp(s). Recheck for squareness.

Step 4   Take a roughing cut. Clean the workpiece and remove burrs. Check the end for squareness. Adjust the workpiece if it has shifted.

Step 5   Decrease the feed. Increase the speed. Take a finish cut to the layout line or required dimension.

Step 6   Stop the spindle. Square up the opposite end. Follow the same steps for rough and finish milling.

Step 7   Measure the overall length for dimensional accuracy.

HOW TO COMPLETE A MILLING PROCESS

Step 1   Stop the machine.
Step 2   Remove, clean, and burr the workpiece.
Step 3   Permit the splash guards to drain for a minute. Remove the chips, wipe the guards, and remove them.
Step 4   Brush any remaining chips from the cutter, workpiece, and work-holding device onto the table.
Step 5   Use a pan to remove any quantity of chips after the cutting fluid has drained from them for a short time. Use a T-slot cleaner and hard bristle brush. Clean the T-slots and the well area at the ends of the table.
Step 6   Lock the spindle. Loosen the arbor nut with the arbor solid open-end wrench.
Step 7   Loosen the arbor support clamp. Slide the support from the overarm.
Step 8   Remove the nut, bearing sleeve, collars, key, and cutter from the arbor.

CAUTION: Protect the table from damage during this process.

Step 9   Clean all parts. Remove any burrs or nicks. Replace the collars, key, bearing sleeve, and nut on the arbor.
Step 10   Clean and store the cutter so that the teeth are protected.
          Note: Examine the cutter teeth. If its teeth are dulled or damaged, the cutter should be resharpened before storing.
Step 11   Remove the arbor. Store type A and B arbors in vertical racks.
Step 12   Loosen the nuts on the T-bolts of the work-holding device. Clean and examine the tongues and base. Remove any burrs. Store in the machine tender or other special storage compartment.
Step 13   Brush away chips from all other parts of the milling machine.
Step 14   Wipe the surrounding areas free of any cutting fluids.
Step 15   Wash the hands and arms if they have been in contact with the cutting fluid.

APPLICATIONS OF THE SURFACE GAGE

Many jobs require plain milling on a surface that is irregular, warped, or varies in thickness. In such instances, it is impractical to seat the part directly on parallels. Surfaces of castings, forgings, and uneven parts may be positioned (leveled) for machining by using a surface gage. The part may be held in a vise or it may be strapped to the table.

The surface gage may also be used for layouts. Guidelines may be scribed on the sides and ends of the workpiece. These layout lines are particularly helpful for determining whether the rough surface will *clean up* when the final finish cut is taken.

HOW TO USE A SURFACE GAGE

Leveling A Workpiece

Step 1   Wipe the table and surface gage. Carefully rub the palm of the hand over the table area and the base of the surface gage.
Step 2   Adjust the scriber and surface gage spindle. The scriber should be able to reach across each corner of the workpiece.
Step 3   Adjust the scriber so that it barely touches one corner of the work.
Step 4   Move the surface gage to the next corner. Slide the scriber to the work. Hold the base against the table. Swivel the surface gage to move the scriber near or over the work surface.
Step 5   Check the scriber height against the corner height. If it is higher than the scriber, tap the corner to the approximate height of the first corner.
Step 6   Check the height of the other corners. Tap where necessary to correctly position the workpiece.
Step 7   Use shims or jacks to support overhanging surfaces.
Step 8   Tighten the vise jaws.
Step 9   Recheck the height of each corner with the surface gage. Further adjust if needed.
          Note: Chalk sometimes is rubbed on the corners. The scriber is brought over the chalked surface. A light line is scribed.

The process is repeated on the other corners.

Note: A dial indicator may be used on smooth surfaces in place of the scriber. This permits more accurate setting of the work height.

Scribing Layout Lines

Step 1   Chalk or apply a dye to the ends and sides of the workpiece.

Step 2   Set the scriber point as close to the work as possible. The position must permit scribing on all required sides.

Step 3   Set the scriber to the required height (thickness of workpiece). Fig. 42-11 shows one method. The end of a steel rule is placed on a parallel. The scriber is adjusted to the required height (thickness of the workpiece).

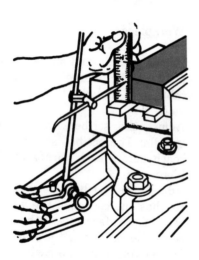

Fig. 42-11. Setting a surface gage to the required thickness of a workpiece.

Step 4   Hold the surface gage base against the table. Slide the surface gage so that the scriber is at an angle to the side to be scribed. Scribe the line as shown in Fig. 42-12.

Note: Care must be taken in handling the surface gage and scribing the line. It is good practice to check the height setting after each line is scribed.

Step 5   Repeat the process on the remaining sides if needed. Transfer the surface gage on the machine table to a place from which the lines may be scribed.

Note: If a side is not accessible, it may be

necessary to reset the spindle and scriber. The first scribed line may serve as a locating line.

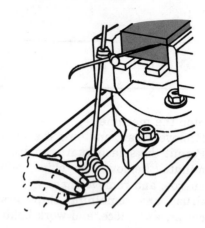

Fig. 42-12. Scribing a line at a required location.

Using Layout Lines To Position A Workpiece

Step 1   Mount the layed out workpiece in a vise, on an angle plate, or in another work-holding setup. Tighten the scriber point just far enough to permit adjustment.

Step 2   Position the scriber point so that it reaches both ends of the layout line.

Note: Fig. 42-13 shows how a workpiece is held and the horizontal line connecting the circles is checked.

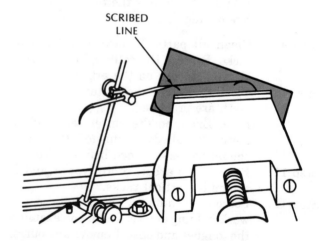

Fig. 42-13. Positioning a layed-out workpiece with a surface gage.

Step 3   Tap the *high end* of the workpiece. When the workpiece is correctly positioned, securely tighten the vise or clamps. Recheck the accuracy of the work position.

SAFE PRACTICES IN PLAIN MILLING SETUPS AND WORK PROCESSES

- Lock the spindle in the off (stopped) position so that it cannot be tripped accidentally during a machine setup.
- Feed the work in climb milling in the same direction in which the cutter rotates.
- Avoid climb milling with a standard miller unless it is rigidly constructed or is equipped with an automatic backlash eliminator.
- Allow enough material for a finish cut so that the cutter teeth can quickly penetrate the surface and remove a fine chip. Otherwise, the resultant burnishing effect will tend to dull the teeth.
- Take a deep first cut below the hard outer scale on castings, forgings, and other irregular or abrasive, rough surfaces.
- Protect the ground surfaces of parallels and vise jaws from being damaged by the rough surfaces of parts.
- Feed the milling cutter into the workpiece by hand. When it is evident that the part will not shift under the cutting force, engage *the power feed*.

- Mill as close as practical to the work-holding device and arbor support arm. The cutting forces should also be directed *toward* the spindle.
- Mill toward the solid vise jaw or the vertical leg of an angle plate.
- Lock the knee and saddle *before* engaging the table feed.
- Stop the machine if any of the common milling problems develop. Check the most probable cause(s). Take the suggested corrective steps.
- Observe standard safety practices for mounting and disassembling the machine and workpiece setups. Carefully clean the machine, work, and tools. Remove burrs. Store each tool and machine part in its correct rack, bin, or other protected location.
- Place rags and oily cloths in a *metal* container. The area around the machine should be dry. After wiping up oil on the floor, use a nonskid compound on the oil spots.

TERMS USED FOR LAYING OUT AND MILLING PLAIN SURFACES

Plain (peripheral milling)	Milling a flat horizontal surface parallel to the axis of the cutter.
Conventional (up) milling	Machining on a horizontal milling machine. Feeding a workpiece into the cutting edge of the rotating milling cutter. Milling a chip form that starts at the depth of the cut. The chip increases to full thickness when the cutting edge cuts through the top surface of the work.
Climb (down) milling	A machine cutting technique. Feeding a workpiece in the same direction that the milling cutter rotates. Milling a chip that is full thickness (equal to the feed) at the beginning of the cut. This diminishes to almost no thickness as the cutter tooth rotates to the depth of the cut.
Teeth marks	Deviations from a perfect plane surface. Crests and hollows corresponding to the feed of the milling cutter in conventional milling. Fine surface indentations resulting from a burnishing rather than a cutting action of a milling cutter.
Automatic backlash eliminator	A design feature of milling machines. A device that automatically compensates for lost motion or play, usually between a feed screw and mating nut or between a rack and pinion. A control feature that makes it possible to climb mill safely.
Cleaning (up) the surface	Machining a surface to just below any low spot or rough surface. Producing a flat plane with a minimum amount of milling.
Seating the work	Firmly securing a workpiece on a parallel, packing block, or other work-positioning device. Tapping the work down so that it rests securely on a parallel surface.
Squaring a block (milling machine)	Milling four sides of a square or rectangle. Milling four adjacent sides at 90° to each other. The opposite sides are parallel along the length.

SUMMARY

- Plain milling cutters are used for producing flat horizontal surfaces on the milling machine.
- Milling process may be performed by conventional (up) milling or climb (down) milling. The latter method is not recommended unless a heavy-duty machine or one equipped with an automatic backlash eliminator is used.
- Regardless of the method, the rotation of the cutter must be correct in relation to the feed.
- Climb milling has the following advantages over conventional milling:
  - Surface burrs are reduced or eliminated
  - The finished surface is almost free of cutter revolution and feed marks
  - Cutting forces are applied against the table or work-holding device
  - Thin-sectioned surfaces may be machined with a greater degree of dimensional accuracy
  - Less power is required.
- Conventional milling is recommended when high-speed steel cutters are to be used on castings, forgings, and other materials with a surface scale.
- Most workpieces that are to be plain milled are held in a standard or universal milling vise, on an angle plate, or in a fixture. Many parts are strapped on parallels directly to the table.
- Workpieces are positioned so that the cutting force is directed toward the *solid* vise jaw, the work stop, the fixture, or the angle plate.
- Time is saved by using the smallest possible cutter diameter.
- Rough and finish cuts should be taken to produce a high-quality, dimensionally-accurate, finished surface.
- From 0.015″ (0.4 mm) to 0.020″ (0.5 mm) is usually allowed for a finish cut. The cutter speed is increased. The feed is decreased.
- The cutter is hand fed for a short distance. This permits checking the rigidity of the entire setup and measuring the workpiece. Power feed is then applied.
- The finish milled surface is held against the ground solid vise jaw. Adjacent machined surfaces are positioned on parallels.
- Chatter, holding to size, burnishing, and quality of surface finish are common workpiece problems. Others—like rapid dulling, burning, and breaking the cutting edges—relate to the cutter. Vibration is a machine problem. Probable causes and corrective steps are provided in table form for easy reference.
- The surface gage may be used to set up the workpiece or to scribe horizontal layout lines.
- Vertical surfaces may be aligned by using a solid-beam steel square or a dial indicator.
- Ends of workpieces may be face, side, or plain milled.
- Ground faces of parallels, vise jaws, and angle plates, and the milled surfaces of the workpiece all must be protected.
- Personal and machine safety rules governing the use of the milling machine, the cutting fluid system, and the cutters must be followed.
- A protective shield or goggles should be worn at all times during any machine process.

UNIT 42 REVIEW AND SELF-TEST

1. Explain why *climb milling* is preferred over *conventional milling* under the following conditions:
   a. Machining deep, narrow slots
   b. Cutting off or machining thin, hard-to-hold workpieces
   c. Eliminating surface burrs and feed marks.
2. Indicate the five major functions the operator performs in setting up a milling machine.
3. List three considerations that guide the operator in selecting the appropriate milling cutter.
4. State what advantages a *vibration damping unit* has over the *standard overarm* in controlling chatter.
5. Identify four practical methods of milling ends square.
6. Tell how the surface gage is used to position a rough-surfaced part that is held in a milling machine vise.
7. Indicate why a deep first cut is taken below the hard outer scale on castings, forgings, and weldments.
8. State two safe practices to observe in climb milling.

Unit 43

FACE MILLING ON THE HORIZONTAL MILLING MACHINE

Three common methods of face milling on the horizontal milling machine are covered in this unit. These include applications of inserted-blade face mills, shell end mills, and solid-shank end mills. Additional technical information about milling cutters and their cutting actions is presented. Step-by-step procedures are given for milling shoulders into the face of a workpiece. The shoulders are produced with shell end mills and solid end mills. Face milling applied to these milling cutters is referred to as *end milling*.

APPLICATIONS OF FACE MILLING CUTTERS AND END MILLS

Solid end mills are generally used to mill flat surfaces that are smaller than 1½" (38 mm) wide. The range of commercial solid end mills is from 1/8" (3 mm) to 2" (50 mm) in diameter. Shell end mills are normally employed to face mill surfaces from approximately 1¼" (32 mm) to 5" (130 mm) wide. Face milling cutters are used in machining surface areas that are wider than 5" (130 mm).

Face Milling Cutter Design Features

The sizes of face milling cutters require that the body be designed of a tough, durable, heat-treated steel. Grooves are cut to accurately position and hold blade inserts. The inserts are made primarily of high-speed steel and cemented carbides. The inserts may be ground, adjusted, and replaced easily and at low cost. This design makes it possible to use the body indefinitely. Fig. 43-1 shows some of the design features. The names of the important ones are listed.

Rake Angles

A face milling cutter is selected in terms of the material to be machined, the nature of the work processes, and the machine setup. Reference is made to manufacturers' tables of rake angles. From these the operator establishes whether to use *positive, zero,* or *negative radial* and *axial rake angles*. The application of a 5° positive and a 5° negative rake

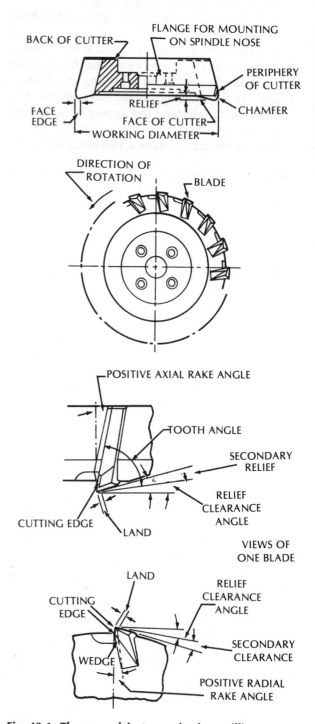

Fig. 43-1. The general features of a face milling cutter.

540

angle are pictured in Fig. 43-2A and B.

Positive rake angles are normally used on high-speed steel cutters. These rake angles are effective for milling tough materials and others that *work harden*. Other advantages of positive rake angles include:

- High quality of surface finish
- Increased cutter life
- Less power required in cutting
- Decreased cutting forces.

Zero and *negative rake angles* are applied to cemented carbide inserts. These rake angles strengthen the cutting edges to withstand heavy impact and other severe cutting conditions. Negative rake angles are not recommended for soft, ductile metals like copper and aluminum or work-hardening materials. Negative rake angles should be used only when the workpiece and machine setup can withstand extreme cutting forces.

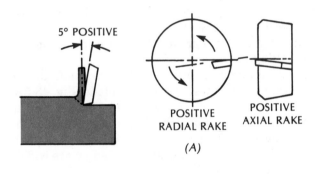

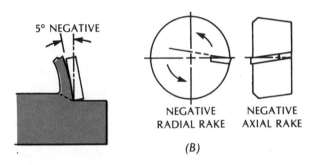

Fig. 43-2. The effects of positive and negative radial rake on cutting.

EFFECT OF LEAD ANGLE AND FEED

Lead Angle

The same principle of lead angle that was explained in lathe work for a single-point cutter applies to face milling. Small lead angles of from 0° to

3° are often used to machine close to a square shoulder. Such lead angles are provided with sufficient clearance to prevent the cutter face from rubbing. With a limited lead angle, Fig. 43-3A shows that a chip thickness of 0.015″ (0.4 mm) is practically

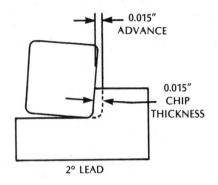

(A) SMALL LEAD ANGLE. FEED AND CHIP THICKNESS ARE EQUAL

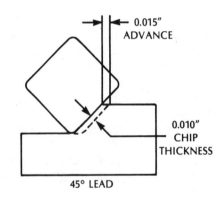

(B) LARGE LEAD ANGLE. SMALLER CHIP THICKNESS, LARGER WIDTH OF CHIP

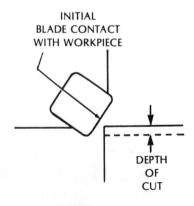

(C) ADVANTAGE OF SIZABLE LEAD ANGLE

Fig. 43-3. The effect of lead angle on chip thickness.

equal to the feed of 0.015″ (0.4 mm).

Using the same depth of cut but increasing the lead angle to 45° changes the chip thickness. Fig. 43-3B illustrates that with a 45° lead angle and a 0.015″ (0.4 mm) feed the chip thickness in this example is 0.010″ (0.25 mm). Note, too, that as the lead angle increases, the chip becomes wider. The steep lead angle limits the depth of cut. For practical purposes, the maximum lead angle on face milling cutters is 30°.

There are a number of advantages to using a sizable lead angle (Fig. 43-3C).

● The cutter contacts the workpiece along the blade rather than at the tip of the cutting edge.

● Cutting forces are applied where the strength of the blade is greater than at the edge.

● Only a partial chip is formed on initial impact. The same thing happens at the end of the chip.

● Cutting a thinner chip thickness adds to effective cutting edge life.

Feed And Cutter Lines

A circular pattern of cutter feed lines may be produced in face milling. *Feed lines* result from:
● The size of the nose radius of the cutter blades
● The coarseness of the feed
● Limited rigidity in the setup
● The nature of the machining process
● The cutter size
● The sharpness of the cutting edges.

An exaggerated rough surface finish is shown in Fig. 43-4A. The ridges produced by the face mill may be reduced by grinding each blade flat (Fig. 43-4B). The flat should be wider than the feed.

FACTORS AFFECTING FACE MILLING PROCESSES

The following major conditions affect dimensional accuracy and quality of surface finish:
● Free play in the table movements
● Eccentricity or cutter run-out. The trailing edges of the cutter produce light cuts that follow the main cut
● Lack of support of the workpiece at the point of cutting action
● Need for more rigid stops. Additional stops may be needed to prevent any movement of the workpiece during cutting
● Position of the table. The table must be positioned as close as possible to the spindle
● Dull cutting edges. These require more power and produce excessive heat
● Cutter revolving at too slow a speed. This causes a buildup on the cutting edge
● Cutter revolving at too high a speed. This produces excessive cutter wear
● Cutter feed is too great. This causes chipping and breaking of the cutting edges
● Positive angle of entry. Contact is made at the cutting tip (weakest area) of the blade (Fig. 43-5A)
● Negative angle of entry (Fig. 43-5B). The initial force is applied along the face and away from the cutting edge. The cutting face is stronger in this area. Zero and negative angles of entry are widely used with cemented carbides
● Failure of the cutting fluid to flood the cutting area. The intermittent cutting process and the speed of carbide face mills produce a *fanning action*. A

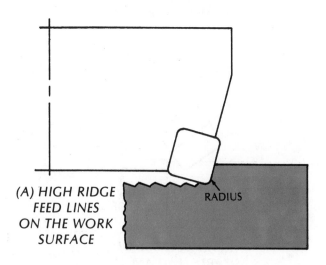

(A) HIGH RIDGE FEED LINES ON THE WORK SURFACE

RADIUS

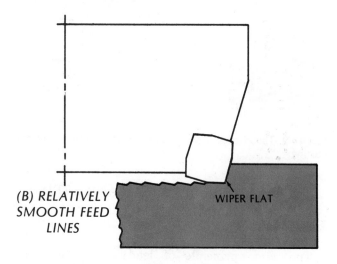

(B) RELATIVELY SMOOTH FEED LINES

WIPER FLAT

Fig. 43-4. The effect of the tooth form on surface finish.

noncontinuous cooling action produces thermal (heat) cracks in the blades. This problem is overcome by forcing the cutting fluid through a fine spray (mist) to continuously reach the cutting edges.

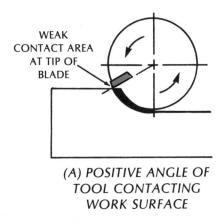

(A) POSITIVE ANGLE OF TOOL CONTACTING WORK SURFACE

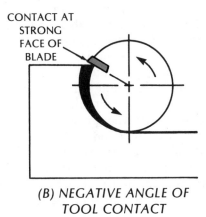

(B) NEGATIVE ANGLE OF TOOL CONTACT

Fig. 43-5. Strong and weak areas of contact between the cutting tool and the workpiece.

GENERAL CONSIDERATIONS IN FACE MILLING

Workpieces are held in such common work-holding devices as a plain or universal vise, an angle plate, or a fixture. Many parts are strapped directly to the milling machine table or a rotary table. Work stops help prevent shifting during cutting.

Solid end mills and shell end mills are usually held in a Type-C arbor. The smaller solid end mills are first mounted in a sleeve, adapter, or collet. The larger face mills have a recessed area on the back side and are screwed into the spindle nose for direct mounting.

In good shop practice, the face milling cutter should be larger than the width of the surface to be milled. The amount varies all the way from 1/4″ to 1″ or more, depending on the width of the surface. Fig. 43-6 shows the relationship between the diameter of a large inserted-tooth face miling cutter and the width of a workpiece.

The rake, clearance angles, and tooth relief are based on the nature of the workpiece and cutter materials, the job requirements, and the machine setup. Machining should take place with maximum rigidity.

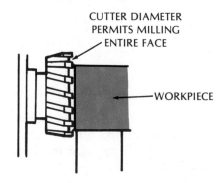

Fig. 43-6. The relationship of an inserted-tooth face milling cutter to the width to be machined.

FACE AND SHOULDER MILLING WITH A SOLID END MILL

Two basic processes may be performed with solid end mills. A plane surface may be milled, or two surfaces may be produced at the same time. The right angle surfaces are referred to as a *step* or *shoulder*. Solid end mills with two or more flutes may be used for face milling. The same cutters produce a shoulder when the face and peripheral teeth are set to cut the two surfaces at one time.

Chip Thickness And Rate Of Feed

Considerable judgement must be exercised by the operator in terms of cutting speeds, cutting feeds, and depth of cut. Table 43-1 provides general recommendations of feed per tooth for different end mill diameters and materials. These are a guide. In actual practice, the feed may be increased or decreased. The feed depends on the usual factors that affect cutter efficiency, dimensional accuracy, surface finish, and the design of the cutting tool itself.

The small sizes of end mills are comparatively fragile. Excessive feeds cause tool breakage and chipping at the edges of the teeth. The rate of feed for

Table 43-1. GENERAL FEEDS FOR END MILLS (0.000″ Feed Per Tooth, HSS)

Diameter of End Mill	Steel				Cast Iron	Nonferrous Metals		
	Low Carbon	High Carbon	Medium Hard Alloy	Stainless		Aluminum	Brass	Bronze
⅛	.0005	.0005	.0005	.0005	.0005	.002	.001	.0005
¼	.001	.001	.0005	.001	.001	.002	.002	.001
⅜	.002	.002	.001	.002	.002	.003	.003	.002
½	.003	.002	.001	.002	.0025	.005	.003	.003
¾	.004	.003	.002	.003	.003	.006	.004	.003
1	.005	.003	.003	.004	.0035	.007	.005	.004
1½	.006	.004	.003	.004	.004	.008	.005	.005
2	.007	.004	.003	.005	.005	.009	.006	.005

solid end mills is calculated by substituting the number of flutes for the number of teeth in the formula. Slower feeds are used for deeper cuts.

Cutting Speeds

Tables of cutting speeds, mentioned previously for other machining processes, apply equally to end mills. The cutting speeds given in such tables are used to establish the spindle and cutter rpm.

Holding Solid End Mills

Taper-shank solid end mills that have a tang are held in a tang drive collet (often called a *sleeve*). A threaded end shank is held in a plain collet adapter. Straight-shank single and double end mills are usually mounted in a spring chuck adapter. The adapters are then secured in a Type-C arbor.

Three cutter-holding adapters are illustrated in Fig. 43-7. The setup for a tang drive is illustrated in (A). The application of the split collet with a threaded end mill is shown in (B). The spring chuck adapter in (C) holds a straight-shank end mill. Fig. 43-8 shows how an end mill adapter, reducing collet, and taper-shank end mill are held for mounting. The setup is locked into the quick-change adapter on the spindle nose by a partial turn of the clamp ring.

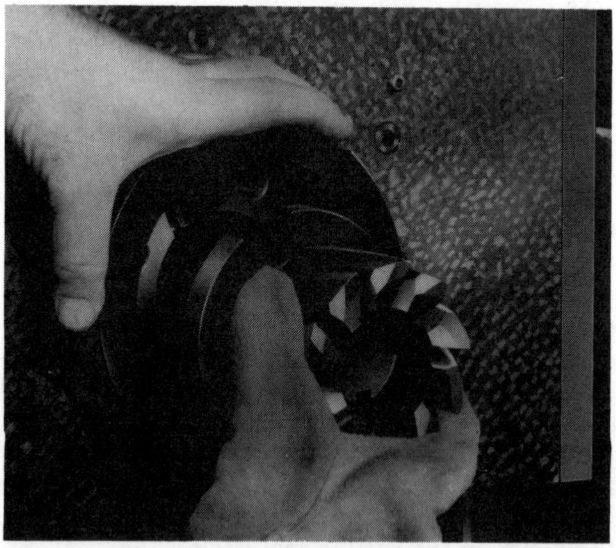

Fig. 43-8. Mounting and locking an end mill and adapter into a quick-change spindle nose adapter. (Courtesy of Cincinnati Milacron)

HOW TO FACE MILL WITH A FACE MILLING CUTTER

Step 1 Read the blueprint or sketch of the part. Determine the job requirements in terms of the work-holding device, type of cutter, spindle speeds, feeds, and cutting fluid.

Step 2 Remove any burrs on the workpiece. Locate it centrally (lengthwise) on the table. Overhang the face to be milled. This may project about 1/4″ beyond the final milled size.

Step 3 Place a stop block(s) against the workpiece to offset the cutting forces.

Step 4 Check the alignment (parallelism) of the face to be milled.

Note: The face may be positioned by measuring the distance it overhangs the

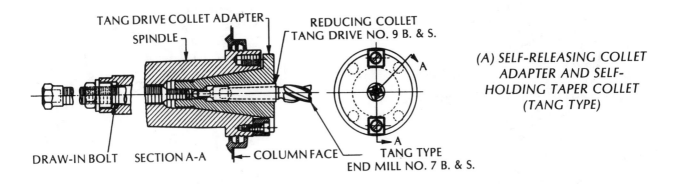

(A) SELF-RELEASING COLLET
ADAPTER AND SELF-
HOLDING TAPER COLLET
(TANG TYPE)

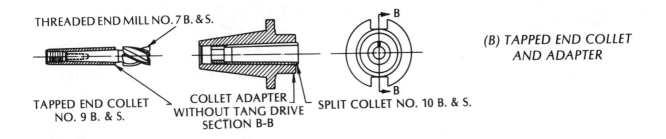

(B) TAPPED END COLLET
AND ADAPTER

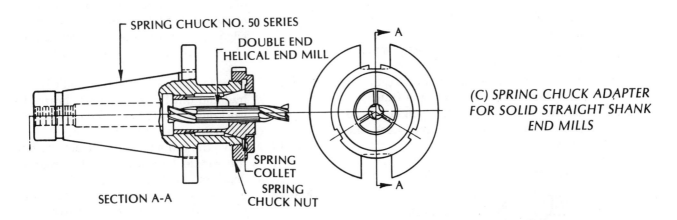

(C) SPRING CHUCK ADAPTER
FOR SOLID STRAIGHT SHANK
END MILLS

Fig. 43-7. Mounting and driving straight-and taper-shank solid end mills. (Courtesy of Cincinnati Milacron)

table. A steel rule or dial indicator may be used, depending on the accuracy required.

Step 5 Tighten the vise, fixture, or clamps to secure the workpiece.

Step 6 Select a cutter. The diameter should measure at least 1" larger than the face to be milled (Fig. 43-9A). Clean the spindle nose (Fig. 43-9B) and the back flange of the cutter.

Step 7 Hold the cutter with both hands and mount it on the spindle nose (Fig. 43-9C). Start the fastening screws and securely tighten them (Fig. 43-9D).

Step 8 Calculate the spindle rpm. Set the speed dial or gear-change levers at the closest lower rpm (Fig. 43-10A). Check the direction of spindle rotation.

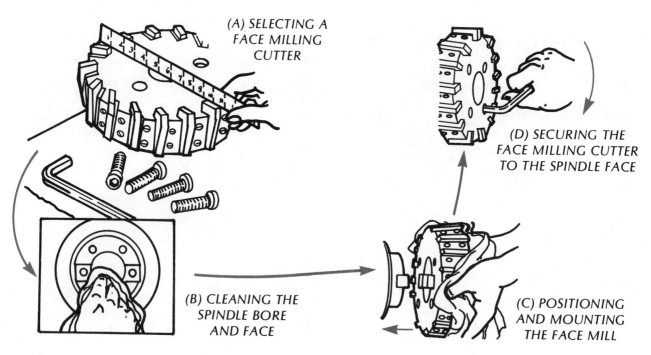

(A) SELECTING A FACE MILLING CUTTER

(B) CLEANING THE SPINDLE BORE AND FACE

(C) POSITIONING AND MOUNTING THE FACE MILL

(D) SECURING THE FACE MILLING CUTTER TO THE SPINDLE FACE

Fig. 43-9. Mounting an inserted-tooth face milling cutter.

Step 9 Calculate the rate of feed. Set the feed dial or feed-change lever to the required feed (Fig. 43-10B).

Step 10 Set the trip dogs. The distance should equal the length of the surface to be milled plus the diameter of the cutter, plus an extra 1/2″ to 1″ at each end.

Step 11 Check the cutting fluid. Position the distributor to flow (or spray on a cemented carbide cutter) the cutting fluid.

Step 12 Center the cutter at the height of the centerline of the workpiece. Lock the knee.

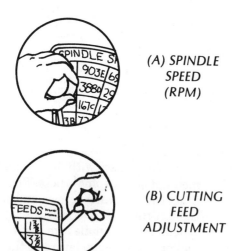

(A) SPINDLE SPEED (RPM)

(B) CUTTING FEED ADJUSTMENT

Fig. 43-10. Setting the speed and feed controls.

Note: If the workpiece is wider than the cutter diameter, the distance of the first cut should be not more than three fourths of the cutter diameter. About one fourth of the cutter should extend above the top edge of the workpiece. The second cut to *clean up the face* must overlap the first cut.

Step 13 Start the spindle. Feed the saddle inward by hand until the cutter just grazes the workpiece. Set the crossfeed micrometer collar at zero. Lock the collar.

Step 14 Move the table to clear the cutter. Turn the saddle handwheel to the required depth of cut. Lock the saddle.

Step 15 Check the cutter rotation. A downward cutting force should be produced.

Step 16 Start the flow of cutting fluid. Feed the face mill by hand. Take a trial cut for a short distance.

Step 17 Stop the spindle and the coolant flow. Check the workpiece for size. Make whatever adjustment is needed.

Step 18 Start the machine and the coolant flow. Engage the power feed. Take the cut across the face of the workpiece.

Step 19 Stop the feed, cutting fluid, and spindle. Return the cutter to the starting position. *Note:* Unlock the knee and move the workpiece up to overlap any second cut that may be required.

Step 20   Increase the speed and decrease the feed for a finish cut. Follow the previous steps to position the cutter for depth, to take a trial cut, to measure, and to mill the finished surface.

Note: The cutter should be changed if it becomes dulled and the required accuracy cannot be held.

HOW TO DISASSEMBLE THE WORK SETUP AND CUTTER

Step 1   Stop the spindle and coolant flow. Lock the spindle.

Step 2   Move the table so that the workpiece and cutter clear each other.

CAUTION: Observe personal, tool, and machine safety procedures in brushing and removing chips and wiping all surfaces.

Step 3   Remove the workpiece. Burr the edges. Move the part to the bench or other work place for further processing.

Step 4   Store all clamps, blocks, bolts, etc., in their proper places.

Step 5   Loosen each screw that holds the face milling cutter to the spindle.

CAUTION: Be sure the cutter is cool enough to be safely handled. Then, place a wiping cloth over the face of the cutter to hold it.

CAUTION: A wooden cradle or tray should be laid on the table under the cutter. A heavy cutter is often placed on a protective tray when it is removed from the spindle nose.

Step 6   Apply a force against the cutter to keep it on the spindle nose. Loosen the fastening screws by hand and remove them. Then, with two hands, move the cutter off the spindle.

Step 7   Clean the cutter. If the teeth need sharpening, report the condition so that the cutter may be reground. Otherwise, return the cutter and screws to the toolroom or store them in an appropriate rack and tray.

Step 8   Wipe the spindle. Return all tools and accessories to their proper storage places.

Step 9   Wipe any oil spots on the floor. Clean the remaining areas of the machine. Apply a corrosion-preventive spray if needed.

Leave the machine clean for the next setup.

HOW TO FACE (END) MILL WITH END MILLS

Face (End) Milling With Solid Or Shell End Mills

Step 1   Set the workpiece in the center area of the table. Provide minimum overhang of the face to be milled. Align the face and secure the workpiece in position.

Step 2   Select the type and largest practical size of end mill that meet the job requirements. Use the appropriate arbor or adapter and split collet, reducing or spring collet, and spring chuck.

Step 3   Secure the Type-C arbor (Fig. 43-11A), adapter, or spring chuck in the spindle nose.

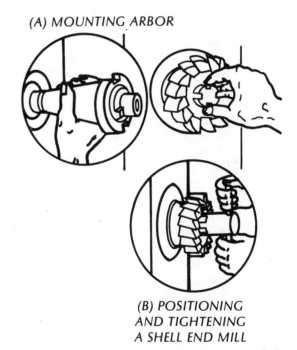

(A) MOUNTING ARBOR

(B) POSITIONING AND TIGHTENING A SHELL END MILL

Fig. 43-11. Mounting a shell end arbor and securing a shell end mill.

Step 4   Assemble the end mill.

Note: A shell end mill is mounted and held directly with a locking screw as shown in (B). A straight-shank solid end mill may be tightened on an adapter or held in a spring collet. Taper-shank solid end mills are held in a tang drive or tapped end collet.

Step 5   Compute the cutter rpm from the known information about the cutting speed and the cutter diameter. Set the spindle speed.

Step 6   Determine the initial feed rate. Set the feed control dial or levers. Position and secure the trip dogs.

Step 7   Position the distributor head of the coolant system so that the required flow will be directed on the workpiece and cutter.

Step 8   Move the face of the end mill so that it is located near the starting place of the cut. Center the cutter at the centerline of the workpiece. Lock the knee.
Note: If two or more cuts are required because the face width is larger than the cutter diameter, follow the same procedures as in face milling.

Step 9   Determine the depth of the cut or cuts. Start the machine. Move the cutting face of the end mill toward the workpiece until it just touches the face.

Step 10   Set the crossfeed micrometer collar at zero. Move the saddle in to the depth of the cut. Lock the saddle.

Step 11   Start the flow of cutting fluid if it is required. Feed the cutter by hand for a short distance. Then move the cutter away from the workpiece.

Step 12   Stop the spindle. Measure the workpiece. Make any necessary adjustments.

Step 13   Start the spindle. Engage the longitudinal power feed. Take the cut.
Note: Some faces are milled to a required length that ends at a shoulder. In such cases, disengage the power feed near the end of the cut. Feed the cutter by hand to the layout line. Lock the table. Unlock the knee. Carefully feed the cutter down, then up. This step removes the small wedge-shaped area. The wedges are produced by the circular form milled by the cutter and the bottom and top edges of the workpiece.

Step 14   Stop the spindle. Return the end mill to the beginning of the cut.

Step 15   Repeat the steps for feeding the cutter to depth, measuring, and taking successive cuts.
Note: Increase the speed and decrease the feed to produce a precision finish cut.

Step 16   Stop the machine. Move the work clear of the cutter. Clean the cutter, workpiece, and machine.

Step 17   Remove the cutter and adapter or arbor. Place them on the machine tender.

Step 18   Disassemble the workpiece and the setup.

Step 19   Burr the workpiece. Return all parts, tools, and accessories to their proper storage places. Leave the machine clean and the floor area free of any spilled cutting fluid.

HOW TO MILL A SHOULDER WITH AN END MILL

Step 1   Select the largest diameter solid or shell end mill that is suitable for the job.
Note: The solid end mill is preferred for small-dimensioned shoulders. A shell end mill is more practical for large shoulders where a heavier cutter is required.

Step 2   Set up the workpiece. Mount the cutter as close to the spindle as practical. Set the trip dogs, speed, feed, and coolant system nozzle.

Step 3   Position the cutter near the workpiece. Start the spindle. Bring the cutter to the workpiece until it just touches the face.

Step 4   Set the cross slide micrometer collar at zero. Move the cross slide in to the depth of the first cut. Lock the saddle.

Step 5   Move the end mill over the workpiece for a short distance. Turn the knee handwheel to bring the workpiece up. Continue until the end mill just skims the surface. Set the knee micrometer collar at zero.

Step 6   Position the workpiece at the start of the cut. Bring the knee up to the depth of the shoulder (Fig. 43-12A).

Step 7   Take a trial cut for a distance equal to at least half the diameter of the end mill (Fig. 43-12B). Stop the spindle. Measure the depth and height (Fig. 43-12C).
Note: The width and depth may be checked against layout lines. Otherwise, a steel rule may be used as illustrated. More accurate measurements are taken with a micrometer depth gage.

Step 8   Stop the spindle at the end of the cut. On long workpieces, use rapid traverse to return the end mill to the starting position.

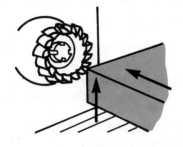

(A) SETTING THE SHELL
END MILL TO THE
REQUIRED DEPTH
AND WIDTH

(B) TAKING A TRIAL
CUT FOR A SHORT
DISTANCE

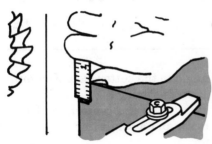

(C) MEASURING
THE DEPTH
AND WIDTH
WITH A
STEEL RULE

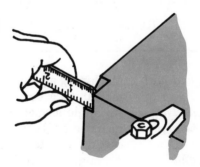

43-12. Trial cut milling a shoulder with a shell end mill.

Step 9   Reset the end mill for other roughing or finish cuts. Lock the saddle and knee in each case. Take a trial cut. Measure the shoulder. Make adjustments. Then take the finish cut.

Step 10 Stop the machine. Follow standard personal, tool, and machine safety practices for disassembling the setup, removing burrs, cleaning up, and properly storing all items (Fig. 43-13).

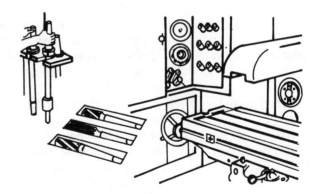

43-13. Returning arbors, adapters, cutters, and work-holding accessories to their proper storage places.

SAFE PRACTICES IN FACE, END, AND SHOULDER MILLING

- Use zero and negative rake angles on cemented carbide insert face mills to mill under heavy impact and severe forces.
- Increase the lead angle of the cutting edge (up to 30°) to reduce chip thickness for heavy feeds.
- Decrease the feed rate and/or grind a wider flat on the cutting edge if feed lines are produced.
- Lock the knee and saddle before face milling.
- Handle the cutter with a cloth, and only when the cutter is cool enough.
- Use a negative angle of entry when the cutting conditions require that the point of impact be in back of the cutting edge.
- Change to a fine spray to deliver cutting fluid continuously over cemented carbide cutter blades and the cutting area.

- Use work stops and heavy-duty strap clamps to rigidly hold a workpiece for deep cuts and coarse feeds.
- Control the cutting force on small-diameter solid end mills. This is to avoid fracture or chipping of the teeth.
- Remove all wiping cloths from the machine before it is placed in operation.
- Place a tray or wooden cradle under large face mills when mounting or dismounting them. This helps prevent damage to the cutter or table.
- Engage the power feed only after it is established that the workpiece is supported rigidly enough to withstand the cutting forces.
- Stop the spindle and lock it during setup, measurement, and disassembling steps.
- Follow standard personal, tool, and machine safety precautions.

TERMS RELATED TO FACE AND END MILLING CUTTERS AND PROCESSES

Face milling; end milling	Machining a plane surface at a right angle to the cutter axis. End milling refers to machining a small-width surface with an end mill.
Positive, zero, and negative radial rake angle	An angle formed by a tooth face in relation to the centerline of the cutter. The included angle formed by the cutter tooth and a centerline that extends from the cutting edge to the center of the cutter.
Positive, zero, and negative axial rake angle	An angle formed by the cutting tooth and the axis of a cutter.
Blades (inserts)	Separate cutter teeth of a face milling cutter. Blades that may be adjusted, reground, or replaced.
Feed and cutter lines	Arc-shaped fine indentations in a face-milled surface. Surface imperfections usually machined when the nose radius and cutting edge are smaller than the rate of feed.
Lead angle effect	Differences in chip thickness and width produced by varying the angle of the cutting edge. The effect on cutter efficiency resulting from changing the lead angle of the cutting edge.
Cutter runout	An eccentric cutter motion. A condition that causes score lines (imperfections) in machining.
Positive and negative angles of entry	A positive or negative angle formed by the face of the cutting tooth and the surface of the workpiece at initial entry.
Mist spray	Flooding a cemented carbide face milling cutter with a fine spray of cutting fluid. A technique of cooling and lubricating a high-speed carbide cutter that otherwise would fan out the cutting fluid.
Overhang	Positioning a workpiece so that the face extends a minimum amount beyond the side of the table. Extending a workpiece a small distance more than the depth of the cuts to be taken.

SUMMARY

- Surfaces are face milled using three common types of milling cutters. Taper- and straight-shank end mills up to 2″ (50 mm) diameter are used to face (end) mill small surfaces. Shell end mills up to 6″ (150 mm) diameter are used to machine surfaces up to about 5″ (125 mm) wide. Inserted-tooth face mills are used for heavy-duty and wide-surface face milling.

- The two basic cutter materials are high-speed steel and cemented carbides.

- Face milling cutters are available with blade inserts. These may be formed with positive, zero, or negative radial and/or axial rakes. Rake is determined by the type of material to be cut, the processes, the work setup, the operating speed, and the feed. Generally, zero and negative rakes are used on cemented carbide inserts for machining under severe conditions.

- The lead angle of the cutting edge affects the chip thickness and cutting efficiency. Lead angles up to 30° are practical.

- Feed and cutter lines may be reduced or practically eliminated. Corrective steps include:
  - Strengthening the setup so that it is more rigid
  - Decreasing the feed or increasing the speed
  - Using a sharp cutter. (Reduce the nose radius and grind the face cutting edge with a flat that it wider than the feed.)

- Use the largest possible diameter of solid end mill. The diameter of a face mill or solid or shell end mill must be larger than the width of the surface to be milled.

- End mills may also be used to cut a shoulder. Care must be taken in setting the cutter for depth and rate of feed. These should not exceed values suitable for the strength of the cutter teeth and body.

- Face milling cutters are mounted directly on the spindle nose. Shell end mills are held on a Type-C arbor. Straight-shank solid end mills are generally chucked in a collet. Taper-shank end mills are first inserted in a tanged or split tapered sleeve (collet).

- Stops should be used to prevent any movement of the workpiece. A trial cut is taken, using hand feed, to test the rigidity of the setup and the cutting action. Then, the power feed is engaged.

- The milling machine operator must make judgments about increasing or decreasing the speed, feed, or depth of cut.

- Special attention must be given to the use of a mist spray. It is needed if the *fanning action* of a cemented carbide cutter prevents the cutting fluid from reaching the cutting teeth and machining area.

- Face milling cutters are heavy and difficult to handle. A wooden cradle is used under the cutter to help in mounting or disassembling.

- Personal, tool, machine, and work safety precautions must be followed. All items should be returned to their proper storage places. The machine must be left clean and ready for the next job.

UNIT 43 REVIEW AND SELF-TEST

1. Give three advantages to using a *steep lead angle* (to a maximum of 30°) on *face milling cutters*.
2. State five factors that affect dimensional and surface finish accuracy when face milling.
3. Identify three *cutter-holding adapters* for straight and/or taper shank *solid end mills*.
4. Give the steps to follow in safely mounting a *face milling cutter*.
5. Tell how to position a face milling cutter to mill an area that is wider than the cutter diameter.
6. Set up the series of steps to follow to *end mill a shoulder* with *a solid or shell end mill*.
7. List three safety precautions to take when disassembling a face milling setup.
8. State two safe practices to follow in end milling with *small diameter end mills*.

Unit 44

SIDE MILLING CUTTER APPLICATIONS

Side milling cutters are widely used on horizontal milling machines. Ends and parallel steps and grooves are produced with various types of side mills. When two parallel surfaces are milled by two cutters mounted on the same arbor, the process is called *straddle milling*. Two or more side milling cutters *(side mills)* may also be mounted with other types of cutters on the same arbor. This *gang* of cutters mills at the same time. The process is known as *gang milling*.

This unit deals with applications of single and multiple side milling cutters, straddle milling, and gang milling.

SETUPS AND APPLICATIONS OF SIDE MILLING CUTTERS

Side milling cutters are of the following three basic designs:
- The *half side milling cutter* has teeth on one side and on the periphery. One, two, or three surfaces may be cut at one time.
- *Full side milling cutters* with staggered teeth are used for milling narrow, deep grooves.
- *Interlocking cutters* are ground with cutting teeth on one side and on the periphery. The combination of two interlocking cutters permits the simultaneous milling of three surfaces, like those of wide grooves.

Half Side Milling Cutter Applications

Side Milling An End Surface Or Step—One of the common operations for a half side milling cutter is milling an end square. The workpiece is usually held in a vise. The vise is accurately aligned parallel to the arbor axis. The half side milling cutter is used with a Type-A short arbor. The diameter of the cutter must permit milling the face in one cut. In other words, the *cutter must clear the work setup* (Fig. 44-1).

The cutter is located and secured on the arbor. As usual, the position of the cutter and the cutting action should take place as close to the spindle as is practical. Care must be taken to tighten the arbor support so that it fits snugly against the end of the

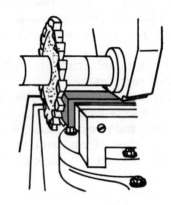

Fig. 44-1. The width of the workpiece must be within the cutting range of the side milling cutter.

arbor. Unless otherwise lubricated, the arbor pilot and side bearing should be oiled before starting any cut.

When the cutter is mounted close to the spindle, or if the workpiece is shorter than the width of the vise, the part must be turned 180° to face the second end. Fig. 44-2 shows how the workpiece is turned and the cutter is set with a steel rule. There are many cases where both ends of the workpiece are milled parallel at the one setting. This setup is described under full side milling processes.

Fig. 44-2. After the workpiece is turned 180°, the cutter is positioned for work length with a rule.

Square Steps Or Shoulders—The half side milling cutter is also used to mill square steps or shoulders. The side teeth mill the vertical face. The peripheral teeth mill the bottom surface. A roughing cut is usually taken, followed by a finish cut. Cutter rpm, cutting speeds, and feeds are determined in the same manner as for plain and face milling.

(A) MILLING FIRST END

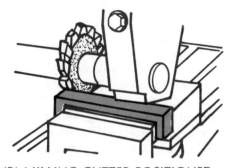

(B) MILLING CUTTER POSITIONED TO MILL SECOND END

Fig. 44-3. A full side mill adapted to mill two ends of a workpiece.

Full Side Milling Cutters

Milling Two End Surfaces—The full side milling cutter is more versatile than the half side mill. With it the setup for milling one end of a workpiece can also be used to mill the second end at the one setting. The cutter in this instance is located farther along the arbor support to permit machining with each cutting face. Fig. 44-3A shows a full side mill positioned to mill one end of the workpiece. After this cut, the cutter is moved to the other end (Fig. 44-3B).

The cutter may be set with a steel rule. The part also may be machined to a more precise linear measurement by positioning the cutter with the aid of the cross slide micrometer collar. The distance the work is moved is equal to the required dimension plus the width of the cutter. It is good practice to move the workpiece beyond this distance.

Backlash is taken out of the cross feed screw. The saddle is then moved back to within 0.004″ to 0.006″ (0.1 mm to 0.2 mm) of the required linear dimension. The saddle is locked. A trial cut is taken. The length is measured by micrometer. This procedure helps correct any machining error that may be produced by the cutter.

Milling A Step—A half side or full side milling cutter may be used to mill a step. The cutter is positioned to cut the vertical face to a required size. The saddle is locked at this position. The knee is then raised until the cutter is set at depth. The knee is also locked. The settings may be made by using the micrometer collars or according to layout lines on the workpiece. The setup is rechecked before any cut is taken. The arbor at final depth of cut must clear the workpiece, strap clamps, T-bolts and nuts, or any other projection.

The diameter of the cutter depends on the depth of the step. As a rule, the smallest diameter of cutter that provides adequate clearance to perform the operation is selected.

Milling A Groove—The sides and depth of a groove are usually layed out. The workpiece is then secured in a vise or strapped directly on the table. A full side milling cutter is used because groove milling requires cutting on the two sides and face of the cutter. As explained before, a staggered tooth side milling cutter is used for machining a deep groove.

If the groove is wider than the cutter, it is roughed out. From 0.010″ to 0.015″ (0.2 to 0.4 mm) is left on all sides for a finish cut. Climb milling is recommended when the cutter groove and cutter size are the same. It is easier to hold to a dimensionally accurate groove by climb milling under these conditions than by conventional milling.

It is important in milling a groove to have an adequate stream of cutting fluid. The chips must be washed out of the teeth and workpiece.

Straddle Milling With Side Milling Cutters

A *straddle milling* setup requires two side milling cutters and spacing collars. Spacing collars that correspond in length to the required dimension of the workpiece feature are placed between the inside hub faces of the cutters (Fig. 44-4). If necessary, a spacing collar may be ground to a needed size.

A typical cutter and work setup for straddle milling both ends of a workpiece is illustrated in Fig. 44-5. The accuracy to which the width is milled depends on how true the cutters run, how well the teeth cut, the removal of chips between the side face

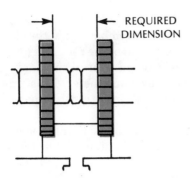

Fig. 44-4. Spacing collars produce the required width between side milling cutters.

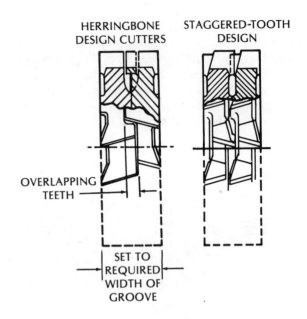

Fig. 44-6. Application of interlocking side milling cutters to machine a groove to an accurate width.

of the cutter and the workpiece, the nature of the cut, the cutting fluid (if required), and the work setup.

Many square, hexagonal, and rectangular parts are machined by straddle milling. This is particularly so with round parts. These are usually held in a chuck on a dividing head. After each cut, the workpiece is accurately positioned by indexing. The positions may also be set when a part is held on a rotary table. The table is rotated to the next angular position.

In some straddle milling operations, like the one illustrated in Fig. 44-4, two steps are milled. The cutters machine to the required width and depth in one cut.

Fig. 44-5. The cutter and work set up for straddle milling two ends of a workpiece.

Interlocking Side Milling Cutters

Interlocking half side milling cutters are used in pairs for milling a wide groove or channel in one operation. The interlocking cutters may be spaced within the limits that the teeth overlap. Interlocking cutters are used to mill a groove or channel to an accurate width.

Fig. 44-6 shows two sets of interlocking side milling cutters. The width is adjusted by adding thin metal collars *(shims)* between the individual cutters. The left-hand and right-hand helical flutes help counteract the cutting forces. Fig. 44-6A shows two cutters of herringbone design. Fig. 44-6B indicates the features of a set of staggered-tooth interlocking side milling cutters.

Gang Milling

Gang milling incorporates a number of different sizes and shapes of milling cutters. These may be mounted together on an arbor to produce a continuous contour. In other instances, there are spaces between the areas that are milled at the same time.

The cutter diameters are selected in terms of the depth of each step or special shape. The spindle speed of a gang milling combination is calculated for the largest diameter of cutter.

Fig. 44-7 shows a gang milling setup and process. Note how close to the spindle the operation is being performed. The table is near the column face. A Type-B overarm and large end bearing provide solid support for the cutters. This is a rigid setup adapted for gang milling.

A gang milling operation may include any combination of different types of cutters. Fig. 44-7 shows a number of different diameters and widths of plain milling cutters and a pair of side milling cutters. Oftentimes, other form milling cutters are included.

Fig. 44-7. A rigid cutter, workpiece, and table set up for gang milling. (Courtesy of Cincinnati Milacron)

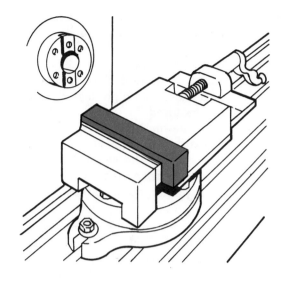

Fig. 44-8. A vise and workpiece set up for milling.

HOW TO MILL ENDS SQUARE

Using One Side Milling Cutter

Step 1   Mount the vise centrally on the milling machine table. Align the solid jaw so that it is parallel to the arbor axis (Fig. 44-8).

Step 2   Support the layed-out workpiece on parallels. Position it so that the end to be milled extends just slightly beyond the depth of cut. Tighten the vise, seating the workpiece on the parallels.

Step 3   Select a side milling cutter. The diameter must be large enough to provide clearance. The arbor bushings, overarm, and the top of the work must clear when the cutter is set to the full depth of cut.

Step 4   Select the sturdiest possible arbor. Mount the arbor.

Step 5   Position the cutter on the arbor close to the spindle. Assemble the collars, key, cutter, bushing, and nut by hand.

Step 6   Move the arbor support into position and clamp it. Use the arbor wrench to securely tighten the arbor nut.
CAUTION: The arbor nut should be tightened only *after* the arbor is supported in the overarm. The spindle should be locked *before* the nut is tightened.

Step 7   Determine the spindle speed. Set the speed control to the required rpm.

Step 8   Determine the feed rate. Set the table feed accordingly.

Step 9   Raise the knee until the cutter clears the bottom of the workpiece. Lock the knee.

Step 10   Move the cutter clear of the workpiece to the point where the cut is to be started. Turn the crossfeed handwheel until the cutter is at the layout line. Lock the saddle.
Note: If a roughing and a finishing cut are to be taken, leave from 0.012″ to 0.016″ (0.3 mm to 0.4 mm) for a finish cut.

Step 11   Feed by hand for a short distance. Then, engage the power feed.

Step 12   Take the finish cut. Stop the spindle. Return the workpiece to the start of the cut.

Step 13   Clean the workpiece and area around the vise. Remove the milling burrs.

Step 14   Turn the workpiece 180°. Seat it on parallels for the next cut.

Step 15   Position the workpiece in relation to the cutter. If there is no layout line, measure the length with a steel rule. Lock the saddle.

Step 16   Start the spindle and the flow of cutting fluid. Take a cut for a short distance.

Step 17   Stop the spindle. Measure the length with a steel rule or micrometer.
Note: Loosen the saddle and make any dimensional adjustment required. Use the micrometer collar on the crossfeed screw to accurately position the workpiece. Clamp the saddle.

Step 18   Continue with the cut. Repeat the cutter-setting steps if a finish cut is to be taken.
Note: Increase the spindle speed and decrease the feed if a high-quality surface finish and high degree of accuracy are required.

Step 19   Stop the machine. Check the length of the workpiece. Remove and burr the part. Disassemble the setup. Clean the machine. Return all tools and accessories to their proper storage areas.

USING TWO SIDE MILLING CUTTERS (STRADDLE MILLING)

Machining Parallel Sides

Step 1   Mount the workpiece in a vise or fixture.

Step 2   Select two half side milling cutters; one right-hand and the other left-hand.
Note: A pair of full side mills may also be used. The diameters must be large enough to mill the ends in one cut and yet clear the work setup and workpiece (Fig. 44-9).

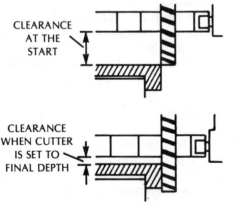

CLEARANCE AT THE START

CLEARANCE WHEN CUTTER IS SET TO FINAL DEPTH

Fig. 44-9. Adequate clearance between the arbor and the workpiece.

Step 3   Mount one cutter on the arbor. Place it as close as possible to the column. Use a key. Check to see that the cutter direction is correct.

Step 4   Make up the required combination of spacing collars, shims, and the arbor support bearing (Fig. 44-10A). The overall length must be the same as the required linear dimension to be milled.
Note: An intermediate arbor support is not always required, nor is it always possible to use one. A short arbor may be used on eight cuts and straddle milling when the cutters are mounted close together. The following steps *do* include the use of the intermediate arbor support.

Step 5   Slide one half the collars over the arbor and key (Fig. 44-10B). Next, add the bearing. This is followed by the remaining spacing collars.

Step 6   Slide the arbor support arm over the arbor. Center it on the bearing. Lock the arm in place (Fig. 44-10C).

Step 7   Slide the remaining collars and the cutter over the arbor and key. The direction of the second cutter must be the same as that of the first.

Step 8   Complete the assembly of the arbor and outer support arm. Securely tighten the arm and the arbor nut (Fig. 44-10D).

Step 9   Determine the correct spindle rpm, rate of feed, and cutting lubricant. Set the machine accordingly.

Step 10   Position the cutters to machine to the layout lines or required dimension. Lock the saddle. Move the cutters so that they clear the bottom of the workpiece. Lock the knee.

Step 11   Start the spindle. Feed by hand. Take a trial cut for a short distance. Stop the spindle and lock it. Measure the workpiece.
Note: If the part is not being straddle milled accurately to the required dimension, it may be necessary to clean the cutter and arbor setup and disassemble it. The spacing collars should be changed to machine to the required width.

Step 12   Restart the spindle. Continue the cut using the power feed.

Step 13   Return the cutters to the starting position. Stop the spindle and lock it. Recheck the finish milled surfaces for width.
Note: If a square, hexagon, or octagon are to be milled, index the workpiece to the next position. Take the second and successive cuts following the same steps. The measurement may be omitted once the cutters are set to the correct width.

Step 14   Stop the machine. Remove the workpiece. File off any milling burrs. Clean the machine. Disassemble the setup. Replace all tools and accessories in their proper storage areas.

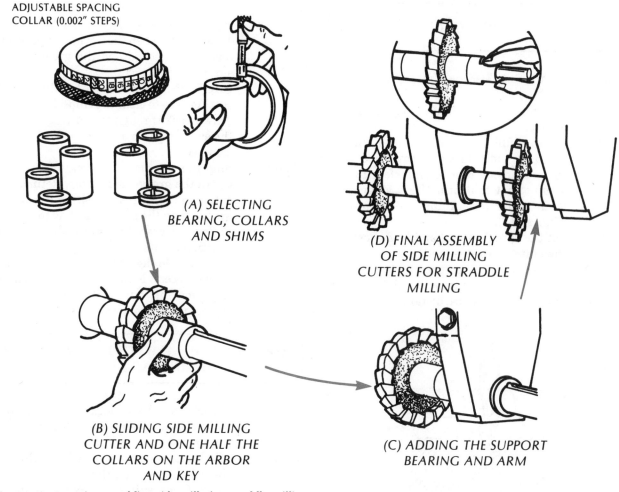

ADJUSTABLE SPACING COLLAR (0.002" STEPS)

(A) SELECTING BEARING, COLLARS AND SHIMS

(B) SLIDING SIDE MILLING CUTTER AND ONE HALF THE COLLARS ON THE ARBOR AND KEY

(C) ADDING THE SUPPORT BEARING AND ARM

(D) FINAL ASSEMBLY OF SIDE MILLING CUTTERS FOR STRADDLE MILLING

Fig. 44-10. Steps in assembling side mills for straddle milling.

Machining Parallel Steps

Parallel steps are milled with cutter, workpiece, and machine setups similar to those used to straddle mill parallel sides. However, if the parallel sides are to be milled to two different depths, the cutter diameters will differ. The smaller diameter cutter is equal to the larger diameter minus one half the difference of the depth dimensions. For example, if 6″ and 4″ diameter cutters are used, there will be a variation of 1″ between the steps that are milled.

Another usual difference is in the way in which the workpiece is held. When deep steps are to be milled, the part is either held in a fixture or clamped on the ends directly on the table. These work setups provide rigid support. At the same time, the straps make it possible for the cutter to mill to the required depth without interference. Fig. 44-11 shows the use of plain (A) and gooseneck straps (B).

Once the cutters and workpiece are set up, the same machining processes are carried on.

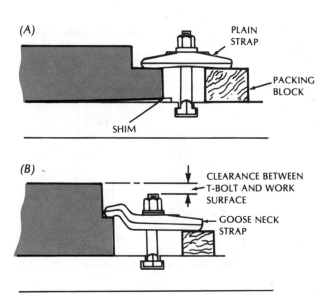

(A)

PLAIN STRAP

PACKING BLOCK

SHIM

(B)

CLEARANCE BETWEEN T-BOLT AND WORK SURFACE

GOOSE NECK STRAP

Fig. 44-11. The use of plain and gooseneck straps for strapping workpieces to the table.

HOW TO MILL GROOVES

Milling With Full Side Milling Cutters

Step 1   Lay out the width and depth of the groove.

Step 2   Select a full side milling cutter that is equal in width to the groove size. The cutter diameter must permit clearing the setup to cut to depth. Use a staggered-tooth cutter for milling a deep groove (Fig. 44-12).

Step 3   Set the spindle speed and feed. Position the cutter according to the layout lines, or measure the required distance (Fig. 44-13). Lock the knee and saddle. Take a trial cut for a short distance.

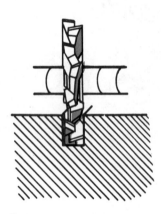

Fig. 44-12. A staggered-tooth, full side milling cutter used for machining a deep groove.

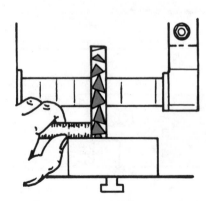

Fig. 44-13. Measuring the location of a slot or groove to position a side milling cutter.

Step 4   Measure the width of the groove. If the cutter is cutting oversize, replace it.

Step 5   Take the cut when the workpiece position has been checked and the groove measurement is correct.
CAUTION: Flood the cutter flutes and cutting area to clear the chips. If they are drawn back through the cutting process, the surface finish and dimension may be changed.

Step 6   Stop the spindle at the end of the cut. Clean the groove and cutter. Return the cutter to the starting position if a wider groove than the cutter width is to be cut. Follow the preceding steps when taking additional cuts.

Step 7   Disassemble the setup at the completion of the process. Follow standard practices for removing burrs, cleaning up, and replacing all tools and accessories.

HOW TO MACHINE WITH INTERLOCKING SIDE MILLING CUTTERS

Step 1   Select a set of interlocking side milling cutters that, when assembled, will cut the groove to the required size.

Step 2   Mount the cutters on an arbor as close to the column as conditions permit.
Note: Place shims between the interlocking faces if the cutters need to be adjusted for size.

Step 3   Position the cutters according to the layout lines on the workpiece.
Note: Sometimes the cutters are positioned by bringing one of the cutter teeth to within 0.001" to 0.002" of a side face of the work. The crossfeed micrometer dial is set. The knee is moved down until the cutter clears the work. The workpiece is then moved to the crosswise position by using the micrometer collar measurement.

Step 4   Move the workpiece up until the teeth just graze the top surface. Set the micrometer collar on the knee. Clear the cutter. Then adjust the cutter for depth of cut. Lock the knee.

Step 5   Set the spindle rpm and feed.
Note: Use the lower rpm and feed ranges. While right- and left-hand helical fluted cutters are used, wide and deep cuts are taken at slower speeds and feeds.

Step 6   Mill the groove. Follow the same procedures for trial cut, measurement, power feeding, and cutting fluid application as used with other milling processes. Deep grooves may require one or more roughing cuts and a finishing cut.

Step 7   Disassemble the setup at the end of the process. Remove any milling burrs. Clean the machine. Return the arbor and parts and cutters and other tools to their proper storage places.

SAFE PRACTICES FOR SIDE MILLING SETUPS AND PROCESSES

- Check for clearance between the arbor collars or overarm and any projecting surfaces on the workpiece or work-holding device.
- Use care in positioning a side milling cutter so that the teeth are not brought against the vise or other holding device.
- Position the cutters and cutting action as close to the column as work conditions permit.
- Tighten the arbor nut only after the overarm is supporting the arbor.
- Use a staggered-tooth full side mill for cutting deep grooves.
- Flood the flutes and area of cutting action to flow the chips away from the cutter and workpiece.

- Stop the spindle and lock it when taking measurements, adjusting the workpiece or cutter, or disassembling the setup.
- Use right-hand and left-hand half side milling cutters for straddle milling. The cutting teeth must face each other. The cutters must travel in the same direction.
- Provide added cutter and work support for wide, deep cutting with interlocking side milling cutters and for multiple-cutter setups for gang milling.
- Observe standard personal, machine, and tool safety practices. These relate to protective goggles; setup, assembly and disassembly procedures; proper clothing; and chip removal and cleanup.

TERMS APPLIED TO SIDE MILLING PROCESSES AND TECHNOLOGY

Straddle milling	The process of milling two parallel surfaces or steps at the same time. The application of two side milling cutters to machine two vertical surfaces at the same time. Milling two parallel steps in one operation.
Clear the setup	Selecting side mills with diameters that are large enough to perform the required operation. Adequate clearance between the arbor, workpiece, and work-holding setups.
Shims (arbor)	Dimensionally accurate, thin metal collars. These permit adjusting the space between cutters so that they mill to a precise dimension. Thin metal collars used alone or with standard-width arbor spacing collars.
Milling a step	The process of milling a vertical face and an adjacent horizontal face. Milling two surfaces at a right angle. One surface is in a vertical plane and the other is in a horizontal plane.
Milling a groove	Milling three surfaces at the same time. The two vertical sides are parallel. They are at right angles to the bottom of the groove.
Parallel steps	Right and left 90° angle steps (shoulders) that are parallel.
Right-hand and left-hand helical flutes (interlocking cutter)	A pair of side milling cutters whose teeth interlock when assembled. A side milling cutter with flutes cut to a right-hand or left-hand helix.
Interlocking faces	The inside face of an interlocking cutter. The face against which a shim may be used to increase the width of a pair of interlocking cutters.
Gang milling	Combining a series of regular and/or special form cutters to mill a surface. The use of multiple cutters on an arbor to simultaneously mill several surfaces.

SUMMARY

- Half side, full side, and interlocking side milling cutters are used to mill one or more surfaces, steps, or a groove.
- The half side milling cutter is applied primarily to the milling of a plane surface or a right-angle step.
- The full side milling cutter is more versatile than the half side milling cutter. The design features permit machining a single face, a right-angle step, or a groove.
- The cutter position for side milling should be as close to the column as practical.
- Layout lines are used to position the cutter and for machining to correct dimension. More accurate machining is done by feeding according to the micrometer collar settings on the cross slide and the knee. Dimensional accuracy is checked with inside, outside, or depth micrometers, as required.
- Trial cuts are taken for a short distance. Dimensions are checked with the spindle stopped. The power feed is engaged only when all conditions are correct.
- Two side milling cutters are required to straddle mill two parallel vertical surfaces or two right-angle steps. Spacing collars of the required width are used between cutters.
- An intermediate arbor support provides rigidity for heavy-duty cutting or when the straddle milling cuts are widely separated.

- A dividing head or rotary table produces the indexing necessary to position a workpiece. Machining square, hexagonal, and octagonal forms are common applications of straddle milling.
- Limited-depth grooves may be milled with full side milling cutters. Climb milling usually produces a more accurately milled narrow groove.
- Deep groove milling requires a staggered-tooth full side milling cutter. The chips must be flowed out of the flutes and away from the cutting area.
- Interlocking side milling cutters may be adjusted for width. Shims are placed between the inner faces to obtain the required dimension. The right- and left-hand helical flutes produce ideal cutting conditions.
- Plain mills, side mills, and form cutters may be combined on an arbor. Such a gang milling setup produces multiple forms at one time. The spindle rpm for gang milling is set for the largest diameter cutter in the series.
- The rigidity of the workpiece and cutter setups, the cutting action, and the appropriate cutting fluid are safe-operating-condition check points.
- Personal hygiene in handling cutting fluids and safe practices in working around moving machinery must be observed.
- Care must be taken to safely and properly remove chips and clean work surfaces. All tools should be cleaned and stored properly.

UNIT 44 REVIEW AND SELF TEST

1. List the major applications of (a) *half side milling cutters* and (b) *full side milling cutters*.
2. Differentiate between milling with *staggered-tooth interlocking side milling cutters* and *gang milling*.
3. Indicate the advantage of *straddle milling* parallel sides compared to machining each side with a single side milling cutter.
4. Tell how to machine parallel steps at different heights with a straddle milling setup.
5. State why the *full side milling cutter flutes and cutting areas* are flooded with cutting fluid when milling a deep groove.
6. Name two precautions to take in straddle milling.
7. List three safe practices to follow in mounting side milling cutters and cutter setups for gang milling.

SECTION THREE

THE

DIVIDING HEAD: PRINCIPLES AND APPLICATIONS

The dividing head is a precision mechanism that is attached to the milling machine. It is used for accurately locating (indexing) workpiece features in relation to one another about a common axis. The dividing head performs this function by rotating the workpiece the required angular distances (degrees) about an axis. This section describes the principles of and procedures for performing direct and simple indexing with the dividing head.

Unit 45

THE DIVIDING HEAD: DIRECT AND SIMPLE INDEXING

The need for dividing a circle and using angular dimensions to produce parts and mechanisms has existed for many centuries. Over 300 years ago a dividing engine was developed to more accurately cut clock gears. The manufacture of interchangeable parts, their increasing complexity, and the continuing demand for ever-higher levels of accuracy brought on a demand for practical devices to machine to precise angular dimensions.

Two basic precision mechanisms were developed to meet these needs. One was the rotary (*circular milling*) *table*. The second was the *dividing head*. These are called *indexing devices*. The rotary head and the dividing head are used to locate (*index*) one surface or angular dimension in an exact relationship with another. These mechanisms have accessories that permit them to also serve as work-holding devices.

RANGES AND TYPES OF INDEXING DEVICES

Direct Indexing Devices

A very simple form of indexing is carried on daily with standard and compound vises. Many workpieces are indexed for a number of different cuts that are at an angle to each other by simply swivelling the vise jaw section. The angle setting is read directly from the graduations on the vise.

A collet index fixture like the one shown in Fig. 45-1 is practical for direct indexing. The two base faces are machined at a right angle. These serve as bases for mounting the fixture horizontally or vertically. Workpieces are held in collets and are indexed directly.

Special milling fixtures are designed with ratchet positioning devices. Once the workpiece is machined on one face, the position is unlocked. The nest in which the workpiece is held is then moved to the next angular setting.

General Precision Dividing Devices

The *circular milling table*, also referred to as the *rotary table*, is a more precise indexing device. The

Fig. 45-1. A direct-index collet indexing fixture. (Courtesy of Hardinge Brothers, Inc.)

different models range from the line-graduated rotary table for direct indexing, to extremely precise inspection devices. Power feed, footstock, and other attachments are available.

There are other *ultraprecise* and *optical rotary tables*. These are applied, primarily, in high-precision jig boring and inspection processes and for the *calibration* (*standards setting*) of master tools. Precision indexing attachments are available to measure within *one tenth* (1/10) *second of arc*. This is 1/36,000 part of one degree.

The most practical, everyday dividing device that is used in shops and laboratories is called a *dividing head*. This is used to divide the circle to mill one surface in relation to another or for machining a number of surfaces around a workpiece. Like the rotary table, the dividing head is manufactured in a number of models. These vary in versatility and degree of precision.

Some workpieces require the machining of equal spaces around the periphery, like four, six, or eight. Such combinations may be positioned easily by using a *simple index head*. The workpiece is mounted in a chucking device. The chuck is connected directly to a spindle. The spindle has a direct-mounted *dial plate*. The work is turned (positioned) to a required location according to the divisions on the dial plate.

The Universal Dividing Head

The regular *universal dividing head* is a versatile, practical, and widely used dividing device. By means of a gear train it is possible to divide a circle into thousands of combinations of equal and unequal parts. Precision linear and circular dimensions are involved. The dividing head turns the workpiece the various distances required to perform many milling operations. These require the use of plain, face, side milling, sawing, slotting, keyway, form, and other commercial cutters. Descriptions, applications, and processes were covered for plain, face, and side milling cutters in earlier units. The other cutters, plus fluting and gear cutters, are treated in the advanced machine technology text. The more advanced *differential indexing* and milling of helical forms are related to these cutters.

Advanced Movements And Precision Dividers

When cuts require a helix angle, it is necessary to swivel the milling machine table to an angle. The cutter is then advanced (fed) into the workpiece according to a fixed lead. This requires the use of another model called the *universal spiral dividing head*.

This head provides a rotary movement of the workpiece. This complements the longitudinal travel of the table. The movement is accomplished by a gear train. This is located between the dividing head spindle and the table lead screw. In other words, the universal spiral dividing head uses gearing to control the turning of the workpiece in relation to the table movement. This type of dividing head is used where helical splines, flutes, spiral teeth, and similar forms are to be milled. Examples of these applications are the milling of flutes on taps, drills and reamers; the cutting of spiral and angular helical gears; and the machining of straight or tapered parts that require the generation of a form at a helix angle.

The *wide range divider* in Fig. 45-2 permits indexing from 2 to 400,000 divisions. This divider has an index plate, an index crank, and sector arms. These parts are mounted in front of the large index plate of the universal dividing head. An additional ratio of 100:1 is obtained. The combination of the original 40:1 ratio (40 turns of the index crank to turn the workpiece one complete revolution) and the 100:1 ratio increases the ratio to 4,000:1. This means it takes 4,000 turns of the index crank to rotate the divider spindle (and therefore the workpiece) one

Fig. 45-2. A wide-range divider adapter on a universal dividing head (range: 2 to 400,000 divisions). (Courtesy of Cincinnati Milacron)

revolution. The accuracy, as stated earlier, is within 1/10 second of arc.

Although dividing heads differ in design and construction, the underlying technology and the fundamental processes are common.

The remainder of this unit deals with the major components of the universal dividing head and its use in direct and simple indexing. A few divider-related mathematical formulas also are presented. This technical information is followed by step-by-step procedures for setting up and operating the dividing head to perform common milling processes.

DESIGN AND FUNCTIONS OF THE DIVIDING HEAD, FOOTSTOCK, AND CENTER REST

Major Parts Of The Dividing Head

A dividing head unit consists of three major components: *dividing head*, *footstock*, and *center rest* (Fig. 45-3). The subassemblies of the dividing head include the *housing*, *swivel block*, and *spindle mechanism*.

The *housing* serves two main functions. (1) *As a base*, the entire head may be bolted to the table. (2) *As a bearing*, the clamping straps permit the swivel block to be moved to an angle. The spindle axis of some makes may be moved through a 145° arc. The swivel block may be positioned at any angle from 5° below a horizontal position to 50° beyond a vertical position.

The angular setting of the swivel block (and spindle axis) is often referred to as the *position of the (dividing) head*. Graduations are usually cut in the swivel block. These are matched with other graduations on an index plate on the housing. The swivel block is set to a required angular position and locked.

The swivel block houses the spindle, the gearing,

Fig. 45-3. The major components of a dividing head unit. (Courtesy of Cincinnati Milacron)

and the indexing mechanism. The sectional drawing in Fig. 45-4 shows many construction features of the dividing head. Some of the principal parts are identified. These are referred to in the following brief description of how the dividing head works.

The spindle (A) has an attached *worm wheel* (B). This meshes with a *worm* (C) on a *trunion* (D). The worm may be engaged or disengaged by turning an *eccentric collar* (E). In a disengaged position, the spindle may be turned by hand. The workpiece may be setup to a required angle by *direct indexing*.

When the worm is engaged, its movement is controlled by the gears between the trunion (D) and the *index crank shaft* (G). As the index crank (H) is moved a complete and/or partial turn, the spindle moves in a *fixed relationship*. This means that the spindle movement is proportional to the *index ratio* between the index crank and the spindle. The most commonly used gear ratio is 40:1.

The dividing head spindle is hollow ground. Generally, the nose has an internal Brown and Sharpe standard taper. This is a self-locking taper. Centers and other mating tapered tools and parts may be mounted in the spindle. The nose is machined on the outside to position and hold chucks and other accessories.

The precise positioning of the workpiece is done by turning the index crank. The partial movement is indicated by the *index plate* (I) and the *index pin* (J). Additional information about these design features is presented later.

Dividing Head Accessories

Driver, Center, And Dog—The three commonly used

driving accessories include the *driver*, milling machine *dog*, and the *dividing head chuck*. Many workpieces are milled between centers. This requires the part to be mounted on the dividing head (spindle) center and the footstock center. A driver is secured to the spindle center.

The slot in the driver receives the tail end of a milling machine dog. The dog is attached to one end of the workpiece. A screw in the driver is tightened against the tail. This provides positive no-slip drive of the workpiece in relation to any movement of the spindle.

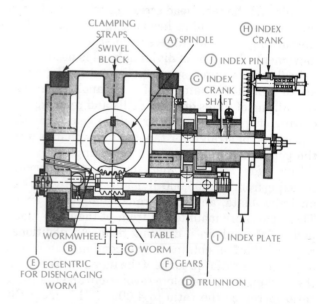

Fig. 45-4. A sectional drawing of a universal dividing head showing its principal parts. (Courtesy of Cincinnati Milacron)

Dividing Head Chuck—Other workpieces are conveniently positioned and held in a *dividing head chuck*. Some are of the *single-step* type. Other designs include conventional *three-step jaws*. Fig. 45-5 shows a common single-step milling machine chuck (A) and one with reversible jaws (B). The combination of *inside and outside jaws* permits chucking a wide range of work diameters and depths.

The chucks are usually of the self-centering type. The chuck itself is fitted with an *adapter plate* on the back side. The plate permits the chuck to be centered accurately and to be secured on the dividing head spindle.

Common work setups are made with the chuck and dividing head positioned horizontally, vertically, or at an angle. Some workpieces are held by the chuck alone. Others are also supported by the footstock.

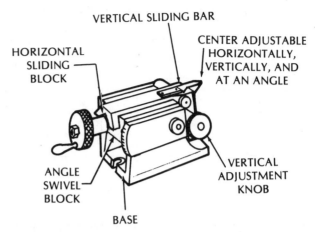

Fig. 45-6. A footstock with vertical, horizontal, and angular positioning parts.

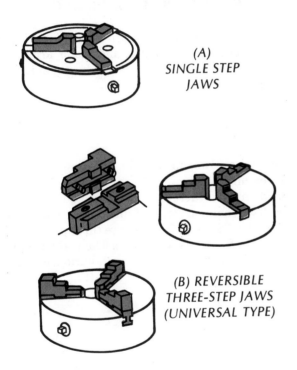

Fig. 45-5. General-purpose milling machine dividing head chucks.

The Footstock (Tailstock)

The terms *footstock* and *tailstock* are used interchangeably. The footstock (Fig. 45-6) serves the following two basic functions:

(1) To support one end of a workpiece

(2) To position the workpiece in a horizontal plane or at an angle.

The footstock consists of a frame and a number of moveable parts. The base of the frame is fitted with

tongues. These permit aligning the tailstock with one of the table slots. A *horizontal sliding block* mounted in the frame permits the *center* to be adjusted toward or away from the dividing head.

A *vertical sliding bar* makes it possible to raise or lower the center. This permits horizontal alignment or offsetting of the footstock center for dividing head work. Angle settings are required to position tapered work surfaces.

The housing for the horizontal and vertical sliding bars may also be moved. The horizontal axis of the footstock center may be changed to a smaller angle position. This makes it possible to properly seat the footstock center in a center hole. Angle graduations on one side of the frame and an index line on the housing are used to make the angular adjustment.

The Center Rest

The *milling machine center rest* serves the same functions as a lathe steady-rest. It is used principally for milling operations on long workpieces or machine setups where the cutting force would otherwise cause the workpiece to bend away from the cut.

The center rest consists of a base and a V-center. This may be adjusted vertically. One or more centers may be positioned along the workpiece, depending upon the number of places where support is needed. Once the center rest is correctly located and adjusted for height, a locking screw is turned to hold the center at the set position.

DIRECT INDEXING

Direct indexing is also called *rapid indexing* or *quick indexing*. When direct indexing is to be used on a universal dividing head, the worm is disengaged

Table 45-1. HOLES IN INDEX PLATE AND DIVISIONS FOR DIRECT INDEXING

Number of Holes in Circle	Divisions Directly Indexed from Holes in Index Plate													
24	2	3	4	—	6	8	—	—	12	—	—	24	—	—
30	2	3	—	5	6	—	—	10	—	15	—	—	30	—
36	2	3	4	—	6	—	9	—	12	—	18	—	—	36

from the worm and wheel by an eccentric device. This makes it possible for the spindle and *direct indexing plate* to be turned freely. This plate is mounted on the spindle. It usually has three circles of holes on the back side: 24, 30, and 36 (Fig. 45-7A). The range of direct divisions is limited to any number that is divisible into 24, 30, or 36. The divisions are shown in Table 45-1. There are additional holes around the rim of the plate. These act as guides to rapidly locate some of the smaller divisions.

A plunger and pin (Fig. 45-7B) are mounted on a bracket. The bracket is attached to the swivel block. The pin may be adjusted to fit the holes in any one of the circles. The workpiece is moved to the required fractional part of the revolution (Fig. 45-8). The spindle is then clamped at this setting by moving the spindle lock. At the end of the milling operation, the spindle is unlocked. The plunger pin is released. The plate is then turned by hand until the plunger pin is aligned with the next required hole. The spindle is then locked in this position.

SIMPLE (PLAIN) INDEXING

The workpiece is positioned in *simple indexing* by using the crank, index plate, and sector arms. The dividing head spindle is connected by engaging the worm and worm wheel.

Index Plates

The *index plates* for simple indexing, like those used for direct indexing, have a number of circles divided into a different number of equally spaced holes. Two common series of index plate hole circles are listed in Table 45-2. There are three plates in the *Brown* and *Sharpe* (B&S) *series*. The standard plate for Cincinnati heads has a combination of eleven different hole circles on each side (Fig. 45-9).

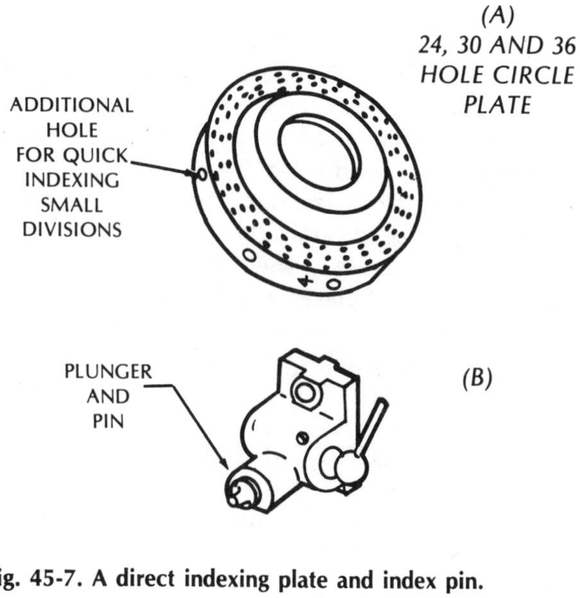

(A)
24, 30 AND 36 HOLE CIRCLE PLATE

ADDITIONAL HOLE FOR QUICK INDEXING SMALL DIVISIONS

PLUNGER AND PIN

(B)

Fig. 45-7. A direct indexing plate and index pin.

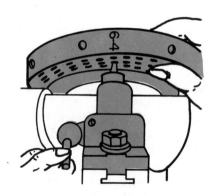

Fig. 45-8. Positioning the index plate and pin in direct indexing.

Table 45-2. HOLES IN CIRCLES ON STANDARD INDEX PLATES

Brown and Sharpe Index Plates	
#1	15-16-17-18-19-20
#2	21-23-27-29-31-33
#3	37-39-41-43-47-49

Standard Cincinnati Index Plate	
First Side	24-25-28-30-34-37 38-39-41-42-43
Reverse Side	46-47-49-51-53-54 57-58-59-62-66

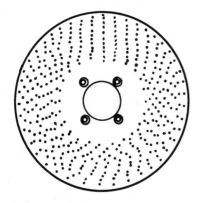

Fig. 45-9. Eleven circles of indexing holes on one manufacturer's standard index plate.

Functions Of The Sector Arms, Index Crank, And Index Pin

Sector arms are used with index plates. The sector arms eliminate the need for counting the number of holes each time a new setting is required. Sector arms are adjustable radial arms. The arms are set and locked in position with the required number of holes between them. In operation, one sector arm is brought against the plunger (index) pin. The spindle is unlocked. The pin is withdrawn and the index crank is turned carefully. When the pin is aligned with the required hole that is positioned at the second sector arm, it drops into the hole. The spindle is then locked and the milling operation is performed.

Design features are included to permit the index plate position to be changed when a cutter or the workpiece is being reset. Sometimes, the adjustment is made by moving the index plate. On other models, the index pin position may be adjusted so that the pin enters the hole in the required circle.

SIMPLE INDEXING CALCULATIONS

The standard gear ratio on the universal dividing head is 40:1. When a fractional part of a revolution is needed, the number of turns of the crank is equal to 40 divided by the number of required divisions. For example, each tooth in a 20-tooth gear is indexed by turning the crank 40 divided by 20, or two complete turns.

Sometimes, angular dimensions are given on a drawing in place of a number of required divisions. With angular divisions, the number of degrees in a circle is divided by the required angular measurement.

Example: The surfaces of a workpiece are to be milled to an included angle of 60°. The number of required divisions must first be determined.

$$\text{Required Divisions} = \frac{360°}{\text{Degrees in Required Angle}}$$
$$= \frac{360}{6}$$
$$= 6.$$

The 6 is then substituted in the formula:
$$\text{Turns of Crank} = \frac{40}{\text{Required Divisions}}$$
$$= \frac{40}{6}$$
$$= 6\ 2/3$$

This means that the index crank must be turned 6 complete revolutions and 2/3 of another revolution.

The 2/3 of a revolution may be indexed by selecting a circle of holes that is divisible by 3. The 15- and 18-hole circles on B&S plate number 1 may be used. The sector arms are set for 10 holes on the 15-hole circle or 12 holes on the 18-hole circle. (It should be noted that in extremely precise angular indexing, the highest degree of accuracy is achieved by using the plate with the greatest number of holes that will accommodate the required partial turn. In this example, the Cincinnati plate with 66 holes may be used.)

Continuing with the example: Each side (60° angle) may be positioned for milling by first unlocking the spindle clamp. The index crank is then turned 6 complete revolutions plus 10 holes on the 15-hole circle. The crank movement is continued slowly until the plunger pin drops into the hole at the sector arm.

If the index arm is turned too far, it first must be turned at least 1/2 revolution in the opposite direction. The index arm is then turned again in the original direction until the index hole is reached.

This step is important because is removes backlash that produces inaccuracies.

There is another one-step method of calculating the indexing to obtain angular dimensions. Since it takes 40 turns of the index crank to turn the spindle one revolution,

$$\text{one turn of the crank} = \frac{360°}{40}$$
$$= 9°.$$

This means a 9° movement of the spindle for each turn of the index crank. Therefore, to calculate the required indexing in the previous example, divide the 60° by 9°. Again, 6 2/3 turns of the crank handle are needed to index for each 60° angle.

Not all indexed workpieces require that the index arm be turned a complete revolution or more. *Example:* Assume that a part is to be machined with 47 divisions.

$$\text{Turns of Crank} = \frac{40}{\text{Required Divisions}}$$
$$= \frac{40}{47} \text{ of a turn.}$$

In this case, the 47-hole circle plate is used. The sector arms are set to 40 holes on the 47-hole circle. Each division is indexed by turning the index crank the distance between the sector arms, or 40/47 of a revolution.

Tables and charts are provided with dividing heads. Because simple indexing is easy to compute, the technical information is most useful in compound indexing. The tables give the number of required divisions. These are followed by the number of whole and partial turns of the index crank, the circle to use, and the number of holes to which the sector arms are set.

HOW TO SET UP THE DIVIDING HEAD AND FOOTSTOCK

Setting Up The Dividing Head

Step 1 Clean the machine and the base of the dividing head. Examine and remove any burrs.

Step 2 Place the dividing head so that its weight and the milling operation can be carried on as near to the center of the table as practical. Use T-bolts and clamp the dividing head.
Note: The index crank should be on the operating side of the table.

Step 3 Position the dividing head at the required milling angle. Loosen the clamping straps and position the swivel block. The angle setting is read on the graduations of the swivel block and index plate on the dividing head body. Fig. 45-10A shows the reading for a horizontal setting. The dividing head may also be held in a vertical position (Fig. 45-10B).
Note: If there is a center in the dividing head, use a *knockout bar* to free the tapers. This is done while the spindle is in the horizontal position.

Step 4 Select the appropriate work-holding device. A center and driver plate are secured to the spindle for work between centers.
Note: If a chuck is to be used, remove the center. Clean the spindle nose and the back plate on the chuck. A wooden cradle should be placed under a heavy chuck. The chuck and block are moved up to the spindle nose. The chuck is positioned on the spindle and locked securely.

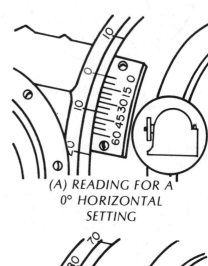

(A) READING FOR A 0° HORIZONTAL SETTING

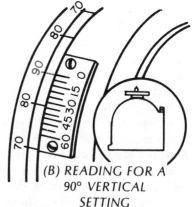

(B) READING FOR A 90° VERTICAL SETTING

Fig. 45-10. Positioning the dividing head.

Setting Up The Footstock

Step 1 Clean the base of the footstock and the table. Remove any burrs.

Step 2 Place one T-head bolt in the center slot. Position it at the place where the footstock is to be located.

Step 3 Locate the footstock, at a distance from the dividing head center. The distance must permit the center to be adjusted to the work length. Tighten the two T-bolts to secure the footstock.

Step 4 Check the swivel block index line and the angle graduations on the frame. The zero lines should coincide for a horizontal setting.
Note: Use a test bar if the centers must be aligned to a greater accuracy. The horizontal alignment may then be checked with an indicator mounted on a surface gage.

Step 5 Check the height of the center. The zero mark on the vertical sliding bar and the zero index line on the front face of the frame should be aligned.

HOW TO PREPARE THE DIVIDING HEAD FOR SIMPLE INDEXING

Changing The Index Plate

Step 1 Loosen the nut on the index crank shaft. Remove the washer and index crank (Fig. 45-11A).

Step 2 Slip off the collar and sector arms (Fig. 45-11B).

Step 3 Remove the index plate screws. Take off the index plate.

Step 4 Screw the new index plate back in position.

Step 5 Replace the collar, sector arms, and the index crank. Select the circle of holes that will produce the required fractional part of a turn.

Setting The Sector Arms

Step 1 Position the index pin in one of the holes. Tighten the crank arm.
CAUTION: The index pin should fit easily into any one of the index holes in the circle when the crank arm is tightened.

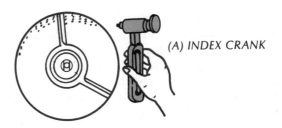

(A) INDEX CRANK

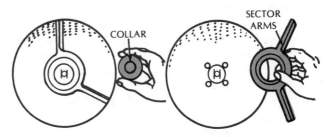

(B) COLLAR AND SECTOR ARMS

Fig. 45-11. Removing the indexing unit parts.

Step 2 Position one sector arm with the beveled edge against the index pin.

Step 3 Count off the required number of holes on the correct circle. Adjust the second arm to accommodate this number of holes. Tighten the lock screw for the two sector arms. The arrangement of the two sector arms and the index pin is shown graphically in Fig. 45-12.
CAUTION: The circle and the number of holes are rechecked before any indexing is done. The *number of open holes* must equal the number of holes required for the partial turn of the index crank.

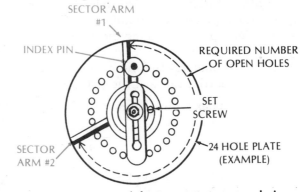

Fig. 45-12. Arrangement of the two sector arms and pin.

HOW TO DO SIMPLE INDEXING

The dividing head extends the range of machining processes. It is used after the craftsperson has gained experience in setting up cutters, workpieces, and carrying on a number of different milling processes. These generally include plain milling, side and straddle milling, angle cutting, slotting, and keyway cutting. The dividing head provides a combination work-holding and accurate-positioning device for carrying on these and other milling processes.

Full Turns

The following steps are taken in simple indexing of a workpiece by full turns of the index crank.

Step 1	Use any circle of holes on the index plate for the complete turns of the index crank. Position the sector arms to zero in one hole or any index plate.
Step 2	Loosen the spindle lock. Withdraw the index pin. Turn the index crank the required number of turns.
Step 3	Lock the dividing head spindle. Take the milling cut. Return the milling cutter to the start of the next cut.
Step 4	Position the workpiece for the second cut. Unlock the dividing head spindle before indexing. Lock the spindle after indexing. **CAUTION:** The crank handle must always be turned in the same direction.
Step 5	Take the cut. *Note:* These steps are repeated until the workpiece has been turned a complete revolution.

Full And/Or Partial Turns

Example: A part is to be indexed for milling graduations around the circumference. These are 60° apart. Set up the dividing head to index each graduation.

Step 1	Compute the number of whole and partial turns of the index crank.

$$\text{Index Crank Turns} = \frac{\text{Required Degrees}}{9°}$$

$$= \frac{60°}{9°}$$

$$= 6\ 2/3.$$

Step 2	Select an index plate with a circle of holes divisible by 3. *Note:* For this example, assume a 24-hole circle plate is already mounted. Use this plate.
Step 3	Set the index pin so that it is aligned with one of the holes. Adjust and set the index crank.
Step 4	Move one sector arm until the beveled edge touches the index pin.

Note: The beveled edge should face in the direction in which the index crank is to be turned.

Step 5	Loosen the lock screw on the sector arms. Count off 16 holes in the 24-hole circle. Start counting with the first hole next to the index pin. This setting is shown in Fig. 45-13.

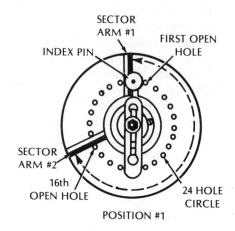

Fig. 45-13. Setting the sector arms.

Step 6	Set the second sector arm close to the 16th hole. Tighten the set screw so that the arms will not move. **CAUTION:** Recheck the number of open holes. In this example there must be 16.
Step 7	Lock the dividing head spindle. Proceed with the milling operation. Return the cutter to the starting position. Index for the next cut.

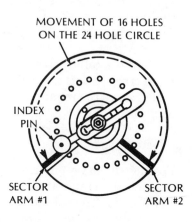

Fig. 45-14. Indexing a partial turn.

Step 8   Revolve the sector arm (#1 in Fig. 45-14) until it touches the index pin. Unlock the dividing head spindle.

Step 9   Move the index crank 16 holes on the 24-hole circle. It is good practice to bring the index pin almost to the hole and tap it gently by hand until it drops into position. Lock the dividing head spindle. **CAUTION:** If the index crank is turned too far, turn the crank in the reverse direction for about half a turn. Then, *come forward again* to take up the back lash. Turn until the index pin drops into the 16th hole.

Step 10   Repeat the steps to unlock the spindle, position the sector arms, move the index crank 16 holes, and lock the dividing head spindle between each cutting step. Continue until the 24 graduations are indexed.

SAFE PRACTICES IN SETTING UP AND USING DIVIDING HEADS AND ACCESSORIES

- Clean the table and base of the dividing head and footstock. Remove any burrs. Check to see that the tongues fit accurately into the table slots.
- Position the dividing head as near the center of the table as is practical.
- Tighten the milling machine dog securely to the driver plate.
- Check the alignment of the dividing head and footstock centers for milling parallel and taper surfaces and for proper seating in the center holes.
- Count the required number of holes on an index plate circle by starting with the first open hole past the index pin.

- Lock the index head spindle after each indexing and before the milling process is started.
- Remove any backlash if the index crank is moved beyond the correct hole.
- Recheck the sector arms to be sure they are held securely.
- Observe personal, machine, tool, and work safety precautions in carrying out the milling processes.

TERMS USED WITH THE DIVIDING HEAD AND INDEXING PRACTICES

Direct indexing   The process of positioning a part for an angular dimension by direct movement of a spindle. Rotating a workpiece a required angular distance without any additional gearing. Positioning a workpiece by using a plate with equally spaced holes in a circle.

Simple (plain) indexing   Dividing a circle into any required number of parts in preparation for inspection, layout, or a machining process. Using a universal dividing head, generally with a gear ratio of 40:1, to index a required division.

Dividing machines   Devices used in inspection, layout, and machining to accurately position a workpiece for angular divisions. The collet index fixture, index head, rotary table, and the universal dividing head are common examples of dividing machines.

40:1 ratio of the dividing head   The ratio between the index crank shaft and the spindle. The ratio of the gears in a dividing head. The number of turns (40) an index crank must be moved to turn a dividing head spindle one complete revolution. A commonly used ratio (40:1) on most indexing heads.

Universal dividing head   An indexing device, usually with a 40:1 ratio. A device that permits a part to be indexed directly or by controlling the full and/or partial turns of the index crank.

Spiral milling	The generation of a helical angle to the axis of a workpiece. Milling of angular channels, flutes, and teeth. Simultaneously turning the workpiece and feeding it into a revolving milling cutter.
Driver	A forked plate that may be attached to a dividing head center. A plate that receives and holds a milling machine dog for positive (no-slip) turning of the workpiece.
Dividing head chuck	Usually a three-jaw universal chuck mounted on the spindle nose of the dividing head.
Footstock	A centering device for workpieces that require support on a centered end.
Index plate	A circular disk with a series of holes equally spaced around a circle. A graduated plate with a series of holes that is used on a dividing head to accurately position the spindle.
Sector arms	Two narrow, beveled arms that are set a required number of holes apart. Arms that simplify the movement and location of the index pin for each division setting.
Index pin and plunger	A sliding pin mounted on the index crank. A pin that may be positioned by a plunger to permit turning the index crank. Once the exact number of whole and partial turns is reached, the pin drops into the required hole on the index plate.

SUMMARY

- The rotary (circular milling) table and the universal dividing head are two basic precision devices for dividing a circle. Other direct indexing devices are also used.

- Indexing devices extend the applications of fundamental milling processes. Plain, face, and side milling, and keyway cutting, slotting, sawing, and form milling cutters and processes are all used in direct, simple, and compound (*differential*) indexing.

- The universal dividing head generally has a 40:1 ratio between the crank shaft and the spindle.

- Intermediate angular dimensions are obtained by using index plates. Different numbers of holes, equally spaced on a series of circles, control the partial distance a workpiece is to be turned.

- The spindle movement of a dividing head is controlled by the whole and partial turns of an index crank. This value is computed by dividing 40 by the required number of divisions. The fractional part is computed by dividing a hole circle number by the denonimator. The result is then multiplied by the numerator.

- When the dimension is given in degrees, the index crank movement is calculated by dividing the required number of degrees by 9°.

- The dividing head normally may be positioned horizontally, vertically, and at any angle within a 145° range.

- The worm gears of a universal dividing head are disengaged for direct indexing. The spindle is always locked after indexing. This ensures a rigid support.

- The dividing head, footstock, and center rest are used in combination to support long workpieces.

- A milling machine dog and driver are used to rotate a workpiece between centers. Universal three-jaw chucks, attached by an adapter plate to the spindle, provide a fast method of holding a workpiece.

- Sector arms are adjustable to represent any angular division. Sector arms are set a required number of holes apart on a specific circle.

- The three index plates in the Brown and Sharpe system and the single reversible Cincinnati plate make it possible to accurately index the full range of divisions required in general machining.

- Once a part is held securely and positioned, the milling machine processes are carried on following the procedures covered in previous units.

- Care must be taken to remove any backlash in indexing. If the index crank is turned too far, turn it back one-half revolution. Then, turn in the original direction until the pin drops into the correct hole.

- Obtain help to position any heavy dividing head on the table. Use a cradle to mount a heavy chuck.

- Follow standard safety practices for the care of the machine and accessories during setting up, operation, and disassembling. Personal safety precautions must be observed, including those about lifting heavy objects and wearing safety goggles.

UNIT 45 REVIEW AND SELF-TEST

1. Distinguish between a *general* and a *universal spiral dividing head*.
2. Tell how an indexing range up to 400,000 divisions and an accuracy of 1/10 second of arc are possible with a *wide-range divider*.
3. Indicate what movements are possible with the *horizontal sliding block* and the *vertical sliding bar* of the *footstock*.
4. Differentiate between *direct indexing* and *plain indexing*.
5. Explain how to eliminate *backlash* caused by turning the index crank of a dividing head too far.
6. List the steps to follow in (a) changing the *index plate* and (b) *setting the sector arms* on a dividing head for simple indexing.
7. Indicate three safety precautions to take in setting up and using a dividing head and accessories.

APPENDIXES

Table A-1. DECIMAL EQUIVALENTS OF FRACTIONAL, WIRE GAGE, LETTER, AND METRIC SIZE DRILLS

Decimal	Inch	Wire	mm.	Decimal	Inch	Wire	mm.	Decimal	Inch	Wire and Letter	mm.	Decimal	Inch	Letter	mm.	Decimal	Inch	mm.
.0217			.55	.0938	3/32			.2040		6		.3445			8.75	.7500	3/4	
.0225		74		.0945			2.4	.2047			5.2	.3465			8.8	.7656	49/64	
.0236			.6	.0960		41		.2055		5		.3480		S		.7677		19.5
.0240		73		.0965			2.45	.2067			5.25	.3504			8.9	.7812	25/32	
.0250		72		.0980		40		.2087			5.3	.3543			9.	.7874		20.
.0256			.65	.0984			2.5	.2090		4		.3580		T		.7969	51/64	
.0260		71		.0995		39		.2126			5.4	.3583			9.1	.8071		20.5
.0276			.7	.1015		38		.2130		3		.3594	23/64			.8125	13/16	
.0280		70		.1024			2.6	.2165			5.5	.3622			9.2	.8268		21.
.0292		69		.1040		37		.2188	7/32			.3642			9.25	.8281	53/64	
.0295			.75	.1063			2.7	.2205			5.6	.3661			9.3	.8438	27/32	
.0310		68		.1065		36		.2210		2		.3680		U		.8465		21.5
.0312	1/32			.1083			2.75	.2244			5.7	.3701			9.4	.8594	55/64	
.0315			.8	.1094	7/64			.2264			5.75	.3740			9.5	.8661		22.
.0320		67		.1100		35		.2280		1		.3750	3/8			.8750	7/8	
.0330		66		.1102			2.8	.2284			5.8	.3770		V		.8858		22.5
.0335			.85	.1110		34		.2323			5.9	.3780			9.6	.8906	57/64	
.0350		65		.1130		33		.2340		A		.3819			9.7	.9055		23.
.0354			.9	.1142			2.9	.2344	15/64			.3839			9.75	.9062	29/32	
.0360		64		.1160		32		.2362			6.	.3858			9.8	.9219	59/64	
.0370		63		.1181			3.	.2380		B		.3860		W		.9252		23.5
.0374			.95	.1200		31		.2402			6.1	.3898			9.9	.9375	15/16	
.0380		62		.1221			3.1	.2420		C		.3906	25/64			.9449		24.
.0390		61		.1250	1/8			.2441			6.2	.3937			10.	.9531	61/64	
.0394			1.	.1260			3.2	.2460		D		.3970		X		.9646		24.5
.0400		60		.1280			3.25	.2461			6.25	.4040		Y		.9688	31/32	
.0410		59		.1285		30		.2480			6.3	.4062	13/32			.9843		25.
.0413			1.05	.1299			3.3	.2500	1/4	E		.4130		Z		.9844	63/64	
.0420		58		.1339			3.4	.2520			6.4	.4134			10.5	1.0000	1	
.0430		57		.1360		29		.2559			6.5	.4219	27/64					
.0433			1.1	.1378			3.5	.2570		F								
.0453			1.15	.1405		28		.2598			6.6							
.0465		56		.1406	9/64			.2610		G								
.0469	3/64			.1417			3.6	.2638			6.7							
.0472			1.2	.1440		27		.2656	17/64									
.0492			1.25	.1457			3.7	.2658			6.75							
.0512			1.3	.1470		26		.2660		H								
.0520		55		.1476			3.75	.2677			6.8							

Handwritten circled group numbers across the top: ① ③ ⑤ ⑦ ⑨

Decimal	Inch	Wire	mm.	Decimal	Inch	Wire	mm.	Decimal	Inch	Wire and Letter	mm.	Decimal	Inch	Letter	mm.	Decimal	Inch	mm.
.0059		97	.15	.0532			1.35	.1495		25		.2717			6.9	.4331		11.
.0063		96	.16	.0550		54		.1496			3.8	.2720		I		.4375	7/16	
.0067		95	.17	.0551			1.4	.1520		24		.2756			7.	.4528		11.5
.0071		94	.18	.0571			1.45	.1535			3.9	.2770		J		.4531	29/64	
.0075		93	.19	.0591			1.5	.1540		23		.2795			7.1	.4688	15/32	
.0079		92	.20	.0595		53		.1562	5/32			.2810		K		.4724		12.
.0083		91	.21	.0610			1.55	.1570		22		.2812	9/32			.4844	31/64	
.0087		90	.22	.0625	1/16			.1575			4.	.2835			7.2	.4921		12.5
.0091		89	.23	.0630			1.6	.1590		21		.2854			7.25	.5000	1/2	
.0095		88	.24	.0635		52		.1610		20		.2874			7.3	.5118		13.
.0098			.25	.0650			1.65	.1614			4.1	.2900		L		.5156	33/64	
.0100		87		.0669			1.7	.1654			4.2	.2913			7.4	.5312	17/32	
.0102			.26	.0670		51		.1660		19		.2950		M		.5315		13.5
				.0689			1.75	.1673			4.25	.2953			7.5	.5469	35/64	
.0105		86		.0700		50		.1693			4.3	.2969	19/64			.5512		14.
.0106			.27	.0709			1.8	.1695		18		.2992			7.6	.5625	9/16	
.0110		85	.28	.0728			1.85	.1719	11/64			.3020		N		.5709		14.5
.0114			.29	.0730		49		.1730		17		.3032			7.7	.5781	37/64	
.0115		84		.0748			1.9	.1732			4.4	.3051			7.75	.5906		15.
.0118			.30	.0760		48		.1770		16		.3071			7.8	.5938	19/32	
.0120		83		.0768			1.95	.1772			4.5	.3110			7.9	.6094	39/64	
.0122			.31	.0781	5/64			.1800		15		.3125	5/16			.6102		15.5
.0125		82		.0785		47		.1811			4.6	.3150			8.	.6250	5/8	
.0126			.32	.0787			2.	.1820		14		.3160		0		.6299		16.
.0130		81	.33	.0807			2.05	.1850		13	4.7	.3189			8.1	.6406	41/64	
.0134			.34	.0810		46		.1870			4.75	.3228			8.2	.6496		16.5
.0135		80		.0820		45		.1875	3/16			.3230		P		.6562	21/32	
.0138			.35					.1890		12	4.8	.3248			8.25	.6693		17.
.0145		79		.0827			2.1	.1910		11		.3268			8.3	.6719	43/64	
.0156	1/64			.0847			2.15	.1929			4.9	.3281	21/64			.6875	11/16	
.0158			.4	.0860		44		.1935		10		.3307			8.4	.6890		17.5
.0160		78		.0866			2.2	.1960		9		.3320		Q		.7031	45/64	
.0177			.45	.0886			2.25	.1969			5.	.3347			8.5	.7087		18.
.0180		77		.0890		43		.1990		8		.3386			8.6	.7188	23/32	
.0197			.5	.0906			2.3	.2008			5.1	.3390		R		.7283		18.5
.0200		76		.0925			2.35	.2010		7		.3425			8.7	.7344	47/64	
.0210		75		.0935		42		.2031	13/64			.3438	11/32			.7480		19.

Table A-2. CONVERSION OF METRIC (mm) TO INCH STANDARD UNITS OF MEASURE

mm Value	(Inch) Decimal Equivalent	mm Value	(Inch) Decimal Equivalent	mm Value	(Inch) Decimal Equivalent	mm Value	(Inch) Decimal Equivalent
.01	.00039	.34	.01339	.67	.02638	1	.03937
.02	.00079	.35	.01378	.68	.02677	2	.07874
.03	.00118	.36	.01417	.69	.02717	3	.11811
.04	.00157	.37	.01457	.70	.02756	4	.15748
.05	.00197	.38	.01496	.71	.02795	5	.19685
.06	.00236	.39	.01535	.72	.02835	6	.23622
.07	.00276	.40	.01575	.73	.02874	7	.27559
.08	.00315	.41	.01614	.74	.02913	8	.31496
.09	.00354	.42	.01654	.75	.02953	9	.35433
.10	.00394	.43	.01693	.76	.02992	10	.39370
.11	.00433	.44	.01732	.77	.03032	11	.43307
.12	.00472	.45	.01772	.78	.03071	12	.47244
.13	.00512	.46	.01811	.79	.03110	13	.51181
.14	.00551	.47	.01850	.80	.03150	14	.55118
.15	.00591	.48	.01890	.81	.03189	15	.59055
.16	.00630	.49	.01929	.82	.03228	16	.62992
.17	.00669	.50	.01969	.83	.03268	17	.66929
.18	.00709	.51	.02008	.84	.03307	18	.70866
.19	.00748	.52	.02047	.85	.03346	19	.74803
.20	.00787	.53	.02087	.86	.03386	20	.78740
.21	.00827	.54	.02126	.87	.03425	21	.82677
.22	.00866	.55	.02165	.88	.03465	22	.86614
.23	.00906	.56	.02205	.89	.03504	23	.90551
.24	.00945	.57	.02244	.90	.03543	24	.94488
.25	.00984	.58	.02283	.91	.03583	25	.98425
.26	.01024	.59	.02323	.92	.03622	26	1.02362
.27	.01063	.60	.02362	.93	.03661	27	1.06299
.28	.01102	.61	.02402	.94	.03701	28	1.10236
.29	.01142	.62	.02441	.95	.03740	29	1.14173
.30	.01181	.63	.02480	.96	.03780	30	1.18110
.31	.01220	.64	.02520	.97	.03819		
.32	.01260	.65	.02559	.98	.03858		
.33	.01299	.66	.02598	.99	.03898		

Table A-3. CONVERSION OF FRACTIONAL INCH VALUES TO METRIC UNITS OF MEASURE

Fractional Inch	mm Equivalent	Fractional Inch	mm Equivalent	Fractional Inch	mm Equivalent	Fractional Inch	mm Equivalent
1/64	0.397	17/64	6.747	33/64	13.097	49/64	19.447
1/32	0.794	9/32	7.144	17/32	13.494	25/32	19.844
3/64	1.191	19/64	7.541	35/64	13.890	51/64	20.240
1/16	1.587	5/16	7.937	9/16	14.287	13/16	20.637
5/64	1.984	21/64	8.334	37/64	14.684	53/64	21.034
3/32	2.381	11/32	8.731	19/32	15.081	27/32	21.431
7/64	2.778	23/64	9.128	39/64	15.478	55/64	21.828
1/8	3.175	3/8	9.525	5/8	15.875	7/8	22.225
9/64	3.572	25/64	9.922	41/64	16.272	57/64	22.622
5/32	3.969	13/32	10.319	21/32	16.669	29/32	23.019
11/64	4.366	27/64	10.716	43/64	17.065	59/64	23.415
3/16	4.762	7/16	11.113	11/16	17.462	15/16	23.812
13/64	5.159	29/64	11.509	45/64	17.859	61/64	24.209
7/32	5.556	15/32	11.906	23/32	18.256	31/32	24.606
15/64	5.953	31/64	12.303	47/64	18.653	63/64	25.003
1/4	6.350	1/2	12.700	3/4	19.050	1	25.400

Table A-4. EQUIVALENT METRIC AND ENGLISH STANDARD UNITS OF MEASURE

Metric Unit of Measure		Equivalent English Standard Unit of Measure
1 millimeter	mm	0.03937079″
1 centimeter	cm	0.3937079″
1 decimeter	dm	3.937079″
1 meter	m	39.37079″
		3.2808992′
		1.09361 yds.
1 decameter	dkm	32.808992′
1 kilometer	km	0.6213824 mi.
1 square cm	cm²	0.155 sq. in.
1 cubic cm	cm³	0.061 cu. in.
1 liter	l	61.023 cu. in.
1 kilogram	kg	2.2046 lbs.

English Standard Unit of Measure	Metric Equivalent Unit of Measure
1 inch	25.4 mm or 2.54 cm
1 foot	304.8 mm or 0.3048 m
1 yard	91.14 cm or 0.9114 m
1 mile	1.609 km
1 square inch	6.452 sq. cm.
1 cubic inch	16.393 cu. cm.
1 cubic foot	28.317 l
1 gallon	3.785 l
1 pound	0.4536 kg

Table A-5. MACHINABILITY, HARDNESS, AND TENSILE STRENGTH OF COMMON STEELS AND OTHER METALS AND ALLOYS

SAE Number	AISI Number	Tensile Strength (psi)	Hardness Brinell	Machinability Rating (Percent)
Carbon Steels				
1015	C1015	65,000	137	50
1020	C1020	67,000	137	52
×1020	C1022	69,000	143	62
1025	C1025	70,000	130	58
1030	C1030	75,000	138	60
1035	C1035	88,000	175	60
1040	C1040	93,000	190	60
1045	C1045	99,000	200	55
1095	C1095	100,000	201	45
Free-Cutting Steels				
×1113	B1113	83,000	193	120-140
1112	B1112	67,000	140	100
....	C1120	69,000	117	80
Manganese Steels				
×1314	....	71,000	135	94
×1335	A1335	95,000	185	70
Nickel Steels				
2315	A2317	85,000	163	50
2330	A2330	98,000	207	45
2340	A2340	110,000	225	40
2345	A2345	108,000	235	50
Nickel-Chromium Steels				
3120	A3120	75,000	151	50
3130	A3130	100,000	212	45
3140	A3140	96,000	195	57
3150	A3150	104,000	229	50
3250	....	107,000	217	44
Molybdenum Steels				
4119	....	91,000	179	60
×4130	A4130	89,000	179	58

SAE Number	AISI Number	Tensile Strength (psi)	Hardness Brinell	Machinability Rating (Percent)
Molybdenum Steels (continued)				
4140	A4140	90,000	187	56
4150	A4150	105,000	220	54
×4340	A4340	115,000	235	58
4615	A4615	82,000	167	58
4640	A4640	100,000	201	69
4815	A4815	105,000	212	55
Chromium Steels				
5120	A5120	73,000	143	50
5140	A5140		174-229	60
52100	E52101	109,000	235	45
Chromium-Vanadium Steels				
6120	A6120		179-217	50
6150	A6150	103,000	217	50
Other Alloys and Metals				
Aluminum (11S)		49,000	95	300-2,000
Brass, Leaded		55,000	RF 100	150-600
Brass, Red or Yellow		25-35,000	40-55	200
Bronze, Lead-Bearing		22-32,000	30-65	200-500
Cast Iron, Hard		45,000	220-240	50
Cast Iron, Medium		40,000	193-220	65
Cast Iron, Soft		30,000	160-193	80
Cast Steel (0.35 C)		86,000	170-212	70
Copper (F.M.)		35,000	RF 85	65
Low-Alloy, High-Strength Steel		98,000	187	80
Magnesium Alloys				500-2,000
Malleable Iron				
Standard		53-60,000	110-145	120
Pearlitic		80,000	180-200	90
Pearlitic		97,000	227	80
Stainless Steel -(12% Cr F.M.)		120,000	207	70

Table A-6. RECOMMENDED SAW BLADE PITCHES, CUTTING SPEEDS, AND FEEDS FOR POWER HACKSAWING FERROUS AND NONFERROUS METALS*

Material		Pitch (Teeth per Inch)	Cutting Speed (Feet per minute)	Feed (Force in pounds)
Ferrous Metals				
Iron	cast	6 to 10	120	125
	malleable	6 to 10	90	125
	drill rod	10	90	125
	forging stock, mild	3, 4, or 6	120	125
	alloy	4 to 6	90	125
	carbon tool	6 to 10	75	125
Steel	high-speed	6 to 10	60	125
	machine	6 to 10	120	150
	stainless	6 to 10	75	125
	structural	6 to 10	120	125
	pipe	10 to 14	120	125
	tubing — thick wall	6 to 10	120	90
	tubing — thin wall	14	120	60
Nonferrous Metals				
Aluminum	alloy	4 to 6	150	60
	heat treated	4 to 10	120	60
	soft	4 to 6	180	60
Brass	hard	10 to 14	90	60
	free machining	6 to 10	150	60
Bronze	castings	4-6 to 10	110	125
	manganese	6 to 10	80	60
Copper	bars	3, 4, or 6	130	150
	tubing	10	120	60

*Note: Coarser pitches are used for sizes over two inches and for thicker wall sections.

**DRILL WITH FLAT
FACE ON CUTTING
LIPS***

Table A-7. RECOMMENDED DRILL POINT AND LIP CLEARANCE ANGLES FOR SELECTED MATERIALS

Workpiece Material		Recommended	
		Included Drill Point Angle	Lip Clearance Angle
• General-purpose, plain carbon steels • Gray cast iron	• Annealed alloy steels • Medium-hardness pearlitic malleable iron	118°	8-12°
• Stainless steel • Heat-treated steel • Hard-to-cut materials	• Alloy steel • Drop forgings	125°	12°
• Soft cast iron • Aluminum • Wood	• Ferritic, malleable iron • Plastics • Hard Rubber	90° to 100°	15°
• Soft aluminum • Die castings	• Plastics • Wood	60°	20°
• Soft and medium copper	• Bronze* • Brass*	100° to 118°	15°
• Magnesium alloys*		60° to 118°	10°
*• 7 to 13 percent manganese steel • Armor plate	• Tough alloy steels • Extra hard materials	136° to 150°	10°

*Grind flat face on cutting lips

Table A-8. CONVERSION OF CUTTING SPEEDS AND RPM FOR ALL MACHINING OPERATIONS (Metric and Inch Standard Diameters)

CUTTING SPEEDS																		
Meters per Minute*	6.1	9.1	12.2	15.2	18.2	21.3	24.4	27.3	30.4	38	45.6	60.8	91.2	121.6	152	182.4	243.2	304
Surface Feet per Minute (sfpm)	20	30	40	50	60	70	80	90	100	125	150	200	300	400	500	600	800	1000

Diameter		REVOLUTIONS PER MINUTE																	
mm	***Inches**																		
1.6	1/16	1222	1833	2445	3056	3667	4278	4889	5500	6112	7639	9167	—	—	—	—	—	—	—
3.1	1/8	611	917	1222	1528	1833	2139	2445	2750	3056	3820	4584	6112	9167	—	—	—	—	—
4.7	3/16	407	611	815	1019	1222	1426	1630	1833	2037	2546	3056	4074	6112	8149	—	—	—	—
6.3	1/4	306	458	611	764	917	1070	1222	1375	1528	1910	2292	3056	4584	6112	7639	9167	—	—
7.8	5/16	244	367	489	611	733	856	978	1100	1222	1528	1833	2445	3667	4889	6112	7334	9778	—
9.4	3/8	204	306	407	509	611	713	815	917	1019	1273	1528	2037	3056	4074	5093	6112	8149	—
10.9	7/16	175	262	349	437	524	611	698	786	873	1091	1310	1746	2619	3492	4365	5238	6985	8731
12.7	1/2	153	229	306	382	458	535	611	688	764	955	1146	1528	2292	3056	3820	4584	6112	7639
15.6	5/8	122	183	244	306	367	428	489	550	611	764	917	1222	1833	2445	3056	3667	4889	6112
18.8	3/4	102	153	204	255	306	357	407	458	509	637	764	1019	1528	2037	2546	3056	4074	5093
21.9	7/8	87	131	175	218	262	306	349	393	437	546	655	873	1310	1746	2183	2619	3492	4365
25.4	1	76	115	153	191	229	267	306	344	382	477	573	764	1146	1528	1910	2292	3056	3820
28.1	1 1/8	68	102	136	170	204	238	272	306	340	424	509	679	1019	1358	1698	2037	2716	3395
31.7	1 1/4	61	92	122	153	183	214	244	275	306	382	458	611	917	1222	1528	1833	2445	3056
34.8	1 3/8	56	83	111	139	167	194	222	250	278	347	417	556	833	1111	1389	1667	2222	2778
38.1	1 1/2	51	76	102	127	153	178	204	229	255	318	382	509	764	1019	1273	1528	2037	2546
41.0	1 5/8	47	71	94	118	141	165	188	212	235	294	353	470	705	940	1175	1410	1880	2351
44.2	1 3/4	44	66	87	109	131	153	175	196	218	273	327	437	655	873	1091	1310	1746	2183
47.3	1 7/8	41	61	82	102	122	143	163	183	204	255	306	407	611	815	1019	1222	1630	2037
50.8	2	38	57	76	96	115	134	153	172	191	239	286	382	573	764	955	1146	1528	1910
	2 1/2	31	46	61	76	92	107	122	138	153	191	229	306	458	611	764	917	1222	1528

*The meters per minute and mm diameters are approximately equivalent to the corresponding sfpm and the diameters given in inches

Table A-9. RECOMMENDED CUTTING SPEEDS FOR DRILLING WITH HIGH-SPEED DRILLS*

Material		Hardness (Bhn)	Cutting Speed (fpm)
Plain Carbon Steels			
AISI-1019, 1020		120-150	80-120
1030, 1040, 1050		150-170	70-90
1060, 1070, 1080, 1090		170-190	60-80
		190-220	50-70
		220-280	40-50
		280-350	30-40
		350-425	15-30
Alloy Steels			
AISI-1320, 2317, 2515		125-175	60-80
3120, 3316, 4012, 4020		175-225	50-70
4120, 4128, 4320, 4620		225-275	45-60
4720, 4820, 5020, 5120		275-325	35-55
6120, 6325, 6415, 8620		325-375	30-40
8720, 9315		375-425	15-30
Alloy Steels			
AISI-1330, 1340, 2330		175-225	50-70
2340, 3130, 3140, 3150		225-275	40-60
4030, 4063, 4130, 4140		275-325	30-50
4150, 4340, 4640, 5130		325-375	25-40
5140, 5160, 5210, 6150		375-425	15-30
6180, 6240, 6290, 6340			
6380, 8640, 8660, 8740,			
9260, 9445, 9840, 9850			
Stainless Steel			
Standard Grades	Austenitic, Annealed	135-185	40-50
	Cold-Drawn, Austenitic	225-275	30-40
	Ferritic	135-185	50-60
	Martensitic, Annealed	135-175	55-70
		175-225	50-60
	Tempered, Quenched	275-325	30-40
		375-425	15-30
Free Machining Grades	Annealed, Austenitic	135-185	80-100
	Cold-Drawn, Austenitic	225-275	60-90
	Ferritic	135-185	100-200
	Martensitic, Annealed	135-185	100-130
	Martensitic, Cold-Drawn	185-240	90-120
	Martensitic, Tempered	275-325	50-60
	Quenched	375-425	30-40

Material	Hardness (Bhn)	Cutting Speed (fpm)
Tool Steels		
Water Hardening	150-250	70-80
Cold Work	200-250	20-40
Shock Resisting	175-225	40-50
Mold	100-150	60-70
	150-200	50-60
High-Speed Steel	200-250	30-40
	250-275	15-30
Gray Cast Iron	110-140	90-140
	150-190	80-100
	190-220	60-80
	220-260	50-70
	260-320	30-40
Malleable Iron		
Ferritic	110-160	120-140
Pearlitic	160-200	90-110
	200-240	60-90
	240-280	50-60
Aluminum Alloys		
Cast Non-Heat Treated		200-300
Heat Treated		150-250
Wrought Cold Drawn		150-300
Heat Treated		140-300
Brass, Bronze (Ordinary)		150-300
Bronze (High-Strength)		30-100

*Use cutting speeds of 50% to 65% of those given in the table for high-speed steel reamers

Table A-10. GENERAL NC TAP SIZES AND RECOMMENDED TAP DRILLS

NC STANDARD THREADS		
Size (Outside Diameter inch)	Threads per Inch	Tap Drill Size* (75% Thread Depth)
¼	20	No. 7
⁵/₁₆	18	"F"
⅜	16	⁵/₁₆
⁷/₁₆	14	"U"
½	13	²⁷/₆₄
⁹/₁₆	12	³¹/₆₄
⅝	11	¹⁷/₃₂
¹¹/₁₆	11N.S.	¹⁹/₃₂
¾	10	²¹/₃₂
¹³/₁₆	10N.S.	²³/₃₂
⅞	9	⁴⁹/₆₄
¹⁵/₁₆	9N.S.	⁵³/₆₄
1	8	⅞
1⅛	7	⁶³/₆₄
1¼	7	1⁷/₆₄
1⅜	6	1¹³/₆₄
1½	6	1¹¹/₃₂
1⅝	5½N.S.	1²⁹/₆₄
1¾	5	1³⁵/₆₄
1⅞	5N.S.	1¹¹/₁₆
2	4½	1²⁵/₃₂

*Nearest commercial drill size to produce a 75% thread depth

Table A-11. GENERAL NF TAP SIZES AND RECOMMENDED TAP DRILLS

NF STANDARD THREADS		
Size (Outside Diameter inch)	Threads Per Inch	Tap Drill Size* (75% Thread Depth)
¼	28	No. 3
⁵/₁₆	24	"I"
⅜	24	"Q"
⁷/₁₆	20	"W"
½	20	²⁹/₆₄
⁹/₁₆	18	³³/₆₄
⅝	18	³⁷/₆₄
¹¹/₁₆	11N.S.	¹⁹/₃₂
¾	16	¹¹/₁₆
¹³/₁₆	10N.S.	²³/₃₂
⅞	14	¹³/₁₆
¹⁵/₁₆	9N.S.	⁵³/₆₄
1	12	⁵⁹/₆₄
1	14N.S.	¹⁵/₁₆
1⅛	12	1³/₆₄
1¼	12	1¹¹/₆₄
1⅜	12	1¹⁹/₆₄
1½	12	1²⁷/₆₄

*Nearest commercial drill size to produce a 75% thread depth

Table A-12. GENERAL MACHINE SCREW TAP SIZES AND RECOMMENDED TAP DRILLS

MACHINE SCREW THREADS		
Screw Gage Number	Threads per Inch	Tap Drill Size* (75% Thread Depth)
0	80	$\frac{3}{64}$"
1	72	#53
1	64	#53
2	64	#50
2	56	#51
3	56	#46
3	48	$\frac{5}{64}$
4	48	#42
4	40	#43
4	36	#44
5	44	#37
5	40	#39
6	40	#33
6	32	#36
8	36	#29
8	32	#29
10	32	#21
10	24	#25
12	28	#15
12	24	#17
14	24	#7
14	20	#10

*Nearest commercial drill size to produce a 75% thread depth

Table A-13. GENERAL METRIC STANDARD TAP SIZES AND RECOMMENDED TAP DRILLS

METRIC STANDARD THREADS				
Metric Thread Size (Nominal Outside Diameter and Pitch) (mm)	Recommended Tap Drill Size			
	Metric Series Drills		Inch Series Drills	
	Size (mm)	Equivalent (inch)	Nominal size	Diameter (inch)
M 1.6 × 0.35	1.25	.0492	- - -	- - -
M 2 × 0.4	1.60	.0630	#52	.0635
M 2.5 × 0.45	2.05	.0807	#45	.0820
M 3 × 0.5	2.50	.0984	#39	.0995
M 3.5 × 0.6	2.90	.1142	#32	.1160
M 4 × 0.7	3.30	.1299	#30	.1285
M 5 × 0.8	4.20	.1654	#19	.1660
M 6 × 1	5.00	.1968	#8	.1990
M 8 × 1.25	6.80	.2677	"H"	.2660
M 10 × 1.5	8.50	.3346	"Q"	.3320
M 12 × 1.75	10.25	.4035	$\frac{13}{32}$	.4062
M 14 × 2	12.00	.4724	$\frac{15}{32}$	.4688
M 16 × 2	14.00	.5512	$\frac{35}{64}$	.5469
M 20 × 2.5	17.50	.6890	$\frac{11}{16}$	.6875
M 24 × 3	21.00	.8268	$\frac{53}{64}$	.8281
M 30 × 3.5	26.50	1.0433	1-$\frac{3}{64}$	1.0469
M 36 × 4	32.00	1.2598	1-¼	1.2500

Table A-14. BASIC DIMENSIONS COARSE-THREAD SERIES (UNC AND NC)

Sizes	Threads Per Inch	Basic Major Diameter (")	Basic Pitch Diameter (")	Minor Diameter		Lead Angle at Basic Pitch Diameter	
				External Threads (")	Internal Threads (")	Deg. (°)	Min. (')
1 (.073)*	64	0.0730	0.0629	0.0538	0.0561	4	31
2 (.086)	56	0.0860	0.0744	0.0641	0.0667	4	22
3 (.099)*	48	0.0990	0.0855	0.0734	0.0764	4	26
4 (.112)	40	0.1120	0.0958	0.0813	0.0849	4	45
5 (.125)	40	0.1250	0.1088	0.0943	0.0979	4	11
6 (.138)	32	0.1380	0.1177	0.0997	0.1042	4	50
8 (.164)	32	0.1640	0.1437	0.1257	0.1302	3	58
10 (.190)	24	0.1900	0.1629	0.1389	0.1449	4	39
12 (.216)*	24	0.2160	0.1889	0.1649	0.1709	4	1
¼	20	0.2500	0.2175	0.1887	0.1959	4	11
⁵⁄₁₆	18	0.3125	0.2764	0.2443	0.2524	3	40
⅜	16	0.3750	0.3344	0.2983	0.3073	3	24
⁷⁄₁₆	14	0.4375	0.3911	0.3499	0.3602	3	20
½	13	0.5000	0.4500	0.4056	0.4167	3	7
⁹⁄₁₆	12	0.5625	0.5084	0.4603	0.4723	2	59
⅝	11	0.6250	0.5660	0.5135	0.5266	2	56
¾	10	0.7500	0.6850	0.6273	0.6417	2	40
⅞	9	0.8750	0.8028	0.7387	0.7547	2	31
1	8	1.0000	0.9188	0.8466	0.8647	2	29
1⅛	7	1.1250	1.0322	0.9497	0.9704	2	31
1¼	7	1.2500	1.1572	1.0747	1.0954	2	15
1⅜	6	1.3750	1.2667	1.1705	1.1946	2	24
1½	6	1.5000	1.3917	1.2955	1.3196	2	11
1¾	5	1.7500	1.6201	1.5046	1.5335	2	15
2	4½	2.0000	1.8557	1.7274	1.7594	2	11
2¼	4½	2.2500	2.1057	1.9774	2.0094	1	55
2½	4	2.5000	2.3376	2.1933	2.2294	1	57
2¾	4	2.7500	2.5876	2.4433	2.4794	1	46
3	4	3.0000	2.8376	2.6933	2.7294	1	36
3¼	4	3.2500	3.0876	2.9433	2.9794	1	29
3½	4	3.5000	3.3376	3.1933	3.2294	1	22
3¾	4	3.7500	3.5876	3.4133	3.4794	1	16
4	4	4.0000	3.8376	3.6933	3.7294	1	11

*Secondary sizes

Table A-15. BASIC DIMENSIONS FINE-THREAD SERIES (UNF AND NF)

Sizes	Threads Per Inch	Basic Major Diameter (")	Basic Pitch Diameter (")	Minor Diameter		Lead Angle at Basic Pitch Diameter	
				External Threads (")	Internal Threads (")	Deg. (°)	Min. (')
0 (.060)	80	0.0600	0.0519	0.0447	0.0465	4	23
1 (0.73)*	72	0.0730	0.0640	0.0560	0.0580	3	57
2 (.086)	64	0.0860	0.0759	0.0668	0.0691	3	45
3 (.099)*	56	0.0990	0.0874	0.0771	0.0797	3	43
4 (.112)	48	0.1120	0.0985	0.0864	0.0894	3	51
5 (.125)	44	0.1250	0.1102	0.0971	0.1004	3	45
6 (.138)	40	0.1380	0.1218	0.1073	0.1109	3	44
8 (.164)	36	0.1640	0.1460	0.1299	0.1339	3	28
10 (.190)	32	0.1900	0.1697	0.1517	0.1562	3	21
12 (.216)*	28	0.2160	0.1928	0.1722	0.1773	3	22
¼	28	0.2500	0.2268	0.2062	0.2113	2	52
⁵/₁₆	24	0.3125	0.2854	0.2614	0.2674	2	40
⅜	24	0.3750	0.3479	0.3239	0.3299	2	11
⁷/₁₆	20	0.4375	0.4050	0.3762	0.3834	2	15
½	20	0.5000	0.4675	0.4387	0.4459	1	57
⁹/₁₆	18	0.5625	0.5264	0.4943	0.5024	1	55
⅝	18	0.6250	0.5889	0.5568	0.5649	1	43
¾	16	0.7500	0.7094	0.6733	0.6823	1	36
⅞	14	0.8750	0.8286	0.7874	0.7977	1	34
1	12	1.0000	0.9459	0.8978	0.9098	1	36
1⅛	12	1.1250	1.0709	1.0228	1.0348	1	25
1¼	12	1.2500	1.1959	1.1478	1.1598	1	16
1⅜	12	1.3750	1.3209	1.2728	1.2848	1	9
1½	12	1.5000	1.4459	1.3978	1.4098	1	3

*Secondary sizes

Table A-16. PERCENT OF THREAD DEPTH FOR RECOMMENDED METRIC AND INCH STANDARD TAP DRILLS FOR METRIC TAP SIZES M1.6 to M27

| Metric Tap Size | Recommended Metric Standard Drill | | | | Closest Recommended Inch Standard Drill | | | |
	Drill Size (Millimeters)	Inch Equivalent	Probable Hole Size (Inches)	Percent of Thread	Drill Size	Inch Equivalent	Probable Hole Size (Inches)	Percent of Thread
M1.6×0.35	1.25	0.0492	0.0507	69	—	—	—	—
M1.8×0.35	1.45	0.0571	0.0586	69	—	—	—	—
M2×0.4	1.60	0.0630	0.0647	69	#52	0.0635	0.0652	66
M2.2×0.45	1.75	0.0689	0.0706	70	—	—	—	—
M2.5×0.45	2.05	0.0807	0.0826	69	#46	0.0810	0.0829	67
M3×0.5	2.50	0.0984	0.1007	68	#40	0.0980	0.1003	70
M3.5×0.6	2.90	0.1142	0.1168	68	#33	0.1130	0.1156	72
M4×0.7	3.30	0.1299	0.1328	69	#30	0.1285	0.1314	73
M4.5×0.75	3.70	0.1457	0.1489	74	#26	0.1470	0.1502	70
M5×0.8	4.20	0.1654	0.1686	69	#19	0.1660	0.1692	68
M6×1	5.00	0.1968	0.2006	70	#9	0.1960	0.1998	71
M7×1	6.00	0.2362	0.2400	70	15/64	0.2344	0.2382	73
M8×1.25	6.70	0.2638	0.2679	74	17/64	0.2656	0.2697	71
M8×1	7.00	0.2756	0.2797	69	J	0.2770	0.2811	66
M10×1.5	8.50	0.3346	0.3390	71	Q	0.3320	0.3364	75
M10×1.25	8.70	0.3425	0.3471	73	11/32	0.3438	0.3483	71
M12×1.75	10.20	0.4016	0.4063	74	Y	0.4040	0.4087	71
M12×1.25	10.80	0.4252	0.4299	67	27/64	0.4219	0.4266	72
M14×2	12.00	0.4724	0.4772	72	15/32	0.4688	0.4736	76
M14×1.5	12.50	0.4921	0.4969	71	—	—	—	—
M16×2	14.00	0.5512	0.5561	72	35/64	0.5469	0.5518	76
M16×1.5	14.50	0.5709	0.5758	71	—	—	—	—
M18×2.5	15.50	0.6102	0.6152	73	39/64	0.6094	0.6114	74
M18×1.5	16.50	0.6496	0.6546	70	—	—	—	—
M20×2.5	17.50	0.6890	0.6942	73	11/16	0.6875	0.6925	74
M20×1.5	18.50	0.7283	0.7335	70	—	—	—	—
M22×2.5	19.50	0.7677	0.7729	73	49/64	0.7656	0.7708	75
M22×1.5	20.50	0.8071	0.8123	70	—	—	—	—
M24×3	21.00	0.8268	0.8327	73	53/64	0.8281	0.8340	72
M24×2	22.00	0.8661	0.8720	71	—	—	—	—
M27×3	24.00	0.9449	0.9511	73	15/16	0.9375	0.9435	78
M27×2	25.00	0.9843	0.9913	70	63/64	0.9844	0.9914	70

Formulas for Metric Tap Drill Size and Percent of Thread

(M in the tap size is the nominal thread size in millimeters)

$$\text{Drilled Hole Size*} = \text{Basic major diameter} - \frac{\%\ \text{Thread} \times \text{Pitch*}}{76.980}$$

$$\text{Percent of Thread} = \frac{76.980}{\text{Pitch*}} \times (\text{Basic Major Diameter*} - \text{Drilled Hole Size*})$$

*In mm

Table A-17. SCREW THREAD FORMS AND FORMULAS (Unified And American Standard)

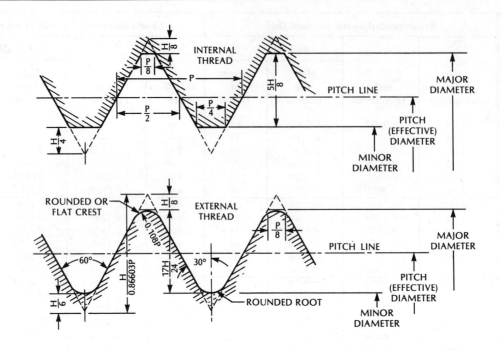

Formulas for Basic Dimensions of Unified and American Standard Unified Systems

H = (height of sharp V-thread)
= 0.86603 × pitch

Pitch	$= \dfrac{1}{\text{Number of Threads per Inch}}$	Crest truncation external thread	$= 0.10825 \times \text{pitch} = \dfrac{H}{8}$
Depth, external thread	$= 0.61343 \times \text{pitch}$	Crest truncation, internal thread	$= 0.21651 \times \text{pitch} = \dfrac{H}{4}$
Depth, internal thread	$= 0.54127 \times \text{pitch}$	Root truncation, external thread	$= 0.14434 \times \text{pitch} = \dfrac{H}{6}$
Flat at crest, external thread	$= 0.125 \times \text{pitch}$	Root truncation, internal thread	$= 0.10825 \times \text{pitch} = \dfrac{H}{8}$
Flat at crest, internal thread	$= 0.250 \times \text{pitch}$	Addendum, external thread	$= 0.32476 \times \text{pitch}$
Flat at root, internal thread	$= 0.125 \times \text{pitch}$	Pitch diameter, external and internal	= Major Diameter − 2 addendum (addendum external thread)

Table A-18. HOLE SIZES FOR TAPPING THREADS TO DIFFERENT DEPTHS AND LENGTHS: Unified and American National Coarse (UNC, NC) and Fine (UNF, NF) Series

UNC, NC	UNF, NF	Thread Size	Hole Size for Different Thread Depth Percentages <1							Recommended Limits for Hole Size Based on Lengths <2 of Thread Engagement — Percentage of Diameter							
										33%		33 to 66%		66% to 1½x		1½ to 3x	
			88⅓%	75%	70%	65%	60%	55%	50%	Min.	Max.	Min.	Max.	Min.	Max.	Min.	Max.
	x	0-80	.0465	.0479	.0486	.0494	.0502	.0510	.0519	.0465	.0500	.0479	.0514	.0479	.0514	.0479	.0514
x		1-64	.0561	.0578	.0588	.0599	.0609	.0619	.0629	.0561	.0599	.0585	.0623	.0585	.0623	.0585	.0623
	x	1-72	.0580	.0595	.0604	.0613	.0622	.0631	.0640	.0580	.0613	.0596	.0629	.0602	.0635	.0602	.0635
x		2-56	.0667	.0686	.0698	.0710	.0721	.0732	.0744	.0667	.0705	.0686	.0724	.0699	.0737	.0699	.0737
	x	2-64	.0691	.0708	.0718	.0729	.0739	.0749	.0759	.0691	.0724	.0707	.0740	.0720	.0753	.0720	.0753
x		3-48	.0764	.0788	.0801	.0815	.0828	.0841	.0855	.0764	.0804	.0785	.0825	.0805	.0845	.0806	.0846
	x	3-56	.0797	.0816	.0828	.0840	.0851	.0862	.0874	.0797	.0831	.0814	.0848	.0831	.0865	.0833	.0867
x		4-40	.0849	.0877	.0893	.0909	.0926	.0942	.0958	.0849	.0894	.0871	.0916	.0894	.0939	.0902	.0947
	x	4-48	.0894	.0918	.0931	.0945	.0958	.0971	.0985	.0894	.0931	.0912	.0949	.0931	.0968	.0939	.0976
x		5-40	.0979	.1007	.1023	.1039	.1056	.1072	.1088	.0979	.1020	.1000	.1041	.1021	.1062	.1036	.1077
	x	5-44	.1004	.1029	.1044	.1059	.1073	.1087	.1102	.1004	.1041	.1023	.1060	.1042	.1079	.1060	.1097
x		6-32	.1042	.1076	.1096	.1117	.1137	.1157	.1177	.1042	.1091	.1066	.1115	.1091	.1140	.1115	.1164
	x	6-40	.1109	.1137	.1153	.1169	.1185	.1201	.1218	.1109	.1148	.1128	.1167	.1147	.1186	.1166	.1205
x		8-32	.1302	.1336	.1356	.1377	.1397	.1417	.1437	.1302	.1345	.1324	.1367	.1346	.1389	.1367	.1410
	x	8-36	.1339	.1370	.1388	.1406	.1424	.1442	.1460	.1339	.1377	.1359	.1397	.1378	.1416	.1397	.1435
x		10-24	.1449	.1495	.1522	.1549	.1576	.1602	.1629	.1449	.1502	.1475	.1528	.1502	.1555	.1528	.1581
	x	10-32	.1562	.1596	.1616	.1637	.1657	.1677	.1697	.1562	.1601	.1581	.1621	.1601	.1641	.1621	.1661
x		12-24	.1709	.1755	.1782	.1809	.1836	.1862	.1889	.1709	.1758	.1733	.1782	.1758	.1807	.1782	.1831
	x	12-28	.1773	.1813	.1836	.1859	.1882	.1905	.1928	.1773	.1815	.1794	.1836	.1815	.1857	.1836	.1878
x		¼-20	.1959	.2012	.2046	.2078	.2111	.2143	.2175	.1959	.2013	.1986	.2040	.2013	.2067	.2040	.2094
	x	¼-28	.2113	.2153	.2176	.2199	.2222	.2245	.2268	.2113	.2152	.2131	.2171	.2150	.2190	.2169	.2209
x		5/16-18	.2524	.2584	.2620	.2656	.2692	.2728	.2764	.2524	.2577	.2551	.2604	.2577	.2630	.2604	.2657
	x	5/16-24	.2674	.2720	.2747	.2774	.2801	.2827	.2854	.2674	.2714	.2694	.2734	.2714	.2754	.2734	.2774
x		⅜-16	.3073	.3142	.3182	.3223	.3263	.3303	.3344	.3073	.3127	.3101	.3155	.3128	.3182	.3155	.3209
	x	⅜-24	.3299	.3345	.3372	.3399	.3426	.3452	.3479	.3299	.3336	.3314	.3354	.3332	.3372	.3351	.3391
x		7/16-14	.3602	.3680	.3726	.3772	.3819	.3865	.3911	.3602	.3660	.3630	.3688	.3659	.3717	.3688	.3746
	x	7/16-20	.3834	.3888	.3921	.3953	.3986	.4018	.4050	.3834	.3875	.3855	.3896	.3875	.3916	.3896	.3937
x		½-13	.4167	.4251	.4301	.4351	.4401	.4450	.4500	.4167	.4225	.4196	.4254	.4226	.4284	.4255	.4313
	x	½-20	.4459	.4513	.4546	.4578	.4611	.4643	.4675	.4459	.4498	.4477	.4517	.4497	.4537	.4516	.4556
x		9/16-12	.4723	.4814	.4868	.4922	.4976	.5030	.5084	.4723	.4783	.4753	.4813	.4783	.4843	.4813	.4873
	x	9/16-18	.5024	.5084	.5120	.5156	.5192	.5228	.5264	.5024	.5065	.5045	.5086	.5065	.5106	.5086	.5127
x		⅝-11	.5266	.5365	.5424	.5483	.5542	.5601	.5660	.5266	.5328	.5298	.5360	.5329	.5391	.5360	.5422
	x	⅝-18	.5649	.5709	.5745	.5787	.5817	.5853	.5889	.5649	.5690	.5670	.5711	.5690	.5730	.5711	.5752
x		¾-10	.6417	.6526	.6591	.6656	.6721	.6785	.6850	.6417	.6481	.6449	.6513	.6481	.6545	.6513	.6577
	x	¾-16	.6823	.6892	.6932	.6973	.7013	.7053	.7094	.6823	.6866	.6844	.6887	.6865	.6908	.6886	.6929
x		⅞-9	.7547	.7668	.7740	.7812	.7884	.7956	.8028	.7547	.7614	.7580	.7647	.7614	.7681	.7647	.7714
	x	⅞-14	.7977	.8055	.8101	.8147	.8194	.8240	.8286	.7977	.8022	.8000	.8045	.8023	.8068	.8045	.8090
x		1-8	.8647	.8783	.8864	.8945	.9026	.9107	.9188	.8647	.8722	.8684	.8759	.8722	.8797	.8760	.8835
	x	1-12	.9098	.9188	.9242	.9296	.9350	.9404	.9459	.9098	.9148	.9123	.9173	.9148	.9198	.9173	.9223
x		1⅛-7	.9704	.9859	.9951	1.0044	1.0137	1.0229	1.0322	.9704	.9790	.9747	.9833	.9789	.9875	.9832	.9918
	x	1⅛-12	1.0348	1.0439	1.0493	1.0547	1.0601	1.0655	1.0709	1.0348	1.0398	1.0373	1.0423	1.0398	1.0448	1.0423	1.0473
x		1¼-7	1.0954	1.1109	1.1201	1.1294	1.1387	1.1479	1.1572	1.0954	1.1040	1.0997	1.1083	1.1039	1.1125	1.1082	1.1168
	x	1¼-12	1.1598	1.1689	1.1743	1.1797	1.1851	1.1905	1.1959	1.1598	1.1648	1.1623	1.1673	1.1648	1.1698	1.1673	1.1723
x		1⅜-6	1.1946	1.2127	1.2235	1.2343	1.2451	1.2559	1.2667	1.1946	1.2046	1.1996	1.2096	1.2046	1.2146	1.2096	1.2196
	x	1⅜-12	1.2848	1.2939	1.2993	1.3047	1.3101	1.3155	1.3209	1.2848	1.2898	1.2873	1.2923	1.2898	1.2948	1.2923	1.2973
x		1½-6	1.3196	1.3377	1.3485	1.3593	1.3701	1.3809	1.3917	1.3196	1.3296	1.3246	1.3346	1.3296	1.3396	1.3346	1.3446
	x	1½-12	1.4098	1.4189	1.4243	1.4297	1.4351	1.4405	1.4459	1.4098	1.4148	1.4123	1.4173	1.4148	1.4198	1.4173	1.4223

1 Thread tests show that general strength requirements are met at 55 to 75% of thread depth.

2 A thread depth of 50% is generally satisfactory when the length of threads is one and one half or more times the outside thread diameter

Table A-19. RECOMMENDED CUTTING SPEEDS FOR REAMING COMMON MATERIALS
(High-Speed Steel Machine Reamers)

Material		Cutting Speed Range	
		fpm	m/min*
Steel	Machinery .2C to .3C	50-70	15-22
	Annealed .4C to .5C	40-50	12-15
	Tool, 1.2C	35-40	10-12
	Alloy	35-40	10-12
	Automotive forgings	35-40	10-12
	Alloy 300-400 brinell	20-30	6-9
	Free machining stainless	40-50	12-15
	Hard stainless	20-30	6-9
Monel Metal		25-35	8-12
Cast iron	Soft	70-100	22-30
	Hard	50-70	15-22
	Chilled	20-30	6-9
	Malleable iron	50-60	15-18
Brass/bronze	Ordinary	130-200	40-60
	High tensile	50-70	15-22
Bakelite		70-100	20-30
Aluminum and its alloys		130-200	40-60
Magnesium and its alloys		170-270	50-80

*m/min: meters per minute

Table A-20. RECOMMENDED CUTTING SPEEDS FOR GENERAL MACHINE TAPPING*

Material to be Tapped			Range of Cutting Speeds (fpm)
Steels	Low carbon (to .25%C)		40 to 80
	Medium carbon (.30 to .60% C)	Annealed	30 to 60
		Heat treated (220 to 280 Bhn)	20 to 50
	Tool, high carbon and high-speed steels		20 to 40
	Stainless		5 to 35
Cast Iron	Gray		40 to 100
	Malleable	Ferritic	80 to 120
		Pearlitic	40 to 80

Material to be Tapped			Range of Cutting Speeds (fpm)
Non-ferrous Metals	Aluminum		50 to 200
	Brass	General	50 to 200
		Manganese	30 to 60
		Naval	80 to 100
	Bronze	Phosphor	30 to 60
		Tobin	80 to 100
	Zinc die castings		60 to 150
Non-metallic Materials	Bakelite		50 to 100
	Plastics	Thermo-Plastics	50 to 100
		Thermo-setting	
	Hard rubber		50 to 100

*The cutting speed to use within a recommended low to high range is affected by the factors and guidelines described for machine tapping

Table A-21. RECOMMENDED CUTTING SPEEDS AND FEEDS (IN SFPM) FOR SELECTED MATERIALS AND LATHE PROCESSES (High-Speed Steel Cutting Tools and Knurls)

Material		Turning, Boring, Facing				Threading	Parting	Knurling
		Rough Cut		Finish Cut				
		rpm	feed (″)	rpm	feed (″)			
Ferrous	Cast Iron	70	.015 to .025	80	.005 to .012	20	80	30
	Machine Steel	100	.010 to .020	110	.003 to .010	40	80	30
	Tool Steel	70	.010 to .020	80	.003 to .010	25	70	30
	Stainless Steel	75-90	.015 to .030	100-130	.005 to .010	35	70	20
NonFerrous	Aluminum	200	.015 to .030	300	.005 to .010	50	200	40
	Brass	180	.015 to .025	220	.003 to .010	45	150	40
	Bronze	180	.015 to .025	220	.003 to .010	45	150	40

Table A-22. RECOMMENDED CUTTING SPEEDS (IN SFPM) FOR CARBIDE AND CERAMIC CUTTING TOOLS

	Material	Carbide Cutting Tools (sfpm)		Ceramic Cutting Tools (sfpm)	
		Rough Cuts	Finish Cuts	Rough Cuts	Finish Cuts
NonFerrous	Cast Iron	200-250	350-450	200-800	200-2000
	Machine Steel	400-500	700-1000	250-1200	400-1800
	Tool Steel	300-400	500-750	300-1500	600-2000
	Stainless Steel	250-300	375-500	300-1000	400-1200
Ferrous	Aluminum	300-450	700-1000	40-2000	600-3000
	Brass	500-600	700-800	400-800	600-1200
	Bronze	500-600	700-800	150-800	200-1000

Table A-23. RECOMMENDED CUTTING SPEEDS AND FEEDS FOR VARIOUS DEPTHS OF CUT ON COMMON METALS (Single-Point Carbide Cutting Tools)

Metal	Depth of Cut (inches)	Feed per Revolution (inches)	Cutting Speed (sfpm)
Aluminum	.005-.015	.002-.005	700-1000
	.020-.090	.005-.015	450-700
	.100-.200	.015-.030	300-450
	.300-.700	.030-.090	100-200
Brass Bronze	.005-.015	.002-.005	700-800
	.020-.090	.005-.015	600-700
	.100-.200	.015-.030	500-600
	.300-.700	.030-.090	200-400
Cast iron (gray)	.005-.015	.002-.005	350-400
	.020-.090	.005-.015	250-350
	.100-.200	.015-.030	200-250
	.300-.700	.030-.090	75-150
Machine Steel	.005-.015	.002-.005	700-1000
	.020-.090	.005-.015	550-700
	.100-.200	.015-.030	400-550
	.300-.700	.030-.090	150-300
Tool Steel	.005-.015	.002-.005	500-700
	.020-.090	.005-.015	400-500
	.100-.200	.015-.030	300-400
	.300-.700	.030-.090	100-300
Stainless Steel	.005-.015	.002-.005	375-500
	.020-.090	.005-.015	300-375
	.100-.200	.015-.030	250-300
	.300-.700	.030-.090	75-175

Table A-24. RECOMMENDED RAKE ANGLES FOR SELECTED FERROUS AND NONFERROUS MATERIALS

	Material	Hardness (Bhn*)	High-Speed Steel Cutting Tools		Carbide Cutting Tools	
			Back Rake Angle, deg.	Side Rake Angle, deg.	Back Rake Angle, deg.	Side Rake Angle, deg.
Steel	Plain Carbon	100 to 200	5 to 10	10 to 20	0 to 5	7 to 15
		200 to 300	5 to 7	8 to 12	0 to 5	5 to 8
		300 to 400	0 to 5	5 to 10	−5 to 0	3 to 5
		400 to 500	−5 to 0	−5 to 0	−8 to 0	−6 to 0
	Alloy	100 to 200	5 to 10	10 to 16	0 to 5	5 to 15
		200 to 300	5 to 7	8 to 12	0 to 5	5 to 8
		300 to 400	0 to 5	5 to 10	−5 to 0	3 to 5
		400 to 500	−5 to 0	−5 to 0	−6 to 0	−6 to 0
	Stainless: Ferritic, Austenitic	130 to 190	5 to 7	8 to 10	0 to 5	5 to 7
	Martensitic	130 to 220	0 to 5	5 to 8	−5 to 5	6 to 15
Aluminum	Nonheat-Treated		10 to 20	30 to 35	0 to 15	15 to 30
	Heat-treated		5 to 12	15 to 20	0 to 5	8 to 15
Magnesium			5 to 10	10 to 20	0 to 5	10 to 20
Iron	Gray Cast	100 to 200	5 to 10	10 to 15	0 to 5	6 to 15
		200 to 300	5 to 7	5 to 10	0 to 5	5 to 7
		300 to 400	−5 to 0	−5 to 0	−6 to 0	−6 to 0
	Malleable Ferritic	110 to 160	5 to 15	12 to 20	0 to 10	7 to 15
	Pearlitic	160 to 200	5 to 8	10 to 12	0 to 5	5 to 8
		200 to 280	0 to 5	−5 to 8	−5 to 5	−5 to 8
Brass	Free cutting		−5 to 5	0 to 10		
	Red, Yellow, Naval		−5 to 5	−5 to 5	−5 to 5	−5 to 5
Bronze	Hard phosphor		−5 to 0	−6 to 3	−5 to 0	−5 to 5
	Manganese		−5 to 5	−5 to 5	−5 to 5	−5 to 5

*Note: The rake angles and hardness values are listed in ascending order, like 5° to 10° and 100 Bhn to 200 Bhn. Actually, the cutting tools are ground to the smaller and even negative angles for the harder (higher) Bhn values. For example, the recommended back rake angle for a high-speed steel cutter to machine plain carbon steel of Bhn 100 hardness is 10°; for Bhn 200, the angle is 5°

Table A-25. AVERAGE CUTTING TOOL ANGLES AND CUTTING SPEEDS FOR SINGLE-POINT, HIGH-SPEED STEEL CUTTING TOOLS

	Material to be Machined	Side [1] Relief	End [1] Relief	Side Rake [2]	True Back [3] Rake	Suggested Cutting Speeds [4]	
						sfpm	m/min
Steel	Free-machining	10°	10°	10°-22°	16°	160-350	50-110
	Low-carbon steel (.05%-.30%C)	10°	10°	10°-14°	16°	90-100	27-30
	Medium-carbon (.30%-.60%C)	10°	10°	10°-14°	12°	70-90	21-27
	High-carbon tool steel (.60-1.70%C)	8°	8°	8°-12°	8°	50-70	15-21
	Tough alloy	8°	8°	8°-12°	8°	50-70	15-21
	Stainless	8°	8°	5°-10°	8°	40-70	12-21
	Stainless, free-machining	10°	10°	5°-10°	16°	80-140	24-42
Cast Iron	Soft	8°	8°	10°	8°	50-80	15-24
	Hard	8°	8°	8°	5°	30-50	9-15
	Malleable	8°	8°	10°	8°	80-100	24-30
Non-Ferrous	Aluminum	10°	10°	10°-20°	35°	200-1500	60-460
	Copper	10°	10°	10°-20°	16°	100-120	30-36
	Brass	10°	8°	0°	0°	150-300	45-90
	Bronze	10°	8°	0°	0°	90-100	27-30
Plastics	Molded	10°	12°	0°	0°	150-300	45-90
	Acrylics	15°	15°	0°	0°	60-70	18-21
	Fiber	15°	15°	0°	0°	80-100	24-30

[1] The side- and end-relief angles are usually ground to the same angle for general machining operations. Average 8° end-relief and 10° side-relief angles are standard for turning most metals. The end- and side-relief angles for shaper and planer cutting tools generally range from 3° to 5°.

[2] The smaller angle is used without a chip breaker; the larger angle, with a chip breaker.

[3] The rake angles are measured from horizontal and vertical planes.

[4] The slower speeds are used for roughing cuts, machining dry. The faster speeds are used for finishing cuts, using cutting fluids. Cutting speeds for cast alloy bits are 50% to 70% faster; for cemented carbide tips and inserts, two to four times faster.

Table A-26. RECOMMENDED CUTTING SPEEDS OF COMMON CUTTER MATERIALS FOR MILLING SELECTED METALS

Workpiece Material		Milling Cutter Material				
		High-Speed Steel	Super High-Speed Steel	Stellite	Tantalum Carbide	Tungsten Carbide
		Cutting Speeds (sfpm)				
Nonferrous	Aluminum	500-1000	. . . .	800-1500	. . . .	1000-2000
	Brass	70-175	. . . .	150-200	. . . .	350-600
	Hard Bronze	65-130	. . . .	100-160	. . . .	200-425
	Very Hard Bronze	30-50	. . . .	50-70	. . . .	125-200
Cast Iron	Soft	50-80	60-115	90-130	. . . .	250-325
	Hard	30-50	40-70	60-90	. . . .	150-200
	Chilled		30-50	40-60	. . . .	100-200
	Malleable	70-100	80-125	115-150	. . . .	350-370
Steel	Soft (low-carbon)	60-90	70-100	. . . .	150-250	300-550
	Medium-carbon	50-80	60-90	. . . .	125-200	225-400
	High-carbon	40-70	50-80	. . . .	100-175	150-250
	Stainless	30-80	40-90	. . . .	75-200	100-300

Table A-27. RECOMMENDED FEED PER TOOTH (0.000″) FOR BASIC MILLING MACHINE PROCESSES
(High-Speed Steel And Cemented Carbide-Tipped Cutters)

Material	High-Speed Steel Milling Cutters					
	Face mills	Spiral mills	Side and Slotting mills	End mills	Form cutters	Saws
Aluminum, soft bronze and brass	0.022	0.017	0.013	0.011	0.006	0.005
Medium-hard bronze and brass, cast iron (soft)	0.018	0.014	0.011	0.009	0.005	0.004
Malleable iron, cast iron (medium-hard)	0.015	0.012	0.009	0.008	0.005	0.004
AISI X-1112 steel, cast iron (hard)	0.013	0.010	0.008	0.006	0.004	0.003
AISI 1020 steel, AISI X-1335 steel	0.011	0.009	0.007	0.005	0.004	0.003
AISI 1045 steel, cast steel	0.009	0.007	0.006	0.005	0.005	0.003
Alloy steel	0.008	0.006	0.005	0.004	0.003	0.002
Medium	0.007	0.005	0.004	0.004	0.002	0.002
Tough	0.005	0.004	0.003	0.003	0.002	0.0015
Hard and stainless	0.006	0.005	0.004	0.003	0.002	0.0015
Cemented Carbide-Tipped Milling Cutters						
Aluminum, soft bronze and brass	0.020	0.016	0.012	0.010	0.006	0.005
Medium hard bronze and brass, cast iron (soft)	0.012	0.010	0.007	0.006	0.004	0.003
Malleable iron, cast iron (medium hard)	0.016	0.013	0.010	0.008	0.005	0.004
AISI 1020 steel, AISI X-1335 steel	0.016	0.013	0.009	0.008	0.005	0.004
AISI 1045 steel, cast steel	0.014	0.011	0.008	0.007	0.004	0.004
Hard and stainless steels	0.010	0.008	0.006	0.005	0.003	0.003

Table A-28. RECOMMENDED CUTTING FLUIDS FOR FERROUS AND NONFERROUS METALS

		FERROUS METALS				NONFERROUS METALS	
Group →		I	II	III	IV	V	VI
Machinability:* →		Above 70%	50-70%	40-50%	Below 40%	Above 100%	Below 100%
Materials →		Low-Carbon Steels High-Carbon Steels		Stainless Steels		Aluminum and Alloys Brasses and Bronzes	
		Malleable Iron Cast Steel Stainless Iron	Cast Iron	Ingot Iron Wrought Iron	Tool Steels High-Speed Steels	Magnesium and Alloys Zinc	Copper Nickel Inconel Monel
Severity	Type of Machining Operation						
(Greatest) 1	Broaching; internal	Em Sul	Sul Em	Sul Em	Sul Em	MO Em	Sul ML
2	Broaching; surface	Em Sul	Em Sul	Sul Em	Sul Em	MO Em	Sul ML
	Threading; pipe	Sul	Sul ML	Sul	Sul		Sul †
3	Tapping; plain	Sul	Sul	Sul	Sul	Em Dry	Sul ML
	Threading; plain	Sul	Sul	Sul	Sul	Em Sul	Sul †
	Gear shaving	Sul L	Sul L	Sul L	Sul L		
4	Reaming; plain	ML Sul	ML Sul	ML SuL	ML Sul	ML MO Em	ML MO Sul
	Gear cutting	Sul ML Em	Sul	Sul	Sul ML		Sul ML
5	Drilling; deep	Em ML	Em Sul	Sul	Sul	MO ML Em	Sul ML
6	Milling plain	Em ML Sul	Em	Em	Sul	Em MO Dry	Sul Em
	multiple cutter	ML	Sul	Sul	Sul ML	Em MO Dry	Sul Em
	Boring; multiple-head	Sul Em	Sul HDS	Sul HDS	Sul Em	K Dry Em	Sul Em
7	Multiple-spindle automatic screw machines and turret lathes: drilling, forming, turning, reaming, cutting-off, tapping, threading	Sul Em ML	Sul Em ML	Sul Em ML HDS	Sul ML Em HDS	Em Dry ML	Sul
8	High-speed, light-feed automatic screw machines: drilling, forming, tapping, threading, turning, reaming, box milling, cutting off	Sul Em ML	Sul Em ML	Sul Em ML	Sul ML Em	Em Dry ML	Sul

(Table A-28 continued)

		FERROUS METALS				NONFERROUS METALS	
Group →		I	II	III	IV	V	VI
Machinability:* →		Above 70%	50-70%	40-50%	Below 40%	Above 100%	Below 100%
Materials →		Low-Carbon Steels High-Carbon Steels		Stainless Steels		Aluminum and Alloys Brasses and Bronzes	
		Malleable Iron Cast Steel Stainless Iron	Cast Iron	Ingot Iron Wrought Iron	Tool Steels High-Speed Steels	Magnesium and Alloys Zinc	Copper Nickel Inconel Monel
Severity	**Type of Machining Operation**						
	Drilling	Em	Em	Em	Em Sul	Em Dry	Em
9	Planing, shaping	Em Sul ML	Em Sul ML	Sul Em	Em Sul	Em Dry	Em
	Turning; single-point tool, form tools	Em Sul ML	Em Sul ML	Em Sul ML	Em SuL ML	Em Dry ML	Em Sul
(Least) 10	Sawing; circular, hack	Sul ML Em	Sul Em ML	Sul Em ML	Sul Em ML	Dry MO Em	Sul Em ML
	Grinding; plain	Em	Em	Em	Em	Em	Em
	form (thread, etc.)	Sul	Sul	Sul	Sul	MO Sul	Sul

SYMBOL	CUTTING FLUID
K	Kerosene
L	Lard Oil
MO	Mineral oils
ML	Mineral-lard oils
Sul	Sulphurized oils, with or without chlorine
Em	Soluble or emulsifiable oils and compounds
Dry	No cutting fluid needed
HDS	Heavy duty soluble oil

*Machinability rating based on 100% for cold drawn Bessemer screw stock (B1112)
†Palm oil is frequently used to thread copper.
Compiled from *Metals Handbook, Machinery's Handbook,* and *AISI Steel Products Manual.*

Table A-29. NATURAL TRIGONOMETRIC FUNCTIONS

Angle	Sine	Cosine	Tangent	Angle	Sine	Cosine	Tangent
1°	.0175	.9998	.0175	46°	.7193	.6947	1.0355
2°	.0349	.9994	.0349	47°	.7314	.6820	1.0724
3°	.0523	.9986	.0524	48°	.7431	.6691	1.1106
4°	.0698	.9976	.0699	49°	.7547	.6561	1.1504
5°	.0872	.9962	.0875	50°	.7660	.6428	1.1918
6°	.1045	.9945	.1051	51°	.7771	.6293	1.2349
7°	.1219	.9925	.1228	52°	.7880	.6157	1.2799
8°	.1392	.9903	.1405	53°	.7986	.6018	1.3270
9°	.1564	.9877	.1584	54°	.8090	.5878	1.3764
10°	.1736	.9848	.1763	55°	.8192	.5736	1.4281
11°	.1908	.9816	.1944	56°	.8290	.5592	1.4826
12°	.2079	.9781	.2126	57°	.8387	.5446	1.5399
13°	.2250	.9744	.2309	58°	.8480	.5299	1.6003
14°	.2419	.9703	.2493	59°	.8572	.5150	1.6643
15°	.2588	.9659	.2679	60°	.8660	.5000	1.7321
16°	.2756	.9613	.2867	61°	.8746	.4848	1.8040
17°	.2924	.9563	.3057	62°	.8829	.4695	1.8807
18°	.3090	.9511	.3249	63°	.8910	.4540	1.9626
19°	.3256	.9455	.3443	64°	.8988	.4384	2.0503
20°	.3420	.9397	.3640	65°	.9063	.4226	2.1445
21°	.3584	.9336	.3839	66°	.9135	.4067	2.2460
22°	.3746	.9272	.4040	67°	.9205	.3907	2.3559
23°	.3907	.9205	.4245	68°	.9272	.3746	2.4751
24°	.4067	.9135	.4452	69°	.9336	.3584	2.6051
25°	.4226	.9063	.4663	70°	.9397	.3420	2.7475
26°	.4384	.8988	.4877	71°	.9455	.3256	2.9042
27°	.4540	.8910	.5095	72°	.9511	.3090	3.0777
28°	.4695	.8829	.5317	73°	.9563	.2924	3.2709
29°	.4848	.8746	.5543	74°	.9613	.2756	3.4874
30°	.5000	.8660	.5774	75°	.9659	.2588	3.7321
31°	.5150	.8572	.6009	76°	.9703	.2419	4.0108
32°	.5299	.8480	.6249	77°	.9744	.2250	4.3315
33°	.5446	.8387	.6494	78°	.9781	.2079	4.7046
34°	.5592	.8290	.6745	79°	.9816	.1908	5.1446
35°	.5736	.8192	.7002	80°	.9848	.1736	5.6713
36°	.5878	.8090	.7265	81°	.9877	.1564	6.3138
37°	.6018	.7986	.7536	82°	.9903	.1392	7.1154
38°	.6157	.7880	.7813	83°	.9925	.1219	8.1443
39°	.6293	.7771	.8098	84°	.9945	.1045	9.5144
40°	.6428	.7660	.8391	85°	.9962	.0872	11.4301
41°	.6561	.7547	.8693	86°	.9976	.0698	14.3007
42°	.6691	.7431	.9004	87°	.9986	.0523	19.0811
43°	.6820	.7314	.9325	88°	.9994	.0349	28.6363
44°	.6947	.7193	.9657	89°	.9998	.0175	57.2900
45°	.7071	.7071	1.0000	90°	1.0000	.0000	

Table A-30. TAPERS AND INCLUDED ANGLES

Taper per Foot	Taper per Inch	Included Angle		
		Deg.	Min.	Sec.
1/8	0.010416	0	35	48
3/16	0.015625	0	53	44
1/4	0.020833	1	11	36
5/16	0.026042	1	29	30
3/8	0.031250	1	47	24
7/16	0.036458	2	5	18
1/2	0.416667	2	23	10
9/16	0.046875	2	41	4
5/8	0.052084	2	59	42
11/16	0.057292	3	16	54
3/4	0.06250	3	34	44
13/16	0.067708	3	52	38
7/8	0.072917	4	10	32
15/16	0.078125	4	28	24
1	0.083330	4	46	18
1 1/4	0.104666	5	57	48
1 1/2	0.125000	7	9	10
1 3/4	0.145833	8	20	26
2	0.166666	9	31	36
2 1/2	0.208333	11	53	36
3	0.250000	14	15	0
3 1/2	0.291666	16	35	40
4	0.333333	18	55	28
4 1/2	0.375000	21	14	2
5	0.416666	23	32	12
6	0.500000	28	4	2

Table A-31. AMERICAN STANDARD MACHINE TAPERS (MORSE*)

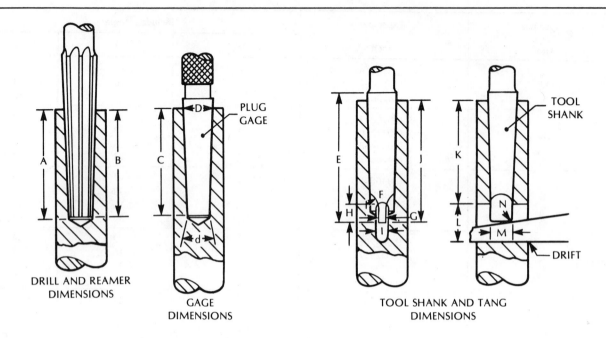

DRILL AND REAMER
DIMENSIONS

GAGE
DIMENSIONS

TOOL SHANK AND TANG
DIMENSIONS

Number of Taper	Taper		Diameter		Depth			Shank			Tang					Tang Slot			Number of Drift
	per inch	per foot	small end (d)	gage line (D)	drilled hole (A)	reamed hole (B)	plug depth (C)	Number of Taper	whole length (E)	depth (J)	thickness (G)	length (C)	radius (F)	diameter (M)	radius (N)	width (I)	length (L)	to end of socket	
0†	0.052000	0.62400	0.252	0.356	$2^{1}/_{16}$	$2^{1}/_{32}$	2	0†	$2^{11}/_{32}$	$2^{7}/_{32}$	0.156	$^{1}/_{4}$	$^{5}/_{32}$	$^{15}/_{64}$	$^{3}/_{64}$	0.166	$^{9}/_{16}$	$1^{15}/_{16}$	0†
1	0.049882	0.59858	0.369	0.475	$2^{3}/_{16}$	$2^{5}/_{32}$	$2^{1}/_{8}$	1	$2^{9}/_{16}$	$2^{7}/_{16}$	0.203	$^{3}/_{8}$	$^{3}/_{16}$	$^{11}/_{32}$	$^{3}/_{64}$	0.213	$^{3}/_{4}$	$2^{1}/_{16}$	1
2	0.049951	0.59941	0.572	0.700	$2^{21}/_{32}$	$2^{39}/_{64}$	$2^{9}/_{16}$	2	$3^{1}/_{8}$	$2^{15}/_{16}$	0.250	$^{7}/_{16}$	$^{1}/_{4}$	$^{17}/_{32}$	$^{1}/_{16}$	0.260	$^{7}/_{8}$	$2^{1}/_{2}$	2
3	0.050196	0.60235	0.778	0.938	$3^{5}/_{16}$	$3^{1}/_{4}$	$3^{3}/_{16}$	3	$3^{7}/_{8}$	$3^{11}/_{16}$	0.312	$^{9}/_{16}$	$^{9}/_{32}$	$^{23}/_{32}$	$^{5}/_{64}$	0.322	$1^{3}/_{16}$	$3^{1}/_{16}$	3
4	0.051938	0.62326	1.020	1.231′	$4^{3}/_{16}$	$4^{1}/_{8}$	$4^{1}/_{16}$	4	$4^{7}/_{8}$	$4^{5}/_{8}$	0.469	$^{5}/_{8}$	$^{5}/_{16}$	$^{31}/_{32}$	$^{3}/_{32}$	0.479	$1^{1}/_{4}$	$3^{7}/_{8}$	4
5	0.052626	0.63151	1.475	1.748	$5^{5}/_{16}$	$5^{1}/_{4}$	$5^{3}/_{16}$	5	$6^{1}/_{8}$	$5^{7}/_{8}$	0.625	$^{3}/_{4}$	$^{3}/_{8}$	$1^{13}/_{32}$	$^{1}/_{8}$	0.635	$1^{1}/_{2}$	$4^{15}/_{16}$	5
6	0.052137	0.62565	2.116	2.494	$7^{13}/_{32}$	$7^{21}/_{64}$	$7^{1}/_{4}$	6	$8^{9}/_{16}$	$8^{1}/_{4}$	0.750	$1^{1}/_{8}$	$^{1}/_{2}$	2	$^{5}/_{32}$	0.760	$1^{3}/_{4}$	7	5
7	0.052000	0.62400	2.750	3.270	$10^{5}/_{32}$	$10^{5}/_{64}$	10	7	$11^{5}/_{8}$	$11^{1}/_{4}$	1.125	$1^{3}/_{8}$	$^{3}/_{4}$	$2^{5}/_{8}$	$^{3}/_{16}$	1.135	$2^{5}/_{8}$	$9^{1}/_{2}$	

*Morse dimensions are identical with the American Standard Machine Tapers
†The #0 taper is not listed in the American Standard Machine Tapers

Table A-32. JARNO TAPERS

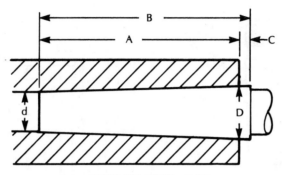

TAPER PER FOOT = 0.600

Number of Taper	Diameter of taper		Depth of Taper (A)	Length of Shank (B)	Clearance (C)
	Large End (D)	Small End (d)			
1	0.125	0.1	0.5	9/16	1/16
2	0.250	0.2	1.0	1 1/8	1/8
3	0.375	0.3	1.5	1 5/8	1/8
4	0.500	0.4	2.0	2 3/16	3/16
5	0.625	0.5	2.5	2 11/16	3/16
6	0.750	0.6	3.0	3 3/16	3/16
7	0.875	0.7	3.5	3 11/16	3/16
8	1.000	0.8	4.0	4 3/16	3/16
9	1.125	0.9	4.5	4 11/16	3/16
10	1.250	1.0	5.0	5 1/4	1/4
11	1.375	1.1	5.5	5 3/4	1/4
12	1.500	1.2	6.0	6 1/4	1/4
13	1.625	1.3	6.5	6 3/4	1/4
14	1.750	1.4	7.0	7 1/4	1/4
15	1.875	1.5	7.5	7 3/4	1/4
16	2.000	1.6	8.0	8 3/8	3/8
17	2.125	1.7	8.5	8 7/8	3/8
18	2.250	1.8	9.0	9 3/8	3/8
19	2.375	1.9	9.5	9 7/8	3/8
20	2.500	2.0	10.0	10 3/8	3/8

Table A-33. BROWN AND SHARPE TAPERS*

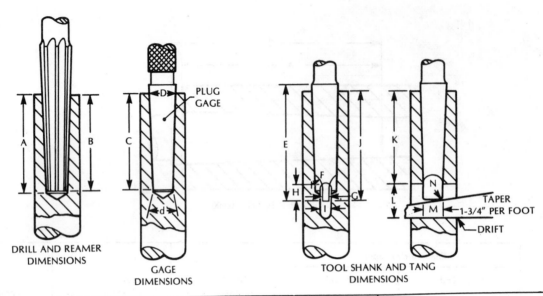

DRILL AND REAMER DIMENSIONS

GAGE DIMENSIONS

TOOL SHANK AND TANG DIMENSIONS

Number of Taper	Taper per Foot	Diameter at Small End (d)	Plug			Tang Slot to End of Socket (K)	Tool Shank Whole Length (E)	Tang Seat		Arbor Tongue Tang				
			Brown and Sharpe Standard	Milling Machine Standard	Miscellaneous			Length (L)	Width (I)	Length (G)	Diameter (M)	Thickness (H)	Radius (F)	Radius (N)
1	0.50200	0.20000	$^{15}/_{16}$	. . .	. . .	$^{15}/_{16}$	$1^3/_{16}$	$^3/_8$	0.135	$^3/_{16}$	0.170	$^1/_8$	$^3/_{16}$	0.030
2	0.50200	0.25000	$1^3/_{16}$	. . .	. . .	$1^{11}/_{64}$	$1^1/_2$	$^1/_2$	0.166	$^1/_4$	0.220	$^5/_{32}$	$^3/_{16}$	0.030
3	0.50200	0.31250	$1^1/_2$	. . .	. . .	$1^{15}/_{32}$	$1^7/_8$	$^5/_8$	0.197	$^5/_{16}$	0.282	$^3/_{16}$	$^3/_{16}$	0.040
			. . .	. . .	$1^3/_4$	$1^{23}/_{32}$	$2^1/_8$	$^5/_8$	0.197	$^5/_{16}$	0.282	$^3/_{16}$	$^3/_{16}$	0.040
			. . .	. . .	2	$1^{31}/_{32}$	$2^3/_8$	$^5/_8$	0.197	$^5/_{16}$	0.282	$^3/_{16}$	$^3/_{16}$	0.040
4	0.50240	0.35000	. . .	$1^1/_4$	. . .	$1^{13}/_{64}$	$1^{21}/_{32}$	$^{11}/_{16}$	0.228	$^{11}/_{32}$	0.320	$^7/_{32}$	$^5/_{16}$	0.050
			$1^{11}/_{16}$	. . .	. . .	$1^{41}/_{64}$	$2^3/_{32}$	$^{11}/_{16}$	0.228	$^{11}/_{32}$	0.320	$^7/_{32}$	$^5/_{16}$	0.050
5	0.50160	0.45000	. . .	$1^3/_4$	. . .	$1^{11}/_{16}$	$2^3/_{16}$	$^3/_4$	0.260	$^3/_8$	0.420	$^1/_4$	$^5/_{16}$	0.060
			. . .	. . .	2	$1^{15}/_{16}$	$2^7/_{16}$	$^3/_4$	0.260	$^3/_8$	0.420	$^1/_4$	$^5/_{16}$	0.060
			$2^1/_8$	. . .	. . .	$2^1/_{16}$	$2^9/_{16}$	$^3/_4$	0.260	$^3/_8$	0.420	$^1/_4$	$^5/_{16}$	0.060
6	0.50329	0.50000	$2^3/_8$	. . .	. . .	$2^{19}/_{64}$	$2^7/_8$	$^7/_8$	0.291	$^7/_{16}$	0.460	$^9/_{32}$	$^5/_{16}$	0.060
7	0.50147	0.60000	. . .	. . .	$2^1/_2$	$2^{13}/_{32}$	$3^1/_{32}$	$^{15}/_{16}$	0.322	$^{15}/_{32}$	0.560	$^5/_{16}$	$^3/_8$	0.070
			$2^7/_8$	. . .	. . .	$2^{25}/_{32}$	$3^{13}/_{32}$	$^{15}/_{16}$	0.322	$^{15}/_{32}$	0.560	$^5/_{16}$	$^3/_8$	0.070
			. . .	3	. . .	$2^{29}/_{32}$	$3^{17}/_{32}$	$^{15}/_{16}$	0.322	$^{15}/_{32}$	0.560	$^5/_{16}$	$^3/_8$	0.070
8	0.50100	0.75000	$3^9/_{16}$	. . .	. . .	$3^{29}/_{64}$	$4^1/_8$	1	0.353	$^1/_2$	0.710	$^{11}/_{32}$	$^3/_8$	0.080
9	0.50085	0.90010	. . .	4	. . .	$3^7/_8$	$4^5/_8$	$1^1/_8$	0.385	$^9/_{16}$	0.860	$^3/_8$	$^7/_{16}$	0.100
			$4^1/_4$	. . .	. . .	$4^1/_8$	$4^7/_8$	$1^1/_8$	0.385	$^9/_{16}$	0.860	$^3/_8$	$^7/_{16}$	0.100

Number of Taper	Taper per Foot	Diameter at Small End (d)	Plug Brown and Sharpe Standard	Plug Milling Machine Standard	Plug Miscellaneous	Tang Slot to End of Socket (K)	Tool Shank Whole Length (E)	Tang Seat Length (L)	Tang Seat Width (I)	Arbor Tongue Tang Length (G)	Arbor Tongue Tang Diameter (M)	Arbor Tongue Tang Thickness (H)	Arbor Tongue Tang Radius (F)	Radius (N)
10	0.51612	1.04465	5	...	...	$4^{27}/_{32}$	$5^{23}/_{32}$	$1^{5}/_{16}$	0.447	$^{21}/_{32}$	1.010	$^{7}/_{16}$	$^{7}/_{16}$	0.110
			...	$5^{11}/_{16}$	...	$5^{17}/_{32}$	$6^{13}/_{32}$	$1^{5}/_{16}$	0.447	$^{21}/_{32}$	1.010	$^{7}/_{16}$	$^{7}/_{16}$	0.110
			...	...	$6^{7}/_{32}$	$6^{1}/_{16}$	$6^{15}/_{16}$	$1^{5}/_{16}$	0.447	$^{21}/_{32}$	1.010	$^{7}/_{16}$	$^{7}/_{16}$	0.110
11	0.50100	1.24995	$5^{15}/_{16}$	...	...	$5^{25}/_{32}$	$6^{21}/_{32}$	$1^{5}/_{16}$	0.447	$^{21}/_{32}$	1.210	$^{7}/_{16}$	$^{1}/_{2}$	0.130
			...	$6^{3}/_{4}$	...	$6^{19}/_{32}$	$7^{15}/_{32}$	$1^{5}/_{16}$	0.447	$^{21}/_{32}$	1.210	$^{7}/_{16}$	$^{1}/_{2}$	0.130
12	0.49973	1.50010	$7^{1}/_{8}$	$7^{1}/_{8}$	...	$6^{15}/_{16}$	$7^{15}/_{16}$	$1^{1}/_{2}$	0.510	$^{3}/_{4}$	1.460	$^{1}/_{2}$	$^{1}/_{2}$	0.150
			...	...	$6^{1}/_{4}$									
13	0.50020	1.75005	$7^{3}/_{4}$	...	...	$7^{9}/_{16}$	$8^{9}/_{16}$	$1^{1}/_{2}$	0.510	$^{3}/_{4}$	1.710	$^{1}/_{2}$	$^{5}/_{8}$	0.170
14	0.50000	2.00000	$8^{1}/_{4}$	$8^{1}/_{4}$	...	$8^{1}/_{32}$	$9^{5}/_{32}$	$1^{11}/_{16}$	0.572	$^{27}/_{32}$	1.960	$^{9}/_{16}$	$^{3}/_{4}$	0.190
15	0.50000	2.25000	$8^{3}/_{4}$	...	...	$8^{17}/_{32}$	$9^{21}/_{32}$	$1^{11}/_{16}$	0.572	$^{27}/_{32}$	2.210	$^{9}/_{16}$	$^{7}/_{8}$	0.210
16	0.50000	2.50000	$9^{1}/_{4}$	...	...	9	$10^{1}/_{4}$	$1^{7}/_{8}$	0.635	$^{15}/_{16}$	2.450	$^{5}/_{8}$	1	0.230
17	0.50000	2.75000	$9^{3}/_{4}$											
18	0.50000	3.00000	$10^{1}/_{4}$											

*All dimensions are in inches

Table A-34. JACOBS TAPERS

Taper Dimensions				
Jacobs Taper	Large Diameter	Small Diameter	Length of Taper	Taper per Foot
0	0.2500	0.2284	$\frac{3}{16}$	0.5915
1	0.3840	0.3334	$\frac{21}{32}$	0.9251
2	0.5590	0.4876	$\frac{7}{8}$	0.9786
2 short	0.5488	0.4876	$\frac{3}{4}$	0.9786
3	0.8110	0.7641	$1\frac{7}{32}$	0.6390
4	1.1240	1.0372	$1\frac{21}{32}$	0.6289
5	1.4130	1.3161	$1\frac{7}{8}$	0.6201
6	0.6760	0.6241	1	0.6229
33	0.6240	0.5605	1	0.7619
E	0.7886	0.7472	$\frac{51}{64}$	0.6240

Table A–35. STANDARD TAPER PINS (Diameters And Drill Sizes)

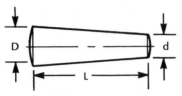

TAPER = 0.250" PER FOOT
= 0.02083" PER INCH

D = LARGE DIAMETER
d = SMALL DIAMETER
L = LENGTH
(d) = (D)-(L + 0.02083)

NUMBER	7/0	6/0	5/0	4/0	3/0	2/0	0	1	2	3	4	5	6	7	8	9	10	11
(D) DIAMETER AT LARGE END																		
	0.0625	0.078	0.094	0.109	0.125	0.141	0.156	0.172	0.193	0.219	0.250	0.289	0.341	0.409	0.492	0.591	0.707	0.857
LENGTH — (d) DIAMETER OF SMALL END OF PIN (AND DRILL SIZE)																		
¼	0.0573 (54)	0.0728 (50)																
⅜	0.0547 (55)	0.0702 (51)	0.0864 (45)															
½	0.0521 (56)	0.0676 (52)	0.0836 (46)	0.0986 (41)	0.1146 (34)	0.1306 (30)	0.1456 (9/64)	0.1616 (5/32)										
⅝	0.0495 (56)	0.0650 (52)	0.0810 (5/64)	0.0960 (3/32)	0.1120 (7/64)	0.1280 (1/8)	0.1430 (9/64)	0.1590 (23)										
¾	0.0469 (56)	0.0624 (53)	0.0784 (48)	0.0934 (43)	0.1094 (36)	0.1254 (31)	0.1404 (29)	0.1564 (24)	0.1774 (11/64)	0.2034 (8)	0.2344 (1)							
⅞		0.0598 (54)	0.0758 (49)	0.0908 (43)	0.1068 (37)	0.1228 (31)	0.1378 (29)	0.1538 (25)	0.1748 (18)	0.2008 (9)	0.2318 (1)							
1		0.0572 (54)	0.0732 (50)	0.0882 (44)	0.1042 (38)	0.1202 (32)	0.1352 (30)	0.1512 (26)	0.1722 (19)	0.1982 (10)	0.2292 (2)	0.2682 (G)	0.3202 (0)					
1⅛				0.0856 (45)	0.1016 (39)	0.1176 (33)	0.1326 (30)	0.1486 (27)	0.1696 (19)	01956 (11)	0.2266 (2)	0.2656 (G)	0.3176 (5/16)					
1¼				0.0830 (46)	0.0990 (41)	0.1150 (33)	0.1300 (1/8)	0.1460 (9/64)	0.1670 (20)	0.1930 (3/16)	0.2240 (7/32)	0.2630 (F)	0.3150 (N)	0.3830 (3/8)				
1⅜					0.0964 (3/32)	0.1124 (7/64)	0.1274 (1/8)	0.1434 (9/64)	0.1644 (20)	0.1904 (3/16)	0.2214 (3)	0.2604 (F)	0.3124 (N)	0.3804 (3/8)				
1½					0.0938 (43)	0.1098 (36)	0.1248 (31)	0.1408 (29)	0.1618 (5/32)	0.1878 (14)	0.2188 (3)	0.2578 (¼)	0.3098 (N)	0.3778 (U)	0.4608 (29/64)			
1¾						0.1045 (38)	0.1195 (32)	0.1355 (30)	0.1565 (24)	0.1825 (16)	0.2135 (4)	0.2525 (D)	0.3045 (19/64)	0.3725 (U)	0.4555 (7/16)			
2						0.0993 (41)	0.1143 (34)	0.1303 (1/8)	0.1513 (26)	0.1773 (11/64)	0.2203 (15/64)	0.2473 (C)	0.2993 (M)	0.3673 (23/64)	0.4503 (7/16)	0.5494 (17/32)		
2¼								0.1251 (31)	0.1461 (27)	0.1721 (19)	0.2031 (8)	0.2421 (B)	0.2941 (L)	0.3621 (T)	0.4451 (7/16)	0.5442 (17/32)		
2½								0.1199 (32)	0.1409 (29)	0.1669 (20)	0.1979 (10)	0.2369 (15/64)	0.2889 (9/32)	0.3569 (S)	0.4399 (27/64)	0.5390 (17/32)	0.6540 (41/64)	
2¾									0.1357 (30)	0.1617 (5/32)	0.1927 (3/16)	0.2317 (1)	0.2837 (J)	0.3517 (11/32)	0.4347 (27/64)	0.5338 (33/64)	0.6488 (41/64)	